V. Z. Aladjev

Computer Algebra Systems: A New Software Toolbox for Maple

Series: Computer Mathematics

Fultus™ Books

Computer Algebra Systems: A New Software Toolbox for Maple

by

V. Z. Aladjev

Series: Computer Mathematics

ISBN 1-59682-000-4

Published by Fultus Corporation

Corporate Web Site: http://www.fultus.com
Fultus eLibrary: http://elibrary.fultus.com
Online Book Superstore: http://store.fultus.com
Writer web site: http://writer.fultus.com/aladjev/
email: production@fultus.com

Table of Contents

This monograph is dedicated to my wife Galina,

daughter Svetlana,

grandsons Arthur and Kristo!

From the Author

This monograph is a fully revised edition of our previous monographs whose pressrun was completely sold out. The book presents the *Library* for the modern computer algebra system *Maple* of releases 6 - 9.5, extending its basic tools. Source codes of the *Library* tools introduce the reader into both effective and *non-standard* programming technique in the *Maple* package. The reader should have a background in the *Maple* environment of any release.

Alongside with it, the monograph considers a lot of essential disadvantages, defects and errors of the package. In addition, a series of them are eliminated by the facilities represented in the present monograph whereas onto others an accent has been done with the purpose to draw developers attention at creation of new releases of the package. Unfortunately, up till now the old tendency is kept - the new *Maple* releases are not free from many old disadvantages, defects and errors, adding their set by new ones.

The tools represented in the book increase the range and efficiency of use of the *Maple* package on *Windows* platforms 95/98/98SE/ME/NT/XP/2000/2003 owing to the innovations in three basic directions: (1) *elimination of a series of basic defects and shortcomings*, (2) *extending of capabilities of a series of standard tools*, and (3) *replenishment of the package by new means which increase capabilities of its program environment, including the means which improve the level of compatibility of releases* 6 - 9/9.5. The basic attention is devoted to additional tools created in the process of practical use and testing the *Maple* package of releases 4 - 9 which by some parameters extend essentially the opportunities of the package and facilitate the work with it. The considerable attention is also devoted to the means providing package compatibility of releases 6 - 9/9.5. The experience in the use of given software has confirmed its high operational characteristics at the usage of *Maple* in numerous appendices.

The above software has been organized into the user *Library* whose a current version contains tools (more than 570 *procedures* and *program modules*) which are oriented onto wide enough sphere of computing and information processing. The *Library* is structurally similar to the main *Maple* library and is supplied with the advanced Help system about the tools located in it. In addition, the *Library* is logically connected with the main *Maple* library, providing access to the tools contained in it similarly to the package tools. The simple guide describes the installation of the *Library* at presence on PC with one of the *Windows* platforms 95/98/98SE/ME/NT/XP/2000/2003 of the installed *Maple* package of releases 6 - 9/9.5.

The *Library* is designed for a wide audience of experts, teachers, post-graduates and students of natural-science professions who use *Maple* of releases 6 - 9/9.5 on the above *Windows* platforms in their own professional work. The *Library* contains well-designed software (a set of procedures and program modules), which supplements well the already available *Maple* software with the orientation towards the widest circle of the *Maple* users, greatly enhancing its usability and effectiveness. Our experience reveals that the use of the *Library* essentially extends opportunities of *Maple* of releases 6 - 9/9.5, simplifying the programming of various practical problems in its environment. This *Library* will present special interest above all for those who use *Maple* not only as a highly intellectual calculator but also as environment for programming of different problems in one's own professional activity. This book will be interesting for scientists, researchers, candidates for doctor's or master's degree in the field of computer algebra systems, and experts and researchers in other areas of physics and mathematics who use computer algebra systems in own professional activity. All *Library* tools can be viewed and modified by users. Therefore, it is useful to learn the *Maple* programming language so that you can modify existing *Maple* code to produce customized means. At last, source codes of the *Library* tools introduce the reader into both effective and non-standard programming technique in *Maple*, better revealing subtleties of the *Maple* programming language. All this allows to hope, that the represented book and the software attached to it will appear useful enough to broad audience of the *Maple* users, both for the beginners and the skilled ones.

Archive with the *Library* versions for *Maple* of releases **6, 7, 8** and **9/9.5** additionally contains the set of *mws*-files composing Help database for the *Library* and the set of *mws*-files with the *Maple* documents containing source codes (for series of them, the different implementations are given) of the *Library* tools with examples of them appendices both in usual and especial situations. The archive contains the detailed guide on installation of the *Library* for the above *Maple* releases; in addition, two installation modes are supported – *self-acting* on the basis of special *mws*-document and *manual* by means of simple copying of datafiles (from the unpacked archive into your *Maple* file system). The installed *Library* is logically linked with the main *Maple* library with the higher priority, providing access to all tools contained in it at level of the standard *Maple* tools. At such approach, the appropriate *Library* tools allow both to extend the *Maple* tools of the same name and to eliminate errors contained in them. The archive has size **6.5** MB and can be downloaded from the web sites of the Publisher and Author:

`http://writers.fultus.com/aladjev/source/UserLib6789.zip`

`http://www.aladjev.newmail.ru/UserLib6789.zip`

Our work on computer algebra systems continues, therefore new software of various purpose and oriented on various appendices will appear. The library represented in the present monograph will be constantly updated by the most useful of these means. Therefore, the reader is recommended to check periodically presence on the above-mentioned sites of new versions of the library. Whereas all questions, remarks and offers can be directed to one of the following addresses:

`valadjev@yahoo.com, aladjev@hotmail.com, aladjev@lenta.ru`

We welcome your feedback. Additionally, for direct inquiries, suggestions, or comments you can use

phone +(372) 635 6078 or **fax** +(372) 600 7969

Preface

Computer Algebra (also known as *Symbolic Computation* or *Computational Algebra*) has found application in many fields of science such as mathematics, physics, chemistry, computer science, engineering, education, technology, computational biology, etc. The *computer algebra systems* (CAS) such as *Maple, Reduce, MuPAD, Axiom, Macsyma, Mathematica, Derive, Magma,* and others are becoming more and more popular in teaching, research and industry. The area of symbolic and algebraic computation aims at automation of mathematical computations of all sorts. The resulting computer systems, both experimental and commercial, are powerful tools for scientists, engineers, and educators. This research combines mathematics with advanced computing techniques. CAS is an interdisciplinary area between *Mathematics* and *Computer Science*. Its research focuses on the development of algorithms for performing *symbolic* manipulations with algebraic objects on computers, and design of programming languages and environments for implementing these algorithms.

In a series of our books and papers [1-20,33,39,41-45,47,48], such packages as *Maple, Reduce, MathCAD* and *Mathematica* have been considered. Our experience in the detailed testing and practical use of different mathematical and physical applications of four mathematical packages (*Reduce, Maple, MathCAD* and *Mathematica*) enables us to consider the packages *Maple* and *Mathematica* as undoubted leaders (*on the basis of a generalized index*) among all listed modern tools of computer algebra. Meanwhile, we give preference to the package *Maple* due to a number of strong reasons described in the above books enough in detail.

Computer algebra becomes a rather powerful and useful tool for scientists and experts from various fields. However, manifold applications demand that the essential significant numerical calculations be combined with algebraic ones. With each new release, the package *Maple* meets more and more requirements. The *Maple* package has been widely used not only as a tool of solving mathematical problems. The package allows us to revise approaches to teaching subjects related to mathematics in universities by defining in many cases the methods for a teaching of subjects with the use of PCs to solve mathematical problems for various purposes [11-14,20-28,32,34-36,40-42].

Researchers use well-known *Maple* package as an essential tool when solving problems related to their investigation. The package is ideal for formulating, solving, and exploring different mathematical models. Its symbolic manipulation facilities extend greatly over a range of problems that can be solved with its help. Educators in high schools, colleges, and universities have revitalized traditional curricula by introducing problems and exercises, which use *Maple's* interactive mathematics and physics. Students can concentrate on the more fundamental concepts rather than on tedious algebraic manipulations. Finally, engineers and experts in industries use *Maple* as an efficient tool replacing many traditional resources such as reference books, calculators, spreadsheets and programming languages. These users easily solve mathematical problems, creating projects and consolidating their computations into professional reports.

Maple products embody advanced technologies such as symbolic computation, infinite precision numeric, innovative *Web* components, extensible user-interface technology, and an unrivalled suite of mathematical algorithms for intelligent management of complex mathematics. Over 3 million users benefit from advanced *Maple* technology. Virtually, all major universities and research institutes in the world, including such as MIT, Oxford, Stanford and Waterloo, has adopted *Maple* products to enhance their education and research activities. Waterloo *Maple's* industrial customer base includes Boeing, Bosch, Canon, NASA, etc.

Meanwhile, our operational experience in the period of 1997 – 2004 with *Maple* of releases 4, 5, 6, 7, 8 and 9/9.5 enabled us not only to estimate its advantages compared with other similar mathematical packages, but has also revealed a number of faults and shortcomings which were eliminated by us. Furthermore, *Maple* does not support a number of important procedures of information processing, *symbolic* and *numeric* computing, including the tools of access to datafiles. By operating *Maple* we have developed rather effective tools (*procedures* and *program modules*), largely extending the possibilities of the package. This software has been organized as a *Library* that is structurally similar to the main *Maple* library, and is provided with a rather detailed Help system analogous to *Maple* Help system.

In particular, *Maple* does not provide sufficient compatibility of releases 6, 7, 8 and 9. This fact and the incompatibility of the package found out by us at a level of base platforms – *Windows* 98SE and lower, on the one hand, and *Windows* XP and above, on the other hand, the decision of a compatibility problem for the *Library* means have demanded.

The tools represented in the *Library* increase the range and efficiency of use of the package on the *Windows* platform owing to the innovations in three basic directions: (1) *elimination of a series of basic defects and shortcomings,* (2) *extension of capabilities of a series of standard tools,* and (3) *replenishment of the package by new means which increase the capabilities of its program environment, including the facilities improving the compatibility of releases* 6, 7, 8 *and* 9/9.5. The basic attention is devoted to additional tools created in the process of practical use and testing of the package of releases 4 – 8 which by some parameters considerably extend the capabilities of the package making the work with it much easier; a considerable attention is also devoted to the tools providing package compatibility of releases 6, 7, 8 and 9/9.5. The experience in using the above software for various applications has confirmed its valuable operational characteristics.

It should be noted that a series of our books and papers on *Maple* [11,13-20,22-25,30-33,39,43-45,47], representing tools developed by the authors and containing suggestions on further development of the package encouraged the development of such applications as package modules *FileTools, LibraryTools, ListTools* and *StringTools.* However, the means suggested by us essentially extend the capabilities of the package, which in many cases exceed those of the specified modules.

The above software has been organized into the user *Library,* whose a current version contains tools (more than 570 *procedures* and *program modules*) which are oriented to a wide area of computing and information processing. The *Library* is structurally similar to the main *Maple* library and is supplied with the advanced Help system about the tools located in it. In addition, it is logically connected with the main *Maple* library, providing access to the tools contained in it similarly to the package tools. The simple guide describes the installation of the *Library* at presence in the PC with the above-mentioned *Windows* platform of the installed *Maple* package of releases 6, 7, 8 or 9/9.5.

The package incompatibility found by us as at the level of its releases, and at the level of underlying platforms – *Windows* 98SE and lower, on the one hand, and *Windows* XP and above, on the other hand, shows need for solving a compatibility problem of for the *Library* tools. A current *Library* version contains means (*more than 570 procedures and program modules*), oriented onto the following basic kinds of processing and calculations:

1. General purpose software
2. Software for operation with procedural and modular objects of *Maple*
3. Software for operation with numeric expressions

4. Software for operation with strings and symbolic expressions
5. Software for operation with sets, tables and lists
6. Software to support data structures of a special type
7. Software to support bit-by-bit processing of symbolic information
8. Tools extending graphic possibilities of *Maple* of releases 6 - 9.5
9. Tools that extend and improve the standard tools of *Maple*
10. Software for working with *Maple* datafiles and documents
    10.1. General purpose software
    10.2. Software for operation with *TEXT* datafiles
    10.3. Software for operation with *BINARY* datafiles
    10.4. Software for operation with *Maple* files
    10.5. Some special tools for operation with datafiles
11. Software for solving problems of mathematical analysis
12. Software for solving problems of linear algebra
    12.1. General purpose software
    12.2. Software for operation with *rtable*-objects
13. Software to support simple statistics problems
    13.1. Software for solving problems of descriptive statistics
    13.2. Software for solving problems of regression analysis
    13.3. Software for testing of statistical hypotheses
    13.4. Elements for simple analysis of time (*dynamic*) and variation series
14. Software for operation with the user libraries

The basic innovations of our *Library* with respect of the listed sections that thematically classify the *Library* tools can be briefly characterized as follows (*in brackets, quantity of corresponding procedures is given*):

1. General pupose softare **(15)**:
    - Increaseng comatibility of *Maple* of releases 6 - 9/9.5 concerning its standard functions {***'assign'***, ***'close'***, ***'fclose'***, ***'fremove'***};
    - definition of full paths to the basic subdirectories of the package *Maple*;
    - definition of installed releases of the package *Maple* on your computer;
    - definition of the current release of the package *Maple*;
    - definition of the current version of underlying operational system of a computer, etc.
2. Software for operation with procedural and modular objects of *Maple* **(30)**:
    - output of source code of a program module or package and variables exported by them;
    - a simple method of procedures debugging based on the basis of a method of checkpoints;

- checking the availability in datafiles of the specified procedures and program modules;
- checking the datafiles for presence of incorrect program modules;
- checking the type of a module table;
- dynamic call of the variables exported by a program module;
- checking the availability in a current *Maple* session of the specified procedures, modules or variables;
- definition of formal arguments of a procedure, a program module or package;
- checking onto compatibility of files of type {"**.m**", "**.lib**"} with the current release of *Maple*;
- merging of datafiles created by the standard *Maple* function ***'save'***;
- converting of a program module into equivalent procedure;
- converting of program modules of the *second* type into equivalent modules of the *first* type;
- extraction of procedures and program modules from text datafiles;
- converting of files of the input *Maple* language format into the internal *Maple* format, and vice versa;
- converting of any *Maple*-object into the program module, etc.

3. Software for operation with numeric expressions **(10)**:
   - converting of the floating-point numbers into the stylized standard format;
   - an useful special converting of integers into the list, and vice versa;
   - converting of the numbers presented in a symbolical or string format, into numerical values, etc.
4. Software for operation with strings and symbolic expressions **(57)**:
   - extraction of *Maple* expressions from their symbolical or string representations;
   - a special converting of strings or symbolical expressions containing *blanks*, for an opportunity of their subsequent correct use by the standard *Maple* functions {***'system'***, ***'ssystem'***}.
   - a set of tools to remove strings and symbolic expressions of their components;
   - a set of tools for an analysis of contents of strings and symbolic expressions;
   - the expanded set of tools for converting of strings and symbolical expressions into lists, vectors and numerical values, and vice versa;
   - the expanded search for the patterns containing *wildcard*-symbols;
   - special insertions of symbols into strings and symbolic expressions;
   - definition of multiple entries into strings, symbolic expressions and lists;
   - special methods of searching for strings and symbolic expressions;
   - reducing to the specified multiplicity of entries of symbols or substrings into a string;
   - extraction from strings and symbolical expressions of substrings limited to the specified delimiters;

- the extended set of tools for substitutions into strings and symbolical expressions;
- the extended set of searching tools to in strings and symbolical expressions;
- the extended set of tools for sorting the elements of strings, symbolic expressions and lists;
- simple tools for coding / decoding of text information, etc.

5. Software for operation with sets, tables and lists **(57)**:
   - support of a special algebra with lists and scalars;
   - support of the set-theory operations with lists;
   - special conversions of lists into sets, and vice versa;
   - tools for expansion of sets and lists, and subsequent elements rearrangements;
   - the extended means for substitutions into lists and sets;
   - a set of special facilities for operation with lists and sets;
   - the extended means of analysis of multiple entries of elements into lists;
   - special conversions of lists into vectors;
   - definition of indices of a table by its entry, and the analysis of multiplicity of its entries;
   - removing elements from tables, and providing the output of a table graph with numerical entries;
   - a representation of special kinds of tables;
   - extension of the standard function ***'map'*** onto lists, tables and sets;
   - tools for sorting ordinary and nested lists;
   - dynamic assignment of expressions to elements of a list or set;
   - a converting of tables into lists, and vice versa;
   - reduction of entries into a list of multiple elements, etc.
6. Software to support data structures of a special type **(20)**:
   - support of a new data structure of *dirax*-type of direct access;
   - an extension of the standard function ***`type`*** by a new *dirax*-type, and conversion of its algorithm of check onto *heap*-type;
   - tools for reorganization of data structures of various types {*stack, queue, heap*};
   - the extended check of a *Maple* object onto type {*dirax, list, table, heap, stack, queue*}, etc.
7. Software to support bit-by-bit processing of symbolic information **(7)**:
   - the above tools are not available in all current releases of the package *Maple*, therefore these means of the *Library* are both rather pertinent, and applicable for a wide range of applications. A number of useful procedures created on their basis are represented.
8. Tools extending graphic possibilities of *Maple* of releases 6 - 9.5 **(16)**:
   - a dynamic coloring of frames of the animated 2D-graphs. For support of the dynamics the method of "*disk transits*", a rather useful in the advanced programming in the *Maple* environment is used;

- an extension of the standard functions *'animate'* and *'animate3d'* onto any quantity of parameters of animation. For maintenance of dynamics the method of ***"disk transits"***, rather useful in conditions of advanced programming in *Maple* is used. The given method has been developed by us and successfully used for computers of 3rd generation (IBM/360, IBM/370, etc.) and appeared especially effective with appearance of the personal computers whose development is characterized (including) by enough quick decreasing of time of access to external memory on hard disks;
- an extension of the standard functions ***'smartplot'*** and ***'smartplot3d'***, etc.

9. Tools that extend and improve the standard tools of *Maple* **(86)**:
    - checking the type of expressions, active in a current *Maple* session;
    - essential extensions of the standard functions **'map'**, **'map2'**, **'op'**, **'read'**, **'save'**, **'seq'**, **'curry'** and **'rcurry'**;
    - an essential extension of the standard function ***'assign'*** with elimination of its incompatibility concerning releases 6 – 9.5 of the package *Maple*;
    - an expansion of the standard operators **'and'**, **'or'** and **'xor'** onto any quantity of operands;
    - converting the expressions sequence into string, and checking an expression onto type *'expressions sequence'*;
    - an extension of the standard function **'type'** onto such types as 'binary', 'nestlist', 'boolproc', 'dir', 'file', 'path', 'byte', 'digit', 'letter', 'ssign', 'mod1', 'package', 'sequent', 'setset', 'realnum', 'upper', 'Upper', 'lower', 'Lower';
    - an expansion of the standard function ***'member'*** for checking if an expression belongs to a set, range, string, symbolical expression, list, module, table, procedure, array and *rtable*-object;
    - an useful extension of the standard function ***'convert'***;
    - checking *Maple* expressions onto *emptiness*;
    - elimination of certain defects of the standard function ***'evalf'***;
    - means of increasing the compatibility of releases 6 – 9.5 of the package *Maple*, etc.

10. Software for working with *Maple* datafiles and documents **(151)**

10.1. General purpose software **(80)**:
- deletion of datafiles having an any set of attributes;
- testing of datafiles, logic input/output channels and input/output devices;
- dynamic increase of quantity of accessible logic input/output channels;
- the extended means of closing of all open datafiles and logic input/output channels of a current *Maple* session;
- definition of components composing a filename or full path to it;
- a standardization of path to a datafile or directory that allows to correctly use them by the standard functions {***'system'***, ***'ssystem'***};
- opening of the same datafile on the different logic input/output channels;

- searching for datafiles with *Maple* documents containing the specified context;
- checking of datafiles and directories onto *emptiness*, copying of one catalogue into another;
- renaming of directories and datafiles with retention of their attributes;
- the extended checking of properties of directories and datafiles;
- checking of datafiles types concerning extensions of their names and logical organization {*TEXT, BINARY*};
- an extension of the standard function ***'type'*** onto a new *rlb*-type of datafiles;
- an extension of the standard function ***'convert'*** providing conversion of an arbitrary datafile into a datafile of type {*rlb, TEXT*};
- definition and redefinition of attributes of datafiles and directories;
- checking datafiles onto a mode of their opening;
- checking of a logic input/output channel for the specified open datafile; opening of a datafile on the specified logic input/output channel;
- essential extensions of the standard functions ***'currentdir'***, ***'fopen'***, ***'open'***, ***'mkdir'*** and ***'rmdir'***;
- creation of a chain of directories of any nesting level or an *empty* closed datafile, full path to which has been specified by a chain of subdirectories of any nesting level;
- creation of an accessible *empty* target datafile, etc.

10.2. Software for operation with *TEXT* datafiles **(32)**:

- an essential extension of the standard functions ***'writedata'*** and ***'readdata'***;
- tools of providing of direct access to datafiles of the *TEXT*-type;
- tools of extended processing of datafiles of the *TEXT*-type providing a number of mass operations with text datafiles, etc.

10.3. Software for operation with *BINARY* datafiles **(12)**:

- an effective processing of a situation «*the end of a datafile*»;
- tools for extended processing of datafiles of the *BINARY*-type providing a number of mass operations with binary datafiles, etc.

10.4. Software for operation with *Maple* files **(12)**:

- a special re-structuring of a text datafile created by the standard function ***'save'***;
- checking the availability of a datafile for reading by the standard function ***'read'***;
- providing of compatibility of a datafile of the internal *Maple* format with the current *Maple* release;
- checking the types of all datafiles of the specified directory;
- removal of all results of calculations from the specified *mws*-file;
- calculation of the attribute determining name length of a *Maple* object in files of the internal *Maple* format;
- definition of presence in *mws*-files of links and their types, etc.

10.5. Some special tools for operation with datafiles **(15)**:

- selection of correct *email*-addresses out of datafiles of the ASCII-format;
- a coding / decoding of datafiles of the ASCII-format;
- selection out of a datafile of words satisfying to the specified conditions;
- providing of support of the mechanism of "*disk transits*";
- a specific statistical analysis of text datafiles and strings, etc.

11. Software for solving problems of mathematical analysis **(15)**:

- extension of the ***shift***-operator onto any quantity of leading variables;
- calculation of values of the tabulated functions;
- calculation of partial derivatives in the specified points;
- full integration of algebraic expressions;
- a polynomial data interpolation;
- a dynamic generation of linear constraints for optimization problems;
- calculation of independent variables of algebraic expressions;
- an analysis of algebraic expressions or functions from one independent variable;
- searching for a minimax of algebraic expressions, etc.

12. Software for solving problems of linear algebra **(24)**

12.1. General purpose software **(10):**

- an interactive creation of numerical rectangular matrixes;
- full factorization of a polynomial from one leading variable;
- an useful special sorting of matrixes of types *Maple* and *NAG*;
- a dynamic generation of cyclic constructions "**for_do**" and "**seq**" of any nesting level;
- an extension of the standard function ***'type'*** onto check of types of indices and entries of tables, etc.

12.2. Software for operation with *rtable*-objects **(14)**:

- the extended converting of *Maple*-objects into *NAG*-objects, and vice versa;
- checking *rtable*-objects, active in a current *Maple* session;
- checking identification numbers of the active *rtable*-objects;
- checking the saving of the active *rtable*-objects in a current *Maple* session;
- checking the availability of *rtable*-objects in *mws*-files with *Maple* documents;
- converting of the *mws*-files containing *rtable*-objects;
- restoration of a work history with *rtable*-objects in the previous *Maple* sessions, etc.

13. Software to support simple statistics problems **(32)**:

- checking of a view of distribution generated by the built-in generator **'rand'** of pseudo-random integers;

- a statistical analysis of text datafiles;
- operations with the weighed statistical data;
- creation of single-factor linear and nonlinear models of regression for the specified resultant and factor variables with calculation of the correlation relation and of correlation coefficient, and with graphic representation of both the initial data, and the model of regress on the same 2D-graph;
- parametric (*Fisher* and *Student*) and nonparametric (*Van der Waerden* and *Mann-Whitney*) criteria of checking of the *null*-hypothesis of equality of populations variances;
- elements of the simple analysis of time and variation series, etc.

14. Software for operation with the user libraries **(46)**:
    - checking of a *Maple* object to be a library similar to the main library of the package;
    - selective call of procedures, modules and variables from the specified libraries;
    - maintenance of the simple user libraries structurally distinct from the main *Maple* library;
    - the extended means of maintenance of the libraries similar to the main *Maple* library;
    - an effective method of a saving of procedures and program modules in libraries;
    - facilities for operation with the damaged libraries similar to the main *Maple* library;
    - the extended gathering of statistics on use of library tools, etc.

Taking into account our long-term experience in operation with the *Maple* package of releases 4 - 9.5 and experience of our colleagues from universities and the academic institutes of Lithuania, Latvia, Belarus, Estonia and Russia, it is should be noted, that many of tools (*or their analogues*) of our *Library* are worth to be included into standard deliveries of subsequent *Maple* releases. At present, they are accessible to the *Maple* users as the offered *Library* supporting releases 6 - 9 and functioning on platforms *Windows* 95 and later. *Library* tools in many cases allow us to facilitate programming of various applied problems in the *Maple* environment of releases 6 - 9.5.

It is possible to state, that a series of our books on the *Maple* problems [30-33,39,43-45,47] that represent the means developed by us and contain useful tips on the further development of the package, has encouraged the development of package modules *FileTools*, *LibraryTools*, *ListTools* and *StringTools*. However, in this respect tools represented by us essentially extend capabilities of the package, exceeding those of the specified package modules in many cases.

The *Library* is designed for a wide audience of experts, teachers, post-graduates and students of natural-science professions who use *Maple* of releases 6 - 9.5 on *Windows* platform in their own professional work. The *Library* contains well-designed software (*a set of procedures and program modules*), which supplements well the already available *Maple* software with the orientation towards the widest circle of the *Maple* users, greatly enhancing its usability and effectiveness. Our experience reveals that the use of *Library* provides more opportunities of *Maple* of releases 6 - 9.5, simplifying the programming of various practical problems in its environment. This *Library* will be of special interest above all to those *who* use *Maple* not only as a highly intellectual calculator but also as environment for programming of different problems in their professional activities. The *Library* has been rewarded by "Smart Award" from Smart DownLoads Network.

Furthermore, the presence in the *Library* delivery of the text datafile “ProcUser.txt” with source codes of the *Library* software and *mws*-files with help-pages composing Help database of the *Library* allows us to adapt it to other underlying platforms different from *Windows*. Furthermore, the source codes, using both the effective and the non-standard technique, can serve as an useful enough practical programming guide on the *Maple* language.

In conclusion of the given preamble we shall rather briefly state, addressing to our numerous readers both hereby, and the future ones, the personal opinion on a comparative rating of packages *Maple* and *Mathematica*. Both one and another package have numerous mistakes (*in many cases absolutely inadmissible ones for systems of a similar sort*) to elimination of which developers both *Waterloo Maple* and *Wolfram Research* pays rather small attention. Of commercial reasons unreasonably often are issued the new releases saving former mistakes and containing in a series of cases various exotic luxuries. However, if *Maple* developers in a form of open dialogue with users in any measure try to solve the given problem, the *Wolfram Research* an enough painfully perceives any criticism (*in overwhelming majority a well-founded*) about own product.

On the other hand, *Wolfram Research* conducts rather aggressive marketing politics completely inadequate to quality of its products. That, first of all, explains its temporary quantitative advantages that quickly enough decrease. Comparing responses of users of *Maple* and *Mathematica*, it is possible to ascertain quite unequivocally, that the second ones at use of the package have essentially more problems. If *Waterloo Maple* can improve the marketing politics, it can change essentially a situation on the market of systems of computer algebra and will play a positive role for the given field of modern computer science in view of the more perspective *Maple* technology. In this respect, certain contribution to affair of popularization of the *Maple* was brought also by a series [8-20,22-33,39,41-45,47,48] of our books and papers in *Russian* and *English* against five books on *Mathematica* [1-3,6,7]. Whereas our above-mentioned *Library* allows to expand first of all basic tools of the package *Maple* of releases since the *sixth* one. Furthermore, very expediently a series of tools out of the *Library* to include in a set of standard basic tools of *Maple*, above all, tools such as for operation with file system of a computer, for support of new data types, for operation with program modules, for increase of compatibility level of the releases of *Maple*, etc. Many of the *Library* tools have shown oneself as effective enough means at problems solving in the various fields, by providing in a series of cases a simplification of programming in *Maple*.

The experience gained by the authors in operating and testing of both packages shows that the *Maple* is essentially a more friendly and open system which uses built-in C-like language simplifying employment of the package by the user improving his skills in programming. Whereas *Mathematica* has a little bit archaic and not so graceful language which in many respects is distinct from popular programming languages. Finally, the *Maple* has the more advanced tools (*for example, for solving differential equations with partial derivatives, for developing algorithms for solving the problems, etc.*), providing a wide spectrum of applications in many fields.

Summing up (*in more details a comparative analysis of both packages can be found in our above-mentioned previous books and papers*), we can recommend to the beginner user of systems of computer algebra, the *Maple* package as the most perspective tool in this area of computer science. This is essentially caused also by creative alliance *Waterloo Maple* with the world famous developer of mathematical software – *NAG Ltd.* Moreover, the given package constantly wins back positions of *Mathematica* and starts to dominate in education what is rather essential with orientation toward future prospect.

At last, here pertinently to make a series of important remarks, keeping force during all subsequent statement. Above all, the library whose tools are described below is intended for *Maple* of releases from the **6** to the **9.5** inclusive on operational platforms *Windows* 95/98/ME/XP/NT/2000/2003. However, in the presence of source codes of all these tools in the body text of the book and in electronic form in the attached library the user can enough easily adapt their wholly or selectively on other operational platform. Along with that, the source codes can serve as an enough well praxis at mastering of programming in the *Maple* environment. Many of source codes represented in the book contain both the effective and the non-standard programming technique that are useful both for the beginners and for the experienced *Maple* users.

Our experience of the package use of releases **4 - 9/9.5** on the different operational platforms *Windows* 95/98/ME/XP/NT/2000/2003 speaks about its rather unstable performance relative to the host operating systems and its incompatibility on a series of its basic functional means relative to the releases. These problems are enough widely discussed in our previous books [29-33,43,44]. In particular, at installing of *Maple* of release **8** on platform *Windows* 98SE one of the main package directory receives name "BIN.W9X", whereas for platform *Windows XP Pro* it will has name "BIN.WIN". A reasonable explanation to that fact to find difficultly enough. At the same time, in particular, infringement of similar type of standardization at the organization of a file structure of the package complicates some questions of programming with its use.

In a series of problems dealing with datafiles processing takes place one essential functional incompatibility of *Maple* relative to host operating systems, namely. On platforms *Windows* 95/98/98SE/ME, the access to an idle disk drive in the *Maple* environment of releases **6 - 9** invokes the appropriate erroneous situation that can be handled by the package tools, for example by means of `**try**` clause of the *Maple* language. This mechanism is rather convenient and effective means for goals of such kind.

However, on platforms *Windows* XP/NT/2000/2003 the above-mentioned access to an idle disk drive can initiate the especial situations with diagnostics "**Mserver.exe** - *No disk: There is no disk in the drive. Please insert a disk into drive* **X** ...", where **X** - logical name of the first idle disk drive on the computer. The request demands one of three answers, namely: ***Cancel, Try Again*** or ***Continue.*** In this case you must answer `***Cancel***` or `***Continue***`. The given circumstance causes rather essential inconveniences in many cases. For example, result of the procedure calls `***currentdir***` with the purpose of redefinition of a current directory (*on condition that disk drives* "**A**" *and* "**D**" *are idle*) cause the following erroneous situations and the above especial situation:

> currentdir("A:\\");

"**Mserver.exe** - No disk: There is no disk in the drive. ..."

Error, (in currentdir) file or directory does not exist

> currentdir("D:\\");

"**Mserver.exe** - No disk: There is no disk in the drive. ..."

Error, (in currentdir) permission denied

Meantime, our experience recommends for elimination of the above especial situation at running of *Maple* on platforms *Windows* XP/NT/2000/2003 to insert in idle drives any volumes (CD and/or diskette). The trick allows you to get rid of appearance of the above especial situation.

At last, at *Maple* loading (*starting with the seventh release*) on a computer connected to Internet and presence of a Firewall, for example "Zone Alarm", *Maple* will demand to give access to Internet. At rejection of access, *Maple* is being closed. In our opinion, this is certain infringement of the user rights, related with ensuring of protection of his information. In order to prevent possible undesirable consequences (*Spyware*, etc.) before the *Maple* loading, we recommend to execute one of the following procedures:

(1) to disconnect Internet and to enable access to it upon inquiry of your Firewall;

(2) to disconnect Internet and Firewall.

It is completely insignificant loss, since performance capabilities of *Maple* (*at least now*) in Internet environment are rather limited. Some other remarks and suggestion on the *Maple* use will be presented during the statement.

In summary, it is necessary to note a rather strange tendency to deterioration of some characteristics (as a whole of rather perspective package) from release to a release; about that testifies as our six-year experience of approbation and use of the package of releases **4 - 9,** and results of similar work of a plenty of the *Maple* users in various applied physical and mathematical appendices. Apparently, it would be necessary to speed up not delivery on the market of new releases of the package (*when between releases elapses about a year*), but its versions eliminating the noticed shortcomings (*passing from the version to the version and from release to a release*) and extending possibilities of the most frequently used means instead of adding new releases by additional means that are often sometimes uninteresting, undeveloped and erroneous ones. In the given direction have been undertaken as our efforts, and efforts of a lot of other *Maple* users. The represented monograph summarises our results in the given direction.

However till now, many of basic remarks and suggestion have not found reflection in last releases of the package. Confirming the told, we shall cite only one simple example of incorrect calculation by the ***sum***-function of the package, namely:

> m:= 'm': sum((-m)^n, n = 0 .. infinity); -> 1/(m + 1)

In this example the ***sum***-function returns result on the basis of the following relation:

lim ((-m)^p*m + 1)/(m + 1) for p -> ∞

by assuming values **m < 1** or **p < 0,** whereas **m**-variable has indeterminate value. And it takes place during all releases **5 - 9.** Along with the marked disadvantages, *Maple* is burdened by a lot of other essential disadvantages which procure to the users many inconveniences and demand the primary attention on the part of developers of a package. On the other hand, the package, since **7**-th release, as the extension, has received the **Sockets** module intended for creation of the *Maple* appendices for operation in networks of the TCP/IP-protocol. However, its means are rather primitive for serious appendices.

All this grows out from not entirely thought-out strategy of developers and the raised haste to please market situation that at creation of complex scientific packages is not pertinent and can lead to absolutely opposite results. Meanwhile, in spite of the told, in our opinion, *Maple* up till now continues to remain one of the leaders among modern systems of computer algebra; however, its further destiny in many respects will depend on the strategy chosen by the developers. Unfortunately, existing tendencies yet no especially give hope - intervals between new releases diminish, whereas many essential shortcomings and mistakes are not eliminated. The quantity of censures grows on the part of the *Maple* users.

Chapter 1.
General purpose software

In the given section, the software providing the most general procedures of operation with package *Maple* of releases 6 - 9.5 is represented. Here and further every procedure is being represented in the following aspects: (1) *procedure's name with its brief characteristic,* (2) *procedure call sequence,* (3) *formal procedure parameters,* (4) *description of the procedure,* (5) *procedure source code,* and (6) *most typical examples of its use.* The source codes allow not only easily to immerse them into the *Maple* environment of releases 6 - 9.5 on many computer platforms, but also to use their as an useful enough illustrative material at a mastering of the advanced programming in the package environment. Many of the software, represented here, use useful enough technique and methods in the practical respect and receptions of programming (including and *non-standard* ones) in the package environment, conditioning to them as the applied, and educational interest. Basic innovations of the software can be briefly characterized in the following way. Above all, the most general procedures of operation with *Maple* are intended for receiving information such as: number of a current *Maple* release, full paths to its basic subdirectories, current version of an underlying system, information about installed releases of *Maple,* etc. Similar tools are absent among the standard *Maple* software, however they appear for us useful enough at more advanced operation with the package *Maple.*

ACP - current code table of MS DOS

Call format of the procedure:

ACP({, nnn})

Formal arguments of the procedure:

nnn - (optional) number of a code table

Description of the procedure:

The procedure call **ACP()** returns the number of a current code table of MS DOS as a 3-digit positive integer. While the procedure call **ACP(nnn)** with optional actual **nnn** argument, where **nnn** - the number of the appropriate code table, provides replacement of a current code table of MS DOS. However in the second case beforehand it is necessary to execute the system program **NLSfunc** which loads information for support of national languages. The information on usage of national languages can be found in the MS DOS user's guide.

```
ACP:= proc()
  op([system(cat("CHCP ",`if`(nargs = 0, "",`if`(type(args[1], posint) and length(args[1]) = 3,
  args[1], ERROR(`invalid argument`))), " > $Art.$Kr")), `if`(filepos(`$Art.$Kr`, infinity) = 0,
  [fremove(`$Art.$Kr`), ERROR(`Code page <%1> not prepared for your system`, args[1])],
  [close(`$Art.$Kr`), cat(``, readbytes(`$Art.$Kr`, TEXT,infinity)), fremove(`$Art.$Kr`)])][2])
end proc
```

Typical examples of the procedure use:

> ACP();

Active code page: ***866***

---

com6_9 – a setting of the current *Maple* release

Call format of the procedure:

com6_9()

Formal arguments of the procedure: **no**

Description of the procedure:

In general, the last five *Maple* releases (6 – 9.5) are incompatible by a series of their functionality. Furthermore, some *Maple* functions of access to datafiles have essential shortcomings. The more vital ones from them can be reflected by the following table.

Release	6	7	8	9/9.5
procedures	fremove	fremove	fremove	fremove
-//-	close, fclose	close, fclose	-	-
-//-	assign	-	-	-

The essence of the above shortcomings can be characterized as follows. Above all, in the sixth release the ***`assign`*** procedure does not allow to assign the *NULL* values, causing erratic situations, whereas in the subsequent releases this shortcoming has been eliminated. This is one of the essential reasons of incompatibility of release 6, on the one hand, and the releases 7 – 9.5, on the other hand. In the releases 6 and 7 an attempt to close a non-existing file (I/O functions ***`close`***, ***`fclose`***) also causes the corresponding erratic situation. At last, in all last five releases an attempt to delete a non-existing file (I/O function ***`fremove`***) also causes the corresponding erratic situation. In a series of problems dealing with access to datafiles two last circumstances complicate implementing of programs containing systems of handling of erratic and especial situations at operations with datafiles.

The procedure call **com6_9()** returns the *NULL* value, i.e. nothing, providing elimination of the above shortcomings along with a higher level of compatibility of releases 6 – 9.5. The fragment represented below very well illustrates the told.

```
Com6_9:= () -> null(alias(fremove = delf),`if`(Release() = 6, alias(assign = assign67, close =
  Close, fclose = Close),`if`(Release() = 7, alias(close = Close, fclose = Close), NULL)))
```

Typical examples of the procedure use:

> assign(IAN=NULL); close("grsu"); fclose("grsu"); fremove("grsu"); # ***Maple 6***

Error, (in **assign**) invalid arguments
Error, (in **close**) file or directory does not exist
Error, (in **fclose**) file or directory does not exist
Error, (in **fremove**) file or directory does not exist

> com6_9(); assign(R=NULL); close("grsu"); fclose("grsu"); fremove("grsu"); R;

> com6_9(); assign(G=NULL, S=NULL), close("F"), fremove("F"), G, S; # ***Maple 8***

> com6_9(); assign(G=NULL, S=NULL), close("F "), fremove("F"), G, S; # ***Maple 9***

CDM – basic directories of the package *Maple*
CCM
KCM

Call format of the procedures:

CDM()
CCM()
KCM()

Formal arguments of the procedures: **no**

Description of the procedures:

The call **CDM()** returns the full path to the main directory of the package as a string.
The call **CCM()** returns the full path to the directory containing the package kernel as a string.
The call **KCM()** returns the name of the directory containing the package kernel as a string.

Along with the information of the built-in function **currentdir()** which returns a current *Maple* subdirectory of an underlying OS, in a whole series of cases the operating information about the root directory of the package and/or about directory with the package kernel can be needed. The above simple procedures solve these problems.

The procedure **CCM** has essential enough meaning, ensuring the correct information on a full path to important BIN-subdirectory of *Maple*. Really, at installation of *Maple 8* on platform *Windows* 98SE this subdirectory will have the name "BIN.W9X", whereas on platform *Windows XP Professional* – "BIN.WIN". The given information allows to use more effectively a file structure of *Maple* at programming various appendices.

```
CDM:= proc ()
local cd, k, h, L;
   assign(L = [libname]);
   for k to nops(L) do
      if Search1(Path(L[k]), "\\lib", 'h') and h = [right] then RETURN(L[k][1..-5]) else next end if
   end do
end proc

KCM:= proc () local a;  assign(a = CCM());  a[Search2(a, {"\\",  "/"})[-1]+1 .. -1] end proc

CCM:= proc()
local a, b, h, f;
   assign(f = "$Art_Kr$", b = "bin.w"), system(cat("DIR /A:D ", blank(CDM()), " > ", f));
   do h := readline(f);
      if h = 0 then fremove(f); ERROR("subdirectory BIN of Maple cannot be identified")
      else a := ewsc(Case(h), [b], [1]);
         if a <> [] then RETURN(fremove(f), cat(CDM(), "\\", a[1])) end if
      end if
   end do
end proc
```

Typical examples of the procedures use:

```
> CDM();
                         "C:\Program Files\Maple 8"
> CCM();
                      "C:\Program Files\Maple 9\bin.win"
```

> KCM();

"bin.win"

Dialog – interactive input of information into a current *Maple* session

Call format of the procedure:

Dialog()

Formal arguments of the procedure: **no**

Description of the procedure:

The call **Dialog()** provides interactive input of the necessary information into a current *Maple* session. The input of each separate expression is terminated by the symbol (;), while the input as a whole is terminated by the symbol `#`. The result is returned as the list of the *Maple* expressions.

```
Dialog:= proc()
local a;
  op([assign('a' = proc () local L, R;  while R <> `#` do R := readstat(`In: `); L := L, R end do;
  [L][2 .. nops([L])-1] end proc), a()])
end proc
```

Typical example of the procedure use:

```
> Art_Kr:= Dialog();
   In: 2003;
   In: 61;
   In: 56;
   In: 36;
   In: 40;
   In: 7;
   In: 15;
   In: `#`;
                    Art_Kr := [2003, 61, 56, 36, 40, 7, 15]
> `+`(op(Art_Kr));
                                 2218
```

Imaple – identification of the installed releases of the package *Maple*

Call format of the procedure:

Imaple()

Formal arguments of the procedure: **no**

Description of the procedure:

In a series of cases for more effective operation with the package, the user can has on a PC four last releases 6, 7, 8 and 9 simultaneously. Such need arises, for example, at programming of software compatible relatively to different releases. In our books [29-33,43,44] the principal questions of actual incompatibility of release 6. on the one hand, and releases 7, 8, on the other hand, of the package *Maple* are discussed enough minutely.

The procedure call **Imaple()** returns the table whose indices are *Maple* release identifiers *Maple6, Maple7, Maple8, Maple9* whereas its entries define sequences of full paths to main *Maple* subdirectory "Bin.wnt", "Bin.W9X" or "Bin.win" for the corresponding release. Naturally, the returned table cannot be empty, whereas its empty entries defines absence of installed corresponding *Maple* releases. In addition, it is necessary to note, if in the same *Windows* environment several versions of 6th release or 6th and {7|8|9}th release can be installed, then the 7th and 8th releases do not admit multiple versions.

```
Imaple:= proc()
local a, d, k, L, j, h, f, t;
  assign(a = [Adrive()], f = "$ArtKr.157");
  L := table([Maple6 = NULL, Maple7 = NULL, Maple8 = NULL, Maple9 = NULL]);
  for j in a do system(cat("Dir /S ",j, ":\\ > ", f));
    do h := readline(f)
      if h = 0 then break else h := Case(h) end if;
      if search(h, "directory of ",'t') then d := h[t .. -1]
      elif search(h, "wmaple.exe") then L[Maple6] := L[Maple6], d[14 .. -1]
      elif search(h, "maplew.exe") then L[Maple7] := L[Maple7], d[14 .. -1]
      elif search(h, "maplew8.exe") then L[Maple8] := L[Maple8], d[14 .. -1]
      elif search(h, "cwmaple9.exe") then L[Maple9] := L[Maple9], d[14 .. -1]
      end if
    end do;
    close(f)
  end do;
  fremove(f), eval(L)
end proc
```

Typical example of the procedure use:

> Imaple();

table([*Maple7* = "c:\program files\maple 7\bin.wnt",
Maple8 = "c:\program files\maple 8\bin.win",
Maple9 = "c:\program files\maple 9\bin.win",
Maple6 = "c:\program files\maple 6\bin.wnt"])

It is necessary to note, the 8th release recognizes itself as the 7th release, whereas the 9th release recognizes itself as the 8th release.

Release – check of the current release of the *Maple* package
Release1

Call format of the procedures:

Release({, **'h'**})
Release1({, **'h'**})

Formal arguments of the procedures:

h - (optional) an unevaluated name

Description of the procedures:

The procedure call **Release()** returns the integer **6, 7, 8** or **9** for four last releases of *Maple*; otherwise the value `*Other release*` is returned. If optional unevaluated actual **h**-argument (*name*) is used, via this name the full path to the main directory of the *Maple* package is returned as a *string*.

In contrast to the first procedure, the procedure call **Release1()** returns the integer **4, 5, 6, 7, 8** or **9** for six last releases of the package; otherwise the value *`Other release`* is returned. If optional unevaluated actual **h**-argument (*name*) is used, via this name the full path to the main directory of the package is returned as a *string*.

First of all, the given procedures represent the doubtless interest in connection with absence of full compatibility (both *"from above – down"* and *"from below – upwards"*) of release **6**, on the one hand, and releases **7, 8** and **9**, on the other hand, of the package *Maple;* in a lot of cases the identification of an used current release of *Maple* is required that the above procedures provide.

```
Release:= proc()
local k, cd, L, R;
  assign(cd = currentdir(), L = [libname]);
  for k to nops(L) do
    try currentdir(cat(L[k][1 .. searchtext("/lib", L[k])], "license"))
    catch "file or directory does not exist": NULL
    end try
  end do
  assign(R = readbytes("license.dat", TEXT, infinity)), close("license.dat"),
    `if`(nargs = 1, assign([args][1] = currentdir()[1 .. -9]), NULL), currentdir(cd);
  `if`(search(R, "Maple7With"), 7, `if`(search(R, "MapleWith"), 6,
    `if`(search(R, "Maple8With"), 8, `if`(search(R, "Maple9With"), 9, `Other release`))))
end proc

Release1:= proc()
local k, L;
  proc () local F; F := 60; save F, `$Kr_Art.m` end proc(), `if`(nargs <> 1, 1,
    [assign(L = [libname]), seq(`if`(L[k][-4 .. -1] = "/lib", assign(args[1] = L[k][1 .. -5]), 1),
  k = 1 .. nops(L))]); assign('F' = readbytes(`$Kr_Art.m`,TEXT,infinity)), fremove(`$Kr_Art.m`);
  `if`(F[4] = "5", 5, `if`(F[4] = "4", 4, `if`(member(SN(F[2]), {6, 7}), SN(F[2]), Release())))
end proc
```

Typical examples of the procedures use:

```
> Release();                                                          # release 6
                                      6
> Release('h'),  h;
                     6, "C:\MAPLE 5 6 7\MAPLE_EDU\MAPLE 6 EDU"
> Release1();
                                      6
> Release1('h'),  h;
                     6, "C:\MAPLE 5 6 7\MAPLE_EDU\MAPLE 6 EDU"
> Release()                                                           # release 7
                                      7
> Release('h'),  h;
                     7, "C:\MAPLE 5 6 7\MAPLE_EDU\MAPLE 7 EDU"
> Release1();
                                      7
> Release1('h'),  h;
                     7, "C:\MAPLE 5 6 7\MAPLE_EDU\MAPLE 7 EDU"
> Release();                                                          # release 8
                                      8
```

```
> Release('h'), h;
                    8, "C:\PROGRAM FILES\MAPLE 8"
> Release1();
                                7
> Release1('h'), h;
                    7, "C:\PROGRAM FILES\MAPLE 8"
> Release();                                            # release 9
                                9
> Release('h'), h;
                    9, "C:\Program Files\Maple 9"
> Release1();
                                9
> Release1('h'), h;
                    9, "C:\Program Files\Maple 9"
```

Of the examples, represented above, follows, the results of the calls of procedures **Release()** and **Release1()** for releases **6** and **7** of *Maple* relatively to the returned release number are identical. Whereas for a case of the 8th release the calls of procedures **Release()** and **Release1()** return the value 8 and 7 accordingly. For the last 9th release both calls **Release()** and **Release1()** return the same value 9. The reason of the given fact is the following important circumstance. As it was already noted above, and it has been considered in detail in our books [29-33,43,44], releases 6 and 7 of the package are incompatible *"from above - down"* and *"from below - upwards"* relative to a lot of the important program parameters. In particular, releases 6 and 7 are different relative to internal format. Appreciably the problem of compatibility of these releases is solved by the software represented in this book and other our books [29-33,43,44].

Meanwhile, the last 9th release of the package relative to the program environment inherits all basic characteristics of the previous 8th release, having other numbering (though and in this case the question of full compatibility of releases 7, 8 and 9 should be the subject of separate consideration). Therefore, if the **Release** procedure returns the official number of a current release, then the **Release1** procedure returns the release number responding a current program environment of the package *Maple*.

Already from the previously mentioned, quite pertinently to make an essential conclusion, the developers of the package first should pay attention to elimination of many essential mistakes and defects of basic means of the previous releases, instead of increase of the numbering of its releases in commercial interests. The part of these defects the *above user library* (**AUL**) eliminates; the **AUL** contains means represented in the present book, and is logically linked to the main *Maple* library with the higher priority of the **AUL** at search of software.

RegMW – registration of *Maple* in the *Windows* environment

Call format of the procedure:

RegMW()

Formal arguments of the procedure: **no**

Description of the procedure:

In a number of cases for more effective operation with the *Maple* package it is recommended to provide recording of a full path to its kernel in the file "Win.ini" of the operating shell *Windows*, which is automatically fulfilled at installation of last versions of the package of the 6th release, whereas releases 7, 8 and 9 automatically do not provide such recording. Such recording consists in a putting into the above file of the appropriate record about a full path to the file with the package kernel. The above-mentioned registration for releases 6 – 9 of the *Maple* package is fulfilled by means of the procedure **RegMW().**

The call **RegMW()** provides the registration of *Maple* in the *Windows* environment by means of putting of appropriate record into the above file "Win.ini" with return of the corresponding message. At detection of non-actual records of the release, they are deleted from the "Win.ini" file.

```
RegMW:= proc()
local a, b, wd, h, p, r, w, f, t, z, q, wrn;
  `if`(member(winver()[9 .. -1], {"98", "95", "2000"}), NULL, RETURN(NULL));
  assign(wrn = proc(r) options operator, arrow; WARNING("the release %1 has been
    registered in file <win.ini>", r) end proc);
  assign(wd = SSF(C, "win.ini")[1], r = Release('h'), f = "61", t = cat("\\", KCM(), "\\""));
  `if`(r = 6, assign(p = cat("wmaple.exe=""",h,t)), `if`(r = 7, assign(p = cat("maplew.exe=""",h,t)),
  `if`(r = 8, assign(p = cat("maplew8.exe=""",h,t)), `if`(r = 9, assign(p = cat("maple9.exe=""",h,t)),
  RETURN(`Release is neither 6 nor 7 nor 8 nor 9`))))), assign(w=readbytes(cat(wd,"\\win.ini"),
    TEXT, infinity)), close(cat(wd, "\\win.ini"));
  `if`(r = 6 and search(w, p) or r = 7 and search(w, p) or r = 8 and search(w, p) or r = 9 and
    search(w, p), RETURN(cat("the release ", r, " has been registered in file <win.ini>")), f);
    assign(a = cat(wd, "\\win.Art"), b = cat(wd,"\\win.Kr")), assign(z = fopen(a,WRITE,TEXT));
  do f := readline(cat(wd, "\\win.ini"));
    if f = 0 then close(a, cat(wd, "\\win.ini")); break else writeline(z, f) end if;
   `if`(f = "[programs]", [writeline(z, p), assign(q = 56)], NULL)
  end do;
    if q = 56 then fopen(b, WRITE, TEXT);
      do f := readline(a);
        if f = 0 then close(a,b); break elif search(f,p[1 .. 10]) and f <> p then next else
          writeline(b,f); next end if
      end do
    else assign(('h') = cat(wd, "\\win.ini")), assign(('z') = fopen(h, APPEND, TEXT));
      writeline(z, "[programs]"), writeline(z, p), close(z), RETURN(wrn(r))
    end if;
  fremove(a, cat(wd, "\\win.ini")), system(cat("ren ", b, " win.ini")), RETURN(wrn(r))
end proc
```

Typical example of the procedure use:

> RegMW();

Warning, the release 9 has been registered in file <win.ini>

winver – the current version of an underlying system

Call format of the procedure:

winver()

Formal arguments of the procedure: **no**

Description of the procedure:

The procedure call **winver()** returns the current version of an underlying system as a *string*. In a series of cases, this information can be useful enough.

```
winver:= proc()
local h;
  system(cat("Ver > wv.$$$")), assign(h = readbytes(`wv.$$$`, TEXT, infinity));
  fremove(`wv.$$$`), h[2 .. searchtext("[", h) - 2]
end proc
```

Typical example of the procedure use:

> winver();

"Microsoft Windows XP"

---

intaddr – the current IP addresses for network connections

Call format of the procedure:

intaddr()

Formal arguments of the procedure: **no**

Description of the procedure:

In a series of cases, it is necessary to know how the network TCP/IP-protocol translates IP-addresses into physical network addresses. The procedure call **intaddr()** returns the four-element list whose the first element defines IP of local host, the second – IP of default gateway, the third – physical address and the fourth – the connection type. This procedure can be rather useful in conjunction with the **Sockets** module that supports a series of tools for network communication in *Maple*. The procedure handles basic especial situations.

```
intaddr:= proc()
local a, b, c, d, h;
  assign67(a = [WD(), WS(), "_$Art_Kr$_"], h = NULL);
  if system(cat(a[1], "\\arp.exe  -a  >  ", a[3])) = 0 then b := fopen(a[3], READ)
  elif system(cat(a[2], "\\arp.exe  -a  >  ", a[3])) = 0 then b := fopen(a[3], READ)
  else delf(a[3]), ERROR("Internet address cannot be defined. Please, use other means")
  end if;
  while not Fend(b) do c := readline(b); d := ewsc(c, ["."], [3]);
  if d <> [] then h := h, d end if  end do;
  delf(a[3]), `if`(h = NULL, ERROR("Network connection is disabled"), 8);
  [op(h[1]), op(SLD(Red_n(FNS(c, " ", 3), " ", 2), " "))]
end proc
```

Typical example of the procedure use:

> intaddr();

["82.147.169.173", "82.147.169.1", "00-30-94-36-95-54", "dynamic"]

---

varsort – variables exchanging by their values

Call format of the procedure:

varsort({*args*} {, *sf*})

Formal arguments of the procedure:

args – (optional) a sequence of unevaluated variables, for example, 'x', 'y', 'z', ...
sf – (optional) an ordering function

Description of the procedure:

In a series of cases, it is necessary to execute the variables exchanging by their values. The simplest example - exchange of values for variables in order their values have been at ascending order. For instance, variables **x, y** and **z** having values 62, 57 and 37 should receive values 37, 57 and 62 accordingly. The following procedure **varsort** solves higher problem even.

The procedure call **varsort('x', 'y', 'z', ...)** returns the *NULL* value, i.e. nothing, providing the exchange of variables **x, y, z, ...** by their values at ascending order. In addition, the exchange can hold only for groups of adjacent variables (*in a sequence of actual arguments*) with values of types {*realnum, symbol*} accordingly; undefined variables remain without change and ordering of values of *symbol*-type holds at lexicographically increasing order.

If optional *sf*-argument is given, it is used to define the ordering for a sorting of *realnum*-values; *sf*-argument must define a Boolean-valued function of two arguments. Specifically, **sf(x, y)** should return the *true* value if **x** precedes **y** in the desired ordering. Otherwise, it should return the *false* value. By varying order of actual arguments at the procedure call, one can receive higher features of the above-described variables exchanging by their values. The procedure **varsort** handles basic especial and erroneous situations.

```
varsort:= proc()
local a, b, c, d, k;
  `if`(nargs = 0, RETURN(), assign(c = []));
  d:= proc(x, y)
       try
          if evalf(x) < evalf(y) then true else false  end if
       catch: `if`(map(type, {x, y}, symbol) = {true}, lexorder(x, y), true)
       end try
    end proc;
  a:= [seq(`if`(type(args[k], boolproc), assign('b' = args[k]), `if`(type(args[k], symbol),
     op(args[k], assign('c' = [op(c), k])), NULL)), k = 1 .. nargs)];
  a:= sort(a, `if`(type(b, boolproc), b, d));
  for k to nops(a) do assign(args[c[k]] = a[k]) end do
end proc
```

Typical examples of the procedure use:

```
> varsort(57, 8, 15, 37, 41, 62),  varsort();
> m:=vsv:  n:=avz:  p:=agn:  varsort('m', 'n', 'p'),  m, n, p;
```

$$agn, avz, vsv$$

```
> x:=57:  y:=37:  varsort('x', 'y'),  x, y;
```

$$37, 57$$

```
> a:=8:  b:=15:  c:=57:  varsort('c', 'a', 'b'),  a, b, c;
```

$$15, 57, 8$$

```
> varsort('a', 'c', 'b'),  a, b, c;
```

$$8, 57, 15$$

```
> x:=2004:  z:=60:  h:=300:  y:=1995:  varsort('x', 'z', 'h', 'y'),  x, z, h, y;
```

$$60, 300, 1995, 2004$$

```
> x:=57:  y:=sqrt(3):  varsort('x', 'y'),  x, y;
                                 3^(1/2), 57
> x:=62: y:=57: z:=37: m:=vsv: n:=avz: p:=agn: varsort('x','y','z','m','n','p'), x,y,z,m,n,p;
                          37, 57, 62, agn, avz, vsv
> x:=62: y:=57: z:=37: m:=vsv: n:=avz: p:=agn: varsort('x','z','y','m','p','n'), x,y,z,m,n,p;
                          37, 62, 57, agn, vsv, avz
> sf:= (a, b) -> `if`(map(type, {a, b}, realnum)={true}, `if`(type(a, even), true, false), false):
> x:=57:  y:=8:  z:=sin(15):  h:=1995:  t:=2004:  v:=37/42:  d:= 62:
   varsort('x', 'y', 'z', 'h', 't', 'v', 'd', sf),  x, y, z, h, t, v, d;
                     sin(15), 37/42, 8, 57, 62, 1995, 2004
> sf1:= (a, b) -> `if`(map(type, {a, b}, realnum)={true}, `if`(type(a, odd), true, false), false):
> x:=57:  y:=8:  z:=tru:  h:=1995:  t:=grsu:  v:=37*a:  d:=62:
   varsort('x', 'y', 'z', 'h', 't', 'v', 'd', sf1),  x, y, z, h, t, v, d;
                      grsu, 1995, 62, 37*a, tru, 57, 8
> restart;  x:=art:  y:=sin(t+8):  z:=(a+b)*ln(h):  H:=() -> `*`(args) + `+`(args):
  f2:= (x, y) -> lexorder(op(map(convert, [x,y], string))):  varsort('x', 'z', 'y', 'H', f2),
  x, y, z, H, eval(y);
          (a + b) ln(h), y, art, sin(t + 8), () -> `*`(args) + `+`(args)
> v:=62: g:=57: s:=37:  ar:=42:  art:=15:  kr:=8:  varsort('v','g','s','ar','art','kr',`<`),
  v, g, s, ar, art, kr;
                           8, 15, 37, 42, 57, 62
> restart;  f3:= (x, y) -> `if`(x="" or x=NULL or y="" or y=NULL, false,
  `if`(op(convert(x[1], bytes)) > op(convert(y[1], bytes)), true, false)):
> x:="avz":  y:="vgs":  z:="Kr":  t:="Art":  s:="":  g:=NULL:  h:="rans":  v:="ian":
   varsort('x', 'y', 'z', 't', 's', 'g', 'h', 'v', f3), x, y, z, t, s, g, g, h, v;
          "vgs", "rans", "ian", "", "avz", "Kr", "Kr", "Art", "ian"
```

The **varsort** procedure is rather broadly used in numerous problems dealing with various variables processing. First, the given procedure is useful for organization of a various kind of exchanges by values between variables.

Whereas other procedures are intended for receiving information such as: number of a current *Maple* release, full paths to its basic subdirectories, current version of an underlying system, information about installed releases of *Maple*, etc. Similar tools are absent among the standard *Maple* software, however they appear for us useful enough at more advanced operation with the package *Maple*.

Chapter 2.
Software of operation with procedural and modular objects of *Maple*

In the given section, a group of tools extending the possibilities of the package *Maple* of releases 6 - 9.5 at operation with procedures and program modules is represented. These tools support kinds of processing such as: converting of modules into procedures; testing of presence in files of incorrect modules; check of parameters of procedures and modules; check of activity (availability for direct use) of a procedure or a module; check of type of a modular table; converting of files of the input *Maple* format containing modules; converting of a module of the second type into the first; converting of a file of the input *Maple* format into a file of the internal *Maple* format, and vice versa, etc. The represented tools provide many of the manifold useful operations with procedural and modular objects of *Maple.* These tools are used enough widely at advanced programming of various problems in the *Maple* environment.

---

stpm - output of source code of a module or package and its exports

Call format of the procedure:

stpm(M, d {, C} {, *all*} {, *prn*})

Formal arguments of the procedure:

M - symbol defining name of a module or a package
D - symbol or string defining a directory for target file
C - (optional) a directory with the user library
all - (optional) key word defining a saving mode of module exports
prn - (optional) key word defining a print mode of the source code

Description of the procedure:

A module is a collection of exported procedures and other data accessible to the user of the module, and private (local) data that is not accessible to the user. However, in a series of cases the necessity of access to the source code for a module or a package arises. This accessibility is essential for enhancing existing routines to meet your specific needs or for mastering of programming in the *Maple* environment. The following procedure **stpm** solves the problem to a certain extent.

The procedure call **stpm(M, d)** saves the source code of a module or a package defined by the first actual argument **M** in ASCII file "M.stm" in a directory defined by the second actual argument **d.** In addition, the directory **d** can have any level of nesting and be absent. The procedure saves the principal schema of a module or a package. In addition, the module definition should be evaluated in a current *Maple* session or be in a library, logically linked with the main *Maple* library. Otherwise, the third optional argument **C** allows to define the path to a directory with the user library containing the sought module or package.

The optional key words *`prn`* and *`all`* are coded after the above-mentioned actual arguments and define a saving mode and a print mode of a module source code. The key word *`prn`* provides output of the module source code on the terminal, whereas the key word *`all`* provides a saving of source codes of all module exports in ASCII file "M.all" in a directory defined by the second actual argument **d.** The successful procedure call outputs the appropriate warnings and, possibly, a source module code. The above-mentioned ASCII files can be useful enough for an analysis of a module or a package. The examples represented below illustrate the results of the procedure calls.

```
stpm:= proc(M::symbol, d::{string, symbol})
local f, p, k, h, r, t, w, v, u;
global libname;
  `if`(type(M, {package, `module`}) or 2 < nargs and type(args[3], dir), assign67(w = libname),
     ERROR("<%1> is not a module", M));
  assign(f = cat("\", M, ".stm")), `if`(type(d, dir), assign(p = cat("", d)), assign67(p = MkDir(d)));
  if 2 < nargs and type(args[3], dir) then libname := libname, args[3] end if;
  try u := with(M)
  catch "invalid input: %1 expects": assign67('libname'=w); ERROR("<%1> is not a module", M)
  end try;
  h := (proc() local a; a := `if`(p = NULL, cat("", d, f), cat(p, f)); save M, a; a end proc)();
  WARNING("source code of module <%1> has been saved in file <%2>", M, h),
   assign67('libname' = w);
  if not member(prn, {args}) then NULL
  else print(Sub_st(["\n" = ""], readbytes(h, TEXT, infinity), r)[1]), close(h)
  end if;
  if not member(all, {args}) then NULL
  else
     assign('h' = cat(h[1 .. Search(h, ".")[-1]], all), v = interface(verboseproc));
     interface(verboseproc=3), writeto(h), print(cat("Source code of the module or package: ",M));
     for k to nops(u) do
        print(cat("Source code of the exported variable <", u[k], ">:"));
        writeline("_$$$_", cat("lprint(eval(`", u[k], "`));print();")), close("_$$$_");
        read "_$$$_"
     end do;
     interface(verboseproc = v), writeto(terminal), fremove("_$$$_");
     WARNING("source codes of module exports <%1> have been saved in file <%2>", M, h)
  end if
end proc
```

Typical examples of the procedure use:

> stpm(ACC, "C:/Temp/Modules\\Academy");

Warning, source code of module <ACC> has been saved in file
<C:/Temp/Modules\Academy\ACC.stm>

> stpm(ACC, "C:/Temp/Modules\\Academy", all, prn);

Warning, source code of module <ACC> has been saved in file
<C:/Temp/Modules\Academy\ACC.stm>

"ACC := module () local p, h, n, S, t; export acc, Sr, Ds; description `Evaluation of autocorrelation coefficient (ACC), a mean (Sr) and dispersion (Ds)`; acc := proc (S, t) options operator, arrow; evalf(sum((S[n+1]-ACC:-Sr(S,2,t))*(S[n]-ACC:-Sr(S,2,t)),n = 1..nops(S)-

1)/sqrt(ACC:-Ds(S,1,t)* ACC:-Ds(S,2,t))/(nops(S)-1),t) end proc; Sr:=proc (S,p,t) options operator, arrow; evalf(sum(S[k], k = p .. `if`(p = 1,nops(S)-1,nops(S)))/(nops(S)-1),t) end proc; Ds := proc (S, h, t) options operator, arrow; evalf(sum(S[k]^2,k = h.. `if`(h =1,nops(S)-1, nops(S)))/(nops(S)-1)-ACC:-Sr(S,h,t)^2,t) end proc; end module;"

Warning, source codes of module exports <ACC> have been saved in file <C:/Temp/Modules\Academy\ACC.all>

> stpm(SimpleStat1,"c:/temp/modules\\academy","c:/program files/maple 9/lib/Galina",all);

Warning, source code of module <SimpleStat1> has been saved in file <C:/Temp/Modules\Academy\SimpleStat1.stm>

Warning, source codes of module exports <SimpleStat1> have been saved in file <C:/Temp/Modules\Academy\SimpleStat1.all>

> stpm(RANS_IAN, "C:/Temp/Modules", "C:/program files/maple 9/lib/Galina", all, prn);

Error, (in **stpm**) <RANS_IAN> is not a module

> stpm(LinearAlgebra, "C:/Temp/Modulcs\\Academy", all);

Warning, source code of module <LinearAlgebra> has been saved in file <C:/Temp/Modules\Academy\LinearAlgebra.stm>

Warning, source codes of module exports <LinearAlgebra> have been saved in file <C:/Temp/Modules\Academy\LinearAlgebra.all>

PP – quick print of source procedures codes

Call format of the procedure:

PP(P)

Formal arguments of the procedure:

P - a symbol or a name

Description of the procedure:

The simple procedure **PP(P)** prints a procedure code defined by the actual **P** argument. Result is printed in format "Text Output". The procedure is convenient enough for quick review of both *Maple* procedures and user procedures, and for dragging of a source procedure code into *Maple Input* paragraph. Any valid *Maple* name formed without using *backquotes* (`) is precisely the same as the name formed by surrounding the name with *backquotes*. Therefore, ***Agn*** and ***`Agn`*** both refer to the same name ***Agn***. On a number of reasons, the procedure **PP** prints names of the sought procedures with *backquotes*.

```
PP:=proc(P::{name, symbol})
local a, h, p;
  `if`(type(P, procedure), assign(h = fopen("$agn56$", APPEND, TEXT)),
  ERROR("%1 is not a procedure",P)); writebytes(h, cat("`",cat("", P), "`")), writebytes(h, ":="),
  writebytes(h, convert(eval(P), string)), writebytes(h, ";"); close(h), assign(p = interface(echo)),
  interface(echo = 3), `if`(type(P,protected),[unprotect(P), assign(a = 7)],1);
    read "$agn56$";
    `if`(a = 7, protect(P), NULL), interface(echo = p), fremove("$agn56$")
end proc
```

Typical examples of the procedure use:

> PP(PP);

> \`PP\`:=proc(P::{symbol, name}) local a, h, p; \`if\`(type(P,procedure),assign(h = fopen("$agn56$", APPEND,TEXT)),ERROR("%1 is not a procedure",P)); writebytes(h,cat("",P)), writebytes(h,":="), writebytes(h,convert(eval(P),string)), writebytes(h,";"), close(h); assign(p = interface(echo)), interface(echo = 3), \`if\`(type(P,protected),[unprotect(P), assign(a = 7)],NULL); read "$agn56$"; \`if\`(a = 7,protect(P),NULL), interface(echo = p), fremove("$agn56$") end proc;

> PP(\`type/dirax\`);

> \`type/dirax\`:=proc (L::anything) options operator, arrow; type(L,list) and evalb(L[1] = dirax) and evalb(L[2] = nops(L[3])) end proc;

Apmv – a testing of program *Maple* objects (procedure, module, and variable)

Call format of the procedure:

Apmv(P {, 'h'})

Formal arguments of the procedure:

P - a symbol or a name
h - (optional) an unevaluated name

Description of the procedure:

The simple procedure **Apmv(P {,'h'})** returns the *true* value, if in a current *Maple* session a **P** name is assigned, i.e. its value is different from the name itself. At coding of the second optional **h** argument via it the type of an object defined by the **P** name, is returned, namely: *procedure*, *module* or *variable*.

In addition, it is necessary to have in mind, the procedure call **Apmv(P)** always returns the *true* value for a procedure or program module **P** which are located in libraries logically linked with the main library of the package, automatically adding their names to a sequence of names defined by the call **anames()**. Testing of a variable by means of the **Apmv** procedure demands to code it in unevaluated form, for example **'V'**. With the purpose of an unification, in such form can be coded any actual **P** argument of the procedure.

```
Apmv:= (P::{symbol}) -> `if`(assigned(P), op([true, `if`(nargs =2, assign(args[2] = `if`(type(eval(P),
   procedure), procedure, `if`(type(eval(P), `module`), `module`, variable))), NULL)]), false)
```

Typical examples of the procedure use:

> map(Apmv, ['Apmv', sin, 'ParProc1', Read1]);

[true, true, true, true]

> seq([Apmv(k, 'h'), h], k=[DIRAX, SimpleStat, Read1, Save]);

[true, module], [true, module], [true, procedure], [true, procedure]

> Vgpti:=1985: Apmv('Vgpti', 't'), t;

true, variable

> Apmv('Apmv', 'h'), h;

true, procedure

> AVZ:= 61*x+42/(y+z)-(a+b)/(c-d): Apmv('AVZ', 't'), t;

true, variable

> Apmv(RANS, 'V'), V;

false, V

ChkPnt – a debug method for *Maple* procedures

Call format of the procedure:

ChkPnt(args)

Formal arguments of the procedure:

args - actual arguments

Description of the procedure:

At present, the problem of debugging of software is elaborated enough. Of simple facilities of debugging, we can note methods of breakpoints, tracing etc. The *Maple* package offers a calculation trace method supported by functions ***trace*** and ***debug*** which represent actually the same function of debugging. The built-in ***DEBUG*** function has the effect of a breakpoint in the *Maple* programs. The function is oriented onto debugging of rather complex procedures in interactive mode and is based on the mechanism of breakpoints, admitting usage more than 16 commands controlling the debug process. At last, the utility ***Mint*** analyzes a *Maple* program, and produces a report about possible errors in the source *Maple* file. This utility is intended for independent testing of syntax of the source texts of programs located in datafiles.

As a simple tool of debugging of the *Maple* procedures, a modified method of breakpoints can be used. This method consists in the following. For generation of procedures providing the return of values of the required expressions in the fixed breakpoints the **ChkPnt** procedure, coded at the beginning of body of the tested **Aproc** procedure as **ChkPnt(*args*)**, serves. Next in the necessary places of the tested **Aproc** procedure, the calls of the following format are coded:

chkpntN(_x1, _x2, _x3, ..., _xn, x1, x2, x3, ..., xn)(); **(1)**
chkpntN(_x1, _x2, _x3, ..., _xn, x1, x2, x3, ..., xn)(x1, x2, x3, ..., xn); **(2)**

where: **N** is the number of a breakpoint and **xj (j = 1 .. n)** are names of expressions whose values it is necessary to receive in the given point of the procedure.

The call of the tested procedure **Aproc(*args*)** on a tuple of its basic actual arguments does not include the mechanism of breakpoints, whereas the call **Aproc(*args*, chkpnt=N)** for **N** from range [1 .. 61] provides output of values of the required expressions in a breakpoint with number **N** (*format* **1**) or the output of values with their return (*format* **2**), as it visually illustrates a clear enough fragment represented below.

The represented mechanism of checkpoints allows easily to set breakpoints and to receive in them values of necessary expressions, and also to return these values, what provides simple mechanism of a programmed exit out of any point of the procedure with return of values of the required expressions. The represented mechanism is rather effective in view of structuredness of the *Maple* procedures that do not use the ***goto***-mechanism. The examples represented below enough lucidly illustrate the previously mentioned.

```
ChkPnt:= proc()
local a, k;
  unassign(seq(cat(chkpnt, k), k = 1 .. 61)); assign(a = {seq(`if`(type(args[k], equation),
  args[k], NULL), k = 1 .. nargs)}); seq(`if`(lhs(a[k]) = chkpnt and type(rhs(a[k]), posint),
  RETURN(assign(cat(chkpnt, rhs(a[k]))) = (() -> [WARNING(" in %1 variables %2 have
    the following values %3 ", procname,[seq(args[k], k = 1 .. 1/2*nargs)], [seq(eval(args[k]),
    k = 1/2*nargs+1 .. nargs)]), RETURN(RETURN)])), eval(cat(chkpnt, eval(rhs(a[k]))))),
    NULL), k = 1 .. nops(a))
end proc
```

Typical examples of the procedure use:

```
Aproc:= proc(X::numeric, Y::numeric, Z::numeric)
local a, b, c;
    ChkPnt(args);
    a := evalf(sqrt(X^2+Y^2+Z^2),4);
    chkpnt1(_X,_Y,_Z,_a,X,Y,Z,a)();
    b := evalf(sqrt(X^3+Y^3+Z^3),4);
    chkpnt2(_a,_b,a,b)(a,b);
    c := evalf(sin(a)+cos(b), 4);
    chkpnt3(_a,_b,_c,a,b,c)();
    WARNING("Result: a=%1, b=%2, c=%3", a, b, c)
end proc
> Aproc(61.42, 56.47, 36.67);
    Warning, Result: a=91.14, b=679.0, c=0.8809
> Aproc(61.42, 56.47, 36.67, chkpnt=2);
    Warning, in chkpnt2 variables [_a, _b] have the following values [91.14, 679.0]  91.14, 679.0
> Aproc(61.42, 56.47, 36.67, chkpnt=1);
    Warning, in chkpnt1 variables [_X, _Y, _Z, _a] have the following values [61.42, 56.47, 36.67,
    91.14]
> Aproc(61.42, 56.47, 36.67, chkpnt=3);
    Warning, in chkpnt3 variables [_a, _b, _c] have the following values [91.14, 679.0, 0.8809]
```

This fragment is clear enough and of any special explanations does not demand. It is recommended to pay attention to a reception used for organization of exit out of tested procedure at once after the activated control point and return (*if it is necessary*) of results of calculations. The represented mechanism allows to easily establish control points and to receive in them values of necessary expressions, and also to return these values what provides the simple mechanism of the programmed exit out of any point of procedure with return of values of required expressions. The given mechanism is effective enough because of structuredness of the *Maple* procedures that do not use the ***goto***-mechanism.

---

filePM – testing of presence in datafiles of procedures and program modules

Call format of the procedure:

filePM(F {, h})

Formal arguments of the procedure:

F - symbol or string defining a filename or full path to a *Maple* datafile
h - (optional) a valid *Maple* expression defining an availability mode of procedures and modules

Description of the procedure:

The procedure call **filePM(F)** returns the 2-element sequence of lists containing names accordingly of procedures and program modules saved in a file specified by the actual **F** argument. If the procedure call uses the second optional actual argument **h,** then procedures and modules saved in a file **F** additionally become available in a current session. The procedure supposes that **F** file was created by the standard function ***`save`*** and has the input *Maple* language format or the internal *Maple* format. The procedure **filePM** is used by other procedures and is useful in a series of applications.

```
filePM:= proc(F::file)
local a, b, c, d, h, n, k, t, p, m, m1, omega, nu, r, psi;
  psi := map(anames, {procedure, `module`});
  assign67(a = {}, p = " := proc (", m = " := module ", m1 = "unprotect('", close(F));
  omega:= proc(nu)
    local a, b, k;
      assign(a = [], b = []), seq(`if`(type(nu[k], procedure), assign('a' = [op(a), nu[k]]),
      `if`(type(nu[k], `module`), assign('b' = [op(b), nu[k]]), NULL)), k = 1 .. nops(nu)), a, b
    end proc;
  if Ftype(F) <> ".m" then
    assign('nu' = interface(warnlevel)), interface(warnlevel = 0);
    try parse(readbytes(F, infinity, TEXT), statement); close(F); read F; n := fopen(F, READ)
    catch "incorrect syntax in parse": ERROR("<%1> is not readable file", F)
    catch "attempting to assign to ": NULL
    finally interface(warnlevel = nu)
    end try;
    while not Fend(n) do assign('h' = readline(n)), assign('b' = Search2(h, {m, p, m1}));
      if b = [] then next
      elif search(h, p, 't') and b[1] = t then a := {op(a), cat(``, h[1 .. t - 1])}
      elif h[1 .. 11] = m1 then search(h, "')", 't'); a := {op(a), cat(``, h[12 .. t - 1])}
      elif search(h, m, 't') and b[1] = t then a := {op(a), cat(``, h[1 .. t - 1])}
      end if
    end do
  else
    try read F; p := {"`6", "f*", "RF*"}; r := SD(readline(F)[2]); close(F)
    catch "": ERROR("file <%1> of release <%2> is incompatible with a current
    Maple release", F, r)
    end try;
    do
      assign('h' = readline(F));
      if h = 0 then break else assign('b' = Search2(h, p)) end if;
      if b = [] then next
      elif h[1] = "I" and member(h[b[1] .. b[1] + 1], p) or member(h[b[1] .. b[1] + 2], p) then
        if evalb(Iddn(b[1] - 3) = h[2]) then a := {op(a), cat(``, h[3 .. b[1] - 1])}
        elif evalb(Iddn(b[1] - 4) = h[2 .. 3]) then a := {op(a), cat(``, h[4 .. b[1] - 1])}
        end if
      end if
    end do
  end if;
  Close(F), assign(d = omega(a));
  if nargs = 1 then for k in [op(d[1]), op(d[2])] do
      if member(k, psi) then next
      else try unassign(k) catch "attempting to assign to ": unprotect(k); unassign(k) end try
      end if
    end do
```

```
  end if;
  d
end proc
```

Typical examples of the procedure use:

```
> restart: MM:=module() export x,m; m:=module() export y; y:= () -> `+`(args) end module
  end module: MM1:=module() local F,m; export a; option package; m:=module() export y;
  y:=() -> `+`(args) end module; F:=() -> m:- y(args); a:=F end module: module SVEGAL ()
  export x; x:=() -> `+`(args)/nargs^6 end module: PP:=proc() `+`(args)/nargs^2 end proc:
  PP1:=proc() `+`(args)/nargs end proc: PP2:= proc() `+`(args)/nargs end proc: module GS
  () export x; x:= () -> `+`(args)/nargs^3 end module: avz:=61: agn:=sqrt(x^2+y^2+z^2):
  Art:=x+y: Kr:=a+b*I: save(MM, PP, avz, SVEGAL, GS, agn, PP1, Art, MM1, Kr,
  "C:/RANS/Grodno/Vilnius/Tallinn/Moscow/Art_Kr"); save(MM, PP, avz, SVEGAL, GS,
  agn, PP1, Art, MM1, Kr, "C:/RANS/Grodno/Vilnius/Tallinn/Moscow/Art_Kr.m");

> restart: filePM("C:/RANS/Grodno/Vilnius/Tallinn\\Moscow\\Art_Kr1.m");
                    [PP, PP1], [SVEGAL, MM, GS, MM1]

> filePM("C:/RANS/Grodno/Vilnius/Tallinn\\Moscow\\Art_Kr", 14);
   Error, (in filePM) attempting to assign to `SVEGAL` which is protected
   Error, (in filePM) attempting to assign to `GS` which is protected
                    [PP, PP1], [SVEGAL, MM, GS, MM1]
```

CorMod – testing of presence in datafiles of incorrect program modules CorMod1

Call format of the procedures:

CorMod(F {, 'h'})
CorMod1(F)

Formal arguments of the procedures:

F - symbol or string defining a filename or full path to a datafile
h - (optional) an unevaluated name

Description of the procedures:

The built-in function *`save`* of *Maple* saves program modules incorrectly. The enough detailed discussion of this question and different aspects can be found in our books [9,29-33,43]. For testing of presence in a file created by means of the *`save`* function of program modules incorrect in the indicated sense two procedures **CorMod** and **CorMod1** can be useful enough.

The procedure **CorMod(F {, 'h'})** returns the *false* value at presence in a file specified by the first actual **F** argument of the incorrect program modules (which have indefinite exported variables or generally without exported variables), otherwise the *true* value is returned. The procedure call with the second optional **h** argument provides the return through it of 2-element list that is the first element defines the list of correct modules, whereas the second – of incorrect program modules contained in a tested file **F.** At lack in the tested file of the incorrect program modules, the second argument **h** is returned unevaluated, if before call of procedure it and was such. The procedure supposes what a file of the input *Maple* format contains only modules correct in the mentioned above sense.

The procedure call **CorMod1(F)** returns the *false* value at presence in a file specified by the actual **F** argument of the incorrect program modules (which have indefinite exported variables or generally without exported variables), otherwise the *true* value is returned.

At presence in a tested file of program modules with ***protected***-attribute, the procedure call of **CorMod** or **CorMod1** causes the error with diagnostics "(in **filePM**) attempting to assign to `%1` which is protected", which does not influence on the call result. In a series of cases, *Maple* does not provide handling of the given error inside a procedure. The examples represented below illustrate not only a calls syntax of both procedures, but also the results of their application for a testing of datafiles of both formats (the input *Maple* language format and the inside *Maple* format).

```
CorMod:= proc(F::{string, symbol})
local `1`, `2`, `3`, `4`, `5`, k, j, `6`, `7`, `8`, `9`;
  `if`(cat(" ", F)[-2 .. -1] <> ".m", RETURN(true), [assign(`1` = {anames(`module`)}),
  Read1([F], '`2`'), assign(`3` = `2`[mods1] union `2`[mods2])]),
  `if`(`3` = {}, RETURN(true), assign('`7`' = [], '`8`' = []));
  for k to nops(`3`) do
    try
      assign('`4`'=with(`3`[k]),'`6`'={}), assign('`5`'={seq(cat(``,`4`[j],`(60)`), j=1..nops(`4`))});
      for j to nops(`4`) do
        assign('`9`' = cat("assign('_avz' = ", `4`[j], `(60)):`)), writebytes(`_$1`, `9`);
        (proc() global _avz; close(`_$1`); read `_$1`; fremove(`_$1`) end proc)();
        `if`(convert(_avz, symbol) = `5`[j], assign('`6`' = {false, op(`6`)}),
        assign('`6`' = {true, op(`6`)}))
      end do
    catch "%1 is not a package": assign('`6`' = {false, op(`6`)})
    catch "": assign('`6`' = {false, op(`6`)})
    end try;
    `if`(`6` = {true}, assign('`7`' = [op(`7`), `3`[k]]), assign('`8`' = [op(`8`), `3`[k]]));
    `if`(nargs = 2, assign(args[2] = [[correct, op(`7`)], [invalid, op(`8`)]]), NULL);
    seq(`if`(member(`3`[k], `1`) or ismLib(`3`[k]), NULL, unassign(`3`[k])), k = 1..nops(`3`)),
      `if`(`8` = [], true, false)
  end do
end proc

CorMod1:= proc(F::{string, symbol})
local `0`, `1`, `3`, `4`, `5`, `6`, `7`, `8`, `9`, k, j, p;
description "Procedure returns `true` if a testable file contains correct modules,
     and `false` otherwise";
  `if`(not type(F, file), ERROR("<%1> is not a file", F), assign(`1` = {anames(`module`)}));
  assign(`3` = {op(filePM(F, 147)[2])}), `if`(`3` = {}, RETURN(true), 6);
  for j to nops(`3`) do
    (proc() save args, `_$1$` end proc)(`3`[j]);
    assign('`4`' = readbytes(`_$1$`, TEXT, infinity), '`5`' = ""), fremove(`_$1$`);
    seq(`if`(`4`[k] = "\n", NULL, assign('`5`' = cat(`5`, `4`[k]))), k = 1 .. length(`4`));
    `if`(search(`5`, `end module`), assign('`4`' = Search2(`5`, {` export `, ` end module`})),
    RETURN(true));
    for k from 1/2*nops(`4`) by -1 to 1 do
      assign('`6`' = `4`[k], '`7`' = `4`[nops(`4`) - k + 1]), assign('`8`' = `5`[`6` .. `7` - 1][9 .. -1]),
        assign('`8`' = `8`[1 .. searchtext(";", `8`) - 1]), assign('`0`' = `5`[`6` + 7 .. `7` - 1]);
      `5` := cat(`5`[1 .. `6` - 1], seq(1, k = 1 .. `7` - `6` + 1), `5`[`7` + 1 .. -1]);
```

```
        assign('`9`' = `if`(search(`8`, `,`), SLD(`8`, ", "), [`8`]));
        for p to nops(`9`) do if search(`0`, cat(`9`[p], " := ")) then next else RETURN(false)
          end if
        end do
      end do;
      for k in `3` do
        if member(k, `1`) or ismLib(k) then NULL
        else try unassign(k) catch "attempting to assign to ": unprotect(k); unassign(k) end try
        end if
      end do
    end do;
    true
end proc
```

Typical examples of the procedures use:

```
> MM:=module() export y,m; m:=61: y:=module() export y; y:= () -> `+`(args) end module
  end module: MM1:=module() local F,m; export a; option package; m:=module() export y;
  y:=() -> `+`(args) end module; F:=() -> m:- y(args); a:=F end module: module SVEGAL()
  export x; x:=() -> `+`(args)/nargs^6 end module: PP:=proc() `+`(args)/nargs^2 end proc:
  PP1:=proc() `+`(args)/nargs end proc: PP2:= proc() `+`(args)/nargs end proc: module GS ()
  export x; x:= () -> `+`(args)/nargs^3 end module:
> save(MM, PP, SVEGAL, GS, PP1, MM1, "C:/RANS/Grodno/Vilnius/Art_Kr");
  Read1(["C:/RANS/Grodno/Vilnius/Art_Kr"], 'h'), eval(h);
    Error, (in unknown) attempting to assign to `SVEGAL` which is protected
    Error, (in unknown) attempting to assign to `GS` which is protected
        table([procs = {PP, PP1}, vars = {}, mods1 = {MM, MM1}, mods2 = {SVEGAL, GS}])
> CorMod("C:/RANS/Grodno/Vilnius/Art_Kr", 't'), t;
                                true, t
> save(MM, PP, SVEGAL, GS, PP1, MM1, "C:/RANS/Grodno/Vilnius/Art_Kr.m");
  Read1(["C:/RANS/Grodno/Vilnius/Art_Kr.m"], 'h'), eval(h);
        table([procs = {PP, PP1}, vars = {}, mods1 = {MM, MM1}, mods2 = {SVEGAL, GS}])
> CorMod("C:/RANS/Grodno/Vilnius/Art_Kr.m", 't'), t;
                 false, [[correct], [invalid, MM, MM1, SVEGAL, GS]]
> map(CorMod1, ["C:/Temp/Vilnius.m", "C:/Temp/Tallinn"]);
    Error, (in filePM) attempting to assign to `SVEGAL` which is protected
    Error, (in filePM) attempting to assign to `GS` which is protected
                              [false, true]
```

DefOpt – validation of optional arguments passed to a *Maple* procedure

Call format of the procedure:

DefOpt(*args*)

Formal arguments of the procedure:

args – a sequence of equations defining values for optional arguments

Description of the procedure:

For providing of validation of optional arguments passed to a *Maple* procedure, the function ***ProcessOptions*** of the following calling sequence serves:

ProcessOptions(n, List, Tab1, Tab2)

where: **n** - quantity of the first obligatory arguments which are passed to a procedure and are eliminated from checkout; **List** - a list of optional arguments subjected to checkout; **Tab1** - a table of types and values by default for all optional arguments of the procedure; **Tab2** - a table containing an index for each optional argument which was defined by the user or whose value by default is provided by the **Tab1** table.

The **ProcessOptions** function is used both for check of correctness of optional arguments passed to a procedure, and for assignment to them of values by default in case of their absence at procedure call. It is possible to familiarize oneself with the **ProcessOptions** function in *Help* system or in our books [9-12,29-33,43] in more details. The **DefOpt** procedure allows inside of a procedure easily to generate the above-mentioned **Tab1** table providing assignment of values by default to its optional arguments.

The **DefOpt** procedure returns the **Tab1** table whose subtables content is defined by arguments of two kinds: **p = [{***default value***},** *type*] and **name = {[***value, type*] | [*type*]} which are received at procedure call. The procedure provides a certain level of automation at use of optional arguments, though in a lot of cases their program handling can be done by other methods considered in our books [1-14,29-33,43,44], much more effectively. In particular, precise knowledge of an algorithm subjected to programming in the *Maple* environment essentially simplifies the problem.

```
DefOpt:= proc()
local `0`, `1`, `2`, `3`;
   [assign('`0`' = [unassign('`2`', '`3`')], '`1`' = [],
   '`3`' = (() -> {seq(lhs([args][1][`2`]), `2` = 1 .. nops(args[1]))})),
   seq(`if`(type(args[`2`], equation) and type(lhs(args[`2`]), symbol), assign('`0`' = [op(`0`),
   lhs(args[`2`]) = [op(rhs(args[`2`])), 'makelist']]), `if`(type(args[`2`], equation) and
   type(lhs(args[`2`]), posint), assign('`1`' = [op(`1`), args[`2`]]),
   ERROR("invalid arguments %1", [args]))), `2` = 1 .. nargs), `if`(nops(`1`) < `3`(`1`)[-1] or
   nops(`1`) <> nops(`3`(`1`)), ERROR("invalid positional values %1", `1`), `if`(nops(`0`) <>
   nops(`3`(`0`)), ERROR("invalid options values %1", `0`),
   TABLE([OptionParms = TABLE(`0`), PositionalParms = TABLE(`1`)]))))]
end proc
```

Typical example of the procedure use:

```
> DefOpt(delete=[{list, symbol}], 1=[0, nonnegint], 2=[nonnegint], test= [true, truefalse],
  rans=[{symbol, string,name}], 3=[Kr, float], 4=[Art, list], create=[{symbol, string}],
  5=[Svet, posint], ian=[{list, set, table, listlist, setset}]);

   [table([OptionParms = table([delete = [{list, symbol}, makelist], test = [true, truefalse, makelist],
   rans = [{name, string, symbol}, makelist], create = [{string, symbol}, makelist],
   ian = [{set, table, list, listlist, setset}, makelist]]),
   PositionalParms = table([1 = [0, nonnegint], 2 = [nonnegint], 3 = [Kr, float], 4 = [Art, list],
   5 = [Svet, posint]])])]

> DefOpt(1=[0, positive], 2=[realnum], trg=[{symbol, string}], 3=[tru, float]);

   [table([OptionParms = table([trg = [{string, symbol}, makelist]]),
   PositionalParms = table([1 = [0, positive], 2 = [realnum], 3 = [tru, float]])])]
```

F_alias – check of type of a modular table
M_Type

Call format of the procedures:

F_alias(M)
M_Type(M)

Formal arguments of the procedures:

M - symbol defining a module name

Description of the procedures:

The simple procedure **F_alias(M)** is intended for analysis of the type of a modular table and receiving of functions aliases supported by a module or package **M.** The procedure call **F_alias(M)** returns the list of equations of a kind *`function name = alias`* for the modular table of the *first* type and the message *`M-modular table has the second type`* for a table **M** of the *second* type. The type of a modular table depends on the *Maple* release, as it illustrates a fragment presented below.

The procedure **M_Type(M)** provides analysis of a module **M** relative to a type of data structure corresponding to it. The procedure call returns the *identifier* of structural organization used by a module **M,** namely: *Tab* – a table, *Proc* – a procedure and *Mod* – a module. Of example represented below follows, that from **68** package modules of *Maple 9* **23** modules use tabular organization, **43** use modular organization and only **1** module has been implemented on the basis of procedure organization.

```
F_alias:= proc(M::{package, `module`})
local p, L, T, H;
  assign(T = eval(M), H = [cat(M, `-modular table has the first type`)]);
  null(seq([assign('L' = convert(T[p], string)), `if`(searchtext("[", L) <> 0,
    RETURN(cat(M, `-modular table has the second type`)),
    assign('H' = [op(H), p =substring(L, searchtext("/", L) + 1..length(L) – 3)]))], p =with(M))), H
end proc

M_Type:= proc(M::symbol)
local a, T, omega, nu, t, k;
  assign(a = {anames(procedure)}, omega = interface(warnlevel)), `if`(M = liesymm,
     RETURN(Tab), assign(nu = (h -> interface(warnlevel = h))));
  try t := [exports(M)]; if t = [] then RETURN(WARNING("<%1> is not a module", M))
     else goto(Fin) end if
  catch "wrong number (or type) of parameters in function %1":
     try nu(0); t := with(M)
     catch "%1 is not a package": nu(omega); RETURN(WARNING("<%1> is not a module", M))
     catch "invalid input: %1": nu(omega); RETURN(WARNING("<%1> is not a module or
                 cannot be initialised", M))
     catch "system level initialisation for": nu(omega); RETURN(WARNING("module <%1>
                 cannot be initialised", M))
     end try
  end try;
  for k in t do if member(k, a) then next else unprotect(k); unassign(k) end if end do;
  Fin;
  nu(omega), assign(T = convert(eval(M), string)), `if`(searchtext("table", T, 1 .. 5) = 1, Tab,
  `if`(searchtext("module", T, 1..6) = 1, Mod, `if`(searchtext("proc", T, 1..4) = 1, Proc, NULL)))
end proc
```

Typical examples of the procedures use:

> map(F_alias, [*group, plots*]); **# *Maple of release 6***

[***group***-*modular table has the* **second** *type*, [***plots***-*modular table has the* ***first*** *type*, *animate* = "animate", *animate3d* = "animate3d", *animatecurve* = "animatecurve", *changecoords* = "changecoords", *complexplot* = "complexplot", *complexplot3d* = "complexplot3d", *conformal* = "conformal", *contourplot* = "contplot", *contourplot3d* = "contplot3d", *coordplot* = "coordplot", *coordplot3d* = "coordplot3d", *cylinderplot* = "cylinder", *densityplot* = "densityplot", *display* = "display", *display3d* = "display", *fieldplot* = "fieldplot", *fieldplot3d* = "fieldplot3d", *gradplot* = "gradplot", *gradplot3d* = "gradplot3d", *implicitplot* = "iplot", *implicitplot3d* = "iplot3d", *inequal* = "inequal", *listcontplot* = "listcontplot", *listcontplot3d* = "listcontplot3d", *listdensityplot* = "listdensityplot", *listplot* = "listplot", *listplot3d* = "listplot3d", *loglogplot* = "loglogplot", *logplot* = "logplot", *matrixplot* = "matplot", *odeplot* = "odeplot", *pareto* = "pareto", *pointplot* = "pointplot", *pointplot3d* = "pointplot3d", *polarplot* = "polarplot", *polygonplot* = "polygonplot", *polygonplot3d* = "polygonplot3d", *polyhedra_supported* = "polyhedra_supported", *polyhedraplot* = "polyhedraplot", *replot* = "replot", *rootlocus* = "rootlocus", *semilogplot* = "semilogplot", *setoptions* = "setoptions", *setoptions3d* = "setoptions3d", *spacecurve* = "spacecurv", *sparsematrixplot* = "sparsematrixplot", *sphereplot* = "spherical", *surfdata* = "surfdata", *textplot* = "textplot", *textplot3d* = "textplot3d", *tubeplot* = "tubeplot"]]

> map(F_alias, [*group, plots*]); **# *Maple of release 9***

[[***group***-*modular table has the* ***first*** *type*, *DerivedS* = "Deriv", *LCS* = "", *NormalClosure* = "NormalClos", *RandElement* = "RandElem", *SnConjugates* = "SnConjuga", *Sylow* = "Sy", *areconjugate* = "areconjug", *center* = "cen", *centralizer* = "centrali", *core* = "c", *cosets* = "cos", *cosrep* = "cos", *derived* = "deri", *elements* = "eleme", *groupmember* = "groupmem", *grouporder* = "groupor", *inter* = "in", *invperm* = "invp", *isabelian* = "isabel", *isnormal* = "isnor", *issubgroup* = "issubgr", *mulperms* = "mulpe", *normalizer* = "normali", *orbit* = "or", *parity* = "par", *permrep* = "perm", *pres* = "p", *transgroup* = "transgr"], [***plots***-*modular table has the* ***first*** *type*, *animate* = "anim", *animate3d* = "animat", *animatecurve* = "animatecu", *arrow* = "ar", *changecoords* = "changecoo", *complexplot* = "complexp", *complexplot3d* = "complexplo", *conformal* = "confor", *conformal3d* = "conforma", *contourplot* = "contp", *contourplot3d* = "contplo", *coordplot* = "coordp", *coordplot3d* = "coordplo", *cylinderplot* = "cylin", *densityplot* = "densityp", *display* = "disp", *display3d* = "disp", *fieldplot* = "fieldp", *fieldplot3d* = "fieldplo", *gradplot* = "gradp", *gradplot3d* = "gradplo", *graphplot3d* = "graphplo", *implicitplot* = "ip", *implicitplot3d* = "iplo", *inequal* = "ineq", *interactive* = "interact", *listcontplot* = "listcontp", *listcontplot3d* = "listcontplo", *listdensityplot* = "listdensityp", *listplot* = "listp", *listplot3d* = "listplo", *loglogplot* = "loglogp", *logplot* = "logp", *matrixplot* = "matrixp", *odeplot* = "odep", *pareto* = "par", *plotcompare* = "plotcomp", *pointplot* = "pointp", *pointplot3d* = "pointplo", *polarplot* = "polarp", *polygonplot* = "polygonp", *polygonplot3d* = "polygonplo", *polyhedra_supported* = "polyhedra_suppor", *polyhedraplot* = "polyhedrap", *replot* = "rep", *rootlocus* = "rootlo", *semilogplot* = "semilogp", *setoptions* = "setopti", *setoptions3d* = "setoption", *spacecurve* = "spacec", *sparsematrixplot* = "sparsematrixp", *sphereplot* = "spheri", *surfdata* = "surfd", *textplot* = "textp", *textplot3d* = "textplo", *tubeplot* = "tubep"]]

> map(M_Type, [PolynomialTools, ExternalCalling, LinearFunctionalSystems, plottools, MatrixPolynomialAlgebra, ScientificErrorAnalysis, CodeGeneration, LinearOperators, CurveFitting, Sockets, Maplets, Student, Matlab, Slode, LibraryTools, Spread, codegen, context, finance, genfunc, LinearAlgebra, geom3d, group, linalg, padic, plots, process, simplex, student, tensor, MathML, Units, DEtools, diffalg, Domains, stats, GaussInt, Groebner, LREtools, PDEtools, Ore_algebra, algcurves, orthopoly, combinat, combstruct, difforms, geometry, inttrans, networks, numapprox, numtheory, powseries, sumtools,

ListTools, RandomTools, RealDomain, SolveTools, StringTools, XMLTools, SumTools, Worksheet, OrthogonalSeries, FileTools, RationalNormalForms, ScientificConstants, TypeTools, VariationalCalculus]);

Warning, module <Matlab> cannot be initialised
--------------------- Domains version 1.0 -------------------------------
Initially defined domains are Z and Q the integers and rationals
Abbreviations, e.g. DUP for DenseUnivariatePolynomial, also made
[Mod, Mod, Mod, Tab, Mod, Mod, Mod, Mod, Mod, Mod, Mod,Mod, Tab, Mod, Mod, Mod, Tab, Mod, Tab, Mod, Tab, Mod, Tab, Mod, Tab, Mod, Tab, Tab, Tab, Mod, Mod, Tab, Tab, Tab, Proc, Mod, Mod, Tab, Tab, Mod, Tab, Tab, Mod, Tab, Tab, Tab, Tab, Tab, Mod, Mod, Mod, Mod, Mod, Mod, Mod, Mod, Mod, Mod, Mod, Mod, Mod, Mod, Mod, Mod, Mod, Mod]

> Mulel(%);

[[*Proc*, 35, 1], [*Mod*, 1, 43], [*Tab*, 4, 22]]

Concerning the diagnostic message about module ***Matlab*** it is necessary to note, it is result not of absence of the given module, but is result of unsuccessful attempt of its initialization, namely: the package ***Matlab*** (for support of the interface with which the given module is intended) has not been installed in our *Maple* environment by reason of disuse of ***Matlab.***

In connection with the procedure **M_Type** it is necessary to do one extremely important note having the direct attitude to a package organization in the *Maple* environment. Many of the packages that are delivered with *Maple* export the functions, which redefine the analogous standard functions or functions exported by other packages. For example, the package **linalg** redefines standard procedures ***'norm'*** and ***'trace'.*** Such approach in some cases does not allow to use simultaneously some packages or a package together with standard or user means. In many cases the given problem is solved by dot calls of functions of a kind **<*package*>[<*function*>](*args*),** however it is not always convenient. All the time it is necessary to keep in mind an opportunity of crossing of activated package means with the built-in or standard *Maple* means.

Meanwhile, and more serious situation takes place, namely: the package **liesymm** exports procedure ***'close'*** which redefines the built-in function ***'close',*** that seriously complicates use of a package **liesymm** together with means of access to datafiles. The similar situation, in our opinion, is inadmissible and urgently demands to use various names for the built-in and standard means, and packages means; i.e. intersection of names sets of means of all *Maple* levels (*kernel*, *libraries* and *packages*) should be empty set. Similar standardization will allow to essentially simplify the programming in many cases. For this purpose a change of the organization of packages is not required - it is enough to add to names of the means exported by a package the package name as a prefix, for example.

Gener – a call mechanism of *Maple* procedures

Call format of the procedure:

Gener(F, d, Fm, n)

Formal arguments of the procedure:

F - symbol or string defining a filename or full path to it
D - symbol or string defining key word ***`same`***, a filename or full path to it
Fm - symbol or list defining types of read-in data
n - a positive integer defining read-in data quantity (*posint*)

Description of the procedure:

The procedure **Gener(F, d, Fm, n)** has four formal arguments whose purpose consists in the following: **F** - qualifier of a datafile with a procedure name, **d** - qualifier of a datafile with actual arguments for the procedure, **Fm** - a type of actual arguments {may be single for all or by list of types for the first **min(nops(Fm), n)** read-in data), and **n** - a quantity of the read-in data.

The data in a text datafile **d** are saved in an arbitrary format and are parted by symbols {" ", "\n", "\t"} (*blanks*, *line feeds*, and *horizontal tabs*); at reading they are interpreted by the **Gener** procedure as a sequence of actual arguments passed to a procedure whose name is situated at the beginning of a text datafile **F.** If the second actual **d** argument is specified as **`*same*`**, the procedure supposes that a text datafile heading specified by the first **F** argument has the following view:

<*procname*>{` `|`\n`|`\t`}**<*args1*>**{` `|`\n`|`\t`}**<*args2*>**{` `|`\n`|`\t`} **...** {` `|`\n`|`\t`}**<*argsp*>**

and a ***procname*** procedure is available in a current *Maple* session, where ***argsk*** **(k = 1 .. p)** - a sequence of actual arguments passed to the ***procname*** procedure.

The **Gener** procedure provides a call mechanism of the user procedures, when a procedure name has been saved in a text datafile, whereas actual arguments passed to it, in a text datafile **d.** The procedure handles all basic erratic and especial situations.

```
Gener:= proc(F::file, d::{symbol, file}, Fm::{symbol, list(symbol)}, n::posint)
local a, b, c, nu, omega, k, t, ER;
  omega:=(x,y,z) -> ERROR("the %-1 read-in value <%3> has a type different from <%2>",x,y,z);
  ER := () -> [interface(warnlevel = nu), ERROR(args)];
  `if`(type(d, file), NULL, `if`(eval(d) = same, assign(t = 7), ERROR("2nd argument should be
  a file or `same`"))); assign(a = readbytes(`if`(t = 7, F, d), TEXT, infinity), c = {" ", "\n", "	"}),
     close(`if`(t = 7, F, d)), assign(nu = interface(warnlevel)), interface(warnlevel = 0);
  `if`(a=0,ER("file <%1> is empty", `if`(t=7,F,d)),7), `if`(Search2(a,c) <> [], assign('a'=SLS(a,c)),
    `if`(t <> 7, assign('a'=[a]), ER("structure of file <%1> is invalid",F))), assign('a'=map(came,a));
  if t <> 7 then b := cat(``, readline(F)); close(F)
  else `if`(1 < nops(a), assign('b' = cat(``, a[1]), 'a' = a[2..-1]), ER("actual arguments are absent"))
  end if;
  `if`(type(b, procedure), NULL, ER("<%1> is not a procedure", b)); `if`(nops(a) < n,
  ER("quantity <%1> of read-in data is less of the given quantity <%2>", nops(a), n), 7);
  if type(Fm, symbol) then for k to n do if type(a[k], Fm) then next else omega(k, Fm, a[k])
       end if end do
  else for k to min(nops(Fm), n) do if type(a[k], Fm[k]) then next else omega(k, Fm[k], a[k])
       end if
     end do
  end if;
  apply(b, op(a[1 .. n])), interface(warnlevel = nu)
end proc
```

Typical examples of the procedure use:

> SR:= () -> `+`(args)/nargs: Gener("C:/Temp/Function", "C:/Temp/Family.txt", integer, 6);

Error, (in **omega**) the 3rd read-in value <40> has a type different from <integer>

> SR:= () -> `+`(args)/nargs: Gener("C:/Temp/Function", "C:/Temp/Family1.txt", integer, 6);

107/3

> SR:= () -> `+`(args)/nargs: Gener("C:/Temp/Function", "C:/Temp/Family1.txt", integer, 7);

Error, (in **Gener**) quantity <6> of read-in data is less of the given quantity <7>

> SR:= () -> `+`(args)/nargs: Gener("C:/Temp/Function", "C:/Temp/Family2.txt", [float, fraction, integer], 6);

26.53191490

> SR:=() -> `+`(args)/nargs: Gener("C:/Temp/Function", "C:/Temp/Family3.txt",realnum,6);

1/6*14^(1/2)+1/6*ln(7)+23.03191490

> Gener("C:/Temp/Grodno.rlb.txt", `same`, realnum, 6);

1/6*14^(1/2)+1/6*ln(7)+23.03191490

> Gener("C:/Temp/Paskula.txt", `same`, realnum, 7);

Error, (in **ER**) <RANS_IAN> is not a procedure

> Gener("C:/Temp/Tallinn.txt", `same`, realnum, 7);

Error, (in **Gener**) file <C:/Temp/Tallinn.txt> is empty

---

dcemod – a dynamic call of a module export

Call format of the procedure:

dcemod(M, ***ex*** **{,** ***args*****})**

Formal arguments of the procedure:

M - symbol defining a module name
ex - symbol defining a name of a variable exported by the module **M**
args - (optional) an arguments sequence passed to the module export *ex*

Description of the procedure:

The *Maple* calls mechanism does not allow to use the dynamically generated names for a variable exported by a module. The following simple example well illustrates the told, namely:

> 3*SimpleStat:- SR([61,56,36,40,7,14],1,10);

107

> M:=SimpleStat: ex:=SR: 3*M:- ex([61,56,36,40,7,14],1,10);

Error, module does not export `**ex**`

However, in a lot cases this circumstance is a rather essential shortcoming leading to complication of programming of problems dealing with modules. The procedure **dcemod** lifts this restriction.

The procedure **dcemod(M,** ***ex*** **{,** ***args*****})** has at least two formal arguments whose purpose consists in the following: **M** - a module name, ***ex*** - name of a variable exported by the module **M,** and ***args*** - optional arguments passed to an export ***ex.*** If the procedure call contains only two actual arguments, then the call ***ex*****()** is supposed. Successful procedure call **dcemod(M,** ***ex*** **{,** ***args*****})** returns a result of call **M:-** ***ex*****({*****args*****}).** The procedure handles all basic erratic and especial situations and in numerous appendices has shown itself as a very useful tool.

```
Dcemod:= proc(M::symbol, ex::symbol)
local a;
global _v61g56;
  `if`(nargs < 2, ERROR("quantity of arguments should be more than 1 but has been received
  <%1>", nargs), `if`(type(M, `module`), `if`(member(ex, {exports(M)}),
```

```
   assign67(a = "_$EE_", _v61g56 = NULL), ERROR("<%1> does not export <%2>", M, ex)),
   ERROR("<%1> is not a module", M))), null(writeline(a, cat("`_v61g56`:=", M, "[", ex, "](",
   `if`(2 < nargs, seqstr(args[3 .. nargs]), ``), "):"))),
   (proc() close(a); read a; fremove(a) end proc)(), _v61g56, unassign('_v61g56')
end proc
```

Typical examples of the procedure use:

```
> dcemod(ACC);
    Error, (in dcemod) quantity of arguments should be more than 1 but has been received <1>
> dcemod(RANS, ian);
    Error, (in dcemod) <RANS> is not a module
> dcemod(ACC, ian);
    Error, (in dcemod) <ACC> does not export <ian>
> dcemod(ACC, Ds, [61,56,36,40,7,14], 1, 10);
                                        360.400000
> dcemod(ACC, Ds);
    Error, (in ACC:-Ds) ACC:-Ds uses a 3rd argument, t, which is missing
> 3*dcemod(SimpleStat, SR, [61,56,36,40,7,14], 1, 10);
                                          107
> M:=SimpleStat:  ex:=SR:  M:- ex([61,56,36,40,7,14],1,10);
> M:- ex([61,56,36,40,7,14],1,10);
    Error, module does not export `ex`
```

---

howAct - check of availability of a procedure, program module or variable

Call format of the procedure:

howAct(P {, 'h'})

Formal arguments of the procedure:

P - symbol defining name of a procedure, module or variable
h - (optional) an unevaluated name

Description of the procedure:

The simple procedure **howAct(P {, 'h'})** returns the *true* value, if a procedure, module or variable specified by the first actual **P** argument is available in a current session (is accessible for direct use); otherwise the *false* value is returned.

If the second optional **h** argument has been coded, the procedure through this argument returns the *true* value if a tested object is in a library logically linked with the main *Maple* library, otherwise the *false* value is returned. Testing of a variable by means of the **howAct** procedure demands to code it in unevaluated form, for example **'V'**.

The **howAct** procedure allows differentially to test a domain of activation of an object (a current session and/or a library); in addition, for the library objects, the pair {*true*, *true*} is always returned, because such objects are always available in a current *Maple* session, if a library containing them, is logically linked with the main package library.

```
howAct:= (P::symbol) -> [`if`(member(P,{anames()}), true, false), `if`(nargs = 2,
    assign(args[2] = ismLib(P)), NULL)][1]
```

Typical examples of the procedure use:

> avz:=61: [howAct(MkDir, 'h'), h], [howAct(SVEGAL, 't'), t], [howAct('avz', 'p'), p];

[*true, true*], [*false, false*], [*true, false*]

ismLib – check of availability of a procedure or program module

Call format of the procedure:

ismLib(P {, 'h'} {, `$`})

Formal arguments of the procedure:

P - symbol defining name of a procedure or program module
H - (optional) an unevaluated name
`$` - (optional) key symbol

Description of the procedure:

The procedure **ismLib(P {, 'h'} {, `$`})** returns the *true* value, if a procedure or program module whose name is specified by the first actual **P** argument is in one of libraries logically linked with the main package library, including main library also; otherwise the *false* value is returned.

At coding of the second optional **h** argument through it the full path to the first library in a chain of libraries defined by the ***libname*** variable of the package, in which a program object, indicated by the **P** argument has been detected, is returned.

At last, at coding of the optional symbol **`$`** as the last argument, the procedure **ismLib** supports a searching of the required procedure or program module in all libraries reflected in the predefined *Maple* variable ***libname,*** excluding the main package library.

```
ismLib:= proc(P::{name, symbol})
local sl, sm, k, j, p, h, LIB;
  assign(h = map(CF, [libname])), `if`(not member(`$`, {args}), assign(LIB = h),
     assign(LIB = [seq(`if`(h[p][-4 .. -1] = "\\lib", NULL, h[p]), p = 1 .. nops(h))]));
  for k to nops(LIB) do
    assign('sl' = march(list, LIB[k])), assign('sm' = map2(op, 1, sl));
    if member(cat("", P), {seq(sm[j][1 .. -3], j = 1 .. nops(sm))}) then
       RETURN(true,
      `if`(2 <= nargs, assign(args[2] = LIB[k]), NULL))
    else NULL
    end if
  end do;
  false
end proc
```

Typical examples of the procedure use:

> ismLib(ParProc1, 'h'), h;

true, "c:\program files\maple 9\lib\userlib"

> ismLib(Read1, 'h'), h, ismLib(MkDir, 't'), t;

true, "c:\program files\maple 9\lib\userlib", *true*, "c:\program files\maple 9\lib\userlib"

```
> ismLib(assign, 'p', `$`), p;  ismLib(assign, 'p'), p;
                              false, p
                 true, "c:\program files\maple 9\lib"
```

IO_proc – check of a procedure to be an *iolib* function

Call format of the procedure:

IO_proc(P)

Formal arguments of the procedure:

P - symbol or positive integer defining name or number of an iolib function accordingly

Description of the procedure:

The procedure call **IO_proc(P)** returns the *number* of an ***iolib*** function specified by its **P** name, or the *name* of an ***iolib*** function specified by its **P** number; otherwise, an appropriate error arises. The procedure does not provide check of ***iolib*** functions of the **process** package of *Maple 6* on *Windows* platform, because the package is supported only under *UNIX* and *UNIX*-compatible operating systems.

```
IO_proc:= proc(P::{posint, symbol})
local a, k, p, T_iolib, omega, h;
  assign(T_iolib = table([currentdir = 27, fclose = 7, close = 7, feof = 21, fflush = 23, fprintf = 9,
     fremove = 8, fscanf = 11, getenv = 28, iostatus = 13, march = 31, mkdir = 29, open = 20,
     readbytes = 4, readline = 2, rmdir = 30, sprintf = 10, sscanf = 12, writebytes = 5, exec = 17,
     block = 19, fork = 16, kill = 22, launch = 36, pclose = 7, pipe = 15, popen = 14, wait = 18,
     "process:-exec" = 17, "process:-block" = 19, "process:-fork" = 16, "process:-kill" = 22,
     "process:-launch" = 36, "process:-pclose" = 7, "process:-pipe" = 15, "process:-popen" = 14,
     "process:-wait" = 18])), assign(omega = (() -> `if`(nargs = 1, args, [args])));
  `if`(type(P, posint), `if`(member(P, {2, 4, 5, 36, p $ (p = 27 .. 31), k $ (k = 7 .. 23)}),
     map(convert, omega(RTab(T_iolib, P)), symbol),
     ERROR("<%1> is invalid number for an iolib Maple function", P)), [`if`(search(P, ":-"),
     assign(h = convert(P, string)), assign(h = P)), assign(a = T_iolib[h]), `if`(type(a, posint),
     RETURN(a), ERROR("<%1> is not an iolib function", P))])
end proc
```

Typical examples of the procedure use:

```
> IO_proc(MkDir); IO_proc(RANS); IO_proc(process:- exec);
   Error, (in IO_proc) <MkDir> is not an iolib function
   Error, (in IO_proc) <RANS> is procedure of a package or program module or is not a procedure
                                        17
> map(IO_proc, [close,mkdir,iostatus,8,27,readbytes,13,21,7,writebytes,8]);
     [7, 29, 13, fremove, currentdir, 4, iostatus, feof, [fclose, process:-pclose, pclose, close], 5, fremove]
> for k in [delf, Fend,15] do try IO_proc(k) catch : WARNING(lasterror): next end try;
  IO_proc(k) end do;
   Warning, <Currentdir> is not an iolib function
   Warning, <Fend> is not an iolib function
                              [pipe, process:- pipe]
> IO_proc(256);
   Error, (in IO_proc) <256> is invalid number for an iolib Maple function
```

ParProc – check of parameters of a procedure, program module or package
ParProc1
Sproc

Call format of the procedures:

ParProc(P)
ParProc1(P {, 'h'})
Sproc(P {, 'h'})

Formal arguments of the procedures:

P - symbol defining name of a procedure, program module or package
h - (optional) an unevaluated name

Description of the procedures:

The important element of work with procedures and program modules is check of their parameters (arguments, local and global variables, options, the exported variables, etc.). The procedures represented below solve this problem to a certain extent.

The procedure call **ParProc(P)** returns the 1-dimensional column-array whose elements have the following view:

<type of parameter> = (<a sequence of values>)

for a procedure specified by its name **P** (a library procedure or procedure activated in a current session). For the built-in functions and functions of the I/O library, their numbers are returned. If the argument **P** specifies name of a program module (active or library module), the list of variables exported by the program module is returned.

The **ParProc1** procedure represents an useful extension of the previous procedure, providing the more detailed analysis of parameters of procedures and program modules. Similarly, to the previous, the procedure call **ParProc1(P)** returns a 1-dimensional column-array of values of parameters for a procedure or program module whose name is specified by the first actual **P** argument. If the second optional **h** argument has been coded, then through it the set of full paths to libraries containing the given program object is returned. Additionally, the appropriate warning about location of sought object is output.

At last, the procedure call **Sproc(P)** returns the *true* value if a procedure or program module specified by the actual **P** argument is located in a library logically linked with the main *Maple* library, including the main library also; otherwise the *false* value is returned. If the second optional **h** argument has been coded and the procedure call returns the *true* value, then through **h** the 3-elements list or 4-elements list of the following kinds are returned, namely:

[*Proc*, {*built-in* | *iolib*}, <*a function number*>, <*full path to the Maple library*>]
[{*Proc* | *Mod* | *package*}, {*Maple* | *User* | *Maple&User*}, <*set of full paths to libraries*>]

The first element of both lists defines the type of a program object {*Proc* - procedure | *Mod* - module | *package* - package}, whereas the others are defined by the implementation kind of a program object, namely: *built-in*, *iolib* or *library*. For the first two kinds the 4-elements list is returned whose the second element defines the implementation kind {*built-in* | *iolib*}, the third element - a function number, and the fourth - full path to the main *Maple* library. Whereas for all implementation kinds the procedure call returns the 3-elements list whose the second element defines the type of library containing a program object {*Maple* – the main *Maple* library | *User* – the user library | *Maple&User* – the main *Maple* and user libraries}, the third element - set of full paths to libraries with a specified program object.

```
ParProc:= proc(P::symbol)
local k, L, R, h;
option system, remember, `Copyright International Academy of Noosphere - Tallinn - 2000`;
  L := [Arguments, locals, `options`, rem_tab, `description`, globals, lex_tab, exit_type];
  assign(R = Release1()), `if`(P = randomize, goto(A), NULL), `if`(type(P, `module`),
    RETURN(cat(P, ` is module with exports`), [exports(P)]), `if`(type(P, procedure),
    assign(h = Builtin(P)), ERROR("<%1> is not a procedure and not a module", P)));
  if type(h, integer) then RETURN(`builtin function`, h)
  else try h := IO_proc(P); if type(h, integer) then RETURN(`iolib function`, h) end if
    catch "": NULL end try
  end if;
  A;
  array(1 .. add(`if`(op(k, eval(P)) = NULL, 0, 1), k = 1 .. R + 1), 1 .. 1,
    [seq(`if`(op(k, eval(P)) <> NULL, [L[k] = op(k, eval(P))], NULL), k = 1 .. R + 1)])
end proc

ParProc1:= proc(M::{procedure, `module`})
local a, b, c, d, p, h, t, z, cs, L, R, N, omega, nu;
global `_62`, ParProc, Sproc;
  unassign(`_62`), assign(L= [exports,locals,globals,`options`,`description`], nu = "AVZ42_61");
  assign(omega=interface(warnlevel), cs =((k::integer, s::symbol) -> [assign('d' =[op(k, eval(s))]),
     assign('p' = [seq(convert(d[v], string), v = 1 .. nops(d))]), assign('t' = [seq(convert(
     substring(p[v], searchtext(":-", p[v]) + 1 .. -1), symbol), v = 1 .. nops(p))]), op(t)]));
  assign(N = (() -> `if`(rhs(op(args)) = NULL, NULL, args))), `if`(nargs = 1, NULL,
     [WARNING("%1 %2 %3", whattype(eval(M)), M, `if`(Sproc(M, 'h') = false,
     op([`has been activated in a current session`, assign(z = 14)]), cat(`is in library `,
     convert(h, name)))), `if`(z <> 14, assign(args[2] = h), NULL)]);
  if type(M, procedure) then RETURN(eval(ParProc(M)))
  else
    assign(a = cat(" `_62` := ", convert(eval(M), string))), interface(warnlevel = 0);
    assign(b = map2(Search, a, [`module`, `export`, `(`])); assign('c' = cat(a[1 .. b[1][1] - 1],
    "proc", a[b[3][1] .. b[2][1] - 1], a[searchtext(";", a, b[2][1] .. -1) + b[2][1] .. b[1][-1] - 1],
      "proc;")), writebytes(nu, c), close(nu);
    (proc() read nu; fremove(nu) end proc)(), assign('a' = ParProc(`_62`)),
      interface(warnlevel = omega);
    assign(R = map(N, [[L[1] = op(cs(1, M))], [L[2] = op(cs(3, M))], [L[3] = op(cs(6, `_62`))],
      [L[4] = op(3, eval(`_62`))], [L[5] = op(5, eval(`_62`))]])), unassign('`_62`'),
      RETURN(array(1 .. nops(R), 1 .. 1, R))
  end if
end proc

Sproc:= proc(P::symbol)
local k, L, R, h, s, omega;
  `if`(type(P, procedure) or type(P, `module`), assign(L = [libname], R = {},
     s = CF(cat(CDM(), "\\lib"))), RETURN(false)), assign(omega = (proc(R, s)
    local a, b;
      assign(a = map(CF, R), b = `if`(type(P, package), package, `if`(type(P, procedure),
      Proc, Mod)));
      if {s} = a then b, Maple elif member(s, a) then b, `Maple&User` else b, User end if
    end proc));
```

```
  `if`(type(P, builtin), RETURN(true, `if`(1 < nargs, assign(args[2] =
      [Proc, builtin, Builtin(P), s]), 7)), 7);
  try assign(h = IO_proc(P)),
      `if`(type(h, integer), RETURN(true, `if`(1 < nargs, assign(args[2] = [Proc, iolib, h, s]), 7)), 7)
  catch "": seq(`if`(search(convert(march(list, L[k]), string),cat("["", P, ".m",")),
      assign('R' = {op(R), L[k]}), NULL), k = 1 .. nops(L))
  end try;
      `if`(R = {}, RETURN(false), RETURN(true, `if`(nargs = 2,
      assign([args][2] = [omega(R, s), R]), NULL)))
end proc
```

The combined use of the procedures **ParProc1** and **Sproc** provides the more differentiable localization of the program objects in the *Maple* environment (a library and/or a current session).

Typical examples of the procedures use:

> ParProc(MkDir), ParProc(came); map(ParProc, ['add', march, goto, iostatus, seq]);

matrix([[*Arguments* = (F::{*string*, *symbol*})], [*locals* = (cd, r, k, h, z, K, L, Lambda, t, d, omega, omega1, u, f, s, g, v)]]), matrix([[*Arguments* = (E::*anything*)], [*locals* = (f, h)], [*globals* = (__Art_Kr_)]])

[*builtin function*, 138, *iolib function*, 31, *builtin function*, 182, *iolib function*, 13, *builtin function*, 253]

> ParProc(DIRAX);

DIRAX is module with exports, [*new, replace, extract, empty, size, reverse, insert, delete, sortd, printd, conv*]

> ParProc(StatLib);

[*Arguments* = (L::{*string*, *symbol*})],
[*locals*=(*a, k, h, S, P, K, G, H, t, pf, unp, u, V, T,R, W, Z,calls1,bytes1,depth1,maxdepth1, time1*)],
[*globals* = (*profile_maxdepth, profile_calls, profile_bytes, profile_depth, profile_time, $Art_Kr, calls2, bytes2, depth2, maxdepth2, time2*)]])

> ParProc(process);

process is module with exports , [*popen, pclose, pipe, fork, exec, wait, block, kill, launch*]

> ParProc1(ParProc1, 'h'), h;

Warning, procedure **ParProc1** is in library {"c:/program files/maple 9/lib/userlib"}

matrix([[*Arguments* = (M::{*procedure, module*})], [*locals* = (a, b, c, d, p, h, t, z, cs, L, R, N, omega, nu)], [*options* = (*system, remember, Copyright International Academy of Noosphere - Tallinn -2000*)], [*globals* = (*61, ParProc, Sproc*)]]), {"c:/program files/maple 9/lib/userlib"}

> ParProc1(Sockets, 't'), t;

Warning, module **Sockets** is in library {"C:\Program Files\Maple 9/lib"}

[*exports* = (socketID, Open, Close, Peek, Read, Write, ReadLine, ReadBinary, WriteBinary, Server, Accept, Serve, Address, ParseURL, LookupService, GetHostName, GetLocalHost, GetLocalPort, GetPeerHost, GetPeerPort, GetProcessID, HostInfo, Status, Configure, pexports)]
[*locals* = (defun, trampoline, soPath, solib, passign, setup, finalise)]
[*options* = (package, noimplicit, unload = finalise, load = setup)]
[*description* = ("package for connection oriented TCP/IP sockets")]
[*package*, Maple, {"C:\Program Files\Maple 8/lib"}]

```
> Art_Kr:= () -> `+`(args)/nargs:  6*Art_Kr(61,56,36,40,7,14);
                                   214
> ParProc1(Art_Kr, 'p'),  p;
   Warning, procedure Art_Kr has been activated in a current session
                       [options = (operator, arrow)], p
> Sproc(goto, 's'),  s;
              true, [Proc, built-in, 193, "c:\program files\maple 9\lib"]
> Sproc(writebytes, 'h'),  h;
              true, [Proc, iolib, 5, "c:\program files\maple 9\lib"]
> Sproc(assign, 'h'),  h;
              true, [Proc, Maple, {"C:\Program Files\Maple 9/lib"}]
> Sproc(MkDir, 'h'),  h;
              true, [Proc, User, {"c:/program files/maple 9/lib/userlib"}]
> Sproc(RANS_IAN_RAC_REA, 'z'),  z;
                                false, z
> Sproc(SimpleStat, 'p'),  p;
              true, [package, User, {"c:/program files/maple 9/lib/userlib"}]
> Sproc(LinearAlgebra, 'p'),  p;
              true, [package, Maple, {"C:\Program Files\Maple 9/lib"}]
> Sproc(`type/package`, 'h'),  h;
true, [Proc, Maple&User, {"C:\Program Files\Maple 9/lib", "c:/program files/maple 9/lib/userlib"}]
```

---

relml – check of compatibility of files {".m", ".lib"} with the current *Maple* release

Call format of the procedure:

relml(F)

Formal arguments of the procedure:

F - symbol or string defining a filename or full path to a datafile of type {**".m"**, **".lib"**}

Description of the procedure:

The datafiles of types {**".m"**, **".lib"**} are not fully compatible relative to the releases of *Maple*, namely: a datafile of a type {**".m"**, **".lib"**} created in the environment of a release {6|7|8|9|9.5} is compatible with any release 7 - 9.5, whereas the datafile created in the environment of a release {7|8|9|9.5} is not compatible with *Maple 6*. The procedure relml checks compatibility of such kind.

The successful call **relml(F)** returns the *true* value if and only if a datafile specified by the actual argument **F** has a type {**".m"**, **".lib"**} and is compatible with the current *Maple* release; otherwise, the *false* value is returned. The procedure handles basic especial and erratic situations.

```
relml:= proc(F::file)
local a, r, t;
  `if`(member(Ftype(F), {".lib", ".m"}), assign(a = readbytes(F, 5), r = Release(), t = Ftype(F)),
     [close(F), ERROR("datafile <%1> has a type different from {".m", ".lib"}", F)]), close(F);
  if t = ".lib" then if a[1] <> 1 then ERROR("<%1> is not a Maple library datafile", F) end if
  end if;
  if r = 6 then if t = ".lib" and a[3] <> 54 or t = ".m" and a[2] <> 54 then false else true end if
  else true
  end if
end proc
```

Typical examples of the procedure use:

> relml("C:/temp/rans_ian6.m"), relml("C:/temp/rans_ian7.m"); # *Maple* 6

true, false

> map(relml, ["C:\\Program Files\\Maple 6\\LIB\\Maple.lib", "C:\\Program Files\\Maple 8/LIB/Maple.lib", "C:\\Program Files\\Maple 9\\LIB\\UserLib\\Maple.lib"]);

[*true, false, false*]

> relml("C:/temp/rans_ian6.m"), relml("C:/temp/rans_ian7.m"); # *Maple* 7

true, true

> map(relml, ["C:\\Program Files\\Maple 6\\LIB\\Maple.lib", "C:\\Program Files\\Maple 7/LIB/Maple.lib", "C:\\Program Files\\Maple 9\\LIB\\UserLib\\Maple.lib"]);

[*true, true, true*]

> relml("C:/temp/rans_ian6.m"), relml("C:/temp/rans_ian7.m"); # *Maple* 9

true, true

> map(relml, ["C:\\Program Files\\Maple 6\\LIB\\Maple.lib", "C:\\Program Files\\Maple 7/LIB/Maple.lib", "C:\\Program Files\\Maple 8\\LIB\\UserLib\\Maple.lib"]);

[*true, true, true*]

> relml("C:/Academy/UserLib9/UserLib/VGTU.lib");

Error, (in **relml**) <C:/Academy/UserLib9/UserLib/VGTU.lib> is not a *Maple* library datafile

---

MmF – merge of files created by the `*save*` function of the package mmf

Call format of the procedures:

MmF(F, F1 {, h})
mmf(F, G, H

Formal arguments of the procedures:

F, F1, G - symbols or strings defining names or full paths to the merged datafiles
H - symbol or string defining name or full path to a target *m*-file
h - (optional) an arbitrary valid *Maple* expression

Description of the procedures:

The successful procedure call **MmF(F1, F2)** returns the *NULL* value, i.e. nothing, providing merge of datafiles (within definitions of program modules and procedures composing their) specified by their qualifiers **F1** and **F2**, and which represent the result of saving of the program objects by means of the built-in function *`save`*. Such merge result updates the first file **F1** with preservation of its format. In addition, the files of different formats can be united, namely: the input *Maple* language format or internal *Maple* format.

The procedure call **MmF(F1, F2, h)** with the optional third argument **h** provides removing out of a current session of those objects, which have been activated as a result of merge of datafiles **F1** and **F2**, and which earlier either have not been made active, or do not belong to libraries logically linked with the main *Maple* library. The procedure has a series of rather useful appendices. In particular, on the basis of the **MmF** procedure an algorithm of converting of datafiles of the input *Maple* language format into the internal *Maple* format is being enough easily implemented.

In addition, it is necessary to have in mind an important circumstance linked with impossibility of correct saving of definitions of program modules in *m*-files (the internal *Maple* format) by means of the *`save`* function. Therefore, if the merged datafiles contain program modules in the internal *Maple* format which were saved by means of the function *`save`*, then and the **MmF** procedure will incorrectly fulfill the merging. With this purpose, the procedure outputs the appropriate warning. The procedure **MmF** handles the basic erratic and especial situations.

At last, the **mmf** procedure is intended for merge of two compatible *m*-files with saving of the result in a new *m*-file. The procedure call **mmf(F, G, H)** merges two *m*-files **F** and **G** compatible relative to the *Maple* releases with return of full path to a target *m*-file **H** and with output of an appropriate warning. If a target *m*-file does not exist, it will be created. The merged *m*-files must have different names or full paths to them and must be created by the compatible *Maple* releases, otherwise errors arise. The procedure call handles all basic erratic and especial situations and does not affect the state of a current *Maple* session. The procedure has a series of rather useful appendices. In particular, on the basis of the **mmf** procedure the procedure **save3** has been implemented which correctly saves program modules in datafiles of the internal *Maple* format.

```
MmF:= proc(F1::file, F2::file)
local act, a, b, c, d, k, Art, Kr, omega;
  omega:= proc(f)
    local nu;
      assign(nu = interface(warnlevel)), interface(warnlevel = 0);
      try parse(readbytes(f, TEXT, infinity), 'statement')
      catch "incorrect syntax in parse:": close(f); interface(warnlevel = nu), RETURN(false)
      end try;
      close(f), interface(warnlevel = nu), true
    end proc;
  `if`(map(Ftype, {F1, F2}) = {".m"} or Ftype(F1) = ".m" and omega(F2) or Ftype(F2) = ".m"
     or map(omega, {F1, F2}) = {true}, assign(act = {anames(procedure), anames(`module`)},
     Art = [], Kr = []), ERROR("one or both files %1 are not readable", {F1, F2}));
  `if`(Rmf(F1) and Rmf(F2), assign(a = [Read1([F1], 'b'), b, Read1([F2], 'c'), c]),
    ERROR("the merged m-files should correspond to a current Maple release"));
  d := {seq(op([op(a[k][procs]), op(a[k][mods1]), op(a[k][mods2])]), k = [1, 2])};
  seq(`if`(type(eval(d[k]), procedure), assign('Art' = [op(Art), d[k]]),
    `if`(type(eval(d[k]), `module`), assign('Kr' = [op(Kr), d[k]]), 1)), k = 1 .. nops(d)),
    `if`(Kr <> [], WARNING("uncorrect saving of modules is highly probable!"), 6);
  `if`(cat(" ", F1)[-2 .. -1] <> ".m" or Kr = [], (proc() save args, F1 end proc)(op(d)),
    (proc() modproc(op(Kr)); save args, F1 end proc)(op(Art), op(Kr)));
  if 1 < nargs then for k in d do
```

```
      if not (member(k, act) or ismLib(k)) then unprotect(k); unassign(k); next end if end do
   end if
end proc

mmf:= proc(F::file, G::file, H::{string, symbol})
local a, b, f;
   `if`(map(IsFtype, {F, G}, ".m") = {true}, NULL, ERROR("at least one file has type different
    from '.m'")); `if`(CF(F) = CF(G), ERROR("files are identical"), assign(a = map(readbytes,
    [F, G], TEXT, 2))); close(F, G), `if`(SD(a[1][2]) = SD(a[2][2]), NULL,
      ERROR("m-files inhere in incompatible releases")); `if`(cat("  ", H)[-2 .. -1] <> ".m",
      assign(f = cat(H, ".m")), assign(f = H)), assign('a' = interface(warnlevel));
   if not type(f, file) then interface(warnlevel = 0); MkDir(f, 1); interface(warnlevel = a) end if;
   assign('a' = readbytes(F, infinity), 'b' = readbytes(G, infinity)), close(F, G);
   writebytes(f, [op(a[1 .. -2]), 44, 10, op(b[6 .. -1])]), close(f); WARNING("merge result of two
   compatible files of the internal Maple format is in file <%1>", f), f
end proc
```

Typical examples of the procedures use:

```
> MM:=module()  export x,m;  m:=module()  export y;  y:=() -> `+`(args) end module end
  module: MM1:=module() local F,m; export a; option package; m:=module() export y;
  y:=() -> `+`(args) end module; F:=() -> m:- y(args); a:=F end module: module SVEGAL()
  export x; x:=() -> `+`(args)/nargs^6 end module: PP:=proc() `+`(args)/nargs^2 end proc:
  PP1:=proc() `+`(args)/nargs end proc: PP2:=proc() `+`(args)/nargs end proc: module GS ()
  export x; x:=() -> `+`(args)/nargs^3 end module: save(MM, PP, "C:/RANS/Temp/File1");
  save(GS, PP1, "C:/RANS/Temp/File1.m"); Read1(["C:/RANS/Temp/File1"], 'h'), eval(h);
                    table([mods2 = {}, mods1 = {MM}, vars = {}, procs = {PP}])
> Read1(["C:/RANS/Temp/File1.m"], 'h'), eval(h);
                    table([mods2 = {GS}, mods1 = {}, vars = {}, procs = {PP1}])
> MmF("C:/RANS/Temp/File1", "C:/RANS/Temp/File1.m");
   Warning, uncorrect saving of modules is highly probable!
> Read1(["C:/RANS/Temp/File1"], 'h'), eval(h);
   Error, (in unknown) attempting to assign to `GS` which is protected
                 table([mods2 = {GS}, mods1 = {MM}, vars = {}, procs = {PP1, PP}])
> map(type, [PP, PP1], procedure), map(type, [MM, GS], `module`);
                              [true, true], [true, true]
> restart: MmF("C:/RANS/Temp/File1", "C:/RANS/Temp/File1.m", 61);
   Error, (in unknown) attempting to assign to `GS` which is protected
   Warning, uncorrect saving of modules is highly probable!
> map(type, [PP, PP1], procedure), map(type, [MM, GS], `module`);
                              [true, false], [false, true]
> F:= "C:/rans/academy/vilnius\\Res.m": G:= "C:/rans/academy/vilnius\\Res1.m":
> mmf(F, G, "C:/RANS/IAN\\Academy/Grodno\\grsu");
   Warning, merge result of two compatible files of the internal Maple format is in file
   <C:/RANS/IAN\Academy/Grodno\grsu.m>
                        "C:/RANS/IAN\Academy/Grodno\grsu.m"
```

```
> read("C:/RANS/IAN\\Academy/Grodno\\grsu.m");
> mmf(F, F, "C:/RANS/IAN\\Academy/Grodno\\grsu");
    Error, (in mmf) files are identical
> mmf(F, "C:/temp/aaa.m", "C:/RANS/IAN\\Academy/Grodno\\grsu");
    Error, (in mmf) m-files inhere in incompatible releases
> mmf("C:/temp/aaa.m", "C:/archive/abs.txt", "C:/RANS/IAN\\Academy/Grodno\\grsu");
    Error, (in mmf) at least one file has type different from '.m'
```

---

mod_proc – converting of a program module into procedure
ModProc
modproc

Call format of the procedures:

mod_proc(M {, F})
ModProc(M)
modproc(*args* {, `$`})

Formal arguments of the procedures:

M – symbol defining name of a program module
F – (optional) symbol or string defining a filename or full path to it
args – a sequence of program modules names
`$` – (optional) key symbol defining a saving mode of result

Description of the procedures:

The package has a rather developed mechanism of profiling of procedures - of main program objects of the *Maple* language. Between that, for program modules the given method of a profiling in its direct view is inapplicable. In our books [8,10-12,29-33,43] we have offered other profiling method, whose essence consists in replacement of a program module by the equivalent procedure. For automation of the given process, it is possible to use two rather useful procedures **mod_proc** and **ModProc**.

With the purpose of the equivalent transformation of a program module (whose definition has been evaluated in a current session) into the procedure, the **mod_proc** procedure can be substantially useful. This procedure returns a procedure with name **proc_N** which is created on the basis of the given program module, where **N** is name of a source module named by any permissible way.

The **mod_proc(M, F)** procedure admits up to two actual arguments. In addition, the first obligatory actual **M** argument specifies name of a program module whose definition has been evaluated in a current session. In this case the procedure call **mod_proc(M)** returns the procedure, which is fully equivalent to the initial program module **M.** Whereas the procedure call with two actual arguments along with the above result saves the generated procedure in a file, the full path to which or its name is specified by the second actual **F** argument.

Because of the above-mentioned converting of a program module into the procedure the exported variables of the module, become the local variables of the procedure, which are being returned during its execution. In addition, the call of the created procedure without arguments returns a sequence of local variables (*exported variables of the source module*). The access to these variables is provided by a way visually represented by examples below. In particular, the **mod_proc** procedure seems as a rather useful facility for a deriving and saving of the source texts of the package modules. However, for modules the clause `**with**` should be previously executed.

The **ModProc(M)** procedure uses only one actual **M** argument – name of an evaluated program module, returning the equivalent procedure. Single restriction for the procedure is the necessity of absence for the converted module **M** of the nested modules. In a case of detection of such module, the error arises. Successful execution of the procedure returns the procedure with name of a view **mp_M** and with output of the appropriate warning. The exported variables of the program module are being put into ***global*** declaration of the external procedure – of the result of converting of the **M** module. The procedure created on the basis of a program module can be profiled by the standard method. The call **mp_M()** provides access to variables exported by an initial **M** module.

In our books [8,10-12,29-33,43,44] the question of incorrect saving of program modules in files of the internal *Maple* format (*m*-files) has been considered in detail enough. We have offered a method of a converting of program modules into equivalent procedures, which next are correctly saved in the files of both formats by means of the standard function `***save***` of the package. This method is based on the **modproc** procedure represented below.

As the actual arguments the **modproc** procedure admits a sequence of names of program modules and, possibly, key symbol **`$`**, which defines a saving mode of the result in files of both formats. The procedure call

modproc(M1, M2, ..., ..., `$`, ..., Mk)   (p=1..k)

returns the *NULL* value, i.e. nothing, providing in a current session the converting of **Mp** program modules into **Mp** procedures which are correctly saved by means of the `***save***` function in files of both *Maple* formats. At presence among the actual arguments of key symbol **`$`**, the converted **Mp** procedures are saved in a current directory in files of both formats with names **`_$$$`** and **`_$$$.m`** accordingly. In addition, the program modules of the second type are converted into modules of the first type without attribute *`protected`*. Because of execution, the procedure can output the appropriate warnings.

Syntax of the first access to the exported variables of such converted program module differs a little from the standard and has the following view:

Name(): Name:- ***exportX***

where ***Name*** – name of a program module and ***exportX*** is any variable exported by it. All subsequent accesses to such module in a current session use syntax adopted in *Maple*. The above said also concerns use of program modules, saved thus in the *m*-files, after reading of an *m*-file containing them, by means of the `***read***` function. The **modproc** procedure is a rather useful facility at making a various sort of software whose algorithm supposes a saving of program modules in files of the internal *Maple* format (*m*-files). The examples represented below enough lucidly illustrate both the results of usage of the above procedures, and algorithms used by them.

```
mod_proc:= proc(M::`module`)
local alpha, beta, m, n, s, t, f, f1, h, loc, ex, e, T, R, z, v, w;
  e:= (a, b, c) -> op([assign('h' = searchtext(c, a, b .. length(a)) + b - 1), h]);
  assign(s = cat(M, " := module "), f = cat([libname][1][1 .. 2], "\\$Sveta"));
  T:= (proc(x) eval(M); save M, x; readbytes(x, TEXT, infinity) end proc)(f);
  close(f), assign(R = map2(Search, T, ["module", "local", "export"]));
  m := cat("proc_", T[searchtext(s, T) .. R[1][1] - 1], "proc() ");
  assign(z = cat(M, ":-"), n = "", ex = T[R[3][1] + 7 .. e(T, R[3][1], ";")]),
    `if`(search(T[1 .. R[3][1]], "local", 't'), assign('m' = cat(m, T[R[2][1] .. e(T, R[2][1], ";") - 1],
    ",", ex)), assign('m' = cat(m, "local ", ex))), assign('m' = cat(m, T[e(T, R[3][1], ";") + 1 ..
    R[1][-1] - 5], ex, " end proc;"));
  seq(assign('n' = cat(n, `if`(m[k] = " " and m[k + 1 .. k + 2] = ":-", "", m[k]))), k = 1 .. length(m));
  assign(loc = Search(n, z)), `if`(loc = [], assign(v = n), [assign(alpha = n[1 .. loc[1] - 1],
```

```
    beta = n[loc[-1] + length(z) .. length(n)]), assign(v = cat(alpha, seq(n[loc[k] + length(z) ..
    loc[k + 1] - 1], k = 1 .. nops(loc) - 1), beta))]);
  (proc(x, y) (proc(z, q) writebytes(q, z), close(q); read q end proc)(x, y),
      `if`(search(y, "\\$Sveta"), fremove(y), fremove(f)), eval(cat(proc_, M))
    end proc)(v, `if`(nargs = 2, args[2], f))
end proc

ModProc:= proc(M::`module`)
local G, N, T, z, t, W, h, F;
  assign(F = cat([libname][1][1 .. 2], `/$`)), (proc(x) save M, x end proc)(F);
  W:= proc(S, K)
    local a, b, c, p, h;
      [assign('a' = [], 'b' = length(S), 'c' = length(K), 'p' = 1), seq(`if`(search(S[p .. b], K, 'h'),
      assign('a' = [op(a), p + h - 1], 'p' = h + p), RETURN(a)), k = 1 .. length(S))]
    end proc;
  assign(T = readbytes(F, TEXT, infinity)), close(F), assign(N = W(T, "module")), unassign('t'),
   `if`(2 < nops(N), ERROR("module <%1> contains nested modules", M),
    assign(h = cat(mp_, M)));
  `if`(search(T, cat("unprotect('", M, "');")), assign('T' = T[length(M) + 16 .. N[2] + 6]), NULL);
  N := map(searchtext, ["module", "export", "global", "end module"], T);
  `if`(N[2] = 0, WARNING("module <%1> has not of the exported names", M), NULL);
  assign(G = cat("mp_", T[1 .. N[1] - 1], "proc")), assign(z = length(T));
  `if`(N[2] = 0, assign('G' = cat(G, T[searchtext("()", T) .. N[4] + 3], "proc;")),
    [search(T[N[2] .. length(T)], "proc", 'z'), `if`(search(T[N[2] .. z + N[2] - 1], "global", 't'),
    assign('G' = cat(G, T[searchtext("()", T) .. N[2] - 1], "global", T[N[2] + 6 ..
    searchtext(";", T[N[2] .. N[2] + t - 1]) + N[2] - 2], ",", T[N[2] + t + 6 .. N[4] + 3], "proc;")),
    assign('G' = cat(G, T[searchtext("()", T) .. N[2] - 1], "global", T[N[2] + 6 .. N[4] + 3],
   "proc;")))]), writeline(F, G), close(F), (proc(y) read y end proc)(F), fremove(F);
  eval(h), WARNING("module <%1> has been converted into procedure with name <%2>", M, h)
end proc

modproc:= proc()
local k, h, t, M, gs;
  assign(gs = (proc(s, n)
    local h;
      `if`(search(s, cat(" ", n, " ()"), 'h'), cat(s[1 .. h - 1], s[h + length(n) + 1 .. -1]), s) end proc)),
    assign(M = []), `if`(nargs = 0, RETURN(NULL), seq(`if`(args[k] = `$`, NULL, `if`(
    type(eval(args[k]), `module`), [assign('M' = [op(M), args[k]]), unprotect(args[k])],
    WARNING("<%1> is not a program module", args[k]))), k = 1 .. nargs));
  `if`(M = [], ERROR("converting has not been done - program modules are absent"), NULL);
  for k to nops(M) do
    assign('h' = gs(convert(eval(M[k]), string), M[k])), assign('h' = cat("", M[k],
      " := proc() RETURN(assign(", M[k], " = ", h, ")) end proc;")), writebytes("_$#@61", h),
        close("_$#@61");
    (proc() read "_$#@61"; fremove("_$#@61") end proc)()
  end do;
  `if`(member(`$`, {args}), null([(proc() save args, `_$$$`; save args, `_$$$.m` end proc)(op(M)),
     WARNING("the converted program modules have been saved in files <%1> and <%2>",
     cat(CDM(), `/_$$$`), cat(CDM(), `/_$$$.m`))]), NULL)
end proc
```

Typical examples of the procedures use:

```
> module M1() local a,b,c,A; global d; option package; description "RANS_IAN"; export
  sr, Dis; A:= module() export h; h:=() -> `+`(args)/nargs end module; sr:=() -> A:- h(args);
  Dis:=() -> add((args[k]-sr(args))^2, k=1..nargs)/(nargs-1); end module;

  module M1 () local a, b, c, A; export sr, Dis; global d; option package; description "RANS_IAN";
                                          end module

> mod_proc(M1);

                proc()
                local a, b, c, A, sr, Dis;
                global d;
                option package;
                description "RANS_IAN";
                A := module() export h; h := () -> `+`(args)/nargs end module;
                sr := () -> A:- h(args);
                Dis := () -> add((args[k] - sr(args))^2, k = 1 .. nargs)/(nargs - 1);
                sr, Dis
                end proc

> 30*seq(proc_M1()[k](42,47,67,62,89,96), k=1..2);

                                 2015, 14249

> mod_proc(M1, "C:/RANS/Temp/Art.txt"); restart; read("C:/RANS/Temp/Art.txt");

                proc()
                local a, b, c, A, sr, Dis;
                global d;
                option package;
                description "RANS_IAN";
                   A := module() export h; h := () -> `+`(args)/nargs end module;
                   sr := () -> A:- h(args);
                   Dis := () -> add((args[k] - sr(args))^2, k = 1 .. nargs)/(nargs - 1);
                   sr, Dis
                end proc
                proc_M1 := proc()
                local a, b, c, A, sr, Dis;
                global d;
                option package;
                description "RANS_IAN";
                   A := module() export h; h := () -> `+`(args)/nargs end module;
                   sr := () -> A:- h(args);
                   Dis := () -> add((args[k] - sr(args))^2, k = 1 .. nargs)/(nargs - 1);
                   sr, Dis
                end proc

> 6*proc_M1()[1](42,47,67,62,89,96), 30*proc_M1()[2](42,47,67,62,89,96);

                                 403, 14249

> Kr:= module() local a; global c; export x,y,z; assign(x=(() -> `+`(args)/ nargs), z=(()-> [args,
  nargs])); y:= proc() local k; c*sum(args[k]^2, k=1..nargs)/x(args) end proc end module:
  ModProc(Kr);
```

Warning, module <Kr> has been converted into procedure with name <mp_Kr>

```
proc()
local a;
global x, y, z, c;
   x := () -> `+`(args)/nargs;
   y := proc() local k; c*sum(args[k]^2, k = 1 .. nargs)/x(args) end proc;
   z := () -> [args, nargs]
end proc
```

> c:=2003: mp_Kr(): 3*x(61,56,36,40,7,14), 103*y(42,61), z(45,67,78);

107, 21972910, [45, 67, 78, 3]

> module M1() local a,b,c; global d; option package; description "RANS_IAN"; export sr, Dis,A; A:= module() export h; h:=() -> `+`(args)/nargs end module; sr:=() -> A:- h(args); Dis:=() -> add((args[k]-sr(args))^2,k=1..nargs)/(nargs-1); end module: Kr:=module() local a; global c; export x,y,z; assign(x=(()->`+`(args)/nargs), z=(()->[*args,nargs*])); y:=proc() local k; c*sum(args[k]^2, k=1..nargs)/x(args) end proc end module: modproc(Kr,`$`,M1);

Warning, the converted program modules have been saved in files <C:\Program Files\Maple 9/_$$$> and <C:\Program Files\Maple 9/_$$$.m>

> M1(): 6*M1:- sr(61,56,36,40,7,14), 30*M1:- Dis(61,56,36,40,7,14);

214, 14192

> restart: read(`_$$$.m`): M1(): 6*M1:- sr(61,56,36,40,7,14), 30*M1:- Dis(61,56,36,40,7,14);

214, 14192

> c:=107: Kr(): 6*Kr:- x(61,56,36,40,7,14), 71*Kr:- y(61,56,36,40,7,14);

214, 2129574

In particular, the **mod_proc** procedure seems as a rather useful tool for receiving and saving the source codes of package modules. For example, by means of constructions of the following general view:

with(*pacman*): interface(*verboseproc*=3): mod_proc(*pacman*); **6 - 8**
with(*PackageManagement*): interface(*verboseproc*=3): mod_proc(*PackageManagement*); **9/9.5**

the specified information can be received for the package modules providing a set of utilities for interactive management by modules for releases 6 - 8 and 9/9.5 of *Maple.*

---

mod21 – converting of a program module of the *second* type into the *first* type

Call format of the procedure:

mod21(M {, h})

Formal arguments of the procedure:

M - symbol defining a program module of arbitrary type
H - (optional) symbol or string defining a directory

Description of the procedure:

A series of inconveniences of operation with program modules of the second type, which in a series of cases require the more complex algorithms of their handling at software making, bring up a problem of program converting them into modules of the first type. In addition, a program module of the *first* type is understood as the module named by construction of view **"Name:= module () ... "**, whereas a program module of the *second* type is characterized by determinative construction of view **"module Name () ... "**. Some useful considerations on use of the program modules of the second type can be found in our books [12, 29-33,43]. Rather simple, but useful enough procedure **mod21** solves the given problem.

The procedure call **mod21(M)** returns the *NULL* value, i.e. nothing, providing a converting in a current session of a program module of the second type specified by the actual argument **M** into the equivalent program module of the first type of the same name. If the procedure call **mod21(M, h)** uses the second optional argument **h**, the argument is considered as a directory in which should be saved a datafile "M.mod1" with definition of the converted program module **M** (if the **h** directory is absent, then it will be created with arbitrary nesting level). In this case, the procedure call returns full path to a datafile with the saved converted program module **M.** A datafile with the saved program module has the input *Maple* language format. The examples below illustrate the said.

```
mod21:= proc(M::{`module`})
local h, t, x, y, nu;
  `if`(type(M, mod1), RETURN(NULL), assign(h = convert(eval(M), string),
  nu = interface(warnlevel))); search(h, " () ", 't'), writebytes("$V+S+G$",
    cat("", M, " := module", h[t .. -1], ":"));
  (proc(x, y) close(x); unprotect(y); read x; fremove(x) end proc)("$V+S+G$", M);
  if 1 < nargs and type(eval(args[2]), {string, symbol}) then
    if type(args[2], dir) then h := cat(args[2], "\\", M, ".mod1"); save M, h; h
    else interface(warnlevel = 0), assign('h' = cat(MkDir(args[2]), "\\", M, ".mod1"));
      (proc() interface(warnlevel = nu); save M, h end proc)(), h
    end if
  end if
end proc
```

Typical examples of the procedure use:

```
> module SVEGAL() export x,y; x:=() -> `+`(args)/nargs; y:=() -> [args, nargs] end module:
> type(SVEGAL, mod1), mod21(SVEGAL), type(SVEGAL, mod1), SVEGAL:- x(61,56,36);
                                   false, true, 51
> module Art_Kr () export SR, DS; SR:=() -> `+`(args)/nargs; DS:=() -> sqrt(sum((args[k] -
  SR(args))^2, k=1..nargs)/(nargs-1)) end module: type(Art_Kr, mod1), mod21(Art_Kr,
  "C:/RANS/IAN/RAC/REA"), type(Art_Kr, mod1);
                     false, "C:/RANS/IAN/RAC/REA\Art_Kr.mod1", true
```

mpl_proc – extracting of procedures and program modules from text datafiles

Call format of the procedure:

mpl_proc(F, R {, `Mod`} {, 'h'})

Formal arguments of the procedure:

F – symbol or string defining a filename or full path to a text datafile
R – symbol or string defining a filename or full path to a target datafile
Mod – (optional) key word defining search mode of program modules
h – (optional) an assignable name

Description of the procedure:

An *mws*-file is the most widespread form of the *Maple* documents, which has structure acceptable for all releases and platforms, and text format, that provides both mobility, and transfer simplicity by email. Along with documents of the given format have circulation (truth, much less wide) documents of the *mpl*-format, which also have text format and contain the source codes of program modules and *Maple* procedures, but their structure is different from the first one. Mainly, these files are represented by the so-called *Share*-library that was delivered with *fifth* release of *Maple*. The procedure **mpl_proc** can appear as a rather useful tool for operation both with *mpl*-files, and with a series of other text files containing definitions of program modules and *Maple* procedures. In addition, in general case an arbitrary text datafile can be as actual value for the first argument of the procedure.

The procedure **mpl_proc(F, R)** on the basis of a text datafile specified by the first actual **F** argument creates a text datafile specified by the second actual **R** argument, which will contain definitions of program modules or procedures, contained in the first one. At coding of the third optional **h** argument through it, the list of names of program modules or procedures saved in the **R** datafile is returned. In addition, the names are returned as strings, because non-empty intersections of them with names, in particular, of predetermined *Maple* variables are possible (see the fourth example below).

The procedure **mpl_proc** is oriented onto processing of text datafiles having a format, similar to the format of types {*mpl, mi*} of the *Share* library; however, it can be applied successfully enough and for text datafiles if the procedures definitions have standard notation, namely:

Name:= proc() ..... end proc

i.e. procedures should be the named *Maple* procedures, whereas program modules should have the first type, i.e. they should be named as follows:

Name:= module () ..... end module

If the procedure call uses the optional key word **`Mod`**, the procedure realizes search of program modules of the first type in a datafile **F**, otherwise the search of the named *Maple* procedures is done. The successful procedure call returns the full path to a target datafile **R** with the extracted procedures or program modules, otherwise the *NULL* value is returned with output of an appropriate warning, or an error arises. The procedure has a series of useful enough appendices.

```
mpl_proc:= proc(F::file, G::{string, symbol})
local a, b, c, k, j, h, p, L, r, s, f, t, omega, nu, z, psi, x, y, g, v, epsilon;
  assign(t = interface(warnlevel), epsilon = {args} minus {F, G, Mod}), interface(warnlevel = 0);
  try `if`(type(F, rlb), assign(s = "", z = [], h = Red_n(Sub_st([" " = " "], readbytes(F, TEXT,
  infinity), b)[1], " ", 2)), RETURN(interface(warnlevel=t), ERROR("<%1> is not a text file", F)))
  catch "invalid subscript selector": close(F); h := Red_n(readbytes(F, TEXT, infinity), " ", 2)
  finally interface(warnlevel = t)
  end try;
  `if`(member(Mod, {args}), assign(x = {":=module(", ":= module(", ":=module (",
    ":= module ("}, v = "program modules of the 1st type", y = {"end module"},
    g = " end module;"), assign(x = {":=proc(", ":= proc(", ":=proc (", ":= proc ("},
    y = {"end proc"}, g = " end proc;", v = "the named Maple procedures"));
  psi:= proc(s)
    local k, h;
      for k from length(s) by -1 to 1 do
```

```
        if s[k] = " " then next else h := s[1 .. k]; break end if end do;
      for k to length(h) do if h[k] = " " then next else RETURN(h[k .. -1]) end if end do
    end proc;
  r:= proc(s, n)
    local k;
      for k from n - 1 by -1 to 1 do
        if member(s[k], {";", "\n", ":", "  "}) then RETURN(s[k + 1 .. n])
        else next
        end if
      end do;
      RETURN(s[1 .. n])
    end proc;
  interface(warnlevel = t), close(F), assign(L = extrS(h, x, y, c));
  if not Empty(c) then for k to nops(L) do
      assign('p' = r(h, c[k][1])), assign('z' = [op(z), psi(p[1 .. -2])]);
      assign('s' = cat(s, p, L[k], `if`(member(Mod, {args}), " end module;", " end proc;"), "\n"))
    end do
  else RETURN(WARNING("file <%1> does not contain %2", F, v))
  end if;
  if type(G, file) then f := G; writebytes(f, s), close(f)
  else interface(warnlevel = 0); f := MkDir(G, 1); writebytes(f, s), close(f)
  end if;
  try interface(warnlevel = 0), parse(readbytes(f, TEXT, infinity), 'statement');
    interface(warnlevel = t), close(f)
  catch "incorrect syntax": interface(warnlevel = t), close(f); WARNING("<%1> can contain
    syntax errors", f)
  catch "attempting to assign": interface(warnlevel = t), close(f)
  end try;
  f, `if`(epsilon = {}, NULL, `if`(type(eval(epsilon[1]), assignable(symbol)), assign(epsilon[1]=z),
    WARNING("argument should be assignable symbol but has received - %1",
    whattype(eval(epsilon[1])))))
end proc
```

Typical examples of the procedure use:

> mpl_proc("C:/rans/ian/rac/rea/Edge.mpl", "C:/RANS/Edge.proc");

"c:\rans\edge.proc"

> mpl_proc("C:/rans/ian/rac/rea/ShapeObj.mpl", "C:/RANS/ShapeObj.proc");

"C:/RANS/shapeobj.proc"

> mpl_proc("C:/rans/ian/rac/rea/Graph.mi", "C:/RANS/Graph.proc");

"C:/RANS/graph.proc"

> mpl_proc("C:/rans/ian/rac/rea/graph.mi", "C:/RANS/graph.proc", h), h;

"C:/RANS/graph.proc", ["Vertices", "Edges", "Order", "Size", "AddEdge"]

> mpl_proc("C:/rans/ian/rac/rea/shapeobj.mpl", "C:/RANS/shapeobj.proc", p), p;

"C:/RANS/shapeobj.proc", ["point", "segment", "len", "midpoint", "circle", "square", "triangle", "area", "circumference"]

> mpl_proc("C:/rans/ian/rac/rea/graph.mi", "C:/RANS/graph.proc", 147);

Warning, argument should be assignable symbol but has received - integer

"C:/RANS/graph.proc"

> mpl_proc("C:/rans/ian/rac/rea/graph.mi", "C:/RANS/graph.mod", \`Mod\`);

Warning, file <C:/rans/ian/rac/rea/graph.mi> does not contain program modules of the 1st type

> mpl_proc("C:/rans/ian/rac/rea/shapeobj.mpl", "C:/RANS/shapeobj.mod", \`Mod\`, d), d;

Warning, <C:/RANS/shapeobj.mod> can contain syntax errors

"C:/RANS/shapeobj.mod", ["ShapeObjects"]

> mpl_proc("c:/temp/rans/Grodno.txt", "C:/Temp/RANS/Grodno.proc");

Warning, file <c:/temp/rans/Grodno.txt> does not contain the named Maple procedures

> mpl_proc("c:/temp/rans/Art.txt", "C:/Temp/RANS/Art.mod", \`Mod\`);

Warning, file <c:/temp/rans/Art.txt> does not contain program modules of the 1st type

---

_mMfile – converting of a datafile of input *Maple* language format into internal *Maple* format, and vice versa

Call format of the procedure:

_mMfile(F)

Formal arguments of the procedure:

F - string or symbol defining a filename or full path to a datafile

Description of the procedure:

The procedure call **_mMfile(F)** returns the full path to a target datafile, providing conversion of a datafile of the input *Maple* language format, specified by the actual **F** argument, into a target datafile of the internal *Maple* format with the same name with adding of extension ".m", and vice versa. However, it is necessary to have in mind, what in a result of such conversion in the output file will be absent the definitions of variables, which can be in an initial datafile **F**.

Thus, the given procedure supports conversion of datafiles at a level of definitions of program modules and procedures composing them. In addition, also it is necessary to have in mind, what on a lower level the **_mMfile** procedure uses the built-in *\`save\`* function which does not support correct saving in a datafile of the internal *Maple* format of program modules, therefore at conversion into the internal *Maple* format the **_mMfile** procedure converts all program modules, contained in a source datafile **F**, into procedures, appropriate to them.

The procedure running is accompanied by output of information warnings of view "*Warning, incorrect saving of modules is highly probable!*" which are generated by the used procedure ***Save2*** and which do not influence the result correctness. Whereas, the returned full path to a target datafile is intended for support of computer-aided programs running mode in the *Maple* environment.

Henceforth at reading of such datafiles by means of the *\`read\`* function the given circumstance supposes change of syntax of the first access in a current session to the program modules contained in these datafiles. The given syntax is rather simple and can be found below at description of procedures of the *Save* group. The procedure in a series of cases allows to eliminate the defects inherent to the built-in *`save`* function of the *Maple* package. The **_mMfile** can seem as an useful facility, in particular, at creation of the user libraries with organization different from the standard one. In a series of cases, such libraries can be more effective.

```
_mMfile:= proc(F::{string, symbol})
local `_$`, `_###`, `_@@@`, _111, _2, `_act$`, k;
  `if`(type(eval(F), file), `if`(isRead(F), `if`(cat(" ", F)[-2 .. -1] = ".m", NULL, sMf(F)),
    RETURN(false)), ERROR("<%1> is not a datafile", F)), assign(`_act$` = {anames()}),
    Read1([F], '`_###`'), assign(_111 = [], _2 = [], `_$` = {op(`_###`[mods2]),
    op(`_###`[mods1]), op(`_###`[procs])});
  map(unprotect, map(op, [`_###`[mods1], `_###`[mods1]]));
  `if`(cat(" ", F)[-2 .. -1] = ".m", assign(`_@@@` = cat("", F, ".txt")),
    assign(`_@@@` = cat("", F, ".m")));
  seq(`if`(type(eval(`_$`[k]), procedure), assign('_111' = [op(_111), `_$`[k]]),
    `if`(type(eval(`_$`[k]), `module`), assign('_2' = [op(_2), `_$`[k]]), NULL)),
    k = 1 .. nops(`_$`)), `if`(`_@@@`[-2 .. -1] <> ".m" or _2 = [],
    (proc() save args, `_@@@`; `_@@@` end proc)(op(`_$`)),
    (proc() Save2(op(`_$`), `_@@@`) end proc)()), unassign('`###`',
    seq(`if`(member(`_$`[k], `_act$`) or ismLib(`_$`[k]), NULL, `_$`[k]), k = 1 .. nops(`_$`)))
end proc
```

Typical examples of the procedure use:

> MM:=module() export x,m; m:=module() export y; y:=() -> \`+\`(args) end module end module: MM1:= module() local F,m; export a; option package; m:=module() export y; y:=() -> \`+\`(args) end module; F:=() -> m:- y(args); a:=F end module: SVEGAL:= module () export x; x:=() -> \`+\`(args)/nargs^6 end module: PP:=proc() \`+\`(args)/nargs^2 end proc: PP1:=proc() \`+\`(args)/nargs end proc: PP2:=proc() \`+\`(args)/nargs end proc: GS:= module () export x; x:=() -> \`+\`(args)/nargs^3 end module:

> save(MM, MM1, SVEGAL, GS, PP, PP1, "C:/RANS/IAN/RAC/REA/Art_Kr.m");

> Read1(["C:/RANS/IAN/RAC/REA/Art_Kr.m"], 'h'), eval(h);

table([*mods2* = {}, *vars* = {}, *mods1* = {*SVEGAL, GS, MM1, MM*}, *procs* = {*PP, PP1*}])

> _mMfile("C:/RANS/IAN/RAC/REA/Art_Kr.m");

"c:/rans/ian/rac/rea/art_kr.m.txt"

> restart; Read1(["C:/RANS/IAN/RAC/REA/Art_Kr.m"], 'h'), eval(h);

table([*mods1* = {*SVEGAL, GS, MM, MM1*}, *vars* = {}, *procs* = {*PP, PP1*}, *mods2* = {}])

> restart: MM:=module() export x,m; m:=module() export y; y:=() -> \`+\`(args) end module end module: MM1:=module() local F,m; export a; option package; m:=module() export y; y:=() -> \`+\`(args) end module; F:=() -> m:- y(args); a:=F end module: SVEGAL:= module () export x; x:=() -> \`+\`(args)/nargs^6 end module: PP:=proc() \`+\`(args)/nargs^2 end proc: PP1:=proc() \`+\`(args)/nargs end proc: PP2:=proc() \`+\`(args)/nargs end proc: GS:=module () export x; x:=() -> \`+\`(args)/nargs^3 end module:

> save(MM, MM1, SVEGAL, GS, PP, PP1, "C:/RANS/IAN/RAC/REA/Art_Kr.txt");

> _mMfile("C:/RANS/IAN/RAC/REA/Art_Kr.txt");

Warning, uncorrect saving of modules is highly probable!
Warning, uncorrect saving of modules is highly probable!

```
Warning, uncorrect saving of modules is highly probable!
Warning, uncorrect saving of modules is highly probable!
Warning, uncorrect saving of modules is highly probable!
Warning, uncorrect saving of modules is highly probable!
                              "c:\rans\ian\rac\rea\art_kr.txt.m"
> restart; read("C:/RANS/IAN/RAC/REA/Art_Kr.txt.m");
> SVEGAL(): 23328*SVEGAL:- x(61,56,36,40,14,7), 3*PP1(61,56,36,40,14,7);
                                        107, 107
```

Remember_T – an update of the *remember*-tables

Call format of the procedure:

Remember_T(NP)

Formal arguments of the procedure:

NP - a procedure name

Description of the procedure:

If the ***remember*** option has been specified in the options field of a *Maple* procedure, then at the end of each procedure call, an entry is made in its *remember*-table that records the result for the specified arguments. Subsequent procedure calls with the same arguments simply retrieve the result out of the *remember*-table. Use of the given mechanism requires the certain memory expenditures, however allows to obtain an essential saving of time at various cyclic calculations on identical values of actual arguments of the procedures. For providing of operation with the *remember*-tables, the package offers a series of useful functional facilities. Here for these purposes the **Remember_T** procedure is offered. The **Remember_T** procedure from indefinite quantity of formal arguments provides a rather effective modification of a *remember*-table by editing its indices and entries.

```
Remember_T:= proc(NP::{name, symbol})
local k, p, h, G, F, S;
global __T;
  if not type(NP, procedure) then ERROR("<%1> is not a procedure", NP)
  elif nargs < 2 then RETURN(op(4, eval(NP)))
  else assign('__T' = op(4, eval(NP)), F = cat([libname][1][1 .. 2], `/$$$`))
  end if;
  for k from 2 to nargs do assign('G' = "", 'S' = "");
    if whattype(args[k]) = `=` then  __T[op(lhs(args[k]))] := rhs(args[k])
    else
      for p to nops(args[k]) - 1 do G := cat(G, convert(args[k][p], string), ",") end do;
      G := cat("'__T[", cat(G, convert(args[k][nops(args[k])], string)),"]';");
      assign('S' = convert(__T[args[k]], string)), assign('S' = cat(S[1 .. 4], S[6 .. -2]));
      assign('G' = cat(cat(S, ":="), G)), writebytes(F, convert(G, bytes)), close(F);
      read F;
      __T[seq(args[k][p], p = 1 .. nops(args[k]))] := %, fremove(F)
    end if
  end do;
  unassign('__T'), op(4, eval(NP))
end proc
```

The first actual **NP** argument of the procedure **Remember_T(NP)** specifies a procedure name whose *remember*-table should be updated. The remaining actual arguments of the procedure are coded in the following rather simple view:

(1) [x1, x2, ... , xn] = <*value*> or **(2) [x1, x2, ... , xn]**

where the first format specifies necessity of replacement of an entry for index **(x1,x2,...,xn)** of a *remember*-table onto the given *value* or inserting into the table of a new index, if the indicated are absent; whereas the second format specifies exclusion of the given index out of the table.

The procedure call returns the *remember*-table updated according to settings for its indices and entries, specified by actual arguments starting with the second. If the procedure call contains only one actual argument - a procedure name, a current *remember*-table of a procedure **NP** is returned. In addition, in case of empty *remember*-table the *NULL* value is returned, i.e. nothing. Examples of the procedure calls **Remember_T** illustrate the update results of *remember*-table of the procedure ***Art_Kr*** from point of view of the above operations, namely: **(1)** replacement of elements of the *remember*-table, **(2)** insertion of new elements into the *remember*-table and **(3)** deletion of elements out of the *remember*-table.

In many cases, the **Remember_T** procedure appears as a rather useful tool, above all, at necessity of the more effective use of the memory at operation with recursive procedures and with often-used procedures in cyclic calculations.

Typical examples of the procedure use:

```
> Art_Kr:=proc() options operator,arrow,remember;  evalf(sum(args[k],k=1..nargs)/nargs,7)
  end proc:  Remember_T(Art_Kr);
> [Art_Kr(3,10,32,37,52,57),Art_Kr(96,89,67,62,47,42),Art_Kr(30,3,99,250,520,5),Art_Kr(47,
  35,14,6,20,2), Art_Kr(1,2,3,4,5,6), Art_Kr(6,34,12,89,96,13)]; op(4,eval(Art_Kr));
                [31.83333, 67.16667, 151.1667, 20.66667, 3.500000, 41.66667]
    table([(30, 3, 99, 250, 520, 5) = 151.1667, (47, 35, 14, 6, 20, 2) = 20.66667, (3, 10, 32, 37, 52, 57) =
    1.83333, (1, 2, 3, 4, 5, 6) = 3.500000, (96, 89, 67, 62, 47, 42) = 67.16667, (6, 34, 12, 89, 96, 13) =
    41.66667])
> Remember_T(Art_Kr,[3,10,32,37,52,57]=2003,[96,89,67,62,47,42]=1995,[31,3,99,42]=[7,14],
  [1,2,3,4,5,6], [47,35,14,6,20,2]=`1995_2003`, [30,3,99,250,520,5]);
    table([(47, 35, 14, 6, 20, 2) = 1995_2003, (3, 10, 32, 37, 52, 57) = 2003, (31, 3, 99, 42) = [7, 14],
    (96, 89, 67, 62, 47, 42) = 1995, (6, 34, 12, 89, 96, 13) = 41.66667])
> Remember_T(Art_Kr);
    table([(47, 35, 14, 6, 20, 2) = 1995_2003, (3, 10, 32, 37, 52, 57) = 2003, (31, 3, 99, 42) = [7, 14],
    (96, 89, 67, 62, 47, 42) = 1995, (6, 34, 12, 89, 96, 13) = 41.66667])
```

The above-mentioned examples of concrete application of the **Remember_T** procedure are clear enough and do not demand any special explanations.

---

sMf – converting of files of the input *Maple* language format containing modules

Call format of the procedure:

sMf(F {, G})

Formal arguments of the procedure:

F - symbol or string defining a filename or full path to a datafile
G - (optional) symbol or string defining a filename or full path to a target datafile

Description of the procedure:

A series of inconveniences of operation with program modules of the second type, which in a series of cases require the more complex algorithms of their handling at implementing of software, bring up the problem of software conversion of them into program modules of the first type. In addition, a program module of the first type is understood as a module named by construction of the next view **"Name:= module () ... "**, whereas a module of the second type is characterized by the determinative construction of view **"module Name () ... "**. Some useful considerations about use of program modules of the second type can be found in our books [29-33,43,44]. The useful enough procedure **sMf** solves the given problem on the level of datafiles.

The procedure call **sMf(F {, G})** returns the full path to a target datafile and provides conversion of a source datafile of the input *Maple* language format, specified by the actual **F** argument and obtained by means of the ***`save`*** function, into a datafile of the same format, in which all program modules of the second type have been converted into program modules of the first type with *protected*-attribute. The procedure call with sole actual **F** argument fulfils update of a source datafile in situ, whereas the procedure call with two arguments provides saving of the result of converting of an **F** datafile into a datafile, specified by the second optional actual **G** argument.

The examples, represented below, illustrate not only syntax of both methods of procedure call, but also the results of its application for conversion of datafiles of the input *Maple* language format which contain program modules of the second type. The results of conversion are tested by the **Read1** procedure until and after of application of the **sMf** procedure to the created target file of the input *Maple* language format. The **sMf** procedure is being successfully used in many appendices dealing with datafiles of the input *Maple* language format that contain the program modules of the above two types.

```
sMf:= proc(F::{string, symbol})
local a, omega, k, h, n, t, p, v, name, w, y, s, R, x, G, f, nu, z;
  `if`(not type(F, file), ERROR("<%1> is not a datafile", F), `if`(type(F, rlb) and
    Ftype(F) <> ".m", assign67(h = readbytes(F, TEXT, infinity), t = "unprotect('", R = "",
    close(F), f = F), ERROR("<%1> is not a text datafile or has the internal Maple format", F)));
  assign(omega = writebytes, nu = (x -> interface(warnlevel = x)), z = (x -> convert(x, string)));
  do
    if search(h, t, 'p') then
      search(h[p .. -1], "');", 'v'), assign('name' = h[p + 11 .. p + v - 2]);
      assign('s' = length(name), 'y' = cat(t[3 .. -1], name, "');"));
      assign('h' = h[1 .. p + v + s + 13], 'w' = h[p + v + s + 14 .. -1]);
      h := cat(h, `if`(w[1 .. s] = name, w[s + 1 .. -1], w));
      assign('x' = Search(h, y)), assign('x' = x[2] + length(y) - 1), assign('R' = cat(R, h[1 .. x]));
    if length(h) - x <= 25 then R := cat(R, h[x+1..-1]); break else h := h[x+1..-1]; next end if
    else R := cat(R, h); break
    end if
  end do;
  if nargs = 1 then close(f); omega(f, R); close(f); z(f)
  else f := args[2];
    if type(f, file) then close(f); omega(f, R); close(f); z(f)
    else a := interface(warnlevel); nu(0); f := MkDir(f, 1); nu(a); omega(f, R); close(f); z(f)
    end if
  end if
end proc
```

Typical examples of the procedure use:

```
> MM:=module() export x,m; m:=module() export y; y:=() -> `+`(args) end module end
  module: MM1:= module() local F,m; export a; option package; m:=module() export y;
  y:=() -> `+`(args) end module; F:=() -> m:- y(args); a:=F end module: module SVEGAL ()
  export x; x:=() -> `+`(args)/nargs^6 end module: PP:=proc() `+`(args)/nargs^2 end proc:
  PP1:=proc() `+`(args)/nargs end proc: PP2:=proc() `+`(args)/ nargs end proc: module GS
  () export x; x:=() -> `+`(args)/nargs^3 end module:
> save(MM,PP,SVEGAL,GS,PP1,MM1,`c:/temp/avz`); Read1([`c:/temp/avz`], 'h'), eval(h);
    Error, (in unknown) attempting to assign to `SVEGAL` which is protected
    Error, (in unknown) attempting to assign to `GS` which is protected
         table([mods1 = {MM1, MM}, mods2 = {SVEGAL, GS}, procs = {PP, PP1}, vars = {}])
> sMf("C:/Temp/AVZ", "C:/RANS/Vilnius"); Read1(["C:/RANS/Vilnius"], 't'), eval(t);
                              "C:/RANS/Vilnius"
       table([mods1 = {SVEGAL, GS, MM1, MM}, mods2 = {}, procs = {PP, PP1}, vars = {}])
> sMf("C:/Temp/AVZ", "C:/Academy"); Read1(["C:/_Academy"], 'p'), eval(p);
                                "c:\_academy"
       table([mods1 = {SVEGAL, GS, MM1, MM}, mods2 = {}, procs = {PP, PP1}, vars = {}])
> sMf("C:/Temp/AVZ"); Read1(["C:/Temp/AVZ"], 'p'), eval(p);
                                "C:/Temp/AVZ"
       table([mods1 = {SVEGAL, GS, MM1, MM}, mods2 = {}, procs = {PP, PP1}, vars = {}])
```

Examples of the above fragment illustrate not only syntax of both methods of the procedure calls, but also results of its application for conversion of datafiles of the input *Maple* language format which contain program modules of the *second* type.

VisM – visualization of a program module

Call format of the procedure:

VisM(M)

Formal arguments of the procedure:

M – a program module name

Description of the procedure:

In contrast to procedures the mechanism of program modules provides the higher level of abstracting, by allowing to `*hide*` from the user the whole groups of subject-linked procedures, modules and data. Meantime, in many cases the necessity of analysis of organization of a program module arises what the simple procedure **VisM(M)** provides by return of source code of a program module **M** in the *string* format.

```
VisM:= proc(M::`module`)
  (proc() local a; a := convert([13], bytes); save M, "$VGS$" end proc)();
  cat(op(subsLS(["\n" = NULL], convert(readbytes("$VGS$", TEXT, infinity), list)))),
     fremove("$VGS$")
end proc
```

Typical examples of the procedure use:

```
> module M1() local a,b,c,A,B; global d; option package; description "RANS"; export sr, Dis; A:=module() export h; h:=()->`+`(args)/nargs end module; sr:=() -> A:-h(args); B:= "RANS\nIAN": Dis:=() -> add((args[k]-sr(args))^2, k=1..nargs)/(nargs-1); end module;
```

module M1 () local a, b, c, A, B; export sr, Dis; global d; option package; description "RANS"; end module

```
> VisM(M1);
```

"unprotect('M1'); M1 := module M1 () local a, b, c, A, B; export sr, Dis; global d; option package; description \"RANS\"; A := module () export h; h := proc () options operator, arrow; `+`(args)/nargs end proc; end module; B := \"RANS\\nIAN\"; sr := proc () options operator, arrow; M1:-A:-h(args) end proc; Dis := proc () options operator, arrow; add((args[k]-M1:-sr(args))^2,k = 1 ..nargs)/(nargs-1) end proc; end module; protect('M1');"

`convert/module` – converting of an *Maple* object into program module

Call format of the procedure:

convert(A, `*module*` {, F})

Formal arguments of the procedure:

A – an assigned name
F – (optional) symbol or string defining a filename or full path to a target datafile

Description of the procedure:

In a series of cases, the representation of an evaluated *Maple* object in a program module form is advisable. For that the **`convert/module`** procedure represented below can be useful enough.

The procedure call **convert(A, `*module*`)** returns the name of a program module - result of conversion of an assigned name specified by the actual argument **A,** i.e. is supposed that **A <> eval(A).** Any valid *Maple* expression can be assigned to the name **A.** In addition, if **A** defines a variable, then it should be coded in unevaluated form, i.e. **'A'.** The returned module is available in a current session only.

The returned module name has the following view `*m_A*`. If the procedure call **convert(A, `*module*`, F)** contains the optional third **F** argument defining a filename or full path to a datafile, the result of conversion is saved in this datafile in the input *Maple* language format. In this case the procedure call returns the **2**-element sequence whose the first element defines the name of the returned program module, whereas the second defines full path to a target datafile with the saved module.

After the procedure completion or read of a datafile containing program modules saved by the above manner, the access to their exported names has the following format: *m_A:- A*{|(*args*)}. The examples represented below illustrate the procedure appendices for conversion of the *Maple* objects of different types into program modules with saving of them in datafiles and without saving.

```
`convert/module`:= proc(A::anything)
local a, b, p, omega, f, r, t, nu, x, y;
  `if`(type(A, assignable), 7, RETURN(WARNING("illegal use of object <%1> as a name", A)));
  assign(p = interface(warnlevel), omega = (x -> interface(warnlevel = x)),
  b = cat("m_", A), f = "$avz$");
  if type(A, `module`) then b := A; r := 7; nu := y -> `if`(cat(" ", y)[-2 .. -1] = ".m", y[1 .. -3], y)
  else a := cat(b, ":=module() export ", A, "; ", A, ":=", convert(eval(A), string), "; end module;")
```

```
  end if;
  if nargs = 1 then t := 147
  elif type(eval(args[2]), {string, symbol}) then
    if type(args[2], file) then f := args[2] else omega(0); f := MkDir(args[2], 1); omega(p)
    end if
  else t := 14; WARNING("<%1> cannot be as a filename or full path to a datafile", args[2])
  end if;
  if r <> 7 then writebytes(f, a), close(f); read f else f := nu(f); save A, f end if;
  cat(``, b), `if`(t = 147 or t = 14, fremove(f), f)
end proc
```

Typical examples of the procedure use:

```
> Art:=14: convert('Art', `module`, "C:/RANS/IAN/Art.mod");
                              m_Art, "C:/RANS/IAN/Art.mod"
> with(m_Art), m_Art:- Art;
                                     [Art], 14
> agn:=() -> `+`(args)/nargs: convert(agn, `module`, "C:/RANS/IAN/AGN.mod");
                              m_agn, "C:/RANS/IAN/AGN.mod"
> with(m_agn), 3*m_agn:- agn(61,56,36,40,7,14);
                                     [agn], 107
> convert(61, `module`, "C:/RANS/IAN/Res.mod");
   Warning, illegal use of object <61> as a name
> Module:=module() export sr; sr:=() -> `+`(args)/nargs: end module;
                        Module := module () export sr; end module
> convert(Module, `module`, "C:/RANS/IAN/M.mod");
                              Module, "C:/RANS/IAN/M.mod"
> with(Module), 3*Module:- sr(61, 56, 36, 40, 7, 14);
                                     [sr], 107
> [convert(Module, `module`, "C:/RANS/IAN/M.m")], [convert(Module, `module`)];
                        [Module, "C:/RANS/IAN/M"], [Module]
> restart; read("C:/RANS/IAN/Art.mod"); read("C:/RANS/IAN/AGN.mod");
  read("C:/RANS/IAN/M.mod");
                        m_Art := module () export Art; end module
                        m_agn := module () export agn; end module
                        Module := module () export sr; end module
> map(with, [m_Art, m_agn, Module]), m_Art:- Art, 3*m_agn:- agn(61, 56, 36, 40, 7, 14),
  3*Module:- sr(61, 56, 36, 40, 7, 14);
                           [[Art], [agn], [sr]], 14, 107, 107
```

Tools represented in this chapter essentially extend the *Maple* facilities at operating with procedural and modular objects both by simplifying programming of a series of problems in the package environment and by uncovering better the nature of these objects. A series of them represent the undoubted interest both for creation of the concrete applications that widely use procedures and the program modules of both types, and for creation of software of wider assignment.

Chapter 3.
Software of operation with numeric expressions

Means of this section extend the possibilities of the *Maple* package at operation with numeric expressions. In particular, these means provide useful procedures such as converting of a floating-point number into symbolic customary format; special converting of integers into lists, and vice versa; converting of an integer into symbolic format with delimiter; converting of a number represented by a format {*string, symbol*} into fraction, etc. This software is represented by useful procedures such as **NDLN, _0N, mod3, NDLN, NDT, SSN** and **NumOfString.**

_0N – converting of a floating-point number into symbolic customary format

Call format of the procedure:

_0N(n)

Formal arguments of the procedure:

n - a floating-point number

Description of the procedure:

At return of a result in the form of a number smaller than **1**, a zero (**0**) in front of the decimal point is absent; in a series of cases it does not correspond to the requirements of designing of final outputs. The similar situation takes place and at converting of an integer into the floating-point number; in this case after the decimal point the zero is absent. Moreover, the scientific notation of numbers contains the **e**-exponent. For sufficing the customary number format for floating-point numbers the **_0N** procedure can be useful enough.

The procedure call **_0N(n)** returns the floating-point number defined by the actual **n** argument in the symbolic format customary for final outputs; in addition, the result has ***symbol***-type. For backward converting of the above customary format into the floating-point number the relation **SN(cat("", _0N(n)))** can be used (see the second example below). The procedure supports all last five releases of the *Maple* package, namely: **5, 6, 7, 8** and **9/9.5.**

```
_0N:= proc(x)
local a, b, c, d, k, z;
  `if`(member(x, {0.}), RETURN(`0.0`), `if`(member(x, {-0.}), RETURN(`-0.0`), NULL)), `if`(
  not type(x, float), convert(x, symbol), [`if`(x < 0, assign(a = convert(abs(x), string), z = `-`),
  assign(a = convert(x, string), z = ``)), `if`(abs(x) < 1, `if`(search(a, "e", 'b'), RETURN(cat(z, `0.`,
  cat(seq(`0`, k = 1 .. SN(a[b + 2 .. -1])), a[2 .. b - 1]))), RETURN(cat(z, `0`, a))), `if`(a[-1] = ".",
  RETURN(cat(z, a, `0`)), `if`(search(a, "e", 'b'), RETURN(assign(c = SN(a[b + 1 .. -1]),
  d = length(a[2 .. b - 1]))), `if`(c = d, cat(z, a), `if`(d < c, cat(z, a[2 .. b - 1], seq(`0`, k = 1 .. c - d),
  `.0`), 13))), RETURN(cat(z, a)))))])
end proc
```

Typical examples of the procedure use:

> map(_0N, evalf([6/14,61,42/47,89/96]));

[0.4285714286, 61.0, 0.8936170213, 0.9270833333]

> map(SN, map2(cat, "", %));

[0.4285714286, 61., 0.8936170213, 0.9270833333]

> _0N(evalf(-6.424767628996*10^13)), _0N(evalf(sqrt(2)/3)), _0N(Float(14,-45)), _0N(-2.3e4);

-64247676290000.0, 0.4714045206, 0.00000000000000000000000000000000000000000014, -23000.0

---

mod3 – computations over integers modulo m

Call format of the procedure:

mod3(*expr*, m {, 'h'})

Formal arguments of the procedure:

expr - an algebraic expression
m - a nonzero integer
h - an unevaluated name

Description of the procedure:

The procedure call **mod3(*expr*, m)** returns the result of evaluation of an expression defined by the first actual ***expr*** argument, over the integers modulo **m,** being an analog to the standard operator **`mod`.** If an unevaluated name has been coded as the third optional **h** argument, then via **h** is returned a **p** number satisfying the equality ***expr* = m*p + `mod`(*expr*, m).** In a series of cases, this procedure allows to simplify programming of the problems dealing with integers processing.

```
mod3:= (expr::integer, m::posint) -> op([expr mod m, `if`(nargs = 3, assign(args[3] = trunc(expr/m)),
        NULL)])
```

Typical examples of the procedure use:

> [mod3(1942,7,'h'), h], [mod3(2003,14, 'h'), h], [mod3(14061942,61, 'V'), V];

[3, 277], [1, 143], [39, 230523]

> mod3(194219471967198919981963,2003, 'f'), f;

1953, 9696428954927554667O

NDLN – a special converting of integers into lists, and vice versa

Call format of the procedure:

NDLN(L)

Formal arguments of the procedure:

L - a positive integer, or a list whose elements are decimal digits

Description of the procedure:

The procedure call **NDLN(L)** returns the result of converting of a list defined by the actual **L** argument into decimal number composed from the list elements. Whereas the procedure call **NDLN(L)** returns the result of converting of a decimal positive integer defined by the actual **L** argument, into the list whose elements are decimal digits of the positive integer **L.**

```
NDLN:= proc(N::{nonnegint, list(digit)})
local a;
  `if`(type(N, nonnegint), [seq(SN(convert(convert(N, string), list)[a]), a = 1 .. `if`(N = 0, 1,
  length(N)))], add(N[a]*10^(nops(N) - a), a = 1 .. nops(N)))
end proc
```

Typical examples of the procedure use:

> map(NDLN, [61,[6,1,3,3,4],56,1995,[2,0,0,3],140642,0,[0],[1,9,9,5]]);

[[6, 1], 61334, [5, 6], [1, 9, 9, 5], 2003, [1, 4, 0, 6, 4, 2], [0], 0, 1995]

> map(NDLN, [61,[6,1,3,3,4],7,1995,[2,0,0,3],140642,14,[7,9],[5,1,9,8,5]]);

[[6, 1], 61334, [7], [1, 9, 9, 5], 2003, [1, 4, 0, 6, 4, 2], [1, 4], 79, 51985]

NDT – converting of an integer into symbolic format with delimiter

Call format of the procedure:

NDT(N, n, D)

Formal arguments of the procedure:

N - an integer
n - an integer **> 0**
D - a string or a symbol (*delimiter*)

Description of the procedure:

The procedure call **NDT(N,n,D)** returns the result of converting of an integer into a special symbolic format; in this format the tuples of **n** numerals of the integer defined by the first actual **N** argument are parted (from left to right) by a symbol or a string defined by the third actual **D** argument. In addition, a type of the third actual **D** argument defines the type of returned result.

Whereas, the procedure call **NDT(N, n, D)** at the condition `` `length(N) <= n` `` is equivalent to the procedure call **convert(N, whattype(D)).** This procedure seems as an useful enough facility at programming of a number of problems dealing with symbolic and integer-valued calculations.

```
NDT:= proc(N::integer, n::posint, D::{string, symbol})
local a, b;
  `if`(length(N) <= n, RETURN(convert(N, whattype(D))), assign(a = cat("", N)));
  cat(seq(`if`(type(b/n, integer), cat("", D, a[b]), a[b]), b = -length(N) .. -1));
  convert(cat(`if`(N < 0, "-", ""), %[`if`(length(N) mod n = 0, length(D) + 1, 1) .. -1]),
  whattype(D))
end proc
```

Typical examples of the procedure use:

> NDT(99140642290747230467,3, " "), NDT(99140642290747230467,3,` `);

"99 140 642 290 747 230 467", *99 140 642 290 747 230 467*

> NDT(-99140642290747230467,6, "** "), NDT(99140642290747230467,42, "@@ ");**

"-99****140642****290747****230467", "99140642290747230467"

> NDT(0,6, "*** "), NDT(140642,2,`-`), NDT(+14062003,4,`***`);**

"0", *14-06-42*, *1406***2003*

SSN – converting of a string or symbol into fraction

Call format of the procedure:

SSN(S)

Formal arguments of the procedure:

S - a string or symbol

Description of the procedure:

The procedure call **SSN(S)** returns the result of converting of a string or symbol defined by the actual argument **S** which contains a rational number with sign or without, into the reduced fraction. While the call **SSN("")** or **SSN(``)** returns *zero* value. The procedure handles the basic especial and erroneous situations.

```
SSN:= proc(S::{name, string, symbol})
local a, b, c, d, e, f, g, h, p, q, r, k;
  op([assign('a' = convert(S,string), 'b' = (c -> add((op(convert(c[k],bytes))-48)*10^(length(c)-k),
  k = 1 .. length(c))), 'SSN_proc' = (proc() ERROR(`invalid argument`) end proc)),
  `if`({".", "-", "+", "/", seq(convert(k, string), k = 0..9)} intersect convert(a, 'set') <> convert(a,
  'set'), SSN_proc(), `if`(member(2, map(nops, map2(Search, a, {"-", "+", "/"}))), SSN_proc(),
  assign('h' = 1))), assign('d' = (e -> `if`(e[1] = "+", 1, `if`(e[1] = "-", -1, 1))*[assign('f' =
  `if`(member(e[1], {"-", "+"}), e[2 .. -1], e)), assign('g' = searchtext(".", f)), assign('p' = `if`(g <> 0,
  [assign('h' = 10^(length(f) - g), 'f' = cat(f[1 .. g - 1], f[g + 1 .. -1])), assign('f' = convert(b(f),
  string))], f)), b(f)/h])), `if`(search(a, "/", 'r'), assign('q' = a[1 .. r - 1], 's' = a[r + 1 .. -1]), NULL),
  `if`(type(r, integer), op(d(q))/op(d(s)), op(d(a))), unassign(seq(convert(k, symbol), k = 0 .. 11))])
end proc
```

Typical examples of the procedure use:

```
> map(SSN, ["42/61",`-7/14`,"+47/56",`61`,"","-1995/2003",`17/69`]);
                    [42/61, -1/2, 47/56, 61, 0, -1995/2003, 17/69]
> map(SSN, ["42/61",`-6/14`,"+47/35",`56`,"",`-5/9`,"-1995/2003"]); SSN("2.56/1.0/61");
                    [42/61, -3/7, 47/35, 56, 0, -5/9, -1995/2003]
Error, (in SSN_proc) invalid argument
```

NumOfString – converting of a string or symbol into numeric list

Call format of the procedure:

NumOfString(S, d)

Formal arguments of the procedure:

S - a string or symbol
d - string, symbol, a list or a set of symbols and/or strings defining delimiters of numbers

Description of the procedure:

The procedure call **NumOfString(S, d)** returns the list of numerical values contained in a string or a symbol **S** and parted by characters, defined by delimiter(s) **d.** Numbers included in the string/character can be of type {*integer, float, fraction*} with sign or without sign. The procedure handles basic especial and erroneous situations. The procedure has many useful appendices.

```
NumOfString:= proc(S::{string, symbol}, d::{set({string, symbol}), list({string, symbol}),
                        string, symbol})
local a, s, t;
  assign(t = interface(warnlevel)), interface(warnlevel = 0);
  if type(d, {string, symbol}) then a := d; s := cat("", S)
  else a := d[1]; s := SUB_S(map(`=`, [op({op(d)} minus {a})], a), cat("", S))
  end if;
  try map(SN, SLD(Red_n(s, a, 2), a)), interface(warnlevel = t)
  catch "Invalid value": interface(warnlevel = t);
    ERROR(`Argument <%1> contains values of types different from
    {integer, float, fraction}`, S)
  end try
end proc
```

Typical examples of the procedure use:

> L:="42 61 47 56 67 36 89 14.7 96 5 46.5 -95.99 42.47 8.09 56/61": NumOfString(L, " ");

[42, 61, 47, 56, 67, 36, 89, 14.7, 96, 5, 46.5, -95.99, 42.47, 8.090, 56/61]

> NumOfString(` 42/61 +95.02 -7.14 2003 19.05 -7/14 36/20 21/47 -25.02`, ` `);

[42/61, 95.020, -7.14, 2003, 19.050, -1/2, 9/5, 21/47, -25.020]

> NumOfString("625 rans 7742815", {" ", "\t", "\n", c});

Error, (in **NumOfString**) Argument <625 rans 7742815> contains values of types different from {integer, float, fraction}

> NumOfString(" 42/62 \n +95.02 a -7.14 2004 \t 19.05 -7/14 b 37/20 c 21/47 -25.02", {" ", a, b, c, "\t", "\n"});

[21/31, 95.020, -7.14, 2004, 19.050, -1/2, 9/5, 21/47, -25.020]

> NumOfString(" 42/62ab\n +95.02 abc -7.14 2003 \t\t\t 19.05 -7/14 bac 36/20 c 21/47 -25.02 \t\n", {" ", a, b, c, "\t", "\n"});

[21/31, 95.020, -7.14, 2004, 19.050, -1/2, 9/5, 21/47, -25.020]

Procedure **ExprOfString** presented in the following chapter generalizes the above **NumOfString** procedure to case of more common *Maple* types, different from *numeric*. Tools of this chapter provide additional facilities simplifying operation with numeric expressions. These tools have confirmed own efficiency in a lot of appendices dealing with the numeric expressions.

Chapter 4. Software of operation with strings and symbolic expressions

In the given section, the tools extending possibilities of the package *Maple* of releases 6, 7, 8 and 9/9.5 at operation with expressions of type {*string, symbol*} are considered. These tools provide a number of useful procedures such as special kinds of converting; comparison of strings or/and symbols; case sensitive pattern searching; exhaustive substitutions into strings or symbols; inversion of symbols, strings or lists; reducing of multiplicity of entries of a symbol into a string; identification of entries of special symbols into a string; and others. In a series of cases, these tools simplify operation with *Maple* objects of type {*string, symbol, name*} in the *Maple* environment.

---

came – extracting of the *Maple* expressions from their symbolic representations

Call format of the procedure:

came(E)

Formal arguments of the procedure:

E – an arbitrary valid *Maple* expression

Description of the procedure:

At a manipulating and evaluations with symbolic expressions in a lot cases a problem of expressions extract from their symbolic representations arises. For instance, from the symbolic expression "x*sin(y) + y*cos(x)" we wish to extract the expression in "in pure form", i.e. x*sin(y) + y*cos(x). For this purpose, the following procedure **came(E)** can be useful enough.

The procedure call **came(E)** returns the result of extract of a syntactically correct expression from its symbolic representation, i.e. from the expression **E** of a type {*symbol, string*}, above all. The expression **E** of other type is returned without processing, excluding its simplification. The procedure provides a handling of cases of syntax incorrectness of the extracted expressions. The examples below illustrate some results of the procedure calls.

```
came:= proc(E::anything)
local f, h;
global __Art_Kr_;
  try `if`(args = NULL or not type(E, {string, symbol}), RETURN(args),
        assign(h = convert(E, string), f = "$61$"));
     `if`(nops(Search(h, " ")) = length(h), RETURN(), parse(h, statement))
  catch: ERROR("incorrect syntax in expression <%1>", E)
  end try;
```

```
  writeline(f, cat("__Art_Kr_:=", h, ":")), close(f);
  (proc(x) read f; fremove(f) end proc)(f), __Art_Kr_, unassign('__Art_Kr_')
end proc
```

Typical example of the procedure use:

> map(came, [`61`, "47.56", `7/14`, "x*sin(x)", `-36.67`, "7+14*I", 42, (a+b)/(c-d)]);

[61, 47.56, 1/2, x*sin(x), -36.67, 7+14*I, 42, (a+b)/(c-d)]

> map(came, [``, "", `  `, "   ", *NULL*]); came("(sin(x)+cos(y)/(x-y^2)");

[]

Error, (in **came**) incorrect syntax in expression <(sin(x)+cos(y)/(x-y^2)>

It is necessary to note, for converting of symbolical representations of values of the *numeric*-type into their numerical equivalent, we recommend to use the **came** procedure having considerably greater responsiveness relative to the **SN** procedure. Besides, the first procedure provides correct converting of symbolical representations of complex numbers also.

frss – a special fragmentation of a string or symbol

Call format of the procedure:

frss(S, n)

Formal arguments of the procedure:

S - a string or symbol
n - a positive integer (*posint*)

Description of the procedure:

In many problems of symbolical processing, a problem of fragmentation of string or symbolical value on pieces of the given length arises. The following simple enough and effective procedure **frss** solves the given problem. The procedure call **frss(S, n)** returns the list whose elements are the pieces received by means of fragmenting of an **S** string or symbol on **n** parts. A type of the pieces is defined by type of the **S** argument. If a fragmentation is not applicable to **S**, the *empty* set is returned, i.e. []. The examples, represented below, clearly illustrate the principle of such fragmentation. This procedure is useful enough tool for many problems of symbolic processing.

```
frss:= proc(S::{string, symbol}, n::posint)
local a, b, c, k;
  assign(a = [], k = 1, c = cat("", S));
  do b := c[k .. k + n - 1]; a := [op(a), b];
    if length(b) < n then break else k := k + n end if
  end do;
  map(convert, `if`(a[-1] = "", a[1 .. -2], a), whattype(S))
end proc
```

Typical example of the procedure use:

> frss(`RANSIAN RAC REA 19952004 TallinnGrodno`, 4);

[*RANS, IAN , RAC , REA , 1995, 2004, Tal, linn, Grod, no*]

> map(frss, ["", ``, "   ", `   `, "123456789", `123456789`], 3);

[[], [], ["   "], [   ], ["123", "456", "789"], [*123, 456, 789*]]

blank – a special converting

Call format of the procedure:

blank(S)

Formal arguments of the procedure:

S - a string, symbol or a name

Description of the procedure:

At use of the built-in functions ***system*** and ***ssystem*** in a number of cases, the necessity of passing to them of strings with *blank* symbols arises. In addition, a condition consists in the following - these functions should consider the passed string as a substring containing *blank* symbols. The given condition is required, in particular, at execution of a number of commands of the MS DOS via the above built-in functions. The simple procedure **blank** solves this problem. The procedure call **blank(S)** returns the result of converting of a string or symbol defined by the actual **S** argument into a special representation of *string*-type. The example, represented below, clearly illustrates the principle of such conversion.

```
blank:= (S::{name, string, symbol}) -> `if`(search(S," "),cat("\"",S,"\""),cat("",S))
```

Typical example of the procedure use:

```
> map(blank,[`a v z`,"agn",`Art  Kr`,"2 0 0 3",`10 09 03`,"RANS  IAN"]);
        ["\"a v z\"", "agn", "\"Art  Kr\"", "\"2 0 0 3\"", "\"10 09 03\"", "\"RANS  IAN\""]
```

This procedure is useful enough tool, in particular, for standardization of file qualifiers.

Case – a case converting

Call format of the procedure:

Case(S {, t})

Formal arguments of the procedure:

S - a string or symbol
t - (optional) a type of case (*lower, upper*)

Description of the procedure:

The procedure call **Case(S)** returns the result of converting of a string or symbol defined by the actual **S** argument into *lowercase*. If is coded the second optional **t** argument (**t** out of {*lower, upper*}), the procedure call **Case(S, t)** returns the result of converting of a string or symbol defined by the actual **S** argument into a case defined by the second actual **t** argument. This procedure in a number of cases appears the more convenient than analogous package tools. The procedure has rather high responsiveness and its usage in procedures, above all, in cyclic constructions is very effective. It has been approved in many appendices.

```
Case:= proc(S::{string, symbol})
local k;
  convert(convert(subs(`if`(nargs = 2 and args[2] = upper, Resl, eval)
  ({seq(65 + k = 97 + k, k = 0 .. 25), seq(192 + k = 224 + k, k = 0 .. 31), 168 = 184}),
  convert(cat("", S), bytes)), bytes), whattype(S))
end proc
```

Typical examples of the procedure use:

> map(Case,[AVZ,TaLliNn,RaNs,IAN, "Lasnamae-2003"]), Case(`1995 Beginning – 2003 eND`);

[avz, tallinn, rans, ian, "lasnamae-2003"], *1995 beginning – 2003 end*

> map(Case, [AVZ,TaLliNn, RaNs, IAN, "Lasnamae-2003"], upper);

[AVZ, TALLINN, RANS, IAN, "LASNAMAE-2003"]

> map(Case, [AVZ, TaLliNn, RaNs, IAN, "Lasnamae-2003"], lower);

[avz, tallinn, rans, ian, "lasnamae-2003"]

CompStr – comparison of strings and/or symbols

Call format of the procedure:

CompStr(S, C {, 'h'})

Formal arguments of the procedure:

S - a string or symbol
C - a string or symbol
H - (optional) an unevaluated name

Description of the procedure:

The procedure call **CompStr(S, C)** returns the *true* value if and only if the both actual arguments **S, C** reduced to common *string*-type are sensitively identical; otherwise, the *false* value is returned. In any case, if actual arguments **S** and **C** are various in lengths, the appropriate warning is printed.

If an unevaluated name has been coded as the third optional actual **h** argument, then via this name the more precise definitions are returned, namely. If the actual arguments **S** and **C** are various in lengths, then via actual **h** argument the 2-element integer list is returned; the first and second elements of the list define lengths of the actual arguments **S** and **C** accordingly. Otherwise, through actual **h** argument the sequence of 3-element lists is returned; the first and second elements of each such list define distinguishable symbols of the actual arguments **S** and **C** accordingly, while the third integer element defines positions of these symbols in the first two actual arguments.

```
CompStr:= proc(S::{string, symbol}, C::{string, symbol})
local a, b, c, s, k, omega;
  assign(a = length(S), b = length(C), s = cat("", S), c = cat("", C)), `if`(s = c, RETURN(true),
    `if`(a <> b, RETURN(false, `if`(nargs = 3, assign(args[3] = [a, b]), NULL),
    WARNING("Various lengths of strings/symbols <%1> and <%2>", S, C)), 6));
  omega := seq(`if`(s[k] = c[k], NULL, [s[k], c[k], k]), k = 1 .. a);
  if omega = [] then true elif nargs = 3 then false, assign(args[3] = omega) else false end if
end proc
```

Typical examples of the procedure use:

> G:="abcdefghrstwhkt": S:="abckefgnrspmnbftde": CompStr(G,S), CompStr(G,S,'h'), h;

Warning, Various lengths of strings/symbols <abcdefghrstwhkt> and <abckefgnrspmnbftde>
Warning, Various lengths of strings/symbols <abcdefghrstwhkt> and <abckefgnrspmnbftde>

false, false, [15, 18]

> G:=`abcdefghrstw`: S:= "abckefgnrspm": CompStr(G,S,'v'), v;

false, ["d", "k", 4], ["h", "n", 8], ["t", "p", 11], ["w", "m", 12]

redss – a reducing of strings or symbols

Call format of the procedure:

redss(S, d, p)

Formal arguments of the procedure:

S – a string or symbol
d – a delimiter (may be *string* or *symbol* of length **1**)
p – a positive integer (*posint*)

Description of the procedure:

The procedure call **redss(S, d, p)** returns the result of reducing of a string or symbol specified by the actual **S** argument by means of deletion out of **S** substrings parted by delimiters **d** whose length is more than **p.** A type of the first argument defines a type of the returned result. If a delimiter **d** does not belong to **S** and **length(S) > p** or all above-mentioned substrings have length **> p** then the *NULL* value is returned, i.e. nothing. The procedure is a rather useful tool at a symbolic processing.

```
redss:= proc(S::{string, symbol}, d::{string, symbol}, p::posint)
local a, k;
  assign(a = SLD(S, d)), `if`(a = [] and p < length(S), RETURN(NULL), assign('a' =
  [seq(`if`(p < length(a[k]), NULL, a[k]), k = 1 .. nops(a))])), `if`(a = [], NULL, clsd(a, ","))
end proc
```

Typical examples of the procedure use:

> redss(\`aaa,bbb,xyzr,ccc,aaaaa,111,22222,77777,666,aaaa,xxx,yyyy,zzzz,hhh\`, ",", 3);

aaa,bbb,ccc,111,666,xxx,hhh

> redss("aaa,bbb,xyzr,ccc,aaaaa,111,22222,77777,666,aaaa,xxx,yyyy,zzzz,hhh", ",", 3);

"aaa,bbb,ccc,111,666,xxx,hhh"

> map(redss,["",\`\`,"111","111,222","1111","333",\`111,222\`,"1995,2003","A,V,Z"],",",3);

["", \`\`, "111", "111,222", "333", *111,222*, "A,V,Z"]

\`convert/lowercase\` – converting of a symbol or a string into lowercase \`convert/uppercase\`

Call format of the procedures:

convert(S, *lowercase*)
convert(S, *uppercase*)

Formal arguments of the procedures:

S – a string or symbol to be convert

Description of the procedures:

The procedure call **convert(S, *lowercase*)** returns the result of converting of a string or a symbol defined by the first actual **S** argument into *lowercase*. Whereas the procedure call **convert(S, *uppercase*)** returns the result of converting of a string or a symbol defined by the first actual **S** argument into *uppercase*. The procedure has rather high responsiveness and its usage in procedures, above all, in cyclic constructions is very effective. It has been approved in many appendices.

```
`convert/lowercase`:= proc(S::{name, string, symbol})
local k;
  convert(convert(subs({seq(65 + k = 97 + k, k = 0 .. 25), seq(192 + k = 224 + k, k = 0 .. 31),
  168 = 184}, convert(cat("", S), bytes)), bytes), whattype(S))
end proc

`convert/uppercase`:= proc(S::{string, symbol})
local k, h, t, G, R;
  assign(G = "", R = convert(S, string));
  for k to length(R) do G := cat(G, op([assign('t' = op(convert(R[k], bytes))),
     `if`(t <= 255 and 224 <= t or t <= 122 and 97 <= t, convert([t - 32], bytes),
     `if`(t = 184, "Ë", R[k]))]))
  end do;
  convert(G, whattype(S))
end proc
```

Typical examples of the procedures use:

```
> map(convert,[`C:\\E_Books`,"C:\\E_Books\\Mail13.txt"], lowercase);
                              [c:\e_books, "c:\e_books\mail13.txt"]
> map(convert,[`C:\\E_Books`,"C:\\E_Books\\Mail13.txt"], uppercase);
                              [C:\E_BOOKS, "C:\E_BOOKS\MAIL13.TXT"]
```

CS – allocation of symbols composing a string

Call format of the procedure:

CS(S {, 'h'})

Formal arguments of the procedure:

S - a string or symbol
h - (optional) an unevaluated name

Description of the procedure:

The procedure call **CS(S)** returns the allocation of symbols composing the analyzed string or symbol, defined by the actual **S** argument, in the context of the Roman alphabet (*Roman*), the Russian alphabet (*Russian*), the Arabic numerals (*decimal*) and special symbols **{~ ! @ # $ % ^ & * ( ) _ + {} : " <> ? | ` - = [ ] ;',./\}**. Depending on the symbolic composition of the string **S**, the call **CS(S)** returns the result in the form of the list whose elements belong to the set {*Roman, Russian, Decimal, Special*}. The procedure call **CS("")** or **CS(``)** returns the *empty* set, i.e. {}. The procedure call **CS("", 'h')** or **CS(``, 'h')** through the second actual **h** argument returns the *NULL* value, i.e. nothing.

If the second optional actual **h** argument has been coded, via **h** the full allocation of symbols composing the analyzed string or symbol defined by the first actual **S** argument, is returned in the form of the sequence of nested lists. The first element of each list of such sequence is element out of the above set {*Roman, Russian, Decimal, Special*}, identifying the type of symbols found in the **S** string. Whereas the remaining elements of the list is **2**-element sublists, whose the first element is the found symbol of the type, defined by the first element of the list, and the second element is its position in the **S** string. The procedure appears as rather useful tool for analysis of textual information situated both in strings/symbols and text datafiles in the context of symbols of types {*Roman, Russian, Decimal, Special*}.

```
CS:= proc(S::{string, symbol})
local a, b, c, d, h, k, p, s, Ro, Ru, De, Sp, Res;
  assign(a = [], b = [], c = [], d = [], s = cat("", S)), seq([assign('h' = op(convert(s[k], bytes))),
    `if`(member(h, {seq(p, p = 65 .. 122)} minus {95, 96}), assign('a' = [op(a), [cat("", s[k]), k]]),
    `if`(member(h, {seq(p, p = 192 .. 255)}), assign('b' = [op(b), [cat("", s[k]), k]]),
    `if`(member(h, {seq(p, p = 48 .. 57)}), assign('c' = [op(c), [cat("", s[k]), k]]),
     assign('d' = [op(d), [cat("", s[k]), k]]))))], k = 1 .. length(S));
  Ro, Ru, De, Sp := [Roman, op(a)], [Russian, op(b)], [Decimal, op(c)], [Special, op(d)];
  Res := [seq(`if`(nops(k) = 1, NULL, k[1]), k = [Ro, Ru, De, Sp])]; Res,
`if`(nargs = 2, assign(args[2] = op([seq(`if`(nops(k) = 1, NULL, k), k = [Ro,Ru,De,Sp])])), NULL)
end proc
```

Typical examples of the procedure use:

```
> GS:= "~AGN%$+61#{RANS!5-IAN-14}Artur/МАНУР_<Кристо/>& `PTP\ (2003)-1995$
Гродно 62 Tallinn 62 $ Lasnamae Galina Art_Kr":  CS(GS, 'Z'),  Z;
```

```
[Roman, Russian, Decimal, Special], [Roman, ["A", 2], ["G", 3], ["N", 4], ["R", 12], ["A", 13], ["N", 14],
["S", 15], ["I", 19], ["A", 20], ["N", 21], ["A", 26], ["r", 27], ["t", 28], ["u", 29], ["r", 30], ["T", 76], ["a",
77], ["l", 78], ["l", 79], ["i", 80], ["n", 81], ["n", 82], ["L", 89], ["a", 90], ["s", 91], ["n", 92], ["a", 93],
["m", 94], ["a", 95], ["e", 96], ["G", 98], ["a", 99], ["l", 100], ["i", 101], ["n", 102], ["a", 103], ["A", 105],
["r", 106], ["t", 107], ["K", 109], ["r", 110]], [Russian, ["М", 32], ["А", 33], ["Н", 34], ["У", 35], ["Р",
36], ["К", 39], ["р", 40], ["и", 41], ["с", 42], ["т", 43], ["о", 44], ["Р", 50], ["Т", 51], ["Р", 52], ["Г", 66],
["р", 67], ["о", 68], ["д", 69], ["н", 70], ["о", 71]], [Decimal, ["6", 8], ["1", 9], ["5", 17], ["1", 23], ["4",
24], ["2", 54], ["0", 55], ["0", 56], ["3", 57], ["1", 60], ["9", 61], ["9", 62], ["5", 63], ["6", 73], ["2", 74],
["6", 84], ["2", 85]], [Special, ["~", 1], ["%", 5], ["$", 6], ["+", 7], ["#", 10], ["{", 11], ["!", 16], ["-", 18], ["-
", 22], ["}", 25], ["/", 31], ["_", 37],["<", 38],["/", 45], [">", 46],["&", 47], [" ", 48], ["`", 49], ["(", 53],
[")", 58], ["-", 59],["$", 64], [" ", 65], [" ", 72], [" ", 75], [" ", 83], [" ", 86], ["$", 87], [" ", 88], [" ", 97], [" ",
104], ["_", 108]]
```

---

conSV – a converting of strings and symbols into vectors, and vice versa
conSA – a converting of strings and symbols into *rtable*-arrays

Call format of the procedures:

conSV(S)
conSA(S1 {, d})

Formal arguments of the procedures:

S - a string, symbol or vector to be converted
S1 - a string or symbol to be converted
d - (optional) a string, symbol or set or list of strings and/or symbols defining delimiter

Description of the procedures:

Maple undergoes certain difficulties at processing of list data structures that cause erratic situations with diagnostic "*assigning to a long list, please use arrays*" at use of lengthy enough lists. In particular, in a lot cases a converting of long strings into lists is needed. However, in the subsequent such lists can cause the above erratic situation. The procedure call **conSV(S)** returns the result of conversion of a string or symbol specified by the actual argument **S** into vector, and vice versa. A type of S defines type of elements of the returned vector, whereas conversion of a vector **S** returns a *string* if majority of **S** elements has *string*-type and a *symbol* otherwise. Vectors are indexed analogously to lists, supporting the same processing algorithms and guarding against the above erratic situations. In particular, the procedure has many rather useful appendices at strings processing.

At last, the procedure call **conSA(S1)** returns a 1-dimensional array of *rtable*-type, whose elements are the symbols composing the source string **S1.** Whereas the procedure call **conSA(S1, d)** returns a 1-dimensional array of *rtable*-type, whose elements are substrings of the source string **S1,** parted by delimiters **d.** As delimiters can act both separate symbols, and strings, and also their set or list. If the second optional argument **d** belongs to the set {``, "", {}, []} or its delimiters do not belong to the source string **S1,** the procedure call **conSA(S1, d)** is equivalent to a call **conSA(S1).** If the first actual argument **S1** is an *empty* symbol or string, i.e. {``, ""}, the procedure call returns it without processing. Procedure handles the basic special and erroneous situations. Both procedures have a lot of useful appendices dealing with the *Maple* expressions of types {*string, symbol*}.

```
conSV:= proc(S::{vector, string, symbol})
local a, b, k, h;
  if type(S, vector) then
    assign(a = rhs(op(2, S))), assign(b = map(type, eval(S), string)), memberL(true, eval(b), 'h');
    convert(cat(seq(convert(S[k], string), k = 1 .. a)), `if`(nops(h) < 1/2*a, symbol, string))
  else vector([seq(convert(cat("", S)[k], whattype(S)), k = 1 .. length(S))])
  end if
end proc

conSA:= proc(S::{string, symbol})
local a, b, c, d, m, k, r, p, t, h, g, z, x, y;
  z := (x, y) -> convert(x, whattype(y));
  if nargs = 1 or 1 < nargs and member(args[2], {{}, [], ``, ""}) then
    if member(S, {``, ""}) then RETURN(S) end if;
    assign(a = cat("", S), m = length(S)); c := Array(1 .. 1, 1 .. m);
    for k to m do c[1, k] := z(a[k], S) end do; c
  else
    if type(args[2], {list({string, symbol}), set({string, symbol}), string, symbol}) then
      d:= args[2]; a := cat("", S); b:= convert([1], bytes); r:= `if`(type(d, {set, list}), {op(d)}, {d})
    else ERROR("the second argument is invalid; it should be {symbol, string, list({symbol,
      string}), set({symbol, string})} but has received <%1>", args[2])
    end if;
    `if`(member(S, {``, ""}), RETURN(S), [assign(g = interface(warnlevel)),
      interface(warnlevel = 0)]);
    a := FNS(Red_n(SUB_S([seq(r[k] = b, k = 1 .. nops(r))], a), b, 2), b, 3);
    assign(interface(warnlevel = g), m = length(a));
    h:= proc(p, t)
      local k;
        for k to nops(p) - 1 do c[1, t + k] := z(a[p[k] + 1 .. p[k + 1] - 1], S) end do; c
      end proc;
    if not search(a, b) then c := Array(1 .. 1, 1 .. m);
      for k to m do c[1, k] := z(a[k], S) end do; c
    else
      assign(p = Search2(a, {b}), 'm' = length(a)); c := Array(1 .. 1, 1 .. nops(p) + 1, []);
      c[1, 1] := z(a[1 .. p[1] - 1], S); c[1, nops(p) + 1] := z(a[p[-1] + 1 .. -1], S); h(p, 1)
    end if
  end if
end proc
```

Typical examples of the procedures use:

> conSV("avzagnvsvvaartkr"), conSV(`avzagnvsvvaartkr`), conSV(vector(["a", "b", "c", "d", "e", "f", "g", "h", a, v, z, 61, a, g, n, 61, Sv, 36, Art, 15, Kr, 7, Ar, 41]));

"a", "v", "z", "a", "g", "n", "v", "s", "v", "v", "a", "a", "r", "t", "k", "r"],
[a, v, z, a, g, n, v, s, v, v, a, a, r, t, k, r], *abcdefghavz61agn61Sv36Art15Kr7Ar41*

> map(conSV, [``, "", `   `, "   ", vector([]), vector([7,14,"a","v","z",2003]), agn, "vsv"]);

[[], [], [` `, ` `, ` `], [" ", " ", " "], "", "714avz2003", [a, g, n], ["v", "s", "v"]]

> conSA("$RANS$$IAN%%RAC&REA%", [`$`, "%", "&"]);

["RANS" "IAN" "RAC" "REA"]

> interface(rtablesize = infinity); map(conSA, ["", "   ", RANSIANRACREA]);

["", [" " " " " "], [*R, A, N, S, I, A, N, R, A, C, R, E, A*]]

> map(conSA, ["", "   ", RANSIANRACREA]);

["", [" " " " " "], ["R" , "A" , "N" , "S" , "I" , "A" , "N" , "R" , "A" , "C" , "R" , "E" , "A"]]

> conSA("194214062004", [57, 62]);

Error, (in **conSA**) the second argument is invalid; it should be {symbol, string, list({symbol, string}), set({symbol, string})} but has received <[57, 62]>

---

swmpat – searching of patterns containing *wildcard* characters swmpat1

Call format of the procedures:

swmpat(S, m, p, d {, **h**} {, `*insensitive*`})
swmpat1(S, P, t {, `*insensitive*`})

Formal arguments of the procedures:

- **S** – string or symbol
- **m** – string or symbol defining a pattern containing *wildcard* characters
- **p** – a list of positive integers defining multiplicities of *wildcard* characters
- **P** – a list of strings or symbols defining a search pattern
- **d** – string or symbol defining a *wildcard* character
- **h** – (optional) an assignable name
- **t** – an unevaluated name
- **`*insensitive*`** – (optional) key word defining a search mode

Description of the procedures:

In a series of cases, a problem of searching in strings or symbols of patterns containing wildcard characters arises. For that, the procedure **swmpat** seems a rather useful tool.

The procedure call **swmpat(S, m, p, d** {, **h**}) returns the *true* value if and only if a string or symbol specified by the actual **S** argument contains entries of substrings matching a pattern **m** with *wildcard* characters specified by the fourth argument **d,** whereas the third actual argument **p** specifies a list of multiplicities of the corresponding entries of *wildcard* characters **d** into a pattern **m.**

For example, the triple **<"a*b*c", [4, 7], "*">** defines the pattern **m**="a****b*******c". The procedure call **swmpat(S,m,[4,7],"*")** defines the fact of entries into a string **S** of non-overlapping substrings which have view of the pattern **m** with arbitrary symbols instead of all entries of *wildcard* character

"*". If the procedure call **swmpat(S, m, p, d {, h})** has used the optional fifth argument **h** and the *true* value has been returned, then through **h** the nested list is returned whose 2-element sublists define the first and ultimate positions of non-overlapping substrings of **S** which are matching a pattern **m.** In addition, if a pattern **m** does not contain *wildcard* characters, then through **h** the integer list is returned whose elements define the first positions of non-overlapping substrings of **S** that are matching the pattern **m.** If the procedure call returns the *false* value, then through **h** the *empty* list is returned, i.e. [].

If the fourth actual argument **d** specifies a string or symbol of length more than **1,** the first its character is used as a *wildcard* character. If a list **p** has fewer elements than quantity of entries of wildcard characters into a pattern **m,** then the redundant entries will receive the multiplicity **1.**

In contrast to the previous procedure, the procedure call **swmpat1(S, P, t)** returns the *true* value if and only if a string or symbol specified by the actual **S** argument contains entries of substrings specified by the actual argument **P** under the condition that their order corresponds to their order in a list specified by the **P** argument. Through the third actual **t** argument, the nested list is returned whose elements define integer lists defining the initial positions of substrings of **P** list that form the allowed entries into the **S** string or symbol. We recommends to notice on an usage of the ***disk transits*** method for organization of a cyclic construction of type "**for ... do**" with undefined nesting level in the procedure implementation.

By default, the both procedures **swmpat** and **swmpat1** do sensitive searching; that is, upper/ lower case differences are considered. If procedure call of any of these procedures uses the optional key word *`insensitive`* then insensitive searching is done; that is, upper/lower case differences are ignored. The procedures **swmpat** and **swmpat1** have many useful appendices in problems of processing of strings and symbols.

```
swmpat:= proc(S::{string, symbol}, m::{string, symbol}, p::list(posint), d::{string, symbol})
local a, b, c, C, j, k, h, s, s1, m1, d1, v, r, res, nu, n, omega, t, epsilon, x, y;
  assign67(c = {args} minus {S, m, p, insensitive, d}, s = convert([7], bytes), y = args);
  C:= (x, y) -> `if`(member(insensitive, {args}), Case(x), x);
  if not search(m, d) then h := Search2(C(S, args), {C(m, args)});
    if h <> [] then RETURN(true, `if`(c = {}, NULL, `if`(type(c[1], assignable), assign(c[1] = h),
  WARNING("argument %1 should be symbol but has received %2", c[1], whattype(eval(c[1]))))))
    else RETURN(false)
    end if
  else
    assign(nu = ((x, n) -> cat(x $ (b = 1 .. n))), omega = (t -> `if`(t = 0, 0, 1)));
    epsilon:= proc(x, y) local k; [seq(`if`(x[k] = y[k], 0, 1), k = 1 .. nops(x))] end proc
  end if;
  assign(s1 = cat("", S), m1 = cat("", m), d1 = cat("", d)[1], v = [], h = "", r = [], res = false, a = 0);
  for k to length(m1) do
    try
      if m1[k] <> d1 then h := cat(h, m1[k]); v := [op(v), 0]
      else a := a + 1; h := cat(h, nu(s, p[a])); v := [op(v), 1 $ (j = 1 .. p[a])]
      end if
    catch "invalid subscript selector": h := cat(h, s); v := [op(v), 1]; next
    end try
  end do;
  assign('h' = convert(C(h, args), list), 's1' = convert(C(s1, args), list)), assign(t = nops(h));
  for k to nops(s1) - t + 1 do
```

```
      if epsilon(s1[k .. k + t - 1], h) = v then res:= true; r:= [op(r), [k, k + t - 1]]; k:= k + t + 1 end if
   end do;
   res, `if`(c = {}, NULL, `if`(type(c[1], assignable), assign(c[1] = r), WARNING(
      "argument %1 should be symbol but has received %2", c[1], whattype(eval(c[1])))))
end proc

swmpat1:= proc(S::{string, symbol}, P::list({string, symbol}), h::evaln)
local a, b, c, d, k, p, s, t, v;
global `00`, `07`;
   `if`(length(S) < length(cat(op(P))), RETURN(false), assign('`00`' = [], '`07`' = [], v = [], t = []));
   `if`(member(insensitive, {args}), assign(s = Case(S), p = map(Case, P)), assign(s = S, p = P));
   for k in p do `00` := [op(`00`), Search2(s, {k})] end do;
   for k to nops(`00`) - 1 do v:=[op(v),`00`[k][cat(1, k)] + length(p[k]) - 1 - `00`[k+1][cat(1, k+1)]]
   end do;
   for k to nops(`00`) do t := [op(t), `00`[k][cat(1, k)]] end do;
   assign(a = "", b = "", c = "", d = [seq(cat(1, k), k = 1 .. nops(`00`))]); b := cat(" if type(",
   convert(v, string), ",list(negative)) then `07`:=[op(`07`),", convert(t, string), "] end if: ");
   for k to nops(`00`) do a:= cat(a, "for `1", k, "` to nops(`00`[", k, "]) do: "); c:= cat(c, " end do:")
   end do;
   (proc() writebytes("_$#@_", cat(a, b, c)); close("_$#@_"); read "_$#@_"; fremove("_$#@_")
      end proc)();
   `if`(`07` = [], false, op([true, assign('h' = `07`)])), unassign('`00`', '`07`', op(d))
end proc
```

Typical examples of the procedures use:

```
> S:= "avz1942agn1947art1986kr1996svet1967art1986kr1996svet": m:= "a*1986*svet":
  p:=[2, 6]: swmpat(S, m, p, "*", z), z;
```

true, [[15, 31], [36, 52]]

```
> swmpat(S, "art1986kr", [7, 14], "*", r), r;
```

true, [15, 36]

```
> swmpat(S, m, p, "*", z);
```

Warning, argument [[15, 31], [36, 52]] should be symbol but has received list

true

```
> S:= "avz1942agn1947Art1986kr1996Svet1967Art1986kr1996Svet": m:= "a*1986*svet":
  p:=[2, 6]: swmpat(S, m, p, "*", a), a;
```

false, []

```
> S:= "avz1942agn1947Art1986Kr1996Svet1967Art1986Kr1996Svet": m:= "a*1986*svet":
  p:=[2, 6]: swmpat(S, m, p, "*", b, `insensitive`), b;
```

true, [[15, 31], [36, 52]]

```
> swmpat(S, "art1986kr", [7, 14], "*", t), t;
```

false, t

```
> swmpat(S, "art1986kr", [7, 14], "*", `insensitive`, t), t;
```

true, [15, 36]

> swmpat1("agnc1942avz1947agn1967vsvxArtzKr", [avz,agn,vsv,Art,Kr], h), h;

true, [[9, 16, 23, 27, 31]]

> swmpat1("agnc1942avz1947agn1967vsvxArtzKr", [avz,agn,vsv,art,kr], p), p;

false, p

> swmpat1("agnc1942avz1947agn1967vsvxArtzKr", [avz,agn,vsv,art,kr], p, `insensitive`), p;

true, [[9, 16, 23, 27, 31]]

> swmpat1("c1234aa2aa345bb4bb567ccxycczz2003zz", [aa,bb,cc,zz], h), h;

true, [[6, 14, 22, 28], [6, 14, 22, 34], [6, 14, 26, 28], [6, 14, 26, 34], [6, 17, 22, 28], [6, 17, 22, 34], [6, 17, 26, 28], [6, 17, 26, 34], [9, 14, 22, 28], [9, 14, 22, 34], [9, 14, 26, 28], [9, 14, 26, 34], [9, 17, 22, 28], [9, 17, 22, 34], [9, 17, 26, 28], [9, 17, 26, 34]]

> swmpat1(`c1234aa2aa345bb4bb567ccxycczz2003zz`, [aa,"bb",cc,"zz"], z), z;

true, [[6, 14, 22, 28], [6, 14, 22, 34], [6, 14, 26, 28], [6, 14, 26, 34], [6, 17, 22, 28], [6, 17, 22, 34], [6, 17, 26, 28], [6, 17, 26, 34], [9, 14, 22, 28], [9, 14, 22, 34], [9, 14, 26, 28], [9, 14, 26, 34], [9, 17, 22, 28], [9, 17, 22, 34], [9, 17, 26, 28], [9, 17, 26, 34]]

DT – a stylized representation of date and time

Call format of the procedure:

DT(S, D, R)

Formal arguments of the procedure:

S - a string or symbol representing date or time
D - a delimiter of components composing date or time
R - a key word defining type of information {**`date`** | **`date1`** | **`time`**}

Description of the procedure:

The procedure call **DT(S, D, R)** returns the result of converting of a string or symbol **S**, which represents date or time with **D** delimiter of components, into the *stylized* format. The third actual **R** argument defines type (*date* or *time*) of information represented by the first actual **S** argument. In addition, a date depending on a value {**`date`** | **`date1`**} is represented in two different formats whose view is easily seen from examples represented below. The returned result has type corresponding to type of the first actual **S** argument. The **DT** procedure handles all basic especial situations.

```
DT:= proc(S::{string, symbol}, D::{string, symbol}, R::symbol)
local d, t, h, omega;
  `if`(search(S, D), assign(t = table([1 = hours, 2 = minutes, 3 = seconds]),
     d = table([1 = January, 2 = February, 3 = March, 4 = April, 5 = May, 6 = June, 7 = July,
     8 = August, 9 = September, 10 = October, 11 = November, 12 = December]),
     omega = (s -> `if`(SD(cat("", s)) < 15, 2000 + SD(cat("", s)), s))), RETURN(S)),
  `if`(length(D) <> 1, goto(ERR), `if`(Search(S, D) = [], goto(ERR),
  `if`(not belong(R, {time, date, date1}), goto(ERR), assign(h = map(SN, SLD(S, D))))));
  `if`(R = date and belong(h[1], 1 .. 31) and belong(h[2], 1 .. 12) and belong(h[3], 0 .. 9999),
    RETURN(convert(cat("", h[1], " ", d[h[2]], " ", omega(h[3])), whattype(S))),
  `if`(R = date1 and belong(h[2], 1 .. 31) and belong(h[1], 1 .. 12) and belong(h[3], 0 .. 9999),
```

```
    RETURN(convert(cat("", d[h[1]], " ", h[2], ", ", omega(h[3])), whattype(S))),
  `if`(R = time and belong(h[1], 0 .. 24) and belong(h[2], 0 .. 59) and `if`(nops(h) = 3,
    belong(h[3], 0 .. 59), true), RETURN(convert(cat("", h[1], " ", t[1], " ", h[2], " ", t[2],
    `if`(nops(h) = 3, cat(" ", h[3], " ", t[3]), ``)), whattype(S))), goto(ERROR1))));
  ERR;
  ERROR("invalid assignment of actual arguments at call %1", 'procname(args)');
  ERROR1;
  ERROR("invalid actual <%1> argument as a date or a time", S)
end proc
```

Typical examples of the procedure use:

> DT(`14/06/42`, "/", time), DT("23:04:1967", `:`, date), DT("14/06/42", "/", time);

14 hours 6 minutes 42 seconds, "23 April 1967", "14 hours 6 minutes 42 seconds"

> DT(`23:04:1967`, ":", date), DT(`14/06/42`, `/`,time), DT(`23:04:1967`, `:`, date);

23 April 1967, 14 hours 6 minutes 42 seconds, 23 April 1967

> DT(`01-22-2004`, "-", date1), DT(`01/22/04`, "/", date1), DT("01:22:04", ":", date1);

January 22, 2004, January 22, 2004, "January 22, 2004"

> DT(`33:09:2003`, `:`, date);

Error, (in **DT**) invalid actual <33:09:2003> argument as a date or a time

---

ewsc – selection from a string or symbol of words with given contexts

Call format of the procedure:

ewsc(S, C, N {, 'h'})

Formal arguments of the procedure:

S - a string or symbol to be searched
C - list of symbols and/or strings defining contexts (*patterns*)
N - list of positive integers or symbols ***`infinity`*** defining multiplicities of appropriate contexts
H - (optional) an assignable name

Description of the procedure:

Under the term "word", we will understand any parts of a string or symbol parted by *blanks*. In particular, a string coincides with word if it does not contain *blanks*.

The procedure call **ewsc(S, C, N)** returns the list containing words of a string or symbol defined by the first actual **S** argument if these words contain entries of patterns defined by the second actual **C** argument with **N** multiplicities appropriate to them. In addition, if the symbol ***`infinity`*** has been coded then the appropriate word is chosen at any quantity (*at least* **1**) of entries of the corresponding pattern. Words are returned as *strings*. Otherwise, the *empty* list is returned, i.e. []. The **ewsc** procedure does *sensitive* patterns searching; that is, upper/lower case differences are essential.

If the optional fourth **h** argument has been coded, then through it the result of deletion of found words from **S** is returned. The given procedure is helpful enough at processing of strings and symbols. Above all, the procedure is rather effective for extraction of words with the given context out of expressions of types {*symbol, string*}.

```
ewsc:= proc(S::{string, symbol}, C::list({string, symbol}), N::list({posint, infinity}))
local a, Res, k, j, t, h, s, z;
  assign(Res = [], z = 0, s = cat("", S)), `if`(Empty(s), RETURN(Res), `if`(nops(C) = nops(N),
    `if`(search(s, " "), assign(a = SLD(Red_n(s, " ", 2), " ")), assign(a = [s])),
    ERROR("discrepancy of dimensions of the second and third arguments")));
  for k to nops(a) do
    for j to nops(C) do t := nops(Search2(a[k], {C[j]}));
      if t = N[j] or t>0 and N[j]=infinity then z := 7 elif 0 < t then h := 62 end if
    end do;
    if z <> 7 or h = 62 then z := 15; h := 57 else Res := [op(Res), a[k]]; z := 15; h := 57 end if
  end do;
  if nargs = 4 and type(args[4], symbol) then assign(args[4] = sub_1([seq(Res[k] = NULL,
    k = 1 .. nops(Res))], S))
  end if;
  Res
end proc
```

Typical examples of the procedure use:

```
> ewsc("agn12   vsv21 Gal art1c2 21Kr Vic Svet19672   1942", [`1`,`2`], [1,1], 'w'),  w;
                ["agn12", "vsv21", "art1c2", "21Kr", "Svet19672", "1942"], "    Gal   Vic    "
> ewsc("  SH  temp a:\\temp  c:/temp\\temp", [`temp`], [infinity], 'a'),  a;
                           ["temp", "a:\temp", "c:/temp\temp"], "  SH     "
> ewsc(`rans   ian Tallinn art1c2 21Kr Grodno SveGal`, [`an`, `2`], [1, 1], 'h'),  h;
                      ["rans", "ian", "art1c2", "21Kr"],     Tallinn   Grodno SveGal
```

---

cdt – a current date

Call format of the procedure:

cdt()

Formal arguments of the procedure: **no**

Description of the procedure:

The procedure call **cdt()** returns a current date in the *stylized* string format whose view is easily seen from an example represented below. Date is returned as a *string*. In particular, this procedure can be rather useful at creation of *Maple* documents demanding the stylized appearance.

```
cdt:= proc()
local a, b, c, d, k;
  assign(b = "_$62$_", c = ["/", ".", ":", "-"]), system(cat("Date /T > ", b)),
     assign(a = readline(b)), delf(b);
  for k to 4 do b := ewsc(a, [c[k]], [2]); if b <> [] then d := c[k]; break end if end do;
  if b = [] then ERROR("incorrect date format") else DT(b[1], d, date1) end if
end proc
```

Typical examples of the procedure use:

```
> cdt();
                                   "May 30, 2004"
```

EntS – identification of entries of a symbol into string

Call format of the procedure:

EntS(S, t)

Formal arguments of the procedure:

S - string or symbol
t - string or symbol of length **1**

Description of the procedure:

The procedure call **EntS(S, t)** returns the sequence of lists whose first elements define positions of entries of a symbol defined by the second actual **t** argument into a string defined by the first actual **S** argument, whereas the second elements define their multiplicities. At lack of entries of **t** symbol into string **S** the *NULL* value is returned. In addition, the **EntS** procedure uses method of sensitive searching; that is, upper/lower case differences are essential. The given procedure represents rather useful tool of processing of strings and symbols.

```
EntS:= (S::{string, symbol}, a::{string, symbol}) -> `if`(search(S, a), KL1(Search(S, a)), NULL)
```

Typical examples of the procedure use:

```
> EntS(`aaaaaaaa6aaaaaagnaaaa14aaaaa61aaahhhaaaartaakraaa`, `a`);
                [1, 8], [10, 6], [18, 4], [24, 5], [31, 3], [37, 4], [43, 2], [47, 3]
```

---

sident – a special converting of a symbol or string into list

Call format of the procedure:

sident(S {, T})

Formal arguments of the procedure:

S - a string or symbol
T - (optional) an element or a subset of the set {*digit, letter, ssign, false*}

Description of the procedure:

The procedure call **sident(S)** returns the list whose elements are formed as continuous substrings consisting of symbols of the same type ***`digit`***, ***`letter`***, ***`ssign`*** or others.

Whereas the procedure call **sident(S, T)** returns the list whose elements are formed as continuous substrings of **S** string that consist of symbols of type defined by the second optional **T** argument. As a value of actual **T** argument can be element or a subset of the set {*digit, letter, ssign, false*}; in addition, the *false*-value defines a type different from types {*digit, letter, ssign*}. The procedure call **sident(S, {*digit, letter, ssign, false*})** is equivalent to the procedure call **sident(S).** Thus, by means of the optional **T** argument a converting mode can vary within wide enough range.

```
sident:= proc(S::{string, symbol})
local k, L, z, P, T;
  `if`(nargs = 1, NULL, `if`(nargs = 2 and belong(args[2], {false, ssign, digit, letter}),
    assign(T = args[2]), ERROR(`sident expects its 2nd optional argument to be an element
    or a subset of the set {digit, letter, ssign, false}, but received <%1>`, args[2])));
  assign(z = "", L = [], P = cat("", S, " "));
```

```
  for k to length(P) - 1 do
    z := cat(z, `if`(nargs = 1, P[k], `if`(belong(stype(P[k]), T), P[k], goto(Cycle))));
    if stype(P[k]) = stype(P[k + 1]) then next else L := [op(L), z]; z := "" end if;
    Cycle;
  end do;
  assign('L' = [op(L), z]), map(convert, `if`(L[-1] = "", L[1 .. -2], L), whattype(S))
end proc
```

Typical example of the procedure use:

> S:="#$%@1942$$Academy&Hooсфера*1995-2003": sident(S);

["#$%@", "1942", "$$", "Academy", "&", "Hooсфера", "*", "1995", "-", "2003"]

> sident(S, {ssign, digit, false});

["#$%@", "1942", "$$", "&", "Hooсфера", "*", "1995", "-", "2003"]

FNS – truncation of a string or symbol

Call format of the procedure:

FNS(S, K, M {, `insensitive`})

Formal arguments of the procedure:

S – a string or symbol
K – a string or symbol
M – a truncation mode; it should be from the set **{1, 2, 3}**
insensitive – (optional) word defining a case of comparison of substrings in string

Description of the procedure:

In a series of problems of symbolic processing the necessity of truncation of a string or a symbol, starting with the first entry of a substring (subsymbol), different from the given string (symbol), up to the end of the initial string (symbol) arises. In addition, truncation can be done (**1**) from beginning of string, (**2**) from the end of string and (**3**) from both ends of string. In particular, similar necessity at truncation in a string of the leading *blanks* arises. For these purposes, the **FNS** procedure can be useful enough.

The procedure call **FNS(S, K, M)** returns the result of truncation of a string or a symbol, defined by the first actual **S** argument, on the basis of deletions of substring or subsymbol, defined by the second actual **K** argument by the above manner and mode, defined by the third actual argument **M.** If **M=1** the truncation is done from beginning of the string, if **M=2** the truncation is done from end of the string, and if **M=3** the truncation is done from both ends of the string. If length of **S** less, than length of **K,** or length of **K** is zero, then the initial **S** expression is returned without processing.

If the fourth optional ***`insensitive`*** argument has been coded then a searching of **K** in **S** is done in insensitive case; otherwise sensitive case is used. This procedure is useful enough tool at symbolic processing.

```
FNS:= proc(S::{string, symbol}, K::{string, symbol}, M::{1, 2, 3})
local d, omega, omicron;
  assign(d = `if`(member(insensitive, [args]), 1, 0)); omicron:= (d, c) -> `if`(d = 1, Case(c), c);
  omega:= proc(S, K, M::{1, 2})
    local a, b, c, t;
```

```
      `if`(search(omicron(d, S), omicron(d, K)), assign(a = cat("", S), b = length(K),
        c = [left, right]), RETURN(S));
      do
        if Search1(omicron(d, a), omicron(d, K), 't') and belong(c[M], t) then
          `if`(M = 1, assign('a' = a[b + 1 .. -1]), assign('a' = a[1 .. -b - 1]))
        else RETURN(a)
        end if
      end do
    end proc;
  convert(`if`(M = 1, omega(S, K, 1), `if`(M = 2, omega(S, K, 2),
    omega(omega(S, K, 1), K, 2))), whattype(S))
end proc
```

Typical examples of the procedure use:

> FNS(`99999999999999RANS_IAN_RAC_REA=61`, "9", 1);

RANS_IAN_RAC_REA=61

> FNS(`ttttttttRANS_IAN-14tttttt`,"TtT", 3), FNS(`ttttttttRANS_IAN-61tttttt`, "TtT", 3, `insensitive`);

ttttttttRANS_IAN-14tttttt, RANS_IAN-61

GG – reducing of a string with converting into list

Call format of the procedure:

GG(S, b, c, Z)

Formal arguments of the procedure:

S – a string or symbol
b – a string of length **1** (*marker*)
c – a string of length **1** (*new marker*)
Z – an unevaluated name (through **Z** the sought list is returned)

Description of the procedure:

In many cases at operation with string structures as a separate procedure the following rather typical problem appears. There is some string **S** of a view **S:="X1X2...Xk...Xn"**, where **Xk** belongs to some alphabet **A** and **X1, Xn** belong to **A\{b}**. It is necessary to reduce all entries of the **b** symbol (string of length **1**) into **S** string to a single multiplicity with the subsequent replacement of each such symbol onto the given **c** symbol (string of length **1**). Simultaneously, it is necessary to create a list; by elements of such list, the segments of **S**-string parted by entries of the b symbol would appear. The problem of reducing of some list of names parted by *blank* into a structure, suitable for single-element processing, often arises as a problem of such type. The **GG** procedure well answers the similar problems.

The procedure call **GG(S, b, c, Z)** returns the result **S1** of reducing of length of substrings composed from a symbol, defined by the second actual **b** argument, to **1** in a string defined by the first actual **S** argument, with posterior replacement of this symbol onto new symbol defined by the third actual **c** argument (*new marker*).

Through the fourth actual **Z** argument the sought list is returned; elements of the list are formed as segments of **S1** string parted by new **c** marker. If the sought **b** symbol is absent in string **S**, the *false* value is returned. In addition, the **GG** procedure uses method of sensitive searching; that is, upper/lower case differences are essential.

```
GG:= proc(L::{name, string, symbol}, b::string, c::string, Z::evaln)
local f, s, k, p, h, z, K, G;
  `if`(search(L, b), assign(s = "", f = 0, G = convert(L, string)), RETURN(false));
  for k to length(G) do s := cat(s, `if`(substring(G, k) = b and f = 0, op([assign('f' = 1), c]),
    `if`(substring(G, k) = b and f = 1, NULL, op([substring(G, k), assign('f' = 0)]))))
  end do;
  [assign(K = [0, seq(`if`(s[h] = c, h, NULL), h = 1 .. length(s)), length(s) + 1]),
    assign(Z = [seq(substring(s, z), z = [(K[p] + 1 .. K[p + 1] - 1) $ (p = 1 .. nops(K) - 1)])]),
    RETURN(s)]
end proc
```

Typical examples of the procedure use:

> LBZ:=`RAC RAN REA RANS 1995 2003`: GG(LBZ, " ", "|",Q), Q;

"RAC|RAN|REA|RANS|1995|2003", ["RAC", "RAN", "REA", "RANS", "1995", "2003"]

> LBZ:=`RAC RAN REA RANS 1995 2003`: GG(LBZ, "_", "*",h), h;

false, h

HS_1D – 1-dimensional Cellular Automata (*Homogeneous Structures*)

Call format of the procedure:

HS_1D(S, n, LTF, M)

Formal arguments of the procedure:

S - a string or symbol
n - a positive integer
LTF - a set or list
M - a string or symbol

Description of the procedure:

The **HS_1D(S, n, LTF, M)** procedure simulates the dynamics of one-dimensional *homogeneous structures* (*Cellular Automata*) at a level of their local transition functions. It is possible to familiarize oneself with exposition of such type of formal parallel information processing systems in our books [1-3]. The procedure has four formal arguments, of which three **S**, **n** and **LTF** define accordingly the following parameters of a structure: an initial finite configuration, a neighborhood size and a local transition function defined by a list or a set of the equations of the view **x1x2 ... xn = x1'**. Whereas the fourth actual **M** argument defines a symbol-*marker* providing an opportunity of extracting from string of the final configuration (the result of evaluation of the next configuration from the initial configuration).

The fragment represents an example of application of the procedure **HS_1D** for generation of one-dimensional binary configurations with the elementary neighborhood index and local transition function of the view **x1x2=x1+x2 mod 2** with respect to the primitive initial configuration **s=`1`**. The sequence of configurations from 36 elements has been obtained. The given procedure allows to investigate many interesting dynamic aspects of the classical one-dimensional homogeneous structures, making a basis for algorithms of generation of sequences of the finite configurations for them. Experiments with the procedure have revealed a few interesting results.

```
HS_1D:= proc(S::{string, symbol}, n::posint, LTF::{set(`=`), list(`=`)}, M::{string, symbol})
local a, b, c, d, f, k, L;
  assign(L = []), seq(assign('L' = [op(L), cat("", lhs(LTF[k])) = cat("", rhs(LTF[k]))]),
  k = 1 .. nops(LTF)); f:= (m, n) -> {seq(length(n(m[k])), k = 1 .. nops(LTF))};
  `if`(1 < nops(f(L, lhs)), goto(E1), `if`(op(f(L, rhs)) = 1, NULL, goto(E2)));
  `if`(n <> length(lhs(L[1])), goto(E3), assign(b = cat("", seq(M, k = 1 .. n - 1)),
     d = (h -> `if`(h = [], 0, h[-1])))); a := [seq(lhs(L[k]) = rhs(L[k]), k = 1 .. nops(L))];
  c := cat(seq(subs(a, cat(b, S, b)[k .. k + n - 1])[1], k = 1 .. length(S) + n - 1));
  RETURN(convert(c[d(Search(c, M)) + 1 .. -1], whattype(S)));
  E1;
  ERROR(`invalid lengths of left members in definition of local transition function`);
  E2;
  ERROR(`invalid lengths of rigth members in definition of local transition function`);
  E3;
  ERROR(`length of HS template conflicts with the local transition function`)
end proc
```

Typical examples of the procedure use:

```
> LTF:= [`00`=`0`,`10`=`1`,`01`=`1`,`11`=`0`,`#1`=`1`,`1#`=`1`]:  s:=`1`:  seq(op([assign('`s`'=
  HS_1D(s, 2, LTF, `#`)), s]), k=1 .. 36);
```

11, 101, 1111, 10001, 110011, 1010101, 11111111, 100000001, 1100000011, 10100000101, 111100001111, 1000100010001, 11001100110011, 101010101010101, 1111111111111111, 10000000000000001, 110000000000000011, 1010000000000000101, 11110000000000001111, 100010000000000010001, 1100110000000000110011, 10101010000000001010101, 111111110000000011111111, 1000000010000000100000001, 11000000110000001100000011, 101000001010000010100000101, 1111000011110000111100001111, 10001000100010001000100010001, 110011001100110011001100110011, 1010101010101010101010101010101, 11111111111111111111111111111111, 100000000000000000000000000000001, 1100000000000000000000000000000011, 10100000000000000000000000000000101, 111100000000000000000000000000001111, 1000100000000000000000000000000010001

The procedure has been used for experiments with one-dimensional binary cellular automata.

Ins – insertion of symbols or strings into a string or symbol

Call format of the procedure:

Ins(S, G, L)

Formal arguments of the procedure:

S - a string or symbol
G - a string or symbol
L - a string of length **1**, list of integers, or set of symbols or/and strings

Description of the procedure:

The procedure call **Ins(S,G,L)** returns the result of insertion of a string or a symbol defined by the second argument **G** into a string or a symbol defined by the first actual **S** argument on the base of positions defined by the third actual **L** argument. If a string of length 1 has presented actual L argument, then insertion is fulfilled after each entry of **L** symbol into the **S** string.

If actual **L** argument is presented by a list of integers then insertion is fulfilled after each position defined by elements of the **L** list. At last, if actual **L** argument is presented by a set of symbols and/or strings, then insertion is fulfilled after each entry of symbols defined by elements of the **L** set. In many appendices dealing with processing of strings or/and symbols this procedure seems as useful enough tool.

```
Ins:= proc(S::{string,symbol}, G::{string,symbol}, L::{string, list(integer), set({string,symbol})})
local s, k, V;
  `if`(type(L, set({string,symbol})), assign(V = Search2(S,L)), assign(V=L)), assign(s=cat("", S)),
  `if`(type(V, list) and length(S) < sort(V)[-1], ERROR(`invalid 3rd argument <%1>`, L),
  convert(cat(seq(`if`(member(`if`(type(V, list), k, s[k]), {op(V)}),
  cat(s[k], G), s[k]), k = 1 .. length(S))), whattype(S)))
end proc
```

Typical examples of the procedure use:

> Ins("19422002602501995613"," - ","0"), Ins("1942200260250"," aaa ","0");

"194220 - 0 - 260 - 250 - 1995613", "194220 aaa 0 aaa 260 aaa 250 aaa "

> Ins("1940022002600250"," == ",{"00"}), Ins(`1942200260250`," - ",[4,8,10]);

"1940 == 0220 == 0260 == 0250", `1942 - 2002 - 60 - 250`

> Ins("19422002602501995613"," - ","Q"), Ins("19422002602501995613"," * ",{`4`,"6",`1`});

"19422002602501995613", "1 * 94 * 220026 * 02501 * 9956 * 1 * 3"

> Ins("635607820026025019956 13"," $$ ",{"9", `4`, "8", `2`, `6`});

"6 $$ 356 $$ 078 $$ 2 $$ 002 $$ 6 $$ 02 $$ 5019 $$ 9 $$ 56 $$ 13"

Insert – insertion of a pattern into a string or symbol

Call format of the procedure:

Insert(S, P, V)

Formal arguments of the procedure:

S - a string or symbol
P - a string or symbol (*pattern*)
V - a string or symbols

Description of the procedure:

The procedure call **Insert(S, P, V)** returns the result of insertion of a string or a symbol defined by the third argument **V** into a string or symbol defined by the first actual **S** argument instead of the first entry of a pattern defined by the second actual **P** argument. The returned result receives a type analogous to a type of the first actual **S** argument of the procedure. If the sought pattern **P** is absent in **S** string, the *false* value is returned. In addition, the procedure uses method of insensitive searching; that is, upper/lower case differences are ignored.

```
Insert:= proc(A::{string, symbol}, V::{string, symbol}, Z::{string, symbol})
local a, b;
  assign(a = searchtext(V, A)), `if`(a = 0, RETURN(false), assign(b = length(V))),
  RETURN(convert(cat(substring(A, 1..a - 1), Z, substring(A, a + b .. length(A))), whattype(A)))
end proc
```

Typical examples of the procedure use:

> Insert("ArtKristoSvetGal", `Art`, "14"), Insert(tthhhzz, "hh", `2003`);

"14KristoSvetGal", *tt2003hzz*

> S:= "AvzbbbddcccttttrrrAVZhhhdddAVz": Insert(S, `AVZ`, "*");**

"***bbbddcccttttrrrAVZhhhdddAVz"

Mulel – multiplicity of entries of symbols

Call format of the procedure:

Mulel(L)

Formal arguments of the procedure:

L - a string, symbol or list

Description of the procedure:

In many problems of operation with structures of types {*symbol, string, name*} the problem of definition of multiplicity of symbols composing an explored structure arises. The following **Mulel** procedure solves this problem for structures even of a wider set of the types, namely: *list, symbol, string* and *name*.

The procedure call **Mulel(L)** returns the nested list, each element of which represents the 3-element list, namely: the first element defines a symbol entry into structure defined by the actual **L** argument; the second element defines its position in the structure, at last, the third element defines the respective *multiplicity* of entry of the symbol into **L**-structure.

```
Mulel:= proc(L::{list, name, string, symbol})
local a, b, c, d, k, p;
   [assign(a = convert(L, list)), assign(b = {op(a)}), assign(c = [0 $ (k = 1 .. nops(b))],
   d = [0 $ (k = 1 .. nops(b))]), add(add(`if`(b[k] = a[p], [assign('c[k]' = c[k] + 1),
   `if`(d[k] = 0, assign('d[k]' = p), NULL)], 1), k = 1 .. nops(b)), p = 1 .. nops(a)),
   [seq([b[k], d[k], c[k]], k = 1 .. nops(b))]][2]
end proc
```

Typical example of the procedure use:

> Mulel("akbaackd RANS ddppe IAN fcsdfh 2003 hhewa");

[["A", 11, 2], [" ", 9, 6], ["R", 10, 1], ["N", 12, 2], ["S", 13, 1], ["2", 32, 1], ["3", 35, 1], ["0", 33, 2], ["a", 1, 4], ["k", 2, 2], ["b", 3, 1], ["c", 6, 2], ["d", 8, 4], ["p", 17, 2], ["e", 19, 2], ["I", 21, 1], ["f", 25, 2], ["s", 27, 1], ["h", 30, 3], ["w", 40, 1]]

N_Cont – a pattern search in a string or symbol
sextr
sextr1
sext

Call format of the procedures:

N_Cont(*string, pattern, mode*)
sextr(S, p, T)
sextr1(S, p, T, T1)
sext(S, a, b, U {, n})

Formal arguments of the procedures:

pattern – a string or name to be searched for
string – a string or name to be searched
mode – an integer defining a search mode of the pattern (***mode***={**1** | **2**})
S – a string or symbol
p – a string or symbol
T, T1 – a valid type
a, b – delimiters (may be *symbols, strings* or *integers* from the range **0 .. 255**)
U – a set or list of *symbols, strings* or *integers* from the range **0 .. 255**)
n – (optional) a positive integer (*posint*)

Description of the procedures:

The procedure **N_Cont** searches for a *pattern* in a *string*, looking for exact matches. The **N_Cont** procedure does sensitive or insensitive searching depending on a value of the third actual *mode*-argument; in addition, for sensitive mode upper/lower case differences are essential, whereas for insensitive mode these differences are inessential. If the *pattern* is not found in the *string* for any searching *mode*, the procedure returns the *empty* list, i.e. []. If the *pattern* is found, the procedure returns an integer two-element list; the first element of the list defines quantity of entries of the *pattern* into the *string*, whereas the second element is the integer list defining the initial positions of all entries of the *pattern* into the *string*.

The procedure call **sextr(S, p, T)** returns the list containing substrings of a string or a symbol defined by the first actual **S** argument that represent the patterns beginning with a prefix **p** and extending up to the first entry of symbol of a type defined by the third actual **T** argument. At missing of such entries the empty list is returned, i.e. []. The procedure seems as useful enough tool, in particular, at processing of program modules in the user libraries and in *m*-files.

The procedure call **sextr1(S, p, T, T1)** returns the list containing substrings of a string/symbol defined by the first actual **S** argument which represent the patterns of **T**-type symbols beginning with prefix **p** and extending up to the first entry of symbol of a type defined by the fourth actual **T1** argument. At missing of such entries the empty list is returned, i.e. []. Similarly to the previous procedure the **sextr1** seems as useful enough tool, in particular, at processing of program modules in the user libraries and in *m*-files.

Finally, the procedure call **sext(S, a, b, U)** returns the sequence of non-overlapping substrings of **S** string which are limited by delimiters **`a`** and **b.** Symbols of the returned substrings should belong to a set or a list defined by the actual **U** argument. The first **S** argument may define a datafile out of which strings are extracted in accord with the above algorithm. The optional fifth **n** argument defines length of readable segment of the datafile **S** (by default the whole file is supposed). Unsuccessful procedure call returns an unevaluated source string **S** or causes an appropriate erratic situation. The procedures of this group are rather useful at expressions processing of types {*symbol*, *string*}.

```
N_Cont:= proc(S::{string, symbol}, G::{string, symbol}, K::{1, 2})
local n, p, d, X, L;
  `if`(search(Case(S), Case(G)), assign({L = [], p = 0, n = 1, d = 1}), RETURN([]));
  `if`(args[3] = 1, assign(X = searchtext), assign(X = SearchText));
  do n := X(G, S, d .. length(S));
    if n <> 0 then d := d + n; p := p + 1; L := [op(L), d - 1] else break end if end do;
  `if`([p, L] = [0, []], [], [p, L])
end proc
```

```
sextr:= proc(S::{string, symbol}, a::{string, symbol}, T::type)
local k, p, R, L, G;
  assign(R = cat("", S), G = [], L = Search(S, a));
  for k to nops(L) do for p from L[k] + length(a) to length(R) do
      if type(R[p], T) then G := [op(G), R[L[k] .. p - 1]]; break end if
    end do
  end do;
  G
end proc

sextr1:= proc(S::{string, symbol}, a::{string, symbol}, T1::type, T2::type)
local k, p, R, L, G;
  assign(R = cat("", S), G = [], L = Search(S, a));
  for k to nops(L) do for p from L[k] + length(a) to length(R) do
      if type(R[p], T1) then next
      elif type(R[p], T2) then G := [op(G), R[L[k] .. p - 1]]; break
      else break
      end if
    end do
  end do;
  G
end proc

sext:= proc(S::{string, symbol}, a::{integer, string, symbol}, b::{integer, string, symbol},
          U::{set({integer, string, symbol}), list({integer, string, symbol})})
local h, k, j, p, omega, omega2, omega3, Omega, s, s1, R, u;
  omega := proc()
    local k, R;
      `if`(nargs = 0, RETURN({}), assign67(R = NULL));
      for k to nargs do
      if type(args[k], {string,symbol}) then R:= R, op(map(convert, convert(args[k], set), string))
       elif type(args[k], integer) and belong(args[k], 0..255) then R:= R, convert([args[k]], bytes)
        else ERROR("the 4th argument is wrong")
        end if
      end do;
      R
    end proc;
  assign67(omega2 = h2, omega3 = h3, p = [], s = `if`(type(S, file), readbytes(S, TEXT,
     `if`(nargs = 5 and type(args[5], posint), args[5], infinity)), cat("", S)), u = {}, R = NULL),
     `if`(type(S, file), close(S), NULL); seq(assign(cat(h, k) = `if`(type(args[k], integer),
     `if`(belong(args[k], 0 .. 255), convert([args[k]], bytes),
    ERROR("the 2nd or 3rd argument is wrong - %1", [a, b])), args[k])), k = 2 .. 3),
   `if`(map2(search, s, {h2, h3}) = {true}, assign(h = Search2(s, {h2, h3})),
     RETURN(S, assign('h2' = omega2, 'h3' = omega3)));
  for k to nops(h) - 1 do
    if s[h[k]] <> h2 then next
    else for j from k + 1 to nops(h) do
        if s[h[j]] = h3 then p := [op(p), [h[k], h[j]]]; k := j + 1; j := k else next end if
      end do
    end if
```

```
  end do;
  assign('h2' = omega2, 'h3' = omega3), `if`(p = [], RETURN(S), NULL);
  h:= [seq(`if`(p[k][2] - p[k][1] = 1, NULL, p[k]), k = 1 .. nops(p))]; Omega:= {omega(op(U))};
  for k to nops(h) do
    s1 := {omega(s[h[k][1] + 1 .. h[k][2] - 1])};
    if s1 intersect Omega = s1 then
      R := R, s[h[k][1] + 1 .. h[k][2] - 1] end if
  end do;
  R
end proc
```

Typical examples of the procedures use:

```
> S:="AvzbbAvzbdAvzdccAvzcAvzttAvztrAvzrAvzrAvZhAvzhhAvzdddAVzAvzAVZAvz":
  S1:=`AvzbAvzbbdAvzdcAvzcAVzcAvzttAVztAvzrAvzrAvzrAvZhhhdddAVzAVzAVzAVz`:
> N_Cont(S,`Avz`,1), N_Cont(S1,`Avz`,1), N_Cont(S, "Avz",2); N_Cont(S, "Avz",3);
        [15, [1, 6, 11, 17, 21, 26, 31, 35, 39, 43, 48, 54, 57, 60, 63]], [15, [1, 5, 11, 16, 20, 24, 29, 33,
            37, 41, 45, 54, 57, 60, 63]], [12, [1, 6, 11, 17, 21, 26, 31, 35, 43, 48, 57, 63]]
    Error, invalid input: N_Cont expects its 3rd argument, K, to be of type {1, 2}, but received 3
> S:=`aaa:-300&bbb:-1942c:-61*%hhh:-56g#:-56%:-300G:-hhh:-`: sextr(S,`:-`, {letter, ssign});
                 [":-300", ":-1942", ":-61", ":-56", ":-56", ":-300", ":-"]
> sextr(S,`:-`,{ssign}), sextr(S,`Q`,{ssign}), sextr1(S,`:-`, {digit}, {letter}), sextr1(S,`Z`, {digit},
  {letter});
      [":-300", ":-1942c", ":-61", ":-56g", ":-56", ":-300G", ":-hhh"], [], [":-1942", ":-56", ":-300", ":-"], []
> x, y:= op(map(convert, [[14], [7]], bytes)): S:=cat(ab,x,y,x,avz,y,x,agn,y,x,cc,y,zz,x,art,y,z,x,
  kristo, y, hhpt): sext(S,14,7, {avz,"gnkris",111,116});
                          "avz", "agn", "art", "kristo"
> writebytes("C:/Temp/Test.txt",S): All_Close(); sext("C:/Temp/Test.txt",14,7,
  {avz, "gnkris", 111, 116});
                          "avz", "agn", "art", "kristo"
```

Pnd – check of a string, symbol or integer to be palindrome

Call format of the procedure:

Pnd(S {, 'h'}{, `*insensitive*`})

Formal arguments of the procedure:

S – a string, symbol, list or integer
h – (optional) an unevaluated name
insensitive – (optional) key word defining insensitive mode of palindrome check

Description of the procedure:

The procedure call **Pnd(S)** returns the *true* value, if a string, symbol, integer or list defined by the actual **S** argument is a *palindrome*; otherwise, the *false* value is returned. The *palindrome* is a string that is symmetric with respect to own center, for example "abcdefghgfedcba". A list defined by the actual **S** argument should has as own elements the integers, symbols or strings. Otherwise, the erratic situation arises.

If as the second optional **h** argument an unevaluated name has been coded, then via this name the 2-element nested list is returned. In addition, sublists of the list define the numbers of positions and symbols in which the property be palindrome is broken for the first time. The name **h** is returned unevaluated, if the first actual **S** argument defines a *palindrome*. At last, if the optional key word **`*insensitive*`** has been coded as the third actual argument, then the procedure uses insensitive mode of the *palindrome* check.

```
Pnd:= proc(S::{integer, string, symbol, list({integer, string, symbol})})
local a, b, c, k, epsilon, kappa, s;
  epsilon:= (h, c) -> `if`(member(insensitive, {op(h)}), Case(c), c);
  kappa:= c -> `if`(member(whattype(S), {string, symbol}), convert(c, whattype(S)),
    `if`(type(S, integer), SN(c), c)); `if`(type(S, list), assign(s = cat(op(S))), assign(s = S));
  assign(a = convert(s, string), b = length(s)), `if`(mul(`if`(epsilon([args], a[k+1]) = epsilon([args],
    a[b - k]), 1, _0N(0, `if`(1 < nargs, assign(args[2] = [[k + 1, kappa(a[k + 1])],
    [b - k, kappa(a[b - k])]]), 0))), k = 0 .. trunc(1/2*b) - 1) = 1, true, false)
end proc
```

Typical examples of the procedure use:

> Pnd(abcdefghgfedcba), Pnd(abcdefghgfedcba,'t'), t, Pnd("abcdefeedcba"); restart;

true, true, t, false

> Pnd(`abcdefeedcba`,'w'), w, Pnd([a,"BT",c,5,e,f,z,h,z,f,e,5,c,"TB",a]), Pnd([a,"BT",c,5,e, f,z,h,z,f,e,5, c,"BT",a],'x'), x;

false, [[6, f], [7, e]], *true*, *false*, [[3, "T"], [15, "B"]]

Pos – identification of non-overlapping entries of a pattern into a string

Call format of the procedure:

Pos(s, S {, `*insensitive*`})

Formal arguments of the procedure:

S – a string or symbol
s – a string or symbol (*pattern*)
insensitive – (optional) key word defining the insensitive mode of pattern search

Description of the procedure:

The procedure call **Pos(s, S)** returns the list whose elements define positions of non-overlapping entries of a pattern defined by the first **s** argument into a string defined by the second actual argument **S.** In the absence of entries of **s** into string **S** the initial **S** value is returned. In addition, the **Pos** procedure uses method of insensitive searching of the s pattern in the **S** string, if the optional third `***insensitive***` argument has been coded, otherwise sensitive searching is used. The procedure **Pos** is used by a series of other procedures for similar purposes.

```
Pos:= proc(s::{string, symbol}, S::{string, symbol})
local a, b, c, k, L, n;
  `if`(nargs = 3 and args[3] = insensitive, assign(a = searchtext), assign(a = SearchText));
  `if`(a(s, S) <> 0, assign(L = [], b = cat("", S), c = cat(seq(convert([1], bytes),
   k = 1 .. length(s)))), RETURN(S));
  do n := a(s, b);
```

```
    if n <> 0 then L:= [op(L), n]; b:= cat(b[1 .. n - 1], c, b[n + length(s) .. -1]) else RETURN(L)
    end if
  end do
end proc
```

Typical examples of the procedure use:

> s:=Kr: S:= `96AGArtKrKrAGSvIANKrKrKr6AGKrAG`: s1:=KR: Pos(s, S), Pos(s1, cat("", S));

[8, 10, 19, 21, 23, 28], "96AGArtKrKrAGSvIANKrKrKr6AGKrAG"

> Pos(s1, cat("", S), insensitive);

[8, 10, 19, 21, 23, 28]

Red_n – reducing of multiplicity of entries of symbols or substrings into a string

Call format of the procedure:

Red_n(S, G, N)

Formal arguments of the procedure:

S - a string or symbol
G - a string or symbol of length **>= 1;** or their list
N - a positive integer (*posint*) or list of positive integers

Description of the procedure:

The procedure call **Red_n(S, G, N)** returns the reducing result of multiplicity of entries of symbols or strings specified by the second actual **G** argument into a string or symbol specified by the first actual **S** argument, to quantity not greater, than the third argument **N** specifies. In particular, if **N** = {**1**|**2**}, a **G** string/symbol is deleted from the string **S** or is remained with a single multiplicity accordingly. In addition, a type of the returned result corresponds to a type of an initial actual **S** argument. The **Red_n(S, G, N)** procedure does sensitive searching; that is, upper/lower case differences are essential. If **G** symbol does not belong to the **S** string, the **Red_n** procedure returns the first actual argument without processing with output of the appropriate warning. The null length of the second actual **G** argument causes error.

If the second and the third actual arguments specify lists between which one-to-one correspondence takes place, then the reducing is done over all elements of list **G** with appropriate multiplicities from list **N.** If ***nops***(G) **>** ***nops***(N), the last ***nops***(G) **–** ***nops***(N) elements of **G** will have multiplicity **1** in the returned result. If **G** is a list and **N** is a positive integer, the all **G** elements receive the same multiplicity **N.** At last, if **G** is a symbol or string and **N** is a list, then **G** receive multiplicity **N[1]** in the returned result. The procedure **Red_n** is a rather useful tool in processing of strings and symbols in symbolic problem solving.

```
Red_n:=proc(S::{string,symbol},G::{string,symbol,list({string,symbol})},N::{posint,list(posint)})
local k, h, Lambda, z, g, n;
  if type(G, {string, symbol}) then g := G; n := `if`(type(N, posint), N, N[1]) else h := S;
    for k to nops(G) do
      try n := N[k]; h := procname(h, G[k], n)
      catch "invalid subscript selector": h := procname(h, G[k], `if`(type(N, list), 2, N))
      catch "invalid input: %1 expects": h := procname(h, G[k], `if`(type(N, list), 2, N))
```

```
      end try
    end do;
    RETURN(h)
  end if;
  `if`(length(g) < 1, ERROR("length of <%1> should be more than 1", g),
     assign(z = convert([2], bytes)));
  Lambda:= proc(S, g, n)
    local a, b, h, k, p, t;
      `if`(search(S, g), assign(t = cat(convert([1], bytes) $ (k = 1 .. n - 1))),
        RETURN(S, WARNING("substring <%1> does not exist in string <%2>", g, S)));
      assign(h = "", a = cat("", S, t), b = cat("", g $ (k = 1 .. n)), p = 0);
      do seq(assign('h' = cat(h, `if`(a[k .. k + n - 1] = b, assign('p' = p + 1), a[k]))),
          k = 1 .. length(a) - n + 1);
        if p = 0 then break else p := 0; a := cat(h, t); h := "" end if end do;
      h
    end proc;
  if length(g) = 1 then h := Lambda(S, g, n, g) else
    h := Subs_All(z = g, Lambda(Subs_All(g = z, S, 2), z, n, g), 2)
  end if;
  convert(h, whattype(S))
end proc
```

Typical examples of the procedure use:

> Red_n(`aaaaaaaa6aaaaaagnaaaa14aaaaaaaahhhaaaartaakr`, `a`, 5);

aaaa6aaaagnaaaa14aaaahhhaaaartaakr

> Red_n("aaaaaaaa6aaaaaagnaaaa14aaaaaaaahhhaaaartaakr", `a`, 5);

"aaaa6aaaagnaaaa14aaaahhhaaaartaakr"

> Red_n("aaaaaaaa6aaaaaagnaaaa14aaaaaaaahhhaaaartaakr", `A`, 5);

Warning, substring <A> does not exist in string
<aaaaaaaa6aaaaaagnaaaa14aaaaaaaahhhaaaartaakr>

"aaaaaaaa6aaaaaagnaaaa14aaaaaaaahhhaaaartaakr"

> Red_n("aaccccbccccccccccccbcccccccckcccckccccccccccc", "cccc", 2);

"aaccccbccccbcccckcccckccccccc"

> Red_n("iii","hhh",2), Red_n("iii","iii",2), Red_n("iii","iii",1), Red_n("iii","iiii",1);

Warning, substring <hhh> does not exist in string <iii>
Warning, substring <iiii> does not exist in string <iii>

"iii", "iii", "", "iii"

> Red_n("avzransian", "", 2);

Error, (in **Red_n**) length of <> should be more than 1

> Red_n("1112222333334444445555556666666", [`1`,`2`,`3`,`4`,`5`,`6`], [2,3,4,5,6,7]);

"12233344455555666666"

> Red_n("1112222333334444445555556666666", [`1`,`2`,`3`,`4`,`5`,`6`], 2);

"123456"

> Red_n("111222233333444444555555566666666", [`1`,`2`,`3`,`4`,`5`,`6`], [1,1,1,1,1,1]);

""

> Red_n("111222233333444444555555566666666", [`1`,`2`,`7`,`4`,`5`,`6`], [2,3,4,5,6]);

Warning, substring <7> does not exist in string <1223333344444455555556666666666>

"122333334444555556"

---

sspos – substitutions of symbols or strings into the given positions of a string or symbol

Call format of the procedure:

sspos(S, p, s)

Formal arguments of the procedure:

S - a string or symbol
p - a list of positive integers
s - a list of symbols and/or strings

Description of the procedure:

The procedure call **sspos(S, p, s)** returns the result of substitutions of symbols or strings specified by the third actual **s** argument into a string or symbol specified by the first actual **S** argument into its positions specified by the second actual **p** argument. In addition, a type of the returned result corresponds to a type of an initial actual **S** argument. If the second and the third actual arguments specify lists between which one-to-one correspondence takes place, then all **S** elements in positions **p** are replaced by the corresponding elements from a list **s**; otherwise, the procedure executes all allowable substitutions, not causing error. The procedure **sspos** is a rather useful tool in processing of strings and symbols in symbolic problem solving.

```
sspos:= proc(S::{string, symbol}, p::list(posint), s::list({string, symbol}))
local a, k;
  a := convert(cat("", S), list);
  for k to nops(p) do
     try a := subsop(p[k] = s[k], a)
     catch "invalid subscript selector": next
     catch "improper op or subscript selector": next
     end try
  end do;
  convert(cat(op(a)), whattype(S))
end proc
```

Typical examples of the procedure use:

> sspos(`aaabbbxyzcccdddeeehhh`, [3,6,9,15,20], [`1`,`2`,`3`,"4","5"]);

aa1bb2xy3cccdd4eeeh5h

> sspos("aaabbbxyzcccdddeeehhh", [3,6,9,15,20], ["4","5"]);

"aa4bb5xyzcccdddeeehhh"

> sspos("ab", [3,6,9], [`1`,`2`]), sspos("",[3,6,9], [`1`,`2`]);

"ab", ""

> sspos(`aaabbbxyzcccdddeeehhh`, [3,6,9,7,15,20], [`1995`,`222`,`3`,`RANS`,"42","2003"]);

aa1995bb222RANSy3cccdd42eeeh2003h

---

extrS – extracting out of a symbol or string of substrings limited by the given delimiters

Call format of the procedure:

extrS(S, L, R {, 'h'})

Formal arguments of the procedure:

S - a string or symbol
L - a set of symbols and/or strings
R - a set of symbols and/or strings
h - (optional) an assignable name

Description of the procedure:

The procedure call **extrS(S, L, R)** returns the list of substrings limited from the left by delimiters specified by the actual **L** argument and from the right by delimiters specified by the actual **R** argument, excluding delimiters themselves. The substrings are extracted out of a string or symbol specified by the actual **S** argument; they can contain only non-intersecting similar substrings or be free from them. In addition, a type of the returned result corresponds to a type of an initial actual **S** argument. The intersection of the second and third arguments should be the *empty* set, otherwise error arises. If the procedure call uses the fourth optional argument **h,** then through it the nested list is returned whose sublists define the initial and end positions of the extracted substrings in an initial **S** string/symbol. At absence of the substrings satisfying to the above conditions, the procedure returns the *empty* list, i.e. []; in this case through **h** argument the *empty* list is returned also. The procedure handles the basic erratic and especial situations. The procedure **extrS** is a rather useful tool in processing of strings and symbols in symbolic problem solving; for example, during a work with text datafiles containing the definitions of procedures.

```
extrS:= proc(S::{string, symbol}, L::set({string, symbol}), R::set({string, symbol}))
local a, b, k, h, res, nu, omega;
  `if`(map2(cat, "", L) intersect map2(cat, "", R) = {}, assign(h = S, res = [], b = convert([7],
     bytes)), ERROR("intersection of the second and third arguments should be the empty set"));
  nu:= proc(h, d)
    local k; [seq(h[d[k][1] + 1 .. d[k][2] - 1], k = 1 .. nops(d))] end proc;
  omega:= z -> `if`(map(Empty, {op(z)}) = {false}, SLj(z, 1), z);
  do a := map2(Search2, h, [L, R]);
    if a[1] = [] and a[2] = [] or a[1] = [] or a[2] = [] then
      try RETURN(nu(S, omega(res)), `if`(3 < nargs and type(eval(args[4]), symbol),
         assign(args[4] = omega(res)), NULL))
      catch "invalid input: %1": RETURN(h, `if`(3 < nargs and type(eval(args[4]), symbol),
         assign(args[4] = omega(res)), NULL))
      end try
    end if;
    for k in a[2] do
      if a[1][-1] < k then res := [op(res), [a[1][-1], k]]; h := sspos(h, [a[1][-1], k], [b, b]); break
      else h := sspos(h, [a[1][-1]], [b])
      end if
```

```
    end do
  end do
end proc
```

Typical examples of the procedure use:

> extrS("aaa(a(aa)a(ccagnccavzccc)bbbb)b", {"("}, {")"});

["a(aa)a(ccagnccavzccc)bbbb", "aa", "ccagnccavzccc"]

> extrS("aaa(a(aa)a(ccccccc(bbbb{b", {"("}, {")"});

["aa"]

> extrS("aaa(a(aa)a(cSvetccArtccKrcc)bbbb)b", {"(", a}, {")", b});

["(a(aa)a(cSvetccArtccKrcc)bbbb)", "a(aa)a(cSvetccArtccKrcc)bbbb", "(aa)a(cSvetccArtccKrcc)bbb", "aa)a(cSvetccArtccKrcc)bb", "a)a(cSvetccArtccKrcc)b", "", "(cSvetccArtccKrcc)", "cSvetccArtccKrcc"]

> extrS("aaa(a(aIANa)a(ccRANScc)bbbb)b", {"(", c}, {")", b});

["a(aIANa)a(ccRANScc)bbbb", "aIANa", "ccRANScc)bbb", "cRANScc)bb", "RANScc)b", "c)", ""]

> extrS("cc123aa456xx789aa1995-2003xxdd", {"aa"}, {"xx"});

["a456", "a1995-2003"]

> extrS("cc123aa456yy789aa101112yydd", {"aa"}, {"xx"});

[]

> extrS(`cc123aa456yy789aa101112yydd`, {"aa"}, {"xx"});

[]

> extrS(`cc123aa456yy789aa101112yydd`, {"aa"}, {"xx"}, z), z;

[], []

> extrS("cc123aayy45aa6yy789aa101112yyddaa2003yy", {"aa"}, {"yy"}, 'p'), p;

["a", "a6", "a101112", "a2003"], [[6, 8], [12, 15], [20, 28], [32, 38]]

Rlss – inversion of symbols, strings or lists
Rssl
Rev

Call format of the procedures:

Rlss(S)
Rssl(S)
Rev(G)

Formal arguments of the procedures:

S - a string, symbol or list
G - a string, symbol or an integer >= 0

Description of the procedures:

The procedure call **Rlss(S)** returns the result of inversion of elements of a string, a symbol or a list defined by the actual **S** argument. The **Rssl** procedure is functionally analogous to the **Rlss** procedure; however, it uses more perfect mechanism of localization of index variable. Starting with the *seventh* release, the procedure **Reverse(L)** of module **ListTools** is intended for inversion of lists, however this procedure yields to the above procedures **Rlss** in time characteristics and has the smaller functionalities.

The procedure call **Rev(G)** returns the result of inversion of elements of a string, a symbol or an integer defined by the **G** argument. The procedure has been programmed by the effective single-string extracode. The above three procedures have the important appendices at programming a various sort of problems of symbolic processing, by illustrating a number of useful receptions of programming, including nonstandard ones also.

```
Rlss:= proc(L::{list, string, symbol})
local k;
  `if`(type(L, list), RETURN([seq(L[-k], k = 1 .. nops(L))]),
     convert(cat(op(procname(convert(convert(L, string), list)))), whattype(L)))
end proc

Rssl:= proc(S::{list, string, symbol})
local k;
  `if`(type(S, list), [S[nops(S) - 'k'] $ ('k' = 0 .. nops(S) - 1)], convert(
  cat(convert(cat("", S), list)[length(S) - 'k'] $ ('k' = 0 .. length(S) - 1)), whattype(S)))
end proc

Rev:= proc(S::{nonnegint, name, string, symbol})
local a, b, k;
  assign(a =convert(S, string), b= ""), seq(assign('b' = cat(b, a[length(S) - k+1])), k= 0..length(S)),
  `if`(member(whattype(S), {string, symbol}), convert(b, whattype(S)), SN(b))
end proc
```

Typical examples of the procedures use:

> map(Rlss, [[67,62,89,96,42,47], "RANS",IAN,Artur,Kristo]), Rlss("1942200260250");

[[47, 42, 96, 89, 62, 67], "SNAR", *NAI*, *rutrA*, *otsirK*], "0520620022491"

> Rlss(\`19422003602501995613\`), Rlss([1,9,42,200,260,250,1995,613,2003]);

31659910520630022491, [2003, 613, 1995, 250, 260, 200, 42, 9, 1]

> map(Rssl, [[67,62,89,96,42,47], "RANS",IAN,Art,Kristo]), map(Rssl,[[1,2,3,5,5,7,8,10,11], Vilnius, "Tallinn", Grodno]);

[[47, 42, 96, 89, 62, 67], "SNAR", *NAI*, *trA*, *otsirK*], [[11, 10, 8, 7, 5, 5, 3, 2, 1], *suinliV*, "nnillaT", *ondorG*]

> n:=10^5: L:=[seq(k,k=1..n)]: with(ListTools): t:=time(): Rlss(L): time()-t; t:=time(): Reverse(L): time()-t; **#** ***Maple*** **8**

0.350
0.352

> n:=10^6: L:=[seq(k,k=1..n)]: with(ListTools): t:=time(): Rlss(L): time()-t; t:=time(): Reverse(L): time()-t;

3.605
3.615

> n:=10^7: L:=[seq(k,k=1..n)]: with(ListTools): t:=time(): Rlss(L): time()-t; t:=time(): Reverse(L): time()-t;

154.692
374.228

> map(Rev, [140619422003,290719472003,23041967,13021989,5081998]);

[300224916041, 300274917092, 76914032, 98912031, 8991805]

RS – exhaustive substitutions into a string or symbol

Call format of the procedure:

RS(*subst, string* {, `*insensitive*`})

Formal arguments of the procedure:

subst – an equation defining a substitution; both members of the equation should be by strings or symbols of length **1**
string – a string or a name to be processed
insensitive – an insensitive mode of substitutions

Description of the procedure:

The procedure call **RS(*subst, string*)** returns the result of substitutions of the right members of equations, defined by the first actual *subst* argument, instead of all entries into a string or symbol defined by the second *string* argument, of the left members of the equations. A type of the returned result corresponds to a type of the second actual *string* argument. The procedure call on the empty string/symbol {""/``} returns the *empty* string/symbol {""/``}.

By default, the **RS(*subst, string*)** procedure does sensitive searching; that is, upper/lower case differences are essential. If the optional actual `*insensitive*` argument has been coded, then the **RS(*subst, string, insensitive*)** procedure does insensitive searching; that is, upper/lower case differences are inessential. If none of the left members of substitutions of the list *subst* does not belong to the *string*, the procedure returns the initial *string* without processing.

```
RS:= proc(L::list(equation), S::{string, symbol})
local a, b, c, k, h, X;
  `if`(nargs = 3 and args[3] = insensitive, assign(X = searchtext), assign(X = SearchText));
  [assign(a = "", b = convert(S, string)), seq([seq(assign('a' = cat(a, `if`(X(lhs(L[h]), b[k]) = 1,
     rhs(L[h]), b[k]))), k = 1 .. length(b)), assign('b' = a),
     assign('c' = a), assign('a' = "")], h = 1 .. nops(L)), convert(c, whattype(S))][-1]
end proc
```

Typical examples of the procedure use:

```
> RS([`*`=`_`,`=`=`-`,`#`=`$`,G=T], `IAN*1995=2003*RANS; #GRU#`);
                         IAN_1995-2003_RANS; $TRU$
> RS([`*`=`_`,`=`=`-`,`#`=`$`,g=T], "IAN*1995=2003*RANS; #GRU#");
                         "IAN_1995-2003_RANS; $GRU$"
> RS([`V`=`G`,`|`=`T`,"w"=`b`], "IAN*1995=2003*RANS;  #GRU#", insensitive);
                         "IAN*1995=2003*RANS;  #GRU#"
```

---

sstr – exhaustive substitutions into a string or symbol instead of its characters

Call format of the procedure:

sstr(L, S)

Formal arguments of the procedure:

L – a list of equations defining substitutions; the left members of the equations should be by strings or symbols of length **1**
S – a string or symbol to be processed

Description of the procedure:

The procedure call **sstr(L, S)** returns the result of substitutions of the right members of equations specified by the first actual **L** argument, instead of all entries into a string or symbol specified by the second **S** argument, of the left members of the equations; in addition, the left members of the equations should have length **1**, i.e. be characters. The **sstr** procedure does sensitive searching; that is, upper/lower case differences are essential. A type of the returned result corresponds to a type of the second **S** argument. At a detecting of especial situations (the *empty* **L** argument and/or **S**, etc.) the procedure call returns the initial **S** argument without processing with output of an appropriate warning. The **sstr** procedure allows to replace characters composing a string or symbol by substrings of length **>=1** and/or to delete characters out of strings or symbols.

```
sstr:= proc(L::{list(equation), set(equation)}, S::{string, symbol})
local a, k, j, h, g, p, u, z;
  `if`(L = [] or L = {} or Sts(S, " "), RETURN(S, WARNING("invalid arguments - %1 and/or
  %2", L, S)), assign67(h = [], z = cat("", S), a = interface(warnlevel),
     g = eval(conSV(S)), u = convert([1], bytes)));
  for k to nops(L) do
     if 1 < length(lhs(L[k])) then WARNING("the left member of substitution %1 is not
        a character, the substitution has been ignored", L[k])
     else h := [op(h), lhs(L[k]) = `if`(rhs(L[k]) = NULL, u, rhs(L[k]))]
     end if
  end do;
  for k to nops(h) do p := Search2(z, {lhs(h[k])});
     if p = [] then next else for j to nops(p) do g[p[j]] := rhs(h[k]) end do end if
  end do;
  interface(warnlevel=0),convert(Red_n(conSV(eval(g)),u,1),whattype(S)),interface(warnlevel=a)
end proc
```

Typical examples of the procedure use:

> sstr([a=NULL, b=61, c=56, d=714, e=avz, f=NULL], "abcdafdbfcedecdeefabcd");

"61567147146156avz714avz56714avzavz6156714"

> sstr([a=NULL, b=61, c=56, d=714, e=avz, f=NULL], \`abcdafdbfcedecdeefabcd\`);

61567147146156avz714avz56714avzavz6156714

> sstr([], "avzransianghtyrserwrtyuimnvgf");

Warning, invalid arguments - [] and/or avzransianghtyrserwrtyuimnvgf

"avzransianghtyrserwrtyuimnvgf"

> sstr([a=NULL, b=61, c=56, d=714, e=avz, fa=NULL], "abcdafdbfcedecdeefabcd");

Warning, the left member of substitution fa = (NULL) is not a character, the substitution has been ignored

"6156714f71461f56avz714avz56714avzavzf6156714"

S_D – identification of entries of the special symbols into a string

Call format of the procedure:

S_D(S, B {, 'h'})

Formal arguments of the procedure:

S - a string or symbol
B - an assignable name
H - (optional) an unevaluated name

Description of the procedure:

The procedure call **S_D(S, B)** returns the *true* value, if a string or symbol specified by the first actual **S** argument, contains symbols {`+`, `-`, `.`, `/`, `^`, `*`, 0..9}, i.e. symbols used for forming of arithmetical expressions. In the absence of entries of the above symbols into the string **S** the *false* value is returned.

If the procedure call **S_D(S, B)** returns the *true* value, then via the second actual **B** argument the sequence of 2-element lists is returned. The first element of these lists defines one of the above symbols {`+`, `-`,`.`, `/`, `^`, `*`, `0`..`9`}, whereas the second element defines its position in the **S** string. Otherwise, through the second actual **B** argument the *NULL* value is returned, i.e. nothing.

If the procedure call uses the third optional actual **h** argument, through it the *true* or the *false* value is returned accordingly depending on presence or absence in the **S** string of symbols different from the above symbols. In particular, the procedure is rather useful at processing of expressions of formats {*symbol, string*} that contain the numerical expressions.

```
S_D:= proc(C::{string, symbol}, B::evaln)
local a, b, c, k, p;
  assign(b = false, c = false), assign('B' = op({seq(op([assign('a' = op(convert(cat("", C)[k],
  bytes))), `if`(member(a, [42, 43, 47, 94, (45 + p) $ ('p' = 0 .. 12)]), [assign('b' = true), `if`(a = 43,
  `+`, `if`(a = 45, `-`, `if`(a = 47, `/`, `if`(a = 46, `.`, `if`(a = 94, `^`, `if`(a = 42, `*`, a - 48)))))), k],
  assign('c' = true))]), k = 1 .. length(C))})), `if`(nargs = 3, assign(args[3] = c), NULL), b
end proc
```

Typical examples of the procedure use:

> G:= "61*A*VZ+56*AGN=7*Kr-14Art^12Sv*34Arne/39.Email": S_D(G,Z), Z;

true, [`*`, 3], [2, 28], [1, 2], [`*`, 5], [`-`, 20], [1, 21], [`^`, 26], [1, 27], [`.`, 41], [6, 10], [`*`, 11], [9, 40], [7, 16], [`*`, 31], [`*`, 17], [`+`, 8],[5, 9], [`/`, 38], [3, 39], [3, 32], [4, 33], [6, 1], [4, 22]

> S_D(G, Z, 'h'), h;

true, true

> S_D("RANS_IAN_RAC_REA=1995-2003", p, 'r'), p, r;

true, [2, 23], [1, 18], [9, 19], [9, 20], [5, 21], [`-`, 22], [0, 24], [0, 25], [3, 26], *true*

> G1:=`61A\*VZ+56AGN=7Kr-14Art^12Sv\*34Arne/39.Email`: S_D(G1,P,'t'), P, t;

true, [1, 2], [6, 1], [1, 18], [`*`, 4], [`+`, 7], [5, 8], [6, 9], [7, 14], [`-`, 17], [4, 19], [`^`, 23], [1, 24], [2, 25], [`*`, 28], [3, 29], [4, 30], [`/`, 35], [3, 36], [9, 37], [`.`, 38], *true*

> S_D("V61G56S36A40Art14Kr7=1995-2003", p, 'r'), p, r;

true, [4, 17], [1, 22], [6, 2], [1, 3], [5, 5], [6, 6], [3, 8], [6, 9], [4, 11], [0, 12], [1, 16], [7, 20], [9, 23], [9, 24], [5, 25], [`-`, 26], [2, 27], [0, 28], [0, 29], [3, 30], *true*

> S_D("RANS+62AVZ=8Kr-15Art^37Sv*42Arne/39.Vgtu\Grsu", P, 't'), P, t;

true, [+, 5], [^, 21], [*, 26], [/, 33], [3, 34], [9, 35], [., 36], [6, 6], [2, 7], [8, 12], [-, 15], [1, 16], [5, 17], [3, 22], [7, 23], [4, 27], [2, 28], *true*

Search – case sensitive and insensitive patterns searching
Search1
Search2
Search3

Call format of the procedures:

Search(string, pattern)
Search1(string, pattern {, 'h'})
Search2(string, L)
Search3(string, **P** {, \`insensitive\`})

Formal arguments of the procedures:

pattern – a string or symbol to be searched for
string – a string or symbol to be searched
h – (optional) an unevaluated name
L – a set or list of strings or symbols
P – a set of patterns (symbols and/or strings)
\`insensitive\` – key word defining a search mode

Description of the procedures:

The following group of three highly effective procedures provides several helpful methods of a sensitive searching; that is, upper/lower case differences are essential. In a series of cases, these procedures are more preferable than similar standard tools of the package and essentially extend the last ones.

The **Search** procedure searches for a *pattern* in a *string* by looking for exact matches. The **Search** procedure does sensitive searching; that is, upper/lower case differences are essential. If the *pattern* is not found in the *string*, then the procedure returns the *empty* list, i.e. []. If the *pattern* is found, the procedure returns the integer list that indicates the initial positions of all entries of the *pattern* into the *string*.

In contrast to the **Search** procedure, the **Search1** procedure returns the *true* value, if it finds at least one entry of a *pattern* into a *string*, otherwise the *false* value is returned. At finding of the *pattern* in the *string*, the **Search1** procedure via the third optional **h** argument returns a mode of entries of the *pattern* into the *string*, namely: *coincidence*, *left* (from the very outset of a string), *inside* and *right* (in the end of a string). A combination of these basic characteristics allows to receive the full enough picture about entries of a *pattern* into a *string*; the given possibility is very useful at solution of many problems related with strings processing.

The **Search2** procedure extends the **Search** procedure, providing a search in a *string* of a set or list of patterns defined by the second actual **L** argument. The procedure returns the integer list of ordered first positions of entries of patterns from **L** into the *string*. If the *pattern* is not found in the *string*, the procedure returns the *empty* list, i.e. []. In particular, the given procedure is helpful enough at processing of the full paths to datafiles or directories of a **PC** file system.

At last, the **Search3(*string*, P)** procedure provides search in a *string* of a *patterns* set specified by the actual **P** argument, considering the multiplicity of non-overlapping entries into the *string*. The procedure returns the nested list whose elements are lists. The first element of the lists defines the corresponding *patterns* from **P** in a format identical to a format of the *string*, whereas the other elements are 2-element integer lists whose the first element defines position of a *pattern* in the *string* and the second – its multiplicity. If the **P** patterns are not found in a *string*, the procedure returns the *empty* list, i.e. [].

By default, the **Search3** procedure does sensitive *patterns* searching; that is, upper/lower case differences are essential. If the optional actual *`insensitive`* argument has been coded, then the procedure does insensitive searching; that is, upper/lower case differences are inessential. The given procedure is helpful enough at processing of strings and symbols. The procedures of this group are widely enough used in many procedures represented in the monograph, essentially simplifying programming of numerous applied problems also.

```
Search:= proc(S::{name, string, symbol}, K::{name, string, symbol})
local a, b, c, d, p, h, t;
  assign(a = cat("", S, " "), b = [], d =length(S), p =length(K), h=1), seq(`if`(search(a[h..d], K, 't'),
  assign('b' = [op(b), h + t - 1], 'h' = h + t), RETURN(b)), c = 1 .. length(a))
end proc

Search1:= proc(S::{string, symbol}, K::{string, symbol})
local h;
  assign(h = Search(S, K)), `if`(h = [], false, `if`(h[1] = 1 and length(S) = length(K), op([true,
  `if`(nargs = 3, assign(args[3] = [coincidence]), NULL)]), `if`(h[1] = 1 and
  h[-1] + length(K) - 1 = length(S), op([true, `if`(nargs = 3, assign(args[3] = [left, `if`(2 < nops(h),
  inside, NULL), right]), NULL)]), `if`(h[1] = 1, op([true, `if`(nargs = 3, assign(args[3] = [left,
  `if`(1 < nops(h), inside, NULL)]), NULL)]), `if`(h[-1] + length(K) - 1 = length(S), op([true,
  `if`(nargs = 3, assign(args[3] = [`if`(1 < nops(h), inside, NULL), right]), NULL)]),
  op([true, `if`(nargs = 3, assign(args[3] = [inside]), NULL)]))))))
end proc

Search2:= proc(S::{string, symbol}, K::{set, list})
local k, L;
  L := {}; seq(assign('L' = {op(Search(S, K[k])), op(L)}), k = 1 .. nops(K)); [op(L)]
end proc

Search3:= proc(S::{string, symbol}, P::set({string, symbol}))
local a, b, c, d, s, p, k, j, t, z, omega, nu, psi;
  assign(psi = (t -> `if`(member(insensitive, {args}), Case(t), t)), nu = []);
  assign(s = cat("", S, convert([1], bytes)), p = map(convert, P, string)), assign(b = length(s));
  for z to nops(p) do assign('a' = 0, 'k' = 1, 'd' = length(p[z]), 'omega' = []);
    if not search(psi(s, args), psi(p[z], args)) then NULL
    else omega := [op(omega), convert(p[z], whattype(S))];
      for k to length(s) do
        if not (psi(s[k .. k + d - 1], args) = psi(p[z], args)) then next
        else assign('c' = k, 'a' = a + 1);
          for j from c + d by d to b do
            if psi(s[j .. j + d - 1], args) = psi(p[z], args) then a := a + 1
            else omega := [op(omega), [k, a]]; k := j; a := 0; break
            end if
          end do
        end if
      end do
    end if;
    nu := [op(nu), omega]
  end do;
  [seq(`if`(nu[k] = [], NULL, nu[k]), k = 1 .. nops(nu))]
end proc
```

Typical examples of the procedures use:

> Search(`RANS5IAN6RANSRANS=SRANS`, RANS), Search(`AAAAAAAAAAAAA`,AAA);

[1, 10, 14, 20], [1, 2, 3, 4, 5, 6, 7, 8, 9, 10, 11]

> Search1("agnavzKr_Artagnagn",`agn`,'t'), t, Search1("VagnavzKr_Artagnagn",`agn`,'t'),t;

true, [*left*, *inside*, *right*], *true*, [*inside*, *right*]

> Search1(`VagnavzKr_Artagnagn`,"Art",'h'), h, Search2(`C:\\Temp/RANS\\IAN/lbz.rtf`, {`.`,`/`,`\\`});

true, [*inside*], [3, 8, 13, 17, 21]

**> Search2(`{HYPERLNK 17 "type[name]" 2 "type[name]" "" }{TEXT 277 1 "," }
{HYPERLNK 17 "type[symbol]" 2 "type [symbol]" "" }{TEXT 278 1 "," }{TEXT -1 1 " " }
{HYPERLNK 17 "substring" 2 "substring" "" }{TEXT 276 2 ", " }{HYPERLNK 17 "cat"
2 "cat" "" }{TEXT 280 1 "," } {TEXT -1 1 " " }`,[`{`,`}`, `HYPERLNK`]);**

[1, 2, 45, 46, 62, 64, 65, 113, 114, 130, 131, 146, 147, 148, 189, 190, 207, 208, 209, 238, 239, 255, 257, 272]

> Search3("123avzavzavzavz333agnagnagnKrKrArtArtArt", {"avz","agn",Kr,Art});

[["Kr", [28, 2]], ["Art", [32, 3]], ["avz", [4, 4]], ["agn", [19, 3]]]

> Search3(`123avzavzavzavz333agnagnagnKrKrArtArtArt`, {"avz","agn",Kr,Art});

[[*Kr*, [28, 2]], [*Art*, [32, 3]], [*avz*, [4, 4]], [*agn*, [19, 3]]]

> Search3("123avzavzavzavz333agnagnagnKrKrArtArtArt", {"avz","agn",Kr7,Art14});

[["avz", [4, 4]], ["agn", [19, 3]]]

> Search3(`123avzavzavzavz333agnagnagnkrkrartartart`, {"avz","agn",Kr,Art});

[[*avz*, [4, 4]], [*agn*, [19, 3]]]

> Search3(`123avzavzavzavz333agnagnagnkrkrartartart`,{"avz","agn",Kr,Art},insensitive);

[[*Kr*, [28, 2]], [*Art*, [32, 3]], [*avz*, [4, 4]], [*agn*, [19, 3]]]

> Search3("123avzavzavzavz333agnagnagnKrKrArtArtArtavzavz", {"avz","agn",Kr,Art});

[["Kr", [28, 2]], ["Art", [32, 3]], ["avz", [4, 4], [41, 2]], ["agn", [19, 3]]]

> Search3("1995aaaaaaaa8aaaaaaa7aaaaaa6aaaaa5aaaa4aaa3aa2a2003", {a,b,c});

[["a", [5, 8], [14, 7], [22, 6], [29, 5], [35, 4], [40, 3], [44, 2], [47, 1]]]

SL – converting of an expression into list of strings or symbols

Call format of the procedure:

SL(*expr* {, `*symbol*`})

Formal arguments of the procedure:

expr - a valid *Maple* expression
symbol - optional argument

Description of the procedure:

The procedure call **SL(*expr*)** returns the result of converting of an arbitrary *Maple* expression into list of strings or symbols where each string represents a symbol composing the expression specified by the actual *expr* argument. If the second optional *`symbol`* argument has been coded then the list of symbols is returned, otherwise the list of strings of length **1** is returned. The given procedure can be useful at symbolical processing of the expressions.

```
SL:= proc(S::anything)
local a, b, k;
  assign67(a = convert(S, string), b = `if`(nargs = 2 and args[2] = symbol, ``, NULL));
  [seq(cat(b, a[k]), k = 1 .. length(a))]
end proc
```

Typical examples of the procedure use:

> SL(sin(7*x+14*y)/(Art+Kr)+Sv*Gal^2-Vic/Art, symbol);

[s, i, n, (, 7, *, x, +, 1, 4, *, y,), /, (, A, r, t, +, K, r,), +, S, v, *, G, a, l, ^, 2, -, V, i, c, /, A, r, t]

> SL(sin(6*x+14*y)/(Art+Kr)+Sv*Gal^2-Vic/Art);

["s", "i", "n", "(", "6", "*", "x", "+", "1", "4", "*", "y", ")", "/", "(", "A", "r", "t", "+", "K", "r", ")", "+", "S", "v", "*", "G", "a", "l", "^", "2", "-", "V", "i", "c", "/", "A", "r", "t"]

SLS – converting of a string or symbol into list or set
SLD

Call format of the procedures:

SLS(S, d {, h})
SLD(S, d1)

Formal arguments of the procedures:

S - a string or symbol
d - a delimiter; a set or list of symbols or strings of length **>= 1**
d1 - a delimiter; a symbol or string of length **>= 1**
h - (optional) a type of the returned result (may be *`list`* or *`set`*)

Description of the procedures:

The procedure call **SLS(S, d)** returns the list or set formed by segments of a string specified by the first actual **S** argument; these segments are formed by partition of the **S** string by delimiters, namely by elements of a list or a set defined by the second actual **d** argument. In addition, a type of the returned result corresponds to a type of the optional third actual **h** argument allowing only two values *`set`* and *`list`*. The **SLS(S, d)** procedure does sensitive searching; that is, upper/lower case differences are essential. If the **d** delimiters do not belong to **S,** the procedure returns the first actual argument **S** without processing with output of appropriate warning.

In contrast to the **SLS,** the highly effective procedure **SLD(S, d1)** admits a string or a symbol of an arbitrary finite length as the delimiter **d1** at the parting of **S** string before its converting into list. Similarly to the first, the procedure call **SLD(S, d1)** returns the list formed by segments of a string, specified by the first actual **S** argument; these segments are formed by partition of the **S** string by a delimiter, namely by a string or a symbol specified by the second actual **d1** argument. In addition, a type of elements of the returned list corresponds to a type of the first actual **S** argument. If the **d1** delimiter does not belong to **S,** the procedure call returns the list **[S]**. Whereas in other especial situations the *empty* list is returned, i.e. []. Both procedures are widely enough used in many procedures represented in the monograph, essentially simplifying programming of numerous applied problems also.

```
SLS:= proc(S::{string, symbol}, s::{list({string, symbol}), set({string, symbol})})
local a, b, c, d, g, k, Z, omega, nu;
  nu:= () -> WARNING("delimiters %1 are absent in the initial string", {op(s)} minus omega);
  `if`(nargs = 2, assign(c = list), `if`(member(args[3], {set, list}), assign(c = args[3]),
    ERROR("3rd argument should be `list` or `set` but has received <%1>", args[3])));
  assign(omega = {seq(`if`(search(S, s[k]), s[k], NULL), k = 1 .. nops(s))}), `if`(omega = {},
    RETURN(nu(), S), `if`(omega = {op(s)}, NULL, nu())), `if`(1 < nops(omega),
    assign(a = [Sub_st([seq(omega[k] = omega[1], k = 2 .. nops(omega))], cat("", S), Z)][1]),
  assign(a = cat("", S))), assign('a' = Red_n(a, omega[1], 2)), assign(b = Search2(a, {omega[1]}));
  convert(map(convert, [`if`(b[1] = 1, NULL, a[1 .. b[1] - 1]), seq(a[b[k] + 1 .. b[k + 1] - 1],
  k = 1 .. nops(b) - 1), `if`(b[nops(b)] = length(a), NULL, a[b[nops(b)] + 1 .. -1])], whattype(S)), c)
end proc
```

```
SLD:= proc(S::{string, symbol}, D::{string, symbol})
local k, p, h, n, d;
  `if`(search(S, D), assign(p = cat("", S), n = length(D), d = cat("", D)) , RETURN([S])),
    assign('p' = Red_n(p, d, 2));
  `if`(p = d or Sts(p, d[1]) and Sts(d, p[1]) and n < length(p) and not type(length(p)/n, integer),
    RETURN([]), assign('p' = `if`(p[1 .. n] = d and p[-n .. -1] = d, p[n + 1 .. -1 - n],
    `if`(p[1 .. n] = d, p[n + 1 .. -1], `if`(p[-n .. -1] = d, p[1 .. -1 - n], p)))));
  `if`(search(p, d), assign(h = Search(p, d)), RETURN([convert(p, whattype(S))])),
    seq(`if`(h[k + 1] - h[k] = 1, assign('h[k + 1]' = 0), 0), k = 1 .. nops(h) - 1);
  [convert(p[1 .. h[1] - 1], whattype(S)), seq(convert(p[h[k] + n .. h[k + 1] - 1], whattype(S)),
    k = 1 .. nops(h) - 1), convert(p[h[nops(h)] + n .. -1], whattype(S))]
end proc
```

Typical examples of the procedures use:

```
> restart: SLS("aa42aaaba47a67aa89abaa14b7a62", {a,b,c,p});
   Warning, delimiters {c, p} are absent in initial string
                 ["42", "47", "67", "89", "14", "7", "62"]
> SLS(`aa42aaaba47a67aa89abaa14b7a62`, {a,b});
                 [42, 47, 67, 89, 14, 7, 62]
> SLS("1, 2, 3, 4, 5, 6, 7, 8, 9, 10, 11, 12, 13, 14", [","]);
     ["1", " 2", " 3", " 4", " 5", " 6", " 7", " 8", " 9", " 10", " 11", " 12", " 13", " 14"]
> SLS("aa42aaaba47a67aa89abaa14b7a62", [a, b, c, p], set);
   Warning, delimiters {p, c} are absent in the initial string
                 {"62", "42", "47", "7", "67", "89", "14"}
> SLS("aa42aaaba47a67aa89abaa14b7a62", [a, b, c, p], art_kr);
   Error, (in SLS) 3rd argument should be `list` or `set` but has received <art_kr>
> SLS("aa42aaaba47a67aa89abaa14b7a62", [x, y, z, h, p], set);
   Warning, delimiters {x, y, p, h, z} are absent in the initial string
                 "aa42aaaba47a67aa89abaa14b7a62"
> SLD("61 56 36 40 7 14"," "), SLD("aa42aaaa47a67aa89aaa96a62", b);
          ["61", "56", "36", "40", "7", "14"], "aa42aaaa47a67aa89aaa96a62"
> SLD("babbcbb",bb), SLD(`babbcbb`,b), SLD("bbbbb",bbb), SLD("bbcbbcbbcb",bbc);
                 ["ba", "c"], [a, c], [], ["b"]
```

Slss – a sorting of symbols, strings or lists

Call format of the procedure:

Slss(S {, *sf*})

Formal arguments of the procedure:

S - a string, symbol or list
sf - (optional) a sorting function

Description of the procedure:

The procedure call **Slss(S)** returns the result of sorting of elements of a string, a symbol or a list defined by the actual **S** argument. The second optional actual *sf* argument is a Boolean procedure of two arguments that defines some sorting algorithm. By default, for the second argument is supposed the value accepted for standard function **`sort`** of the package *Maple*.

```
Slss:= L::{list, string, symbol} -> `if`(type(L, list), RETURN(sort(L, `if`(nargs = 2,
   args[2], NULL))), convert(cat(op(procname(convert(convert(L, string), list)))), whattype(L)))
```

Typical examples of the procedure use:

> map(Slss, [[67,62,89,96,42,47],"RANS",IAN,Artur,Kristo]), Slss("1942200360250");

[[42, 47, 62, 67, 89, 96], "ANRS", *AIN*, *Arrtu*, *Kiorst*], "0000122234569"

> Slss([1,9,42,200,260,250,1995,613,2003]), map(Slss, [RANSREA, IANRAC],`lexorder`);

[1, 9, 42, 200, 250, 260, 613, 1995, 2003], [*AAENRRS*, *AACINR*]

> map(Slss, [[1,2,3,7,4,5,5,7,61,8,9,10,11,12,13,14,56,36,40], Grodno, Vilnius, "Tallinn", Moscow]);

[[1, 2, 3, 4, 5, 5, 7, 7, 8, 9, 10, 11, 12, 13, 14, 36, 40, 56, 61], *Gdnoor*, *Viilnsu*, "Taillnn", *Mcoosw*]

> map(Slss, [[q,w,e,r,t,y,u,i,o,p,a,s,d,f,g,h,j,k,l,z,x,c,v,b,n,m], Lasnamae, RANS_IAN, Noosphere]);

[[a, b, c, d, e, f, g, h, i, j, k, l, m, n, o, p, q, r, s, t, u, v, w, x, y, z],*Laaaemns*, *AAINNRS_*, *Neehooprs*]

SN – converting of symbolic representation of a number into its numeric value

Call format of the procedure:

SN(*str*)

Formal arguments of the procedure:

str - a numeric value or symbolic representing of a numeric value {may be *integer, float, fraction, symbol, string*}

Description of the procedure:

The procedure call **SN(*str*)** returns the numeric value defined by the actual ***str*** argument; this argument is a numeric value or symbolic representation {*string* | *symbol*} of a number with sign or without of a type {*integer* | *float* | *fraction*}. The procedure call on the empty string/symbol or on a string/symbol consisting only of *blanks* returns *zero*.

The given procedure is dispatcher of four special procedures **SN_6, SN_7, SN_8** and **SN_9** that are oriented onto releases **6, 7, 8** and **9** accordingly. In addition, implementations of procedures **SN_7, SN_8** and **SN_9** are identical, and are different from implementation of procedure **SN_6**.The procedure **SN** calls one of the these four procedures depending on a current release of *Maple*. Such approach has been used in connection with some incompatibility taking place for the last four releases **6, 7, 8** and **9** of *Maple*. The compatibility problems of the above releases are discussed in our books [12,13,29-33,43,44] enough in detail. The procedure **SN** plays important enough role in many of the problems of symbolic processing in the *Maple* environment. The **came** procedure extends the procedure **SN** possibilities and has the larger responsiveness.

```
SN:= proc(S::{numeric, string, symbol})
local s;
  assign(s = convert(S, string)), `if`(belong({op(convert(s, bytes))}, {32}), RETURN(0),
  convert(cat("SN_", Release()), symbol)(s))
end proc

SN_6:= proc(S::string)
local k, `1`, `2`, `3`, `4`, `5`, p;
  [`if`(S[1] = "+", assign('`1`' = S[2 .. length(S)]), assign('`1`' = S)), `if`(`1`[length(`1`)] = "/",
  assign('`1`' = cat(`1`, "1")), NULL), `if`(`1`[length(`1`)] = ".", assign('`1`' = cat(`1`, "0")),
  NULL), `if`(`1`[1] = ".", assign('`1`' = cat("0", `1`)), NULL), `if`(`1`[1 .. 2] = "-.", assign('`1`' =
  cat("-0", `1`[2 .. length(`1`)])), NULL), mul(`if`(member(`1`[p], ["/", ".", "-", seq(convert(k,
  string), k = 0 .. 9)]), 1, ERROR("Invalid value <%1>", `1`)), p = 1 .. length(`1`))*1(unassign('`2`',
  '`3`', '`5`'))*`if`(`1`[1] = "-", -1(assign(`3` = 2)), 1(assign(`3` = 1)))*1(assign(`2` = (`5` ->
  add((op(convert(`5`[k], bytes)) - 48)*10^(length(`5`) - k), k = 1 .. length(`5`))))) * `if`(search(`1`,
  "/", '`4`'), `2`(`1`[`3` .. `4` - 1])/`2`(`1`[`4` + 1 .. length(`1`)]), `if`(search(`1`, ".", '`4`'),
  `2`(`1`[`3`..`4` - 1]) + evalf(`2`(`1`[`4` + 1..length(`1`)])/10^(length(`1`) - `4`), length(`1`) - `4`),
  `2`(`1`[`3` .. length(`1`)]))), unassign(seq(convert(k, symbol), k = 1 .. 5))][1]
end proc

SN_9:= proc(S::string)
local k, `1`, `2`, `3`, `4`, `5`, p;
  op([`if`(S = "1" or 1 < length(S) and S[-1] = "1" and {op(convert(S[1 .. -2], bytes))} = {48},
  RETURN(1), NULL), op(1, `if`(S[1] = "-", -1, 1)*[`if`(S[1] = "+", assign('`1`' = S[2..length(S)]),
  assign('`1`' = S)), `if`(`1`[length(`1`)] = "/", assign('`1`' = cat(`1`, "1")), NULL),
  `if`(`1`[length(`1`)] = ".", assign('`1`' = cat(`1`, "0")), NULL), `if`(`1`[1] = ".", assign('`1`' =
  cat("0", `1`)), NULL), `if`(`1`[1 .. 2] = "-.", assign('`1`' = cat("-0", `1`[2 .. length(`1`)])), NULL),
  mul(`if`(member(`1`[p], ["/", ".", "-", seq(convert(k, string), k = 0 .. 9)]), 1,
  ERROR("Invalid value <%1>", `1`)), p = 1 .. length(`1`))*1(unassign('`2`', '`3`', '`5`'))*
  `if`(`1`[1] = "-", -assign('`3`' = 2), assign('`3`' = 1))*assign(`2` = (`5` -> add((op(convert(`5`[k],
  bytes)) - 48)*10^(length(`5`) - k), k = 1..length(`5`))))*`if`(search(`1`, "/", '`4`'), `2`(`1`[`3`..`4` -
  1])/`2`(`1`[`4` + 1 .. length(`1`)]), `if`(search(`1`, ".", '`4`'), `2`(`1`[`3` .. `4` - 1]) + evalf(`2`(`1`[
  `4` + 1 .. length(`1`)])/10^(length(`1`) - `4`), length(`1`) - `4`), `2`(`1`[`3` .. length(`1`)]))),
  unassign(seq(convert(k, symbol), k = 1 .. 5))][1])])
end proc
```

Typical examples of the procedure use:

```
> map(SN, ["+2003", "-1995", "19.89", "19/99", "+56/61", "-19.42", "6/14", "-56/47", "61.42", "4120",
  "004261", "-0061/56", "+/47", "", "  ", "-3.03", `  `]);
    [2003, -1995, 19.89, 19/99, 56/61, -19.42, 3/7, -56/47, 61.42, 4120, 4261, -61/56, 0, 0, 0, -3.030, 0]
> map(SN, [`+2003`,`-1995`,`19.89`,`19/99`,`+56/61`,`-19.42`,`6/14`,`-56/47`,`61.42`,`4120`,`004261`,
  `-0061/56`,`+/47`,"","  ",`-3.03`,`  `]);
    [2003, -1995, 19.89, 19/99, 56/61, -19.42, 3/7, -56/47, 61.42, 4120, 4261, -61/56, 0, 0, 0, -3.030, 0]
```

```
> map(SN, [0,+2003,-1995,19.89,19/99,+56/61,-19.42,7/14, -56/47,61.42,4120, 004261, -0061/56,
  +1/47,-3.03]);
    [0, 2003, -1995, 19.89, 19/99, 56/61, -19.42, 1/2, -56/47, 61.42, 4120, 4261, -61/56, 1/47, -3.030]
```

Sts – check of a string or symbol to be homogeneous

Call format of the procedure:

Sts(S, s)

Formal arguments of the procedure:

S – a string or symbol
s – a symbol or string of length **1**

Description of the procedure:

The procedure call **Sts(S, s)** returns the *true* value if and only if a string or a symbol specified by the first actual **S** argument is concatenation of the same string or symbol of length **1** specified by the second actual **s** argument. Otherwise, the *false* value is returned. This procedure is very useful for a series of problems relating to the strings processing.

```
Sts:= proc(S::{string, symbol}, s::{string, symbol})
local a;
   op(1, [assign(a = ""), seq(assign('a' = cat(a, s)), k = 1 .. length(S)),
        `if`(convert(S, string) = a, true, false)])
end proc
```

Typical examples of the procedure use:

```
> Sts("      "," "), Sts("5555",`5`), Sts(`55515`,"5"), Sts(RANS_IAN,`A`), Sts("======", `=`);
                                  true, true, false, false, true
```

Subs_All – exhaustive substitutions into a string or symbol
Subs_all1
Subs_all2

Call format of the procedures:

Subs_all1(*subst, string* {, h})
Subs_all2(*subst, string* {, h})
Subs_All(*subst, string,* r {, h})

Formal arguments of the procedures:

subst – an equation defining substitution; both members of the equation may be by strings or symbols
string – a string or a name to be processed
r – a substitution mode (may be **1** or **2**)
h – (optional) key word *`insensitive`* defining case-insensitive search

Description of the procedures:

The following group of three highly effective procedures provides several helpful methods of sensitive exhaustive substitutions into a string or a symbol instead of the given contexts; that is, upper/lower case differences are essential. In a series of cases these procedures are more preferable than similar standard tools of the package and essentially extend the last ones.

The procedure call **Subs_all1(*subst, string*)** returns the result of substitution of the right member of an equation, defined by the first actual *subst* argument, instead of all entries into a string or symbol, defined by the second *string* argument, of the left member of this equation; in addition, the crossing entries are processed also. A type of the returned result corresponds to a type of the second actual argument. The **Subs_all1** procedure does sensitive searching; that is, upper/lower case differences are essential. If pattern **lhs(*subst*)** does not belong to the *string*, the procedure returns the *string* without processing and prints the appropriate warning.

In contrast to the **Subs_all1,** the procedure call **Subs_all2(*subst, string*)** returns the result of substitution of the right member of an equation, defined by the first actual *subst* argument, instead of all entries into a string or a symbol, defined by the second *`string`* argument, of the left member of this equation; in addition, after each done substitution the appropriate entry of the left member of the *subst* equation is deleted from a current state of the processed *string***.** The **Subs_all2** procedure does sensitive searching also; that is, upper/lower case differences are essential. If the pattern **lhs(*subst*)** does not belong to the *string*, the procedure returns the *string* without processing and prints the appropriate warning. If any of the above two procedures uses as the optional actual **h** argument the key word ***`insensitive`***, then it supports case-insensitive search of the left members of substitutions.

At last, the **Subs_All** procedure has been programmed by an one-string extracode and it carries the dispatcher's functions providing switch of calls of procedures **Subs_all1** or **Subs_all2** depending on the required mode of substitutions **r={1|2}**. If the procedure uses as the optional actual **h** argument the key word ***`insensitive`***, then it supports case-insensitive search of the left members of substitutions. The above procedures seem as useful tools at operation with strings and symbols. The many procedures represented in the monograph essentially use the above procedures.

```
Subs_All:= (E::equation, S::{name, string, symbol}, r::{1, 2}) -> cat(Subs_all, r)(args[1 .. 2],
            `if`(nargs = 4, args[4], NULL))

Subs_all1:= proc(E::equation, S::{name, string, symbol})
local a, l, r, d, t, omega;
  omega:= (s, v) -> `if`(member(insensitive, v), Case(s, lover), s);
  `if`(search(omega(S, {args}), omega(lhs(E), {args})), assign(a = cat("", S), l = lhs(E),
    r = rhs(E), d = length(lhs(E))), RETURN(S));
  do
    if search(omega(a, {args}), omega(l, {args}), 't') then a := cat(a[1 .. t - 1], r, a[t + 1 .. -1])
    else RETURN(convert(a, whattype(S)))
    end if
  end do
end proc

Subs_all2:= proc(E::equation, S::{name, string, symbol})
local a, l, r, d, t, omega;
  omega:= (s, v) -> `if`(member(insensitive, v), Case(s, lover), s);
  `if`(search(omega(S, {args}), omega(lhs(E), {args})), assign(a = cat("", S), l = lhs(E), r = rhs(E),
    d = length(lhs(E))), RETURN(S));
  do
    if search(omega(a, {args}), omega(l, {args}), 't') then a := cat(a[1 .. t - 1], r, a[t + d .. -1])
    else RETURN(convert(a, whattype(S)))
    end if
  end do
end proc
```

Typical examples of the procedures use:

> Subs_all1("aaa"="s","aaa19aaa42aaaaaaa7aaaa14aaaahaaaaaaa");

"saa19saa42sssssaa7ssaa14ssaahsssssaa"

> Subs_all2("aaa"="s","aaa19aaa42aaaaaaa7aaaa14aaaahaaaaaaa");

"s19s42ssa7sa14sahssa"

> Subs_All("aaa"="s","aaa19aaa42aaaaaaa7aaaa14aaaahaaaaaaa", 1);

"saa19saa42sssssaa7ssaa14ssaahsssssaa"

> Subs_All("aaa"="s","aaa19aaa42aaaaaaa7aaaa14aaaahaaaaaaa", 2);

"s19s42ssa7sa14sahssa"

> Subs_all1("aAa"="s","aaa19aaa42aaaaaaa7aaaa14aaaahaaaaaaa");

"aaa19aaa42aaaaaaa7aaaa14aaaahaaaaaaa"

> Subs_all1("aAa"="s","aaa19aaa42aaaaaaa7aaaa14aaaahaaaaaaa", `insensitive`);

"saa19saa42sssssaa7ssaa14ssaahsssssaa"

> Subs_all2("aAA"="s","aaa19aaa42aaaaaaa7aaaa14aaaahaaaaaaa", `insensitive`);

"s19s42ssa7sa14sahssa"

SUB_S – substitution of symbols or strings into a string or symbol

Call format of the procedure:

SUB_S(L, S {, `*sensitive*`})

Formal arguments of the procedure:

S – a string or symbol
L – a list of equations defining the rules of substitutions
sensitive – (optional) key word defining the case-sensitive search mode

Description of the procedure:

The procedure call **SUB_S(L,S)** returns the result of execution of substitutions from a list specified by the first actual argument **L**, into a string or symbol specified by the second actual **S** argument; in addition, the next substitution of the list **L** is executed till its full exhaustion in the **S** string. A type of the returned result corresponds to a type of the second actual **S** argument. If the substitutions **L** are not applicable to the **S** string, then the initial string **S** is returned without any change. At coding of the third optional argument `*sensitive*` the procedure supports the case-sensitive search, otherwise case-insensitive.

```
SUB_S:= proc(L::list(equation), S::{string, symbol})
local k, d, G, h, P, R, l;
  assign(l = [seq(convert(lhs(L[k]), string) = convert(rhs(L[k]), string), k = 1 .. nops(L))],
    R = cat("", S)), `if`(nargs = 3 and args[3] = sensitive,
        assign(P = SearchText), assign(P = searchtext));
  for k to nops(L) do assign('d' = P(lhs(l[k]), R), 'h' = lhs(l[k]), 'G' = rhs(l[k]));
    if d <> 0 then R := cat(substring(R, 1 .. d - 1), G, substring(R, d + length(h) .. length(R)));
      k := k - 1 end if
```

```
  end do;
  convert(R, whattype(S))
end proc
```

Typical examples of the procedure use:

> SUB_S([ab=cdgh, aa=hhsg, cd=cprhkt], "aababccabaacdcdaavrhcdab");

"acprhktghcprhktghcccprhktghhhsgcprhktcprhkthhsgvrhcprhktcprhktgh"

> SUB_S([Ab=cdgh, aa=hhsg ,Cd=cprhkt], "aababccabaacdcdaavrhcdab");

"acprhktghcprhktghcccprhktghhhsgcprhktcprhkthhsgvrhcprhktcprhktgh"

> SUB_S([Ab=cdgh, Aa=hhsg, Cd=cprhkt], `AababccabaaCdcdaavrhcdAb`, sensitive);

hhsgbabccabaacprhktcdaavrhcdcdgh

Sub_st – exhaustive substitutions into a string or symbol

Call format of the procedure:

Sub_st(E, S, R {, `*insensitive*`})

Formal arguments of the procedure:

E – a list of equations defining a substitutions system; both members of substitutions equations have a type {*string, symbol*}
S – a string or symbol to be processed
R – an assignable name
Insensitive – (optional) key word defining the case-insensitive search mode

Description of the procedure:

The procedure call **Sub_st(E, S, R)** returns the result of substitutions of the right members of equations specified by the first actual **E** argument, instead of all entries into a string or symbol specified by the second argument **S**, of the left members of substitutions equations; in addition, the crossing entries are processed also. A type of the returned result corresponds to a type of the second actual **S** argument.

The result of processing of **S** by means of the substitutions **E** is only the first element of the returned 2-element sequence, whereas its second element is an integer vector which defines the number of applications to **S** of the corresponding substitutions out of **E.** If the left members of the substitutions of **E** do not belong to **S** or at least one substitution initiates infinite process, the procedure prints the appropriate warnings.

The substitution rules are given by a list of equations of the following view [Y1 = X1, Y2 = X2, ..., Yn = Xn]; where for both members of the equations the values of a type {*string*|*symbol*} are supposed. The algorithm of applying of substitutions system **E** consists in the following: the first substitution of **E** is applied to **S** until its full exhaustion in **S,** then the same operation is done with the second substitution of **E** and so on until full exhaustion of **E.**

At coding of the fourth optional argument `*insensitive*` the procedure supports the case-insensitive search, otherwise case-sensitive. At last, via the third actual **R** argument is returned the following value: (**1**) the *true*, if the processing of **S** has been completed successfully, (**2**) result of processing of **S** until detecting of a substitution leading to an infinite process (*cycling*) and (**3**) the *false* otherwise. The given procedure generalizes the above procedure **Subs_All**, and has numerous appendices at expressions processing of types {*symbol, string*}.

```
Sub_st:= proc(E::list(equation), S::{string, symbol}, R::evaln)
local k, h, G, v, omega, p;
  assign(p = {args}, omega = ((s, v) -> `if`(member(insensitive, v), Case(s, lover), s)));
  `if`(nops(E) = 0, WARNING("Substitutions system <%1> is absent", E),
    assign(G = cat("", S), R = true, v = array(1 .. nops(E), [0 $ (k = 1 .. nops(E))])));
  `if`(Search2(omega(S, p), {seq(omega(lhs(E[k]), p), k = 1 .. nops(E))}) = [],
    RETURN(assign('R' = false), WARNING("Substitutions system %1 is not applicable to
    <%2>", E, S)), NULL);
  for k to nops(E) do
    `if`(omega(lhs(E[k]), p) = omega(rhs(E[k]), p), [assign('R' = G), RETURN(WARNING(
      "Substitution <%1> generates an infinite process", E[k]))], assign('v[k]' = 0));
    while search(omega(G, p), omega(lhs(E[k]), p)) do
      assign('h' = searchtext(omega(lhs(E[k]), p), omega(G, p)), 'v[k]' = v[k] + 1);
      G := cat(G[1 .. h - 1], rhs(E[k]), G[h + length(lhs(E[k])) .. length(G)])
    end do
  end do;
  convert(G, whattype(S)), evalm(v)
end proc
```

Typical examples of the procedure use:

```
> S:= "ARANS95IANGIANRANS95RAEIAN99RACREAIANRANSIANR99ANSRANS":
     Sv:=["RANS"="Art","IAN"=" ","95"="S", "99"= "Kr"]: Sub_st(Sv,S,Z);
                  "AArtS G ArtSRAE KrRACREA Art RKrANSArt", [4, 5, 2, 2]
> Sv:=["rans"="Art","ian"=" ","95"="S", "99"= "Kr"]: Sub_st(Sv,S,Z);
    "ARANSSIANGIANRANSSRAEIANKrRACREAIANRANSIANRKrANSRANS", [0, 0, 2, 2]
> Sv:=["rans"="Art","ian"=" ","95"="S", "99"= "Kr"]: Sub_st(Sv,S,Z, `insensitive`);
                  "AArtS G ArtSRAE KrRACREA Art RKrANSArt", [4, 5, 2, 2]
> Sv1:=[ "RANS"="RANS","IAN"=" ","95"="S", "99"="Kr"]: Sub_st(Sv1,S,Z), Z;
   Warning, Substitution <RANS = RANS> generates an infinite process
       "ARANS95IANGIANRANS95RAEIAN99RACREAIANRANSIANR99ANSRANS"
```

---

sub_1 – substitution of symbols or strings into a string or symbol

Call format of the procedure:

sub_1(E, S {, `*insensitive*`})

Formal arguments of the procedure:

S – a string or symbol
E – an equation or a list of equations defining the rules of substitutions
insensitive – (optional) key word defining the case-sensitive search mode

Description of the procedure:

The procedure call **sub_1(E, S)** returns the result of execution of a substitution or substitutions of a list specified by the first actual argument **E**, into a string or a symbol specified by the second actual **S** argument; in addition, each substitution is executed only for the first entry of its left member into **S** string. A type of the returned result corresponds to a type of the second actual **S** argument. If no substitutions **E** are applicable to **S** string, then the initial **S** string is returned without any change. At coding of the third optional argument *`insensitive`* the procedure supports the case-insensitive search, otherwise case-sensitive.

```
sub_1:= proc(s::{equation, list(equation)}, S::{string, symbol})
local k, xi, r, t, m;
  `if`(member(insensitive, {args}), assign(m = 6), assign(m = 14));
  xi:= (s, S) -> `if`(search(`if`(m = 0, Case(S, lover), S), `if`(m = 6, Case(lhs(s), lover), lhs(s)), 't'),
    convert(cat(cat("", S)[1 .. t - 1], rhs(s), cat("", S)[t + length(lhs(s)) .. -1]), whattype(S)), S);
  `if`(type(s, equation), xi(s, S), [assign(r = S), seq(assign('r' = xi(s[k], r)), k = 1 .. nops(s)),
    RETURN(r)])
end proc
```

Typical examples of the procedure use:

```
> sub_1(ab=`[cdgh]`,"aaababccabaacdcdaaasvrhcdabaa");
                     "aa[cdgh]abccabaacdcdaaasvrhcdabaa"
> sub_1([ab=`[cpgh]`,aa=7,cd=14],"aaababccabaacdcdaaasvrhcdabaa");
                     "7[cpgh]abccabaa14cdaaasvrhcdabaa"
> sub_1([ab=`[cpgh]`,AA=7,Cd=14],"aaababccabaacdcdaaasvrhcdabaa");
                     "aa[cpgh]abccabaacdcdaaasvrhcdabaa"
> sub_1([ab=`[cpgh]`,AA=7,Cd=14],"aaababccabaacdcdaaasvrhcdabaa", `insensitive`);
                     "7[cpgh]abccabaa14cdaaasvrhcdabaa"
```

Suffix – check of suffix of a string or symbol

Call format of the procedure:

Suffix(G, B, h::*evaln* {, `*insensitive*`} {, `*del*`})

Formal arguments of the procedure:

G – a string or symbol
B – a symbol or string
h – a name
insensitive – (optional) key word defining the case-sensitive search mode
del – (optional) key word defining deletion mode of a suffix

Description of the procedure:

In a context of *suffixed*-type, the **Suffix** procedure useful at operation with string structures presents the undoubted interest. The procedure call **Suffix(G,B,h)** returns the *true* value, if the first actual **G** argument of a type {*symbol* | *string*} begins from a value specified by the second actual **B** argument of the same permissible type; otherwise, the *false* value is returned.

At return of the *true* value via the third actual **h** argument of the procedure the result of deletion of **B**-suffix out of **G** expression is returned. The expression returned via the third argument **h** of the procedure, has the same type, as an expression specified by the first actual **G** argument. At coding of the optional key word **`*insensitive*`** the procedure supports the case-insensitive search, otherwise case-sensitive. Whereas, at coding of the optional key word **`*del*`**, at return of the *true* value via the third actual **h** argument of the procedure the result of deletion of all multiple entries of **B**-suffix out of **G** expression is returned.

```
Suffix:= proc(G::{name, string, symbol}, B::{name, string, symbol}, S::evaln)
local a, b, n, m, t, p;
  assign(a = cat("", G), b = cat("", B), n = length(G), m = length(B)), `if`(3 < nargs and
    member(insensitive, {args[4 .. -1]}), assign('a' = Case(a), 'b' = Case(b)), NULL);
  if n < m or not search(b, a[1 .. m]) then false elif not member(del, {args[4 .. -1]}) then true,
assign('S' = convert(cat("", G)[m + 1 .. -1], whattype(G))) else
    do if procname(a, b, t) then a := t; p := length(t) else break end if end do;
    true, assign('S' = convert(cat("", G)[-p .. -1], whattype(G)))
  end if
end proc
```

Typical examples of the procedure use:

> Suffix(RANS_RAC_IAN_REA, RANS, Q), Q, Suffix(RANS_RAC_IAN_REA, rans, R), R;

true, *_RAC_IAN*, *false*, *R*

> Suffix("RANS_RAC_IAN_REA", rans, R, insensitive), R;

true, "_RAC_IAN_REA"

> Suffix("agnagnSveta", "agn", c), c, Suffix("agnagnSveta","agn", b, del), b;

true, "agnSveta", *true*, "Sveta"

> Suffix("aGnAgnSvetaV","Agn", p), p, Suffix(`aGnAgnSvetaV`,"Agn", t, del, insensitive), t;

false, p, *true*, SvetaV

Type_D – check of entry of decimal digits into a string or symbol

Call format of the procedure:

Type_D(S)

Formal arguments of the procedure:

S – a string or symbol

Description of the procedure:

The procedure call **Type_D(S)** returns the *false* value if a string or a symbol specified by the actual **S** argument contains no decimal digits, otherwise two-element sequence of a view "*true*, [[d1,n1], [d2,n2],..., [dp,np]]" is returned where the second element is a nested integer list. The elements of sublists of the second element of the sequence define a decimal digit and its position in the string **S** accordingly. In particular, the procedure is rather useful at processing of expressions of formats {*symbol*, *string*} that contain the numerical expressions.

```
Type_D:= proc(C::{string, symbol})
local a, k, p, t;
  assign(a =cat("", C), t=[]), op({seq(`if`(member(op(convert(a[k], bytes)), [(48+p) $ ('p' = 0..9)]),
  assign('t' = [op(t), [k, SD(a[k])]]), NULL), k = 1 .. length(a))}), `if`(t = [], false, op([true, t]))
end proc
```

Typical examples of the procedure use:

**> Type_D(`RANS-2003-IAN`), Type_D("rans-rac-rea-ian"),
Type_D("avz61agn56Kr7Art14SvArne");**

true, [[6, 2], [7, 0], [8, 0], [9, 3]], *false*, *true*, [[4, 6], [5, 1], [9, 5], [10, 6], [13, 7], [17, 1], [18, 4]]

XSN – an integer code of a string or symbol; a simple text decoder
decode

Call format of the procedures:

XSN(S)
decode(T)

Formal arguments of the procedures:

S - a string, symbol or an integer **>= 0**
T - a string defining text for decoding

Description of the procedures:

In a series of symbolic processing problems the use of concept of an integer code, which describes a string or symbol, seems an useful enough index. At such approach, each string (symbol) has an unique integer code, but not vice versa; i.e. not each integer represents a string (symbol).

The procedure call **XSN(S)** on a symbolic value (*string* or *symbol*), defined by the actual **S** argument, returns the own integer code, while the call **XSN(S)** on an integer **S** returns the string, appropriate to **S** as its integer index. For arbitrary string (symbol) **S** the relation **XSN(XSN(S)) = S** takes place with an accuracy to a symbolic type {*symbol, string*}. This procedure is an useful enough tool in some problems that combine the symbolic and numeric algorithms.

In particular, on the basis of procedures **XSN** and **xpack** a simple enough procedure **decode(T)** can be implemented which provides a coding of some **T** text with subsequent decoding of it according to the following relation \`**decode(T) ≡ decode(decode(T), p)**\`, where **p** - an arbitrary valid *Maple* expression. These procedures have a series of useful enough appendices.

```
XSN:= proc(S::{string, symbol, nonnegint})
local a, b, k, n;
  assign(b = [``, `0`, `00`], n = length(S) mod 3);
  if not type(S, integer) then
  assign(a=convert(cat("", S), bytes)), SN(cat("", seq(cat(b[4 - length(a[k])], a[k]), k=1..nops(a))))
  else assign('n' = cat("", `if`(n = 0, S, `if`(n = 1, cat(b[3], S), cat(b[2], S)))));
    cat(seq(convert([SN(n[3*k + 1 .. 3*k + 3])], bytes), k = 0 .. 1/3*length(n) - 1))
  end if
end proc

decode:= T::{string, symbol} -> `if`(nargs = 1, xpack(XSN(T)), XSN(xpack(T)))
```

Typical examples of the procedures use:

```
> A:= XSN(`AVZ_AGN+Art+Kr, Sv, Arn=61`);
   A := 65086090095065071078043065114116043075114044083118044065114110061054049
> XSN(A);
                    "AVZ_AGN+Art+Kr, Sv, Arn=61"
> B:= XSN("aa42aaaa47a67aa89aaa96a62");
   B := 97097052050097097097097052055097054055097097056057097097097057054097054050
> XSN(B);
                    "aa42aaaa47a67aa89aaa96a62"
> C:= XSN(`Effective Operation in the Package Maple 6/7 - 14.06.2002`);
   C := 69102102101099116105118101032079112101114097116105111110032105110032116\
   41010320800970991070971031010320770971121081010320540470550320450320490520460\
   48054046050048048050
```

> XSN(C);

"Effective Operation in the Package Maple 6/7 - 14.06.2002"

> XSN(XSN(2003)), XSN(XSN("RANS_IAN_REA_RAC-Tallinn-2003"));

2003, "RANS_IAN_REA_RAC-Tallinn-2003"

> decode(`Systems of Computer Algebra: A New Software Toolbox for Maple - Series: Computer Mathematics. The second edition.– Tallinn, 2003 - 2004, ISBN 9985-9277-8-8`);

"éxw{wÇgv÷{i‡wv†˜l×wv÷xw× | vwzi†Ëvçivvþw ¦ ýkæ˜l¶˜mçgw˜n—wv‡ | wöýw§gi†êwwwvæþ w\w†i‡hwwzi†Ýo×xvçgi†«i†évwzv ·gw¶¾i†Íwwow‡}wÇgw ¦ ˜mÖýwÇjvwoo× | v¶ÿw¶¬i†êv§g i‡{vvÿ\wvvf˜vwfv · | v ·wwf¬j¶˜n ¦ ývçnv ·vwfªi†¶jæ®kv˜j¶˜kf®jæ¸j ¦ ˜m–élÆÞi†½kÖ¼k– «kÖ¶k¶»j¶¼j¶¼"

> decode(%, 7);

"Systems of Computer Algebra: A New Software Toolbox for Maple - Series: Computer Mathematics. The second edition.– Tallinn, 2003 - 2004, ISBN 9985-9277-8-8"

Qsubstr – quantity of substrings (subsymbols) in a string (symbol)

Call format of the procedure:

Qsubstr(S, d)

Formal arguments of the procedure:

S - a string or symbol
d - a marker {*string* | *symbol* | *set* | *list*}

Description of the procedure:

The procedure call **Qsubstr(S, d)** returns the quantity of substrings (*subsymbols*) of a string (*symbol*) **S** parted by **d**-markers; both the separate strings (*symbols*) and their list or a set as **d**-argument can appear. The given procedure represents a rather useful tool of processing of strings and symbols.

```
Qsubstr:=proc(S::{string,symbol}, d::{string,symbol, set({string,symbol}), list({string,symbol})})
local k, N, G, L;
  assign(G = cat("", S), N = 1), assign(L = [op(Search2(G, {op(d)})), length(S)]);
  if L = [] then N
  else
    for k to nops(L) - 1 do if 1 < L[k + 1] - L[k] then N := N + 1 end if end do; `if`(N = 0, 1, N)
  end if
end proc
```

Typical examples of the procedure use:

> Kr:= "12b345aaa aaa123456a666b6666666bggggggbgggggga1111b11a1111b11aabbb": Qsubstr(Kr, `b`);

7

> Art:= "aa245@876asdf@987654#aaa1256#a6666666#6bggggg@gggggа1111@11a111aa@": Qsubstr(Art, {`#`, `@`});

8

> Qsubstr("\tRANS\tIAN\tRAC\tREA\1995\t2004\t", "\t");

6

ExprOfString – extracting from string or symbol of correct *Maple* expressions

Call format of the procedure:

ExprOfString(S)

Formal arguments of the procedure:

S - a string or symbol

Description of the procedure:

The procedure call **ExprOfString(S)** returns the sequence of correct *Maple* expressions contained in a string or symbol **S** and parted by *tabs*. Usage of tabs as a delimiter allows to essentially extend presentable class of correct *Maple* expressions. The incorrect expressions are ignored; at absence of correct expressions, the procedure call returns the *NULL* value, i.e. nothing.

```
ExprOfString:= proc(S::{string, symbol})
local a, b, c, k;
  assign(c = interface(warnlevel)), interface(warnlevel = 0);
  assign67(a = SLD(Rcd_n(cat("", S), "\t", 2), "\t"), b = NULL);
  for k to nops(a) do try b := b, came(a[k]) catch "incorrect": next end try end do;
  interface(warnlevel = c), b
end proc
```

Typical examples of the procedure use:

```
> ExprOfString("""A B""\t0/0\t57\t\t 37\t(a+b)/(c-d)\tx+8\ty/15\t""manur""\tRANS");
```

"A B", 57, 37, (a + b)/(c – d), x + 8, y/15, "manur", *RANS*

```
> ExprOfString("\t\t 0/0 \t\t 300/0 (a+b)/(c- \t\t \t a/(8-8)");
> ExprOfString("\t `SveGal` \t 0/0 \t\t 300/60 \t (a+b)*(c-d) \t\t`RANS`\t");
```

SveGal, 5, (a + b) (c - d), *RANS*

Resources of the given chapter provide a lot of useful additional possibilities for expressions processing of types (*string, symbol, name*} which essentially extend resources of a package in the given direction. It is especially actual, that the above-mentioned types of data structures form the *Maple* environment base. Usage of the given resources at programming a lot of the manifold tasks has confirmed high enough their operating performances. A whole series of these tools have been used at implementation in the *Maple* environment of algorithms of many procedures represented in this monograph.

Chapter 5. Software of operation with sets, tables and lists

This section represents tools that extend the possibilities of *Maple* at operation with objects of types {*list, set, table*}. The list structures play an extremely important role, defining the ordered sequences of elements. Since the *sixth* release, a possibility of substantial extending of operations with the list structures had appeared. As an example having the interesting practical appendices, we consider definition of algebra on a set of all lists having the same length. Algebraic operations the corresponding procedures, presented below, provide. A series of procedures of the section supports useful kinds of processing such as: a special converting of lists into sets, and vice versa; operation with rarefied lists; dynamic assignment of values to elements of a list or a set; evaluation of indices of a table over its entry; representation of a special type of tables; special kinds of exhaustive substitutions into lists or sets; a series of important kinds of sorting of nested lists, and also many others. The given tools are rather useful at operation with objects of the above types in the *Maple* environment.

\`&+\` – an algebra with lists and scalars – the package module AlgLists
\`&-\`
\`&*\`
\`&/\`
\`&^\`

Call format of the procedures:

AlgLists[\`&+\`](L, L1)
AlgLists[\`&-\`](L, L1)
AlgLists[\`&*\`](L, L1)
AlgLists[\`&/\`](L, L1)
AlgLists[\`&^\`](L, L1)

Formal arguments of the procedures:

L, L1 - a list or scalar

Description of the procedures:

The list structures play an extremely important role, defining the ordered sequences of elements. Since the sixth release a possibility to essentially extend operations with the list structures had appeared. As an example having the interesting practical applications, we shall consider definition of algebra on a set of all lists having the same length. With the purpose of the better visualization and without loss of generality, we shall consider the algebra with lists on examples of lists of a fixed length. The operators supporting the above algebra with lists and scalars are implemented as the procedures composing the package module **AlgLists.** The one-time operator's calls have one of kinds **AlgLists:- <*operator name*>(*args*)** or **AlgLists[<*operator name*>](*args*),** whereas the call **with(AlgLists)** returns the list of procedures, providing their availability in a current session.

Operator **L &+ L2 {`&+`(L, L2)}** defines operation of *addition* of two lists **L** and **L2** of the same length; in general, any combinations of lists and/or scalars can be as operands for the operator.

Operator **L &– L2 {`&-`(L, L2)}** defines the operation of *subtraction* of two lists **L** and **L2** of the same length; in general, any combinations of lists and/or scalars can be as operands for the operator.

Operator **L &* L2 {`&*`(L, L2)}** defines operation of *product* of two lists **L** and **L2** of the same length; in general, any combinations of lists and/or scalars can be as operands for the operator.

Operator **L &/ L2 {`&/`(L, L2)}** defines operation of *division* of two lists **L** and **L2** of the same length; in general, any combinations of lists and/or scalars can be as operands for the operator.

Operator **L &^ L2 {`&^`(L, L2)}** defines operation of *raise to the power* with two lists **L** and **L2** of the same length; in general, any combinations of lists and/or scalars can be as operands for the operator.

Definitions of all operators **{&+, &-, &*, &/, &^}** are identical. All operations with lists are element-wise made and allow to program the algorithms in a series of cases more compactly and transparently mathematically; consequently, in a number of cases a considerable decrease of temporary costs of execution of mathematical algorithms concerning usage of traditional approaches has been observed.

The first example of a fragment given below represents definitions of two concrete lists **L, L2,** and two special lists - *zero* list (**L0**) and *unity* list (**L1**). The second example demonstrates the operations of *addition* and *subtraction* for data structures of such type (lists of identical length). While the remaining examples illustrate the other operations of the algebra **&*** (*product*), **&/** (*division*) and **&^** (*exponentiation*). The algebra of lists has been used in a series of appendices, above all, at processing of list data structures. As structures of this type are basic *Maple* structures, the algebra of lists can be useful in many appendices.

```
`&*`:= proc(L::{list, scalar}, L1::{list, scalar})
local k, RT, OP, omega;
   assign(omega = convert(cat("", procname)[-1], symbol),
   OP = (() -> `if`(type(args[1], scalar), scalar, list)));
   RT := table([(list, list) = [seq(omega(L[k], L1[k]), k = 1 .. nops(L1))], (scalar, list) =
      [seq(omega(L, L1[k]), k = 1 .. nops(L1))], (list, scalar) =
      [seq(omega(L[k], L1), k = 1 .. nops(L))], (scalar, scalar) = omega(L, L1)]);
   `if`(OP(L) = list and OP(L1) = list and nops(L) <> nops(L1),
      ERROR(`lists of different length`), RT[OP(L), OP(L1)])
end proc
```

```
`&*`:= proc(L::{list, scalar}, L1::{list, scalar})
local k, RT, OP, omega;
   assign(omega = convert(cat("", procname)[-1], symbol),
   OP = (() -> `if`(type(args[1], scalar), scalar, list)));
   RT := table([(list, list) = [seq(omega(L[k], L1[k]), k = 1 .. nops(L1))],
      (scalar, list) = [seq(omega(L, L1[k]), k = 1 .. nops(L1))],
      (list, scalar) = [seq(omega(L[k], L1), k = 1 .. nops(L))], (scalar, scalar) = omega(L, L1)]);
   `if`(OP(L) = list and OP(L1) = list and nops(L) <> nops(L1),
      ERROR(`lists of different length`), RT[OP(L), OP(L1)])
end proc
```

```
`&+`:= proc(L::{list, scalar}, L1::{list, scalar})
local k, RT, OP, omega;
  assign(omega = convert(cat("", procname)[-1], symbol),
  OP = (() -> `if`(type(args[1], scalar), scalar, list)));
  RT := table([(list, list) = [seq(omega(L[k], L1[k]), k = 1 .. nops(L1))],
    (scalar, list) = [seq(omega(L, L1[k]), k = 1 .. nops(L1))],
    (list, scalar) = [seq(omega(L[k], L1), k = 1 .. nops(L))], (scalar, scalar) = omega(L, L1)]);
  `if`(OP(L) = list and OP(L1) = list and nops(L) <> nops(L1),
    ERROR(`lists of different length`), RT[OP(L), OP(L1)])
end proc

`&-`:= proc(L::{list, scalar}, L1::{list, scalar})
local k, RT, OP, omega;
  assign(omega = convert(cat("", procname)[-1], symbol),
  OP = (() -> `if`(type(args[1], scalar), scalar, list)));
  RT := table([(list, list) = [seq(omega(L[k], L1[k]), k = 1 .. nops(L1))],
    (scalar, list) = [seq(omega(L, L1[k]), k = 1 .. nops(L1))],
    (list, scalar) = [seq(omega(L[k], L1), k = 1 .. nops(L))], (scalar, scalar) = omega(L, L1)]);
  `if`(OP(L) = list and OP(L1) = list and nops(L) <> nops(L1),
    ERROR(`lists of different length`), RT[OP(L), OP(L1)])
end proc

`&/`:= proc(L::{list, scalar}, L1::{list, scalar})
local k, RT, OP, omega;
  assign(omega = convert(cat("", procname)[-1], symbol),
  OP = (() -> `if`(type(args[1], scalar), scalar, list)));
  RT := table([(list, list) = [seq(omega(L[k], L1[k]), k = 1 .. nops(L1))],
    (scalar, list) = [seq(omega(L, L1[k]), k = 1 .. nops(L1))],
    (list, scalar) = [seq(omega(L[k], L1), k = 1 .. nops(L))], (scalar, scalar) = omega(L, L1)]);
  `if`(OP(L) = list and OP(L1) = list and nops(L) <> nops(L1),
    ERROR(`lists of different length`), RT[OP(L), OP(L1)])
end proc

`&^`:= proc(L::{list, scalar}, L1::{list, scalar})
local k, RT, OP, omega;
  assign(omega = convert(cat("", procname)[-1], symbol),
  OP = (() -> `if`(type(args[1], scalar), scalar, list)));
  RT := table([(list, list) = [seq(omega(L[k], L1[k]), k = 1 .. nops(L1))],
    (scalar, list) = [seq(omega(L, L1[k]), k = 1 .. nops(L1))],
    (list, scalar) = [seq(omega(L[k], L1), k = 1 .. nops(L))], (scalar, scalar) = omega(L, L1)]);
  `if`(OP(L) = list and OP(L1) = list and nops(L) <> nops(L1),
    ERROR(`lists of different length`), RT[OP(L), OP(L1)])
end proc
```

Typical examples of the procedures use:

```
> with(AlgLists);
```

[&*, &+, &-, &/, &^]

```
> L:= [a,b,c,d,e,g,h,i,j,k]:  L2:= [e,f,g,h,p,s,z,t,h,p]:  L0:= [0$nops(L)]:  L1:= [1$nops(L)]:
```

Addition and subtraction of lists of the same length; **L0** – zero list and **L1** – unity list

> L &+ L2, L &- L2, a &+ L, L &- b;

[a+e, b+f, c+g, d+h, e+p, g+s, h+z, i+t, j+h, k+p], [a-e, b-f, c-g, d-h, e-p, g-s, h-z, i-t, j-h, k-p], [2*a, a+b, a+c, a+d, a+e, a+g, a+h, a+i, a+j, a+k], [a-b, 0, c-b, d-b, e-b, g-b, h-b, i-b, j-b, k-b]

Product of two lists of the same length or scalars

> L &* L2, \`&*\`(L,L2);

[a e, b f, c g, d h, e p, g s, h z, i t, j h, k p], [a e, b f, c g, d h, e p, g s, h z, i t, j h, k p]

> L0 &* L0, L0 &* L1, L1 &* L1, x &* b, L &* x;

[0, 0, 0, 0, 0, 0, 0, 0, 0, 0], [0, 0, 0, 0, 0, 0, 0, 0, 0, 0], [1, 1, 1, 1, 1, 1, 1, 1, 1, 1], x b, [a x, x b, c x, d x, e x, g x, h x, i x, j x, k x]

Division of two lists of the same length

> L &/ L2, x &/ L2, x &/ y, L &/ L1;

[a/e, b/f, c/g, d/h, e/p, g/s, h/z, i/t, j/h, k/p], [x/e, x/f, x/g, x/h, x/p, x/s, x/z, x/t, x/h, x/p], x/y, [a, b, c, d, e, g, h, i, j, k]

Operation of raise to the power with two lists of the same length

> L &^ L2, a &^ L2, a &^ b;

[a^e, b^f, c^g, d^h, e^p, g^s, h^z, i^t, j^h, k^p], [a^e, a^f, a^g, a^h, a^p, a^s, a^z, a^t, a^h, a^p], a^b

---

set operations with lists – the package module SoLists
\`union\` – list union operation
\`intersect\` – list intersection operation
\`minus\` – list difference operation
sublist – subset operation

Call format of the procedures:

SoLists:- \`union\`(L1 {, L2, L3, ...})
SoLists:- \`intersect\`(L1 {, L2, L3, ...})
SoLists:- \`minus\`(L1 {, L2})
SoLists:- sublist(L1 {, L2})

Formal arguments of the procedures:

L1, L2, L3, ... - lists

Description of the procedures:

The **\`union\`**, **\`intersect\`**, **\`minus\`** and **sublist** procedures are used for the list union, intersection, difference, and sublist operations. The **\`union\`** and **\`intersect\`** procedures are **n**-ary, whereas the **\`minus\`** and **sublist** procedures are binary or unary. The procedures have been implemented as exports of the **SoLists** module.

If any argument is not a list, an error is returned. The procedure names **\`union\`**, **\`intersect\`** and **\`minus\`** must be enclosed in backquotes since they are keywords. In contrast to operations of the same name for sets, these operations save order of lists elements. The above procedures are useful in many problems dealing with the list data structures. As list structures and set structures are not equivalent relative to saving of elements order, the set operations with lists can be useful in many appendices dealing with lists processing.

```
`union`:= proc()
local a, b, c, k;
   assign(a = nargs), `if`(a = 0, ERROR("actual arguments are absent"),
      gelist([seq(type(args[k], list), k = 1 .. a)], b)); assign(c = SLj(b, 1)[1]), `if`(c[1], NULL,
      ERROR("arguments with numbers %1 are not lists", c[2 .. -1])); redL(map(op, [args]))
end proc

`intersect`:= proc()
local a, b, c, k, j, t, R;
   assign(a = nargs), `if`(a = 0, ERROR("actual arguments are absent"),
      gelist([seq(type(args[k], list), k = 1 .. a)], b)); assign(c = SLj(b, 1)[1]), `if`(c[1], NULL,
   ERROR("arguments with numbers %1 are not lists", c[2 .. -1]));
   `if`(map(Empty, {args}) <> {false}, RETURN([]), `if`(nargs = 1, RETURN(redL(args[1])),
   NULL)); assign('a' = redL(args[1]), R = [], t = 14), assign('b' = nops(a));
   for k to b do
      for j from 2 to nargs do
         if not member(a[k], args[j]) then t := 7; break end if end do;
      `if`(t = 7, NULL, assign('R' = [op(R), a[k]])), assign('t' = 14)
   end do;
   redL(R)
end proc

`minus`:= proc(L::list, N::list)
local a, b, c, k, j;
   if L = [] and N = [] then [] elif N = [] then redL(L)
   else assign(a = redL(N), b = [], c = redL(L));
      for k to nops(a) do for j to nops(c) do
            if c[j] = a[k] then b := [op(b), j] end if end do end do;
      subsop(seq(b[k] = NULL, k = 1 .. nops(b)), c)
   end if
end proc

sublist:=(L::list, N::list) -> `if`(L = [] and N = [], true, `if`(L = [], true, `if`(N = [],false, belong(L, N))))
```

<u>Typical examples of the procedures use:</u>

> L1:=[a,b,c,d,d,a,c]: L2:=[z,y,a,b,c,y,c,a]: L3:=[a,b,c,d,d,a,c,h]: with(SoLists);

Warning, the protected names intersect, minus and union have been redefined and unprotected

[intersect, minus, sublist, union]

> `union`(L1,L2), `union`([],L2), `union`(L1,[]), `union`([],[]), `union`(L1,L2,L3);

[a, b, c, d, z, y], [z, y, a, b, c], [a, b, c, d], [], [a, b, c, d, z, y, h]

> `minus`(L1,L2), `minus`([],L2), `minus`(L1,[]), `minus`([],[]);

[d], [], [a, b, c, d], []

> `intersect`(L1,L2), `intersect`([],L2),`intersect`(L1,[]),`intersect`([],[]), `intersect`(L1,L2,L3);

[a, b, c], [], [], [], [a, b, c]

> sublist(L1,L3), `sublist`([],L3), `sublist`(L1,[]), `sublist`([],[]), `sublist`(L1,L2);

true, true, false, true, false

Cls – a special converting of lists into sets, and vice versa

Call format of the procedure:

Cls(L)

Formal arguments of the procedure:

L – a list or a set

Description of the procedure:

The data structures of a type {*list, set*} are enough widely used by *Maple*, however in contrast to lists, whose elements are strictly ranked, the elements of sets are ranked according to the agreements of the package, causing the certain difficulties at the problem solving, which use the order of elements of data structures. For a possibility of restoration of order of list elements after its converting into a set the **Cls** procedure, represented below, can be useful enough.

If the actual **L** argument is a list, then the procedure call **Cls(L)** returns the result of converting of a list into a set **S** organized by the special manner. Whereas the procedure call **Cls(S)** with a set of such organization as the actual argument returns the initial list **L**, i.e. with preservation of elements order of the initial list and entries of multiple elements. This procedure has been successfully used in a series of appendices dealing with list data structures of special organization.

```
Cls:= proc(G::{set, list})
local a, b, c, d, k, p, t, H, N;
  H:= h -> `if`(N(op(0, op(2, h))) = op(op(1, h)), true, false);
  N:= s -> op([assign('t' = convert(s, bytes)), add((t[i] – 48)*10^(nops(t) – i), i = 1 .. nops(t))]);
  `if`(type(G, 'set'), op([assign(c = add(`if`(patmatch(G[k], [a::integer]@(b::function)), 0, 1),
    k = 1 .. nops(G))), `if`(c <> 0, G, [assign(d = [k $ (k = 1 .. nops(G))]), seq(`if`(H(G[p]),
    assign('d'[op(op(G[p])[1])] = op(op(G[p])[2])), NULL), p = 1 .. nops(G)), op(d)])]),
    {seq([k]@(convert(k, symbol)(G[k])), k = 1 .. nops(G))})
end proc
```

Typical examples of the procedure use:

> L:= [a,61,Art^14,h,sin(x),[v,t],1/b,0.42,a*x+d,agn,Kr/(Art+18),a*x^2+ b*x+c,sqrt(2),n..m, {x,y,z}, RANS/(ian+rac+rea), assign(Sr = (() -> `+`(args)/nargs))];

$$L := \left[a, 61, Art^{14}, h, \sin(x), [v, t], \frac{1}{b}, 0.42, a\,x+d, agn, \frac{Kr}{Art+18}, a\,x^2+b\,x+c, \sqrt{2}, n\,..\,m, \{x, y, z\}, \frac{RANS}{ian+rac+rea}\right]$$

> L1:= Cls(L);

$$L1 := \{[11]@\left(11\left(\frac{Kr}{Art+18}\right)\right), [1]@(1(a)), [2]@(2(61)), [3]@(3(Art^{14})), [4]@(4(h)), [5]@(5(\sin(x))), [6]@(6([v, t])), [7]@\left(7\left(\frac{1}{b}\right)\right), [8]@(8(0.42)), [9]@(9(a\,x+d)), [10]@(10(agn)), [12]@(12(a\,x^2+b\,x+c)), [13]@(13(\sqrt{2})), [14]@(14(n\,..\,m)), [15]@(15(\{x, y, z\})), [16]@\left(16\left(\frac{RANS}{ian+rac+rea}\right)\right)\}$$

> Cls(L1);

$$\left[a, 61, Art^{14}, h, \sin(x), [v, t], \frac{1}{b}, 0.42, a\,x+d, agn, \frac{Kr}{Art+18}, a\,x^2+b\,x+c, \sqrt{2}, n\,..\,m, \{x, y, z\}, \frac{RANS}{ian+rac+rea}\right]$$

> L2:= Cls([[a1,b1,c1,d1,e1], [a2,b2,c2,d2,e2], [a3,b3,c3,d3,e3], [a4,b4,c4, d4,e4], [a5,b5,c5,d5,e5]]);

L2 := {@([1],1([a1, b1, c1, d1, e1])), @([2], 2([a2, b2, c2, d2, e2])), @([3], 3([a3, b3, c3, d3, e3])), @([4], 4([a4, b4, c4, d4, e4])), @([5], 5([a5, b5, c5, d5, e5]))}

> Cls(L2);

[[a1, b1, c1, d1, e1], [a2, b2, c2, d2, e2], [a3, b3, c3, d3, e3], [a4, b4, c4, d4, e4], [a5, b5, c5, d5, e5]]

> L3:= Cls([a,b,c,a,a,d,f,d,b,a,h,j,k,p,u,x,x,x]);

L3 := {@([17],17(x)), @([1],1(a)), @([2],2(b)), @([3],3(c)), @([4],4(a)), @([5],5(a)), @([6],6(d)), @([7],7(f)), @([8],8(d)), @([9],9(b)), @([10],10(a)), @([11],11(h)), @([12],12(j)), @([13],13(k)), @([14],14(p)), @([15],15(u)), @([16],16(x)), @([18],18(x))}

> Cls(L3);

[a, b, c, a, a, d, f, d, b, a, h, j, k, p, u, x, x, x]

The examples represented above enough lucidly illustrate the principle used for organization of the above-mentioned special converting of lists into sets, and vice versa.

---

clsd – a special converting of lists and sets into strings

Call format of the procedure:

clsd(L, d)

Formal arguments of the procedure:

L - a list or a set
d - a symbol or string defining filler between elements

Description of the procedure:

The simple **clsd** procedure is intended for a special converting of lists and sets into strings. The call **clsd(L, d)** returns the string formed by means of concatenation of elements of a set or a list defined by the actual argument **L** into string with insert between elements of filler defined by the second actual **d** argument. A type of the first element of the actual **L** argument defines a type of the returned result. The call **clsd([], d)** or **clsd({}, d)** returns the empty list ([]) or empty set ({}) accordingly. In particular, the given procedure is useful at solution of many problems, linked with organization of access to datafiles. Between procedures **CF, CFF** and **clsd** an useful relation takes place, namely:

CF(clsd(CFF(F), d)) = CF(F)

where **F** is a path and **d** belongs to the set {"\\", "/"}. In addition, the **CF** procedure converts the directory delimiter "/" into delimiter "\\" that is acceptable for standard procedures ***`system`*** and ***ssystem***.

```
clsd:= proc(L::{set, list}, d::{string, symbol})
local k, h;
  `if`(Empty(L), RETURN(L), seq(assign('h' = cat("", h, convert(L[k], string), d)), k=1..nops(L)));
  convert(h[2 .. -length(d) - 1], whattype(L[1]))
end proc
```

Typical examples of the procedure use:

> clsd(["C:", "Program Files", "Maple 9", "bin.win", "xxx", "yyy"], "\\");

"C:\Program Files\Maple 9\bin.win\xxx\yyy"

> F:= "C:/aaa/bbb/ccc\\ddd/ppp/hhh.txt": CF(clsd(CFF(F),"/")) = CF(F);

"c:\aaa\bbb\ccc\ddd\ppp\hhh.txt" = "c:\aaa\bbb\ccc\ddd\ppp\hhh.txt"

> clsd([V,61,G,56,A,40,S,36,Ar,14,Kr,7], `=&=`);

"V=&=61=&=G=&=56=&=A=&=40=&=S=&=36=&=Ar=&=14=&=Kr=&=7"

> map(clsd, [[], {}, ["a",v,z], {a,g,n}], ".");

[[], {}, "a.v.z", *g.a.n*]

FLL – an expansion of lists

Call format of the procedure:

FLL(L, p, m, ah)

Formal arguments of the procedure:

L - a list
p - an integer
m - an integer
ah - an expression

Description of the procedure:

The procedure call **FLL(L, p, m, ah)** returns the result of expansion of a list **L** by a sublist consisting from **m** elements **ah** after position **p** of the initial **L** list. If **p <= 0** the list is expanded to the left, if **p >= nops(L)** – to the right, otherwise the list **L** is separated by the sublist after **p** position. Along with common purpose, the given procedure seems convenient enough in problems of bit-by-bit symbolic processing.

```
FLL:= proc(L::list, p::integer, m::posint, a::anything)
local h, k;
  assign(h = [a $ (k = 1 .. m)]), `if`(p <= 0, [op(h), op(L)], `if`(nops(L) <= p, [op(L), op(h)],
  [op(L[1 .. p]), op(h), op(L[p + 1 .. -1])]))
end proc
```

Typical examples of the procedure use:

> FLL([67,62,89,96,42,47], 3, 6, `Art`), FLL([67,62,89,96,42,47], 0, 6, `Art`);

[67, 62, 89, *Art, Art, Art, Art, Art, Art*, 96, 42, 47],[*Art, Art, Art, Art, Art, Art*, 67, 62, 89, 96, 42, 47]

---

subsLS – substitutions into lists and sets

Call format of the procedure:

subsLS(S, L)

Formal arguments of the procedure:

L - a list or a set
S - a list of substitutions of view **a=b**

Description of the procedure:

The simple **subsLS** procedure is intended for providing of substitutions into lists and sets instead of their elements. The procedure call **subsLS(S, L)** returns the result of substitutions into a list or set specified by the actual **S** argument of the right members of substitutions from a list specified by the second actual argument **L**, namely: all elements of **S** which are identical to the first entries of the left members of the substitutions from **L** are replaced by their right members. In addition, if the right member has the *NULL* value, an appropriate element of **S** is omitted. The procedure has a series of rather useful appendices at operation with sets, lists, symbols and strings.

```
subsLS:= proc(S::list(equation), L::{set, list})
local omega, k, x, y;
  omega:= (x, y) -> op([seq(`if`(x = lhs(y[k]), RETURN(rhs(y[k])), NULL), k = 1..nops(y)), x]);
  map(omega, L, S)
end proc
```

Typical examples of the procedure use:

> L:= [RANS,IAN,RAC,REA,Art,Kr,RANS,2003,1995,a,b,c,d,2003,x,y,z,1995]:

> subsLS([RANS=61,IAN=56,Kr=Art,Art=Kr,REA=NULL,b=Sv,1995=2003,y=G], L);

[61, 56, RAC, Kr, Art, 61, 2003, 2003, a, Sv, c, d, 2003, x, G, z, 2003]

> S:= {RANS,IAN,RAC,REA,Art,Kr,RANS,2003,1995,a,b,c,d,2003,x,y,z,1995}:

> subsLS([RANS=61,IAN=56,Kr=Art,Art=Kr,REA=NULL,b=Sv,1995=2003,y=G], S);

{56, 61, 2003, a, x, G, d, c, Kr, Art, RAC, Sv, z}

---

subSL - substitutions into lists instead of their sublists

Call format of the procedure:

subSL(S, L {, \`*sensitive*\`})

Formal arguments of the procedure:

L – a list to be processed
S – a list of substitutions of view **[a, b, ...]=[x, y, ...]**
\`*sensitive*\` – (optional) key word defining a search mode at substitutions

Description of the procedure:

The procedure **subSL** is intended for a providing of substitutions into lists instead of their sublists. The procedure call **subSL(S,L)** returns the result of substitutions into a list specified by the second actual **L** argument of the right members (*sublists*) of substitutions from a list specified by the first actual argument **S**, namely: all non-overlapping sublists of **L** which are identical to the left members (*lists*) of the substitutions from **S** are replaced by their right members (*lists*). If the procedure call uses the optional key word **\`*sensitive*\`**, then at substitutions the case-sensitive mode of search is used, otherwise the case-insensitive mode. The procedure has a series of rather useful appendices at operation with lists.

```
subSL:= proc(S::list({equation(list)}), L::list)
local a, b, c, k, n, t, f;
  `if`(Empty(S), RETURN(L), assign(n = map(nops, [S, L]), c = map(convert, L, string)));
  assign67(t = `if`(2 < nargs and args[3] = sensitive, sensitive, NULL)), unassign('_61_');
  a := [seq(map(convert, lhs(S[k]), string) = map(convert, rhs(S[k]), string), k = 1 .. n[1])];
  assign(b = [seq(LSL(lhs(a[k])) = LSL(rhs(a[k])), k = 1 .. n[1])], 'c' = LSL(c), f = "_$&$_");
  c := sstr(["`" = NULL], cat("_61_:=", convert(map(convert, LSL(SUB_S(b, c, t)), symbol),
       string), ":"));
  (proc() writebytes(f, c); close(f); read f; delf(f) end proc)(), _61_, unassign('_61_')
end proc
```

Typical examples of the procedure use:

> M:=matrix([[1995,2003],[RANS,IAN]]): T:=table([V=61,G=56,S=36]): L:=[1995,art,kr, rans,42, G,art+kr,Sv*Ar,M,A,V,ian,grsu,TRU,Vgtu];

L := [1995, art, kr, rans, 42, G, art+kr, Sv*Ar, M, A, V, ian, grsu, TRU, Vgtu]

> subSL([[42,G,Art+Kr,Sv*Ar,M]=[61,300,RANS], [V,IAN,Grsu]=[M,ln(y),T]], L);

[1995, art, kr, rans, 61, 300, RANS, A, M, ln(y), T, TRU, Vgtu]

> map(eval, %);

$$\left[1995, art, kr, rans, 61, 300, RANS, A, \begin{bmatrix} 1995 & 2003 \\ RANS & IAN \end{bmatrix}, \ln(y), \mathrm{table}([V = 61, G = 56, S = 36]), TRU, Vgtu\right]$$

> subSL([[42,G,Art+Kr,Sv*Ar,M]=[61,300,RANS],[V,IAN,grsu]=[M,ln(y),T]], L, `sensitive`);

[1995, art, kr, rans, 42, G, art+kr, Sv*Ar, M, A, V, ian, grsu, TRU, Vgtu]

> map(eval, %);

$$\left[1995, art, kr, rans, 42, G, art + kr, Sv\ Ar, \begin{bmatrix} 1995 & 2003 \\ RANS & IAN \end{bmatrix}, A, V, ian, grsu, TRU, Vgtu\right]$$

> subSL([[42]=[61],[V]=[M,ln(y),T]], L, `sensitive`);

[1995, art, kr, rans, 61, G, art+kr, Sv*Ar, M, A, M, ln(y), T, ian, grsu, TRU, Vgtu]

> subSL([[V]=61, [G]=[M,T]], L, `sensitive`);

Error, invalid input: **subSL** expects its 1st argument, S, to be of type list({equation(list)}), but received [[V] = 61, [G] = [M, T]]

---

perml – a permutation of lists elements

Call format of the procedure:

perml(L, N)

Formal arguments of the procedure:

L - a list
N - a list of non-negative integers

Description of the procedure:

The **perml** procedure is intended for permutation of lists elements with possibility of deletion of their elements. The procedure call **perml(L, N)** returns the permutation result of a list elements specified by the actual argument **L** in conformity with an integer list specified by the actual argument **N.** The lengths of lists **L** and **N** should be identical and the **N** elements are integers from range **0..nops(L)** which define the positions of permutation and/or deletion of the **L** elements, namely: if **N[k]=0** then out of the returned list an element **L[k]** is deleted, otherwise element **L[k]** in the returned list will has position **L[N[k]]** with correction for length reduction of the returned list because of elements deletion. The procedure has many appendices in the problem solving dealing with lists, sequences, strings and symbols, and with the list structures, in general. It is especially actual, that the above-mentioned types of data structures form the *Maple* environment base. Usage of the given resources at programming a lot of the manifold tasks has confirmed high enough their operating performances. A whole series of these tools have been used at implementation in the *Maple* environment of algorithms of many procedures represented in this monograph.

```
perml:= proc(L::list, N::list(nonnegative))
local a, b, k, P;
  assign(a = convert([0], bytes), b = nops(L)), assign(P = [a $ (k = 1 .. b)]);
  if belong({op(N)} minus {0}, {k $ (k = 1 .. b)}) and b = nops(N) then
    for k to b do if N[k] = 0 then P[k] := a else P[N[k]] := L[k] end if end do;
      subsLS([a = NULL], P)
  else ERROR("lists lengths mismatching or invalid elements of the second list")
  end if
end proc
```

Typical examples of the procedure use:

```
> perml([a,b,c,d,7,8,9,1995,2003,61,h,r,t,f], [7,5,1,2,3,4,6,14,10,8,11,13,12,9]);
                  [c, d, 7, 8, b, 9, a, 61, f, 2003, h, t, r, 1995]
> perml([a,b,c,d,7,8,9,1995,2003,61,h,r,t,f], [7,5,1,0,3,4,6,0,10,8,11,0,12,9]);
                  [c, 7, 8, b, 9, a, 61, f, 2003, h, t]
> perml([a,b,c,d,7,8,9,1995,2003,61,h,r,t,f], [0,5,1,0,3,0,6,0,0,8,0,0,12,9]);
                  [c, 7, b, 9, 61, f, t]
```

---

insL – an expansion of lists

Call format of the procedure:

insL(L, p, z)

Formal arguments of the procedure:

L – a list
p – a set of nonnegative integers
m – a valid *Maple* expression

Description of the procedure:

The procedure **insL** similarly to the previous procedure **FLL** gives one more useful expansion of lists. The procedure call **insL(L, p, z)** returns the result of expansion of a list **L** by an expression **z** or its elements after the **L** positions specified by a set **p.** If **z** is a list and the relation **nops(z) = nops(p1)** takes place, where **p1** – a subset of set **p** whose elements belong to range **0..nops(L),** then the **z** elements will be inserted into list **L** after its elements with positions from **p1.** Otherwise, a sequence of actual arguments, starting with **z,** will be inserted into list **L** after its elements with positions from **p1.** At the impossibility of such expansion, the source **L** list is returned. Examples of fragment represented below very clearly illustrate results of the procedure application. Along with common purpose, the given procedure seems convenient enough in problems of bit-by-bit symbolic processing and strings processing.

```
insL:= proc(L::list, p::set(nonnegative), z::anything)
local a, b, c, k, n;
  assign(a = nops(L), n = nops(p)), assign(b = {seq(`if`(a < p[k], NULL, p[k]), k = 1 .. n)});
  if b = {} then RETURN(L)
  else assign('n' = nops(b)), `if`(type(z, list) and n = nops(z), 14, assign(c = 7));
     [op(L[1 .. b[1]]), `if`(c = 7, args[3 .. nargs], z[1]), seq(op([op(L[b[k] + 1 .. b[k + 1]]),
     `if`(c = 7, args[3 .. nargs], z[k + 1])]), k = 1 .. n – 1), op(L[`if`(b[n] = a, 0, b[n] + 1) .. -1])]
  end if
end proc
```

Typical examples of the procedure use:

```
> insL([a,b,c,d,e,f,g,h], {0,2,3,5,9,11}, [x,y,z,t]);
                    [x, a, b, y, c, z, d, e, t, f, g, h]
> insL([a,b,c,d,e,f,g,h], {0,2,3,5,9,11}, avz);
                    [avz, a, b, avz, c, avz, d, e, avz, f, g, h]
> insL([a,b,c,d,e,f,g,h], {p$('p'=1..8)}, "$");
                    [a, "$", b, "$", c, "$", d, "$", e, "$", f, "$", g, "$", h, "$"]
> insL([a,b,c,d,e,f,g,h], {p$('p'=1..8)}, [(p)$('p'=1..8)]);
                    [a, 1, b, 2, c, 3, d, 4, e, 5, f, 6, g, 7, h, 8]
> insL([a,b,c,d,e,f,g,h], {p$('p'=0..8)}, [(p)$('p'=0..8)]);
                    [0, a, 1, b, 2, c, 3, d, 4, e, 5, f, 6, g, 7, h, 8]
> insL([a,b,c,d,e,f,g,h], {p$('p'=0..8)}, 1,2,3);
        [1, 2, 3, a, 1, 2, 3, b, 1, 2, 3, c, 1, 2, 3, d, 1, 2, 3, e, 1, 2, 3, f, 1, 2, 3, g, 1, 2, 3, h, 1, 2, 3]
> insL([a,b,c,d,e,f,g,h], {2,4,6}, [61], 2003, {56});
                [a, b, [61], 2003, {56}, c, d, [61], 2003, {56}, e, f, [61], 2003, {56}, g, h]
> insL([a,b,c,d,e,f,g,h,p,t,r,u], {2,4,6,8,10,12}, [V,G,S,Art,Kr,Arn]);
                    [a, b, V, c, d, G, e, f, S, g, h, Art, p, t, Kr, r, u, Arn]
> insL([a,b,c,d,e,f,g,h,p,t,r,u], {0,2,4,6,8,10,12}, [V,G,S,Art,Kr,Arn,F]);
                    [V, a, b, G, c, d, S, e, f, Art, g, h, Kr, p, t, Arn, r, u, F]
> insL([a,b,c,d,e,f,g,h,p,t,r,u], {0,2,4,6,8,10,12}, [V,G,S,Art,Kr,Arn,F]);
                    [V, a, b, G, c, d, S, e, f, Art, g, h, Kr, p, t, Arn, r, u, F]
> insL([a,b,c,d,e,f,g,h,p,t,r,u], {}, [V,G,S,Art,Kr,Arn,F]);
                    [a, b, c, d, e, f, g, h, p, t, r, u]
> insL([a,b,c,d,e,f,g,h,p,t,r,u], {}, []);
                    [a, b, c, d, e, f, g, h, p, t, r, u]
> insL([a,b,c,d,e,f,g,h,p,t,r,u], {3,5,7}, []);
                    [a, b, c, [], d, e, [], f, g, [], h, p, t, r, u]
> insL([a,b,c,d,e,f,g,h,p,t,r,u], {14,7,5}, [avz,agn]);
                    [a, b, c, d, e, avz, f, g, agn, h, p, t, r, u]
> insL([a,b,c,d,e,f,g,h,p,t,r,u], {14,7,5,3,10}, [avz,agn]);
            [a, b, c, [avz, agn], d, e, [avz, agn], f, g, [avz, agn], h, p, t, [avz, agn], r, u]
```

InvL – inversion of lists
InvList

Call format of the procedures:

InvL(L)
InvList(L)

Formal arguments of the procedures:

L - a list

Description of the procedures:

The procedure call **InvL(L)** returns the result of *inversion* of elements of a list specified by the actual **L** argument. Analogously to the above procedure, the procedure call **InvList(L)** returns the result of *inversion* of elements of a list defined by the actual **L** argument. However, the **InvL** procedure applies the more perfect algorithm, allowing to gain time essentially at inversion of the long enough lists.

```
InvL:= proc(L::list) local k; [seq(L[nops(L) - k], k = 0 .. nops(L) - 1)] end proc
InvList:= L::list -> foldl((a, b) -> [b, op(a)], [], op(L))
```

Typical examples of the procedures use:

> map(InvL, [[67,62,89,96,42,47], ["RANS",IAN,Artur,Kristo]]);

[[47, 42, 96, 89, 62, 67], [*Kristo*, *Artur*, *IAN*, "RANS"]]

> map(InvList, [[67,62,89,96,42,47], ["RANS",IAN,Artur,Kristo]]);

[[47, 42, 96, 89, 62, 67], [*Kristo*, *Artur*, *IAN*, "RANS"]]

KL - clusters of an integer list
KL1

Call format of the procedures:

KL(L)
KL1(L)

Formal arguments of the procedures:

L - an integer list

Description of the procedures:

In respect to a sorted integer list, a new concept of *"cluster"* is introduced. Under the *cluster,* we shall understand a sublist of an integer list whose elements constitute a sequence of numbers going one after another, taking into consideration their sign; i.e. the *cluster* itself represents a continuous segment of the integer series. A series of problems of processing of the list and string structures with success can use the given *cluster* concept; see our books [29-33,43,44].

The procedure call **KL(L)** returns the allocation of clusters in an integer list defined by the actual **L** argument, as a sequence of **2**-element lists; the first element of each such list defines an initial position of the cluster in list **L**, whereas the second its element defines length of the cluster itself. In addition, the procedure takes into consideration the clusters having at least two elements only. If list **L** contain no clusters the *NULL* value is returned, i.e. nothing.

The **KL1(L)** procedure is similar to the previous procedure but with two differences, namely: (**1**) clusters of any length are taken into consideration, and (**2**) if list **L** contain no clusters the *empty* list is returned, i.e. []. The given procedures represent useful enough tools in a series of problems of processing of symbols and strings.

```
KL:= proc(l::{list(integer)})
local k, p;
```

```
  `if`(l = [], RETURN(NULL), `if`(l = [seq(l[1] + p, p = 0 .. nops(l) - 1)], RETURN(l), KL1(l)));
  seq(`if`([%][k][2] = 1, NULL, [%][k]), k = 1 .. nops([%]))
end proc

KL1:= proc(l::{list(integer)})
local a, b, k, n;
  `if`(l = [], RETURN([]), assign(a = cat("", seq(`if`(1 < abs(l[n + 1] - l[n]), cat(l[n], `#`),
  cat(l[n], `|`)), n = 1 .. nops(l) - 1), l[-1]))), assign(b = SLD(a, `#`)),
  seq([SN(b[n][1 .. searchtext(`|`, b[n]) - 1]), nops(Search(b[n], `|`)) + 1], n = 1 .. nops(b))
end proc
```

Typical examples of the procedures use:

> KL([1,4,5,6,7,8,9,10,13,14,15,42,47,59,60,61,62,63,70,71,80,81,82]);

[4, 7], [13, 3], [59, 5], [70, 2], [80, 3]

> KL([1,4,6,8,10,13,15,42,47,59,60,63,70,80,82]), `|`, KL([]), `|`, KL([-2,-1,0,1,2,3,4,5]);

[59, 2], |, |, [-2, -1, 0, 1, 2, 3, 4, 5]

> KL([-1,0,1,3,4]), `|`, KL([1,3,5,7]), `|`, KL([3,5,6,8,9,10,11,13]);

[-1, 3], [3, 2], |, |, [5, 2], [8, 4]

> KL1([1,4,6,8,10,13,15,42,47,59,60,63,70,80,82]);

[1, 1], [4,1], [6,1], [8, 1], [10,1], [13,1], [15, 1], [42, 1],[47, 1], [59,2], [63, 1], [70, 1], [80,1], [82,1]

Lrare – check of an integer list to be as the rarefied

Call format of the procedure:

Lrare(L)

Formal arguments of the procedure:

L – an integer list

Description of the procedure:

In many cases at operation with values of *string*-type, the problem of definition of degree of distribution density of the numbers of positions of entries of some pattern arises. Let us term a list of the numbers of positions of entries by the *rarefied*, if pairwise elements of the list differ more, than on **1.** For testing of the fact of *rarity* of such list the procedure **Lrare** can be useful enough. The procedure call **Lrare(L)** returns the *true* value, if an integer list, defined by the actual **L** argument, is the *rarefied* and the *false* value otherwise. The **Lrare** procedure represents particular interest in many of the appendices operating lists of different nature.

```
Lrare:= proc(L::{list(integer)})
local a, b;
  assign(a = sort([op(L)])), `if`({seq(`if`(a[b + 1] - a[b] = 1, false, true), b = 1 .. nops(a) - 1)} =
    {true} or nops(a) = 1, true, false)
end proc
```

Typical examples of the procedure use:

> map(Lrare, [[1,2,3,4,5,6],[1,3,5,7,9], [6,13,55,35,39], [42,47,62,67,89,96], [7,14,36,40,56,61]]);

[*false*, *true*, *true*, *true*, *true*]

LSL – converting of a string or symbol into list, and vice versa

Call format of the procedure:

LSL(L)

Formal arguments of the procedure:

L - a list of integers, strings and/or symbols; a symbol or string

Description of the procedure:

The package has the advanced enough tools for work with string structures relative to list structures, therefore in a number of problems it is expedient to convert a list into string for execution of necessary processing with the subsequent inverse converting. In this respect, the procedure **LSL** is represented as useful enough tool. The procedure has been implemented by one-line extracode; it works on the basis of the *switch*, and converts a list into string, and vice versa.

The procedure call **LSL(L)** returns the result of converting of a list, defined by the actual **L** argument, into string. On the other hand, if the actual **L** argument of the procedure call **LSL(L)** defines a symbol or a string then the list of symbols or strings is returned accordingly. The returned symbol/string contains the symbols **convert([k], bytes) (k=1 .. 3)** as delimiters of elements of the source list L which depend on their types (*integer, symbol, string*).

```
LSL:= proc(G::{string, symbol, list({integer, string, symbol})})
local a, b, k, c, g;
  assign(g = map(convert, [[1], [2], [3]], bytes)), assign(a = table([integer = g[1],
  symbol = g[2], string = g[3]])), `if`(type(G, list), RETURN(`if`(G = [], "", cat("", seq(cat(G[b],
  a[whattype(G[b])]), b = 1 .. nops(G))))), RETURN(`if`(G = `` or G = "", [],
  [assign(c = Search2(cat("", G), {g[k] $ (k = 1 .. 3)})), `if`(c = [], G, [assign('c' = [0, op(c)]),
  RETURN([seq(`if`(G[c[k]] = g[1], SN(G[c[k - 1] + 1 .. c[k] - 1]), `if`(G[c[k]] = g[2],
  cat(``, G[c[k - 1] + 1 .. c[k] - 1]), G[c[k - 1] + 1 .. c[k] - 1])), k = 2 .. nops(c))])])])))
end proc
```

Typical examples of the procedure use:

> map(LSL, [[67,62,89,-96,42,47], ["RANS", IAN, Artur, Kristo], [1995, -2003]]);

["67□62□89□-96□42□47□", "RANS□IAN□Artur□Kristo□", "1995□-2003□"]

> map(LSL, %);

[[67, 62, 89, -96, 42, 47], ["RANS", *IAN*, *Artur*, *Kristo*], [1995, -2003]]

---

memberL – an expanded check of entries into a list or vector

Call format of the procedure:

memberL(z, L {, 'h'})

Formal arguments of the procedure:

L - a list or vector
z - an expression
h - (optional) an unevaluated name

Description of the procedure:

The procedure call **memberL(z, L)** returns the *true* value, if an expression **z** is element of a list or vector **L**, otherwise the *false* value is returned. On the other hand, if the optional third argument **h** has been coded and the expression **z** belongs to **L**, then via **h** the list of positions (indices) of entries of a **z** into **L** is returned. In a series of cases, the procedure is useful enough at operation with list structures and vector structures of the different types.

```
memberL:= proc(a::anything, L::{vector, list})
local b, k, G;
   assign(b = `if`(type(L, list), nops(L), rhs(op(2, L))));
   `if`(not belong(a, L), RETURN(false), RETURN(assign(G = []), true, `if`(nargs = 2, NULL,
      null([seq(`if`(L[k] = a, assign('G' = [op(G), k]), NULL), k = 1 .. b), assign(args[3] = G)]))))
end proc
```

Typical example of the procedure use:

```
> L:= [S,G,Art,61,61,ian,61,Kr,Sv,61,56,61,36,61,2003,61,ian,61,61]: memberL(61,L,'h'), h,
  memberL(61,vector([56,61,61,7,14,36,61,rans,61,h,61,300,61,42,61,2004,61,61]),'h'), h;
                true, [4, 5, 7, 10, 12, 14, 16, 18, 19], true, [2, 3, 7, 9, 11, 13, 15, 17, 18]
```

gelist – an analysis of multiple list elements
inel

Call format of the procedures:

gelist(L {, 'h'})
inel(L {, P})

Formal arguments of the procedures:

L - a list
h - (optional) an unevaluated name
P - (optional) a procedure from one argument

Description of the procedures:

The procedure call **gelist(L)** returns the nested list of 2-element sublists whose the first element defines an element of a list specified by actual **L** argument whereas the second element defines its multiplicity. If the procedure call uses the second optional argument, then through **h** the nested list is returned also. However, in contrast to the above list its sublists have indefinite quantity of elements, namely: their first element is the same as in the first case, whereas the others define positions of entries of the corresponding element into an **L** list.

The procedure **inel** implemented on the basis of other algorithm in a sense extends the previous procedure. The procedure call **inel(L)** returns the sequence of 2-element lists whose the first element defines an element of a list specified by actual **L** argument whereas the second element defines the set of positions of elements *matched* to it (*coincidence of elements*). If the procedure call uses the second optional argument - a procedure **P** from one argument, then the *matching* is considered relative to the elements **map(F, L).** The both procedures are rather useful in a number of appendices dealing with the list data structures.

```
gelist:= proc(L::list)
local a, b, c, d, k, j, p, n, m, g, omega, h, r, t, psi;
   `if`(L = [], RETURN(L), assign(d = [], p = 1, h = [], n = nops(L), m = nops(L) + 1));
   for k to n do h := [op(h), `if`(type(L[k], rtable), [cat(convert(L[k], string), "_$r"), k], [L[k], k])]
   end do;
   assign(b = SLj(h, 1), g = (x -> convert(x, string))), assign(a = array(1 .. m, [b[j] $ (j = 1 .. n),
```

```
    [infinity, infinity]]));
  omega:= x -> [`if`(Search1(g(x[1]), "_$r", 't') and t = [right], came(x[1][1..-4]), x[1]),nops(x)-1];
  psi:= x -> [`if`(Search1(g(x[1]), "_$r", 't') and t = [right], came(x[1][1..-4]), x[1]), op(x[2..-1])];
  _avz4262300;
  for k from p to m do c := a[k][2];
    for j from k + 1 to m do
      if a[j][1] = a[k][1] then c:= c, a[j][2] else p:= j; d:= [op(d), [a[k][1], c]]; goto(_avz4262300)
      end if
    end do
  end do;
  `if`(1 < nargs and type(args[2], symbol), assign(args[2] = map(psi, d)), NULL), map(omega, d)
end proc

inel:= proc(L::list)
local k, h, n, R, omega;
  assign(n = nops(L), h = `if`(1 < nargs and type(eval(args[2]), procedure) and nops([op(1,
  eval(args[2]))]) = 1, map(args[2], L), L)), assign(R = table([seq(h[k] = {}, k = 1 .. n)]),
    omega = (y -> [L[y[1]], y]));
  for k to n do R[h[k]] := {k, op(R[h[k]])} end do;
  assign('R' = [op(map(omega, map(op, [entries(R)])))]), assign('n' = nops(R));
  assign('R' = SLj([seq([op(R[k]), R[k][2][1]], k = 1 .. n)], 3)), seq(R[k][1 .. 2], k = 1 .. n)
end proc
```

Typical example of the procedures use:

> gelist([y,y,z,a,b,A,c,a,c,b,A,a,x,x,a,z,a,c,A,x,y,y,a], 'h'); h;

[[A, 3], [a, 6], [b, 2], [c, 3], [x, 3], [y, 4], [z, 2]], [[A, 6, 11, 19]
[a, 4, 8, 12, 15, 17, 23], [b, 5, 10], [c, 7, 9, 18], [x, 13, 14, 20], [y, 1, 2, 21, 22], [z, 3, 16]]

> {gelist([[a,b],[x,y],[a,b],[x,y]])}, {gelist([])}, {gelist([`1995_2003`])};

{[[[a, b], 2], [[x, y], 2]]}, {[]}, {[[*1995_2003*, 1]]}

> A:=Array([[S,G],[Art,Kr]]): V:=Matrix([[7,14],[36,40]]): V1:=Vector[row]([1995,2003]): c:=Vector([42,61]): gelist([y,y,z,a,b,A,c,a,c,b,A,a,V,x,x,a,z,V1], 'p');

$$\left[\left[\begin{bmatrix} S & G \\ Art & Kr \end{bmatrix}, 2\right], \left[\begin{bmatrix} 7 & 14 \\ 36 & 40 \end{bmatrix}, 1\right], \left[\begin{bmatrix} 42 \\ 61 \end{bmatrix}, 2\right], [[1995, 2003], 1], [a, 4], [b, 2], [x, 2], [y, 2], [z, 2]\right]$$

> p;

$$\left[\left[\begin{bmatrix} S & G \\ Art & Kr \end{bmatrix}, 6, 11\right], \left[\begin{bmatrix} 7 & 14 \\ 36 & 40 \end{bmatrix}, 13\right], \left[\begin{bmatrix} 42 \\ 61 \end{bmatrix}, 7, 9\right], [[1995, 2003], 18], [a, 4, 8, 12, 16], [b, 5, 10], [x, 14, 15], [y, 1, 2], [z, 3, 17]\right]$$

> L:=[a,A,A,kr,A,`14`,Art,c,d,D,C,Kr,`7`,A,x,kr,X,`7`,X,`14`,art,Z,Art,Kr,`14`]: inel(L, Case);

[a, {1, 2, 3, 5, 14}], [kr, {4, 12, 16, 24}], [14, {6, 20, 25}], [Art, {7, 21, 23}], [c, {8, 11}],
[d, {9, 10}], [7, {13, 18}], [x, {15, 17, 19}], [Z, {22}]

> Function:= (x) -> cat(convert(x, string)): inel(L, Function);

[a, {1}], [A, {2, 3, 5, 14}], [kr, {4, 16}], [14, {6, 20, 25}], [Art, {7, 23}], [c, {8}], [d, {9}],
[D, {10}], [C, {11}], [Kr, {12, 24}], [7, {13, 18}], [x, {15}], [X, {17, 19}], [art, {21}], [Z, {22}]

Resl – an inversion of lists

Call format of the procedure:

Resl(L)

Formal arguments of the procedure:

L – a list or a set whose elements have the type {`**equation**` | `=`}

Description of the procedure:

The procedure call **Resl(L)** returns the result of interchange of left and right members of elements of a list or a set specified by the actual **L** argument. This procedure is useful enough tool in many important tasks of data processing in the *Maple* environment.

```
Resl:= proc(L::{'set'(`=`), list(`=`)})
local k, G;
  G := [seq(rhs(L[k]) = lhs(L[k]), k = 1 .. nops(L))]; `if`(type(L, list), G, {op(G)})
end proc
```

Typical examples of the procedure use:

> Resl([V=61, G=56, Sv=36, Arn=40, Art=14, Kr=7, 1995=2003, RANS=IAN, RAC=REA]);

[61 = *V*, 56 = *G*, 36 = *Sv*, 40 = *Arn*, 14 = *Art*, 6 = *Kr*, 2003 = 1995, *IAN* = *RANS*, *REA* = *RAC*]

> Resl({V=61, G=56, Sv=36, Arn=40, Art=14, Kr=7, 1995=2003, RANS=IAN, RAC=REA});

{14 = *Art*, 36 = *Sv*, *REA* = *RAC*, 2003 = 1995, *IAN* = *RANS*, 7 = *Kr*, 40 = *Arn*, 56 = *G*, 61 = *V*}

---

nlcvector – a converting of lists into vectors

Call format of the procedure:

nlcvector(L)

Formal arguments of the procedure:

L – a list whose elements have an arbitrary *Maple* type

Description of the procedure:

Maple has certain difficulties at processing of lengthy enough lists with recommending to convert their into the corresponding *array*-structures, for example, *vectors*. However, if a list contains elements of *list*-type, i.e. list is a nested list (*nestlist*), the use for these purposes of the procedure `***convert***` can result in a violation of internal list structure, or to errors, as the examples represented below illustrate. Meantime, in a series of cases the saving of internal list structures is necessary.

The procedure call **nlcvector(L)** returns the result of conversion of an arbitrary list specified by the actual argument **L** into vector with saving of internal structure of the source list. This procedure is useful enough tool in many important tasks of list processing in the *Maple* environment. Especially, it concerns the case when a problem deals with list data structures with large number of elements. In this case, erratic situations are rather probable.

```
nlcvector:= proc(L::list)
local a, b, k;
  assign(b = nops(L)), `if`(type(L, nestlist), [assign(a = vector(b)),
  seq(assign('a'[k] = L[k]), k = 1 .. b), RETURN(eval(a))], convert(L, vector))
end proc
```

Typical examples of the procedure use:

> convert([[a],b,[c],2003,7,[14],sin(x),1995,[2003],agn], vector);

Error, (in index/fill) initialization of arrays must be done with lists

> nlcvector([[a],b,[c],2003,7,[14],sin(x),1995,[2003],agn]);

[[a], b, [c], 2003, 7, [14], sin(100), 1995, [2003], agn]

> convert([[a,b,c],[x,y,z],[1995,2003,300],[rans,grsu,vgtu]], vector);

$$\begin{bmatrix} a & b & c \\ x & y & z \\ 1995 & 2003 & 300 \\ rans & grsu & vgtu \end{bmatrix}$$

> nlcvector([[a,b,c],[x,y,z],[1995,2003,300],[rans,grsu,vgtu]]);

[[a, b, c], [100, y, z], [1995, 2003, 300], [rans, grsu, vgtu]]

> convert([61,56,36,40,7,14,rans], vector), nlcvector([61,56,36,40,7,14,rans]);

[61, 56, 36, 40, 7, 14, rans], [61, 56, 36, 40, 7, 14, rans]

RTab – determination of indices of a table by its entry

Call format of the procedure:

RTab(T, *expr*)

Formal arguments of the procedure:

T – a table
expr – a valid *Maple* expression

Description of the procedure:

The simple **RTab** procedure is intended for operation with objects of *table*-type, by allowing by a value of entry of a table to obtain a sequence of its indices corresponding to the indicated entry. The procedure call **RTab(T, *expr*)** returns the sequence of **T**-table indices, to which an entry ***expr*** corresponds. At absence of the given entry the *NULL* value is returned, i.e. nothing. The given procedure is useful at solution of many problems, linked with processing of tables in the *Maple* environment.

```
RTab:= proc(T::table, a::anything)
local k, h, L;
  assign(h = op(2, eval(T)), L = []); null(add(`if`(rhs(h[k]) = a,
     _1N(0, assign('L' = [op(L), lhs(h[k])])), 0), k = 1 .. nops(h))), op(L)
end proc
```

Typical examples of the procedure use:

> C:= table([x=b,y=b,z=c,w=b]): T:=table([avz=61,agn=56,Kr=61,vsv=36,arn=40,Art=61]):

> [RTab(C, b)], [RTab(T, 61)], [RTab(C, z)], [RTab(T, 2003)];

[x, y, w], [*avz*, *Kr*, *Art*], [], []

getable – an analysis of multiple table entries

Call format of the procedure:

getable(T {, h})

Formal arguments of the procedure:

T – a table
h – (optional) a valid *Maple* expression

Description of the procedure:

The procedure call **getable(T)** returns the nested list of 2-element sublists whose the first element defines an entry of a table specified by actual **T** argument whereas the second element defines its multiplicity. If the procedure call uses the second optional argument (any valid *Maple* expression), then the nested list is returned also. However, in contrast to the above list its sublists have indefinite quantity of elements, namely: their first element is the same as in the first case, whereas the others define indices for the corresponding entry of a **T** table. The procedure is a rather useful in a number of appendices dealing with the tabular data structures.

```
getable:= proc(T::Table)
local a, b, k, t, omega, x;
  `if`(Empty(T), RETURN([]), assign(a = op(2, eval(T))));
  assign(b = [seq(rhs(a[k]), k = 1 .. nops(a))], omega = (x -> [x[1], seq(lhs(a[k]), k = x[2..-1])]));
  gelist(b, 't'), `if`(nargs = 1, NULL, RETURN(map(omega, t)))
end proc
```

Typical example of the procedure use:

> T:=table([a=x,b=y,c=y,d=x,e=y,h=x,x1=7,y1=14,x2=7,y2=14,x3=7,a=14,b=7]);

T := table([x1 = 7, y1 = 14, c = y, x2 = 7, d = x, y2 = 14, e = y, x3 = 7, h = x, a = 14, b = 7])

> getable(T), getable(T,h);

[[7, 4], [14, 3], [x, 2], [y, 2]], [[7, x1, x2, x3, b], [14, y1, y2, a], [x, d, h], [y, c, e]]

> M:=Matrix([[7,14], [36,40]]): V:=Vector([61,56,36,40,7,14]): G:=table([a=M,v=61,b=V, g=56, c=M, s=36, d=V, h=M, t=V]);

$$G := \text{table}([t = \begin{bmatrix} 61 \\ 56 \\ 36 \\ 40 \\ 7 \\ 14 \end{bmatrix}, a = \begin{bmatrix} 7 & 14 \\ 36 & 40 \end{bmatrix}, v = 61, b = \begin{bmatrix} 61 \\ 56 \\ 36 \\ 40 \\ 7 \\ 14 \end{bmatrix}, g = 56, c = \begin{bmatrix} 7 & 14 \\ 36 & 40 \end{bmatrix}, s = 36, d = \begin{bmatrix} 61 \\ 56 \\ 36 \\ 40 \\ 7 \\ 14 \end{bmatrix}, h = \begin{bmatrix} 7 & 14 \\ 36 & 40 \end{bmatrix}])$$

> getable(G), getable(table([a=b])), getable(table([a=b]), 14);

$$\left[[36, 1], [56, 1], [61, 1], \left[\begin{bmatrix} 61 \\ 56 \\ 36 \\ 40 \\ 7 \\ 14 \end{bmatrix}, 3 \right], \left[\begin{bmatrix} 7 & 14 \\ 36 & 40 \end{bmatrix}, 3 \right] \right], [[b, 1]], [[b, a]]$$

> getable(G, 7), getable(table([]));

$$\left[[36, s], [56, g], [61, v], \left[\left[\begin{matrix}61\\56\\36\\40\\7\\14\end{matrix}\right], t, b, d\right], \left[\left[\begin{matrix}7 & 14\\36 & 40\end{matrix}\right], a, c, h\right]\right], [\ ]$$

deltab – deletion of table elements

Call format of the procedure:

deltab(T, r, a)

Formal arguments of the procedure:

T - a table
r - a deletion mode; may be {**0, 1, 2**}
a - a valid *Maple* expression

Description of the procedure:

Entries can be removed from a table by assigning a table entry to its own name. Thus, **Tb[Kr]:= 'Tb[Kr]'** removes the element of a table **Tb** whose index is **Kr.** The procedure **deltab** provides the more wide functions for deletion of table elements.

The procedure call **deltab(T, r, a)** returns the *NULL* value, i.e. nothing, providing deletion out of a table specified by the first actual argument **T** of all elements which have a match relatively to an expression `**a**` depending on a deletion mode **r**, namely: **r=0** - equality of indices, **r=1** - equality of entries, and **r=2** - presence of any of the above equalities. The procedure is a rather useful in a number of appendices dealing with the tabular data structures.

```
deltab:= proc(T::table, r::{0, 1, 2}, a::anything)
local b, c, d, k, p, n;
   assign(b = op(2, eval(T)), d = op(1, eval(T)), c = ((x, y) -> `if`(x(y) = args[3 .. -1], NULL, y)));
   assign67(n = nops(b), p = args[3 .. -1]), assign('T' = table(d, [seq(`if`(r = 0, c(lhs, b[k], p),
      `if`(r = 1, c(rhs, b[k], p), `if`(c(rhs, b[k], p) = NULL or c(lhs, b[k], p) = NULL,
      NULL, b[k]))), k = 1 .. n)]))
end proc
```

Typical example of the procedure use:

> T:= table([x=a, y=b, z=Art, Art=15, Kr=7]);

T := table([z = Art, Art = 15, Kr = 7, x = a, y = b])

> deltab(T,0, Art), eval(T); deltab(T,1,a), eval(T); deltab(T,2, Art), eval(T);

table([z = Art, Kr = 7, x = a, y = b])
table([z = Art, Kr = 7, y = b])
table([Kr = 7, y = b])

> T1:= table(symmetric, [(y,x)=avz,(y,z)=G, (Art,Kr)=[15,7], sv=(Art,Kr)]);

T1 := table(*symmetric*, [(y, z) = G, (x, y) = avz, sv = (Art, Kr), (Art, Kr) = [15, 7]])

> T1[Art,Kr], T1[Kr,Art],T1[x,y], T1[y,x], T1[y,z], T1[z,y], T1[sv];

[15, 7], [15, 7], avz, avz, G, G, Art, Kr

> **deltab(T1,0,(y,x)), eval(T1); deltab(T1,1,[15,7]), eval(T1); deltab(T1,2,(Art,Kr)), eval(T1);**

table(*symmetric*, [(y, z) = G, (x, y) = avz, sv = (Art, Kr), (Art, Kr) = [15, 7]])
table(*symmetric*, [(y, z) = G, (x, y) = avz, sv = (Art, Kr)])
table(*symmetric*, [(y, z) = G, (x, y) = avz])

plotTab – plot of a table with entries of *realnum*-type

Call format of the procedure:

plotTab(T, {, *popts*})

Formal arguments of the procedure:

T – a table with entries of *realnum*-type
popts – (optional) plot options

Description of the procedure:

The **plotTab(T)** procedure is intended for data plotting of a table specified by the first actual **T** argument. The argument should define a table with entries of *realnum*-type. The procedure call **plotTab(T)** returns the data plot of a **T**-table whose entries define data for **X**-axis and indices define data for **Y**-axis. Whereas the table indices may be arbitrary valid *Maple* expressions, they are sorted lexicographically and are printed in a view of the equations list that defines the one-to-one correspondence between the table entries and points of **X**-axis. The optional actual arguments beginning with the second one specify a sequence of plot options. The given procedure is useful at solution of a series of problems linked with processing of tables in the *Maple* environment.

```
plotTab:= proc(T::table)
local _, b, c, k, p, f;
  `if`(ttable(T, realnum), assign67(_ = map(op, {indices(T)}), b = map(evalf, eval(T)),
    f = TIMES, BOLD), ERROR("table <%1> has non-realnum entries", T)),
    print([seq(k = _[k], k = 1 .. nops(_))]); assign(p = nops(_)),
    assign(c = plots[listplot]([seq(b[_[k]], k = 1 .. p)], `if`(nargs = 1, NULL, args[2 .. -1])));
  plots[display]([c, seq(plottools[line]([k, 0], [k, b[_[k]]], color = red), k = 1 .. p)], xtickmarks = p,
    labels = ["entries", "indices"], labelfont = [f, 11], axesfont = [f, 10])
end proc
```

Typical examples of the procedure use:

> **Fam:= table([V=61/6, S=36, G=56, Ar=40, Art=14, Kr=sqrt(50)]): plotTab(Fam, thickness=2, color=green);**

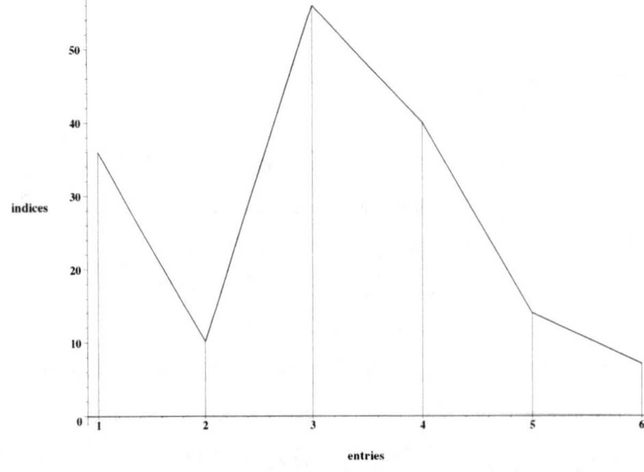

[1 = G, 2 = S, 3 = Art, 4 = Kr, 5 = V, 6 = Ar]

```
> Fam1:= table([V=61, S=36, G=56, Ar=40, Art=14, Kr="14_7"]): plotTab(Fam1, thickness=2,
  color=green, axesfont=[TIMES,BOLD,10]);
```

Error, (in **plotTab**) table <Fam1> has non-realnum entries

Search_D – search of words in dictionaries
Search_D1
Search_D2

Call format of the procedures:

Search_D(D, W)
Search_D1(D, W {, `*insensitive*`})
Search_D2(Dic, W)

Formal arguments of the procedures:

D – a dictionary defined by a list of strings or symbols
W – a sought word defined by a symbol or a string
insensitive – (optional) key word defining the search mode of words in the dictionary
Dic – a table whose entries are words of dictionary

Description of the procedures:

Search is one of the most mass procedures at symbolical processing. The procedures, presented below, provide effective enough search of words in dictionaries defined by a list. In case of detection of a sought word, defined by the second actual **W** argument in dictionary defined by the first actual **D** argument, the procedure call **Search_D(D,W)** returns the list of a view [*true*, <*location of the word in the dictionary*>], otherwise the *false* value is returned.

The **Search_D1** procedure is similar to the above procedure **Search_D,** but in contrast to the second one, it allows to do both the sensitive (*by default*) and the insensitive search of a word **W** in dictionary **D.** A search mode is defined by the third optional argument `***insensitive***`.

At last, the highly effective procedure **Search_D2(Dic, W)** (in contrast to the above ones) is based on a tabular structure of a dictionary defined by the first actual **Dic** argument. Words of such dictionary are defined as indices of a table and can be of type {*string, symbol*}, while its entries can define some semantics of the words and be of any valid *Maple* type.

In case of detection of a sought word, defined by the second actual **W** argument, in dictionary, defined by the first actual **Dic** argument, the procedure call **Search_D2(Dic, W)** returns the list of a view [*true*, <*semantics of a word in the dictionary*>], otherwise the *false* value is returned. The procedure provides only the sensitive search in a **Dic** dictionary. It is necessary to pay attention, for organization of the dictionaries and some other data structures providing a direct access to the own elements can be used. A discussion of the similar questions can be found in our books [29-33,43].

```
Search_D:= proc(Dictionary::list, Word::{name, string, symbol})
local k, G, Flag;
  G:= proc(S::{name, string, symbol})
    local k, L, H;
      assign({H = ``, L = convert(S, 'bytes')});
      for k to nops(L) do H := cat(H, L[k]) end do;
      op(sscanf(H, %d))
    end proc;
```

```
  for k to nops(Dictionary) do if G(Word) = G(Dictionary[k]) then RETURN([true, k])
     else next
     end if
  end do;
  `if`(Flag = 59, NULL, false)
end proc

Search_D1:= proc(Dic::list({name, string, symbol}), W::{name, string, symbol})
local a, b, k, omega;
  omega:= (a, b) -> `if`(member(insensitive, a), Case(b), b);
  op({seq(`if`(omega({args}, cat("", W)) = omega({args}, cat("", Dic[k])), RETURN([true, k]),
     false), k = 1 .. nops(Dic))})
end proc

Search_D2:= proc(Dic::table, W::{name, string, symbol})
local Res;
  Res := Dic[cat(``, W)]; `if`(type(Res, indexed), `if`(type(Dic[cat("", W)], indexed), false,
     [true, Dic[cat("", W)]]), [true, Res])
end proc
```

Typical examples of the procedures use:

> F:= [IAN,Svet,Art,RAC,Kr,Vic,Gal,Arn,RANS]: Search_D(F,Kr), Search_D(F,Avz);

$[true, 5], \; false$

> Search_D1(F,Kr), Search_D1(F,KR), Search_D1(F,KR, `insensitive`);

$[true, 5], \; false, \; [true, 5]$

> k:=5: n:=10^k: H:=[seq(cat(avz,k),k=1..n)]: t:=time(): Search_D1(H,avz99999): time()-t;

120.742

> Dic := table([RANS = [2000-2002], IAN = 1995, Svet = 1967, Art = 1989, RAC = 1994, Kr = 1998, Vic = 1942, Gal = 1947, Arn = 1962]): Search_D2(Dic,Art), Search_D2(Dic,"Art");

$[true, 1989], [true, 1989]$

> k:= 5: n:= 10^k: H:=table([seq(cat(avz,k)=k, k=1..n)]): t:=time(): Search_D2(H,avz99999): time()-t;

0.001

---

tabar – representation of a special type of tables

Call format of the procedure:

tabar(T, C1, C2 {, n})

Formal arguments of the procedure:

T - a table
C1 - a string or symbol
C2 - a string or symbol
n - (optional) a value of *numeric* type

Description of the procedure:

The **tabar** procedure is intended for visual representation of a special type of a table specified by the first actual argument. Indices of such table are values of a type {*symbol*, *string*}, and entries –

values of *numeric*-type. The procedure call **tabar(T, C1, C2)** returns the (Nx2)-array, whose the first column defines the indices of a table **T**, and the second column defines entries of **T**; in addition, the first row of the returned array defines titles of columns indicated by the actual arguments **C1** and **C2** accordingly; in addition, in general case the relation **N=nops(op(op(T))) + 1** takes place. The returned array represents the result of sorting of the initial table **T** in decreasing order of values of its entries; in addition, in case of equal values of entries the indices are sorted in lexicographical order. In case of the procedure call **tabar(T, C1, C2, n)** with the fourth optional argument of *numeric* type the initial part of a sought array (values of elements of the second column of this part are not less than value of the fourth actual **n** argument) is returned only. At absence of such rows for an array, the procedure returns the *lack*-value.

The given procedure seems as useful tool, taking into account a wide spectrum of appendices of data structures of *table*-type. In particular, with its help the contents of tables generated by the ***profile*** procedure, intended for a profiling of calls of the indicated procedures is reflected enough visually.

The procedure **tabar** represents a convenient enough tool of generating of special type of tables intended for a visual representation of the information when the first column of the table defines *names* of some objects, whereas the second – *numerical* characteristics corresponding to them.

```
tabar:= proc(T::table, C1::{string, symbol}, C2::{string, symbol})
local a, h, k, p, A, B;
   assign(a = op(2, eval(T))), assign(h = [seq([lhs(a[k]), rhs(a[k])], k = 1 .. nops(a))]);
   assign('h' = SLj(h, 2, `>`)), assign(A = array(1 .. nops(h) + 1, 1 .. 2, []));
   assign(A[1, 1] = cat(``, C1), A[1, 2] = cat(``, C2));
   seq(assign('A'[k + 1, 1] = cat(``, h[k][1]), 'A'[k + 1, 2] = h[k][2]), k = 1 .. nops(h));
   `if`(nargs = 3, eval(A), [`if`(not type(args[4], numeric),
      ERROR("Invalid 4th argument <%1> - should be numeric", args[4]),
      seq(`if`(A[k, 2] < args[4], 1, assign('p' = k)), k = 1 .. nops(h) + 1)),
      assign(B = array(1 .. `if`(type(p, symbol), RETURN(lack), p), 1 .. 2, [])),
      seq(assign('B'[k, 1] = A[k, 1], 'B'[k, 2] = A[k, 2]), k = 1 .. p), RETURN(eval(B))])
end proc
```

Typical examples of the procedure use:

> FF:=table([Galina=57,Victor=62, Sveta=37, Arne=40, Artur=15, Kristo=7]): tabar(FF, Name,Age);

$$\begin{bmatrix} Name & Age \\ Victor & 62 \\ Galina & 57 \\ Arne & 40 \\ Sveta & 37 \\ Artur & 15 \\ Kristo & 7 \end{bmatrix}$$

Below in section **14**, the appendices of the **tabar** procedure will be illustrated by examples of reflection of contents of the tables generated by the standard ***profile*** procedure used for a profiling of the procedures calls from the user library logically linked with the main *Maple* library. The **tabar** procedure is successfully used in numerous appendices.

mapTab – an extending of the standard procedure `*map*` onto tables, sets and lists
mapLS

Call format of the procedures:

mapTab(F, T, n {, x1, x2, ..., xn})
mapLS(F, L, n {, x1, x2, ..., xn})

Formal arguments of the procedures:

F – symbol defining name of a procedure
T – a table
L – a list or a set whose elements are relations or arbitrary *Maple* expressions
n – an integer defining a processing mode (may be **0, 1** or **2** only)
x1, x2, ..., xn – (optional) arguments passed to the **F** function since the second one

Description of the procedures:

In the package *Maple* the **map** and **map2** procedures providing the cyclic application of a required procedure to each operand of a given expression are enough widely used. We additionally have introduced two useful procedures **map3** and **mapN.** The following procedure **mapTab** extends the standard **map** procedure onto tables, namely: the procedure allows to apply a procedure both to indices and to entries of a table.

The procedure call **mapTab(F, T, n {, x1, x2, ..., xn})** returns the result of application of a **F** function from arguments **<y, x1, ..., xn>** to indices or/and entries **y** of a **T** table depending on a **n** value (**0** – only indices, **1** – only entries, and **2** – indices and entries).

The procedure call **mapLS(F, L, n {, x1, x2, ..., xn})** returns the result of application of a **F** function from arguments **<y, x1, ..., xn>** to elements of a list or set **L.** In addition, if a **L** element has *relation*-type {`<`, `<=`, `<>`, `=`}, then **F** function is applied to its left and/or right members depending on a **n** value (**0** – only left members, **1** – only right members, and **2** – left and right members). The examples of appendices of procedures **mapTab** and **mapLS** well illustrate the told. The procedures **mapTab** and **mapLS** extend functional possibilities of the above procedures **`map`, map2, map3** and **mapN.**

```
mapTab:= (F::symbol, T::table, n::{0, 1, 2}) -> table(mapLS(F, op(eval(T)), n,
            `if`(nargs = 3, NULL, args[4 .. -1])))

mapLS:= proc(F::symbol, L::{set, list}, n::{0, 1, 2})
local a, b, h, p, x, y, r, omega;
  assign(p = nops(L), h = ((x, y) -> convert(x, whattype(y))), omega = whattype, r = relation);
  if 3 < nargs then a := args[4 .. -1] else a := NULL end if;
  if n = 1 then h([seq(op([assign('b' = L[k]), `if`(type(b, r), omega(b)(lhs(b), F(rhs(b), a)),
      F(b, a))]), k = 1 .. p)], L)
  elif n = 0 then h([seq(op([assign('b' = L[k]), `if`(type(b, r), omega(b)(F(lhs(L[k]), a), rhs(L[k])),
     F(L[k], a))]), k = 1 .. p)], L)
  else h([seq(`if`(type(L[k], r), map(F, L[k], a), F(L[k], a)), k = 1 .. p)], L)
  end if
end proc
```

Typical examples of the procedures use:

```
> T:= table([a=b,c=d,t=b,omega=nu]); mapTab(H,T,1,x,y,z);
```

T := table([a = b, omega = nu, c = d, t = b])
table([a = H(b, x, y, z), omega = H(nu, x, y, z), c = H(d, x, y, z), t = H(b, x, y, z)])

```
> mapTab(H,T,0,x,y,z);
            table([H(c,x,y,z) = d, H(t,x,y,z) = b, H(a,x,y,z) = b, H(omega,x,y,z) = nu])
> mapTab(H,T,2,x,y,z);
        table([H(c,x,y,z) = H(d,x,y,z), H(t,x,y,z) = H(b,x,y,z), H(a,x,y,z) = H(b,x,y,z),
                              H(omega,x,y,z) = H(nu,x,y,z)])
> F:= table([V=62,G=57,S=37,Art=15,Kr=8,Arn=41]): mapTab(convert, F, 2, string);
        table(["S" = "37", "V" = "62", "Art" = "15", "Kr" = "8", "G" = "57", "Arn" = "41"])
> mapLS(F, [a=b,c=d,e=h,g=h], 1, x,y,z);
                 [a = F(b,x,y,z), c = F(d,x,y,z), e = F(h,x,y,z), g = F(h,x,y,z)]
> mapLS(F, [a=b,c=d,e=h,g=h], 0, x,y,z);
                 [F(a,x,y,z) = b, F(c,x,y,z) = d, F(e,x,y,z) = h, F(g,x,y,z) = h]
> mapLS(F, [a=b,c=d,e=h,g=h], 2, x,y,z);
    [F(a,x,y,z) = F(b,x,y,z), F(c,x,y,z) = F(d,x,y,z), F(e,x,y,z) = F(h,x,y,z), F(g,x,y,z) = F(h,x,y,z)]
> mapLS(F, {a=b,c<d,e<=h,A,g<>h,V,Z}, 2,x,y,z);
      {F(c,x,y,z) < F(d,x,y,z), F(e,x,y,z) <= F(h,x,y,z), F(g,x,y,z) <> F(h,x,y,z), F(V,x,y,z),
                      F(a,x,y,z) = F(b,x,y,z), F(A,x,y,z), F(Z,x,y,z)}
> mapLS(F, {a=b,c=d,e=h,g=h,V>1995,S<=2004,Art<>Kr}, 1,x,y,z);
        {g = F(h,x,y,z), 1995 < F(V,x,y,z), e = F(h,x,y,z), c = F(d,x,y,z), a = F(b,x,y,z),
                         S <= F(2004,x,y,z), Art <> F(Kr,x,y,z)}
> mapLS(F, [a=b,c=d,e<>h,g=h,Art>Kr,G<>S], 0,x,y,z);
   [F(a,x,y,z) = b, F(c,x,y,z) = d, F(e,x,y,z) <> h, F(g,x,y,z) = h, F(Kr,x,y,z) < Art, F(G,x,y,z) <> S]
```

InvT – interchange of indices and entries of a table

Call format of the procedure:

InvT(T)

Formal arguments of the procedure:

T – a table

Description of the procedure:

The procedure call **InvT(T)** returns the result of interchange of indices and entries of a table specified by the actual **T** argument. Whereas a table can has multiple entries, the interchange of its indices and entries is generally invalid. However, in a series of cases the given table operation seems useful enough. The procedure is used in the **Iddn1** procedure for change of view of a table being returned by it. However, the basic procedure destination lies in sphere of work with the *Maple* objects of type *`table`*. The procedure call for a table with multiple entries causes erratic situation.

```
InvT:= proc(T::table)
local a, k;
  `if`(nops(op(2, eval(T))) <> nops(map(op, {entries(T)})), ERROR("table <%1> cannot be
       correctly inverted", T), op([assign(a = op(eval(T))),
       table([seq(rhs(a[k]) = lhs(a[k]), k = 1 .. nops(a))])]))
end proc
```

Typical examples of the procedure use:

> T:= table([V=62, G=57, S=37, Art=15, Kr=8, Ar=40]): InvT(T), InvT(InvT(T));

table([8 = Kr, 15 = Art, 37 = S, 40 = Ar, 57 = G, 62 = V]), table([S = 37, G = 57, Art = 15, Kr = 8, Ar = 40, V = 62])

> T1:= table([a=b, c=d, h=b, t=d, z=d, p=b]): InvT(T1);

Error, (in **InvT**) table <T1> cannot be correctly inverted

---

sls – a special sorting of lists

Call format of the procedure:

sls(L, h {, \`<\` {, \`*insensitive*\`}})

Formal arguments of the procedure:

L – a list whose elements are strings or symbols
h – a pattern (string or symbol)
\`<\` – (optional) a sorting mode
insensitive – (optional) a search case of the pattern

Description of the procedure:

The procedure call **sls(L, h)** returns the result of sorting of a list, defined by the first actual **L** argument, depending on the number of entries into its elements of a pattern, defined by the actual **h** argument. If key symbol **\`<\`** has been coded then **L** is sorted in *ascending* numerical order, otherwise in *descending* numerical order depending on the number of entries of the pattern **h** into its elements (*by default*).

If key word \`***insensitive***\` has been coded, the searching of the pattern **h** in the list elements is done in *insensitive* case, otherwise in *sensitive* case (*by default*). In all cases, indicated above, for the list elements with identical number of entries of the pattern **h** the sorting is done in *lexicographic* order. The given procedure seems as useful tool at processing of many tree data structures, in particular, at processing of a PC file structure.

```
sls:= proc(L::list({string, symbol}), h::{string, symbol})
local G, SF, alpha, omega;
  assign(omega = `if`(member(insensitive, [args]), 1, 0)), assign(alpha = (x -> `if`(omega = 1,
     Case(x), x)));
  SF:= (x, y) -> `if`(nops(Search(alpha(y), alpha(h))) <= nops(Search(alpha(x), alpha(h))),
     true, false);
  assign(G = sort(L, string)), `if`(member(`<`, [args]), Rlss(sort(G, SF)), sort(G, SF))
end proc
```

Typical examples of the procedure use:

> sls([\`Real\`,"IAN",\`RANS\`,\`2003\`,"REA",\`7\`,\`RAAN\`,\`RAC\`,\`14\`,\`ALLA\`,\`1995\`, Tallinn], \`A\`, \`<\`);

[*Tallinn, Real, 7, 2002, 1995, 14,* "REA", *RANS, RAC,* "IAN", *RAAN, ALLA*]

> sls([\`Real\`,"IAN",\`RANS\`,\`2003\`,"REA",\`7\`,\`RAAN\`,\`RAC\`,\`ALLA\`,\`1995\`, Tallinn], \`a\`, \`insensitive\`);

[*ALLA, RAAN,* "IAN", *RAC, RANS,* "REA", *Real, Tallinn, 1995, 2003, 7*]

> sls([\`Real\`,"IAN",\`RANS\`,\`2003\`,"REA",\`7\`,\`RAAN\`,\`RAC\`,\`14\`,\`ALLA\`,\`1995\`, Tallinn],"a");

[*Real, Tallinn, 14, 1995, 2003, 7, ALLA,* "IAN", *RAAN, RAC, RANS,* "REA"]

SLj – a sorting of nested lists

Call format of the procedure:

SLj(L, n {, *sf*})

Formal arguments of the procedure:

L - a nested list
n - a positive integer (*posint*)
sf - (optional) an ordering function

Description of the procedure:

At processing of the multivariate data, represented by *nested* lists (i.e. data structures have the *nestlist*-type), in many cases a problem of sorting of such lists according to the given position of elements of their sublists arises. For solution of similar problems, the **SLj** procedure can be useful enough.

The procedure call **SLj(L, n)** returns the result of sorting of a nested list specified by the first actual **L** argument, according to the given position **n** of elements of its sublists. The third optional actual *sf* argument is Boolean procedure of two arguments defining some ordering algorithm. By default, for the third argument is supposed a value accepted for standard function *`sort`* of the *Maple* package.

```
SLj:= proc(L::nestlist, n::posint)
local a, b, c, k, p, R;
   assign(b = map(nops, {op(L)})[1], c = nops(L)), `if`(b < n, ERROR("invalid 2nd argument
      <%1>; must be >=1 and <=%2", n, b), assign(R = [])), assign(a = sort([op({seq(L[k][n],
      k = 1 .. c)})], `if`(2 < nargs and type(args[3], boolproc), args[3], NULL)));
   for k to nops(a) do for p to c do if a[k] = L[p][n] then R := [op(R), L[p]] end if end do end do;
   R
end proc
```

Typical examples of the procedure use:

```
> K:=[[a,b,c], [g,d,c], [a,g,l], [l,s,a,s], [f,d,k,c], [s,a,d,a,a]]: SLj(K, 2, lexorder), SLj(K,1, lexorder);
     [[s, a, d, a, a], [a, b, c], [g, d, c], [f, d, k, c], [a, g, l], [l, s, a, s]], [[a, b, c], [a, g, l], [f, d, k, c],
     [g, d, c], [l, s, a, s], [s, a, d, a, a]]
> Art:= [["V",42,61,300], ["G",47,56], ["K",7,96], ["Ar",14,89], ["S",36,67,4], ["A",40,62]]: SLj(Art, 2),
   SLj(Art, 3),  SLj(Art, 1, lexorder);
     [["K", 7, 96], ["Ar", 14, 89], ["S", 36, 67, 4], ["A", 40, 62], ["V", 42, 61, 300], ["G", 47, 56]], [["G", 47,
     56], ["V", 42, 61, 300], ["A", 40, 62], ["S", 36, 67, 4], ["Ar", 14, 89], ["K", 6, 96]], [["A", 40, 62], ["Ar",
     14, 89], ["G", 47, 56], ["K", 6, 96], ["S", 36, 67, 4], ["V", 42, 61, 300]]
```

_SL – dynamic assignment of values to elements of a list or set

Call format of the procedure:

_SL(L, *expr* {, p})

Formal arguments of the procedure:

L - a list or a set
expr - an expression
p - a positive integer (*posint*)

Description of the procedure:

The procedure call **_SL(L, *expr*)** returns the *NULL* value, providing the assignment of a value, given by the second actual ***expr*** argument, to all elements of a list or set defined by the first actual **L** argument; the replacement of elements of the initial **L** argument is being done in situ. In addition, assignments are done for the unassigned names only.

If the third optional actual **p** argument has been coded, the procedure call **_SL(L, *expr*, p)** provides the assignment of ***expr*** value to the **p**th element of a list or set **L.** The given procedure allows easily to actualize a mechanism of dynamic assignment of values to variables both outside and in body of a procedure, what essentially facilitates programming of many problems of various purposes.

```
_SL:= proc(L::{set, list}, a::anything)
local k;
  `if`(3 <= nargs, `if`(belong(args[3], 1 .. nops(L)), assign(`if`(type(op(args[3], L), symbol),
  op(args[3], L) = a, NULL)), ERROR(`Invalid 3rd argument`)), assign(seq(`if`(type(eval(L[k]),
  symbol), op(k, L) = a, NULL), k = 1 .. nops(L))))
end proc
```

Typical examples of the procedure use:

> restart; _SL([Art,Kr,V,G,Sv,Arn], 2003); [Art, Kr, V, G, Sv, Arn];

[2003, 2003, 2003, 2003, 2003, 2003]

> Art, Kr:= 14, 7: _SL([Art,Kr,x,y,z,h], 1942), [Art,Kr,x,y,z,h];

[14, 7, 1942, 1942, 1942, 1942]

> k:=61: L:=[y1,y2,y3,y4,y5,y6,y7]: seq(_SL(L,k^k,k),k=1..nops(L)); k,eval([seq(cat(``,y,k), k=1..6)]);

61, [*19421, 19422, 19423, 19424, 19425, 19426*]

> _SL([1942, 1947, X, 1967, 1962, Y, 1989, 1996, Z], 2003), [X, Y, Z];

[2003, 2003, 2003]

SSet – address sorting of list elements

Call format of the procedure:

SSet(*args*)

Formal arguments of the procedure:

args – a sequence of actual arguments

Description of the procedure:

Four functions ***assemble, disassemble, addressof*** and ***pointto*** are known as *`hackware`* tools in *Maple*. These functions provide access to the internal representations of the *Maple* objects and to the addresses pointing to them; these functions are considered enough in detail in Help and in our books [9,29-33]. The following useful procedure **SSet** is based on these built-in *Maple* functions.

The procedure call **SSet(*args*)** returns the set of equations, whose the left members by a value meet to actual *`**args**`* arguments of the procedure, whereas the right members define the numbers of elements corresponding to them, if sequence of ***args*** arguments would be converted into a set. The procedure call **SSet()** returns the *empty* set, i.e. {}. In particular, the analysis of the procedure allows deeper to understand the address sorting of elements for structures of *set*-type used by *Maple* and provides more simple and effective algorithms of operations with the sets.

```
SSet:= proc()
local a, k, p;
  {assign(a = [disassemble(addressof(args))][2 .. nops([args]) + 1]), seq(seq(`if`(a[k] =
  [op({op(a)})][p], args[k] = p, NULL), p = 1 .. nops({op(a)})), k = 1 .. nops(a))}
end proc
```

Typical examples of the procedure use:

```
> restart; E:= 1995,m,g,42,Art,61,G^2, {n,6}, [h,14], Kr, 2003:  SSet(E), {E};
        {42 = 1, 1995 = 3, m = 8, g = 7, Art = 5, 61 = 2, G^2 = 9, {6, n} = 10, [h, 14] = 11, Kr = 6,
              2003 = 4}, {42, 61, 1995, 2003, Art, Kr, g, m, G^2, {6, n}, [h, 14]}
> restart; E1:= sqrt(1),v,z,42,Art,61,56, {k$k=1..6}, [7,14], Kr, sin(6):  SSet(E1), {E1};
        {1 = 1, v = 5, z = 8, 42 = 2, Art = 6, 61 = 4, 56 = 3, {1, 2, 3, 4, 5, 6} = 11, [7, 14] = 10, Kr = 7,
              sin(6) = 9}, {1, 42, 56, 61, v, Art, Kr, z, sin(6), [7, 14], {1, 2, 3, 4, 5, 6}}
> E2:= ln(7), v, z, 42, Art, 61, 56, Kr, sin(6):  SSet(E2);
        {42 = 1, ln(7) = 9, v = 4, z = 7, Art = 5, 61 = 3, 56 = 2, Kr = 6, sin(6) = 8}
```

Sub_list – check of a list to be as sublist

Call format of the procedure:

Sub_list(K, L {, 'h'})

Formal arguments of the procedure:

K, L - lists
h - (optional) an unevaluated name

Description of the procedure:

In many problems of operation with data structures of *list*-type, the necessity of definition of the fact of entry of some list into the given list in the form of a *sublist* arises. This problem successfully is solved by means of the **Sub_list** procedure. The procedure call **Sub_list(K, L)** returns the *false* value, if a list, defined by the first actual argument **K**, is no *sublist* or no *element* of a list **L** specified by the second actual argument, otherwise the *true* value is returned. In addition, if the third optional **h** argument has been coded, then through it the 4-element list of a view **[***sublist*, t, *element*, p**]** is returned, where **t** - number of entries of **op(K)** in the form of a *sublist* and **p** - a list of positions of entries of **K** in the form of *elements* (*numbers of elements*) of **L.** The *empty* list as the actual arguments of the procedure causes the erroneous situation.

```
Sub_list:= proc(K::list, L::list)
local T, k, p, h, n, t, v;
  `if`(K = [] or L = [], ERROR(`at least one of actual arguments is empty list`), assign(t = 0));
  assign(T = table([seq(k = L[k], k = 1 .. nops(L))])), assign(p = [RTab(T, K)]);
  assign(h = map(convert, [K, L], string)), assign(v = h[1][2 .. -2]), assign(n = Search2(h[2], {v}));
  seq(`if`(h[2][n[k] - 1] = "[" and h[2][n[k] + length(v)] = "]", 6, assign('t' = t + 1)), k=1..nops(n));
  `if`(t=0 and p=[], false, op([true, `if`(nargs=3, assign('args'[3] = [sublist,t,element,p]), NULL)]))
end proc
```

Typical examples of the procedure use:

```
> K:= [a,b,c]:  L:= [a,b,c,t,[a,b,c],y,5,a,b,c,6,7,8,9,[a,b,c],0,7,14,a,b,c]:  Sub_list(K,L,'h'), h;
                                true, [sublist, 3, element, [5, 15]]
```

TabList – converting of tables into lists, and vice versa

Call format of the procedure:

TabList(T)

Formal arguments of the procedure:

T – a list or table

Description of the procedure:

If **T** is a table then the procedure call **TabList(T)** returns the nested list of 2-elements sublists; the first element of such sublist represents an *index* of the table **T**, whereas the second – an *entry* corresponding to it. On the other hand, if **T** is a list then the call **TabList(T)** returns the table whose *indices* are positions of elements of **T** while *entries* – the list elements corresponding to them.

```
TabList:= proc(T::{list, table})
local k;
  `if`(type(T, list), eval(table([seq(k = T[k], k = 1 .. nops(T))])),
  [seq([op(indices(T)[k]), op(entries(T)[k])], k = 1 .. nops([indices(T)]))])
end proc
```

Typical examples of the procedure use:

> T:= TabList([67, 62, 89, 96, 42, 47, 37, 40, 15, 7, 62.57]);

T:= table([1 = 67, 2 = 62, 3 = 89, 4 = 96, 5 = 42, 6 = 47, 7 = 37, 8 = 40, 9 = 15, 10 = 7, 11 = 62.57])

> TabList(T);

[[1, 67], [2, 62], [3, 89], [4, 96], [5, 42], [6, 47], [7, 37], [8, 40], [9, 15], [10, 7], [11, 62.57]]

`convert/ssll` – converting into listlist/setset

Call format of the procedure:

convert(R, *ssll*)

Formal arguments of the procedure:

R – a listlist or setset

Description of the procedure:

The procedure **convert(R, *ssll*)** provides a converting of **R** structure of a type {*setset* | *listlist*} into structure of a type {*listlist* | *setset*} accordingly.

```
`convert/ssll`:= proc(S::{listlist, setset})
local k, G;
  [`if`(type(S, setset), [assign(G = []), seq(assign('G' = [op(G), [op(S[k])]]), k = 1 .. nops(S))],
  [assign(G = {}), seq(assign('G' = {{op(S[k])}, op(G)}), k = 1 .. nops(S))]), G][2]
end proc
```

Typical examples of the procedure use:

> S:={{2,3,5}, {1,2,3}, {9,5,3}, {2,3,5}, {9,8,5}, {3,9,7}, {42,47,67},{89,96,62}}; L:=convert(S, ssll);

S := {{2, 3, 5}, {3, 5, 9}, {1, 2, 3}, {5, 8, 9}, {3, 7, 9}, {42, 47, 67}, {62, 89, 96}}
L := [[2, 3, 5], [3, 5, 9], [1, 2, 3], [5, 8, 9], [3, 7, 9], [42, 47, 67], [62, 89, 96]]

> S:= convert(L, ssll);

S := {{2, 3, 5}, {3, 5, 9}, {1, 2, 3}, {5, 8, 9}, {3, 7, 9}, {42, 47, 67}, {62, 89, 96}}

expLS – an expanding of lists or sets

Call format of the procedure:

expLS(G)

Formal arguments of the procedure:

G – a set or a list

Description of the procedure:

The procedure call **expLS(G)** returns the list that is a result of disclosing of all levels of sublists and/or subsets that compose the initial set or list, defined by the actual **G** argument. The procedure is useful enough tool at processing of data structures of a type {*set, list*}.

```
expLS:= proc(G::{list, set})
local k, p, Kr, S;
  assign(p = 0, Kr = G, S = []);
  do
    for k to nops(Kr) do
      if type(Kr[k], {list, set}) then p := p + 1; S := [op(S), op(Kr[k])] else S := [op(S), Kr[k]]
      end if
    end do;
    if p <> 0 then Kr := S; p := 0; S := []; next else RETURN(S) end if
  end do
end proc
```

Typical examples of the procedure use:

> expLS([[a,b,c], [1,2], [3, [42,47], 5, 6], {n,m}]), expLS([[[a,b,c],[1, 2]],[3, [42, 47], 5, 6], [{n,m}, 2003]]);

[a, b, c, 1, 2, 3, 42, 47, 5, 6, m, n], [a, b, c, 1, 2, 3, 42, 47, 5, 6, m, n, 2003]

RedList – reducing of lists
redL
redlt

Call format of the procedures:

RedList(L)
redL(G {, t})
redltL(G, p::evaln {, t})

Formal arguments of the procedure:

L – a list with elements of *realnum*-type
G – a list with elements of any valid *Maple* type
p – an evaluated name
t – an arbitrary valid *Maple* expression

Description of the procedures:

In a number of problems of processing of the numerical data there is the necessity of extracting out of list **L** of data of *realnum*-type of a sublist containing unitary entrances of data with preservation of their entry order into the initial list **L**; in addition, the first entrances of the identical data are saved only. The given problem is solved by the procedure **RedList(L)** that returns the required sublist. The procedure returns the initial list **L** if the list does not contain *multiple* entrances of elements.

In contrast to the above procedure, the **redL(G {, t})** procedure admits arbitrary valid *Maple* expressions as elements of an **G** list, returning the reduced list in the above-mentioned sense if the one actual **G** argument has been coded. At a call with two actual arguments **(G, t)** in the returned list the last entrances of the identical elements are saved only.

At last, the procedure **redlt** provides deletion of *multiple* entrances of elements of an **G** list relative to their types. The procedure call **redlt(G, p)** returns the list containing first entrances of the identical (*relative to their type*) elements. In addition, comparison is done on the basis of standard **`*whattype*`** procedure which returns the "*top level*" data type as determined by the order of precedence of the operators. Through the second actual argument **p,** the procedure call returns the list of types appropriate to elements of the main result with bijective correspondence between both lists. Whereas, at coding of the third optional argument **t,** the procedure call **redlt(G, p, t)** does the above comparison relative to types **`eval(h)`**, where **h** – arbitrary element of the list **G.** The given procedure, in particular, is rather useful mean at creation of procedures with optional key formal arguments. The procedures of this group are useful in many appendices.

```
RedList:= proc(L::list(realnum))
local T, k, S, K, G;
  assign(K = [], G = []), seq(assign('T[L[k]]' = 0), k = 1 .. nops(L));
  for k to nops(L) do T[L[k]] := T[L[k]], k end do;
  assign(S = op(2, eval(T))), seq(assign('K' = SLj([op(K), [rhs(S[k])[2], lhs(S[k])]], 1)),
    k = 1 .. nops(S)); seq(assign('G' = [op(G), K[k][2]]), k = 1 .. nops(K)), G
end proc

redL:= proc(L::list)
local a, k, j, nu, omega, t;
  assign(nu = {op(L)}, omega = nlcvector(L), a = convert([7, 14, 26], bytes));
  for k to nops(nu) do memberL(nu[k], L, 't');
    if nops(t) = 1 then next else for j from 2 to nops(t) do omega[t[j]] := a end do end if
  end do;
  [seq(`if`(omega[k] = a, NULL, omega[k]), k = 1 .. rhs(op(2, eval(omega))))]
end proc

redlt:= proc(L::list, p::evaln)
local a, b, k, t, h;
  assign(a = [], b = []);
  for k to nops(L) do  t := whattype(`if`(2 < nargs, eval(L[k]), L[k]));
    if  not member(t, b) then  a := [op(a), L[k]]; b := [op(b), t]  end if
  end do;
  a, assign(p = redL(b))
end proc
```

Typical examples of the procedures use:

```
> L:=[4.7,4.7,-6,-6,9,-6,42.6,7.03,7/14,7/14,3.02,3.02,4.7,20.03,7/14,3.02,-6,42.6,7.03]: RedList(L);
                    [4.7, -6, 9, 42.6, 7.03, 1/2, 3.02, 20.03]
> L:= [a,IAN,61,19.42,"RANS",61,-8/9,(a+b)/(c-d),"RANS",(a+b)/(c-d),IAN,-8/9,IAN, 56,
  "RANS","RANS",a,61,IAN, (a+b)/(c-d),"RANS",(a+b)/(c-d),56,a];
  L := [a, IAN, 61, 19.42, "RANS", 61, -8/9, (a+b)/(c-d), "RANS", (a+b)/(c-d), IAN, -8/9, IAN, 56,
           "RANS", "RANS", a, 61, IAN, (a+b)/(c-d), "RANS", (a+b)/(c-d), 56, a]
> redL(L);
                 [a, IAN, 61, 19.42, "RANS", -8/9, (a+b)/(c-d), 56]
```

> redL(L, 7);

[a, IAN, 61, 19.42, "RANS", -8/9, (a+b)/(c-d), 56]

> R:=rand(7 .. 14): Lr:=[seq(R(), k=1 .. 10000)]: redL(Lr), redL(Lr, 14);

[9, 8, 10, 12, 7, 13, 11, 14], [14, 10, 8, 11, 7, 12, 13, 9]

> redL([[a],b,[c],[a,b],2003,7,[14], sin(x), [14],1995,[2003], agn, [2003],[a,b],[a,b],[2003],[a]]);

[[a], b, [c], [a, b], 2003, 7, [14], sin(x), 1995, [2003], agn]

> h:= table([]): S:= matrix([]): redlt([H, 15, G, H, G, k, S, u, S, -15], t, 7), t;

[*H*, 15, *G*, *S*], [*table, integer, symbol, array*]

> h:= table([]): S:=matrix([]): redlt([H, 15, G, H, G, k, S, u, S, -15], p), p;

[*H*, 15], [*symbol, integer*]

SortL – sorting of symbolic lists

Call format of the procedure:

SortL(L, n, m {, *sf*})

Formal arguments of the procedure:

L – a list with elements of a type {*string, symbol*}
n – a positive integer (*posint*)
m – a positive integer (*posint*)
sf – (optional) a sorting function (boolean procedure of two arguments)

Description of the procedure:

The procedure **SortL(L,n,m {, *sf*})** sorts a list **L** whose elements have a type {*string, symbol*}, concerning the keys determined by positions **n..m** of each element. The sorting algorithm is defined by the optional fourth *sf*-argument (a sorting function); at absence of the fourth argument, a sorting in the lexicographic order is made. The given procedure seems useful enough in many problems of symbolical processing generalizing standard function **`*sort*`** of the package *Maple.*

```
SortL:= proc(L::list({string, symbol}), n::posint, m::posint)
local G, S, k, h, p;
  `if`(m < n, ERROR(`3rd argument is wrong`), assign(G = [], S = [], p = nops(L)));
  for k to p do h := cat("", L[k]); G := [op(G), [L[k], `if`(length(h) <= n, h,
       `if`(length(h) < m, h[n .. -1], h[n .. m]))]]
  end do;
  assign('G' = SLj(G, 2, `if`(nargs = 4, args[4], lexorder))),
       seq(assign('S' = [op(S), G[k][1]]), k = 1 .. p), S
end proc
```

Typical examples of the procedure use:

> L:= [Victor, Galina, Svetlana, Artur, Kristo, Arne, `19aaa42`, `20bbb03`]: SortL(L, 3, 5);

[*19aaa42, 20bbb03, Victor, Svetlana, Kristo, Galina, Arne, Artur*]

---

ssf – an ordering function for sorting

Call format of the procedure:

ssf(x, y, p, d {, t})

Formal arguments of the procedure:

x – the first element of a sorted pair {*string, symbol*}
y – the second element of a sorted pair {*string, symbol*}
p – substring of the pair elements (*delimiter*)
d – a sorting type (may be **0** or **1** only)
t – (optional) a sorting mode (the value *numeric*)

Description of the procedure:

Generally, the standard ***sort*** function admits use of some ordering function **F** that should be a Boolean-valued function of two arguments. Specifically, **F(x, y)** should return the *true* value if **x** precedes **y** in the desired ordering. Otherwise, it should return the *false* value. For a series of cases of the lists sorting and manipulations with *TEXT* datafiles, the following ordering function can be useful. In particular, the procedure is useful at *TEXT* datafiles processing.

The procedure call **ssf(x, y, p, d {, t})** returns the *true* value if the left (**d=0**) or the right (**d=1**) substring of an **x** string parted by a **p** delimiter precedes the left (**d=0**) or the right (**d=1**) substring of a **y** string parted by a **p** delimiter in the desired ordering. Otherwise, it should return the *false* value. At absence of the fourth actual **t** argument, the lexicographic ordering is made. If the ***`numeric`*** word is coded as the fourth actual argument the numerical ordering is made. The given procedure is used in a series of rather useful procedures represented in the book that deal with symbolic processing and the access to *TEXT* datafiles. Namely, on the basis of the procedure a series of the useful ordering functions can be implemented.

```
ssf:= proc(x::{string, symbol}, y::{string, symbol}, p::{string, symbol}, d::{0, 1})
local a, b, n, nu, k;
  assign(n = map(Search, [x, y], p), nu = length(p)), assign(a = ((d, k) -> `if`(d = 0,
    `if`(n[k][1] = 1, nu + 1, 1), `if`(n[k][1] + nu = length(args[k]) + 1, 1, n[k][1] + nu))),
    b = ((d, k) -> `if`(d = 0, `if`(n[k][1] + nu = length(args[k]) + 1, nu - 1, -1),
    `if`(n[k][1] + nu = length(args[k]) + 1, nu - 1, -1)))), seq(assign(cat(_, k) =
    cat("", args[k])[a(d, k) .. b(d, k)]), k = 1 .. 2);
  `if`(sort(`if`(member(numeric, {args}), map(SD, [_1, _2]), [_1, _2]))[1] =
    `if`(member(numeric, {args}), SD(_1), _1), true, false), unassign('_1', '_2')
end proc
```

Typical examples of the procedure use:

> L:= [Victor, Galina, Svetlana, Artur, Kristo, Arne, `19aaa42`, `20bbb03`]: sort(L);

[19aaa42, 20bbb03, Arne, Artur, Galina, Kristo, Svetlana, Victor]

> L1:= [`Victor###4`, `Galina###1`, `Svetlana###2`, "Artur###7", "Kristo###3", `Arne###8`, `19aaa42###5`, `20bbb03###6`]: sf1:= (x,y) -> ssf(x, y, "###", 1, numeric): sort(L1, sf1);

[*Galina###1, Svetlana###2,* "Kristo###3", *Victor###4, 19aaa42###5, 20bbb03###6,* "Artur###7", *Arne###8*]

> sf2:= (x,y) -> ssf(x, y, "###", 0): sort(L1, sf2);

[*19aaa42###5, 20bbb03###6, Arne###8,* "Artur###7", *Galina###1,* "Kristo###3", *Svetlana###2, Victor###4*]

> sf3:= (x,y) -> ssf(x, y, "#", 0): sort(L1, sf3);

[*19aaa42###5, 20bbb03###6, Arne###8,* "Artur###7", *Galina###1,* "Kristo###3", *Svetlana###2, Victor###4*]

> sf4:= (x,y) -> ssf(x, y, "#", 1): sort(L1, sf4);

[*Galina###1*, *Svetlana###2*, "Kristo###3", *Victor###4*, *19aaa42###5*, *20bbb03###6*, "Artur###7", *Arne###8*]

---

subseqn – exhaustive substitutions into lists or sets

Call format of the procedure:

subseqn(S, L {, r})

Formal arguments of the procedure:

S - a list or a set with elements of a type {`=`, *`equation`*}
L - a list or a set
r - (optional) a type {*relation, range*}

Description of the procedure:

The procedure call **subseqn(S, L {, r})** returns the result of exhaustive substitutions determined by the equations of the form **v=w** from a list or a set **S,** into a list or a set **L.** In addition, substitutions are made as follows: **(1)** the list elements whose type is distinct from a type given by optional third argument **r** {*relation, range*} or whose left members are distinct from **v,** remain without change; **(2)** if the left member of an element of list **L** coincides with **v** and **w** <> *NULL*, the right member of the element of **L** is replaced on **w; (3)** if **w=***NULL* the corresponding element of **L** is removed. The third optional argument **r** defines one of types {`=`, `<`, ` <=`, `<>`, `..`, *range, relation, equation*} or their list or set. At absence of the third argument, any of the above types is supposed.

```
subseqn:= proc(S::{set(`=`), list(`=`)}, L::{set, list})
local k, p, K, G, r;
  assign(K = L, G = [], r = {`..`, `=`, `<`, `<=`, `<>`, equation, relation, range});
  `if`(nargs = 3 and op(map(belong, args[3], r)), assign('r' = args[3]), NULL);
  for k to nops(S) do
    for p to nops(K) do G := [op(G), `if`(not type(K[p], r), K[p], `if`(lhs(S[k]) = lhs(K[p]) and
      rhs(S[k]) <> NULL, whattype(K[p])(lhs(K[p]), rhs(S[k])), `if`(lhs(S[k]) = lhs(K[p]) and
      rhs(S[k]) = NULL, NULL, K[p])))]
    end do;
    assign('K' = G, 'G' = [])
  end do;
  `if`(type(L, list), K, {op(K)})
end proc
```

Typical examples of the procedure use:

> subseqn({a=NULL,v=42,g=47,s=67,h=NULL,v=2003}, [m,v=67,g=23,b,a=61,v=1942,s=36, s=88, h=77, h=2003, a=45]);

[m, v = 2003, g = 47, b, v = 2003, s = 67, s = 67]

> subseqn({a=NULL,v=42,g=47,s=67,h=NULL,v=2003}, {m,v=67,g=23,h..32,s=88,h=77}, {range, equation});

{m, g = 47, s = 67, v = 2003}

> subseqn({a=NULL,v=62,g=57,s=37,t=NULL,v=2004}, [t,v=67,g=23,b,s=88,h=77]);

[t, v = 2004, g = 57, b, s = 37, h = 77]

NDP – evaluation of decimal digits of a *realnum*-number

Call format of the procedure:

NDP(n, p {, 'h'})

Formal arguments of the procedure:

n - an integer of *realnum*-type
p - an integer
h - (optional) an unevaluated name

Description of the procedure:

The procedure call **NDP(n, p)** returns the decimal digit being on the **p**th position of a number **n** of *realmum*-type; in addition, **p < 0** defines number of position to the left of decimal point, whereas **p > 0** defines number of position to the right of decimal point. If the optional third argument **h** has been coded, then via it the nested 2-element list is returned. The first sublist defines decimal digits of the number **n** to the left of decimal point including its sign (-), whereas the second sublist defines decimal digits to the right of decimal point.

```
NDP:= proc(n::realnum, p::integer)
local a, b, c, k, t, m;
   assign(m = convert(_0N(evalf(n)), string), b = []); `if`(m[1] = "-", [assign('m' = m[2 .. -1],
   a = [`-`])], assign(a = [])); assign(c = search(m, ".", 't')), seq(assign('a' = [op(a), SD(m[k])]),
   k = 1 .. t – 1), seq(assign('b' = [op(b), SD(m[k])]), k = t + 1 .. length(m));
   try `if`(nargs = 3, assign('args'[3] = [a, b]), NULL), `if`(p < 0, a[p], `if`(0 < p, b[p],
      ERROR(`2nd argument <%1> is invalid`, p)))
   catch "invalid subscript selector": RETURN(0)
   end try
end proc
```

Typical examples of the procedure use:

> Digits:=12: NDP(-2003*6*14*Pi*Catalan*gamma, -6, 'h'), h;

2, [[-, 2, 7, 9, 4, 6, 4], [9, 6, 9, 6, 2, 2]]

> Digits:=14: NDP(-2003*7*14*Pi*Catalan*gamma, -7, 'h'), h;

-, [[-, 3, 2, 6, 0, 4, 2], [4, 6, 4, 5, 5, 8, 8, 6]]

> evalf(-2003*7*14*Pi*Catalan*gamma);

-326042.46455886

A series of procedures of the present chapter supports useful kinds of processing of *Maple* objects of types {*set, list, table*}. The given tools are rather useful at operation with objects of the above types in the *Maple* environment. Usage of these tools in numerous appendices of different purpose has confirmed their high enough performance, essentially simplifying the programming in a lot of cases. In a series of cases, they have shown oneself as useful enough tools.

Chapter 6. Software to support data structures of a special type

This section contains the extended tools of work with data structures such as *stack, queue, priqueue,* etc. Along with standard data structures of types {*stack, queue, heap, priqueue*}, a new *dirax*-type maintained by program module **DIRAX** has been defined. The module supports data structures of direct access of a *dirax*-type that by a series of parameters essentially extends data structures of types {*stack, heap, queue*}. The data structures of *dirax*-type are founded on a list organization; however allow to address itself immediately to its elements according to their addresses (numbers). Moreover, in contrast to structures of a type {*stack, queue*}, the *dirax* structures are precisely identified by *ispd*-index; a reception, used in the module, allows easily to define precise classification of data structures of a type {*stack, queue, heap, dirax*}. At the same time, the **DIRAX** module provides an effective solution of problems of shared use of data structures of all above-mentioned types. At last, tools of testing of the above data structures types have been essentially precised, because the standard tools have some inaccuracies.

STACK – stack data structures

Call format of the procedure:

STACK(O, *args*)

Formal arguments of the procedure:

O – a name from the set {new, push, top, depth, empty, pop}
args – the rest of actual arguments

Description of the procedure:

The ***stack*** is a data structure widely used in the computer science for storage of objects which should be processed by the principle *`last-in, first-out`*, and for creation on its basis of very reactive programming languages. In addition, any passing of the information does not happen, contents of the register – of the *stack* pointer vary only.

It is known [10-11], the *stack* structure represents a special type of a table what allows to work with it both by tools of the ***stack*** function and by the language tools oriented onto work with the table objects. The **STACK** procedure is equivalent to the ***stack*** function, but has been implemented by exclusively functional tools of the *Maple* language for work with tables. At programming of the procedure, a series of useful nonstandard receptions has been used, allowing to program a series of problems in the *Maple* environment by more mathematically clearly and by more effective during their running. A number of them are approved on many applied and system problems [9,29-32].

The **STACK** procedure admits more than one actual **O** argument – an identifier of a *stack* operation; as the second actual argument the identifier of the *stack* or a sequence of elements of the *stack* can appear, whereas as the third actual argument a separate element of the ***stack*** appears. The set of the operations admitted for the procedure **STACK** as a value for the argument **O** is analogous to the standard ***stack*** procedure, namely: *new, depth, push, top, empty* and *pop.*

```
STACK:= proc(O::name)
local k, St;
  `if`(O = new, `if`(nargs = 1, [assign(St = table([0 = 0])), RETURN(eval(St))],
     [assign(St = table([0 = nargs - 1, seq(k - 1 = args[nargs - k + 2], k = 2 .. nargs)])),
    RETURN(eval(St))]), `if`(type(args[2], table) and indices(args[2])[1] = [0],
    assign(St = args[2]), ERROR(`STACK expects 2nd argument to be of stack-type, but had
    received <%1>`, convert(args[2], symbol))));
  if O = push then [assign('St'[0] = St[0] + 1), assign('St'[St[0]] = args[3]), RETURN(args[3])]
  elif O = top then RETURN(St[St[0]])
  elif O = depth then RETURN(St[0])
  elif O = empty then RETURN(`if`(St[0] = 0, true, false))
  elif O = pop then
    writebytes(`G$`, cat(`St[`, convert(St[0], symbol), `]:='St[`, convert(St[0], symbol), `]':`));
    [close(`G$`), assign('St' = args[2]), assign('k' = St[St[0]], 'St'[0] = St[0] - 1)];
    read `G$`;
    [fremove(`G$`), RETURN(k)]
  elif O <> new then ERROR(`The 1st argument <%1> for STACK is incorrect`, O)
  end if
end proc
```

Typical examples of the procedure use:

> Kristo:= STACK(new, 61,56,36,40,7,14); STACK(depth, Kristo), STACK(push, Kristo, 2003), eval(Kristo), STACK(empty, Kristo), STACK(top, Kristo), STACK(pop, Kristo), eval(Kristo);

Kristo := table([0 = 6, 1 = 14, 2 = 7, 3 = 40, 4 = 36, 5 = 56, 6 = 61])
6, 2003, table([0 = 6, 1 = 14, 2 = 7, 3 = 40, 4 = 36, 5 = 56, 6 = 61, 7 = 2003]), false, 2003, 2003,
table([0 = 6, 1 = 14, 2 = 6, 3 = 40, 4 = 36, 5 = 56, 6 = 61, 7 = 2003])

Queue – queue data structures

Call format of the procedure:

Queue(O, *args*)

Formal arguments of the procedure:

O – a name from the set {new, enqueue, dequeue, inqueue, front, empty, clear, length, reverse}
args – the rest of actual arguments

Description of the procedure:

The ***queue*** – a data structure used widely in the computer science for storage of objects that should be processed by principle *"first in, first out"*. In contrast to the *stack* data structure, a table represents the queue data structure; the order of elements of the table (*queue*) corresponds to the order of elements at its creation. Tools intended for operation with table data structures have implemented the Queue procedure.

The **Queue** procedure admits more than one actual **O** argument – an identifier of *queue* operation; as the second actual argument an identifier of an *queue* or a sequence of elements of an *queue* can appear, whereas as the third actual argument a separate element of an *queue* appears. The set of operations admitted for the **Queue** procedure as a value for the actual argument **O** is analogous to the standard *`queue`* procedure, namely: *new, dequeue, empty, front, length, reverse, clear, enqueue.*

In addition, in addition to standard operations with queues the **Queue** procedure admits a new *`inqueue`* operation, for which the call **Queue(*inqueue*, Q, b)** returns the number in a **Q** queue of the indicated **b** element or at its absence the *NULL* value is returned, i.e. nothing.

```
Queue:= proc(O::name)
local k, h, Q;
  `if`(O = new, `if`(nargs = 1, [assign('Q' = table([0 = 0])), RETURN(eval(Q))],
     [assign(Q = table([0 = nargs - 1, seq(k - 1 = args[k], k = 2 .. nargs)])),
     RETURN(eval(Q))]), `if`(type(args[2], table) and indices(args[2])[1] = [0],
     assign(Q = args[2]), ERROR(`The Queue expects 2nd argument to be of queue-type,
     but had received <%1>`, convert(args[2], symbol))));
  if O = enqueue then [assign('Q'[Q[0]+1] = args[3]), assign('Q'[0] = Q[0]+1), RETURN(args[3])]
  elif O = clear then [assign('k' = Q[0], 'Q' = table([0 = 0])), RETURN(k)]
  elif O = front then RETURN(`if`(Q[0] <> 0, Q[1], `Queue is empty`))
  elif O = length then RETURN(Q[0])
  elif O = reverse then RETURN(table([0 = Q[0], seq(k = Q[Q[0] - k + 1], k = 1 .. Q[0])]))
  elif O = empty then RETURN(`if`(Q[0] = 0, true, false))
  elif O = dequeue then `if`(Q[0] <> 0, RETURN(Q[1], table([0 = Q[0] - 1, seq(k - 1 = Q[k],
     k = 2 .. Q[0])])), RETURN(`Queue is empty`))
  elif O = inqueue then RETURN(seq(`if`(seq(map(rhs, op(eval(Q)))[k], k = 2 .. Q[0] + 1)[h] =
     args[3], h, NULL), h = 1 .. Q[0]))
  elif O <> new then ERROR(`The 1st argument <%1> for Queue is incorrect`, O)
  end if
end proc
```

Typical examples of the procedure use:

```
> Q:= Queue(new, 61,56,36,40,7,14); Q:= Queue(reverse, Q); Q:= Queue(dequeue, Q)[2],
  Queue(empty, Q), Queue(length, Q), Queue(front, Q), Queue(enqueue, Q,99), eval(Q);
```

Q := table([0 = 6, 1 = 61, 2 = 56, 3 = 36, 4 = 40, 5 = 7, 6 = 14])
Q := table([0 = 6, 1 = 14, 2 = 7, 3 = 40, 4 = 36, 5 = 56, 6 = 61])
Q := table([0 = 5, 1 = 7, 2 = 40, 3 = 36, 4 = 56, 5 = 61]), false, 6, 14, 99,
table([0 = 7, 1 = 14, 2 = 7, 3 = 40, 4 = 36, 5 = 56, 6 = 61, 7 = 99]), 6

A rather useful version of *queue* data structures is the *heap*-structure (the *ordered queue*), supported by the built-in ***heap***-function of the package. In this connection, we shall mention only two useful enough procedures **T_SQHT(T)** and **Heap(T,SF)** where the first is the testing procedure and returns a type of structure defined by its first **T**-argument. Whereas the second procedure returns a *heap* data structure received on the basis of reorganization of the structure determined by the first actual **T**-argument of a type{*stack* | *queue* | *heap*}, according to an ordering procedure determined by the second actual **SF**-argument. Hence, the given procedure provides both the reordering of the initial *heap* data structure, and converting of any of the specified data structures into *heap* data structures.

Heap – reorganization of a stack, an queue or a heap

Call format of the procedure:

Heap(T, SF)

Formal arguments of the procedure:

T - a stack, an queue or a heap
SF - a symbol (name of a *ranging function*)

Description of the procedure:

The procedure call **Heap(T, SF)** returns the *heap* data structure obtained on the basis of reorganization of a {*stack* | *queue* | *heap*} data structure, defined by the first actual **T** argument, according to a ranging function defined by the second actual **SF** argument. As the **SF** argument any Boolean function **B(x,y)** is admitted which returns the *true* value if **x** and **y** satisfy to the given order, otherwise the *false* value is returned.

```
Heap:= proc(T::{stack, heap, queue}, SF::{name, symbol})
local k, h;
  table([`<` = SF, 0 = T[0], seq(T[0] - k + 1 = sort([seq(T[h], h= 1 .. T[0])], SF)[k], k = 1 .. T[0])])
end proc
```

Typical examples of the procedure use:

```
> assign(Q = Queue(new, 61,56,36,40,7,14)), assign(H = Heap(Q, `<`)), eval(Q), eval(H),
  type(H, heap);

   table([0 = 6, 1 = 61, 2 = 56, 3 = 36, 4 = 40, 5 = 7, 6 = 14]), table([0 = 6, 1 = 61, 2 = 56, 3 = 40,
                        4 = 36, 5 = 14, 6 = 7, `<` = `<`]), true

> assign(Kr=STACK(new, 61,56,36,40,7,14)), assign(H1=Heap(Kr, `>`)), eval(Kr), eval(H1),
  type(H1, heap);

   table([0 = 6, 1 = 14, 2 = 7, 3 = 40, 4 = 36, 5 = 56, 6 = 61]), table([0 = 6, 1 = 7, 2 = 14, 3 = 36,
                        4 = 40, 5 = 56, 6 = 61, `<` = `>`]), true
```

As shown in our books [12,13], the standard `***type***` function does not provide due testing of data structures of a type {*stack, queue, heap*}. The reason of it is the table organization common for all these structures. Meanwhile, the following simple procedure **T_SQHD** allows to classify the *Maple* tables upon *heap*-structures and {*queue, stack*}-structures, whereas lists - upon *list*-structures and *dirax*-structures, considered below.

T_SQHD - check of an object to have a type dirax, list, table, heap, stack or queue

Call format of the procedure:

T_SQHD(T)

Formal arguments of the procedure:

T - a table or a list

Description of the procedure:

The standard function `***type***` does not provide a proper testing of data structures of a type {*stack, queue, heap*}. The reason is in the table organization common for all these structures. The simple **T_SQHD** procedure allows to classify the *Maple* tables upon *heap* structures and {*stack, queue*} structures, whereas the lists upon *dirax*-structures and *list*-structures. The procedure call **T_SQHD(T)** returns *dirax, list, table, heap, stack* or *queue* depending on a type of data structure defined by the actual **T** argument.

```
T_SQHD:= proc(T::{table, list})
local a;
  `if`(type(T, list), `if`(T[1] = dirax, dirax, list), op([ assign('a' = op(eval(T))), `if`(type(T, stack),
  `if`(lhs(a[-1]) = `<`, heap, `stack or queue`), table)]))
end proc
```

Typical examples of the procedure use:

> A:= [dirax, 6, [V,G,S,Ar,Art,Kr]]: H:= heap[new](lexorder, g,s,v): Kr:= [a,b,c]: L:= stack[new](1,9,9,5): Tab:= table([1=V,2=G,3=S]): map(T_SQHD, [A,Kr,Tab,H,L]);

[*dirax, list, table, heap, stack or queue*]

T_SQHT – check of a table, a stack, an queue or a heap

Call format of the procedure:

T_SQHT(T)

Formal arguments of the procedure:

T - a table, stack, queue or heap

Description of the procedure:

The procedure call **T_SQHT(T)** returns the type of a data structure, defined by the actual **T** argument. The detailed elaboration of a type testing is supported on the level of the types *heap, table* and {*stack, queue*}. In case of a *heap* data structure along with its type a ranging function is returned as the second element of the returned list.

```
T_SQHT:= T::{table, heap, stack, queue} -> `if`(whattype(T[`<`]) = symbol, [heap, T[`<`]],
            `if`(type(T[0], integer), [stack, queue], table))
```

Typical examples of the procedure use:

> assign(Q = Queue(new, 61,56,36,40,6,14)), assign(H = Heap(Q, `<`)), eval(Q), eval(H), type(H, heap);

table([0 = 6, 1 = 61, 2 = 56, 3 = 36, 4 = 40, 5 = 7, 6 = 14]), table([0 = 6, 1 = 61, 2 = 56, 3 = 40, 4 = 36, 5 = 14, 6 = 7, `<` = `<`]), *true*

> assign(Kr=STACK(new, 61, 56, 36, 40, 7, 14)), assign(H1=Heap(Kr, `>`)), eval(Kr), eval(H1), type(H1, heap);

table([0 = 6, 1 = 14, 2 = 7, 3 = 40, 4 = 36, 5 = 56, 6 = 61]), table([0 = 6, 1 = 7, 2 = 14, 3 = 36, 4 = 40, 5 = 56, 6 = 61, `<` = `>`]), *true*

> T:= table([v=61,g=56,s=36,ar=40,art=14,kr=7]): map(T_SQHT, [Q, H, Kr, H1, T]);

[[*stack, queue*], [*heap*, <], [*stack, queue*], [*heap*, >], *table*]

DIRAX – a data structure of direct access

Call formats of the variables exported by the module:

DIRAX:- X(*args*) or **with(DIRAX): X(*args*)**

Formal parameters of the module call:

X - a name of function exported by the **DIRAX** module
args - formal arguments corresponding to a function exported by the module

Description of the module:

The module **DIRAX** supports a new type of data structures of direct access. Along with the standard data structures of types {*stack, queue, heap, priqueue*}, the new *dirax*-type supported by the module **DIRAX** has been defined. The module supports a data structure of direct access of the *dirax*-type, which by a series of positions essentially expands data structures of types {*queue, stack, heap*}. The *dirax* data structure is founded on a list organization however allows to address itself immediately to its elements according to their addresses (*numbers*). Moreover, in contrast to structures of type {*stack, queue*}, the *dirax* structure is precisely identified by the *ispd*-index; the reception, used in the module, allows easily to determine precise classification of data structures of types {*stack, queue, heap, dirax*}. At the same time, the **DIRAX** module provides an effective solution of problems of shared use of data structures of all above-mentioned types.

Such *dirax* structure represents a list structure organized by the special manner, namely:

[*dirax, number of elements*, [*element_1, element_2, ... , element_N*]]

The functional tools of the **DIRAX** module allow to create such structures, to test them, to work with them at a level of a specific element, to sort their, etc. The given **DIRAX** module exports the following twelve functional tools that support a work with *dirax* data structures, namely:

- ***new***($a_1,a_2,..,a_n$) – creation of a new structure *dirax* with elements a_k **(k=1 .. n);**
- ***type***(**D**) – testing of a **D**-object onto its belonging to the *dirax*-type;
- ***replace***(**D,n,a**) – replacement of the **n**th element of *dirax*-structure **D** onto **`a`;**
- ***extract***(**D,n**) – extraction of the **n**th element out of *dirax*-structure **D;**
- ***empty***(**D**) – the *true* value, if *dirax*-structure **D** is empty, and the *false* value otherwise;
- ***size***(**D**) – number of elements of *dirax*-structure **D;**
- ***reverse***(**D**) – inversion of elements of *dirax*-structure **D;**
- ***insert***(**D,n,a**) – insertion of an element **`a`** into *dirax*-structure **D** after its **n**th element;
- ***delete***(**D,n**) – deleting out of *dirax*-structure **D** of its **n**th element;
- ***sortd***(**D,F**) – sorting of *dirax*-structure **D** according to a ranging **F**-function;
- ***printd***(**D**) – return of elements of *dirax*-structure **D;**
- ***conv***(**D,T**) – converting of structure **D** into structure of a type **T=**{*set* | *list*}**.**

The call of any of the above exported function returns the result according to the presented description and any additional comments are not demanded. Along with that, the **DIRAX** module via global variable **`*type/dirax*`** defines in a current session a new *dirax* data type tested by the extended function **`*type*`** of the package *Maple*.

```
DIRAX:=module () local Wrgn: global `type/dirax`: description `Module for support of simple
  data structure with direct access = Tallinn - Vilnius 18.10.2001`: options package,
  `CopyRight © RANS. All rights reserved.`, load = Wrgn:
  export new, replace, extract, empty, size, reverse, insert, delete, sortd, printd, conv:
  `type/dirax`:=(L::anything) -> type(L,list) and evalb(L[1]=`dirax`) and evalb(L[2]=nops(L[3])):
  new:= () -> [`dirax`, nargs, [args]]:
  replace:= proc(L::dirax, n::posint, a::anything) try if 1<=n and n<=L[2] then [L[1], L[2],
    [op(L[3][1..n-1]),a,op(L[3][n+1..L[2]])]] else ERROR("Member") end if
    catch "Member": Wrgn(L,n) end try end proc:
  extract:= proc(L::dirax,n::posint) try if 1<=n and n<=L[2] then L[3][n] else
    ERROR("Member") end if catch "Member": Wrgn(L,n) end try end proc:
  empty:= (L::dirax) -> `if`(L[2]=0, true, false):
  size:= (L::dirax) -> L[2]:
  reverse:= (L::dirax) -> `if`(L[2]<2, L, [L[1], L[2], [seq(L[3][L[2]-k+1], k=1..L[2])]]):
```

```
   delete:= (L::dirax, n::posint) -> `if`(n>=1 and n<=L[2], [L[1], L[2]-1, [seq(`if`(k<>n, L[3][k],
      NULL), k=1..L[2])]], Wrgn(L,n)):
   printd:= (L::dirax) -> L[3]:
   sortd:= (L::dirax, h) -> `if`(L[2] >= 2, [L[1], L[2], sort(L[3], h)], L):
   conv:= (L::dirax, T::type('set', list)) -> convert(L[3], T):
   insert:= (L::dirax, n::posint, a::anything) -> `if`(n>=1 and n<=L[2], [L[1], L[2]+1,
      [op(L[3][1..n]), a, op(L[3][n+1..L[2]])]], Wrgn(L,n)):
   assign('Wrgn'= (() -> `if`(nargs<>2, NULL, WARNING("Element with number %1 in
      dirax-structure %2 does not exist", args[2], args[1])))):
end module;

DIRAX := module ()
local Wrgn;
export new, replace, extract, empty, size, reverse, insert, delete, sortd, printd, conv;
global `type/dirax`;
options package, `CopyRight © RANS. All rights reserved.`, load = Wrgn;
description `Module for support of simple data structure with direct access =
   Tallinn - Vilnius 18.10.2001`;
end module
```

Typical examples of the module use:

```
> D_st:= DIRAX:- new(61, 56, 36, 40, 7, 14, 42, 47, 67, 62, 96, 89);
```

D_st := [*dirax*, 12, [61, 56, 36, 40, 7, 14, 42, 47, 67, 62, 96, 89]]

```
> DIRAX:-extract(D_st,8),DIRAX:-empty(D_st),DIRAX:- extract(D_st,58), DIRAX:-
delete(D_st,6);
```

Warning, Element with number 58 in dirax-structure [dirax, 12, [61, 56, 36, 40, 7, 14, 42,47, 67, 62, 96, 89]] does not exist

47, *false*, [*dirax*, 11, [61, 56, 36, 40, 7, 42, 47, 67, 62, 96, 89]]

```
> DIRAX:- replace(D_st,5, Kristo), DIRAX:- sortd(D_st,`<`), DIRAX:- printd(D_st);
```

[*dirax*, 12, [61, 56, 36, 40, *Kristo*, 14, 42, 47, 67, 62, 96, 89]], [*dirax*, 12, [7, 14, 36, 40, 42, 47, 56, 61, 62, 67, 89, 96]], [61, 56, 36, 40, 7, 14, 42, 47, 67, 62, 96, 89]

```
> DIRAX:- reverse(D_st), DIRAX:- conv(D_st, set), type(D_st, dirax), with(DIRAX);
```

[*dirax*, 12, [89, 96, 62, 67, 47, 42, 14, 7, 40, 36, 56, 61]], {7, 14, 36, 40, 42, 47, 56, 61, 62, 67, 89, 96}, *true*, [*conv, delete, empty, extract, insert, new, printd, replace, reverse, size, sortd*]

```
> with(DIRAX): Svetla:= new(62,57,37,42,8,15,42,47,67,62,96,89): sortd(Svetla, `>`);
```

[*dirax*, 12, [96, 89, 67, 62, 62, 57, 47, 42, 42, 37, 15, 8]]

SQHD - converting of a table or a list into stack, queue, heap or dirax

Call format of the procedure:

SQHD(T, t {, SF})

Formal arguments of the procedure:

T - a table or a list
t - a symbol (name of a type {*stack, queue, heap, dirax*})
SF - (optional) a symbol (*name of a ranging function*)

Description of the procedure:

The useful procedure **SQHD(T, t)** provides a converting of any **T**-table or **T**-vector into data structure of a type **t::**{*stack, queue, heap ,dirax*}. This procedure represents an useful enough tool at operation with many types of data structures.

The procedure call **SQHD(T, t)** returns the result of converting of a table or a list defined by the first actual **T** argument into data structure of a type defined by the second actual **t** argument; in addition for **t** the types {*stack, queue, heap, dirax*} are admitted.

The third optional **SF** argument of the procedure defines a ranging function of elements for the created *heap* structure. As **SF** argument any Boolean function **B(x, y)** is admitted, which returns the *true* value, if **x** and **y** satisfy to the given order, otherwise the *false* value is returned.

The **SQHD** procedure has useful enough appendices at problem solving which deal with data structures of the above types, namely: *stack, queue, heap, dirax, table,* and *list.*

```
SQHD:= proc(T::{table, list}, t::{symbol, name})
local O, k, Wrng;
  `if`(member(t, {heap, stack, queue, dirax}), NULL, ERROR("%1 - inadmissible type", t));
  `if`(type(T, table), assign(O = op(eval(T))), assign(O = nops(T)));
  Wrng:= (T, N) -> RETURN(WARNING("The object %1 is a %2", T, N));
  `if`(type(O, integer), `if`(T[1] = dirax and t = dirax, Wrng(T, t), `if`(T[1] = dirax,
     RETURN(t[new](`if`(t = heap, `if`(nargs = 2, lexorder, args[3]), NULL), op(T[3])))),
     `if`(t <> dirax, RETURN(t[new](`if`(t = heap, `if`(nargs = 2, lexorder, args[3]), NULL),
     op(T))), RETURN([dirax, nops(T), T])))), NULL);
  `if`(type(T, stack) and t = dirax and lhs(O[-1]) = `<`, RETURN([dirax, nops(O) - 2,
     [seq(rhs(O[k]), k = 2 .. nops(O) - 1)]]), `if`(type(T, stack) and t = dirax, RETURN([dirax,
     nops(O) - 1, [seq(rhs(O[k]), k = 2 .. nops(O))]]),
  `if`(t = dirax, RETURN([dirax, nops(O), [seq(rhs(O[k]), k = 1 .. nops(O))]]), NULL)));
  `if`(type(T, stack), `if`(lhs(O[-1]) = `<`, Wrng(T, heap), Wrng(T, `stack or queue`)),
   t[new](`if`(t = heap, `if`(nargs = 2, lexorder, args[3]), NULL), seq(rhs(O[k]), k = 1 .. nops(O))))
end proc
```

Typical examples of the procedure use:

```
> restart; L:=[dirax,6,[V,G,S,Ar,Art,Kr]]: Tab:=table([1=V,2=G,3=S, 4=Ar,5=Art,6=Kr]):
  L1:=[61, 56, 36,40,7,14]: assign(Dx=SQHD(Tab, dirax), H=SQHD(Tab, heap),
  St=SQHD(L, stack), H1=SQHD(L1, heap, `>`));
> SQHD(St, heap), op(map(eval,[Dx,H,St,H1])), type(Dx, dirax), type(H, heap),
  type(St, stack), type(H1, heap);
   Warning, The object St is a stack or queue
        [dirax, 6, [V, G, S, Ar, Art, Kr]], table([0 = 6, 1 = V, 2 = G, 3 = S, 4 = Ar, 5 = Art,
        6 = Kr, `<` = lexorder]), table([0 = 6, 1 = V, 2 = G, 3 = S, 4 = Ar, 5 = Art, 6 = Kr]),
      table([0 = 6, 1 = 7, 2 = 40, 3 = 14, 4 = 61, 5 = 56, 6 = 36, `<` = `>`]), true, true, true, true
```

In view of told and of examples of application of the module **DIRAX** it is easily to convince oneself, what the module really supports the data structure of direct access of *dirax*-type, which by a number of positions essentially extends types {*stack, queue, heap*} of data structures supported by the package. The data structure such as *dirax* is based not on *table*-organization, but on *list*-organization, but allows to work with its elements, providing direct addressing to them by means of their addresses (*numbers*). At the same time, the **DIRAX** module provides an effective decision of questions of sharing of data structures of all types considered above.

At last, the procedures **`type/dirax`** and **`type/heap`** provide correct testing of data structures of a type {*dirax, heap*} with return on them of the *true* value accordingly and the *false* value otherwise. In particular, the package procedure **`type/heap`** incorrectly fulfills testing. The reason of that lies in its erroneous algorithm.

`type/dirax` – check of an object to be the *dirax*-structure

Call format of the procedure:

type(L, *dirax*)

Formal arguments of the procedure:

L – any valid *Maple* expression

Description of the procedure:

The procedure call **type(L, *dirax*)** returns the *true* value if **L** is a *dirax*-structure; otherwise, the *false* value is returned. The procedure provides rigorous testing of *dirax* data structures basing on use of *dirax*-index that identifies list structures of such type. The approach used for identifying of *dirax*-structures provides their strict testing and the use simplicity in contrast to standard data structures.

```
`type/dirax`:= L::anything -> type(L, list) and evalb(L[1] = dirax) and evalb(L[2] = nops(L[3]))
```

Typical examples of the procedure use:

> T:= table([0=6,1=`IAN`,2=`REA`,3=`RAC`,4=`RANS`,5=Art,6=Kr]): H:= heap[new] (lexorder, Kristo, Artur, Sveta,Galina,Arne,Victor): type(T, dirax), type(H, dirax), type(H, heap);

false, false, true

> A:=[dirax, 6,[V,G,S,Ar,Art,Kr]]:Dr:=DIRAX:-new(61,56,36,40,14,6): type(A, dirax),type(Dr,dirax);

true, true

`type/heap` – check of an object to be the *heap*-structure

Call format of the procedure:

type(x, *heap*)

Formal arguments of the procedure:

x – any valid *Maple* expression

Description of the procedure:

The testing of *heap* data structures by means of the standard ***type***-function is not quite reliably, as it illustrates a rather simple example represented below. The reason is in the erroneous algorithm of the standard procedure **`type/heap`.** The error of the standard procedure consists in that, what it does not do a check of the tested table on presence in it of the index **`<`** for a Boolean procedure of ranging of elements of an queue. The following procedure of the same name eliminates this shortage. The procedure call **type(x, *heap*)** returns the *true* value, if **x** is a *heap*-structure; otherwise, the *false* value is returned.

```
`type/heap`:= proc(T::anything)
option `Copyright (c) 2000 by the International Academy of Noosphere. All rights reserved.`;
  `if`(type(T, table) and member([`<`], {indices(T)}), true, false)
end proc
```

Typical examples of the procedure use:

```
> restart; libname:= "C:\\Program Files\\Maple 9/lib": interface(verboseproc=3):
  eval(`type/heap`);                                  # source code of the Maple procedure

    proc(h)
    option `Copyright © 1990 by the University of Waterloo. All rights reserved.`;
      type(h, 'table') and type(h[0], 'integer')
    end proc

> T:= table([0=6, 1=`IAN`, 2=`REA`, 3=`RAC`, 4=`RANS`, 5=Art, 6=Kr]):
  H:= heap[new](lexorder, Kristo, Artur, Sveta, Galina, Arne, Victor): type(T, heap),
  type(H, heap), type(T, table), type(H, table);

                              false, true, true, true
```

DagTag – mapping between IDs of internal representation of data structures and their names

Call format of the procedure:

DagTag()

Formal arguments of the procedure: **no**

Description of the procedure:

In the internal representation of *Maple* objects they are identified by the *object identifiers* (**IDs**) representing the type of the object, and can be expressed as an *integer*, or as a *name*. The valid object identifiers names are depended from the current *Maple* release. The procedure call **DagTag()** returns the table whose indices are positive integers defining numeric object identifiers, whereas its entries define their names. Thus, the returned table defines the mapping between **IDs** of internal representation of data structures and their names. The procedure is a rather useful tool at operation with the internal representations of *Maple* objects, for example, in combination with procedures ***`assemble`*** and ***`dismantle`***, defining a "*hackware*" in *Maple.*

```
DagTag:= proc()
local k, a;
  try assign(a = []), seq(assign('a' = [op(a), k = kernelopts(dagtag = k)]), k = 1 .. 500)
  catch "invalid data structure tag": NULL("End of a representation table of internal
                                     data structures")
  end try;
  eval(table(a))
end proc
```

Typical examples of the procedure use:

```
> DagTag();                                                              # Maple 6

    table([1 = INTNEG, 2 = INTPOS, 3 = RATIONAL, 4 = FLOAT, 5 = HFLOAT, 6 = COMPLEX,
    7 = STRING, 8 = NAME, 9 = MEMBER, 10 = TABLEREF, 11 = DCOLON, 12 = CATENATE,
    13 = POWER, 14 = PROD, 15 = SERIES, 16 = SUM, 17 = ZPPOLY, 18 = FUNCTION,
    19 = UNEVAL, 20 = EQUATION, 21 = INEQUAT, 22 = LESSEQ, 23 = LESSTHAN, 24 = AND,
    25 = NOT, 26 = OR, 27 = EXPSEQ, 28 = LIST, 29 = LOCAL, 30 = PARAM, 31 = LEXICAL,
    32 = PROC, 33 = RANGE, 34 = SET, 35 = TABLE, 36 = RTABLE, 37 = MODDEF,
```

38 = *MODULE*, 39 = *ASSIGN*, 40 = *FOR*, 41 = *IF*, 42 = *READ*, 43 = *SAVE*, 44 = *STATSEQ*,
45 = *STOP*, 46 = *ERROR*, 47 = *TRY*, 48 = *RETURN*, 49 = *BREAK*, 50 = *NEXT*, 51 = *USE*,
52 = *BINARY*, 53 = *HASH*, 54 = *HASHTAB*, 55 = *GARBAGE*, 56 = *FOREIGN*, 57 = *CONTROL*,
58 = *DEBUG*])

> DagTag(); **# *Maple 7 - 9***

table([1 = *INTNEG*, 2 = *INTPOS*, 3 = *RATIONAL*, 4 = *FLOAT*, 5 = *HFLOAT*, 6 = *COMPLEX*,
7 = *STRING*, 8 = *NAME*, 9 = *MEMBER*, 10 = *TABLEREF*, 11 = *DCOLON*, 12 = *CATENATE*,
13 = *POWER*, 14 = *PROD*, 15 = *SERIES*, 16 = *SUM*, 17 = *ZPPOLY*, 18 = *FUNCTION*,
19 = *UNEVAL*, 20 = *EQUATION*, 21 = *INEQUAT*, 22 = *LESSEQ*, 23 = *LESSTHAN*, 24 = *AND*,
25 = *NOT*, 26 = *OR*, 27 = *XOR*, 28 = *IMPLIES*, 29 = *EXPSEQ*, 30 = *LIST*, 31 = *LOCAL*,
32 = *PARAM*, 33 = *LEXICAL*, 34 = *PROC*, 35 = *RANGE*, 36 = *SET*, 37 = *TABLE*, 38 = *RTABLE*,
39 = *MODDEF*, 40 = *MODULE*, 41 = *ASSIGN*, 42 = *FOR*, 43 = *IF*, 44 = *READ*, 45 = *SAVE*,
46 = *STATSEQ*, 47 = *STOP*, 48 = *ERROR*, 49 = *TRY*, 50 = *RETURN*, 51 = *BREAK*, 52 = *NEXT*,
53 = *USE*, 54 = *BINARY*, 55 = *HASH*, 56 = *HASHTAB*, 57 = *GARBAGE*, 58 = *FOREIGN*,
59 = *CONTROL*, 60 = *DEBUG*])

In the present chapter, a new ***dirax***-type maintained by program module **DIRAX** has been defined. The module supports data structures of direct access of *dirax*-type that by a series of parameters essentially extends standard data structures of types {*stack, heap, queue*}. In contrast to data structures of types {*stack, queue*}, the *dirax* structures are precisely identified by ***ispd**-index*; a reception, used in the module, allows easily to define precise classification of data structures of types {*stack, queue, heap, dirax*}. At the same time, the **DIRAX** module provides an effective solution of problems of shared use of data structures of all above-mentioned types. Furthermore, tools of testing of the above data structures types have been essentially precised, because the standard tools have some inaccuracies. In a series of problems of data processing, the *dirax*-type has shown oneself as the most effective data structure.

Chapter 7.
Software to support bit-by-bit processing of symbolic information

In the seventh section, the tools, which support a *bit-by-bit* information processing in the *Maple* environment, are represented. The package does not possess tools of the similar type. The software offered by us is represented by six useful procedures such as **Bit, Bit1, xbyte, xbyte1, xNB** and **xpack.** These procedures serve for *bit-by-bit* information processing, i.e. the user has a possibility to operate with strings or symbols at a level of separate bits composing them.

Bit – **bit-by-bit information processing**
Bit1
xbyte
xbyte1
xNB

Call format of the procedures:

Bit(B, *bits*)
Bit1(B1, *bits*)
xbyte(L)
xbyte1(L1)
xNB(N)

Formal arguments of the procedures:

B – a symbol or a string of length **1,** or an integer **1**-element list
B1 – an integer from the range **0..255**
bits – a leading variable of a polynomial
L – a list from **8** binary digits
L1 – a positive binary number of length **<= 8,** or a list from no more than **8** binary digits
N – a positive integer (*posint*)

Description of the procedures:

The tools represented below serve for a *bit-by-bit* information processing, i.e. the user has a possibility to operate with the symbols or strings on a level of separate bits composing them. The first **Bit** procedure allows effectively enough to execute *bit-by-bit* processing of separate symbols.

The procedure call **Bit(B,** ***bits*)** allows to execute the following basic *bit-by-bit* operations with a symbol, defined by the first actual **B** argument of the procedure (as argument can appear an one-element string or symbol of length 1, and also 1-element list whose element defines a decimal code of a symbol from the range **0 .. 255**):

- (**1**) if the second actual ***bits*** argument has a view **[n1, n2, ...]**, then the procedure call **Bit(B, *bits*)** returns the values of bits of symbol **B**, which are located in its $\mathbf{n_j}$ positions ($\mathbf{n_j}$ belong to the range **1..8**);

- (**2**) if the second actual ***bits*** argument has a view **[n1=b1, n2=b2,...]**, the decimal code of a symbol, obtained by replacement of values of bits of symbol **B** (which are in its positions $\mathbf{n_k}$) onto binary values $\mathbf{b_k}$, is returned ($\mathbf{n_k}$ **= 1 .. 8;** $\mathbf{b_k}$ **= {0|1}**).

- (**3**) if the second actual ***bits*** argument is a name of procedure defined over the lists, the result of application of the procedure to a list of values of all bits of a symbol, defined by the first actual **B** argument, is returned.

In addition, in two last cases the decimal code of a symbol – of the result of processing of the initial symbol, defined by the first actual **B** argument of the procedure, is returned.

The **Bit1(B1, *bits*)** procedure is a modification of the above **Bit** procedure. In the procedure the more effective algorithm of evaluations has been implemented, however a handling of special and erroneous situations are not fulfilled, and as the first actual **B1** argument only the decimal code of a symbol subjected to bit-by-bit processing is admitted. The second ***bits*** argument has remained the same. The given procedure is more reactive and can be successfully used in the cyclic computational constructions.

For providing of *bit-by-bit* symbolic processing two basic procedures **Bit** and **Bit1** represented above are intended. The simple procedure **xbyte(L)** fulfills an inverse function, returning a symbol with the given bit value defined by the actual **L** argument; as a value for actual **L** argument a list from **8** binary digits is used.

Provided that the arbitrary symbol "X" has a *bit-by-bit* representation **B::list(*binary*)**, for the procedures **xbyte** and **Bit** the following defining relations take place:

"X" = xbyte(Bit("X", [k$k=1..8]))   **Bit(xbyte(B), [k$k=1..8]) = B**

The procedure **xbyte1(L)** extends the above procedure ***xbyte*** onto positive binary numbers, i.e. the numbers consisting of numerals {**0,1**} only; for example **11001011.** In this case we can represent a symbol in the form of such binary number of length, not greater than **8;** at a smaller length the procedure supplements a *bit-by-bit* representation at the left by *zeroes***.** Moreover, a binary list of length **< 8,** obtained at the procedure call, is supplemented at the left by *zeroes* up to the standard list of length **8.** Such approach allows in many cases more conveniently to represent the symbols at a level of bits composing them.

The simple procedure **xNB(N)** has been implemented by a one-string extracode and is useful enough at operation with information at the bit level. The procedure call **xNB(N)** returns the *true* value, if a positive integer **N** consists of binary numerals only, otherwise the *false* value is returned. The five procedures **Bit, Bit1, xbyte, xbyte1** and **xNB,** represented above, along with the introduced types {*digit, binary, byte*} create an quite satisfactory basis for operation with symbolic information at the *bit-by-bit* level. The examples represented below illustrate use of all procedures represented above for a *bit-by-bit* information processing in the *Maple* environment.

```
Bit:= proc(B::{string, symbol, list(integer)}, bits::{procedure, list(equation), list(posint)})
local a, b, c, d, h, k, R;
   assign(c = "00000000"), `if`(type(bits, procedure), goto(Evaluation), `if`(member(B, {`0`, "0"}),
     [assign(b = ""), goto(Art)], assign(a = {k $ (k = 1 .. 8)}, h = {[k] $ (k = 1 .. 8)}))),
     `if`(type(B, list) and nops(B) = 1 and member(op(B), {k $ (k = 0..255)}) and nops(bits) <= 8,
     goto(Kr), `if`(type(B, {name,string}) and length(B)=1 and nops(bits)<=8, goto(Kr), goto(E)));
```

```
  Kr;
  `if`(type(bits, list(posint)) and {op(bits)} union a = a or type(bits, list(`=`)) and
     {entries(table(bits))} union {[1], [0]} = {[1], [0]} and {indices(table(bits))} union h = h,
     goto(Evaluation), goto(Error2));
  E;
  ERROR("invalid arguments <%1> or/and 2nd argument <%2>", B, bits);
  Error2;
  ERROR("invalid 2nd argument <%1>", bits);
  Evaluation;
  b := cat("", convert(op(`if`(type(B, list), B, convert(cat("", B), bytes))), binary));
  Art;
  R:= L::list -> convert(SN(cat("", seq(L[k], k = 1 .. nops(L)))), decimal, binary);
  assign(d = cat("", c[1 .. 8 - length(b)], b)), `if`(type(bits, procedure),
   RETURN(R(bits([seq(SN(d[k]), k=1..8)]))), `if`(type(bits, list(posint)), [seq(SN(d[k]), k=bits)],
   `if`(type(bits, list(equation)), [seq(assign('d' = cat(d[1 .. lhs(bits[k]) - 1], rhs(bits[k]),
   d[lhs(bits[k])+1..-1])), k = 1..nops(bits)), RETURN(convert(SN(d), decimal, binary))],NULL)))
end proc
```

```
Bit1:= proc(B::integer, bits::{procedure, list})
local b, d, c, k;
option remember;
  assign(c = (a -> convert(a, decimal, binary)), b = NDLN(convert(B, binary))),
    assign(d = [0 $ (k = 1 .. 8 - nops(b)), op(b)]); `if`(type(bits, procedure),
      RETURN(c(NDLN(bits(d)))), `if`(type(bits, list(posint)), [seq(d[k], k = bits)],
    `if`(type(bits, list(equation)), [seq(assign('d' = [op(d[1 .. lhs(bits[k]) - 1]), rhs(bits[k]),
      op(d[lhs(bits[k]) + 1 .. -1])]), k = 1 .. nops(bits)), RETURN(convert(NDLN(d), decimal,
      binary))], NULL)))
end proc
```

```
xbyte:= proc(L::{list(binary)})
option remember;
  `if`(nops(L) <> 8, ERROR(`argument %1 is invalid`, L),
  convert([convert(SN(cat("", op(L))), decimal, binary)], bytes))
end proc
```

```
xbyte1:= proc(L::{nonnegint, list(binary)})
local k;
option remember;
  `if`(type(L, list), xbyte([0 $ (k = 1 .. 8 - nops(L)), op(L)]), `if`(xNB(L) and length(L) <= 8,
  RETURN(xbyte([0 $ (k = 1 .. 8 - length(L)), op(map(SN, convert(cat("", L), list)))])),
  ERROR(`argument <%1> is invalid`, L)))
end proc
```

```
xNB:= N::nonnegint -> belong({op(map(SN, convert(cat("", N), list)))}, {0, 1})
```

Typical examples of the procedures use:

```
> Bit(`S`, [k$k=1..8]), Bit([61], [1,3,8]), Bit([61], [2=1,4=0,8=0]), Bit("G", [6]), Bit("0", [6=1,7=1,8=1]);
                    [0, 1, 0, 1, 0, 0, 1, 1], [0, 1, 1], 108, [1], 7
> R:=(L::list)->[L[nops(L)-k]$k=0..nops(L)-1]: Bit("G",[k$k=1..8]),Bit("G",R),Bit("S",R),Bit([61], R);
                    [0, 1, 0, 0, 0, 1, 1, 1], 226, 202, 188
```

In particular, the given example illustrates use as the second actual *bits* argument of the **R**-procedure providing the inversion of lists:

> K:=(L)->[L[8],L[2],L[6],L[4]+1 mod 2,L[5]+1 mod 2,L[3],L[7],L[1]]: map(Bit1, [61,56,36,40,7,14], K);

[164, 4, 60, 20, 186, 50]

> map(xbyte, [[1,1,0,1,0,1,0,1], [0,0,1,0,0,1,0,1]]);

["X", "%"]

> B:=Bit("G", [k$k=1..8]); "G"=xbyte(Bit(G,[k$k=1..8])), Bit(xbyte(B), [k$k=1..8])=B;

B := [0, 1, 0, 0, 0, 1, 1, 1]

"G" = "G" [0, 1, 0, 0, 0, 1, 1, 1] = [0, 1, 0, 0, 0, 1, 1, 1]

> map(xbyte1, [101101, 1110101, 11101011, 11011, 1011001, 1001010, 10001001, 10000001, 10101]);

["-", "u", "л", "\e", "Y", "J", "‰", "Ѓ", "□"]

> map(xbyte1, [1010101,11010101,11100011,110110,10001001,10000001,1010101,11111001]);

["U", "X", "г", "6", "‰", "Ѓ", "U", "щ"]

> map(xNB, [10011,11001101,1102011,10101,1001010011101,11100011,110110,10001001]);

[*true*, *true*, *false*, *true*, *true*, *true*, *true*, *true*]

---

xpack – a packing/unpacking of the positive integers

Call format of the procedure:

xpack(N)

Formal arguments of the procedure:

N - a positive integer or string

Description of the procedure:

The **xpack(N)** procedure, working according to the principle of *switch*, provides a packing of positive integers (*with providing of the subsequent restoration*) by means of putting of two decimal numerals in one byte. On a positive integer as the actual **N** argument, the procedure returns the string equivalent of the number **N**, while on the equivalent the number represented by it is being returned. However, the given operation is not reversible, i.e. not each string is an equivalent of some positive integer, as that illustrates the last example of use of the **xpack** procedure of a fragment represented below. Namely, the following relation takes place: **xpack(xpack(N)) = N,** where **N** - a positive integer; this relation, in general, is incorrect for arbitrary string **N**, excluding a case when **N** is a result of the procedure call **xpack(M),** where **M** - a positive integer.

The **xpack(N)** procedure essentially uses the **Bit** and **xbyte** procedures considered above. The examples represented below illustrate use of this procedure for a packing/unpacking of information.

```
xpack:= proc(N::{nonnegint, string})
local a, b, c, d, k, s, p, h, q, Tab_con;
option remember;
  Tab_con := table([0 = [0, 1, 1, 0], 1 = [0, 1, 1, 1], 2 = [1, 0, 0, 0], 3 = [1, 0, 0, 1], 4 = [1, 0, 1, 0],
     5 = [1, 0, 1, 1], 6 = [1, 1, 0, 0], 7 = [1, 1, 0, 1], 8 = [1, 1, 1, 0], 9 = [1, 1, 1, 1]]),
```

```
    assign(q = "", h = [p $ (p = 1 .. 8)]);
  if type(N, integer) then assign67(a = cat("", `if`(type(length(N), even), N, cat(`0`, N))), s = "");
    for k by 2 to length(a) do assign67('b' = SN(a[k .. k + 1])), assign67('c' = trunc(1/10*b));
      assign67('b' = b - 10*c), assign67('d' = [op(Tab_con[c]), op(Tab_con[b])]);
      assign67('s' = cat(s, xbyte(d)))
    end do;
    RETURN(s)
  else
    for k to length(N) do assign67('a' = Bit(N[k], h));
      assign('b' = RTab(Tab_con, a[1 .. 4]), 'c' = RTab(Tab_con, a[5 .. 8]));
      assign67('q' = cat(q, b, c))
    end do;
    RETURN(SN(q))
  end if
end proc
```

Typical examples of the procedure use:

> assign('G' = map(xpack, [0,1942,2004,14062004])), assign('S' = map(xpack, G)), G, S;

["f", "□¨", "†j", "zl†j"], [0, 1942, 2004, 14062004]

> xpack(424706762089961), xpack(xpack(424706762089961));

"jЉЬЦНИпяЗ", 424706762089961

> xpack(xpack("Grodno")), xpack(05240001400453510549041414), xpack(xpack("Tallinn"));

"gvцno", "kЉЬfg¦j№·kЇjzz", "llinn"

> xpack(%[2]), xpack("RANS_IAN"), xpack(8938), xpack(1302198904081996), xpack("yh□пjn□ь");

524000141004535105490414140, 8938, "пћ", "yh□пjn□ь", 1302198904081996

The five procedures **Bit, Bit1, xbyte, xbyte1** and **xNB,** represented above, along with the introduced data types {*digit, binary, byte*} create a quite satisfactory basis for operation with the symbolical information at the bit-by-bit level. In many problems, dealing with the bit-by-bit processing, these tools have shown oneself from the best party. Taking into account the important role of the *bit-by-bit* information processing in the *Maple* environment, it is supposed, the tools represented in the present chapter, should be essentially extended with increase of their reactivity and functionality. It is especially topical, for with growth of numbers of the *Maple*'s releases their responsiveness is being reduced as a whole.

Chapter 8.
Tools extending graphic possibilities of *Maple* of releases 6 – 9.5

In the given section, the tools extending graphic possibilities of the package ***Maple*** of releases **6 – 9.5** are represented. The package not possesses tools of the similar type. The tools offered by us are represented by twelve useful enough procedures having undoubted applied interest.

A_Color – dynamical coloring of frames of the animated 2-dimensional graphs

Call format of the procedure:

A_Color(*Art*)

Formal arguments of the procedure:

Art – a function of one variable

Description of the procedure:

For today, the *Maple* language tools oriented onto work with graphical animated objects have a number of restrictions on dynamic updating of their characteristics. Meanwhile, many of the problems dealing with graphical animated objects suppose an availability of dynamic updating of their characteristics.

In particular, in many cases the necessity of dynamical modification of coloring of a curve subjected to animation by the ***animate*** function arises. The given problem is solved by the useful enough procedure **A_Color.** The procedure call **A_Color(*Art*)** has undefined quantity of the formal arguments and is intended for modification of a coloring of frames of a **2**-dimensional functional dependence subjected to animation. The first actual ***Art*** argument of the procedure defines the **ANIMATE** structure, whereas all its other arguments are coded as a sequence of equations of a view "***colour* = <*colour*>**"**.** In addition, their amount should strictly agree with amount of frames for an object subjected to animation. The successful procedure call **A_Color** returns the animated graphics objects whose frames are not identical, but have a coloring, given by the user. We recommend to pay attention to a reception used for organization of mechanism (so-called mechanism of "***disk transits***") of dynamical modification of coloring of the animated curves in the *Maple* environment.

```
A_Color:= proc(Art::function)
local k, n, m, p, z, h, t, d, v, Kr, LC, LC1, L, L1, Ln, F;
  [interface(labelling = false), assign(F = cat([libname][1][1 .. 2], `/$`))];
  Kr, LC1, n, L, p, h, t, m, L1 := convert(Art, symbol), [], nops([op(op(1, Art))]), [], 1, 0, 0,
     length(convert(Art, symbol)), [];
  LC := [seq(op(op(op(op(1, Art))[k]))[-1], k = 1 .. n)];
```

```
  `if`(nargs < 2, ERROR("Number of real arguments is incorrect: %1", args), NULL);
  `if`(nops(LC) <> nargs - 1, ERROR("Number of Color-arguments is not equal to number of
     frames: %1", args), NULL);
  for k from 2 to nargs do LC1 := [op(LC1), op(convert(args[k], PLOToptions))] end do;
  LC1 := map(convert, LC1, symbol);
  do z := searchtext(COLOUR, Kr, p .. m);
     if z = 0 then break else L := [op(L), z + h]; h := h + z + 3; p := h + 1 end if
  end do;
  for k to nops(L) do h := searchtext(")", Kr, L[k] .. m); L1 := [op(L1), L[k] + h - 1] end do;
  assign('t' = open(F, WRITE)), writebytes(t, cat(`D_R_Col:=`, substring(Kr, 1 .. L[1] - 1)));
  for k to nops(L) - 1 do writebytes(t, LC1[k]); writebytes(t, substring(Kr, L1[k]+1 .. L[k+1] - 1))
  end do;
  writebytes(t, LC1[nargs - 1]); [writebytes(t, cat(substring(Kr, L1[nargs-1]+1..m), `;`)), close(t)];
  (proc(x) read x end proc)(F), [fremove(F), RETURN(D_R_Col)]
end proc
```

Typical example of the procedure use:

```
> Art:= plots[animate](sin(t*x^4), x=-Pi..Pi, t=1..3, frames=7, thickness=2): A_Color(Art,
  color=red, color=green, color=yellow, color= pink, color=gold, color=navy, color=magenta);
```

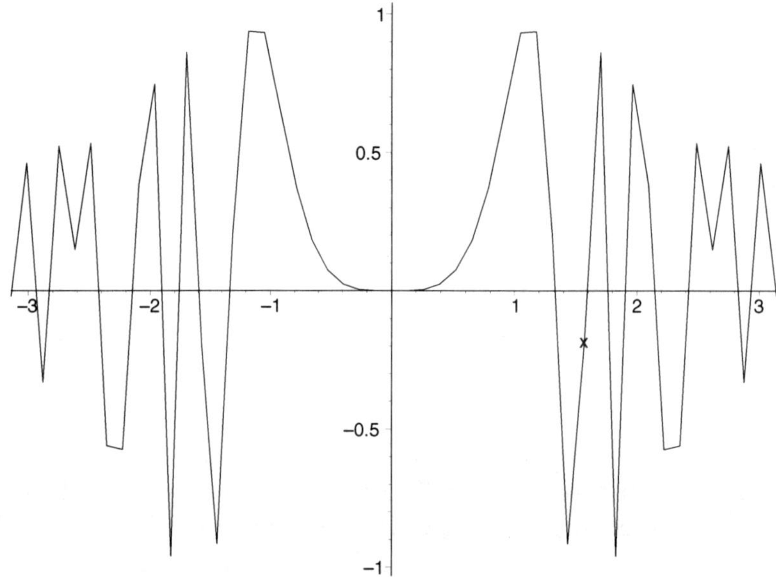

As it was marked above, the procedure arguments determining of frames coloring are set as equations. However, the procedure can be easily extended to a case when instead of equations a set or a list of colors for frames are set; in this case it is possible to define modes of choosing colors for frames, for example, in a pseudorandom way. We leave the given updating of the procedure to the reader as an useful enough exercise.

Pul_Bal – **some useful examples of animation of 2D and 3D graphical objects**

Pul_Bal_Cor
Porshen
Pulsar
Chess
Histo

Call format of the procedures:

Pul_Bal(N)
Pul_Bal_Cor(N)
Porshen(N)
Pulsar(N)
Chess(N)
Histo(m, n, a, b, C)

Formal arguments of the procedures:

N – a positive integer defining quantity of frames taking part in animation
m, n, a, b – integers
L – a list defining a frames colouring

Description of the procedures:

Six examples represented below illustrate a number of receptions of organization of animation of **2D** and **3D** graphical objects in the *Maple* environment of releases **6 – 9.** These receptions can be useful for practical programming of graphical objects in the *Maple* environment.

The procedure call **Pul_Bal(N)** returns the animated **2D** graphical object a *disk*, defined by an appropriate graphical **2D**-primitive of the module **plottools** of the package. The essence of animation consists in cyclic alternation of a sequence of **N** frames, imitating a motion of the object along a sine trajectory, during of which the object cyclically changes own size and color. We recommend to pay attention to a principle of creation of a frames list from **ST** data structure, modified by the ***subs*** function, and onto artificial insertion of the ***COLOR***-option into a graphic *disk* structure for an opportunity of the subsequent modification of a setting of the given option. It is necessary to have in mind, the increasing of **N** value gives rise essential enough increasing of time costs. This note takes place also for the procedures represented below.

The **Pul_Bal_Cor(N)** procedure is an extension of the previous procedure **Pul_Bal** and returns the animated **2D** object, whose essence of animation consists in a motion of a disk centre along a sine curve; in addition, its size and its coloring change cyclically (under the sine law) during such motion. During animation simultaneously take place the static (*graph*) and dynamic (*disk*) components of the graphical object. Quantity of the frames is defined by the actual **N** argument.

The procedure call **Porshen(N)** returns the more complex animated **3D** graphical object consisting of a *torus* and a *cylinder* that are defined by the appropriate graphical **3D** primitives. Quantity of frames is defined by the actual **N** argument. The essence of animation is reduced to an illustration of reciprocal motion (downwards – upwards) of a cylinder via centre of a fixed torus. During animation a coloring of the torus dynamically varies also. By means of the ***display*** function a combining of the stationary graphical structure (*torus*) with the modified structure (*cylinder*) for each frame participating in the process of animation is done.

The procedure call **Pulsar(N)** returns the animated **3D** graphical *sphere* object defined by an appropriate **3D** graphical primitive. The essence of animation consists in cyclic change of sizes of the object (pulsation under the sine law) and its color appearance. As a practically useful, here can appear a reception used for dynamic change of the color characteristic of the object. Quantity of frames is defined by the actual **N** argument.

The **Chess(N)** procedure illustrates making an animated structure based on the graphical ***rectangle*** primitive, on the recursive calls of functions ***seq*** and ***display,*** and on the ***rand*** generator of pseudorandom numbers providing dynamic modification of coloring of square fields of the chessboard. Quantity of frames is defined by the actual **N** argument. Because of animation an effect of *illumination* of squares of the chessboard is reached.

The procedure call **Histo(p,n,a,b,C)** returns the dynamic *histogram* constructed on **(p-1)** equal intervals; height of columns of the diagram is a random function of time with values uniformly distributed on an interval **[a, b].** Actual arguments **n** and **C** define a quantity of animation frames and a color appearance of the *histogram* accordingly. In addition, the relation **nops(C)=p-1** should be fulfilled, and values of elements out of **C** list should belong to interval **[0, 1].** The examples represented below illustrate (*in static form*) use of the above six procedures; whereas their dynamic running is possible only in the *Maple* environment of releases **5 – 9.5.**

```
Pul_Bal:= proc(N::posint)
local x, y, k, r, S, Lf;
  [with(plots, display), with(plottools, disk), assign(S = disk([x, y], r, color = black))];
  Lf := [subs([x = evalf(4*k*Pi/N), y = sin(evalf(4*k*Pi/N)), r = 0.02 + sin(evalf(4*k*Pi/N)),
     COLOUR(RGB, 0, 0, 0)  = COLOUR(RGB, 0.99, 0.42 + k/N, 0.89 – k/N)], S) $ ('k' = 0..N)];
  display(Lf, insequence = true, axes = none, thickness = 2, scaling = constrained)
end proc

Pul_Bal_Cor:= proc(n::posint)
local x, y, k, r, df, ST, Lf, Kor;
  [with(plots, display), with(plottools, disk), assign(Kor, plot(sin(x), x = 0 .. 4*Pi, color = green)),
     assign(ST, disk([x, y], r, color = black)), assign(df, [])];
  Lf := [subs([x = evalf(4*k*Pi/n), y = sin(evalf(4*k*Pi/n)), r = 0.02 + sin(evalf(4*k*Pi/n)),
     COLOUR(RGB, 0, 0, 0) = COLOUR(RGB, 0.98, 0.42 + k/n, 0.89 – k/n)], ST) $ ('k' = 0..n)];
  seq(assign('df' = [op(df), display(Kor, Lf[k])]), k = 1 .. nops(Lf));
  display(df, insequence = true, axes = none, thickness = 2, scaling = constrained)
end proc

Porshen:= proc(N::posint)
local k, z, tor, C, Lf;
  [with(plots, display), with(plottools, torus, cylinder), assign(Lf = [], C = [XYZ, XY, Z,
     ZGREYSCALE, ZHUE]), assign(tor = torus([0, 0, 0], 5, 12))];
  for k from 0 to N do z := 12*sin(evalf(2*k*Pi/N));
     Lf := [op(Lf), display(tor, cylinder([0, 0, z – 16], 3, 32, shading = C[(k mod 5) + 1]))]
  end do;
  display(Lf, insequence = true, orientation = [45, 64], scaling = constrained)
end proc

Pulsar:= proc(N::posint)
local k, r, C, Lf;
  [with(plots, display), with(plottools, sphere), assign(Lf, []), assign(C, [XYZ, XY, Z,
     ZGREYSCALE, ZHUE])];
  for k from 0 to N do r:= 2*sin(evalf(k*Pi/N));
     Lf := [op(Lf), sphere([0, 0, 0], 0.3 + r, shading = C[(k mod 5) + 1])]
  end do;
  display(Lf, insequence = true, thickness = 2, style = PATCHNOGRID, scaling = constrained)
end proc

Chess:= proc(N::posint)
local k, p;
  [map(with, [plottools, plots]), RETURN(display(seq(display(seq(seq(rectangle([k, p], [k-1, p+1],
  color = COLOR(RGB, 1/1000000000000*k*rand(), 1/1000000000000*p*rand(),
  1/1000000000000*k*rand())), k = 1 .. 8), p = 0 .. 7)), k = 1 .. N), insequence = true,
  axes = none, thickness = 2))]
end proc
```

```
Histo:= proc(m::integer, n::integer, a::integer, b::integer, C::list)
local k, p, L;
   [map(with, [plottools, plots]), assign(L, rand(a .. b)),
   RETURN(display(seq(display(seq(rectangle([p, L()], [p + 1, 0],
   color = COLOR(RGB, 0.6*C[p], 0.2*C[p], 0.1*C[p])), p = 1 .. m - 1)), k = 1 .. n),
   insequence = true, axes = none, thickness = 2))]
end proc
```

Typical examples of the procedures use:

> Pul_Bal(32); Pul_Bal_Cor(32);

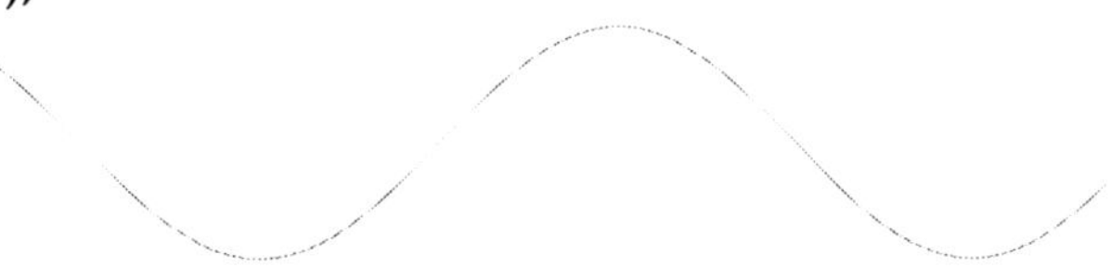

> Porshen(32);

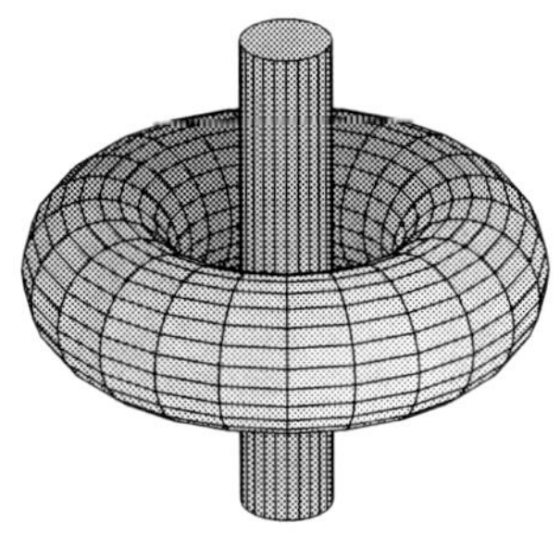

> Pulsar(32); Chess(32);

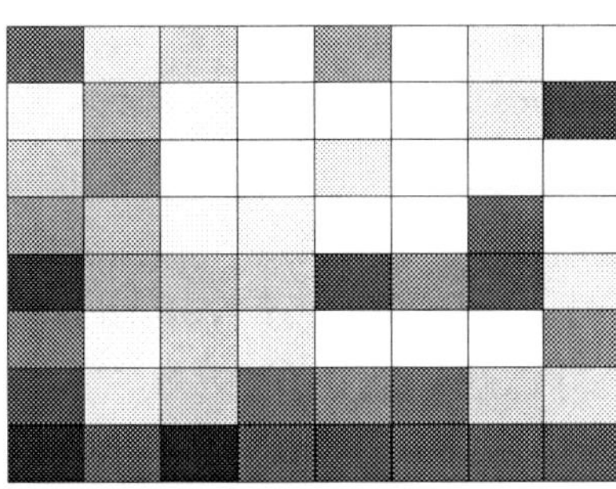

> Histo(7, 18, 3, 10, [0.95*'k'$'k'=1..14]);

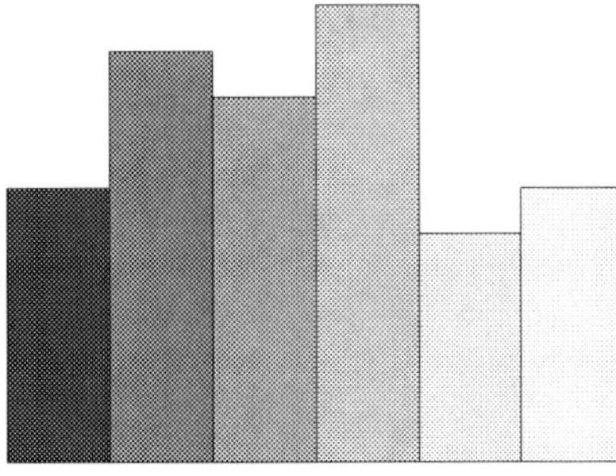

Clr – management by color of the 2D graphics
UserC

Call format of the procedures:

Clr(C)
UserC(UC, x, y, z)

Formal arguments of the procedures:

C - a name defining a color for **2D** graphics
UC - an assignable name defining the user color
x, y, z - a floating-point number from the range **0..1**

Description of the procedures:

Two useful procedures **Clr** and **UserC** provide operations with the ***color***-option of **2D** graphical objects. The procedure call **Clr(C)** returns the element of a graphical structure corresponding to one of the **25** predetermined C-colors of the *Maple* package for **2**-dimensional graphics.

The procedure call **UserC(UC,x,y,z)** returns the *NULL* value, defining the user **UC**-color that can be used along with the predetermined colors of the package for **2**-dimensional graphics. Values of the actual arguments **x, y** and **z** should have *float*-type and be in the range **0 .. 1;** they define quotas of *red, green* and *blue* colors accordingly at compounding of the user color. The given tools extend possibilities of the user over the management of color appearance of the 2-dimensional graphical objects in the *Maple* environment of releases 6 – 9.5.

```
Clr:= proc(C::{name, symbol})
local a, b, c;
  assign('a' = convert(plot([[0, 0]], colour = C), string), 'c' = cat([libname][1][1 .. 2], "/$VGS")),
  search(a, "COLOUR", 'b'), writeline(c, cat("_OOO:=", substring(a, b .. searchtext(")", a,
  b..length(a))+b-1), ":")), close(c), (proc(x) read x end proc)(c), RETURN(fremove(c), _OOO)
end proc

UserC:= proc(C::evaln, x::float, y::float, z::float)
local k;
  `if`({seq(`if`(abs(args[k]) <= 1, 1, 0), k = 2 .. nargs)} = {1}, assign(C = COLOR(RGB, abs(x),
  abs(y), abs(z))), ERROR(`Invalid values %1 for colour`, [args[2 .. nargs]]))
end proc
```

Typical examples of the procedures use:

> Clr(magenta), Clr(red), Clr(green), Clr(blue), Clr(yellow), Clr(brown);

COLOUR(RGB,1.00000000,0.,1.00000000), COLOUR(RGB,1.00000000,0.,0.),
COLOUR(RGB, 0.,1.00000000,0.), COLOUR(RGB,0.,0.,1.00000000),
COLOUR(RGB,1.00000000,1.00000000,0.), COLOUR(RGB,0.64705882,0.16470588,0.16470588)

> UserC(UC,0.62,0.1942,0.2004); UC; plot(sin(x),x= -7*Pi..7*Pi, thickness = 2, colour = UC);

COLOR(RGB, 0.62, 0.1942, 0.2004)

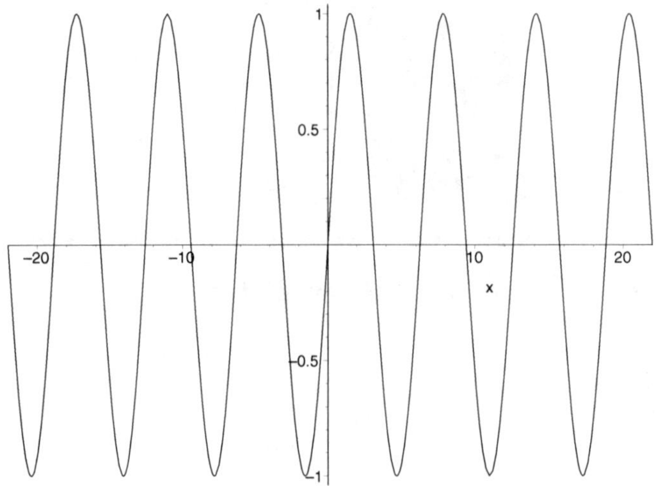

Animate2D – an extending of the standard procedures *animate* and *animate3d* on any finite number of animation parameters
Animate3D

Call format of the procedures:

Animate2D(F1, x, xa {, a {, b {, c ...}}})
Animate3D(F, x, y, xa, ya {, a {, b {, c ...}}})

Formal arguments of the procedures:

F – an algebraic expression **F(x, y, a, b, c, ...)** from two leading variables
F1 – an algebraic expression **F1(x, a, b, c, ...)** from one leading variable
x, y – the leading variables
xa, ya – ranges on **x** and **y**
a, b, c, ... – (optional) numeric variation ranges and/or lists and/or sets for animation parameters

Description of the procedures:

The standard procedure **animate3d** of the package **plots** provides animation of **3D**-functional dependences depending on one animation parameter only. Whereas the procedure **Animate3D** represented below allows any quantity of animation parameters for an algebraic expression **F(x,y,a,b,c, ...)** from two leading variables **{x, y}** and finite quantity of animation parameters **{a, b, c, ...}**. The procedure essentially uses the above-mentioned method of ***"disk transits"***.

The procedure call **Animate3D(F, x, y, xa, ya {, a {, b {, c ...}}})** returns the animated graphic **3D**-object specified by an algebraic expression **F(x, y)** from leading variables **{x, y}** on animation parameters **{a, b, c, ...}** whose variation values are specified by numeric ranges and/or lists and/or sets starting with **6**th actual argument. In addition, the relation **nops({a, b, c, ...}) <= nargs-5** must be; otherwise, an error arises. Furthermore, between a set **A={a, b, c, ...}** and list **B=[args[6 .. (nops(A) + 6]** the one-to-one correspondence is supposed, i.e. for **A[k]** a variation values set is **B[k], k=1 .. nops(A).**

If an expression **F(x, y)** depends on variables **{x, y}** only, the procedure call returns a graph of the **F** expression in specified diapasons **{xa, ya}** on the leading variables. The procedure handles the basic especial and erratic situations.

In view of reduction of dimensionality, the previously mentioned concerning the procedure **Animate3D** in the full measure concerns the procedure **Animate2D** providing animation of **2D** graphic objects, defined by functional dependences from one leading variable, relative to any finite number of the animation parameters. The procedures **Animate2D** and **Animate3D** can be extended onto the wider circle of types of the first argument, using the above method of "*disk transits*".

```
Animate2D:= proc(F::algebraic, x::symbol, xa::range)
local a, b, c, d, n, k;
global __0;
  `if`(map(type, {lhs(xa), rhs(xa)}, numeric) = {true}, NULL, ERROR("3rd argument is invalid"));
  assign(a = Indets(F), n = nargs, '__0' = [], b = "", d = "");
  if not member(x, a) then ERROR("quantity of unknowns is more than 1")
  elif not (nops(a) = 1) and nargs = 3 then ERROR("<%1> contains non-numeric values", F)
  elif n = 3 then RETURN(plot(F, x = xa))
  elif not (nops(a) - 2 <= nargs - 3) then ERROR("quantity of parameters for animation is invalid")
  else a := a minus {x};
    if nargs - 3 < nops(a minus {x}) then ERROR("quantity of parameters for animation is more
      than quantity of variation ranges for them")
    elif {seq(type(args[k],
```

```
      {range(numeric), list(numeric), set(numeric)}), k = 4 .. nargs)} <> {true} then
        ERROR("values of parameters for animation are invalid")
      else
        for k to nops(a) do b := cat(b, "seq(") end do;
        for k to nops(a) do d := cat(d, a[k], "=", convert(args[3 + k], string), "),") end do;
        c := cat("assign('`__0`'=[op(`__0`),", convert(F, string), "]),");
        writeline("_$$$_", cat(b, c, cat(d[1 .. -2], ":"))), close("_$$$_");
        (proc(x) read x; fremove(x) end proc)("_$$$_"), assign('__0' = map(evalf, __0))
      end if
  end if;
  plots[display](map(plot, __0, x = xa), insequence = true), unassign('__0')
end proc
```

```
Animate3D:= proc(F::algebraic, x::symbol, y::symbol, xa::range, ya::range)
local a, b, c, d, n, k;
global __0;
  `if`(map(type, {lhs(xa), rhs(xa), lhs(ya), rhs(ya)}, numeric) = {true}, NULL,
     ERROR("4th and/or 5th argument is invalid"));
  assign(a = Indets(F), n = nargs, '__0' = [], b = "", d = "");
  if not belong({y, x}, a) then ERROR("quantity of unknowns is more than 2")
  elif not (nops(a) = 2) and nargs = 5 then ERROR("<%1> contains non-numeric values", F)
  elif n = 5 then RETURN(plot3d(F, x = xa, y = ya))
  elif not (nops(a) - 2 <= nargs - 5) then ERROR("quantity of parameters for animation is invalid")
  else a := a minus {y, x};
     if nargs - 5 < nops(a minus {y, x}) then ERROR("quantity of parameters for animation is
        more than quantity of variation ranges for them")
     elif {seq(type(args[k], {range(numeric), list(numeric), set(numeric)}), k = 6..nargs)} <> {true}
     then ERROR("values of parameters for animation are invalid")
     else
        for k to nops(a) do b := cat(b, "seq(") end do;
        for k to nops(a) do d := cat(d, a[k], "=", convert(args[5 + k], string), "),") end do;
        c := cat("assign('`__0`'=[op(`__0`),", convert(F, string), "]),");
        writeline("_$$$_", cat(b, c, cat(d[1 .. -2], ":"))), close("_$$$_");
        (proc(x) read x; fremove(x) end proc)("_$$$_"), assign('__0' = map(evalf, __0))
     end if
  end if;
  plots[display](map(plot3d, __0, x = xa, y = ya), insequence = true), unassign('__0')
end proc
```

Typical examples of the procedure use:

```
> p:= evalf(2*Pi):  Animate3D(cos(d*x)*(c*y+a)*sin(b*y),x,y,-p..p,-p..p,[1,2,3],{5,6,7}, 7..15,
  {7,15},[1,5,7]);
```

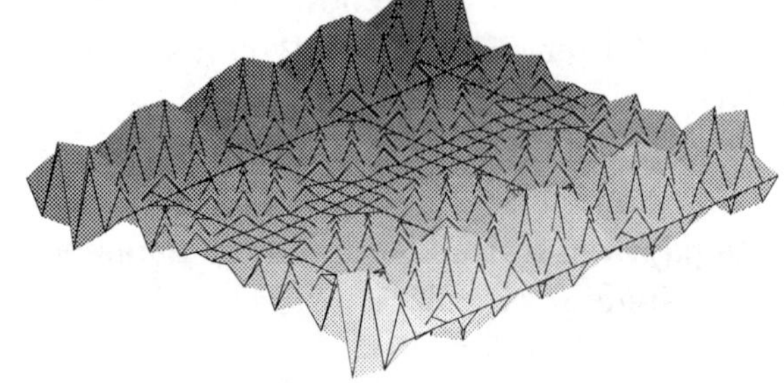

> Animate3D((sin(b*y)-a)^2+(cos(a*x)+b)^2, x, y, -7..7, -7..7, {1,2,3,-3,-2,-1}, -7..7);

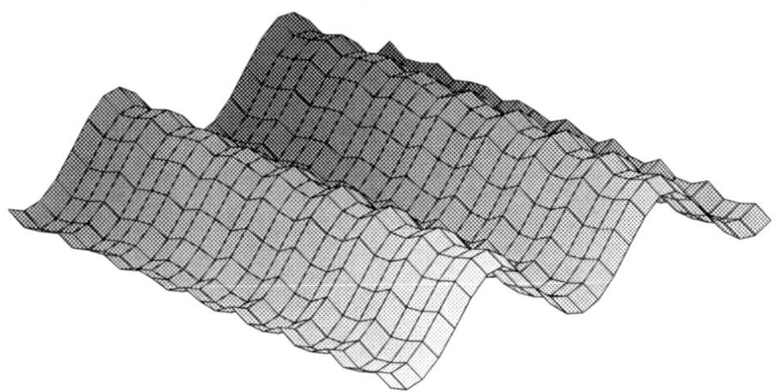

> Animate2D(cos(d*x)*(c*x+a)*sin(b*x), x, -p..p, [1,2,3], {5,6,7}, 7..15, {7,15}, [1,5,7]);

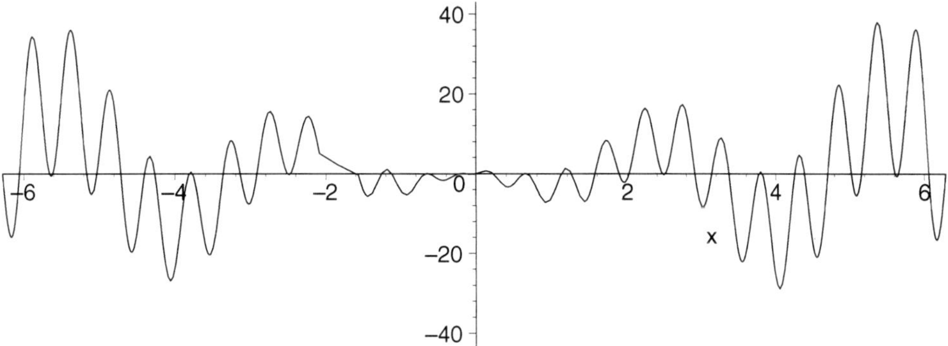

> p:= evalf(2*Pi): Animate2D(cos(d*x)*(c*x+a)*sin(b*x), x, -p..p);

Error, (in **Animate2D**) <cos(d*x)*(c*x+a)*sin(b*x)> contains non-numeric values

> p:= evalf(2*Pi): Animate2D(cos(d*x)*(c*x+a)*sin(b*x), x, -p..p, [1,2,3], {5,6,7}, 7..15);

Error, (in **Animate2D**) quantity of parameters for animation is more than quantity of variation ranges for them

> Animate2D((a+x)*cos(x-a)*(b+x)*sin(x+a), x, -p..p, {1,2,3,-3,-2,-1}, 1..7);

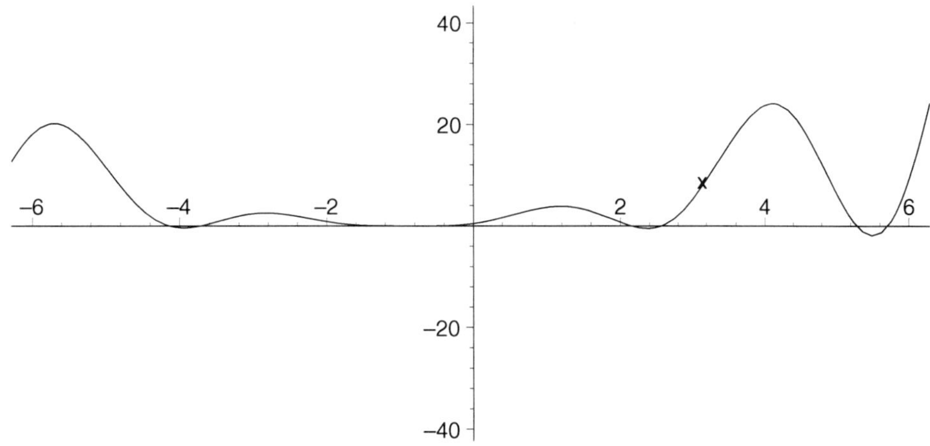

Is_Color – check of colors admitted by the package *Maple*

Call format of the procedure:

Is_Color(n)

Formal arguments of the procedure:

n – a positive integer from the range **1..25**

Description of the procedure:

The procedure call **Is_Color(n)** provides a visual representation of colors that are supported by graphics tools of the package *Maple*. Over an **n** number (**1 <= n <= 25**) the text with indication of a *name* of color, formed by a color corresponding to the **n** number, is output.

```
Is_Color:= proc(n::posint)
local Color;
  `if`(1 <= n and n <= 25,NULL,ERROR("Inadmissible argument n=%1; must be 1<=n<=25",n));
  Color := [aquamarine, black, blue, brown, coral, cyan, gold, gray, green, grey, khaki, magenta,
    maroon, navy, orange, pink, plum, red, sienna, tan, turquoise, violet, wheat, white, yellow];
  plots[textplot]([0,0,cat(`This color is: `,Color[n])],color=Color[n],font=[TIMES, BOLDITALIC,
    14], axes = none)
end proc
```

Typical example of the procedure use:

> Is_Color(7); Is_Color(14); Is_Color(20); Is_Color(25);

This color is: *gold* This color is: *navy* This color is: *tan* This color is: *yellow*

Kr_Mesh – forming the MESH-structure of graphical objects

Call format of the procedure:

Kr_Mesh(F, a, b, c, d, m, n)

Formal arguments of the procedure:

F - a function of two variables
a, b, c ,d - algebraic expressions
m, n - integers

Description of the procedure:

As the graphical *plot3d*-structure has the text format and clear organization, in principle, it can be formed in the environment of a word processor or directly in the environment of a current *Maple*-document. For these purposes can be used useful enough **Kr_Mesh** procedure represented below. For an arbitrary **F** function of two leading variables the **Kr_Mesh(F,a,b,c,d,m,n)** procedure forms a graphical structure of the *Mesh*-type over an area **[a <= x <= b, c <= y <= d]** with strong (**mxn**)-lattice of points **(x, y)**.

The **Kr_Mesh** procedure has at least seven formal arguments from which explicitly appear only the first seven, namely: **F** - identifier of a functional dependence (function), boundary values for an area **[a <= x <= b, c <= y <= d]**, and dimensionality for strong (**mxn**)-lattice of points **(x, y)** of the area.

The actual arguments, since the eighth, should be coded in the form of equations and to define settings for **3D** graphical options, i.e. should have the following view *`option = value`*, for example *color = green*. The procedure returns the graph of a **3D** functional dependence corresponding to the evaluated *Mesh*-structure. In addition, the appearance of resultant graphics object is defined by defaults for graphics options, and by their settings made by means of actual arguments passed to the procedure, since the eighth argument. Description of the procedure algorithm is represented in our books [8,9,29-33], whereas its source text is represented below.

```
Kr_Mesh:= proc(F::{name, symbol}, a::algebraic, b::algebraic, c::algebraic, d::algebraic,
                  m::integer, n::integer)
```

```
local k, p, Mesh, Mesh1;
  `if`(nargs<7, ERROR("Number of real arguments must be < 7 but received %1",[args]), NULL);
  `if`(7 < nargs, `if`(member(false, {seq(type(args[k], equation), k = 8 .. nargs)}),
    ERROR("Options must be coded in equation format but received %1", [seq(args[k],
    k = 8 .. nargs)]), NULL), NULL);
  Assign(Mesh = NULL, Mesh1 = NULL);
  for k from evalhf(a) by evalhf((b – a)/m) to evalhf(b) do
    for p from evalhf(c) by evalhf((d – c)/n) to evalhf(d) do Mesh := Mesh, [k, p, F(k, p)] end do;
    Assign('Mesh1' = Mesh1, [Mesh], 'Mesh' = NULL)
  end do;
  PLOT3D(MESH(evalf([Mesh1], 3)), `if`(7 < nargs, seq(op(convert(args[k],
    PLOT3Doptions)), k = 8 .. nargs), NULL))
end proc
```

Typical example of the procedure use:

> a:= 4*Pi: AG:= (x,y) -> x*sin(y) + y*sin(x): Kr_Mesh(AG,-a,a,-a,a,61,61, color = green, orientation=[45,30], thickness=2);

LGD – detection of the available graphical drivers

Call format of the procedure:

LGD(V)

Formal arguments of the procedure:

V - an assignable name

Description of the procedure:

In the delivered versions and releases of the package *Maple* and depending on an used platform, the set of really supported graphical drivers can be various in the presence of the basic set that is standard for all. The procedure call **LGD(V)** returns the set of graphical drivers supported by both a current release of the package *Maple* and the underlying operating platform, whereas through actual **V** argument the list of all graphical drivers recognized by a current *Maple* release is returned.

```
LGD:= proc(V::evaln)
local k, GD, L;
  interface(warnlevel = 0), plotsetup();
  L, GD := {}, sort([op(map2(op, 1, {indices(`plotsetup/devices`)}))], lexorder);
  for k to nops(GD) do plotsetup(GD[k]); L := {op(L), rhs(plotsetup()[3])} end do;
  plotsetup(inline), interface(warnlevel = 3), assign(V = GD), RETURN(L)
end proc
```

Typical example of the procedure use:

> LGD(Art_Kr), Art_Kr; # *Maple* **6**

{inline, dxf, tek, pov, x11, wmf, ps, bmp, pcx, gif, hpgl, char, jpeg, window}, [PostScript, X11, bmp, char, cps, default, dumb, dxf, gif, hpgl, hplj, inline, jpeg, laserjet, pcx, postscript, pov, ps, tek, tektronix, window, wmf, x11, xwindow]

> LGD(Art_Kr), Art_Kr; # *Maple* **8**

{ps, gdi, hpgl, inline, jpeg, dxf, window, wmf, bmp, x11, tek, char, gif, pcx, pov}, [PostScript, X11, bmp, char, cps, default, dumb, dxf, gdi, gif, hpgl, hplj, inline, jpeg, laserjet, pcx, postscript, pov, ps, tek, tektronix, window, wmf, x11, xwindow]

> LGD(Art_Kr), Art_Kr; # *Maple* **9**

{inline, hpgl, jpeg, gdi, wmf, bmp, pov, dxf, ps, pcx, x11, tek, char, window, gif}, [PostScript, X11, bmp, char, cps, default, dumb, dxf, gdi, gif, hpgl, hplj, inline, jpeg, laserjet, pcx, postscript, pov, ps, tek, tektronix, window, wmf, x11, xwindow]

Polyhedra – check of a polyhedron type

Call format of the procedure:

Polyhedra(n)

Formal arguments of the procedure:

n – a positive integer from the range **1 .. 123** (for releases **6, 7, 8** and **9**)

Description of the procedure:

The procedure call **Polyhedra(n)** provides a visual representation of polyhedrons that are supported by the ***polyhedraplot*** procedure of the *Maple* package. Over a **n** number (**1<=n<=123**) the procedure call **Polyhedra(n)** outputs the polyhedron corresponding to the **n** number; its appearance is defined by default or by settings coded as optional actual arguments behind the first argument **n.** Output of a polyhedron is accompanied by its name.

```
Polyhedra:= proc(n::posint)
local k, L, Opt;
  L := sort([op(plots[polyhedra_supported](n))]);
  `if`(n < 1 or nops(L) < n, ERROR("The first argument <%1> is wrong", n), NULL);
  if 2 <= nargs then Opt := args[k] $ (k = 2 .. nargs) else Opt := style = patch  end if;
  RETURN(plots[polyhedraplot]([0, 0, 0], polytype = L[n], Opt, title = L[n]))
end proc
```

Typical example of the procedure use:

> Polyhedra(61, titlefont= [TIMES, BOLD, 14], lightmodel=light2, thickness=2);

MetagyrateDiminishedRhombicosidodecahedron

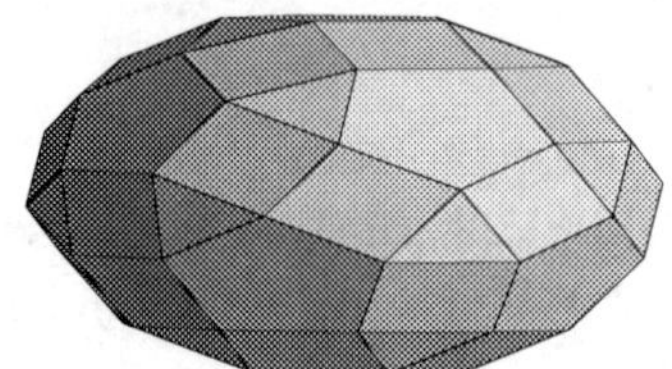

Q2plot – an quick data graph

Call format of the procedure:

Q2plot(X, Y {, x {, y} {, s}})

Formal arguments of the procedure:

X – a list of data along **X**-axis
Y – a list of data along **Y**-axis
x, y – (optional) symbols; labels along axes **X** and **Y** accordingly
s – (optional) a symbol defining style of a data graph

Description of the procedure:

The package gives the developed enough tools for a graphical data representation and functional dependences. However, in many cases it appears by useful to have simple tools of prompt visualization of numerical data, not addressing oneself to the package tools. The **Q2plot** procedure can be as one of such tools.

The first two arguments of the procedure define lists of data values along axes **X** and **Y** accordingly, whereas the optional third and fourth arguments set the labels along these axes. If the optional actual **s** argument has been coded as ***LINE***, straight lines connect the data points; otherwise, the procedure simply outputs the data points.

```
Q2plot:= proc(X::list, Y::list)
local k;
  `if`(nops(X) = nops(Y), PLOT(CURVES([seq([X[k], Y[k]], k = 1 .. nops(X))],
  COLOUR(RGB, 0, 0, 0)), SYMBOL(CIRCLE, 12), STYLE(`if`(member(LINE, [args]), LINE,
  POINT)), THICKNESS(2), AXESLABELS(`if`(2 < nargs and args[3] <> LINE,
  convert(args[3], string), ""), `if`(3 < nargs and args[4] <> LINE, convert(args[4], string), "")),
  VIEW(convert({op(X)}[1], float) .. convert({op(X)}[-1], float), DEFAULT)),
  ERROR(`Disparity of the first two arguments [%1 <> %2]`, nops(X), nops(Y)))
end proc
```

Typical examples of the procedure use:

> X:=[1,2,3,4,5,6]: Y:=[61,56,36,40,14,7]: P:=Q2plot(X,Y,name,age,LINE): P1:=Q2plot(X,Y): plots[display](array(1..2, [P,P1]), axesfont= [TIMES,BOLD,10], thickness=2);

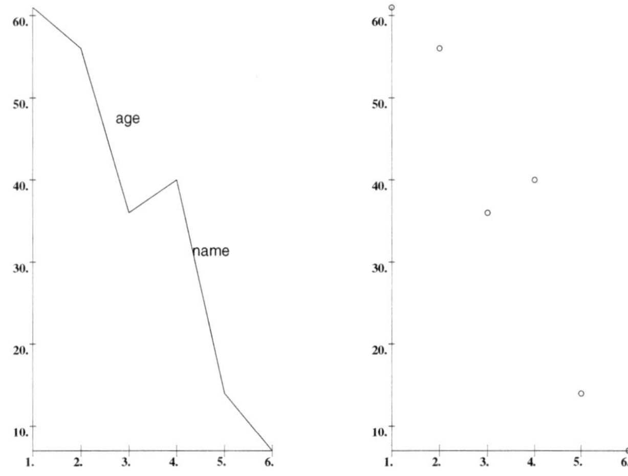

The resulted example, in particular, illustrates an incorrectness of the package at formation of labels of axes of coordinates in a case when the schedule is located in structures of a type {*array, vector, matrix*}. It is necessary to note, the similar note takes place and concerning of a series of other graphic characteristics [11-13,29-33].

Smart – creation of 2-dimensional or 3-dimensional plots

Call format of the procedure:

Smart(*args*)

Formal arguments of the procedure:

args – a sequence of algebraic expressions or equations

Description of the procedure:

Concerning graphic tools of the previous releases of the package *Maple*, its releases 6 – 9 not only differ by substantial improvement of the existing tools, but also they have been extended with a new possibility of fast receiving of so-called intellectual (*smart*) graphs defined by the *Maple* expressions. The given opportunity is supported for the present only at a level of the package shell and at a level of tools of its language by means of only two procedures **smartplot** and **smartplot3d** that have been considered in our books [13,14] in detail enough. Both procedures serve for output of graphs of the dependences determined by algebraic expressions or their set in the united system of the Cartesian coordinates of dimension 2 and 3 accordingly.

The standard procedures **smartplot** and **smartplot3d** are used primarily by the graphical user interface to generate initial plots that can be further tailored via use of interactive controls. Meanwhile, the use of these procedures on expressions/equations from quantity of independent variables greater than dimensionality without one can cause the erroneous situations. Therefore for more effective use of the given procedures is recommended to apply the **Smart** procedure that handles the basic special and erroneous situations, arising at calls of the standard procedures **smartplot** and **smartplot3d.**

The procedure call **Smart(*args*)** outputs the graphs of one or more algebraic expressions or equations defined by the actual ***args*** arguments. Syntax of the actual ***args*** arguments fully corresponds to the syntax of procedures arguments **smartplot** and **smartplot3d.**

```
Smart:= proc()
local k, L, Z;
  if nargs = 0 then RETURN() else L, Z := {}, {} end if;
  for k to nargs do
    if type(args[k], algebraic) then L := {nops(indets(args[k], symbol)), op(L)}
    elif type(args[k], `=`) and type(rhs(args[k]), algebraic) and type(lhs(args[k]), algebraic) then
      Z := {nops(indets(args[k], symbol)), op(Z)}
    else ERROR("Among arguments are invalid types <%1>", args)
    end if
  end do;
  if L minus {0, 1} = {} and Z minus {1, 2} = {} then smartplot(args)
  elif L minus {1, 2} = {} and Z minus {2, 3} = {} then smartplot3d(args)
  else ERROR("Among arguments are invalid expressions <%1>", args)
  end if
end proc
```

Typical examples of the procedure use:

```
> Smart(X^2+b*Y^2+c*Z);  Smart(a<>b, 8*X^2+X*Y+Y^2+Z+H, c<d);
    Error, (in Smart) Among arguments are invalid expressions <X^2+b*Y^2+c*Z>
    Error, (in Smart) Among arguments are invalid types <a <> b>
```

```
> Smart(y^2, x*cos(x), y^2, h, -t);  Smart(x^2+y^2+z^2=20, y^2+x^2);  Smart(x^2+y^2+z^2=20);
```

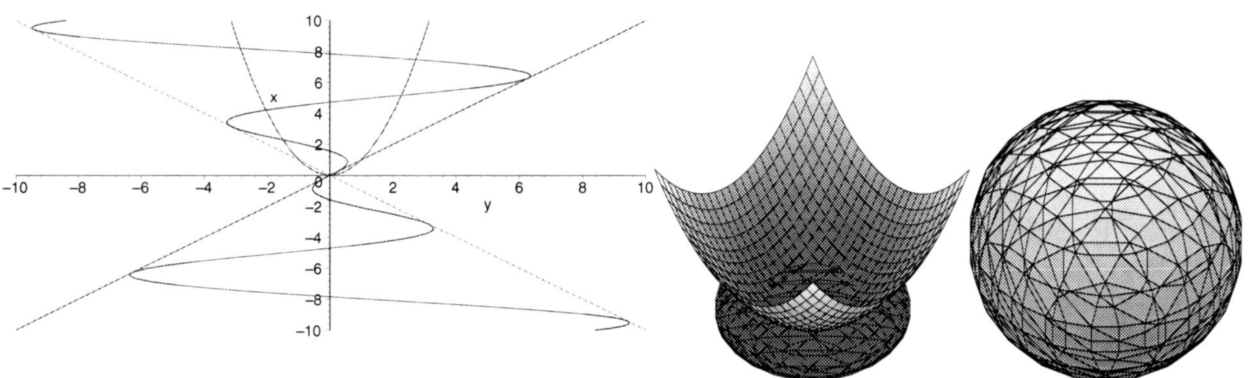

The example represented below illustrates an opportunity of the **Smart** procedure for graphical searching of real roots of the polynomial equations.The given reception appears useful enough at numerical calculation of roots of the equations and their systems when the behavior of curves is complex enough in the area of roots. On the basis of the given approach, it is possible to receive graphically effectively enough the initial values for calculated roots of the equations and their systems.

```
> Smart(x^4-10*x^2+3=x^2);  evalf(solve(x^4-10*x^2+3=x^2,x));
```

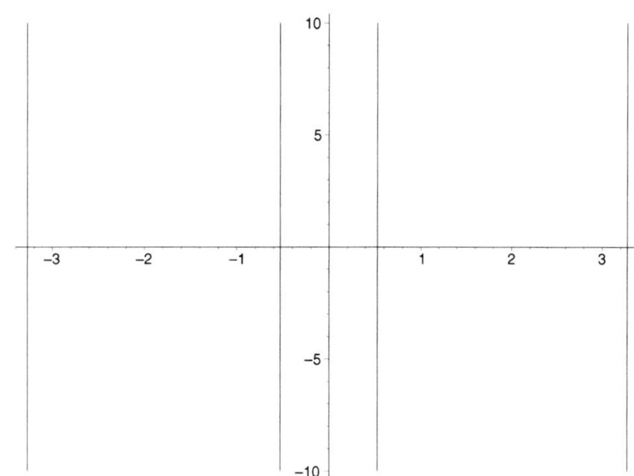

-3.274164512, 3.274164512, -0.5290054300, 0.5290054300

The given graphical approach in a great extent can be useful for solution of the algebraic equations and their systems of more general view.

Smart_Plot - converting of a graphic Smart structure into standard graphic structure

Call format of the procedure:

Smart_Plot(S, P)

Formal arguments of the procedure:

S - a functional dependence
P - an assignable name

Description of the procedure:

As it was already marked, the functions {***smartplot, smartplot3d***} take somewhat special place among graphical tools of the *Maple* language and their sharing with other graphical tools is enough limited. The reason of it is that they do not return standard graphic data structures and are oriented only onto direct visualization of graphic *"sketches"* of functional dependences. The **Smart_plot** procedure represented below solves this problem.

The procedure call **Smart_Plot(S, P)** returns via the second actual **P** argument the result of converting of a *Smart*-structure, defined by the first **S** argument, into the standard **{2D | 3D}** graphic structure. The **Smart_Plot** procedure does not process a *Smart* structure created by the procedure **smartplot{3d}**, if its actual argument is an equation. The **Smart_Plot** procedure provides a convenient tool for support of combined use of both the above *smart* tools and the standard graphical tools of *Maple*.

```
Smart_Plot:= proc(S::function, P::evaln)
local a, Z;
  [assign(Z = convert(S, symbol), a = smartplot), assign('Z' = `if`(searchtext(cat(a, `2D`), Z) <> 0,
    `if`(type(op(S)[1], `=`), RETURN(), plot), `if`(searchtext(cat(a, `3D`), Z) <> 0,
    `if`(type(op(S)[1], `=`), RETURN(), plot3d), RETURN())))];
  `if`(Z = plot, assign('Z' = Z(op(op(S)))), assign('Z' = Z(op(S)[1], op(S)[2] = -10 .. 10,
    op(S)[3] = -10 .. 10))), assign(P = Z)
end proc
```

Typical example of the procedure use:

> SM:= smartplot3d(x^2+y^2): Smart_Plot(SM, PS); PS;

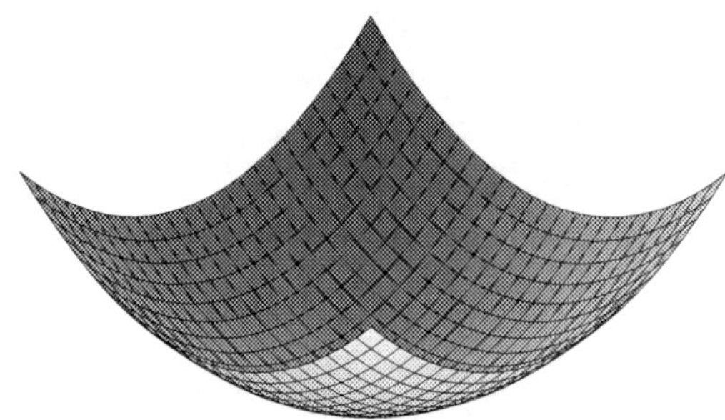

T_Font – restoration of fonts for the graphs titles

Call format of the procedure:

T_Font(ST, U)

Formal arguments of the procedure:

ST - a function defining a graphical structure
U - an equation defining a setting for the *font*-option

Description of the procedure:

In a series of cases because of redefinition of fonts for titles of 3D graphical objects, the losses of a text are possible. The given situation takes place for all releases of the *Maple* package, complicating the text appearance of graphical objects. The **T_Font** procedure represented below eliminates the indicated shortcoming and gives the extended tools of text appearance of the 3D graphical objects.

The first actual **ST** argument of the procedure **T_Font(ST, U)** defines a 3D graphical structure, whereas the second actual **U** argument defines the graphical option ***`titlefont = Value`***. The procedure defines the global *_Art* variable whose value is a 3D graphical structure - a result of the corresponding modification of the initial graphic structure **ST**.

```
T_Font:= proc(ST::function, U::`=`)
local n, m, h, p, Kr, Ln, F;
  assign(Kr = convert(ST, symbol), F = cat([libname][1][1 .. 2], "/$Kr"));
  `if`(nargs <> 2, ERROR("Number of arguments is wrong: %1", nargs),
    `if`(searchtext(PLOT, Kr) = 0 and searchtext(PLOT3D, Kr) = 0, RETURN(),
    `if`(lhs(args[2]) <> titlefont, RETURN(), assign(h = length(Kr)))));
  assign(n = searchtext(TITLE, Kr)), assign(m = n + searchtext(")", Kr, n .. h) - 1);
  p := n + searchtext(FONT, Kr, n .. m) - 1;
  assign(Ln = open(F, WRITE)), writebytes(Ln, cat(`_Art:=`, substring(Kr, 1 .. n - 1)));
  if p = n - 1 then writebytes(Ln, cat(substring(Kr, n .. m - 1), ",", "FONT(",
    substring(convert(rhs(args[2]), symbol), 2 .. -2), ")", substring(Kr, m .. h), ";"))
  else writebytes(Ln, cat(substring(Kr, n .. p + 4), substring(convert(rhs(args[2]), symbol), 2 .. -2),
    substring(Kr, m .. h), ";")), close(Ln)
  end if;
  (proc(x) read x; fremove(x) end proc)(F)
end proc
```

Typical example of the procedure use:

```
> S:= plot3d({x^6+y^6, 61}, x=-7..7, y=-7..7, title = `Artur_Kristo`): T_Font(S, titlefont =
  [HELVETICA, BOLDOBLIQUE, 20]),  _Art;
```

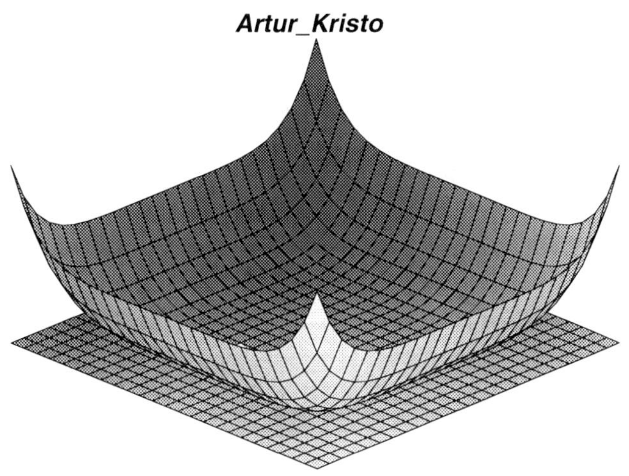

plotvl – an extension of the standard procedure *`plot`*

Call format of the procedure:

plotvl(F, h {, v} {, *opts*} {, *pvl*})

Formal arguments of the procedure:

F - function(s) to be plotted
h - horizontal range
v - (optional) vertical range
opts - (optional) graphical options
pvl - (optional) option defining a setting of vertical lines

Description of the procedure:

In a series of cases the necessity of partition of area of a 2D-graph by means of setting of vertical lines in the given points of abscissa axis arises. The procedure **plotvl** solves this problem.

All procedure arguments (both *obligatory* and *optional* ones) with the exception of ***pvl***-option are similar to the standard `***plot***` procedure, which is intended for creation of a two-dimensional plot of functions. The procedure call **plotvl(F, h, ...)** without optional *pvl*-option is equivalent to the procedure call **plot(F, h, ...)**.

The ***pvl***-option has one of following two coding formats, namely:

pvl = [a, b, c, ...] or **pvl = [[a, b, c, ...],** ***graphical options*]**

where: **a, b, c, ...-** values of *realnum*-type whereas `***graphical options***` are the same as graphical options for the standard procedure `***plot***` but on the assumption that only one function is considered. The points **[a, b, c, ...]** of setting of vertical lines should belong to **h**-interval, otherwise in points outside this interval, the vertical lines will not be set. The procedure handles basic especial and erroneous situations. The **plotvl** procedure has been successfully used in many problems dealing with numerical evaluation of roots of algebraic equations and their systems.

```
plotvl:= proc(F, h)
local a, b, c, d, g, vl, k, z, t, u, Xmin, Xmax, Ymin, Ymax;
  assign67(b = [], t = [], u = NULL);
  vl := proc(L::list(realnum), n::realnum, m::realnum)
    local a, k;
      a := [color = black, linestyle = 1, thickness = 1];
      plots[display](seq(plottools[line]([args[1][k], n], [args[1][k], -abs(m)]),
        k = 1 .. nops(args[1])), `if`(nargs = 3, op(a), args[4 .. nargs]))
    end proc;
  seq(`if`(type(args[k], equation) and lhs(args[k]) = 'pvl', assign('c' = rhs(args[k])),
    assign('b' = [op(b), args[k]])), k = 1 .. nargs);
  if nops(b) = nargs then plot(op(b)) else
    if type(c, list(realnum)) and c <> [] or type(c, nestlist) and type(c[1], list(realnum)) and
    c[1] <> [] then if type(c, list(realnum)) then z := map(evalf, c)
      else z := map(evalf, c[1]); u := op(c[2 .. -1]) end if;
      try g := plot(op(b)); a := convert(g, string); d := Search2(a, {"[[", "]]"});
        for k by 2 to infinity do
          try t := [op(t), op(came(a[d[k] .. d[k + 1] + 1]))]
          catch: break
          end try
        end do;
        assign(Xmin = SLj(t, 1)[1][1], Xmax = SLj(t, 1)[-1][1],
          Ymin = SLj(t, 2)[1][2], Ymax = SLj(t, 2)[-1][2]);
        z := [seq(`if`(belong(z[k], Xmin .. Xmax), z[k], NULL), k = 1 .. nops(z))]
      catch: WARNING("main plot has been defined incorrectly - %1", b);
            RETURN(vl(z, 10, -10, u))
      end try;
      if z = [] then plot(op(b)) else t:= vl(z, Ymax, Ymin, u); plots[display](g, t) end if
    else ERROR("option <pvl> is invalid - %1", c)
    end if
  end if
end proc
```

Typical example of the procedure use:

> f:= [TIMES, BOLD, 11]: plotvl([x^2, x^3, 62*x+57], x=-10..10, thickness=[2,2,2], pvl=[[-5,-9,2,8,62,57],color=blue], color=[red,magenta,brown], labelfont=f,axesfont=f);

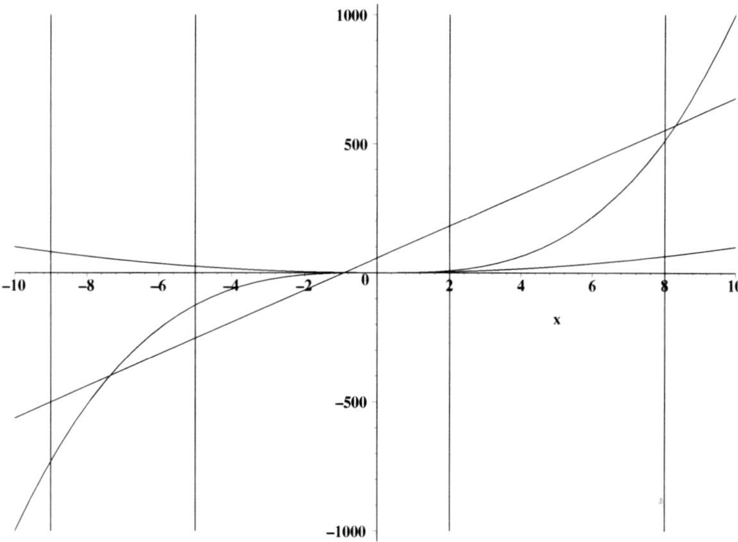


```
> plotvl([x^2, x^3, 62*x+57], x=-10..10, thickness=[2,2,2], color=[green,gold,magenta],
  labelfont=f, axesfont=f);
```

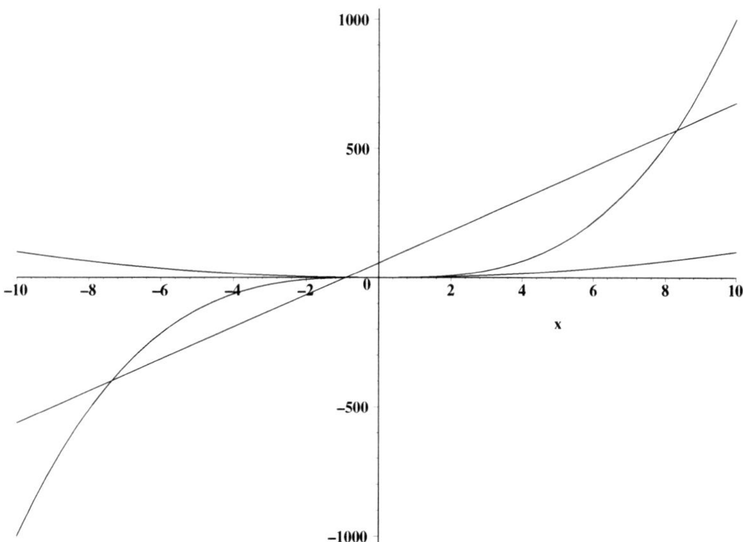

The procedure has many appendices; in particular, the procedure can be rather useful at graphical solution of equations and their systems, for horizontal calibration of 2D-graphs, etc.

In conclusion of the present chapter considering questions of work with graphic objects in the *Maple* environment, quite pertinently to make a few rather essential remarks. Absence of the advanced mechanism of dynamic memory release from "garbage" in a working area of the package leads to ineffective use of the RAM, to rise of swapping intensity and to decrease of reactivity of *Maple*. As a result can be a closing of a current *Maple* session. In addition, after the unforeseen and unauthorized termination of a current session the package nevertheless holds module "**mserver.exe**" in active tasks queue of *Windows* (in particular, the matter concerns *Maple* **8** and *Windows XP Pro*) what is absolutely inadmissible since the subsequent loadings of the package only aggravate the given situation. It is necessary to mark, with growth of number of the *Maple* release is noticeably reduced its reactivity. A layout distortion of the axial labels takes place at output of two-dimensional graphs into *array*-structure. A series of other essential remarks about the *Maple* shortcomings at work with graphical objects can be found in our books [29-31,33,39,43,44]. It demands improvement of the appropriate *Maple* algorithms, above all of algorithms dealing with processing of graphical objects.

frame_n – determination of the given frame of animated graph of functions

Call format of the procedure:

frame_n(d, F, x, {, y} t {, *gopt*} {, *frame* = n})

Formal arguments of the procedure:

d - dimensionality of graphical objects (should be **2** or **3**)
F - function(s) to be animated
x, y - **x** axis range and **y** axis range
t - frame parameter range
gopt - (optional) graphical options admitted by the `*plot*` procedure (for **d=2** only)
frame = n - (optional) option defining return of frame with number **n**

Description of the procedure:

In a series of cases, the problem of determination of *frame* with the given number of an animated graph of functions arises. It is enough essentially, inasmuch as at copying of an animated object its first *frame* is being copied only. The procedure call **frame_n(d, F, x, {, y} t {, *gopt*} {, *frame* = n})** returns the nth *frame* of an animated graph defined by the same factual arguments as for standard `**animate{3d}**` procedure. If the last option **frame=n** has been not coded, the procedure call is equivalent to the call of the standard `**animate{3d}**` procedure. If **d=2,** the procedure call admits optional graphical `***plot***` options. The procedure has many useful appendices.

```
frame_n:= proc(d::{2, 3}, F)
local as;
  if type(args[-1], equation) and lhs(args[-1]) = `frame` then
    if belong(rhs(args[-1]), rhs(args[d + 2])) then
      as := plots[`if`(d = 2, animate, animate3d)](args[2 .. -2]);
      RETURN(plots[display](op(op(as)[1])[rhs(args[-1])]))
    else ERROR("<%1> is inadmissible value for the frame number", rhs(args[-1]))
    end if
  end if;
  plots[`if`(d = 2, animate, animate3d)](args[2 .. -1])
end proc
```

Typical examples of the procedure use:

```
> frame_n(3, cos(t*x)*sin(t*y), x=-Pi .. Pi, y=-Pi .. Pi, t=1 .. 8, frame = 6);
> frame_n(2, {x-x^3/u, sin(u*x), x/u}, x=0 .. Pi/2, u=1 .. 16, color=magenta, frame = 8);
> a:=Pi: frame_n(3,{x^2+t*x*y+y^2,sin(t*x)+cos(t*y)},x=-a..a,y=-a..a,t=1..15,frame=8):
> frame_n(3, {x^2+t*x*y+y^2, sin(t*x)+cos(t*y)}, x=-a..a, y=-a..a, t=1..15, frame=37):
   Error, (in frame_n) <37> is inadmissible value for the frame number
> frame_n(0, {x^2+t*x*y+y^2, sin(t*x)+cos(t*y)}, x=-a..a, y=-a..a, t=1..15, frame=15):
   Error, invalid input: frame_n expects its 1st argument to be of type {2,3}, but received 0
```

In particular, the **frame_n** procedure provides decomposition of an animated {2D|3D}-graph of functions onto frames composing it. In a series of cases this possibility is rather useful, for example, at preparing of the printing materials or for illustration of mechanisms of animation in the *Maple* environment.

In more details it is possible to familiarize oneself with the tools represented in the given section in a series of our books [8-14,29-33,43,44], whereas the source texts of all these tools can serve as first-hand examples of programming of the graphical objects in the *Maple* environment.

Chapter 9.
Tools that extend and improve the standard tools of *Maple*

In the given section, the tools that extend and improve the standard *Maple* tools for releases 6 - 9.5 are represented. These tools are used enough widely both at operation with the *Maple* package in interactive mode and at programming of various problems in its environment. They represent undoubted interest at programming of various problems in the *Maple* environment, both by simplifying the programming and by making it by more clear.

Aobj – testing of types of the active *Maple* expressions

Call format of the procedure:

Aobj()

Formal arguments of the procedure: **no**

Description of the procedure:

The procedure call **Aobj()** returns the table whose *indices* are types of all objects *active* in a current *Maple* session, while its *entries* - a set of identifiers corresponding to them. The **Aobj** procedure seems as useful supplement to the standard tools {*type, whattype, hastype, hasfun,* etc.} of testing of the expressions types.

```
Aobj:= proc()
local k, A, B, x, y;
  assign(A = [anames()]), assign(B = seq([whattype(eval(A[k])), A[k]], k = 1 .. nops(A)));
  assign(x = {seq(B[k][1], k = 1..nops([B]))}), assign(y = table([seq(x[k] = {}, k = 1..nops(x))]));
  for k to nops([B]) do y[B[k][1]] := {op(y[B[k][1]]), B[k][2]} end do;
  assign('y[procedure]' = y[procedure] minus {assign, Aobj, interface}), eval(y)
end proc
```

Typical example of the procedure use:

```
> restart: St:="RANS": sym:=`IAN`: Vic:=19.42: A:=2003: M:=module () end module:
  Proc:=proc() end proc: Mt:=Matrix(): Art_Kr:=7/14: tab:=table(): List:=[]: Set:={}:
  func:=Tallinn(x): Z:=sqrt(61): Aobj();
```

table([*function* = {*type/interfaceargs, func*}, *set* = {*Set*}, *procedure* = {*Matrix, Proc, sqrt*}, *float* = {*Vic*}, *string* = {*St*}, *integer* = {*A, sqrt/primes*}, *symbol* = {*sym*}, *list* = {*List*}, *table* = {*tab*}, *Matrix* = {*Mt*}, ^ = {*Z*}, *fraction* = {*Art_Kr*}, *module* = {*M*}])

Call of this procedure after `*restart*` command allows to obtain both the main objects activated in a current *Maple* session and the package objects accompanying them.

Aproc – generating a curried procedure

Call format of the procedure:

Aproc(F, L, p)

Formal arguments of the procedure:

F - a procedure or a name to be curried
L - an expression sequence of arguments to be curried over
p - a positive integer defining a curing mode

Description of the procedure:

In the package *Maple* 6 for possibility of extension of the list of arguments of a procedure two new procedures ***curry*** and ***rcurry*** have been introduced. As a generalization of these procedures the **Aproc(F, L, p)** procedure providing an extension of formal arguments of a procedure both {to the left|to the right} and within of its formal arguments is offered. The procedure call **Aproc(F, L, p)** returns the result of extension of a procedure defined by the first actual **F** argument by a list of the formal arguments defined by the second actual **L** argument according to the given mode defined by the third actual **p** argument.

The third argument **p** defines the following modes of extension: **0** – extension to the left (analog of the ***curry*** procedure), **-1** – extension to the right (analog of the ***rcurry*** procedure), and in case of other permissible positive integer **p** the extension of arguments is done within after their **p**th argument. For **p** argument the values in the integer range **[-1 .. nops([op(1, eval(F))]) – 1]** are permitted only, otherwise erroneous situation arises. The given procedure seems as useful enough tool for generation of the procedures with the extended list of formal arguments from existing procedures in a context of the user procedures activated in a current *Maple* session.

```
Aproc:= proc(F::symbol, A::list, p::integer)
local a, k, n, m, P, R, S, Z, h, cd;
  assign(S = "", Z = ""), seq(assign('S' = cat(S, A[k], ",")), k = 1 .. nops(A));
  R:= (a, b, h) -> cat(P[1 .. a], `if`(h = -1, ",", ""), S[1 .. length(S) – 1], `if`(h = 0, ",", ""),
     P[b .. length(P)]);
  Release1('h'), assign(cd = currentdir()), currentdir(h), (proc(x) save x, "$Kr$" end proc)(F);
  assign(P = readbytes("$Kr$", TEXT, infinity)), close("$Kr$"), assign(a = [op(1, eval(F))]),
      assign('P' = cat("`ъ`", P[searchtext(":=", P) .. length(P)]));
  assign(n = searchtext("proc (", P) + 5), assign(m = searchtext(")", P, n .. length(P)) + n – 1);
  if nops(a) = 0 then writebytes("$Kr$", R(n, m, 2)), close("$Kr$"); read "$Kr$";
    RETURN(eval(`ъ`), fremove("$Kr$"))
  elif p = 0 or p = -1 then writebytes("$Kr$", `if`(p = 0, R(n, n + 1, 0), R(m – 1, m, -1))),
      close("$Kr$"); read "$Kr$"; RETURN(eval(`ъ`), fremove("$Kr$"))
  elif nops(a) <= p or p < 1 then ERROR(`Invalid third argument <%1>`, p)
  else seq(assign('Z' = cat(Z, a[k], ",")), k = 1 .. p), assign('Z' = cat(Z, S)),
       seq(assign('Z' = cat(Z, a[k], ",")), k = p + 1 .. nops(a)); writebytes("$Kr$", cat(P[1 .. n],
       Z[1 .. length(Z) – 1], P[m .. length(P)])), close("$Kr$"); read "$Kr$";
    RETURN(eval(`ъ`), fremove("$Kr$"), currentdir(cd))
  end if
end proc
```

Typical examples of the procedure use:

```
> Art:= proc() `+`(args)/nargs  end proc: Aproc(Art, [x,y,z,t,v,r,w], 0);
                    proc(x, y, z, t, v, r, w, a, b, c, d) `+`(args)/nargs end proc
```

```
> Aproc(Art, [x,y,z,t,v,r,w], -1);
                 proc(a, b, c, d, x, y, z, t, v, r, w) `+`(args)/nargs end proc
> Aproc(Art, [x,y,z,t,v,r,w], 2);
                 proc(a, b, x, y, z, t, v, r, w, c, d) `+`(args)/nargs end proc
> G:= (a,b,c,d,f,g) -> [args]: Aproc(G, [x,y,z,h], -1), Aproc(G, [x,y,z,h], 0);
          (a, b, c, d, f, g, x, y, z, h) -> [args], (x, y, z, h, a, b, c, d, f, g) -> [args]
> Aproc(G, [x,y,z,t,h,r,w], 2);
                        (a, b, x, y, z, t, h, r, w, c, d, f, g) -> [args]
```

Assign - performing of the assignments
assign6
assign7
assign67

Call format of the procedures:

Assign(*args*)
assign6(*args*)
assign7(*args*)
assign67(*args*)

Formal arguments of the procedures:

args – a sequence of actual arguments

Description of the procedures:

The changes, which touched the standard procedure `***assign***`, in particular, are one of reasons of incompatibility "*from above – downwards*" of release **6** on the one hand and the releases **7 - 9.5** on the other hand. The given problem is essential at translating of software created in environment of the *sixth* release, into environment of releases **7 - 9.5.** This problem is considered in our books [8,9,29-33,43,44] in detail. For its decision, we used two basic approaches. The first has consisted in adaptation of the existent tools for all releases what demanded their reprocessing whereas new tools already beforehand were planned to execution in five last releases. The second approach has consisted in creation of a managing procedure that depending on a current *Maple* release should execute the corresponding procedure, oriented onto one or another release. The procedures, represented below, eliminate the above problem of incompatibility of the releases along with extending of possibilities of the standard procedure `***assign***`, allowing to execute the assignments of sequences of expressions to variables, in particular.

If a sequence of actual arguments at the procedure call **assign7** contains tuples of a view **<a=b, x1, x2, x3, ..., xp>** (at satisfying of the condition **map(type, {b, seq(xj, j=1 .. p)}, `*equation*`) = {*false*}**), then the procedure call returns a result which is analogous to a result of the procedure call **assign(a,b,x1,x2,x3,...,xp).** Furthermore, on actual arguments causing the erratic situations for `***assign***`-procedure, the procedure **assign7** returns the *NULL* value, providing the continuity of computations. In other cases the **assign7** procedure is similar to the standard procedure `***assign***`. Such generalization of the standard function `***assign***` allows, in particular, in one call of the **assign7** procedure to execute assignment of a sequence of values to several variables simultaneously (see examples below). The implementation of the procedure **assign7** has been optimized relatively to conditions of environment of the *seventh* release. The told concerning the procedure **assign7** for release **7** is valid for releases **8** and **9/9.5** of *Maple* also.

Similarly, the **assign6** procedure extends the standard procedure *`assign`* of the **6**th release according to its opportunities. The procedure call **assign6(x{,|=}a,b,c,..)** returns the *NULL* value, providing assignment **x:=[a,b,c,...]**. While the call on actual argument of a view **(x1,x2,…,xn) = (a1,a2,…,an)** also returns the *NULL* value with providing of multiple assignments **xj:=aj (j= 1..n).** The procedure call **assign6(X=NULL)** returns the *NULL* value, providing assignment of the *NULL* value to a variable **X.** In addition, the ending of the call **assign6(X=*NULL*)** by the delimiter **";"** automatically assigns the *NULL* value to **X**, otherwise for that a separate reference to **X** is demanded. While on remaining configurations of actual arguments, the call of procedure **assign6** is similar to the call of the standard procedure *`assign`* of the **6**th release. In addition, the procedure call **assign6** is correct on any tuple of the actual arguments, providing a rather wide set of types and varieties of assignments, along with the continuity of calculations.

In many cases the given properties, by which the represented expansions of the standard procedure *`assign`* are provided, seem useful enough in practical programming. In addition, depending on a release {**6**|**7**, **8**, **9/9.5**} the appropriate procedure **assign**{**6**|**7**} or the procedure manager ***Assign*** irrespectively from a concrete release can be used. The **Assign** procedure (depending on a current *Maple* release) realizes the call of one of the above procedures **assign6** (release **6**) or **assign7** (releases **7**, **8** or **9/9.5**).

At last, for the package of releases **6, 7, 8** and **9/9.5** the procedure **assign67** can be offered as a replacement of the standard procedure *`assign`*. The **assign67** procedure by own opportunities exceeds the standard procedure *`assign`* of the package of releases **6 – 9/9.5,** providing in addition operations such as: multiple assignments of sequences of values to variables, multiple assignments of the *NULL* values, etc. In addition, call of the **assign67** procedure is correct on any tuple of the actual arguments, providing a rather wide set of types and varieties of assignments, along with the continuity of calculations. The given procedure in many cases allows essentially to simplify programming in the *Maple* environment of releases **6 – 9/9.5.** It is best; the **assign67** possibilities can be illustrated on a series of the most typical examples represented below. For example, if the procedure call **assign67(x1,x2,...)** as the actual arguments **xj** contains indefinite variables only, the assignment **x1:= x2,x3,...** is done. While for the procedure call **assign67(x1,x2,…,xk, y=a,...)** the assignments **x1:= *NULL*, … , xk:= *NULL*** are done.

```
Assign:= proc()
local R;
  R := Release(); if R = 6 then assign6(args) elif 7 <= R then assign7(args) else assign(args)
  end if
end proc

assign6:= proc()
local f, k;
  try
    `if`(nargs = 1 and type(eval(args), symbol), RETURN(NULL), `if`(nargs = 1 and
       type(args, `=`) and rhs(args) = NULL, RETURN(null(unassign('`0`'),
      (proc() global `0`; `0` := NULL end proc)(), macro(lhs(args) = `0`))),
       assign(f = (proc() seq(assign(args[1][k], args[2][k]), k = 1 .. nops(args[1])) end proc))));
    `if`(nargs = 1 and type(args, equation) and nops([lhs(args)]) = nops([rhs(args)]),
      f([lhs(args)], [rhs(args)]), `if`(1 < nargs and map(type, {args}, equation) = {true},
      assign(args), `if`(type(args[1], `=`), assign(lhs(args[1]), [rhs(args[1]), args[2 .. -1]]),
      assign(args[1], [args[2 .. -1]]))))
  catch "invalid subscript selector": RETURN(NULL)
  catch "invalid arguments": RETURN(NULL)
```

```
  end try
end proc
assign7:= proc()
local ass, k;
  ass:= proc()
    local a, b, c, k, p;
      try
        `if`(nargs = 1 and type(args, equation) and nops([lhs(args)]) = nops([rhs(args)]),
          RETURN(assign(args)), assign(a = [args], b = []));
        add(`if`(type(a[k], equation), assign('b' = [op(b), k]), 1), k = 1 .. nargs);
        `if`(not type(a[1], equation), RETURN(assign(args)), assign(c = nops(b)));
        seq(assign(lhs(a[b[p]]), rhs(a[b[p]]), op(a[b[p] + 1 .. b[p + 1] - 1])), p = 1 .. c - 1),
          assign(lhs(a[b[-1]]), rhs(a[b[-1]]), op(a[b[-1] + 1 .. nargs]))
      catch "invalid subscript selector": RETURN(NULL)
      catch "invalid arguments": RETURN(NULL)
      end try
    end proc;
  if nargs = 0 or nargs = 1 and not type(args, equation) then RETURN(NULL)
  elif type(args[1], equation) then ass(args)
  else for k to nargs do if not type(args[k], equation) then next else break end if end do
  end if;
  try if k < nargs then ass(args[k .. -1]) else RETURN(NULL) end if
  catch "cannot determine if this expression is true or false": RETURN(NULL)
  end try
end proc
assign67:= proc(X)
local Art, Kr, k;
  `Copyright (C) 2002 by the International Academy of Noosphere. All rights reserved.`;
  Art, Kr := [], [];
  try
    for k to nargs do if type(args[k], equation) then Kr := [op(Kr), k] else NULL end if end do;
    if nargs = 0 then NULL elif Kr = [] and type(X, {symbol, name, `::`, function})
    then X := args[2 .. -1]
    elif nargs = 1 and type(X, equation) and nops([lhs(X)]) = nops([rhs(X)]) then
      seq(procname(op(k, [lhs(X)]), op(k, [rhs(X)])), k = 1 .. nops([lhs(X)]))
    elif nargs = 1 and type(X, {list, 'set'}) then seq(procname(op(k, X)), k = 1 .. nops(X))
    elif type(X, `=`) and Kr = [1] or type(X, {list, 'set'}) then procname(op(X), args[2 .. -1])
    elif Kr <> [] and type(X, {symbol, name, `::`, equation, function}) then
      for k to Kr[1] - 1 do procname(args[k]) end do;
      for k to nops(Kr) - 1 do procname(args[Kr[k] .. Kr[k + 1] - 1]) end do;
      procname(args[Kr[-1] .. -1])
    else procname('Art' = [seq(`if`(type(args[k], name)  or type(args[k], equation) and
      type(lhs(args[k]), name), k, NULL), k = 1 .. nargs)]),
      `if`(Art = [], NULL, procname(args[Art[1] .. -1]))
    end if
  catch "wrong number (or type) of parameters": seq(procname(args[Kr[k]]), k = 1 .. nops(Kr))
  end try;
  NULL
end proc
```

Typical examples of the procedures use:

assign7-procedure for releases **7, 8** and **9/9.5**

> assign7(w, x=a,b,c, y=m, z=n,{a1,a2}, t=h,p, v=b); [x], [y], [z], [t], [v], [w];

[a, b, c], [m], [n, {a2, a1}], [h, p], [b], [w]

> restart: assign7(a,b,c, x=a1,a2,a3, y=b1,b2,b3, z=c1,c2,c3); [a,b,c], [x], [y], [z];

[a, b, c], [a1, a2, a3], [b1, b2, b3], [c1, c2, c3]

> restart: assign7(Art,Kr, v=h,r,z,61,42,2003); [v], [Art], [Kr];

[h, r, z, 61, 42, 2003], [Art], [Kr]

> restart: assign7([assign7('h'=iostatus())]); h;

[0, 0, 7]

> restart; assign7(IAN=*NULL*), assign7(lbz), [IAN], [lbz];

[], [lbz]

> restart: assign7(h,p, x=b,c,v,op({y=d}),e,f); [x], [y], [h,p];

[b, c, v], [d, e, f], [h, p]

assign6-procedure for release **6**

> assign6(G=56, V=61, Ar=40, Kr=7, Art=14, Y=2003); G, V, Ar, Kr, Art, Y;

56, 61, 40, 7, 14, 2003

> assign6((x,y,z,u,t,h)=(a,b,c,d,f,v)); [x, y, z, u, t, h], assign6(Art_Kr), assign6(Q=NULL), Art_Kr; Q;

[a, b, c, d, f, v], *Art_Kr*

> assign6(n=42, m=47, k=67,2003, p=62, d=89, q=96); n, m, k, p, d, q;

[42, *m* = 47, *k* = 67, 2003, *p* = 62, *d* = 89, *q* = 96], *m*, *k*, *p*, *d*, *q*

> assign6(['n'=42, 'm'=47, 'k'=67, 'p'=62, 'd'=89, 'q'=96]); n, m, k, p, d, q;

42, 47, 67, 62, 89, 96

Assign-procedure for releases **6, 7, 8** and **9/9.5**

> Assign(w,x=a,b,c,y=m,z=n,{a1,a2},t=h,p,v=b); [x], [y], [z], [v], [w]; # *release* 7 - 9/9.5

[a, b, c], [m], [n, {a1, a2}], [b], [w]

> Assign(G=56,V=61,Ar=40,Kr=7,Art=14,Y=2003); G, V, Ar, Kr, Art, Y;

56, 61, 40, 7, 14, 2003

> Assign(n=42,m=47,k=67,2003,p=62,d=89,q=96); n, m, [k], p, d, q;

42, 47, [67, 2003], 62, 89, 96

> Assign(w,x=a,b,c,y=m,z=n,{a1,a2},t=h,p,v=b); [x], [y], [z], [v], [w]; # *release* 6

[x], [y], [z], [v], [[x = a, b, c, y = m, z = n, {a2, a1}, t = h, p, v = b]]

> Assign(G=56, V=61, Ar=40, Kr=7, Art=14, Y=2003); G, V, Ar, Kr, Art, Y;

56, 61, 40, 7, 14, 2003

> Assign(n=42, m=47, k=67,2003, p=62, d=89, q=96); n, m, [k], p, d, q;

[42, m = 47, k = 67, 2003, p = 62, d = 89, q = 96], m, [k], p, d, q

assign67-procedure for releases **6, 7, 8** and **9/9.5**

> restart: assign67(n=42, m=47, k=67,2003, p=62, d=89, q=96); n, m, [k], p, d, q;

42, 47, [67, 2003], 62, 89, 96

> assign67(v=7,b,c, x=a1,a2,a3, y=b1,b2,b3, z=c1,c2,c3); [v], [x], [y], [z];

[7, b, c], [a1, a2, a3], [b1, b2, b3], [c1, c2, c3]

> restart; assign67(w=t,h, d=61,42,2003, (T,R,U)=(62,20,1942)); [w], [t], [d], [T, R, U];

[t, h], [t], [61, 42, 2003], [62, 20, 1942]

> assign67('w', 67, h, 'd'=61,42); [w], [h], [d];

[], [], [61, 42]

> assign67([assign67('h'=iostatus())]); h;

[0, 0, 7]

> restart: assign67(7, 14, 'z'=t,h, d=61,42,2003, r={42,61},2003); [z], [d], [r];

[t, h], [61, 42, 2003], [{42, 61}, 2003]

> assign67(7,14, v=h,r,z, 'd'=61,42,2003); [v], [d];

[h, {42, 61}, 2003, t, h], [61, 42, 2003]

> assign67(('q','w','e')=(42, 47)); [q], [w], [e];

[w, e, 42, 47], [w], [e]

> assign67(1,2,3); assign67(); assign(NULL); assign(Z), Z, assign67(Art_Kr=NULL),[Art_Kr];

[]

> assign67(('q','w','e')=(1,2,3)); assign67(('x','y','z')=(61,56,36)), [x,y,z], [q], [w], [e];

[61, 56, 36], [1], [2], [3]

> assign67(V=42,61,G=47,56,S=67,36,Art=89,14,Kr=96,7,Ar=62,40),[V],[G],[S],[Art], [Kr],[Ar];

[42, 61], [47, 56], [67, 36], [89, 14], [96, 6], [62, 40]

> assign67(X1,X2,X3,X4,X5,X6,X7); [X1], [X2], [X3], [X7];

[X2, X3, X4, X5, X6, X7], [X2], [X3], [X7]

> restart; assign67(X1,X2,X3=4,X5,X6,X7); [X1], [X2], [X3], [X7];

[], [], [4, X5, X6, X7], [X7]

We recommend the reader to consider the organization of the above procedures and opportunities, given by them, which are useful enough in applied programming of various types of the problems in the *Maple* environment.

---

andN – extension of the built-in Boolean operators \`and\`, \`or\` and \`xor\`
orN
xorN

Call format of the procedures:

andN(e1, e2, e3, ...)
orN(e1, e2, e3, ...)
xorN(e1, e2, e3, ...)

Formal arguments of the procedures:

ej – an arbitrary valid *Maple* expression

Description of the procedures:

Boolean expressions are formed by using the logical operators **and, or, xor, implies**, and **not,** and relational operators **<, <=, >, >=, =,** and **<>.** The *Maple* package provides the operation with binary Boolean operators **and, or** and **xor** only. New Boolean operators **andN, orN** and **xorN** extend the standard operators **and, or** and **xor** accordingly onto arbitrary finite amount of operands. The procedure call **{and|or|xor}N()** returns the *true* value, whereas the procedure call **{and|or|xor}N(p)** returns an evaluated expression defined by the single actual **p** argument.

```
andN:= () -> foldl(`and`, true, args)

orN:= proc()
local a, b, c, k, p, j;
  `if`(nargs = 0, RETURN(true), assign(p = [(2*k - 1) $ (k = 1 .. ceil(1/2*nargs))]));
  seq(assign(cat(_, k) = args[k]), k = 1 .. nargs); `if`(type(nargs, even), NULL,
    assign(cat(_, nargs + 1) = "_$")); assign(b = cat(seq(cat(convert(eval(cat(_, k)), string),
    " or ", convert(eval(cat(_, k + 1)), string), " or "), k = p))), writebytes("_$$$_",
    cat("`_0`:= ", b[1 .. -`if`(b[-10 .. -1] = " or _$ or ", 11, 5)], ";"));
  (proc() close("_$$$_"); read "_$$$_"; fremove("_$$$_") end proc)(), _0,
    unassign(seq(k, k = [seq(cat(_, j), j = 0 .. nargs + 1)]))
end proc

xorN:= proc()
local a, b, c, k, p, j;
  `if`(nargs = 0, RETURN(true), assign(p = [(2*k - 1) $ (k = 1 .. ceil(1/2*nargs))]));
  seq(assign(cat(_, k) = args[k]), k = 1 .. nargs); `if`(type(nargs, even), NULL,
    assign(cat(_, nargs + 1) = "_$")); b := cat(seq(cat(convert(eval(cat(_, k)), string), " xor ",
    convert(eval(cat(_, k + 1)), string), " xor "), k = p));
  writebytes("_$$$_", cat("`_0`:= ", b[1 .. -`if`(b[-12 .. -1] = " xor _$ xor ", 13, 6)], ";"));
  (proc() close("_$$$_"); read "_$$$_"; fremove("_$$$_") end proc)(), _0,
    unassign(seq(k, k = [seq(cat(_, j), j = 0 .. nargs + 1)]))
end proc
```

Typical examples of the operators use:

> andN(), andN(A), andN(G, S, Art, Kr, Ar, V, Ian, Rans, TRU, Rac, Rea);

true, A, G **and** *S* **and** *Art* **and** *Kr* **and** *Ar* **and** *V* **and** *Ian* **and** *Rans* **and** *TRU* **and** *Rac* **and** *Rea*

> orN(), orN(A), orN(G, S, Art, Kr, Ar, V, Ian, Rans, TRU, Rac, Rea);

true, A, G **or** *S* **or** *Art* **or** *Kr* **or** *Ar* **or** *V* **or** *Ian* **or** *Rans* **or** *TRU* **or** *Rac* **or** *Rea*

> xorN(), xorN(A), xorN(G , S, Art, Kr, Ar, V, Ian, Rans, TRU, Rac, Rea);

true, A, G **xor** *S* **xor** *Art* **xor** *Kr* **xor** *Ar* **xor** *V* **xor** *Ian* **xor** *Rans* **xor** *TRU* **xor** *Rac* **xor** *Rea*

---

belong – test for belonging to a set, list, module, procedure, symbol, string, etc.

Call format of the procedure:

belong(*expr*, V)

Formal arguments of the procedure:

expr – an arbitrary valid *Maple* expression
V – a set, range, string, symbol, list, module, procedure, table, array, Array, Vector, or Matrix

Description of the procedure:

The procedure call **belong(*expr*, V)** returns the *true* value if ***expr*** is the member of a set, module, list, Array, array, procedure, Vector, Matrix or belongs to a string or a symbol defined by the second actual **V** argument; otherwise the *false* value is returned. If **V** is a module or a procedure, the fact of entry of the ***expr*** expression into the module or procedure as the export variables or local and global variables is tested accordingly, whereas in case of a table its entries are tested.

Along with it, the **belong** procedure tests the fact of entry of a list or a set ***expr*** as a sublist or subset of the list or set **V,** and also belonging of elements of a list or a set to the given range. The **belong** procedure essentially extends the built-in `member` procedure of the package *Maple.*

```
belong:= (a::anything, V::{table, procedure, list, `module`, range, string, symbol, set, Matrix,
        Vector, Array, array})
  -> `if`(a = V, true, `if`(member(whattype(eval(V)), {table, Matrix, Array, array, Vector[row],
    Vector[column]}), RETURN(procname(a, convert(V, list))), `if`(type(V, procedure),
    procname(a, map(op, [2, 6], eval(V))), `if`(type(eval(V), symbol) or type(eval(V), string),
    `if`(nops(Search2(V, {op(a)})) = 0, false, true), `if`(type(V, `module`), member(a, V),
    `if`(type(V, range) and type(a, numeric), `if`(a <= rhs(V) and lhs(V) <= a, true, false),
    `if`(type(a, 'set') and type(V, {list, 'set'}), evalb({op(a)} intersect {op(V)} = {op(a)}),
    `if`(type(a, list) and type(V, list), `if`(Sub_list(a, V) = false, false, true), `if`(type(V, range)
    and type(a, {list(numeric), set(numeric)}), `if`(map(procname, {op(a)}, V) = {true}, true,
    false), `if`(type(V, {string, symbol}), `if`(map2(search, V, map(cat, {op(a)}, ``)) = {true},
    true, false), member(a, V)))))))))))
```

Typical examples of the procedure use:

```
> M1:= module() export ex; end module: belong(y, {x, y, z}), belong(y, {x*y, y*z}),
  belong(w, [x, y, w, u]), belong(ex, M1);
```

true, false, true, true

```
> belong(avz, M1), belong(A,RANS), belong(AN, "RANS"), belong(61,42..99);
```

false, true, true, true

```
> belong([42,47],[42,47,67,62,89,96]), belong({42,47},{42,47,67,62,89,96}),belong([42,47,67], 42..99);
```

true, true, true

```
> Proc:= proc() local a; global b; a:= `+`(args); a/nargs end proc: belong([a,b], Proc);
```

true

```
> map2(belong, 47, [{7,47,13}, 42..99, [34,47,59]]), map2(belong,[96,89],[{7,47,13},42..99,[34, [96,89]]]);
```

[*true, true, true*], [*false, true, true*]

```
> belong(Art, `vsv_agn_avz_Art_Kr`), belong([Art,Kr], "vsv_agn_avz_Art_Kr");
```

true, true

```
> m:=matrix(2,3,[a,b,c,42,96,47]): v:=vector([42,G,47,g]): a:=array(1..2,1..3,[[42,47,67], [62,89,
  96]]): t:= table([c=z]): V:=Vector([61,56,36,40,7,14]): V1:=Vector[row]([61,56,36,40,7,14]):
  M:=Matrix(1..2, 1..2, [[T,S],[G,K]]): A:=Array(1..2,1..3,1..2,[]): A[2,2,2]:=42: A[1,2,2]:=47:
  A[2,3,2]:=67: A[2,1,2]:=89: A[2,2, 1]:= 96: belong(47, m), belong(47, a), belong(61, V),
  belong(61, V1), belong(K, M), belong(67, A), belong(z, t);
```

true, true, true, true, true, true, true

Examples of the fragment represented above evidently illustrate the calls results of the **belong** procedure on various tuples of actual arguments. The given procedure in many cases essentially simplifies programming in the *Maple* environment.

Builtin – testing of the built-in functions of the current *Maple* release

Call format of the procedure:

Builtin(*expr*)

Formal arguments of the procedure:

expr – (optional) a positive integer (*posint*), a name or the *NULL* value

Description of the procedure:

The procedure call **Builtin()** without arguments returns the list of equations whose left members define names of the built-in functions, whereas the right members define their decimal numbers. On the other hand, the call **Builtin(N::*symbol*)** returns the number of a built-in function with **N** name, if such exists. In addition, vice versa, the call **Builtin(N::*posint*)** returns the name of a built-in function with the given number **N.** In addition, it is necessary to have in mind; the results of a procedure call depend on the current release of *Maple*.

```
Builtin:= proc()
local a, k, L, A, h, Z, R, G, M;
   assign(a = [seq(convert(k, string), k = 0 .. 9)], R = [], Z = []);
   assign(A = [anames(builtin)]), assign(L = map(convert, map(eval, A), string));
   for k to nops(L) do
      seq(`if`(member(L[k][j], a, 'h'), assign('R' = [op(R), h – 1]), N), j = 1 .. length(L[k]));
      assign('Z' = [op(Z), add(R[k]*10^(nops(R) – k), k = 1 .. nops(R))], 'R' = [])
   end do;
   `if`(1 < nargs, ERROR(`Inadmissible number of arguments; should be <= 1`),
      assign(G = [seq(A[k] = Z[k], k = 1 .. nops(A))]));
   `if`(nargs=0, RETURN(G), `if`(type(args,symbol), seq(`if`(lhs(G[k])=args,RETURN(rhs(G[k])),
      assign('M' = cat(args, ` is not a built-in function`))), k = 1 .. nops(G)), `if`(type(args, posint),
      seq(`if`(rhs(G[k]) = args, RETURN(lhs(G[k])), assign('M' = cat(args, ` is not number of a
      builtin function`))), k = 1 .. nops(G)), NULL))), `if`(length(M) = 1, NULL, RETURN(M))
end proc
```

Typical examples of the procedure use:

> map(Builtin, [222, remove, 144, goto, rhs, \`eval\`]), Builtin(\`evalfh\`); Builtin(); ***# Release 9***

[*max*, 248, *Scale2*, 193, 249, 94], *evalfh is not a built-in function*

[traperror = 104, map2 = 221, normal = 234, indets = 203, \`>=\` = 125, hastype = 195, irem = 209,

==

==,

rtable_is_zero = 254, remove = 248, tcoeff = 279, mod = 227, kernelopts = 212, OrderedNE = 137]

\`convert/list\` – converting of the *Maple* expressions into list or set \`convert/set\`

Call format of the procedures:

convert(R, *list*)
convert(R, *set*)

Formal arguments of the procedures:

R – a range, table, array, Array, vector, Vector, matrix, Matrix, string, symbol, expression

Description of the procedures:

It is necessary to note, what by means of the standard function *`convert`* the expressions of a type {*string, symbol*} formally can be converted into structures of a type {*list* | *set*}. The results of such converting will be formally recognized by the test functions {*whattype, type, typematch*} correctly, however, as a matter of fact, these structures remain as structures of a type {*string, symbol*}, excepting a case of converting of a string into list.

That speaks about necessity of essential revision of the given aspect because of rather frequent necessity of a converting of structures of the indicated types. In full measure, it relates to the converting **`convert/set`** also. The modified procedure **`convert/list`** will correctly convert an array, Array, range, table, vector, Vector, matrix, Matrix, symbol, expression, string specified by the first actual **R** argument, into the list (*list*). While the modified procedure **`convert/set`** will correctly convert an array, Array, range, table, vector, Vector, matrix, Matrix, symbol, expression, string specified by the first actual **R** argument, into the set (*set*). In addition, a construction **{{*expr*} | [*expr*]}** provides conversion of an arbitrary expression ***expr*** into {*set* | *list*} as a whole.

```
`convert/list`:= proc(expr::anything)
local a, k, L, omega;
  omega:= a -> `if`(type(a, equation), rhs(a), a);
  `if`(type(expr, Vector) or type(expr, Matrix), RETURN(map(omega, [op(op(2, expr))])),
    `if`(type(expr, Array), RETURN(map(omega, [op(op(3, expr))])), NULL)),
    `if`(type(expr, list), RETURN(expr), NULL), `if`(type(expr, range) and type(lhs(expr),
    numeric) and type(rhs(expr), numeric), [assign(L = [seq(k, k = lhs(expr) .. rhs(expr))]),
    RETURN(`if`(belong(rhs(expr), L), L, [op(L), rhs(expr)]))], `if`(type(expr, range),
    RETURN([expr]), NULL));
  `if`(type(eval(expr), symbol) or type(expr, string), [seq(substring(expr, k..k), k=1..length(expr))],
    `if`(type(expr, table) or type(expr, vector), [assign(a = [entries(expr)]), seq(op(a[k]),
    k = 1 .. nops(a))], `if`(member(whattype(expr), {list, set}), [op(expr)], [expr])))
end proc
```

```
`convert/set`:= proc(expr::anything)
local a, k, L, omega;
  omega:= a -> `if`(type(a, equation), rhs(a), a);
  `if`(type(expr, Vector) or type(expr, Matrix), RETURN(map(omega, op(2, expr))),
    `if`(type(expr, Array), RETURN(map(omega, op(3, expr))), NULL)), `if`(type(expr, range)
    and type(lhs(expr), numeric) and type(rhs(expr), numeric), [assign(L = {seq(k, k = lhs(expr) ..
    rhs(expr))}), RETURN(`if`(belong(rhs(expr), L), L, {rhs(expr), op(L)}))],
    `if`(type(expr, range), RETURN({expr}), NULL));
  `if`(type(eval(expr), symbol) or type(expr, string), {seq(substring(expr,k..k),k=1..length(expr))},
    `if`(type(expr, table) or type(expr, vector), {op([assign(a = [entries(expr)]), seq(op(a[k]),
    k = 1 .. nops(a))])}, `if`(member(whattype(expr), {list, set}), {op(expr)}, {expr})))
end proc
```

Typical examples of the procedures use:

```
> S:= "RANS_IAN": Z:=`RANS_IAN`: convert(Z, list), convert(S, list), convert(Z, set);

        [R, A, N, S, _, I, A, N], ["R", "A", "N", "S", "_", "I", "A", "N"], {I, S, _, N, A, R}

> B:= array(1..3, 1..6, [[42,47,67,62,89,96], [61,56,36,40,7,14], [1,2,3,4,5,6]]): convert(S, set),
  convert(B, list);

        {"I", "_", "S", "N", "A", "R"}, [7, 1, 42, 2, 6, 96, 62, 4, 89, 5, 14, 3, 47, 61, 36, 67, 40, 56]
```

```
> T:= table([Art=14, Kr=7, V=61, G=56, Sv=36, Ar=40]): convert(B, set), convert(T, set),
  convert(T, list);
        {1, 2, 3, 4, 5, 6, 7, 14, 36, 40, 42, 47, 56, 61, 62, 67, 89, 96}, {7, 14, 36, 40, 56, 61},
                                    [14, 40, 61, 36, 7, 56]
> convert((a+b)/(x+y+z)^2+sin(c+d)-exp(h), set), convert((a+b)/(x+y+z)^2+sin(c+d)-exp(h), list);
    {-2, -1, *, ^, +, a, x, y, exp, sin, b, h, d, c, z}, [+, *, +, a, b, ^, +, x, y, z, -2, sin, +, c, d, *, -1, exp, h]
> R:=19.42..60.61: R1:=42..61: map(convert, [R, R1], list), map(convert, [R, R1], set);
   [[19.42, 20.42, 21.42, 22.42, 23.42, 24.42, 25.42, 26.42, 27.42, 28.42, 29.42, 30.42, 31.42, 32.42, 33.42,
  34.42, 35.42, 36.42, 37.42, 38.42, 39.42, 40.42, 41.42, 42.42, 43.42, 44.42, 45.42, 46.42, 47.42, 48.42, 49.42,
  50.42, 51.42, 52.42, 53.42, 54.42, 55.42, 56.42, 57.42, 58.42, 59.42, 60.42, 60.61], [42, 43, 44, 45, 46, 47, 48,
  49, 50, 51, 52, 53, 54, 55, 56, 57, 58, 59, 60, 61]], [{31.42, 32.42, 33.42, 34.42, 35.42, 30.42, 36.42, 37.42,
  38.42, 39.42, 40.42, 41.42, 42.42, 43.42, 44.42, 45.42, 46.42, 47.42, 48.42, 49.42, 50.42, 51.42, 52.42, 53.42,
  54.42, 55.42, 56.42, 57.42, 58.42, 59.42, 60.42, 19.42, 60.61, 20.42, 21.42, 22.42, 23.42, 24.42, 25.42, 26.42,
  27.42, 28.42, 29.42}, {42, 43, 44, 45, 46, 47, 48, 49, 50, 51, 52, 53, 54, 55, 56, 57, 58, 59, 60, 61}]
> m:=matrix(2,3,[a,b,c,42,96,47]): v:=vector([42,G,47,g]): a:=array(1..2,1..3,[[42,47,67], [62,89,96]]):
  V:= Vector([61,56,36,40,7,14]): M:= Matrix(1..2,1..2, [[T,S],[G,K]]): A:=Array(1..2,1..3,1..2,[]):
  A[2,2,2]:=42:A[1,2,2]:=47:A[2,3,2]:=67:A[2,1,2]:=89:A[2,2,1]:=96: map(convert, [m,v,a,V,M,A], list);
        [[a, b, 42, 47, c, 96], [42, G, 47, g], [42, 47, 62, 96, 67, 89], [61, 56, 36, 40, 7, 14],
                          [T, "RANS_IAN", G, K], [47, 89, 96, 42, 67]]
> map(convert, [1995+2004*I, sqrt(1995+2004*I), a*b*sin(x)*c*(-8)^(1/3)*d*(-27)^(4/3)+
cos(y)/f,sin(x)+cos(y),`RANS IAN RAC REA`], list);
   [[Complex, 1995, 2004], [^, Complex, 1995, 2004, Fraction, 1, 2], [+, *, -27, a, b, sin, x, c, ^, -8,
    Fraction, 1, 3, d, ^, -27, Fraction, 1, 3, *, cos, y, ^, f, -1], [+, sin, x, cos, y], [R, A, N, S, ``,
                              I, A, N, ``, R, A, C, ``, R, E, A]]
```

PSubs – substitutions into the *Maple* expressions

Call format of the procedure:

PSubs(S, E)

Formal arguments of the procedure:

S – (optional) a sequence of equations, the lists and/or sets of equations defining substitutions rules
E – (optional) an arbitrary valid *Maple* expression

Description of the procedure:

As the first optional argument of the **PSubs** procedure can be a sequence, whose elements are separate equations (*substitutions rules*) and their sets or lists. The second optional **E** argument represents a processed expression itself. The procedure call **PSubs(S, E)** returns the result of sequential application of substitution rules, defined by the first actual **S** argument, to an expression, defined by the second actual **E** argument. Simultaneously, the information on application of each of substitution rules is printed. The procedure call **PSubs()** returns the *NULL* value, i.e. nothing; whereas the procedure call **PSubs(E)** returns the initial **E** expression without processing; the rest of situations are identified by the appropriate warnings.

```
PSubs:= proc()
local k, h, L, W;
  `if`(nargs = 0, RETURN(), assign(W = args[nargs], L = 61));
```

```
  for k to nargs - 1 do L := L, `if`(type(args[k], equation), args[k], `if`(type(args[k], set(equation))
     or type(args[k], list(equation)), op(args[k]), RETURN(`Substitution rules are incorrect`)))
  end do;
  `if`(nops([L]) - 1 = nargs, RETURN(WARNING("Substitution rules %1 are not applicable to
     absent expression", [args])), NULL);
  for k from 2 to nops([L]) do h := numboccur(W, lhs(L[k]));
     `if`(h = 0, print(cat(`Rule(`, convert(L[k], symbol), `) is inactive`)),
        [print(cat(`Rule(`, convert(L[k], symbol), `) was applied `,
        convert(h, symbol), ` times`)), assign('W' = subs(L[k], W))])
  end do;
  W
end proc
```

Typical examples of the procedure use:

> PSubs(x=a, {y=b,z=c}, [t=h, k=v], K(x^2+y^2+z^2)/(t+k+sin(x*y-t*k)) - A(x+z+t+k)*(cos(x)+sin(y)+t*h)/(G(x-t+h*k)*V(t+h*k-z)));

Rule(x = a) was applied 5 times
Rule(y = b) was applied 3 times
Rule(z = c) was applied 3 times
Rule(t = h) was applied 6 times
Rule(k = v) was applied 5 times

$$\frac{\mathrm{K}(a^2+b^2+c^2)}{h+v+\sin(a\,b-h\,v)}-\frac{\mathrm{A}(a+c+h+v)\,(\cos(a)+\sin(b)+h^2)}{\mathrm{G}(a-h+h\,v)\,\mathrm{V}(h+h\,v-c)}$$

Empty – testing of emptiness of the *Maple* objects

Call format of the procedure:

Empty(S)

Formal arguments of the procedure:

S – a set, symbol, string, list, listlist, table, matrix, vector, array, Array, Matrix or Vector

Description of the procedure:

The procedure call **Empty(S)** returns the *true* value, if the actual argument **S** defines the *empty* set, list, symbol, string, listlist, matrix, table, vector, array, Array, Matrix or Vector; otherwise, the *false* value is returned.

```
Empty:= S::{listlist, set, Matrix, Vector, matrix, vector, Array, array, list, string, symbol, table}
        -> `if`(member(S, {"", ``}) or expLS(convert(S, list)) = [], true, false)
```

Typical examples of the procedure use:

> v:= vector(6,[]): m:= matrix(7,14,[]): v1:= vector([42,G,47,h]): m1:= matrix(2,3,[a,b,c,42, 96,47]): a:= array(1..2,1..3,[[42,47,67], [62,89,96]]): M:= Matrix(1..2,1..2,[[T,S], [G,K]]): A:= Array(1..2, 1..3,1..2,[]): A[2,2,2]:=42: A[1,2,2]:=47: A[2,3,2]:=67: A[2,1,2]:=89: A[2,2,1]:=96: M1:=Matrix(3,4,[]): U:=table([]): A1:= Array(1..2,1..3,1..6,[]):

> map(Empty, [``,"", [], {}, v, m, v1, m1, a, M, A, M1, U, A1, {[],[],{},{},{{},{[],[],{}},[],[]}}]);

[*true*, *true*, *true*, *true*, *true*, *true*, *false*, *false*, *false*, *false*, *false*, *true*, *true*, *true*, *true*]

Etest – testing entry of a subexpression into an arbitrary *Maple* expression VTest

Call format of the procedures:

Etest(V, H, R)
VTest(V, H, R)

Formal arguments of the procedures:

V - a valid *Maple* expression
H - a valid *Maple* expression
R - an assignable name

Description of the procedures:

The procedure call **Etest(V, H, R)** returns the *true* value, if an expression **H** appears in an expression **V** as an elementary term, otherwise the *false* value is returned. By the *elementary term* is being understood a term **G** for which takes place the following relation **nops({op(G)})=1.** Through the third actual **R** argument of the procedure is returned the set of all elementary terms composing the tested expression **V** without taking of their multiplicity into consideration.

The **VTest(V, H, R)** procedure is analogous to the previous procedure **Etest** with the distinction, what through its third actual **R** argument is returned the list of all elementary terms composing the analyzed expression **V** with taking into consideration of their multiplicity.

```
Etest:= proc(V::anything, H::anything, R::evaln)
local k, S, L, Z;
  [assign(S = {op(V)}), assign(L = {}), assign(Z = {})];
  do
    for k to nops({op(S)}) do
      `if`(nops({op(S[k])}) <> 1, assign('L' = {op(S[k]), op(L)}), assign('Z' = {op(Z), S[k]}))
    end do;
    if L = {} then break else end if;
    [assign('S' = L), assign('L' = {})]
  end do;
  [assign(R = Z), RETURN([false, true][1 + charfcn[Z](H)])]
end proc

VTest:= proc(V::anything, h::anything, r::evaln)
local k, S, L, Z;
  [assign(S = {op(V)}), assign(L = []), assign(Z = [])];
  do
    for k to nops({op(S)}) do
      `if`(nops({op(S[k])}) <> 1, assign('L' = [op(L), op(S[k])]), assign('Z' = [op(Z), S[k]]))
    end do;
    if L = [] then break else end if;
    [assign('S' = L), assign('L' = [])]
  end do;
  [assign(r = Z), RETURN([false, true][1 + charfcn[{op(Z)}](h)])]
end proc
```

Typical examples of the procedures use:

```
> Etest(x*cos(y)*Art+y*(a+exp(x)*Art)/(ln(x)-tan(z))-(Art+Kr)/(Sv+Ar), Art,R),  R;
                true, {-1, Kr, Art, Sv, Ar, x, a, y, cos(y), exp(x), ln(x), tan(z)}
```

Typical examples of the procedure use:

> E:=a*b*sin(x)*c*(-8)^(1/3)*d*(a+b)^(1/3)+cos(y)/f: [evalf(E), Evalf(E)];

$$\left[(1.0000000 + 1.7320508\, I)\, a\, b\, \sin(x)\, c\, d\, (a+b)^{\left(\frac{1}{3}\right)} + \frac{\cos(y)}{f},\ -2.0000000\, a\, b\, \sin(x)\, c\, d\, (a+b)^{\left(\frac{1}{3}\right)} + \frac{\cos(y)}{f}\right]$$

> E:=a*b*sin(x)*c*(-8)^(1/3)*d*(-27)^(4/3)+cos(y)/f: [evalf(E), Evalf(E)];

$$\left[(81.000000 - 140.29611\, I)\, a\, b\, \sin(x)\, c\, d + \frac{\cos(y)}{f},\ -162.00000\, a\, b\, \sin(x)\, c\, d + \frac{\cos(y)}{f}\right]$$

> map(evalf, [(-8)^(1/3),(-8)^(4/3)]), map(Evalf, [(-8)^(1/3),(-8)^(4/3)]);

[1.00000+1.73205 **I**, -8.00000-13.8564 **I**], [-2.00000, 16.0000]

> map(evalf, [(-8)^(1/3),(-8)^(4/3)], 4), map(Evalf, [(-8)^(1/3),(-8)^(4/3)], 4);

[1.00000+1.73205 **I**, 4., -8.00000-13.8564 **I**, 4.], [-2.000, 16.00]

> E:=(a*(-8)^(1/3)+b*(-27)^(5/3))/(c*(-32)^(3/5)-b*(-27)^(1/3)): [evalf(E), Evalf(E)];

$$\left[\frac{(1.0000000 + 1.7320508\, I)\, a + (121.50000 - 210.44417\, I)\, b}{-(2.4721360 - 7.6084521\, I)\, c - (1.5000000 + 2.5980762\, I)\, b},\ \frac{-2.0000000\, a + 243.00000\, b}{-8.0000000\, c + 3.0000000\, b}\right]$$

> E:=(a*(-8)^(1/3)+b*(-27)^(5/3))/(c*(-32)^(3/5)-b*(-27)^(1/3)): [evalf(E, 4), Evalf(E, 4)];

$$\left[\frac{(1.000 + 1.732\, I)\, a + (121.5 - 210.4\, I)\, b}{-(2.472 - 7.608\, I)\, c - (1.500 + 2.598\, I)\, b},\ \frac{-2.000\, a + 243.0\, b}{-8.000\, c + 3.000\, b}\right]$$

> E:= (a*(-8)^(1/3)+b*(-27)^(5/3)+(-a)^(1/3))/(c*(-32)^(3/5)-b*(-27)^(1/3)-(-c)^(1/5)):
[evalf(E, 4), Evalf(E, 4)];

$$\left[\frac{(1.000 + 1.732\, I)\, a + (121.5 - 210.4\, I)\, b + (-1.\, a)^{\left(\frac{1}{3}\right)}}{-(2.472 - 7.608\, I)\, c - (1.500 + 2.598\, I)\, b - 1.\, (-1.\, c)^{\left(\frac{1}{5}\right)}},\ \frac{-2.000\, a + 243.0\, b + (-1.\, a)^{\left(\frac{1}{3}\right)}}{-8.000\, c + 3.000\, b - 1.\, (-1.\, c)^{\left(\frac{1}{5}\right)}}\right]$$

> G:=(a+(-7+(-3)^(1/3))^(1/5))+(b+(-14+(-3)^(1/5))^(1/7)): [evalf(G,4), Evalf(G,4)];

[a+2.510+1.421 **I**+b, a+2.570+1.541 **I**+b]

> Z:= (-27.0)^(1/3): Z1:= (-27)^(1/3): map2(type, Z, [\`^\`, complex]), map2(type, Z1, [\`^\`, complex]);

[*false*, *true*], [*true*, *false*]

IDS – testing entry of a subexpression into a *Maple* expression

Call format of the procedure:

IDS(*expr*)

Formal arguments of the procedure:

expr – an arbitrary valid *Maple* expression

```
> VTest(x*cos(y)*Art+y*(a+exp(x)*Art)/(ln(x)-tan(z))-(Art+Kr)/(Sv+Ar), Art,R), R;
        true, [-1, y, x, cos(y), Art, Art, Kr, -1, a, Sv, Ar, exp(x), Art, ln(x), -1, tan(z)]
```

Evalf – evaluations using floating-point arithmetic

Call format of the procedure:

Evalf(E {, n})

Formal arguments of the procedure:

E - an arbitrary valid algebraic expression
n - (optional) an integer specifying amount of the processed digits

Description of the procedure:

Maple has trouble at evaluation already of a whole series of simple radicals with rational from negative values, if denominator of exponent is odd number. In this case, even the st function *evalf* appears powerless, what the rather simple examples very well illustrate:

```
> evalf((-8)^(1/3)), evalf((-8)^(4/3)), evalf(-8^(4/3)), evalf((8)^(1/3)), evalf(8^(1/3));
        1.00000+1.73205 I, -8.00000-13.8564 I, -16.0000, 2.00000, 2.00000
```

In particular, the given circumstance can serve as a reason of impossibility of output in the environment of the graphical representation of a number of functions inasmuch as at calculat their coordinates the *evalf*-function is essentially used. With the purpose of partial elimina the given defects, we had introduced the procedure **Evalf** allowing to calculate the real va expressions of the above type correctly.

The **Evalf** function just as the *evalf* function evaluates both the real and complex floating numbers, expressions (or subexpressions) which contain constants such as **Pi, exp(1), gamm** also functions such as ***exp, ln, sin, cosh, arctan, GAMMA,*** and ***erf.*** The parameter **n** sets a va the predefined ***Digits***-variable that controls the number of digits that *Maple* uses when calcu with software floating-point numbers. The examples represented below well illustrate diffe between procedure **Evalf** and the built-in function ***evalf.*** However, this procedure does not the above problem as a whole, what is conditioned by the insufficiently advanced mechani types used by *Maple*, complicating an analysis of algebraic expressions. Indeed, it is very diffic agree with the following result:

```
> type((-27)^(1/3), `^`), type((-27.)^(1/3), `^`);
        true, false
```

The last examples of the fragment well illustrate the told. In many cases at usage of floating-p arithmetic, the **Evalf** procedure allows to receive the more acceptable results, if expressions con radicals with rational degrees.

```
Evalf:= proc(E::algebraic)
local a, b, k, omega, r, psi, n;
  n := `if`(nargs = 2 and type(args[2], posint), args[2], NULL);
  omega:=x -> `if`(type(denom(op(x)[2]),odd),evalf(abs(evalf(x)))*sign(evalf(op(x)[1])),evalf(x));
  assign(psi = (x -> op([x, seq(op(k, x), k = 1 .. nops(x))])), a = {}, b = {E});
  do if a = b then break else a := b end if; b := {op(map(psi, b))} end do;
  r := [seq(`if`(type(a[k], radical) and type(op(1, a[k]), numeric), a[k], NULL), k = 1 .. nops(a))];
  `if`(r = [], evalf(E, n), evalf(subs([seq(r[k] = omega(r[k]), k = 1 .. nops(r))], E), n))
end proc
```

Description of the procedure:

The procedure call **IDS(*expr*)** returns the list of parse of an expression, given by the actual ***expr*** argument, in the terms of types of objects accepted in *Maple*. Each of the objects is identified either by a symbolic or by a numeric identifier. The procedure call **IDS()** without arguments returns the list of equations whose left and right members define the numerical and symbolic object identifiers accordingly. This information can depend on the current *Maple* release.

```
IDS:= proc()
local k, d, ids, R;
  ids:= proc()
    local a, k, b, L, R, Q;
      for k to 256 do
        try assign('L' = [op(L), k = kernelopts(dagtag = k)])
        catch "invalid data structure tag": NULL
        end try
      end do;
      `if`(nargs = 0, RETURN(L[2 .. -1]), assign('L' = L[2 .. -1]));
      `if`(member(whattype(args), {fraction, Matrix, Array, series, float, integer, string, symbol,
        Vector[row], exprseq, Vector[column]}) or member(whattype(eval(args)),
        {procedure, array, `module`}), RETURN(args), NULL);
      assign(R =[disassemble(addressof(args))]), assign(Q =[subs({op(L)}, R[1]), op(R[2..-1])]);
      Q := seq(`if`(type(Q[k], symbol), Q[k], pointto(Q[k])), k = 1 .. nops(Q))
    end proc;
  `if`(nargs = 0, RETURN(ids()), NULL);
  `if`(member(whattype(args), {fraction, Matrix, Array, series, float, integer, string, symbol,
    Vector[row], exprseq, Vector[column]}) or member(whattype(eval(args)), {procedure, array,
    `module`}), RETURN(args), assign(d = 59, R = [ids(args)]));
  while d <> 0 do assign('d' = 0),
    assign('R' =[seq(`if`(ids(R[k]) = R[k], R[k], op([assign('d' =d+1), ids(R[k])])), k=1..nops(R))])
  end do;
  R
end proc
```

Typical examples of the procedure use:

> IDS(); IDS(a^2+sqrt(Art^2+Kr^2) – (a+sin(x))/(x-1)+1995*G-a^b+2004*I);
IDS(Art/Kr-1995*G-2004*I);

[1 = *INTNEG*, 2 = *INTPOS*, 3 = *RATIONAL*, 4 = *FLOAT*, 5 = *HFLOAT*, 6 = *COMPLEX*, 7 = *STRING*, 8 = *NAME*, 9 = *MEMBER*, 10 = *TABLEREF*, 11 = *DCOLON*, 12 = *CATENATE*, 13 = *POWER*, 14 = *PROD*, 15 = *SERIES*, 16 = *SUM*, 17 = *ZPPOLY*, 18 = *FUNCTION*, 19 = *UNEVAL*, 20 = *EQUATION*, 21 = *INEQUAT*, 22 = *LESSEQ*, 23 = *LESSTHAN*, 24 = *AND*, 25 = *NOT*, 26 = *OR*, 27 = *XOR*, 28 = *IMPLIES*, 29 = *EXPSEQ*, 30 = *LIST*, 31 = *LOCAL*, 32 = *PARAM*, 33 = *LEXICAL*, 34 = *PROC*, 35 = *RANGE*, 36 = *SET*, 37 = *TABLE*, 38 = *RTABLE*, 39 = *MODDEF*, 40 = *MODULE*, 41 = *ASSIGN*, 42 = *FOR*, 43 = *IF*, 44 = *READ*, 45 = *SAVE*, 46 = *STATSEQ*, 47 = *STOP*, 48 = *ERROR*, 49 = *TRY*, 50 = *RETURN*, 51 = *BREAK*, 52 = *NEXT*, 53 = *USE*, 54 = *BINARY*, 55 = *HASH*, 56 = *HASHTAB*, 57 = *GARBAGE*, 58 = *FOREIGN*, 59 = *CONTROL*, 60 = *DEBUG*]

[*SUM*, *PROD*, a, 2, 1, *PROD*, *SUM*, *PROD*, *Art*, 2, 1, *PROD*, *Kr*, 2, 1, 1/2, 1, *PROD*, *SUM*, a, 1, *FUNCTION*, *sin*, x, 1, 1, *SUM*, x, 1, -1, 1, -1, -1, G, 1995, *POWER*, a, b, -1, *COMPLEX*, 2004, 1]

[*SUM*, *PROD*, *Art*, 1, *Kr*, -1, 1, G, -1995, *COMPLEX*, -2004, 1]

map3 – an extending of the standard procedures `map` and `map2`
map4
map5

Call format of the procedures:

map3(L, P {, 0})
map4(F, R)
map5(X, Y, z, p, t)

Formal arguments of the procedures:

L – a list of symbols
F – a symbol or list of symbols
P – a list, listlist or nestlist
R – a nestlist
0 – (optional) *zero* value
z – a valid *Maple* expression
X, Y – symbols defining functions names
P – a set or list
T – a valid *Maple* expression

Description of the procedures:

In *Maple* the procedures **map** and **map2** provide the cyclic application of a required procedure to each operand of a given expression; they are used enough widely. The following procedures **map3, map4** and **map5** extend the above procedures largely.

The procedure call **map3(L, P)** returns the calculation result of functions whose names are defined by the first actual **L** argument (a list of names), from arguments defined by the second actual argument a **P** list. Whereas the procedure call **map3(L, P, 0)** returns the calculation result of functions whose names are defined by the first actual **L** argument, from arguments defined by a **P** list as a whole for each function. If in the first case between lists **L** and **P** the one-to-one correspondence is absent, then the pairs **{L[k], P[k]}** **[k=1 .. min(nops(L), nops(P))]** are considered only.

The procedure call **map4(F, R)** returns the calculation result of functions whose names are defined by the first actual **F** argument (a symbol or a list of symbols), from arguments defined by the corresponding sublists of the second actual argument **R** (a nestlist). If between actual arguments **F** and **R** the one-to-one correspondence is absent, then the pairs **{F[k], R[k]}** **[k=1 .. min(nops(L), nops(R))]** are considered only. If the first actual argument **F** is a symbol, the procedure call **map4(F, R)** returns the calculation result of an **F** function on the all tuples defined by sublists of **R** or a list **R** as a whole. The procedure **map4** handles the basic erratic situations.

At last, the procedure call **map5(X, Y, z, p, t)** depending on type of the fourth argument **p** returns the set or list of values of a view **Y(m, n)** where **m** possesses all possible values from **map2(X, z, p),** whereas **n** possesses the values from **t;** in addition, if an **j**-element of **t** is a set or list, instead of it the value **op(t[j])** is accepted. Examples of the fragment below allow better to clarify a result kind returned by the procedure call.

```
map3:= proc(L::list(symbol), ex::{list, listlist, nestlist})
local k;
  `if`(nargs = 3 and args[3] = 0, [seq(k(op(ex)), k = L)], [seq(L[k](op(ex[k])),
  k = 1 .. min(nops(L), nops(ex)))])
end proc
```

```
map4:= proc(F::{symbol, list(symbol)}, L::nestlist)
local k, n, p, d, x, v;
  `if`(F = [], ERROR("the first actual argument is empty"), `if`(Empty(L), RETURN(F),
     assign(p = `if`(type(F, symbol), nops(L), nops(F)), n = nops(L))));
  [seq(`if`(type(F, symbol), F, F[k])(seq(x, x = [seq(L[k][j], j = 1 .. nops(L[k]))])),
     k = 1 .. `if`(p <= n, p, n))]
end proc

map5:= proc(F::symbol, G::symbol, x::anything, p::{set, list}, t::anything)
local a, b, k, j;
  assign67(a = map2(F, x, p), b = NULL);
  for k to nops(a) do for j to nops(t) do b := b, G(a[k], `if`(type(t[j], {set, list}), op(t[j]),
        `if`(nops(t) = 1, `if`(type(t, {set, list}), op(t), t), t[j])))
     end do
  end do;
  `if`(type(p, set), {b}, [b])
end proc
```

Typical examples of the procedures use:

> map3([F,G,H], [a,b, [c,d], f,h,g]), map3([F,G,H], [a,b, [c,d] ,f,h,g], 0);

[F(a), G(b), H(c,d)], [F(a,b,[c, d],f,h,g), G(a,b,[c, d],f,h,g), H(a,b,[c, d],f,h,g)]

> map3([F,G,H,S,A,T], [a,b, [c,d], f,h,g]), map3([F,G,H,S,A,T,R,Z], [a,b, [c,d], f,h,g], 0);

[F(a), G(b), H(c,d), S(f), A(h), T(g)], [F(a,b,[c, d],f,h,g), G(a,b,[c, d],f,h,g), H(a,b,[c, d], f,h,g), S(a,b,[c, d],f,h,g), A(a,b,[c, d],f,h,g), T(a,b,[c, d],f,h,g), R(a,b,[c, d],f,h,g), Z(a,b, [c, d],f,h,g)]

> map3([F,G,H,S,A,T,R,Z],[a,b,[c,d],[f,h,g]]), map3([F,G,H,S,A,T,R,Z],[a,b,[c,d],[f,h,g]], 0);

[F(a), G(b), H(c,d), S(f,h,g)], [F(a,b,[c, d],[f, h, g]), G(a,b,[c, d],[f, h, g]), H(a,b,[c, d], [f, h, g]), S(a,b,[c, d],[f, h, g]), A(a,b,[c, d],[f, h, g]), T(a,b,[c, d],[f, h, g]), R(a,b,[c, d], [f, h, g]), Z(a,b,[c, d],[f, h, g])]

> map4(F, [[x,y,z,t], [m,n], [a,b,c,d], [t,u,w,z,s], [x1,x2,x3,x4], [y1,y2,y3], [z1,z2,z3,z4]]);

[F(x,y,z,t), F(m,n), F(a,b,c,d), F(t,u,w,z,s), F(x1,x2,x3,x4), F(y1,y2,y3), F(z1,z2,z3,z4)]

> map4([F,G,H,S,A,V,K],[[x,y,z,t],[m,n], [a,b,c,d],[t,u,w,z,s],[x1,x2,x3,x4],[y1,y2,y3],[z1,z2,z3,z4]]);

[F(x,y,z,t), G(m,n), H(a,b,c,d), S(t,u,w,z,s), A(x1,x2,x3,x4), V(y1,y2,y3), K(z1,z2,z3,z4)]

> map4([F,G,H], [[x,y,z,t], [m,n], [a,b,c,d], [t,u,w,z,s], [x1,x2,x3,x4], [y1,y2,y3], [z1,z2,z3,z4]]);

[F(x,y,z,t), G(m,n), H(a,b,c,d)]

> map4([F,G,H,S,A,V,K], [[x,y,z,t], [m,n], [a,b,c,d]]);

[F(x,y,z,t), G(m,n), H(a,b,c,d)]

> map4([F,G,H,S,A,V,K,T,W,P,D,Z], [[]]);

[F, G, H, S, A, V, K, T, W, P, D, Z]

> map4([F,G,H,S,A,V,K,T,W,P,D,Z]);

Error, (in **map4**) map4 uses a 2nd argument, L (of type nestlist), which is missing

> map5(F, G, 5, [a,b,c], {y,u,t});

[G(F(5, a), y), G(F(5, a), t), G(F(5, a), u), G(F(5, b), y), G(F(5, b), t), G(F(5, b), u), G(F(5, c), y), G(F(5, c), t), G(F(5, c), u)]

> L:= [restart,`done`,`quit`,`stop`]: L:= map5(cat,cat,"",L, [":", ";"]);

L := ["restart:", "restart;", "done:", "done;", "quit:", "quit;", "stop:", "stop;"]

> map5(F, G, 5, [a,b,c], {}), map5(F, G, 5, [], {y,u,t});

[], []

> map5(F, G, 5, [a,b,c], {{y,u,t}, [n,m]});

[G(F(5,a), y, t,u), G(F(5,a), n,m), G(F(5,b), y, t,u), G(F(5,b), n,m), G(F(5,c), y, t,u), G(F(5,c), n,m)]

> map5(F, G, 5, [a,b,c], {y,u,t});

[G(F(5, a), y), G(F(5, a), t), G(F(5, a), u), G(F(5, b), y), G(F(5, b), t), G(F(5, b), u), G(F(5, c), y), G(F(5, c), t), G(F(5, c), u)]

> map5(F, G, 5, [a,b,c], d), map5(F, G, 5, [a,b,c], {d});

[G(F(5, a), d), G(F(5, b), d), G(F(5, c), d)], [G(F(5, a), d), G(F(5, b), d), G(F(5, c), d)]

The procedures **map3, map4** and **map5** essentially extend the standard *Maple* procedures **map** and **map2.**

---

mapN – variation of the given argument of a function or procedure

Call format of the procedure:

mapN(F, p, x1, x2, ..., xn, [a, b, c, ...])

Formal arguments of the procedure:

F – a symbol or a name
p – an integer valid *Maple* expression
xj – arguments of a **F** function
a, b, c, ... – values received by the **p**th argument of the **F** function

Description of the procedure:

In the package *Maple* the ***map*** and ***map2*** procedures providing the cyclic application of a required procedure to each operand of the given expression are enough widely used. In this context the procedure **mapN** seems useful enough.

The procedure call **mapN(F,p,x1,x2, ..., xn, [a,b,c,...])** returns the calculation result of a **F** function from arguments **<x1, x2, ..., xn>** under condition of receiving by its **p**th argument of values from a list or set specified by the last actual argument of the procedure. The procedure seems as useful enough supplement to the above two standard *Maple* procedures.

The tuple **<F, p, x1, x2, ...,xn, [a,b,c,...]>** is coded at call of the given procedure as its actual arguments, where: **F** – a computational function; **p** – number of a formal function argument whose values vary; **x1,x2, ..., xn** – a set of values of actual arguments (the varied argument can be coded by an arbitrary variable) and **[a,b,c,...]** is a list or set of variants for the **p**th argument of the **F** function. The examples of appendices of the procedure **mapN** well illustrates the aforesaid.

```
mapN:= proc(F::symbol, p::posint)
local k;
  `if`(nargs - 3 < p, ERROR(`position number <%1> is invalid`, p),
  `if`(p=1, convert([seq(F(args[-1][k], args[4 .. -2]), k = 1 .. nops(args[-1]))], whattype(args[-1])),
  `if`(p = nargs - 2, cnvtSL([seq(F(args[3 .. -2], args[-1][k]), k = 1 .. nops(args[-1]))],
  whattype(args[-1])), cnvtSL([seq(F(args[3 .. p + 1], args[-1][k], args[p + 3 .. -2]),
  k = 1 .. nops(args[-1]))], whattype(args[-1])))))
end proc
```

Typical examples of the procedure use:

> F:= (x,y,z,r,t,h) -> G(x,y,z,r,t,h): mapN(F, 6, x,y,z,r,t,h, [a,b,c,d,e,f,n,m,p,u]);

[G(x,y,z,r,t,a), G(x,y,z,r,t,b), G(x,y,z,r,t,c), G(x,y,z,r,t,d), G(x,y,z,r,t,e), G(x,y,z,r,t,f), G(x,y,z,r,t,n), G(x,y,z,r,t,m), G(x,y,z,r,t,p), G(x,y,z,r,t,u)]

> P:= proc(x,y,z,t,h,r) `+`(args)/nargs end proc: mapN(P, 3, x,y,z,t,h,r, [a,b,c,d,e]);

[1/6*x+1/6*y+1/6*a+1/6*t+1/6*h+1/6*r, 1/6*x+1/6*y+1/6*b+1/6*t+1/6*h+1/6*r, 1/6*x+1/6*y+1/6*c+1/6*t+1/6*h+1/6*r, 1/6*x+1/6*y+1/6*d+1/6*t+1/6*h+1/6*r, 1/6*x+1/6*y+1/6*e+1/6*t+1/6*h+ 1/6*r]

The **mapN** procedure essentially extends the functional possibilities of the above procedures **map, map2, map3** and **map4.**

---

null – spreading of some of possibilities of *Maple* 6 onto its next releases 7 – 9/9.5
_1N

Call format of the procedures:

null(*expr*)
_1N(n, *expr*)

Formal arguments of the procedures:

expr - a valid *Maple* expression
n - an integer

Description of the procedures:

In our books [29-33] the questions of incompatibility of releases **6** and {**7**|**8**} both *`from above – downwards`* and *`from below – upwards`* are considered enough in detail. Therefore in a lot of cases the appropriate editing of the source codes of programs, prepared and debugged in environment of the **6**th release, before implanting them into environment of releases **7 – 9/9.5** of the package is required. Just such operation has been carried out for tools represented in the present library, by providing their complete compatibility for last four releases of the *Maple* package.

In particular, in the **6**th release the calls **NULL(*expr*)** and **N(*expr*)** return the *NULL* value and **N** accordingly, evaluating an ***expr*** expression at the same time, where **N** – an integer, whereas in the releases **7 – 9/9.5** the same calls return the *NULL* value and **N** accordingly also, however evaluation of the ***expr*** expression is not done. The first two examples below illustrate the previously mentioned. These possibilities allow to create very effective single-string extracodes in environment of the *sixth* release of *Maple* together with simplification of programming of a series of problems. The next two simple procedures, which in many cases solve the problem of compatibility of the releases **6** and {**7**|**8**|**9**|**9.5**} of the *Maple* package, are represented below.

The procedure call **null**(*expr*) returns the *NULL* value, evaluating an *expr* expression at the same time for environment of releases **6 – 9/9.5.** Whereas the procedure call **_1N(n,** *expr*) returns the **n** value, evaluating an *expr* expression at the same time for environment of releases **6 – 9/9.5.** In particular, the returned **n** value can be used in the next calculations as the completion code for choice of a processing path in programs. On the basis of the above two procedures **null** and **_1N,** many procedures using the above properties of the *sixth* release can be relatively easily adapted for the releases **7 – 9/9.5**.

```
null:= () -> NULL
_1N:= (n::nonnegint, Expr::anything) -> n
```

Typical examples of the procedures use:

> [NULL(assign(Art=14)), 7(assign(Kr=96))], [Art,Kr]; *# release 6*

[7], [14, 96]

> [NULL(assign(Art=14)), 7(assign(Kr=96))], [Art,Kr]; *# release 7*

[7], [*Art, Kr*]

> [NULL(assign(Art=14)), 7(assign(Kr=96))], [Art,Kr]; *# release 8*

[7], [*Art, Kr*]

> [NULL(assign(Art=14)), 7(assign(Kr=96))], [Art,Kr]; *# release 9*

[7], [*Art, Kr*]

The examples represented below have been executed analogously in releases **6, 7, 8** and **9/9.5**:

> [null(assign('Art'=14))], [Art], [_1N(7, assign('Art'=14))], [Art];

[], [14], [7], [14]

> [null(assign('Art'=14)), _1N(7, assign('Kr'=96))], [Art, Kr];

[7], [14, 96]

The procedures **null** and **_1N** represented above can be successfully used for suppression of results of intermediate calculations and return of completion codes of calculations by means of the constructions of view **null(*Expr*)** and **_1N(n, *Expr*)** accordingly, where **n** is an integer and ***Expr*** is any valid *Maple* expression. Along with that, these procedures are used in a series of appendices for a forming the return code of kind the *NULL*, i.e. nothing, or the given integer.

Fac – extension of the `*factorial*` function onto values of numeric type

Call format of the procedure:

Fac(n)

Formal arguments of the procedure:

n - an arbitrary *Maple* expression of the *realnum*-type

Description of the procedure:

The procedure call **Fac(n)** returns the factorial of a value **n** of the *realnum*-type defined in this section below.

```
Fac:= proc(n::realnum)
local a;
  `if`(n = 0, 1, `if`(type(n, integer), sign(n)*abs(n)!, op([assign(a = evalf(n)),
  `if`(0 < a, a*GAMMA(a), -abs(a)*GAMMA(abs(a)))])))
end proc
```

Typical examples of the procedure use:

> map(Fac, [7, -6, evalf(8.9!), 8.9, 7/15, -19.42, sqrt(2004)]);

[5040, -720, 0.3224319988e1457431, 289867.7039, 0.8856140385, -0.4254799596e18, 0.4899975454e56]

maxl – maximum of expressions with respect to their length
minl – minimum of expressions with respect to their length
MinMax – minimax of values of actual arguments

Call format of the procedures:

maxl(e1, e2, e3, ...)
minl(e1, e2, e3, ...)
MinMax(e1, e2, e3, ...)

Formal arguments of the procedures:

e1, e2, ... – any valid *Maple* expressions

Description of the procedures:

The calls **maxl(e1, e2, e3, ...)** and **minl(e1, e2, e3, ...)** return the arguments with *maximum* and *minimum* length among all actual arguments accordingly. The procedure call **MinMax(e1,e2,e3,...)** in the general case returns the **2**-element sequence whose the first element represents the **2**-element list with *minimal* and *maximal* values of symbolic actual arguments **e1, e2, e3,...** , while the second element – the **2**-element list with *minimal* and *maximal* values of *realnum* actual arguments **e1, e2, e3,...** . At presence among actual arguments of values of the types different from the above ones, the appropriate warning is output. The *realnum*-type is considered below. Each actual **h** argument that satisfies condition **type(eval(h), {*symbol, realnum*}) -> *false*** initiates the warning with indication about such argument and its position.

```
maxl:= proc()
local p, h, k, sf;
  sf:= (a::symbol, b::symbol) -> `if`(length(a) <= length(b), true, false);
  `if`(nargs = 0, RETURN(), assign(h = sort(map(convert, [args], symbol), sf))),
    assign(p = length(h[-1])), assign('h' = {seq(`if`(length(convert(args[k], symbol)) = p, args[k],
    NULL), k = 1 .. nargs)}), `if`(1 < nops(h), h, op(h))
end proc

minl:= proc()
local p, h, k, sf;
  sf:= (a::symbol, b::symbol) -> `if`(length(a) <= length(b), true, false);
  `if`(nargs = 0, RETURN(), assign(h = sort(map(convert, [args], symbol), sf))),
    assign(p = length(h[1])), assign('h' = {seq(`if`(length(convert(args[k], symbol)) = p, args[k],
    NULL), k = 1 .. nargs)}), `if`(1 < nops(h), h, op(h))
end proc

MinMax:= proc()
local a, k, S, G, Kr;
  `if`(nargs = 0, RETURN(), assign(Kr = (() -> `if`(nargs = 0, NULL, [args[1][1], args[1][-1]]))));
  assign(S = {}, G = {}), seq([assign('a' = evalf(eval(args[k]))), `if`(type(a, symbol) and
    whattype(eval(a)) = symbol, assign('S' = {a, op(S)}), `if`(type(a, realnum),
    assign('G' = {a, op(G)}), WARNING("[%1, %2] has a type different from type
    {symbol, realmum}", args[k], k)))], k = 1 .. nargs);
  Kr(`if`(Empty(S), NULL, [maxl(op(S)), minl(op(S))])), Kr(`if`(Empty(G), NULL, [G[-1], G[1]]))
end proc
```

Typical examples of the procedures use:

> minl(), maxl(), maxl(0.42, `0123456789`, "Tallinn", rans, ian, 7.14, "vilnius", avzransian), minl(61, Tallinn, 7.14, "rans", IAN1,0.42, grodno, avz, 56);

{*avzransian, 0123456789*}, {56,61}

> T:= table([]): M:=matrix([]): MinMax(), MinMax(a,b,h,g,c), MinMax(rans, ian, sqrt(61), ln(56)), MinMax(avz,b,h,c,4.7,4.2,7/8,89,96,T,M,sqrt(147));

Warning, [T,10] has a type different from type {symbol, realmum}
Warning, [M,11] has a type different from type {symbol, realmum}
[{h,a,b,c,g}, {h, a, b, c, g}], [*rans, ian*], [4.025351691, 7.810249676], [*avz,* {h, b, c}],[12.12435566, 4.7]

> MinMax(Grodno,AVZ,Tallinn,4.2,99,95,56,Pi,sqrt(2),14/6,gamma,Catalan,56/61);

[*Tallinn, AVZ*], [0.9180327869, 4.2]

In a series of cases, the above procedures seem as useful enough tools at programming in *Maple.*

OP – extracting of operands of all levels out of a *Maple* expression

Call format of the procedure:

OP(*expr*)

Formal arguments of the procedure:

expr - a valid *Maple* expression

Description of the procedure:

The procedure call **OP(*expr*)** returns the result of extracting of operands of all levels out of an expression specified by the actual ***expr*** argument. The procedure call **OP()** returns the *NULL* value, i.e. nothing.

```
OP:= proc(expr::anything)
local k, h, Op, L, G;
  Op:= proc(ex::anything)
    local a, k;
      assign67(a = NULL, k = 0);
      do
        try a := a, op(k, ex); k := k + 1 catch "improper op or subscript selector": RETURN(a)
        end try
      end do
    end proc;
  `if`(nargs = 0, RETURN(NULL), assign(G = [], L = [Op(expr)], h = 0));
  do
    for k to nops(L) do
      if nops(L[k]) = 1 and not type(L[k], function) then G := [op(G), L[k]]; next
      else G := [op(G), Op(L[k])]; h := 61
      end if
    end do;
    if h = 61 then L := G; G := []; h := 0; next else break end if
  end do;
  op(L)
end proc
```

Typical examples of the procedure use:

> T:= table([Art=14,Kr=6,V=61,G=56,Sv=36,Ar=40]): map(OP,[(a+b)/(x+y+z)^2, T]), [OP()];

[*, +, a, b, ^, +, x, y, z, -2, *symbol, table, list,* =, *Art,* 14, =, V, 61, =, *G,* 56, =, *Sv,* 36,=, *Kr,* 6, =, Ar, 40], []

> OP((a+b)/(x+y+z)^2+sin(c+d)-exp(h)); M:= module() export x; x:=() -> `+`(args)/nargs; end:

+, *, +, a, b, ^, +, x, y, z, -2, *sin*, +, c, d, *, -1, *exp*, h

> P:= proc() local a; global b; a:= () -> `+`(args)/nargs^b; end proc: [OP(P)], [OP(M)];

[*symbol*, *procedure*, a, b], [*symbol*, *module*, x, *moduledefinition*, *thismodule*, x]

> [OP(f(g(a,b), h(c,d,e,j,k,l)))], [OP(series(sin(x), x=2, 6))];

[f, g, a, b, h, c, d, e, j, k, l], [+, x, -2, *sin*, 2, 0, *cos*, 2, 1, *, *Fraction*, -1, 2, *sin*, 2, 2, *, *Fraction*, -1, 6, *cos*, 2, 3, *, *Fraction*, 1, 24, *sin*, 2, 4, *, *Fraction*, 1, 120, *cos*, 2, 5, O, 1, 6]

Read – reading of the *Maple* files

Call format of the procedure:

Read(L)

Formal arguments of the procedure:

L – a list or a set of filenames or/and full paths to them (may be *strings* and/or *symbols*)

Description of the procedure:

The **Read** procedure (unlike "**read**" statement of *Maple*) is used to read one or more files of the *Maple* language format or the internal *Maple* format into a current session. The objects stored in the datafiles defined by a list or a set **L** are read into a current *Maple* session and become *available* for use. Paths or names of the datafiles should have a type {*symbol* | *string*} and to meet the *Maple* agreements. The successful call of the procedure returns the *NULL* value, i.e. nothing, with output of the appropriate warning about the read files.

The procedure handles the erratic and especial situations conditioned by the following reasons, namely: (**1**) absence of files for reading, (**2**) the mode ***"open"*** for a file, and (**3**) non-readability of a file. Except the first case, the procedure omits a file causing an especial situation and outputs the appropriate warning characterizing a detected situation. The examples below illustrate the told.

```
Read:= proc(F::{set({string, symbol}), list({string, symbol})})
local k, p, n, h, t, G, Lambda, w;
  `if`(Empty(F),ERROR("files which should be read are absent"),writeline(`_$_`, "`$ $`:=null();"));
  assign67(close(`_$_`), h = [op(F), `_$_`], w = interface(warnlevel));
  assign(Lambda = (t -> interface(warnlevel = t)), n = nops(h), G = []);
  for k to n do
    if not type(h[k], file) then WARNING("<%1> is not a file", h[k]); next
    elif Ftype(h[k]) = ".m" then read h[k]; G := [op(G), h[k]]; next
    elif IsOpen1(h[k]) then WARNING("file <%1> is open", h[k]); next
    else
      try Lambda(2); parse(readbytes(h[k], TEXT, infinity), statement)
      catch "incorrect syntax in parse": WARNING("file <%1> is not readable", h[k]); next
      finally close(h[k]); Lambda(w)
      end try;
      (proc() read h[k]; `if`(k = n, fremove(`_$_`), assign('G' = [op(G), h[k]])) end proc)()
    end if
  end do;
  unassign('`$ $`'), WARNING("files %1 have been read successfully into a current session", G)
end proc
```

Typical example of the procedure use:

```
> Read([]);
    Error, (in Read) files which should be read are absent
> Read(["C:/Temp/av.txt", "C:/Temp/ag.m", `C:/Temp/rans.m`, "C:/Academy/RANS.m",
  "C:/Academy/ReadMe.htm", "C:/Temp/h.txt", "C:/Temp/Order/ABS.txt"]);
    Warning, file <C:/Temp/h.txt> is open
    Warning, <C:/Temp/rans.m> is not a file
    Warning, file <C:/Academy/ReadMe.htm> is not readable
    Warning, files [C:/Temp/av.txt, C:/Temp/ag.m] have been read successfully into a current
    session
```

Read1 – extended reading of the *Maple* files

Call format of the procedure:

Read1(L {, 'T' {, 'V'}})

Formal arguments of the procedure:

L – a list or a set of filenames or/and full paths to them (may be strings and/or symbols)
T – (optional) a table of names and types of the read *Maple* objects
V – (optional) a table of names and types of those *Maple* objects that were active in a current session until the procedure call **Read1**

Description of the procedure:

The **Read1** procedure (in addition to functions of the above **Read** procedure) is used to obtain names of the *Maple* objects contained in the read files, and their types. The objects stored in the files defined by a list or a set **L** are read into a current *Maple* session and become available for use.

In case of coding only of the second actual **T**-argument through it the table informing about all activated program objects is returned. The objects are presented in the following slits: procedures (*procs*), variables (*vars*), the program modules of the first type (*mods1*) and the second type (*mods2*). In addition, a program module of the *first* type is understood as the module defined by a construction of view **"Name:=module () …"**, whereas a module of the *second* type is characterized by the determinative construction of view **"module Name () … "**. Some useful considerations at use of program modules of the second type can be found in our books [10,29-33,43,44]. Indices of the returned table **T** are the above identifiers of types of program objects, and entries – the sets of names corresponding to them; as that, the examples of appendices of the **Read1** procedure illustrate.

In case of coding of the second and third actual argument through the third **V** argument, the table structurally similar to the table **T** is returned. The **V**-table in the above slits defines the lists of names of those program objects, which before already were active in a current *Maple* session. The given information can be useful at organization of mechanisms of management by a withdrawing of program objects out of a current session.

The **Read1** procedure provides a reading of both closed and opened files of the *Maple* language format and interior format of the package; in addition, open files are being closed. A path or name of the *Maple* file should has a type {*symbol, string*} and to meet the package agreements.

If the **Read1** procedure reads a file of the *Maple* language format containing the program modules of the second type, then for each of them the erroneous situation of a view "Error, (*in unknown*) attempting to assign to ..." is identified that however does not influence the correctness of the returned result. Similar situation takes place and for case of the built-in `**read**` function of the package *Maple.*

The procedure handles the erratic and especial situations conditioned by the following reasons, namely: (**1**) absence of files for reading, (**2**) the mode ***"open"*** for a file, and (**3**) non-readability of a file. Except the first case, the procedure omits a file causing an especial situation and outputs the appropriate warning characterizing a detected situation. The examples below illustrate the told.

```
Read1:= proc(F::{set({string, symbol}), list({string, symbol})})
local k, j, act_tab, rf, v, t, fg, nm, T, U, R;
  `if`(F = [] or F = {}, ERROR("files for reading are absent"), NULL);
  assign(fg = (a::string -> `if`(search(a, "/"), a[2 .. -2], a)), U = {anames()});
  assign(act_tab = table([procs = {}, mods1 = {}, mods2 = {}, vars = {}]),
    T = table([procs = {}, mods1 = {}, mods2 = {}, vars = {}]));
  v := table(["""" = 1, "#" = 2, "$" = 3, "%" = 4, "&" = 5, "'" = 6, "(" = 7, ")" = 8, "*" = 9, "+" = 10,
    "," = 11, "-" = 12, "." = 13, "/" = 14, "0" = 15, "1" = 16, "2" = 17, "3" = 18, "4" = 19, "5" = 20,
    "6" = 21, "7" = 22, "8" = 23, "9" = 24, ":" =25, ";" = 26, "<" = 27, "=" = 28, ">" =29, "?" = 30,
    "@" = 31, "A" = 32, "B" = 33, "C" = 34, "D" = 35, "E" = 36, "F" = 37, "G" = 38, "H" = 39,
    "I" = 40, "J" = 41, "K" = 42, "L" = 43, "M" = 44, "N" = 45, "O" = 46, "P" = 47, "Q" = 48,
    "R" = 49, "S" = 50, "T" = 51, "U" = 52, "V" = 53, "W" = 54, "X" = 55, "Y" = 56, "Z" = 57]);
  rf:= proc(f)
    local _h, _p, _p1, k, _z, _g, _r, _q;
      close(f), assign(_p ={}, _p1={}, _z ={seq(lhs(op(2, eval(v))[k]), k=1..nops([indices(v)]))});
      do _h := readline(f);
        if _h = 0 then break
        else
          if cat(" ", f)[-2 .. -1] = ".m" then
            if _h[1] = "I" and member(_h[2], _z) then
              _p := {op(_p), cat(``, _h[3 .. 2 + eval(v[_h[2]])])}; _p1 := _p
            end if
          else
            if _h[1 .. 11] = "unprotect('" then
              _g := Search(_h, ` `); _p := {op(_p), cat(``, fg(_h[_g[1] + 1 .. _g[2] - 1]))}
            elif evalb(assign('_q' = Search2(_h, {" := mod", " := proc"})), _q <> []) then
              _p := {op(_p), cat(``, fg(_h[1 .. _q[1] - 1]))}
            elif search(_h, ` := `, '_r') then _p1 := {op(_p1), cat(``, fg(_h[1 .. _r - 1]))}
            end if
          end if
        end if
      end do;
      close(f), [_p, _p1]
    end proc;
  for k to nops(F) do
    try
      if not type(F[k], file) then WARNING("<%1> is not a file or does not exist", F[k]); next
     elif not TRm(F[k]) then
        WARNING("file <%1> is not readable by statement <read>",F[k]); next
      else
        if nargs = 1 then read F[k]; next
```

```
          else
            nm := (proc(`_$$$`) read `_$$$`; rf(`_$$$`) end proc)(F[k]);
            for j to nops(nm[1]) do
              if type(eval(nm[1][j]), procedure) and member(nm[1][j], {anames(procedure)})
              then act_tab[procs] := {nm[1][j], op(act_tab[procs])}
              elif type(eval(nm[1][j]), `module`) and member(nm[1][j], {anames(`module`)})
              then
                if type(nm[1][j], mod1) then act_tab[mods1] := {nm[1][j], op(act_tab[mods1])}
                else act_tab[mods2] := {nm[1][j], op(act_tab[mods2])}
                end if
              end if
            end do;
            act_tab[vars] := {op(act_tab[vars]), op(nm[2] intersect {anames()})} minus
              map(op, {act_tab[procs], act_tab[mods1], act_tab[mods2]})
          end if
        end if
      catch "local and global": NULL(`Processing of especial situations with program modules`)
      end try
    end do;
    `if`(member(_Vars_, act_tab[procs]), _Vars_(), NULL), assign(R = act_tab[mods1] union
      act_tab[mods2]), `if`(R <> {}, map(eval, R), NULL);
    `if`(act_tab[procs] <> {}, (proc() try map(act_tab[procs], ``) catch "": NULL end try
      end proc)(), NULL);
    null(`if`(1 < nargs, [assign(args[2] = eval(act_tab)), `if`(nargs =3, [seq(assign('T[k]' = U intersect
      act_tab[k]), k = [procs, mods1, mods2, vars]), assign(args[3] = eval(T))], NULL)], NULL))
end proc
```

Typical examples of the procedure use:

```
> MM:= module() export x,m; m:= module() export y; y:= () -> `+`(args) end module end
  module: MM1:= module() local F,m; export a; option package; m:= module() export y;
  y:=() -> `+`(args) end module: F:=() -> m:-y(args); a:=F end module: module SVEGAL ()
  export x; x:= () -> `+`(args)/nargs end module: PP:= proc() `+`(args)/nargs end proc:
  PP1:=proc() `+`(args)/nargs end proc: PP2:=proc() `+`(args)/nargs end proc: module GS()
  export x; x:= () -> `+`(args)/nargs end module: avz:= 61: agn:=56: vsv:=36: arn:=40:
  Art:=14: Kr:=6: save(avz,agn,vsv,arn,Art,Kr,GS,MM1,PP,MM,PP1,SVEGAL,PP2,
  `C:/Academy/RANS`); save(avz,agn,vsv,arn,Art,Kr,GS, MM1, PP,MM,PP1,SVEGAL,PP2,
  "C:/Academy/IAN.m");
> restart; Read1(["C:/Academy/RANS"], 'h'), eval(h);
  Error, (in unknown) attempting to assign to `GS` which is protected
  Error, (in unknown) attempting to assign to `SVEGAL` which is protected
  table([mods1 = {MM1, MM}, procs = {PP, PP1, PP2}, vars = {avz, agn, vsv, arn, Art, Kr},
  mods2 = {SVEGAL, GS}])
> Read1({"C:/Academy/IAN.m"}, 'h'), eval(h);
  table([mods1 = {MM1, MM}, procs = {PP, PP1, PP2}, vars = {avz, agn, vsv, arn, Art, Kr},
  mods2 = {SVEGAL, GS}])
> Read1({}, 'h'), eval(h);
  Error, (in Read1) files for reading are absent
```

Rfact – check of an arbitrary integer to be the factorial

Call format of the procedure:

Rfact(N {, 't'})

Formal arguments of the procedure:

N - an arbitrary non-negative integer (*nonnegint*)
t - an arbitrary unevaluated name

Description of the procedure:

The procedure call **Rfact(N)** returns the *true* value, if **N** is the factorial and the *false* value otherwise. If optional actual **t**-argument has been coded, through it an integer **n** for which **n! = N** is returned.

```
Rfact:= proc(N::nonnegint)
local a, k, p;
option remember;
  `if`(N <> 0, NULL, RETURN(true, `if`(nargs = 1, 1, assign('args'[2] = 0))));
  for p to N do assign('a' = mul(k, k = 1 .. p)), `if`(nargs = 1, 1, assign('args'[2] = p));
    if a < N then next elif N < a then RETURN(false) else RETURN(true) end if
  end do
end proc
```

Typical examples of the procedure use:

```
> map(Rfact, [62!, 1942, 2004]), Rfact(894!, 'h'), h, map(Rfact, [7!, 1947, 1995]),
  Rfact(87178291200, 't'), t, Rfact(14061942);
```

$$[true, false, false], true, 894, [true, false, false], true, 14, false$$

Save – extended saving of the *Maple* procedures and program modules
Save1
Save2

Call format of the procedures:

Save(P, F)
Save1(P, F)
Save2(P, F)

Formal arguments of the procedures:

P - a list of names of the *Maple* procedures or/and program modules assigned to saving
F - a list of files assigned to receiving of the above *Maple* objects

Description of the procedures:

The **Save** procedure (in contrast to the built-in function `*save*`) is used to save the *Maple* objects (*procedures* or/and *program modules*) defined by the first actual **P** argument in files defined by the second actual **F** argument. Between elements of the lists **P** and **F** should be one-to-one correspondence, namely: the **j**th object of **P** is saved in the **j**th file of **F.** In a case of violation of such coincidence the error is arises. In addition, if path to a receiving file does not exist, then it is created with output of appropriate warning, if it is necessary; this note is valid for all tools of so-called **Save**-group. For these purposes, the mechanism of the **pathtf** procedure is used.

The built-in function `*save*` of *Maple* saves program modules incorrectly. An quite detailed discussion of this question can be found in our books [10,29-33,43,44]. By reason of that the **Save** procedure saves both the procedures and the program modules defined by actual **P** parameter in files of the *Maple* language format, whereas the procedures in files of both formats. If the procedure call **Save** defines a saving of a program module in file of internal *Maple* format (*m*-file), then the module will be saved in file of the input *Maple* language format with main name of the target file and with *txt*-extension. In this case, an appropriate warning is output.

Extremely useful procedure **Save1** essentially extends both the built-in function `*save*`, and the above **Save** procedure. The extension of the procedure allows to save the groups of program objects (procedures and program modules) in separate files. In this case the first actual **P** argument of the procedure call **Save1(P, F)** can be represented by a list, whose elements are the lists or the sets of names of the saved objects (whose definitions were evaluated in a current session or are in a library logically linked with the main *Maple* library), whereas the second argument **F** is analogous to the second argument of the **Save** procedure represented above. The **Save1** procedure in a series of cases essentially simplifies programming problems dealing with organization of storage of procedures and program modules in the peripheral memory. The detailing of description of the **Save1** procedure can be found below in an example illustrating its application.

At last, the **Save2** procedure quite essentially extends the built-in function `*save*` with complete inheriting of syntax of its call. The **Save2** procedure uses the same formal arguments, as the built-in function `*save*`, but in contrast to the second it allows correctly to save in the *m*-files the variables, procedures, and program modules. The procedure call returns full path to a target datafile.

For saving of variables, they at a procedure call should be coded in *single quotes* (for example, **'X'**). The procedure can output the warnings of a view *"Warning, Incorrect saving of modules is very likely!"*, notifying that during a saving the procedure has met program modules, which can be incorrect, for example, variables, exported by them, are not defined. The given warning has rather notifying character, and at a priori correctness of modules it does not influence the correctness of their saving in files of both formats (the input *Maple* language format and the internal *Maple* format). The detailing of description of the **Save2** procedure can be found below in an example illustrating its application.

Thus, the represented group of procedures **Save, Save1, Save2** and **Read1** fulfills a much wider set of functions, than 2 built-in functions `*save*` and `*read*`. The examples of joint appendices of the procedures **Save2, Read1** and `*read*` represented below, illustrate the told above.

```
Save:= proc(P::list({symbol}), F::list({string, symbol}))
local a, b, c, d, k, h;
  `if`(P = [] or F = [], ERROR("one or both actual arguments are empty"), assign(a = nops(P),
    d = nops(F))), `if`(a <> d, ERROR("disparity of quantities of procedures and
    datafiles: %1 <> %2", a, d), 1);
  for k to a do assign('c' = pathtf(F[k]));
    if not type(eval(P[k]), {procedure, `module`}) then
      WARNING("<%1> is not a procedure and not a module", P[k]); next
    else
      if c[-2 .. -1] = ".m" and type(P[k], `module`) then c := cat(c[1 .. -3], ".txt");
        WARNING("module <%1> has been saved in datafile <%2>", P[k], c)
      end if;
      null(iolib(5, "_$$", cat("save(", P[k], ",", """, CF1(c), """, "):")), iolib(7, "_$$")),
        (proc() read "_$$"; iolib(8, "_$$") end proc)()
    end if
  end do
end proc
```

```
Save1:= proc(P::list({symbol, list(symbol), set(symbol)}), F::list({string, symbol}))
local a, d, k, j, h, f, Q, `_$avz_agn_sv_art_kr_arn$`, R;
  `if`(P = [] or F = [], ERROR("one or both actual arguments are empty"),
     assign(a = nops(P), d = nops(F), f = map(pathtf, F)));
  `if`(a <> d, ERROR("disparity of quantities of procedures and datafiles: %1<>%2", a, d), 6);
  `if`(nops(f) <> nops(map(CF, {op(f)}, string)), WARNING("multiple names in %1", f), 1);
  for k to nops(P) do
     assign('R' = [seq(`if`(member(whattype(eval(P[k][j])), {procedure, `module`}), P[k][j],
        null(WARNING("<%1> is unassigned or a variable", P[k][j]))), j = 1 .. nops(P[k]))],
        '`_$avz_agn_sv_art_kr_arn$`' = 1942194767899662);
     save `_$avz_agn_sv_art_kr_arn$`, f[k];
     `if`(cat(" ", f[k])[-2 .. -1] = ".m", seq(`if`(type(eval(R[j]), `module`), modproc(R[j]), NULL),
        j = 1 .. nops(R)), NULL);
     `if`(cat(" ", f[k])[-2 .. -1] <> ".m", seq(`if`(type(eval(R[j]), `module`), mod21(R[j]), NULL),
        j = 1 .. nops(R)), NULL);
     assign('Q' = [seq(cat(`_$ArtKr$`, j), j = 1..nops(R))]), seq(Save([R[j]], [Q[j]]), j =1..nops(R));
     seq(MmF(f[k], Q[j]), j = 1 .. nops(R)), iolib(8, op(Q))
  end do
end proc
```

```
Save2:= proc()
local a, b, c, j, k, G, S, nu, f, omega;
  `if`(nargs < 2, RETURN(NULL), assign(nu = interface(warnlevel),
     omega = (x -> interface(warnlevel = x))));
  if type(args[-1], file) then f := args[-1]; omega(0) elif not type(eval(args[-1]), {string, symbol})
  then ERROR("<%1> cannot specify a datafile", args[-1]) else omega(0); f := MkDir(args[-1], 1)
  end if;
  assign(G = [], S = ""), omega(2), assign(a = {op(SLD(convert('procname(args)', string)[7 .. -1],
     ",")[1 .. -2])}); seq(`if`(type(eval(args[j]), {procedure, `module`}), assign('G' = [op(G),
     args[j]]), NULL), j = 1 .. nargs - 1), assign(b = a minus {op(map(convert, G, string))}),
     `if`(b = {} and G = [], RETURN(omega(nu)), NULL);
  if b <> {} then assign(c = cat("_Vars_", " := proc() global ", seq(cat(b[j], ","), j = 1 .. nops(b)),
     "; assign(", seq(cat(convert(cat("'", b[j], "'") = eval(convert(b[j], symbol)), symbol), `,`),
     j = 1 .. nops(b)), ") end proc:")), seq(assign('S' = cat(S, `if`(member(k, {op(Search2(c, {"'",
     ",;", ",)"}))}), NULL, c[k]))), k = 1 .. length(c)), iolib(5, "_$", S), iolib(7, "_$"),
     (proc() read "_$"; iolib(8, "_$") end proc)()
  end if;
  Save1([[op(G), `if`(b = {}, NULL, _Vars_)]], [f]), omega(nu), f
end proc
```

<u>Typical examples of the procedures use:</u>

```
> P:=() -> `+`(args)/nargs: M:=module() export x;  x:=() -> `+`(args)/nargs end module: F1:=
  `C:/AVZ/rans`: F2:=`C:/AVZ/IAN.m`: F3:=`C:/AVZ/IAN`: Save([P,M,M], [F1,F2,F3]);

    Warning, module <M> has been saved in datafile <c:\avz\ian.txt>
```

Application of the procedure **Save1** for saving of the procedures **PP, PP1, PP2** and program modules **SVEGAL, GS, MM, MM1** of two types in two files of various formats (the input *Maple* format and the internal *Maple* format) with the subsequent check of a saving result by means of the procedure **Read1.** Of this example, the saving principle realized by the **Save1** procedure is well being viewed, namely:

(**1**) the procedures save own type regardless of a type of a receiving file;

(**2**) in case of a receiving file of the input *Maple* format, the program modules of the second type are converted into modules of the first type (*what will provide their subsequent correct use*);

(**3**) in case of receiving file of the input *Maple* format, the program modules of both types are converted into procedures (*what will provide their subsequent correct use by a special syntax of the first access to variables exported by them*).

The first access to variables exported by a module, saved thus, has the following simple view:

Name(): Name:- S(args)

where: ***Name*** – a name of a saved program module, **S** – a variable exported by it, and ***args*** – actual arguments passed to the variable. The following fragment illustrates some appendices of the **Save1** procedure.

```
> MM:= module() export x,m; m:= module() export y; y:=() -> `+`(args) end module end
  module: MM1:= module() local F,m; export a; option package; m:= module() export y;
  y:=() -> `+`(args) end module; F:=() -> m:- y(args); a:=F end module: module SVEGAL()
  export x; x:=() -> `+`(args)/nargs^6 end module: PP:=proc() `+`(args)/nargs^2 end proc:
  PP1:=proc() `+`(args)/nargs end proc: PP2:=proc() `+`(args)/nargs end proc: module GS()
  export x; x:=() -> `+`(args)/nargs^3 end module:
> Save1([[MM,PP,PP2,GS], {MM1,SVEGAL,PP1}], ["C:/temp/VSG", "C:/temp/VSG.m"]);
    Warning, Incorrect saving of modules is very likely!
> restart; Read1(["C:/temp/VSG"], 'h'), eval(h);
              table([procs = {PP, PP2}, mods2 = {}, mods1 = {GS, MM}, vars = {}])
> restart; Read1(["C:/temp/VSG.m"], 'h'), eval(h); A:= 2*6^5:
             table([procs = {SVEGAL, PP1, MM1}, mods2 = {}, mods1 = {}, vars = {}])
> SVEGAL(): A*SVEGAL:- x(61,56,36,40,6,14), 3*A*SVEGAL:- x(42,47,67,62,96,89);
                                      71, 403
> MM1(): MM1:- a(61,56,36,40,6,14), MM1:- a(42,47,67,62,96,89);
                                      213, 403
```

Application of the procedure **Save2** for a saving of the procedures **PP, PP1,** the program modules **GS, MM, SVEGAL, MM1** of two types, and variables {**x, y, x, t, v, u, m**} in a file of the internal *Maple* format (*m*-file) with the subsequent check of a saving result by means of the built-in function **`*read*`**. Of this example, the saving principle realized by the **Save2** procedure, is enough well being viewed, namely:

(**1**) the procedures save own type regardless of a type of a receiving file;

(**2**) the variables are saved in a receiving file as procedures;

(**3**) in case of receiving file of the input *Maple* format, the program modules of the second type are converted into modules of the first type (*what will provide their subsequent correct use*);

(**4**) in case of receiving file of the internal *Maple* format, the program modules of both types are converted into procedures (*what will provide their subsequent correct use by a special syntax of the first access to variables exported by them*).

The first access to variables exported by a program module, saved thus, and next read by the built-in `***read***` function has the following simple view:

Name(): Name:- S(*args*)

where: *Name* - a name of a saved program module, **S** - a variable exported by it, and ***args*** - actual arguments passed to the variable. While the variables, saved by the **Save2** procedure, after reading of *m*-file containing them, become accessible in a current session after the unitized call **_Vars_()**{:|;|,}. While a reading of the objects saved by the **Save2** procedure, by means of the **Read1** procedure already at the first access to them suppose the standard *Maple* syntax. The following fragment illustrates some appendices of the **Save2** procedure.

```
> MM:= module() export x,m;  m:= module()  export y; y:= () -> `+`(args) end module  end
  module:  MM1:= module()  local F,m;  export a; option package; m:= module()  export y;
  y:=() -> `+`(args) end module;  F:=() -> m:- y(args);  a:=F end module: module SVEGAL()
  export x;  x:=() -> `+`(args)/nargs^6 end module:  PP:= proc()  `+`(args)/nargs^2  end proc:
  PP1:=proc() `+`(args)/nargs  end proc: PP2:=proc() `+`(args)/nargs end proc: module GS()
  export x; x:=() -> `+`(args)/nargs^3 end module: x:=61:  y:=56: z:=36: t:=67: v:=89: u:=96:
  m:=6:  Save2('x',MM,'y',MM1,'z',SVEGAL,'t',GS,'v',PP,'u',PP1,'m', "C:/Temp/AVZ.m");
```

Warning, Incorrect saving of modules is very likely!

```
> restart;  Read1(["C:/Temp/AVZ.m"], 'h'),  eval(h);
```

table([*mods1* = {}, *procs* = {*PP1*, *MM*, *MM1*, *GS*, *SVEGAL*, *_Vars_*, *PP*}, *vars* = {}, *mods2* = {}])

```
> [x, y, z, t, v, u, m],  36*PP(42,47,67,62,89,96),  SVEGAL:- `x`(33);
```

[61, 56, 36, 67, 89, 96, 6], 403, 33

```
> restart; read("C:/Temp/AVZ.m"); SVEGAL(): [SVEGAL:-`x`(33)], _Vars_(), x,y,z,t,v,u,m;
```

[33], 61, 56, 36, 67, 89, 96, 6

```
> Save([], []);
```

Error, (in **Save**) one or both actual arguments are empty

---

save1 – an extension of the standard *Maple* function `save`

Call format of the procedure:

save1(N, E, F)

Formal arguments of the procedure:

N - a valid *Maple* expression defining an assignable name or list of such expressions
E - a valid *Maple* expression to be assigned to a name **N** or list of such expressions to be assigned to the names from **N**
F - symbol or string defining filename or full path to a target file

Description of the procedure:

The call **save(n1, n2, ..., nk,** ***fileName*****)** of the standard *Maple* function writes the specified variables **n1, n2, ..., nk** into ***fileName*** as a sequence of assignment statements. If a particular **np** is not assigned a value, the assignment **np:= np** is written. Each argument except the last one is "*evaluated to a name*", while the last argument (a filename or full path to it) is fully evaluated. However, the `***save***` function does not allow to save in a file the dynamically evaluated names, what in a lot cases is essential shortcoming. The following procedure **save1** eliminates this shortcoming.

The procedure call **save1(N, E, F)** writes the specified variables defined by the actual **N** argument into a file defined by the actual **F** argument as a sequence of assignment statements. If a particular element **n** from **N** is not assigned a value, the assignment **n:=n** is written with output of appropriate warning. If a particular element **n** from **N** is protected, the element is ignored with output of appropriate warning. In addition, if all elements from **N** are protected, an erratic situation arises. The successful procedure call returns full path to a saving datafile with **F** fulfilling of the given assignments of an **E** expressions to the corresponding symbols **N** in a current *Maple* session with output of necessary warnings.

This procedure essentially extends the `*save*` possibilities, allowing to save in datafiles of both the input *Maple* language format and internal *Maple* format the assigned variables with dynamically generated names. In a lot tasks it is very important possibility. Furthermore, the procedure supports saving of the *Maple* objects in datafiles with arbitrary paths to them. The examples of a fragment represented below illustrate the most typical appendices of the **save1** procedure.

```
save1:= proc(N::{symbol, list(symbol)}, E::anything, F::{string, symbol})
local a, b, c, k, nu, omega, zeta, psi, r, s, x;
  assign(b = (x -> interface(warnlevel = x)), zeta = "_$Euro$_", nu = "save(",
    s = "symbols %1 are protected");
  if not type(eval(F), {string, symbol}) then ERROR("argument <%1> cannot specify a datafile", F)
  elif not type(F, file) then c := interface(warnlevel); b(0); r := CF1(MkDir(F, 1)); b(c)
  else r := CF1(F)
  end if;
  psi:= proc(f)
    local a, k, p, h;
      `if`(f[-2 .. -1] = ".m", RETURN(), assign(p = fopen(f, READ, TEXT), a = "_$$$_"));
      while not Fend(p) do h := readline(p); writeline(a, `if`(h[-1] <> ";", h, cat(h[1 .. -2], ":")))
      end do;
      null(close(f, a), writebytes(f, readbytes(a, infinity)), close(f), fremove(a))
    end proc;
  omega:= proc(N::{symbol, list(symbol)}, E::anything, F::{string, symbol})
    local k;
      `if`(type(N, symbol), assign('nu' = cat(nu, N, ",")), seq(assign('nu' = cat(nu, N[k], ",")),
        k = 1 .. nops(N))), writeline(zeta, cat(nu, """", r, """", "):")), close(zeta);
      if type(N, symbol) then assign(N = eval(E))
      elif type(E, list) then for k to min(nops(N), nops(E)) do assign(N[k] = eval(E[k])) end do
      else assign(N[1] = eval(E))
      end if;
      (proc(zeta) read zeta; fremove(zeta) end proc)(zeta)
    end proc;
  if type(N, symbol) and type(N, protected) then ERROR("symbol <%1> is protected", N)
  elif type(N, list) then assign(a = []);
    for k to nops(N) do `if`(type(N[k], protected), NULL, assign('a' = [op(a), N[k]])) end do;
    `if`(a = [], ERROR("all symbols %1 are protected", N), op([omega(a, args[2], r), psi(r),
      `if`(nops(a) = nops(N), NULL,WARNING(s, {op(N)} minus {op(a)}))]))
  else omega(args[1 .. 2], r), psi(r)
  end if;
  r, WARNING("the saving result is in datafile <%1>", r)
end proc
```

Typical examples of the procedure use:

```
> xyz, abc, ht, RANS, IAN, G:= 57, 62, 1995, 2004, 19.42, 8.15:
> save(cat(x, y, z), cat(ab, c), ht, RANS, "C:\\Vilnius\\Grodno\\save.txt");
    Error, save can only save names
```

```
> save(xyz, abc, ht, RANS, "C:\\RANS\\aaa\\bbb\\ccc\\ddd/zzz.m");
    Error, could not open `C:\RANS\aaa\bbb\ccc\ddd/zzz.m` for writing
> restart: save1([sin, cat(x,y,z), cat(a,b,c), cat(h,t), convert("RANS", symbol)], [56,61,36],
  "C:\\temp\\aaa\\bbb\\ccc\\zzz.m"); xyz, abc, ht, RANS;
    Warning, unassigned variable `RANS` in save statement
    Warning, symbols {sin} are protected
    Warning, the saving result is in datafile <C:\temp\aaa\bbb\ccc\zzz.m>
                          "C:\temp\aaa\bbb\ccc\zzz.m"
                              56, 61, 36, RANS
> save1(RANS, [`1995_2003`], "C:\\RANS/IAN\\RAC\\REA/aaa.m"); RANS;
    Warning, the saving result is in datafile <C:\\RANS/IAN\\RAC\\REA/aaa.m>
                          "C:\RANS/IAN\\RAC\REA/aaa.m"
                                [1995_2003]
> restart; read("c:\\temp\\aaa\\bbb\\ccc\\zzz.m"); read("c:/RANS/IAN/RAC/REA/aaa.m");
  xyz, abc, ht, RANS;
                           56, 61, 36, [1995_2003]
> save1([x1,x2,x3], (a+b)/(c+d)+sin(x)*cos(y), "C:\\RANS/aaa\\xxx.m"); x1, x2, x3;
    Warning, unassigned variable `x2` in save statement
    Warning, unassigned variable `x3` in save statement
    Warning, the saving result is in datafile <C:\RANS/aaa\xxx.m>
                            "C:\RANS/aaa\xxx.m";
                       (a+b)/(c+d)+sin(x)*cos(y), x2, x3
> save1(ln, (a+b)/(c+d)+sin(x)*cos(y), "C:\\RANS/aaa\\xxx.m");
    Error, (in save1) symbol <ln> is protected
> save1([ln, sin, cos, sqrt], (a+b)/(c+d)+sin(x)*cos(y), "C:\\RANS/aaa\\xxx.m");
    Error, (in save1) all symbols [ln, sin, cos, sqrt] are protected
> save1([sin, ln, cat(x,y,z), cat(a,b,c), cat(h,t), convert("RANS",symbol)], [56,61,36],
  "C:/RANS/IAN/RAC/REA"); xyz, abc, ht, RANS;
    Warning, Path element <rea> has been found, it has been renamed on <_rea>
    Warning, unassigned variable `RANS` in save statement
    Warning, symbols {sin, ln} are protected
    Warning, the saving result is in datafile <C:/RANS/IAN/RAC/REA>
                           "C:/RANS/IAN/RAC/REA"
                              56, 61, 36, RANS
> restart; read("C:/RANS/IAN/RAC/_REA"); xyz, abc, ht, RANS;
                              56, 61, 36, RANS
> save1([G,cat(x,y,z),cat(a,b,c),cat(h,t),convert("RANS", symbol), cat(S,v)],[56,61,36,40,7,14],
  "C:\\Academy"); G, xyz, abc, ht, RANS, Sv;
    Warning, the saving result is in datafile <c:\_academy>
                               "c:\_academy"
                           56, 61, 36, 40, 7, 14
```

```
> restart; read("C:\\_Academy"); G, xyz, abc, ht, RANS, Sv;
                              56, 61, 36, 40, 7, 14
> save1([cat(x,y,z),cat(y,z),cat(z,6)], [proc() `+`(args)/nargs end,56,36], "C:\\TEMP");
   Warning, the saving result is in datafile <c:/_temp>
                                   "c:/_temp"
> restart; read("c:/_temp"); [yz,z6,3*xyz(61,56,36,40,7,14)];
                                  [56, 36, 107]
```

---

save2 – an useful modification of the standard *Maple* functions *`save`* and *`read`*
save3
read3

Call format of the procedures:

save2(S, F)
save3(S, F)
read3(G)

Formal arguments of the procedures:

S - names or symbols; sequence of one or more specified variables
F - symbol or string defining a filename or full path to a target file
G - symbol or string defining a filename or full path to a datafile with saved *Maple* objects

Description of the procedures:

The call **save(n1, n2, ..., nk,** ***fileName*****)** of the standard *Maple* function writes the specified variables **n1, n2, ..., nk** into ***fileName*** as a sequence of assignment statements. If a particular **np** is not assigned a value, the assignment **np:= np** is written. Each argument except the last one is "*evaluated to a name*", while the last argument (a filename or full path to it) is fully evaluated. However, the ***`save`*** function saves names in formats which in a lot cases are rather improper. The following procedure **save2** extends the standard procedure ***`save`***, supporting formats useful in a series of cases.

The procedure call **save2(S, F)** writes the specified variables defined by the actual **S** argument into a file specified by the actual **F** argument as a sequence of assignment statements. If a particular **Si** from **S** is not assigned a value, the assignment **Si:= Si** is written. The successful procedure call returns the *NULL* value with output of a warning about full path to a target datafile.

This procedure allows to save in datafiles of both the input *Maple* language format and internal *Maple* format the assigned variables without the *protected*-attribute. The program modules before saving are converted into modules of the *first* type. In many tasks, it is very important possibility. In particular, the **save2** result is read by the standard procedure ***`read`***; moreover, this datafile allows to receive names of *Maple* objects saved in it, their types without their activation in a current session. Furthermore, the procedure supports saving of the *Maple* objects in datafiles with arbitrary paths to them. In addition, if the standard procedure ***`save`*** demands to code names of the saved constants in backquotes, i.e. <`>, whereas the **save2** procedure demands to code such names as unevaluated expressions, i.e. in single quotes <'>. The examples of a fragment represented below illustrate the most typical appendices of the **save2** procedure.

The procedure call **save3(S, F)** writes the specified variables defined by the actual **S** argument into a file specified by the actual **F** argument as a sequence of assignment statements. If a particular **Si** from **S** is not assigned a value, the assignment **Si:= Si** is written. The successful procedure call returns the *NULL* value with output of a warning about full path to a target datafile, and a mode of its read.

This procedure allows to save in datafiles of both the input *Maple* language format and internal *Maple* format the assigned variables without the *protected*-attribute. The program modules before saving are converted into modules of the *first* type. In a lot tasks it is very important possibility. Furthermore, the procedure supports saving of the *Maple* objects in datafiles with arbitrary paths to them. In addition, if the standard procedure `**save**` demands to code names of the saved constants in backquotes, i.e. <`>, whereas the **save3** procedure (analogously to the above procedure **save2**) demands to code such names as unevaluated expressions, i.e. in single quotes <'>.

In contrast to the standard function `**save**`, the procedure **save3** provides correct saving of program modules in datafiles of the internal *Maple* format. The subsequent read of such *m*-files is supported by both the standard function `**read**` and the procedure **read3** represented below. The procedure call **save3(S, F)** for a datafile **F** of format different from "**.m**" is analogous to use of the standard function `**save**` only that the saved procedures and modules lose the *protected*-attribute and modules additionally are converted into modules of the *first* type. Whereas for case of saving of the assigned variables in datafiles of the internal *Maple* format is used on principle other approach. The approach allows correctly to read *m*-files created by the **save3** by both standard function `**save**` and the procedure **read3.** If a *m*-file created by the **save3** has been read by the standard function `**save**`, then for subsequent activation in a current session of a **M** module contained in the *m*-file it is necessary to evaluate expression `**M()**{:|;}`. That provides use of **M** module by technique accepted in *Maple*.

The procedure **read3** is used to read *Maple* language files into a current *Maple* session. If the filename ends with the characters "**.m**", the file is assumed to be in the internal *Maple* format. The objects stored in the file are read into a current *Maple* session and become available for use. The procedure call **read3(G)** supposes that a *Maple* datafile defined by the actual **G** argument has been previously prepared by the above procedures `**save**`, **save1, save2, save3**, etc. (for *Maple* files of both internal and the input formats) or by a *Windows* application (for *Maple* files of input language format).

The successful procedure call **read3(G)** returns the *NULL* value by activating in a current *Maple* session the objects previously saved in a datafile **G** of the input *Maple* language format or the internal *Maple* format. In addition, an access to these objects is fully analogous to the technique accepted in *Maple***.** The **read3** procedure extends the standard function `**read**`, providing the more advanced handling of both erratic and especial situations. The examples of a fragment represented below illustrate the most typical appendices of the above procedures **save2, save3** and **read3.**

```
save2:= proc()
local b, c, k, r, x, nu;
  assign(b = (x -> interface(warnlevel = x)), x = nargs, nu = (() -> map(unprotect, [args])),
    omega = (x -> `if`(type(x, `module`), `if`(type(x, mod1), NULL, mod21(x)), NULL)));
  if x < 2 then ERROR("quantity of actual arguments must be at least 2, but has been
    received %1", x)
  elif not type(eval(args[-1]), {string, symbol}) then
    ERROR("<%1> cannot specify a datafile", args[-1])
  elif not type(args[-1], file) then c := interface(warnlevel); b(0); r := MkDir(args[-1], 1); b(c)
  else r := args[-1]
```

```
  end if;
  if cat(" ", r)[-2 .. -1] = ".m" then map(omega, [args[1 .. -2]]), nu(args[1 .. -2]);
    (proc(x) save args, r end proc)(args[1 .. -2])
  else map(omega, [args[1 .. -2]]), nu(args[1 .. -2]);
    for k to x – 1 do writeline(r, cat("", args[k], " := ", convert(eval(args[k]), string), ":")) end do
  end if;
  close(r), WARNING("the saving result is in datafile <%1>", r)
end proc

save3:= proc()
local c, d, k, f, m, p, nu, omega, psi, epsilon;
  `if`(nargs < 2, ERROR("quantity of arguments must be at least 2 but has been received %1",
    nargs), 7);
  assign(m=[], p=[], d=nargs-1, f =cat("", args[-1]), omega="_56_.m", nu ="_61_", psi = "_56_");
  assign(c = interface(warnlevel)), interface(warnlevel = 0), seq(`if`(type(args[k], procedure),
    assign('p' = [op(p), unprotect(args[k]), args[k]]), `if`(type(args[k], `module`), [`if`(type(args[k],
    mod1), 7, mod21(args[k])), assign('m' = [op(m), unprotect(args[k]), args[k]])], 7)), k = 1 .. d),
    assign(epsilon = "_42_.m"), `if`(type(f, file), 7, assign('f' = MkDir(f, 1)));
  if m = [] or Ftype(f) <> ".m" then
    (proc(x) save args, f end proc)(args[1 .. -2]), interface(warnlevel = c)
  else
    for k to nops(m) do writeline(nu, cat("", m[k], ":=proc() global ", m[k], "; ", m[k], ":=",
      convert(eval(m[k]), string), " end proc:"))
    end do;
    (proc(x) close(nu); read nu; save args, f end proc)(args[1 .. -2]);
    writeline(psi, Red_n(cat("proc() local __1,", seqstr(op(m)), ";__1:=map(eval,[",
      seqstr(seq(cat(m[k], `()`), k = 1 .. nops(m))),"]): save(__1,"_56_.m") end proc():"), "`", 1));
    (proc() close(psi); read psi; mmf(f, omega,epsilon);
        writebytes(f, readbytes(epsilon, infinity)); close(f) end proc)();
    interface(warnlevel = c), fremove(psi, epsilon, omega, nu)
  end if;
  WARNING("a saving result is in file <%1>, it can be read by functions 'read' or 'read3'", f)
end proc

read3:= proc(F::file)
local a;
  if Ftype(F) = ".m" then
    if Release1() = SD(readbytes(F,TEXT,2)[2]) then close(F); read F; eval(__1); unassign('__1')
    else close(F), ERROR("file <%1> is incompatible with a current Maple release", F)
    end if
  else
    try assign(a = interface(warnlevel)), interface(warnlevel = 0);
      parse(readbytes(F, TEXT, infinity), statement);
      (proc() close(F), interface(warnlevel = a); read F end proc)()
    catch "": close(F); interface(warnlevel = a), ERROR("file <%1> cannot be read by function
      'read'", F)
    end try
  end if
end proc
```

Typical examples of the procedures use:

```
> Art:= () -> `+`(args)/nargs: Kr:= () -> [nargs,`+`(args)/nargs]: protect(Kr); AGN:=56:
  protect('AGN'); Svet:=36: M1:=module () export x; option package; x:=()->`+`(args) end
  module: module M2 () export x; option package; x:=()->`+`(args) end module:
> save2(Art,Kr,'AGN','Svet',M1,M2,PP,Z, "C:/rans/AGN/AVZ/Art_Kr.txt");
    Warning, the saving result is in datafile <C:/rans/AGN/AVZ/Art_Kr.txt>
> save2(Art,Kr,'AGN','Svet',M1,M2,PP,Z, "C:/rans/AGN/AVZ/Art_Kr.m");
    Warning, the saving result is in datafile <C:/rans/AGN/AVZ/Art_Kr.m>
> restart; read("C:/rans/AGN/AVZ/Art_Kr.txt"): AGN, Svet, map(whattype,map(eval,
  [Art, Kr, M1, M2])), map(type, [AGN,Svet,Art,Kr,M1,M2], protected);
```

$$56, 36, [procedure, procedure, module, module], [false, false, false, false, false, false]$$

```
> restart; read("C:/rans/AGN/AVZ/Art_Kr.m"): AGN, Svet, map(whattype,map(eval,
  [Art, Kr, M1, M2])), map(type, [AGN,Svet,Art,Kr,M1,M2], protected);
```

$$56, 36, [procedure, procedure, module, module], [false, false, false, false, false, false]$$

```
> restart; Art:=14: Kr:=7: Svet:=2003: save(`Art`, `Kr`, `Svet`, "C:/rans/family.txt");
  save2(`Art`, `Kr`, `Svet`, "C:/rans/family1.txt");
    Warning, the saving result is in datafile <c:\rans\family1.txt>
> save('Art','Kr','Svet', "C:/rans/fam2.txt"); save2('Art','Kr','Svet', "C:/rans/fam3.txt");
    Error, illegal use of an object as a name
    Warning, the saving result is in datafile <c:\rans\fam3.txt>
> restart; Art:=module() export a,x,y,z,u,t; x:=()->`+`(args)/nargs; y:=()->sqrt(x(args));
  a:=module () export b; b:=()->`+`(args)/nargs end module; z:=()->x(args)+y(args); u:=()->
  sqrt(x(args)+y(args)); t:=proc(x) x^2 end proc end module: eval(delf): agn:=56: avz:=61:
  Vgtu:=module () export a; a:=()->min(args)/max(args) end module: v:=vector([1995,2004]):
  T:=table([x=a,y=b]): m:=matrix([[42,61],[47,56]]): module Kr () export z; z:=()->
  sqrt(sum(args[k]^2, k=1..nargs)) end module:
> save3(Art,Kr,'agn',v,Vgtu,'avz',T,m, "C:/rans/academy/vilnius\\Res.m");
    Warning, a saving result is in file <C:/rans/academy/vilnius\Res.m>, it can be read by
    functions 'read' or 'read3'
> save3(Art,Kr,'agn',v,Vgtu,'avz',T,m, "C:/rans/academy/vilnius\\Res");
    Warning, a saving result is in file <C:/rans/academy/vilnius\Res.m>, it can be read by
    functions 'read' or 'read3'
> restart; read3("C:/rans/academy/vilnius\\Res.m"); agn, avz, map(type, [Art,Kr,Vgtu],
  `module`), map(eval, [v, T, m]);
```

$$56, 61, [false, false, false], \left[[1995, 2004], \text{table}([x = a, y = b]), \begin{bmatrix} 42 & 61 \\ 47 & 56 \end{bmatrix} \right]$$

```
> map(with, [Art,Kr,Vgtu]), map(exports, [Art,Kr,Vgtu]);
    Warning, the name z has been rebound
    Warning, the name a has been rebound
                 [[a, t, u, x, y, z], [z], [a]], [a, x, y, z, u, t, z, a]
> Art:- x(42,47,67,62,89,96), Kr:- z(42,47,67,62,89,96), Vgtu:- a(42,47,67,62,89,96);
                 403/6, 29443^(1/2), 7/16
```

seq1 – generating of the *Maple* expressions sequences
seq2

Call format of the procedures:

seq1(V, p, n, m)
seq2(V, p, n, m)

Formal arguments of the procedures:

V - any valid *Maple* expression
p - a name or symbol - an index variable
n, m - the numerical values

Description of the procedures:

In our previous books [8,10,29-33,43,44] some principal demerits of the built-in function `*seq*` were discussed. In particular, execution of the cyclic constructions of view **seq(F(k), k=1 .. n)** does not cease by **stop**-button of the **GUI** and in case of large enough **n** values can result in "hanging" of the package requiring a **PC** reboot. At the same time, use of values `*infinity*` and **kernelopts(***maximmediate***) = 1073741823** for the actual **n** argument is impossibly, where *maximmediate* - the largest integer that can be represented in immediate-integer form (represented in a word instead of a pointer to the multiword data structures). The following procedures **seq1** and **seq2** eliminates the above demerits of the standard function `*seq*` of the package *Maple* of releases **6 – 9/9.5.**

The procedure call **seq1(V, p, n, m)** corresponds to the call of the built-in function **seq(V, k = n .. m)** provided that the **p** argument defines a **V**-expression indexed by a variable that is coded even at absence of indexing of the **V**-expression. At last, the call **seq2(V, p, n, m)** is analogous to the call of standard procedure **seq(V, p=n..m)** if a value of its actual argument **m** is different from `*infinity*` and **kernelopts(***maximmediate***).** Otherwise, the call **seq2(V,p,n,m)** outputs an appropriate warning and forms the global `*_rseq*` variable whose value is available in a current session after termination of calculations by **stop**-button of the **GUI.** The following examples illustrate successful executions of the procedures **seq1** and **seq2** there, where the built-in function `*seq*` causes erroneous situations.

```
seq1:= proc(V::anything, p::symbol, n::numeric, m::{pos_infinity, numeric})
local k, H;
  for k from n to m do H := H, eval(subs(p = k, V)) end do; H[2 .. nops([H])]
end proc

seq2:= proc(V::anything, p::symbol, n::numeric, m::{pos_infinity, numeric})
local a, k;
global _rseq;
  assign67(a = kernelopts(maximmediate), '_rseq' = NULL, k = n - 1), `if`(not member(m,
     {infinity, a}), RETURN(V $ (p = n .. m)), WARNING(`potential-unlimited cycle`));
  do assign('k' = k+1), `if`(k = a, RETURN(_rseq), assign67('_rseq' = _rseq, eval(subs(p = k, V))))
  end do
end proc
```

Typical examples of the procedures use:

```
> seq(k, k=1..infinity);
    Error, unable to execute seq
> seq(k, k=1..kernelopts(maximmediate));
    Error, object too large
```

> seq1(k, p, 1 ,infinity);

Warning, computation interrupted

> seq1(k, p, 1, kernelopts(maximmediate));

Warning, computation interrupted

> [seq(k,k=4.2..18.7)], [seq1(k,k,4.2,18.7)];

[4.2, 5.2, 6.2, 7.2, 8.2, 9.2, 10.2, 11.2, 12.2, 13.2, 14.2, 15.2, 16.2, 17.2, 18.2], [4.2, 5.2, 6.2, 7.2, 8.2, 9.2, 10.2, 11.2, 12.2, 13.2, 14.2, 15.2, 16.2, 17.2, 18.2]

> seq2(p^2, p, 1, infinity);

Warning, potential-unlimited cycle
Warning, computation interrupted

> _rseq;

1, 4, 9, 16, 25, 36, 49, 64, 81, 100, 121, 144, 169, 196, 225, 256, 289, 324, 361, 400, 441, 484, 529, 576, 625, 676, 729, 784, 841, 900, 961, 1024, 1089, 1156, 1225, 1296, 1369, 1444, 1521, 1600, 1681, 1764, 1849, 1936, 2025, 2116, 2209, 2304, 2401, 2500, 2601, 2704, 2809, 2916, 3025, 3136, 3249, 3364, 3481, 3600, 3721, 3844, 3969, 4096, 4225, 4356, 4489, 4624

> seq2(k*(k^7+2003), k, 1, 17);

2004, 4262, 12570, 73548, 400640, 1691634, 5778822, 16793240, 43064748, 100020030, 214380914, 430005732, 815756760, 1475817098, 2562920670, 4294999344, 6975791492

In addition, the interruptions of the examples **2**, **3** and **5** are caused by an interruption of computations by **stop**-button of the **GUI** (*Graphic User Interface*).

---

seqstr – converting of an expressions sequence into the string

Call format of the procedure:

seqstr(*args*)

Formal arguments of the procedure:

args - an expressions sequence

Description of the procedure:

The procedure call **seqstr(*args*)** returns the result of converting of an expressions sequence, defined by the actual ***args*** arguments, into the *string*.

```
seqstr:= () -> convert([args], string)[2 .. -2]
```

Typical examples of the procedure use:

> seqstr(libname), seqstr(), seqstr(a,b,c,d,e,f,g,h), seqstr(1995,RANS,IAN,2004);

""c:/program files/maple 9/lib/userlib", "C:\Program Files\Maple 9/lib"", "", "a, b, c, d, e, f, g, h", "1995, RANS, IAN, 2004"

Subset – check of a set to be the subset

Call format of the procedure:

Subset(S, S1)

Formal arguments of the procedure:

S, S1 - the compared sets

Description of the procedure:

The procedure call **Subset(S, S1)** returns the *true* value if a set **S** is the subset of a set **S1,** otherwise the *false* value is returned. In a series of cases, this procedure is more convenient than the built-in **subset** operator of *Maple.*

```
Subset:= (a::set, b::set) -> evalb(a  intersect  b = a)
```

Typical examples of the procedure use:

> Subset({a,b,c,y},{a,b,c,m}), {a,b,c,y} subset {a,b,c,m}, Subset({ag,av}, {42,ag,av,95});

false, false, true

---

ttable – check of a *Maple* object upon type `*table*` with entries of the given type

Call format of the procedure:

ttable(T {, Tp})

Formal arguments of the procedure:

T - any valid *Maple* expression
Tp - (optional) an admissible *Maple* type

Description of the procedure:

As is indicated in our books [10-12], the **'indexable'(h)** type does not test, in particular, **h**-types of elements of such indexed structures as a list, an array and a set, while the **'tabular'(h)** type does not test **h**-types of elements of such structures as an array and a table.

In this connection we recommend to use for the types testing of elements of an **A**-structure of a type {*set* | *list* | *array*} the constructions of view ***type*(A, '**{*set* | *list* | *array*}**'(Tp))** accordingly, where **Tp** - an admissible *Maple* type. Whereas for similar testing of elements of a table the **ttable** procedure can be used successfully.

The procedure call **ttable(T)** returns the *true* value, if **T** is a table and the *false* value otherwise. If the second optional **Tp** argument has been coded, then the testing upon the given by it type of entries of the table is done. The *true* value is returned only if **T** is a table and each its entry has a type, indicated by the second actual **Tp** argument. Otherwise, the procedure call **ttable(T,Tp)** returns the *false* value.

```
ttable:= proc(T::anything)
local a, k;
  `if`(type(T, table), `if`(nargs = 2, op([assign(a = op(2, eval(T))), evalb({seq(
     `if`(type(rhs(a[k]), args[2]), true, false), k = 1 .. nops(a))} = {true})]), true), false)
end proc
```

Typical examples of the procedure use:

> T:=table([V,G,S,Art,Kr]): T1:=table(["A","V","P","Art","Kr"]): T2:=table([{},{}]):
T3:=table(["61",56,36,7,14]): T4:=table([61,56,36,14,7]): ttable(T, symbol),
ttable(T1, string), ttable(T2, set), ttable(T3, posint), ttable(T4, posint);

true, true, true, false, true

> map4(ttable, [[T, symbol], [T1, string], [T2, posint], [T3, {string, posint}],[T4, integer]]);

[true, true, false, true, true]

In many cases, it is not necessary to know the exact value of an expression; quite enough to know that the expression belongs to some broad class or group of expressions that share some general properties. These classes or groups are known as ***"types".*** If **T** represents a type, then we will say that an expression is of type **T** if it belongs to the class that **T** represents. For instance, an expression, say, has type *"integer"* if it belongs to the class of expressions denoted by the type named *"integer"* which is the set of all integers with sign or without, including *zero.*

Many *Maple* procedures use types to control by a branching in the different algorithms or to decide whether an expression is a valid input. Basic types and more complicated *"type expressions"* are built up via the grammar from the more primitive ***type*** expressions. A type called ***"type"*** describes the class of expressions that represent valid types. That is, the call **type(T, *type*)** returns the *true* value only when **T** is a valid ***type*** expression, otherwise the *false* value is returned. Thus, the call **type(2004, *posint*)** returns the *true* value, while the call **type(-1995, *posint*)** - the *false* value.

Types may be used to represent expressions that are based on general data structures such as {*list, set, table, matrix, vector*} or may be used to represent mathematical properties such as {*posint, prime, float, fraction*}. The **\`*type*\`** function allows you to test whether a *Maple* expression is of a given type. The procedure call **type(*exp*, t)** returns the *true* value if an expression ***exp*** has a type **t,** and returns the *false* value otherwise. At present, the ***type*** function of the last *ninth* release of the package *Maple* recognizes the set of 182 basic types.

However, in many cases, the standard set of basic types seems insufficient; therefore, within of operation with the package *Maple* of releases **4 – 9/9.5** we had additionally defined a number of new basic types useful enough in practical programming of a lot of problems of various purposes. The description of the given types concludes the present section.

\`type/binary\` – check of a *Maple* object upon type \`*binary*\`

Call format of the procedure:

type(x, *binary*)

Formal arguments of the procedure:

x - any valid *Maple* expression

Description of the procedure:

For many problems the *binary*-type seems as useful enough, which is absent in the package of all current releases. The procedure call **type(x, *binary*)** returns the *true* value, if **x** is a binary number (i.e. takes place the defining relation "**member(x, {0, 1}) -> *true***"), and the *false* value otherwise.

```
`type/binary`:= x::anything -> member(x, {0, 1})
```

Typical examples of the procedure use:

> map(type, [0,1,8,15,sin(0),7^0,round(2004/1942),6-5,2-14^0,sqrt(4)-1,nops({})], binary);

[true, true, false, false, true, true, true, true, true, true, true]

\`type/nestlist\` – check of a *Maple* expression upon type \`*nestlist*\`

Call format of the procedure:

type(expr, *nestlist*)

Formal arguments of the procedure:

expr – an arbitrary *Maple* expression

Description of the procedure:

In a number of problems dealing with structures of the *list*-type, the necessity of use of so-called *nested* lists structures (*nestlist*) arises. Let us say, a list is the *nested* list if at least one its element has the *`list`* type. For check of a list to be the nested list, a new extending of the standard ***type*** procedure of the package serves. The procedure call **type(*expr*, *nestlist*)** returns the *true* value, if a list defined by the first actual ***expr*** argument is the nested list and the *false* value otherwise.

```
`type/nestlist`:= L::anything -> type(L, list) and {op(map(type, L, list))} minus {false} = {true}
```

Typical examples of the procedure use:

> L:=[a,c,[h],x,[a,b,[]]]: type([], nestlist), type([[]], nestlist), type(L, nestlist), type([[[]]], nestlist), type([[61], [56], [36], [7], [14]], nestlist), type([a, b, c, {}], nestlist);

false, true, true, true, true, false

`type/boolproc` – check of a *Maple* procedure to be the Boolean procedure

Call format of the procedure:

type(P, *boolproc*)

Formal arguments of the procedure:

P – a valid *Maple* expression

Description of the procedure:

For many problems the *boolproc*-type seems as useful enough, which is absent in the package of all current releases. The procedure call **type(P, *boolproc*)** returns the *true* value, if **P** is a *Boolean* procedure (i.e. returns the values *`true`* and *`false`* only), and the *false* value otherwise. The procedure is used widely enough by tools represented in the monograph.

```
`type/boolproc`:= proc(P::anything)
local _Art_Kr_, _avz_, omega, nu, z;
option `Copyright International Academy of Noosphere – Tallinn, November, 2003`, remember;
  `if`(type(eval(P), procedure), assign(`type/_avz_` = eval(P)), RETURN(false));
  assign(omega = (x -> interface(warnlevel = x)), z = (() -> unassign('`type/_avz_`')));
  try nu := interface(warnlevel); omega(0); type(_Art_Kr_, _avz_); true, omega(nu), z()
  catch "result from type `%1` must be true or false": RETURN(false, omega(nu), z())
  catch "%1 is not a procedure": RETURN(false, omega(nu), z())
  catch "": RETURN(true, omega(nu), z())
  end try
end proc
```

Typical examples of the procedure use:

> map(type, [IsOpen, `type/file`, PP, isDir, IsFempty, MkDir, lexorder, Empty,ZZ], boolproc);

[true, true, false, true, true, false, true, true, false]

> map(type, [com_exe1, `type/rlb`, `type/byte`, `type/boolproc`, frame_n], boolproc);

[false, true, true, true, false]

`type/dir` – check of a *Maple* expression to be a directory, a path or a file
`type/path`
`type/file`

Call format of the procedures:

type(K, *dir***)**
type(P, *path***)**
type(P, *file***)**

Formal arguments of the procedures:

K - name or full path to a directory or a file; its type should be {*symbol, string*}
P - symbol or string defining full path to a directory or a file

Description of the procedures:

For many of the problems dealing with a **PC** file system the types {*dir, path, file*} which are absent in all releases of the *Maple* package seem as quite useful ones. The following procedures eliminate this shortage.

The procedure call **type(K,** ***dir*)** returns the *true* value if **K** defines a real directory and the *false* value otherwise. In addition, it is necessary to have in view, the return of the value *false* speaks only that the tested directory **K** is not: (**1**) as the last subdirectory in the real chain determined by the actual argument **K,** (**2**) in the subdirectories chains determined by the predetermined variable ***libname,*** (**3**) in the subdirectories chain determined by function **currentdir(),** (**4**) in the subdirectories chains determined by open files, or (**5**) is the main directory of non-active drive. Whereas the given directory, generally speaking, can be as an element of file system of a computer. The procedure seems as a rather important tool in light of file concept of the operating MS-DOS system and of key distinctions in handling of subdirectories and datafiles.

The procedure call **type(K,** ***file*)** returns the *true* value if **K** is a real file and the *false* value otherwise. In addition, it is necessary to have in view, the return of the value *false* speaks only that the tested file **K** is not: (**1**) as the last element in the real chain determined by the actual argument **K,** (**2**) in the main *Maple* directory, (**3**) in the directory containing the *Maple* kernel, (**4**) in the USERS subdirectory, (**5**) among the open files. Whereas the given file, generally speaking, can be as an element of file system of a computer. The procedure seems as a rather important tool in light of file concept of the operating MS-DOS system and of key distinctions in handling of subdirectories and datafiles. Call of this procedure does not change attributes and status {*open* | *close*} of the tested files.

In general case a full path is the chain of elements of file system of the following general view:

***<logical name of a drive>*:{/ | \\}*directory_1*{/ | \\}*directory_2*{/ | \\} ... {/ | \\} *directory_n*{/ | \\}file**

The procedure call **type(P,** ***path*)** returns the *true* value if **P** is the real full path to a directory or a datafile, and the *false* value otherwise. The given tool seems a rather important in light of the file concept of the operating system MS DOS and of key distinctions in handling of directories and datafiles. In many procedures dealing with access to datafiles the procedures of this group play a rather important part, providing strict identification of types of elements of file system of a computer.

```
`type/dir`:= proc(F::anything)
local a, b, c, d, k, u, omega, t, nu, h, p, n;
  `if`(type(eval(F), {name, string, symbol}), NULL, RETURN(false)), assign(h = currentdir());
  omega, u := () -> ERROR("<%1> is invalid path to a directory or a file", F),
     interface(warnlevel);
  `if`(Red_n(cat(" ", F), " ", 2) = " " or member(F, {``, ""}), omega(), NULL),
```

```
    interface(warnlevel = 0);
  assign(a = Red_n(Subs_All("/" = "\\", Case(cat("", F)), 1), "\\", 2),
    nu = (() -> interface(warnlevel = u)));
  if length(a) = 1 then assign('a' = cat(a, ":\\"))
    elif length(a) = 2 and a[2] = ":" then assign('a' = cat(a, "\\"))
  elif length(a) = 3 and a[2 .. 3] = ":\\" or a[2 .. 3] = ":/" then NULL
  end if;
  `if`(a[2] = ":", `if`(member(a[1], map(Case, {Adrive()})), NULL,
    RETURN(nu(), false)), NULL);
  if  length(a) < 4 then  try  p := currentdir(a)
    catch "directory exists and is not empty":  NULL
    catch "file or directory does not exist":  NULL
    catch "permission denied":  currentdir(h); RETURN(nu(), true)
    catch "file I/O error":  currentdir(h); RETURN(nu(), true)
    end try;
    `if`(p = NULL or p = h, RETURN(nu(), null(currentdir(h)), true), NULL)
  end if;
  `if`(a[-1] = "\\", assign('a' = a[1 .. -2]), NULL);
  if a[2 .. 3] <> ":\\" then assign(b = iostatus(), d = CF(currentdir()),
      n = cat("\\", a), c = map(CF, [libname]));
    `if`(search(d, n) or {op(map(search, c, n))} <> {false}, [nu(), RETURN(true)], 158);
    `if`(nops(b) = 3, 1, seq(`if`(search(CF(b[k][2]), cat("\\", a, "\\")),
      [nu(), RETURN(true)], NULL), k = 4 .. nops(b)))
  end if;
  `if`(type(holdof(hold), set(integer)), NULL, assign(t = 7));
  try close(fopen(a, READ))
  catch "file or directory does not exist": `if`(t = 7, holdof(restore, 7), NULL), nu(),
    RETURN(nu(), false)
  catch "file I/O error":  `if`(t = 7, holdof(restore, 7), NULL), nu(), RETURN(true)
  catch "permission denied":  `if`(t = 7, holdof(restore, 7), NULL), nu(), RETURN(true)
  end try;
  `if`(t = 7, holdof(restore, 7), NULL), nu(), false
end proc

`type/file`:= proc(F::anything)
local a, b, k, p, t, t1, t2, t3, h, u, omega;
  `if`(type(eval(F), {name, string, symbol}), assign(b = iostatus()), RETURN(false));
  try CF(F) catch "<%1> is invalid path to a directory or a file": RETURN(false)  end try;
  omega, u:= () -> ERROR("<%1> is invalid path to a directory or a file", F), interface(warnlevel);
  `if`(Red_n(cat(" ", F), " ", 2) = " " or member(F, {"", ``}), omega(), interface(warnlevel = 0));
  assign(a = Red_n(Subs_All("/" = "\\", Case(cat("", F)), 1), "\\", 2)), interface(warnlevel = u);
  `if`(3 < length(a) and a[-1] = "\\", assign('a' = a[1 .. -2]), `if`(length(a) = 2 and a[-1] = ":",
    assign('a' = cat(a, "\\")), NULL));
  `if`(nops(b) = 3, NULL, seq(`if`(CF(a) = CF(b[k][2]), RETURN(true), NULL), k = 4..nops(b)));
  `if`(type(holdof(hold), set(integer)), NULL, assign(t = 7));
  if a[2 .. 3] <> ":\\" then
    try  assign67(h = cat(CCM(), "\\", "maplesys.ini"), Close(h));
      do t1 := readline(h);  if t1[1 .. 5] = "UserD" then t1 := t1[15 .. -1]; Close(h); break end if
      end do;
      t1, t2, t3 := cat(t1, "\\", a), cat(CDM(), "\\", a), cat(CCM(), "\\", a);
```

```
    if {op(map(procname, [t1, t2, t3]))} <> {false} then RETURN(holdof(restore, 7), true) end if;
      `if`(nops(b) = 3, NULL, seq(`if`(Search1(CF(b[k][2]), a, 'p') and p = [right],
        RETURN(holdof(restore, 7), true), 7), k = 4 .. nops(b)))
    catch "file or directory does not exist": RETURN(false)
    end try
  end if;
  try Close(fopen(a, READ))
  catch "file already open": `if`(t = 7, holdof(restore, 7), NULL), RETURN(true)
  catch "file or directory does not exist": `if`(t = 7, holdof(restore, 7), NULL), RETURN(false)
  catch "file I/O error": `if`(t = 7, holdof(restore, 7), NULL), RETURN(false)
  catch "permission denied": `if`(t = 7, holdof(restore, 7), NULL), RETURN(false)
  end try;
  `if`(t = 7, holdof(restore, 7), NULL), true
end proc

`type/path`:= F::anything -> `if`(not type(eval(F), {name, string, symbol}), false,
              `if`(CF(F, string)[2 .. 3] = ":\\", `if`(type(F, dir) or type(F, file), true, false), false))
```

Typical examples of the procedures use:

> map(type, [`C:\\Academy`, "C:\\E_Books\\Mail13.txt", `C:\\TEMP/`, "C:\\"], dir);

[true, false, true, true]

> type("C:/Temp/aaa.bbb", dir); isdir("C:/Temp/aaa.bbb");

false

Error, (in **isdir**) file or directory does not exist

> map(type, [`C:\\Academy`, "bin.win","Program Files", `UserLib`, "LIB", `C:\\TEMP/`, "C:\\E_Books\\Mail13.txt", "C:\\", `Maple 9`, "C:/Archive\\Booklet"], dir);

[true, true, true, true, true, true, false, true, true, true]

> map(type, [A, "C:", "users", LIB, UserLib, HelpBase, MPL9MWS], dir);

[false, true, true, true, true, true, true]

> map(type, [`C:\\TEMP/`, `A:/EBooks`, "a:/avz", "c:/Program files/Maple 9/Users/aaa.mws", "C:/Academy/Books/ReadMe.htm"], path);

[true, false, false, true, true]

> map(type, [`C:/Archive/abs.rtf`, `C:/Autoexec.bat`, "C:/Temp/Re.txt", "c:/Temp/books.7", "C:\\Program Files\\Maple 8\\Users\\ProcLib_6_7_8_9.mws"], file);

[true, true, false, false, true]

> map(type, ["ProcLib_6_7_8_9.mws", `Maplew9.ico`, "helpman.mws", `Excel.txt`, "ICCS_2004.txt"], file);

[true, true, true, true, false]

`type/fpath` – check of a *Maple* expression to be an allowable full path

Call format of the procedure:

type(P, ***fpath*****)**

Formal arguments of the procedure:

P – symbol or string defining *full path* to a directory or datafile

Description of the procedure:

For many problems dealing with a PC file system the types {*dir, path, file*} which are absent in all releases of the *Maple* package seem as quite useful ones. The three previous procedures eliminate this shortage. The following procedure **`type/ftype`** solves similar problem in case of testing of full path to a directory or datafile. In addition, the procedure tests only correctness of syntax without taking real existence of such path. For example, the procedure call

type("G:/aaa\\bbb/ccc\\ddd/salcombe.txt", ***fpath*****)**

returns the true value even if disk drive "G" is idle. In such case, the procedure call outputs the appropriate warning.

The procedure call **type(P,** ***fpath*****)** returns the *true* value if **P** defines an allowable *full path* and the *false* value otherwise. In addition, it is necessary to have in view, the return of the *true* value speaks only, the tested directory **P**, generally speaking, can be as an element of file system of a computer. The procedure well complements the above procedure **`type/path`.** The procedure **`type/ftype`** seems as a rather important tool in many problems of datafiles processing.

```
`type/fpath`:= proc(P::{string, symbol})
local a, f, f1, k, p, w, z, dir, df;
  assign(a = CF2(P), w = interface(warnlevel), z = (x -> interface(warnlevel = x))), z(0);
  `if`(Red_n(a, " ", 1) = "" or length(a) < 3 or a[2 .. 3] <> ":\\", [z(w), RETURN(false)],
    assign(p = 97 .. 122));
  z(w), `if`(member(convert(a[1], bytes), {[k] $ (k = p)}),
    `if`(member(a[1], map(CF2, {Adrive()})), NULL,
    WARNING("<%1> is idle disk drive", Case(a[1], upper))),
    ERROR("<%1> - illegal logical drive name", Case(a[1], upper)));
  f := cat(WT(), "\\_$$$_2004"); f1 := cat(f, a[3 .. -1]);
  df:= proc(x) for k to nops(x) do try rmdir(x[k]) catch: next end try end do end proc;
  try MkDir(f1) catch: dir := [op(InvL(DirF(f))), f]; df(dir); RETURN(false) end try;
  dir := [op(InvL(DirF(f))), f]; df(dir); true
end proc
```

Typical examples of the procedure use:

> type("G:/aaa\\bbb/ccc\\ddd/salcombe.txt", fpath);

Warning, <G> is idle disk drive

true

> type("C:/*aaa/bbb/cc*c/ddvb/hhh\\ttt/ddd", fpath);

false

> type("@:/aaa/bbb/ccc/ddvb/hhh\\dir\\test.rlb", fpath);

Error, (in **type/fpath**) <@> – illegal logical drive name

The given procedures extend the standard ***type***-function of the *Maple* package, providing an opportunity of sufficiently fail-safe check of elements of a **PC** file system to be a *subdirectory*, a *file* or a *path.* In particular, the releases **8** and **9/9.5** release has been ensured by the ***isdir*** function that tests if a file is a directory. However, as is shown by the above examples, this function is less convenient for use than the above procedure **`type/dir`.**

\`type/byte\` – check of a *Maple* expression upon type *\`byte\`*

Call format of the procedure:

type(x, *byte*)

Formal arguments of the procedure:

x – any valid *Maple* expression

Description of the procedure:

For many problems the *byte*-type seems useful enough which is absent in the package *Maple* of all current releases. The procedure call **type(x, *byte*)** returns the *true* value, if **x** is a byte and the *false* value otherwise. According to the given procedure to the *\`byte\`* type belong the strings and symbols of length **1**, positive integers of length not greater than **8** of binary digits, and also the binary lists with elements amount equal **8.**

```
`type/byte`:= B::anything -> `if`(type(eval(B), {string, symbol}) and length(B) = 1 or
    type(B, list(binary)) and nops(B) = 8 or type(B, nonnegint) and length(B) <= 8 and
    xNB(B), true, false)
```

Typical examples of the procedure use:

> map(type, [a, [1,1,0,0,1,0,1,1], 110101, "G", aa, [1,1,0,0,0,1,0,1], 10011110, \`S\`, \`t\`], byte);

[*true, true, true, true, false, true, true, true, true*]

\`type/digit\` – check of a *Maple* expression upon type *\`digit\`*, *\`letter\`* or *\`ssign\`*

\`type/letter\`
\`type/ssign\`
stype

Call format of the procedures:

type(x, *digit*)
type(x, *letter*)
type(x, *ssign*)
stype(S)

Formal arguments of the procedures:

x – any valid *Maple* expression
S – a symbol or string

Description of the procedures:

For a lot of problems the types *\`digit\`*, *\`letter\`* and *\`ssign\`* seems useful enough, missing in the *Maple* package of all current releases. The procedure call **type(x, *digit*)** returns the *true* value, if **x** is a decimal digit (i.e. "**member(x, {0,1, 2, ..., 9, \`0\`, \`1\`,\`2\`, ...,\`9\`,"0","1","2", ...,"9"}) -> *true***"), and the *false* value otherwise. This definition is very useful at operation with expressions of types {*symbol, string*}.

The procedure call **type(x, *letter*)** returns the *true* value, if **x** is a Roman letter and the *false* value otherwise.

The procedure call **type(x, *ssign*)** returns the *true* value, if **x** is a special character of the set {~,!,@,#,$,%, ^, &, *, (,),_,+,\`,-, =, {,},[,],:,;, ",', <, \`,\`, >,., ?,/, |,\}, and the *false* value otherwise.

At last, the procedure call **stype(S)** returns the *type* of a symbol or string defined by the actual **S** argument in the context of types {*letter, digit, ssign*}; in addition, if **length(S) > 1** then a type {*symbol, string*} is returned; if type of **S** is different from the above three types, the *false* value is returned.

```
`type/digit`:= proc(N::anything)
local `0`, k;
  member(N, {seq(op({cat(``, k), cat("", k), k}), k = 0 .. 9)})
end proc

`type/letter`:= proc(a::anything)
local mu;
  assign(mu = convert(a, string)); `if`(length(mu) <> 1, false,
  belong(op(convert(mu, bytes)), 65 .. 90) or belong(op(convert(mu, bytes)), 97 .. 122))
end proc

`type/ssign`:= mu::anything -> search("~!@#$%^&*()_+`-={}[]:";'<>?|,./\", convert(mu, string))
stype:= s::{string, symbol} -> `if`(1 < length(s), op(0, s), `if`(type(s, ssign), ssign,
        `if`(type(s, letter), letter, `if`(type(s, digit), digit, false))))
```

Typical examples of the procedures use:

> map(type, [8,1,15,`3`,56, "4",2, "0",61,`7`,5, round(14/6),(14-6)/2, sqrt(4)], digit);

[true, true, false, true, false, true, true, true, false, true, true, true, true, true]

> map(type, [a,1,H,3,`$`,4,g,0,s,7,A,cat("", round(14/6)),"A",`R`,15,S," "], letter);

[true, false, true, false, false, false, true, false, true, false, true, false, true, true, false, true, false]

> map(type, [a,1,`~`,3,`$`, ";",g,`(`,s,8,A, "%",A,"*","@",`+`,`~`,`=`], ssign);

[false, false, true, false, true, true, false, true, false, false, false,true, false, true, true, true, true, true]

> map(stype, [a,`1`,H, "a1",`3`,`$`,"4",g,`01`,s,`7`,A, "Д",`Ж`,`avz`]);

[letter, digit, letter, string, digit, ssign, digit, letter, symbol, letter, digit, letter, false, false, symbol]

`type/mod1` – testing of a program module type

Call format of the procedure:

type(M, *mod1*)

Formal arguments of the procedure:

M – a program module of any admissible type

Description of the procedure:

The procedure call **type(M, *mod1*)** returns the *true* value, if **M** is a program module of the *first type* and the *false* value otherwise. In addition, the program module of the *first type* is understood as a module named by construction of view **"Name:= module () ... "**, whereas the module of the *second type* is characterized by determinative construction of view **"module Name () ... "**. Some useful enough considerations relative to use of the program modules of the *second* type can be found in our books [8-14,29-33,43,44].

```
`type/mod1`:= M::anything -> `if`(type(M, `module`), `if`(convert(eval(M), string)[1 .. 8] =
              "module (", true, false), false)
```

Typical examples of the procedure use:

> MM:=module() export x,m; m:=module() export y; y:=() -> \`+\`(args) end module end module: module SVEGAL () export x; x:= () -> \`+\`(args)/nargs end module: AG:=module() export s; s:= () -> [[args],nargs] end module: map(type, [MM, SVEGAL, AG], mod1);

[*true*, *false*, *true*]

\`type/package\` – check of a *Maple* object for its belonging to the \`*package*\` type

Call format of the procedure:

type(x, *package*)

Formal arguments of the procedure:

x – any valid *Maple* expression

Description of the procedure:

With respect to previous releases a new \`*package*\` type, intended for testing objects for the given type has been introduced in the 7th release of *Maple*. Meanwhile, both the results of its use and the analysis of the source code of the standard procedure \`***type/package***\` speak about an incorrectness of its definition, as illustrate two first examples of the fragment represented below. Analysis of cause of such incorrectness is considered in our books [29-33,43,44] enough in detail. For elimination of this defect a modification of the \`***type/package***\` procedure of the same name has been defined. Into the basis of the procedure algorithm were put both the conceptual and the intuitive considerations about such object as the *package*.

The procedure call **type(x, *package*)** returns the *true* value, if **x** is a package; otherwise, the *false* value is returned. The given procedure in a series of cases provides effective enough testing of the *Maple* objects upon the \`*package*\` type.

```
`type/package`:= proc(P::anything)
local a, b, c, k, nu;
  assign(c = interface(warnlevel), nu = (h -> interface(warnlevel = h))),
    `if`(type(P, {name, symbol}), NULL, RETURN(false));
  `if`(traperror([assign(a = (`if`(member(Release(), {6, 7, 8}), pacman,
    PackageManagement)):- pexports(P)), assign(b = nops(a))]) = lasterror, RETURN(false),
    [nu(0), `if`(a = [] or add(`if`(eval(P[a[k]]) = a[k], 1, 0), k = 1 .. b) = b or {seq(whattype(a[k]),
    k = 1 .. b)} <> {symbol}, RETURN(nu(c), false), `if`({protected, package, _syslib} intersect
    {attributes(eval(P)), attributes(P)} <> {} or type(P, table), RETURN(nu(c), true),
    RETURN(nu(c), false)))])
end proc
```

Typical examples of the procedure use:

> T:=table([42,47,67,62,89,96,61,56,36]); with(T); # for the \`***type/package***\` procedure in *Maple* **8**

T:= table([1 = 42, 2 = 47, 3 = 67, 4 = 62, 5 = 89, 6 = 96, 7 = 61, 8 = 56, 9 = 36])

Error, (in **pacman:– pexports**) invalid arguments to sort

> T:=table([42,47,67,62,89,96,61,56,36]); with(T); # for the \`***type/package***\` procedure in *Maple* **9**

T:= table([1 = 42, 2 = 47, 3 = 67, 4 = 62, 5 = 89, 6 = 96, 7 = 61, 8 = 56, 9 = 36])

Error, (in **PackageManagement**:-pexports) invalid arguments to sort

> packages(), type(T, package); # for the above two releases of *Maple*

[], *true*

> packages(), type(T, package); # for our procedure *`type/package`* in *Maple* of releases **6 – 9**

[], *false*

> map(type, [T, pacman, PackageManagement, stats, linalg, plots, LinearAlgebra, DIRAX, process, plottools, Sockets], package);

[*false, false, true, true, true, true, true, true, true, true, true*]

---

typeseq – check of a *Maple* object upon type *`expressions sequence`*

Call formats of the procedure:

typeseq(*expr, seqn*)
typeseq(*expr, seqn*(T))

Formal arguments of the procedure:

expr – any valid *Maple* expression
seqn – a key name
T – a sequence of valid *Maple* types

Description of the procedure:

For a name of an inner data structure of the type *`sequence of expressions`* is used the symbol ***exprseq*** not recognized by the ***type***-function of the package *Maple* as a type. For testing of expressions of the *exprseq*-type, the package *Maple* has only ***whattype*** function providing return of a data type of the top level only. However, in many cases it is desirable to have a tool for the more detailed testing of the *expressions sequences.* For these purposes, the following **typeseq** procedure can be a rather useful at programming in the *Maple* environment of a series of problems.

The procedure call **typeseq(*expr, seqn*)** returns the *true* value, if ***expr*** is an expressions sequence of any valid *Maple* type; otherwise the *false* value is returned. The procedure call **typeseq(*expr, seqn*(T))** returns the *true* value, if ***expr*** is an expressions sequence of any *Maple* type defined by the actual **T** value – by a sequence of types; otherwise the *false* value is returned. The given procedure in a series of cases provides effective enough testing of an expressions sequence and types of its components. It allows effectively enough to process the data structures of the ***sequence***-type what is very important in a number of problems of symbolic processing in the *Maple* environment. The given procedure in a series of cases provides effective enough testing of the expressions sequence and types of their components.

```
typeseq:= proc()
local k;
  `if`(nargs <= 2, false, `if`(args[-1] = seqn, true, `if`(op(0, args[-1]) = seqn,
  `if`({seq(type(args[k], {op(args[-1])}), k = 1 .. nargs - 1)} = {true}, true, false),
  ERROR("the last argument <%1> is invalid", args[-1]))))
end proc
```

Typical examples of the procedure use:

> typeseq("Kr",Art,6,14,"rans", seqn(integer, symbol, string)), typeseq(Kr,Art, seqn), typeseq(Kr, seqn), typeseq(G, "Gr", 56, seqn(posint, symbol, string));

true, true, false, true

> **whattype("Kr", Art, 6, 14, "RANS", IAN, Tallinn, Vilnius, Grodno);**

exprseq

> **[typeseq()], typeseq(a,b,6,14, seqn), typeseq("Kr",Art,6,14, "RANS",IAN,Tallinn,Vilnius, Grodno, seqn), typeseq(G,[a],{b},61,7/14, seqn(integer, symbol, list, set, fraction));**

[false], true, true, true

`type/sequent` – check of a *Maple* expression upon type *`sequential`*

Call format of the procedure:

type(x, *sequent*)

Formal arguments of the procedure:

x - any valid *Maple* expression

Description of the procedure:

The package of five last releases untruly declares testability of expressions of the *`string`* type relative to belonging their to the *`sequential`* type, what with all conspicuity illustrates an example below. The procedure **`type/sequent`** allows to eliminate this defect. The procedure call **type(x, *sequent*)** returns the *true* value, if **x** is an object of one of the following types {*list, set, string*}; otherwise, the *false* value is returned.

```
`type/sequent`:= A::anything -> `if`(member(whattype(A), [list, set, string]), true, false)
```

Typical examples of the procedure use:

> **type("That is a string", sequential), map(type, ["That is a string",{a,b,7,96},[c,d,14,89], "RANS_IAN", cat("",Kr_Art), ""], sequent);**

false, [true, true, true, true, true, true]

`type/setset` – check of a *Maple* expression upon type *`setset`*

Call format of the procedure:

type(S, *setset*)

Formal arguments of the procedure:

S - a valid *Maple* expression

Description of the procedure:

In a number of problems dealing with structures of the *`set`* type, the necessity of use of so-called *setsets* structures (analogously to *listlists*) arises. Let us say, a set has the *`setsets`* type if its elements are sets of the same cardinality. For test of a set to be *setset* a new extending of the standard *`type`* procedure of the *Maple* package serves. The procedure call **type(S, *setset*)** returns the *true* value, if a *Maple* expression defined by the first actual **S** argument is the *setset* and the *false* value otherwise. A set of sets which is formed by the above manner is analogous to the type *`list of lists`*.

```
`type/setset`:= proc(L::anything)
local k;
   `if`(type(L, 'set'), `if`(nops(L) <> 0 and nops({seq(`if`(type(L[k], 'set'), nops(L[k]),
        RETURN(false)), k = 1 .. nops(L))}) = 1, true, false), false)
end proc
```

Typical examples of the procedure use:

> map(type, [{{}}, {{},{}}, {{{7}}}, {{a,b}, {c,d}}, {{a,b}, {c}}, {{6,14},{61,56}, {36,40}}], setset);

[*true, true, true, true, false, true*]

`type/realnum` – check of a *Maple* expression upon type `*realnum*`

Call format of the procedure:

type(N, *realnum*)

Formal arguments of the procedure:

N – any valid *Maple* expression

Description of the procedure:

In a series of cases, it is important to determine belonging of an expression to *real* numbers. The procedure call **type(N, *realnum*)** returns the *true* value if and only if **N** is a real number; otherwise, the *false* value is returned.

```
`type/realnum`:= N::anything -> type(evalf(N), float)
```

Typical examples of the procedure use:

> map(type, [sqrt(2),ln(10),Pi/2,sin(14),Catalan,4.2,6/14,61,ln(42.47),gamma], realnum);

[*true, true, true, true, true, true, true, true, true, true*]

`convert/listlist` – conversion of a list, set, setset, array or *rtable* into listlist

Call format of the procedure:

convert(A, *listlist*)

Formal arguments of the procedure:

A – a list, set, setset, array or rtable

Description of the procedure:

In problems of a statistical analysis enough often it is necessary to carry out a conversion of two-dimensional arrays into list structures (*nested lists* or *listlists*), what, in particular, is actual in case of graphic representation of the discrete information. As a rule, for these purposes the standard function `*convert*` is used. However in a series of versions of the 6th release the call **`convert(A::array, listlist)`** causes the erratic situation with diagnostics "Error, in convert/old_array_to_listlist) Seq expects its 2nd argument ... ". For elimination of the given situation a modification of the standard procedure of the same name **`convert/listlist`** can be used.

The modified procedure **`convert/listlist`** will correctly convert a list, set, setset, array or rtable, defined by the first actual argument **A**, into the list of lists (*listlist*). A list of lists, which is formed from an array, is equivalent to the list that would be passed to the ***array***-function to recreate the original array. The examples represented below enough clearly illustrate the results of appendices of the above procedure to the program objects of types {*setset, list, set, array, rtable*}, that use command-line programs of host operating system.

```
`convert/listlist`:= proc(A::{set, array, rtable, list, setset})
local k, j, Artll;
```

```
  Artll:= proc(A::rtable)
    local a, b, c, k, p, L;
      assign(a = `if`(type(A, Array), [op(2, eval(A))], [seq(1..[op(1, A)][k],
      k=1..nops([op(1, A)]))]), c = 0, b = op(`if`(type(A, Array), 3, 2), eval(A))),
      null(seq([assign('L' = [seq(c, p = 1 .. rhs(a[abs(k)]))]), assign('c' = L)], k = -nops(a) .. -1),
      seq(assign('L'[lhs(b[k])] = rhs(b[k])), k = 1 .. nops(b))), L
    end proc;
  `if`(type(A, rtable), Artll(A), `if`(type(A, setset), [seq([op(A[k])], k = 1 .. nops(A))],
   `if`(nops([op(2, eval(A))]) = 1, convert(A, list), [seq([seq(A[k, j], j=1..rhs(op(2, eval(A))[2]))],
   k = 1 .. rhs(op(2, eval(A))[1]))])))
end proc
```

Typical examples of the procedure use:

> A:=array(1..2,1..5,[[42,47,67,62,89],[60,55,35,39,6]]): convert(A, listlist); **#** ***Maple*** **6**

Error, (in convert/old_array_to_listlist) Seq expects its 2nd argument, p, to be of type symbol, but received n=1..6

> A:=Array(1..3,1..3,1..3,[]): SS:={{a,b,c},{1,2,3},{4,5,6}}: L:=[42,47,67]: S:={Sv,Art,Kr,Arn}: A1:=Array(1..2,1..2,1..2,[[[a,b],[c,d]],[[e,f],[42,47]]]): M:= Matrix(6,5,[]): V:= Vector(1..3, [67,89,96]): V1:= Vector[row](1..3,[36,14,7]): Ar:= array(1..2,1..3,[[a,b,c],[42,47,67]]):

> convert(A1, listlist), convert(V, listlist); convert(A, listlist), convert(V1, listlist); convert(L, listlist), convert(Ar, listlist), convert(S, listlist); convert(M, listlist);

[[[a, b], [c, d]], [[e, f], [42, 47]]], [67, 89, 96]
[[[0, 0, 0], [0, 0, 0], [0, 0, 0]], [[0, 0, 0], [0, 0, 0], [0, 0, 0]], [[0, 0, 0], [0, 0, 0], [0, 0, 0]]], [36, 14, 7]
[42, 47, 67], [[a, b, c], [42, 47, 67]], [Arn, Kr, Art, Sv]
[[0, 0, 0, 0, 0], [0, 0, 0, 0, 0], [0, 0, 0, 0, 0], [0, 0, 0, 0, 0], [0, 0, 0, 0, 0], [0, 0, 0, 0, 0]]

\`type/upper\` – check of register of a symbol or string
\`type/lower\`
\`type/Upper\`
\`type/Lower\`

Call format of the procedures:

type(S, *upper*)
type(S, *lower*)
type(S, *Upper*)
type(S, *Lower*)

Formal arguments of the procedures:

S - any valid *Maple* expression of a type {*symbol, string*}

Description of the procedures:

In a series of cases it is important to determine belonging of symbols composing an expression **S** of a type {*symbol, string*} to a register {*upper, lower*}. The procedure call **type(S, *upper*)** returns the *true* value if and only if the Latin letters composing **S** belongs to the *upper*-register; otherwise, the *false* value is returned.

The procedure call ***type*(S, *lower*)** returns the *true* value if and only if the Latin letters composing **S** belongs to the *lower*-register; otherwise, the *false* value is returned.

The procedure call **type(S, *Upper*)** returns the *true* value if and only if **S** is composed of symbols belonging to the *upper*-register of **US** keyboard; otherwise, the *false* value is returned.

The procedure call **type(S, *Lower*)** returns the *true* value if and only if **S** is composed of symbols belonging to the *lower*-register of **US** keyboard; otherwise, the *false* value is returned.

In a series of cases the above four procedures extends admissible types of textual data.

```
`type/upper`:= proc(S::anything)
local a, b, k, n;
  `if`(type(S, {string, symbol}), [assign(a = convert(S, string), b = ""),
    assign(n = length(a))], RETURN(false));
  seq(`if`(type(a[k], letter), assign('b' = cat(b, a[k])), NULL), k = 1 .. n);
  `if`(belong({op(convert(b, bytes))}, {k $ (k = 65 .. 90)}), true, false)
end proc

`type/lower`:= proc(S::anything)
local a, b, k, n;
  `if`(type(S, {string, symbol}), [assign(a = convert(S, string), b = ""),
    assign(n = length(a))], RETURN(false));
  seq(`if`(type(a[k], letter), assign('b' = cat(b, a[k])), NULL), k = 1 .. n);
  `if`(belong({op(convert(b, bytes))}, {k $ (k = 97 .. 122)}), true, false)
end proc

`type/Upper`:= proc(S::anything)
local a, k;
  `if`(type(S,{string,symbol}),assign(a={op(convert(convert(S,string),bytes))}),RETURN(false));
  `if`(belong(a, {58, 60, 94, 95, k $ (k = 62 .. 64), k $ (k = 123 .. 126), k $ (k = 65 .. 90),
  k $ (k = 33 .. 38), k $ (k = 40 .. 43)}), true, false)
end proc

`type/Lower`:= proc(S::anything)
local a, k;
  `if`(type(S,{string,symbol}),assign(a={op(convert(convert(S,string),bytes))}), RETURN(false));
  `if`(belong(a, {39,57,59,61,96,k $ (k = 91..93), k $ (k = 44..56), k $ (k = 97..122)}), true, false)
end proc
```

Typical examples of the procedures use:

```
> type("A~!@#$%^&*()_+B{}:""<>?|C", Upper), type("a`1234567890b-=[];',./\\b", Lower);
                                  true, true
> map(type, ["c",C,G47,"Art",KR], upper), map(type,["c",C,G47,"Art",KR], lower);
          [false, true, true, false, true], [true, false, false, false, false]
```

\`type/complex1\` - check for an *Maple* object of type \`*complex*\`

Call format of the procedure:

type(Z, *complex1*)
type(Z, *complex1*(t))

Formal arguments of the procedure:

Z - a valid *Maple* expression
t - (optional) any valid *Maple* type

Description of the procedure:

Call of the standard procedure **type(Z, *complex*)** should return the *true* value if **Z** is an expression of the form **x + y*I,** where **x** (*if exists*) and **y** (*if exists*) are finite and of type *`realcons`*. In addition, the procedure call **type(Z, *complex*(t))** should return the *true* value if **Re(Z)** (*if exists*) and **Im(Z)** (*if exists*) are both of type **t.** Unfortunately, the standard procedure does not provide correct testing of *Maple* expressions for the above type *`complex`*. The following simple example earnestly confirms the told, namely:

> map(type, [-3.1942, 62, 300, -1995, 8/15], *complex*) -> [*true, true, true, true, true*]

Unfortunately, this serious error takes place for all *Maple* releases starting with the *sixth.* The procedure **`type/complex1`** has the same formal arguments as the standard procedure and eliminates errors inherent to it. Furthermore, it provides the broader check of *Maple* objects of type *`complex`*. However, the procedure does not distinguish numeric types {*fraction, rational*}, identifying them with more common *numeric*-type.

```
`type/complex1`:= proc(Z::anything)
local a, b, c, d, h, t, k, rm, im, sv, tn;
option `Copyright (C) 2004 by the International Academy of Noosphere. All rights reserved.`;
  if map2(search, convert(Z, string), {"I*", "I", "*I"}) = {false} then false
  else assign(h = interface(warnlevel), a = normal(evalf(Z))), interface(warnlevel = 0);
    tn:= proc(x) local a; a:= convert(x, string); `if`(a[-1] = ".", came(a[1..-2]), x) end proc;
    sv:= (s) -> `if`(length(s) > 4 and member(s[1 .. 4], {"+1.*", "-1.*"}), s[5 .. -1], s);
    d := denom(a);
    if d = 1 or Search2(convert(d, string), {"I*", "I", "*I"}) = []
    then a := convert(normal(evalf(a)), string)
    else a := convert(expand(normal(evalf(conjugate(d)*numer(a)))), string)
    end if;
    a := `if`(member(a[1], {"-", "+"}), a, cat("+", a));
    assign67(b = seqstr('procname(args)'), t = NULL);
    b := b[length(cat("`type/convert1`", "(", convert(Z, string))) + 1 .. -2];
    if b = "" then t := realnum
    else t:= came(sub_1([seq(k = "realnum", k = ["fraction", "rational"])], b[2 .. -1]))
    end if;
    c := Search2(a, {"-", "+"});
    if nops(c) = 1 then a:= sub_1(["*I" = NULL, "I*" = NULL], a);
        interface(warnlevel=h), type(tn(came(sv(a))), t)
    else assign(rm = "", im = "", 'b' = [seq(a[c[k] .. c[k + 1] - 1], k = 1 .. nops(c) - 1),
        a[c[-1] .. -1]]);
      for k to nops(b) do
        if Search2(b[k], {"I*", "*I"}) = [] then rm := cat(rm, b[k])
        else im := cat(im, sub_1(["*I" = NULL, "I*" = NULL], b[k]))
        end if
      end do;
      try `if`(im = "", RETURN(false), assign('rm' = came(sv(rm)), 'im' = came(sv(im))))
      catch: RETURN(interface(warnlevel = h), false)
      end try;
      interface(warnlevel = h), `if`(map(type, {tn(rm), tn(im)}, t) = {true}, true, false)
    end if
  end if
end proc
```

Typical examples of the procedure use:

[Standard *Maple* procedure **`type/complex`**

> type((a + I)*(a - a*I + 3*I), complex(algebraic));

true

> type(a+sqrt(8)*I, complex({symbol, realnum}));

true

> map(type, [-3.19421947, 62, 2004, 300, -1995, 8/15,60], complex);

[*true, true, true, true, true, true, true*]

> map(type, [-3.65478, 62, 2004, -1995, 8/15], complex(numeric));

[*true, true, true, true, true*]

> type([a, b]-3*I, complex({realnum, list}));

true

> type((a + I)*(a - a*I + 3*I), complex(algebraic));

true

[Our procedure **`type/complex1`**

> type((a + I)*(a - a*I + 3*I), complex1(algebraic));

true

> type(a+sqrt(8)*I, complex1({symbol, realnum}));

true

> map(type, [-3.65478, 62, 2004, -1995, 8/15], complex1);

[*false, false, false, false, false*]

> map(type, [-3.65478, 62, 2004, -1995, 8/15], complex1(numeric));

[*false, false, false, false, false*]

> type(2004, complex1(numeric)), type([a, b]-3*I, complex1({realnum, list}));

false, true

> type(-3.67548, complex), type(-3.67548, complex1);

true, false

> type(-3.67548, complex), type((3+4*I)/(5+6*I), complex);

true, true

> type(-3.67548, complex1), type((3+4*I)/(5+6*I), complex1);

false, true

> type((a-I)*(a+I), complex(algebraic)), type((a-I)*(a+I), complex1(algebraic));

true, false

> type(8+I^2, complex(algebraic)), type(8+I^2, complex);

true, true

```
> type(8+I^2, complex1(algebraic)), type(8+I^2, complex1);
                              false, false
> map(type, [8*I, a-b*I, -15*I, 62], complex1({even, symbol, negative, posint}));
                          [true, true, true, false]
> map(type, [8*I, a-b*I, -15*I, 62], complex({even, symbol, negative, posint}));
                          [true, true, true, true]
```

In contrast to the standard procedure, the **`type/complex1`** procedure provides more correct check of *Maple* expressions for the *`complex`*-type.

MSK – deleting of unnecessary *"mserver.exe"* out of active tasks manager

Call format of the procedure:

MSK()

Formal arguments of the procedure: **no**

Description of the procedure:

The *Maple* kernel runs as a separate process from the user interface. This allows a single user interface to use multiple kernels in parallel and mixed kernel modes, or multiple user interfaces to use one kernel. The kernel executable, ***"mserver.exe"***, starts automatically in the background when *Maple* starts. It communicates with the user interface over a TCP/IP socket. The socket connection is secure; it is local to the machine running *Maple* and cannot be accessed from other machines on the network.

Absence of the advanced mechanisms of dynamic memory allotment may lead to closing of a current *Maple* session. The given situation enough often arises at working with complex 3D-objects (above all, in *animation* mode) and in a series of multi-stage cyclic symbolic calculations.

In addition, after such unforeseen and unauthorized termination of a current session the package nevertheless holds module ***"mserver.exe"*** in active tasks queue of *Windows* (in particular, the matter concerns *Maple 8* and *Windows XP Pro*) what is absolutely inadmissible since the subsequent loadings of the package only aggravate the above-mentioned situation. In principle, such situation can take place at a series of other emergency completions of *Maple*. Generally, similar situation takes place for any *Maple* release since the *seventh*. To some degree, the following procedure **MSK** allows to solve the given problem, deleting copies of unnecessary ***"mserver.exe"*** out of active tasks manager of *Windows*.

The procedure use supposes, at the present moment in *Windows* environment only one *Maple* session is activated. Furthermore, the availability of one of pairs of the system utilities {***tlist.exe, kill.exe***} and {***tasklist.exe, taskkill.exe***} is expected. The second pair is standardly included in *Windows XP Pro*; at their absence, they can be downloaded into system *Windows* directory from the appropriate *Windows Tools Kit* or from *Internet*. At absence of the above utilities, the procedure call returns the *false* value with output of the appropriate warning.

Successful procedure call **MSK()** returns Process ID of the current active ***"mserver.exe"***, preliminary having removed all unnecessary ones out of active tasks manager (for releases **7 – 9/9.5**), whereas the *NULL* value is returned for release **6**, i.e. nothing. The procedure handles all basic especial and erroneous situations. As a rule, the procedure call **MSK()** is used if in a current *Windows* session the previous *Maple* session for release **7 – 9/9.5** has aborted.

```
MSK:= proc()
local a, b, c, d, f, k, h;
  assign(b = "$ArtKr$", d = [], a = interface(warnlevel)), interface(warnlevel = 0);
  h:= () -> WARNING("utility tasklist or tlist does not exist, please ensure its availability
     or use other tool");
  if evalb(delf(b), system(cat("tasklist.exe > ", b)) = 0) then f:= fopen(b, READ, TEXT);
    while not Fend(f) do c := SLD(Red_n(Case(readline(f)), " ", 2), " ");
      if c[1] = "mserver.exe" then d := [op(d), c[2]] end if
    end do;
    delf(b), assign('c' = "taskkill.exe "), interface(warnlevel = a);
    if 1 < nops(d) then seq(assign('c' = cat(c, cat("/PID ", d[k], " "))), k = 1 .. nops(d) - 1);
      c := cat(c, "/F");
      if system(c) <> 0 then WARNING("one or a few of mserver.exe with PID(s)
          %1 cannot be deleted", op(d[1 .. -2]))
      end if
    end if;
    `if`(d=[], NULL, came(d[-1]))
  elif evalb(delf(b), system(cat("tlist.exe > ", b)) = 0) then f:= fopen(b, READ, TEXT);
    while not Fend(f) do c := SLD(Red_n(Case(readline(f)), " ", 2), " ");
      if c[2] = "mserver.exe" then d := [op(d), c[1]] end if
    end do;
    delf(b), assign('c' = "kill.exe /f "), interface(warnlevel = a);
    if 1 < nops(d) then for k in d[1 .. -2] do
        if system(cat(c, " ", k)) <> 0 then
          WARNING("mserver.exe with PID %1 cannot be deleted")
        end if
      end do
    end if;
    `if`(d=[], NULL, came(d[-1]))
  else delf(b), interface(warnlevel = a), h(), RETURN(false)
  end if
end proc
```

Typical examples of the procedure use:

```
> MSK();                                                              # Maple 6
> MSK();                                                              # Maple 8
                                    1868
> MSK();
   Warning, utility tasklist or tlist does not exist, please ensure its availability or use other tool
                                    false
```

---

com_exe1 – additional support for standard functions `*system*` and `*ssystem*`
com_exe2

Call format of the procedures:

com_exe1(P, L)
com_exe2({S})

Formal arguments of the procedures:

P - value {**0**|**1**} or full path to a program file of type {"**.com**", "**.exe**"}
L - full path to a *Maple* library or to arbitrary element of file system of a computer
S - (optional) set or list of full names of program files of type {"**.com**", "**.exe**"}

Description of the procedures:

The standard functions `*system*` and `*ssystem*` pass command to the host operating system (for example, *Windows, UNIX*) which performs the appropriate function. The result of these functions is the *"return status"* of the command performed. These tools allow to perform a series of important enough procedures, providing the more effective programming of numerous problems of different purpose. In fact, on some operating systems, some operating system commands may not be accessible via this mechanism.

As a rule, the above-mentioned *Maple* functions deal with program files of types {"**.com**", "**.exe**"} that are standardly delivered with host operating system. Meantime, the set of such files is utterly large and it contains whole series of very useful means that are of interest for programming of many problems in the *Maple* environment.

In this connection, there is a serious problem of compatibility of the user software (as a separate, and at a level of the whole libraries) that use the above-mentioned standard means on set of the program files which are not delivered standardly with operating systems. One of the approaches that solves the given problem consists in delivery with your software utilizing the above-mentioned means, of the program files whose guarantee of presence in standard deliveries of host operating systems is absent. The following two procedures **com_exe1** and **com_exe2** in full measure support this approach.

The description of the algorithms used by these procedures, are given relative to our library described in the monograph. However it is easily enough extrapolated and onto other ways of implementation of the given approach. As the first factual **P** argument of the procedure call **com_exe1(P, L)** can be the full path to a program file (*executable in command line mode*) of type {"**.com**", "**.exe**"} or {**0**|**1**}-value defining check mode of global ***_TABprog***-variable created/updated by the procedure **com_exe1.**

The ***_TABprog***-variable defines special tabular object whose indices are full names {`***aaa.com***`, `***bbb.exe***`} of program files whereas its entries are executable modules appropriate to them. Successful procedure **call com_exe1(0, L)** returns the list of full names of programs saved in the ***_TABprog***-table of a *Maple* library whose full path is defined by the second **L** argument. Whereas the procedure call **com_exe1(1, L)** returns the list of full names of programs saved in the ***_TABprog***-table accessible in a current *Maple* session; in addition, as the second **L** argument may be an arbitrary element of file system of a computer. The ***_TABprog***-table of a *Maple* library having the highest priority in the predefined *libname*-variable is accessible (*active*) in a current *Maple* session.

At last, the procedure call **com_exe1(P, L)** updates a *Maple* library whose full path is defined by the second **L** argument by program whose full path is defined by the first **P** argument; in addition, the program should be defined by its full name, for example, "**tlist.exe**". If ***_TABprog***-table is being updated by a program, whose name is the table index, then entry appropriate to it is being substituted for a new program module. Successful procedure call returns the *NULL* value, i.e. nothing, with output of appropriate warning. The procedure handles all basic especial and erroneous situations.

The **com_exe2** procedure, concomitant to the above procedure, is intended for uploading the required program into main *Windows* directory. The procedure call **com_exe2()** uploads all programs defined by the active ***_TABprog***-table into the main *Windows* directory with output of appropriate warning; in addition, if real uploading has been done, the procedure call returns the list of full names of the uploaded programs, otherwise the *NULL* value is returned, i.e. nothing.

The procedure call **com_exe2(S)** uploads programs defined by a set or list **S** of their full names that are in active ***_TABprog***-table into the main *Windows* directory with output of appropriate warning; in addition, if real uploading has been done, the procedure call returns the list of full names of the uploaded programs, otherwise the *NULL* value is returned, i.e. nothing. The procedure handles all basic especial and erroneous situations.

```
com_exe1:= proc(P::{0, 1, path}, L::path)
local a, b, c, f;
global _TABprog, __ABprog;
  if P = 1 then  if type(_TABprog, table) then
        RETURN(map(op, [indices(_TABprog)]))
     else ERROR("_TABprog is inactive in a current Maple session")
     end if
  end if;
  if type(L, dir) and type(L, mlib) then assign(f = "$ArtKr$.m", c = interface(warnlevel)),
     interface(warnlevel = 0)
  else ERROR("<%1> is not a Maple library", L)
  end if;
  if P = 0 then  march(extract, L, `_TABprog.m`, f);
     if type(f, file) then
        (proc(f)  filepos(f, 8), writebytes(f, [95]), close(f);  read f;  delf(f)  end proc)(f);
        RETURN(interface(warnlevel = c),  map(op, [indices(__ABprog)]),
           assign('__ABprog' = '__ABprog'))
     else ERROR("_TABprog is absent in library <%1>", L)
     end if
  elif type(P, file) then  assign(a = Ftype(P), b = CFF(P)[-1])
  else interface(warnlevel = 0), ERROR("the first factual argument must be zero or file,
     but has received - %1", P)
  end if;
  if member(a, {".exe", ".com"}) then _TABprog[cat(``, b)]:= readbytes(P, infinity); close(P)
  else ERROR("filetype must be {`.com`, `.exe`}, but has received - %1", a)
  end if;
  UpLib(L, [_TABprog]), interface(warnlevel = c);
  WARNING("_TABprog of library <%1> has been updated by program <%2>", L, b)
end proc

com_exe2:= proc()
local a, b, c, d, p, k, h;
  if type(_TABprog, table) then assign(a = WD(), d = [], p = [],
     b = map(op, [indices(_TABprog)]))
  else ERROR("_TABprog is inactive in a current Maple session")
  end if;
  if  nargs = 0  then  h := b;  goto(A)
  else
     if  type(args[1], {set({symbol, string}), list({symbol, string})}) then
        if  Empty(args[1])  then  h := map(op, [indices(_TABprog)]);  goto(A)
        else for k in args[1] do
              if member(cat("     ", k)[-4 .. -1], {".exe", ".com"}) then  p := [op(p), cat(``, k)]
```

```
          else WARNING("<%1> is an inadmissible filename", k)
          end if
        end do
      end if
    else ERROR("the first factual argument %1 is invalid", args[1])
    end if;
    h := SoLists:- `intersect`(p, b);
    if p = [] or h = [] then ERROR("programs appropriate for uploading are absent") end if
  end if;
  A;
  for k in h do  c := cat(a, "\\", k);
    if not type(c, file) then writebytes(c, _TABprog[k]); close(c); d:= [op(d), k] end if
  end do;
  if d <> [] then d, WARNING("programs %1 have been uploaded into the main
     Windows directory", d)
  else WARNING("the required programs are already in the Windows directory")
  end if
end proc
```

Typical examples of the procedures use:

> com_exe1("C:\\Windows\\tlist.exe", "C:\\Temp\\UserLib7\\UserLib");

Warning, _TABprog of library <C:\Temp\UserLib7\userlib> has been updated by program <tlist.exe>

> com_exe1(0, "C:\\Temp\\UserLib9\\UserLib");

[*taskkill.exe*, *tlist.exe*, *tasklist.exe*, *kill.exe*]

> com_exe1(1, "C:\\Program Files");

[*taskkill.exe*, *tlist.exe*, *tasklist.exe*, *kill.exe*]

> com_exe2();

Warning, programs [tlist.exe, kill.exe, tasklist.exe, taskkill.exe] have been uploaded into the main Windows directory

[*tlist.exe*, *kill.exe*, *tasklist.exe*, *taskkill.exe*]

> com_exe2({`Tallinn.TRG`, `SveGal.com`});

Warning, <tallinn.trg> is an inadmissible filename

Error, (in **com_exe2**) programs appropriate for uploading are absent

> com_exe2();

Warning, the required programs are already in the Windows directory

Both procedures support a peculiar interface between *Maple* and the set of all programs suitable for execution in command line mode, providing their usage in the *Maple* environment. For support of the approach described above of usage of programs, the ***_TABprog***-table can be as in the user library together with procedures and the program modules, and in the special library containing only this table. It is presupposed, in any case, that such library should be logically linked with the main *Maple* library, i.e. full path to it should be indicated in the predetermined *libname*-variable of the package. In this case, the ***_TABprog***-table and all programs contained in it will available (*active*) in a current *Maple* session. At last, usage of the main *Windows* directory for programs uploading allows to simplify coding of their calls in functions **`*system*`** and **`*ssystem*`**, since this directory, as a rule, is one of basic directories in which programs are being searched by default. The approach described above can be useful enough at implementation of many general-purpose procedures in the *Maple* environment.

---

mwsname – name testing of a current *Maple* document

Call format of the procedure:

mwsname()

Formal arguments of the procedure: **no**

Description of the procedure:

In a series of cases, the problem of name testing of a current *Maple* document arises. The following procedure mwsname solves this problem. Successful procedure call **mwsname()** returns the name of a current *Maple* document as the *symbol*. The procedure implementation essentially uses the approach supported by the previous two procedures. Sole limitation is that a current *Windows* session should contains one *Maple* session only. As a rule, this condition is true.

```
mwsname:= proc()
local a, b, c, t, k;
  VGS;
  assign('c' = "$ArtKr$", 't' = interface(warnlevel)), assign('a' = system(cat("tlist.exe > ", c)));
  if a <> 0 then
    try interface(warnlevel = 0), com_exe2({`tlist.exe`})
    catch "programs appropriate": interface(warnlevel = t), RETURN(delf(c), false)
    end try;
    interface(warnlevel = t), goto(VGS)
  end if;
  assign(b = fopen(c, READ));
  while not Fend(c) do a := readline(c);
    if search(a, " Maple ") then k:= Search2(a, {"[", "]"}); delf(c), assign('a' = a[k[1]..k[-1]]);
      if search(a, "[Untitled", 'c') then RETURN(cat(``, a[c + 1 .. Search(a, ")")[1]]))
      else c:= Search(a, ".mws"); b:= Search(a, "["); RETURN(cat(``, a[b[1] + 1..c[1] + 3]))
      end if
    end if
  end do
end proc
```

Typical example of the procedure use:

> mwsname();

SveGal.mws

The procedure **mwsname** is rather useful at software processing of current *Maple* documents.

In the given chapter, the tools that extend and improve the standard *Maple* tools for releases **6 – 9.5** have been presented. These tools are used enough widely both at operation with the *Maple* package in interactive mode and at programming of various problems in its environment. They represent undoubted interest at programming of various problems in the *Maple* environment, both by simplifying the programming and by making it by more clear. In a series of cases the facilities simplify work with *Maple* after different emergency completions.

Chapter 10.
Software of working with *Maple* datafiles and documents

Being the programming language in the package environment, oriented, first, to symbolic calculations (*computer algebra*) the *Maple* language has the relatively limited opportunities at operation with data that are located in external computer memory. Moreover, in this respect the *Maple* language essentially yields to traditional programming languages such as *C, COBOL, FORTRAN, PL/1, Pascal, ADA, Basic,* etc. At the same time, the *Maple* language, oriented, first of all, onto solution of problems of mathematical character, gives a set of tools for access to datafiles which can quite satisfy a broad enough audience of users of physical and mathematical appendices of the package. In the present section, additional means of access to datafiles are represented, essentially extending opportunities of the package in the given direction. Many of them simplify the programming of many problems dealing with access to datafiles of different purpose.

With all evidence we can assert, that new package modules **FileTools** and **LibraryTools** have been inspired by a series of our books [33,39,43,44] with which the *Maple* developers have been acquainted. However, our set of the similar procedures is considerably more representative and is focused on wider practical use.

10.1. General purpose software

In the given item, the tools of access to the data which have general purpose and are oriented to the widest use in problems of access to the datafiles and to a work with file system of a computer as a whole are represented.

---

delf - deletion of datafiles
delf1

Call format of the procedures:

delf(*files*)
delf1(*files* {, \`\*\`})

Formal arguments of the procedures:

files - one or more file names or their descriptors
\`\*\` - (optional) key symbol defining a removal mode of *read-only* datafiles

Description of the procedures:

The procedure **delf(*files*)** extends the standard procedure ***fremove***, providing the more effective processing of the arising erratic situations at deletion of ***files.*** The procedure call **delf(*files*)** returns the *NULL* value, i.e. nothing. The specified datafiles are closed if they do exist and are open, and then removed. If at the procedure call a directory, non-existing file or descriptor has been coded, the procedure does not cause an erratic situation. This circumstance allows to delete files without processing of the above erroneous situation. If the user does not have the necessary permissions to remove a datafile (a file with attribute *read-only*), an appropriate erratic situation arises.

The procedure **delf1(*files*)** considerably extends the previous procedure, providing the more effective processing of the arising erratic situations at deletion of files. The procedure **delf(*files*)** returns the *NULL* value, i.e. nothing. The specified datafiles are closed if they do exist and are open, and then removed. If at the procedure call a non-existing file or directory has been coded, the procedure does not cause an error situation. This circumstance allows to delete files without processing of the above erroneous situation. If a removed datafile has *read-only* attribute, the procedure requests permission to remove the datafile, an answer different from **{Y | y}** cancels the removal request, otherwise the datafile will be removed. In addition, the coding of the optional key symbol `*` allows automatically to remove all datafiles without any user permission. Both procedures have many appendices at datafiles processing. The examples represented below well illustrate the told.

```
delf:= proc()
local k;
  if nargs = 0 then NULL
  else for k to nargs do
        try iolib(8, args[k])
        catch "file or directory does not exist": NULL
        catch "file descriptor not in use": NULL
        end try
     end do
  end if
end proc

delf1:= proc()
local k, omega, psi, s, t;
  assign(psi = (proc(x) F_atr1(x, []); fremove(x) end proc),
        omega = (proc(t) s := cat("<", t, "> is read-only. Delete?(y/n): ");
        if Case(readstat(s)) = y then true else false end if end proc));
  if nargs = 0 then NULL
  else for k to nargs do
        if args[k] = `*` then next
        elif isflo(args[k], 't') then
           try fremove(t)
           catch "permission denied": if `if`(member(`*`, {args}), true, omega(t)) then psi(t) end if
           catch "file I/O error": if `if`(member(`*`, {args}), true, omega(t)) then psi(t) end if
           end try
        end if
     end do
  end if
end proc
```

Typical examples of the procedures use:

> delf("C:/RANS/Book.doc"); # file has attribute "*read-only*"

Error, (in **delf**) permission denied **or**
Error, (in **delf**) file I/O error

```
> fopen("C:/RANS/abs.txt", READ); delf(0);
                                        0
> delf("C:/RANS/Book.147", "C:/Temp", "C::\\RANS", 0, 1, 2, 5);
> delf1("C:/rans/RANS.txt", "C:/rans/Academy.txt", 2, 4, 5, 6);
   [ <C:/rans/RANS.txt> is read-only. Delete?(y/n): n;
> delf1("C:/rans/RANS.txt","C:/rans/IAN.txt","C:/rans/RAC.txt", 9, 1, 3,7,14, "C:/Temp");
   [ <C:/rans/RANS.txt> is read-only. Delete?(y/n): y;
   [ <C:/rans/IAN.txt> is read-only. Delete?(y/n): n;
   [ <C:/rans/RAC.txt> is read-only. Delete?(y/n): y;
> delf1("C:/rans"), delf1("C:/Archive"), delf1("C:/Order"), delf1("C:/Grodno");
> delf1("C:/rans/1.txt", "C:/rans/abs.txt", `*`, "C:/rans/2.txt", "C:/rans/acri_2004.txt");
> delf1("C:\\Program Files\\Maple 8\\Users\\Tallinn\\SveGal.mws");
   <C:\Program Files\Maple 8\Users\Tallinn\SveGal.mws> is read-only. Delete?(y/n):  y;
```

---

isflo – testing of datafiles and logical I/O channels

Call format of the procedure:

isflo(F {, 'h'})

Formal arguments of the procedure:

F - a filename, its descriptor, or full path to it (may be *string, symbol* or *integer* from range **0..6**)
h - (optional) an assignable name

Description of the procedure:

The procedure call **isflo(F)** returns the *true* value if and only if the first actual **F** argument specifies name of an existing datafile, full path to it, or descriptor of an open datafile; otherwise, the *false* value is returned. Thus, the procedure additionally tests busy condition of logical **I/O** channels. If the optional second argument **h** has been coded, then through it a filename or full path of an existing datafile is returned. The procedure is functionally equivalent to the procedure **`type/file`**, however it allows additionally to receive names or full paths for open datafiles on the basis of their descriptors. The examples represented below well illustrate the told.

```
isflo:= proc(F::{nonnegative, string, symbol})
local a, b, k, omega;
  omega:= proc(a, F)
      for k to nops(a) do if a[k][1] = F then RETURN(a[k][2]) end if end do end proc;
  if type(F, {string,symbol}) and type(F,file) then `if`(1 < nargs, assign(args[2] = F), NULL), true
  elif type(F, nonnegative) and belong(F, 0 .. 6) then a := iostatus();
    if nops(a) = 3 then false
    else b := omega(a[4 .. -1], F); `if`(b = NULL, false, op([`if`(1 < nargs,
        assign(args[2] = b), NULL), true]))
    end if
  else false
  end if
end proc
```

Typical examples of the procedure use:

> fopen("C:/rans/rans.txt",READ), fopen("C:/rans/ian.txt",READ), fopen("C:/rans/abs.txt", READ), fopen("C:/rans/ACRI_2004.txt", READ); iostatus();

[4, 0, 7, [0, "C:/rans/rans.txt", *STREAM*, *FP* = 2013506696, *WRITE*, *TEXT*],
[1, "C:/rans/ian.txt", *STREAM*, *FP* = 2013506728, *WRITE*, *TEXT*],
[2, "C:/rans/abs.txt", *STREAM*, *FP* = 2013506760, *WRITE*, *TEXT*],
[3, "C:/rans/ACRI_2004.txt", *STREAM*, *FP* = 2013506792, *WRITE*, *TEXT*]]

> map(isflo, [0,1,2,3,5,6,7], 't'); seq([isflo(k, 't'), t], k=[0,1,2,3]);

[*true*, *true*, *true*, *true*, *false*, *false*, *false*]

[*true*, "C:/rans/rans.txt"], [*true*, "C:/rans/ian.txt"], [*true*, "C:/rans/abs.txt"],
[*true*, "C:/rans/ACRI_2004.txt"]

> seq([isflo(k, 't'), t], k=["C:/rans/RANS.txt", "C:/rans/IAN.txt", "C:/rans/ABS.txt", "C:/rans/ACRI_2004.txt"]);

[*true*, "C:/rans/RANS.txt"], [*true*, "C:/rans/IAN.txt"], [*true*, "C:/rans/ABS.txt"],
[*true*, "C:/rans/ACRI_2004.txt"]

AFdes – check of the I/O channels to be open

Call format of the procedure:

AFdes({n {, 'h'}})

Formal arguments of the procedure:

n – (optional) a positive integer defining number of **I/O** channel
h – (optional) an assignable name

Description of the procedure:

The procedure call **AFdes()** without actual arguments returns the list of logic numbers of the open **I/O** channels, whereas at coding of a channel number **n** the {*true*|*false*} value is returned, if the given channel {*is busy*|*is not busy*} by a file accordingly. If the second optional actual **h** argument has been coded, then via it the full path to a file ascribed to the **n** channel is returned.

```
AFdes:= proc()
local a, b, c;
  [assign(a = iostatus()), `if`(a[-1] = 7, RETURN(`if`(nargs = 0, [], false)),
  [assign(c=[seq(a[b][1], b=4..nops(a))]), RETURN(`if`(nargs=1 and belong(args, {b $ (b=0..7)}),
  belong(args, c), `if`(nargs = 2 and belong(args[1], c), op([seq(`if`(a[k][1] = args[1],
  RETURN(true, assign(args[2] = a[k][2])), NULL), k = 4 .. nops(a)), false]), c)))])]
end proc
```

Typical examples of the procedure use:

> AFdes(), AFdes(1), AFdes(3), AFdes(6), AFdes(0);

[0, 1, 2, 3], *true*, *true*, *false*, *true*

> seq(fopen(cat("C:/Archive/",H),READ),H=["ABS.rtf","Lectures.rtf","ICCS_2004.rtf"]):
> AFdes(), AFdes(0), AFdes(1), AFdes(2), AFdes(2, 'h'), h;

[0, 1, 2], *true*, *true*, *true*, *true*, "C:/Archive/ICCS_2004.rtf"

sof – status of an open datafile

Call format of the procedure:

sof(h)

Formal arguments of the procedure:

h - a nonnegative integer, symbol or string defining a filename or full path to it

Description of the procedure:

The useful procedure **sof(h)** provides analysis of status of a datafile specified by the actual **h** argument. If **h** defines a logical I/O channel of an open datafile, the procedure call **sof(h)** returns 4-element list whose elements are (**1**) a filename or a full path to it, (**2**) a scanned position, (**3**) an access mode, and (**4**) a type of the file. If **h** defines a filename of an open file or full path to it, the procedure call **sof(h)** also returns 4-element list whose the first element is number of logical I/O channel, while the other elements are the same as for the above case. If **h** defines a closed file, the *false* value is returned. In the other cases, especial or erratic situations arise. The procedure call **soft()** is equivalent to the procedure call **iostatus().** The **sof** procedure relates to general tools of operation with datafiles of different types.

```
sof:= proc(h::{integer, string, symbol})
local a, b, c, k;
  assign(a=iostatus()), `if`(nargs=0,RETURN(a), `if`(nops(a)=3,ERROR("all files are closed"),1));
  `if`(type(h, integer) and belong(h, 0 .. 7), assign(b = 1, c = h), `if`(type(h, file),
    assign(b = 2, c = cat("", h)), ERROR("%1 is not a file or not a I/O channel", h))),
    seq(`if`(a[k][b] = c, RETURN([a[k][2^b mod 3], filepos(h), a[k][-2], a[k][-1]]), NULL),
    k = 4 .. nops(a)), false
end proc
```

Typical examples of the procedure use:

> fopen("C:/RANS/Lectures.rtf", READ, BINARY): iostatus();

[1, 0, 7, [0, "C:/RANS/Lectures.rtf", *STREAM*, *FP* = 2013506696, *WRITE*, *BINARY*]]

> sof("C:/RANS/Lectures.rtf"), sof(0);

[0, 0, *WRITE*, *BINARY*], ["C:/RANS/Lectures.rtf", 0, *READ*, *BINARY*]

In particular, from the first example follows, the *Maple* 9 (*in contrast to the previous releases*) incorrectly indicates status of a datafile open by means of the fopen function with file mode *READ* and file type *BINARY*. In a series of cases, this circumstance may be reason of incorrect access to datafiles if it is based on information of the ***iostatus*** call. In the case we recommend open a datafile by means of the ***readbytes*** function. In this respect, the **sof** well supplements the standard function ***iostatus*** and is useful in many appendices on work with datafiles of different types and purpose.

---

holdof – an extension of the available I/O channels

Call format of the procedure:

holdof(R {, h})

Formal arguments of the procedure:

R - an I/O channels extension mode (may be *`hold`* or *`restore`* only)
h - (optional) a valid *Maple* expression

Description of the procedure:

For the user the *Maple* simultaneously admits no more than **7** open I/O channels. An attempt to open **I/O** channels beyond this limit causes the error with diagnostics "*too many files already open*". However, in a series of cases the necessity of usage of an additional available I/O channel arises. Above all, such necessity arises in some testing procedures that should analyze components of file system of a computer with using for these purposes of a logical I/O channel without change of status of previously open I/O channels whose quantity may be maximally allowable. For providing of such possibility the procedure **holdof** serves.

If the quantity of open I/O channels is less than **7** the procedure call **holdof(*hold*)** returns the list of logical numbers of open I/O channels, otherwise the call returns a logical number of an open I/O channel with saving of a file status open previously on this channel. The successful procedure call allows to use an additional I/O channel for datafiles processing. After termination of a work, the call **holdof(*restore*)** restores a deferred status of a datafile open previously on this I/O channel with return of the *NULL* value, i.e. nothing, and with output of an appropriate warning. If the procedure call **holdof(*restore*, h)** uses the optional actual **h** argument, the warning output is suppressed. Thus, a scheme of the **holdof** use has the following principal view, namely:

holdof(*hold*) => *datafiles processing on a released* I/O *channel* => holdof(*restore*)

The above-mentioned scheme keeps validity only within a current session until execution of the **restart** clause or arising of a serious error of *Maple* or system that require the *Maple* restart or its reload. A fragment represented below clearly illustrates an usage principle of the **holdof** procedure. The given procedure is used in a number of procedures of the present library. Along with it, the procedure is effective enough tool at organization of systems of access to datafiles when simultaneous operation with the large number of files is required. In a number of examples of such type, the procedure well has recommended itself.

```
holdof:= proc(C::symbol)
local a, b, k, psi, j, omega;
global _avzagnartkrarn61;
  psi:= x -> {k $ (k = 0 .. 6)} minus {seq(x[j][1], j = 4 .. nops(x))};
  `if`(member(C, {restore, hold}), assign(a = iostatus()), ERROR("argument should be `hold`
    or `restore` but has received <%1>", C));
  if C = hold then `if`(nops(a) < 10, RETURN(psi(a)), NULL);
    for k from 4 to nops(a) do
      if not member(a[k][2], {"default", "terminal"}) and member(a[k][3], {STREAM, RAW})
          then _avzagnartkrarn61:= filepos(a[k][2]), a[k]; close(a[k][1]), RETURN(a[k][1])
      end if
    end do;
    if type(_avzagnartkrarn61, assignable) then ERROR("too many files already open") end if
  else
    try `if`(not typeseq(_avzagnartkrarn61, seqn), RETURN(),
      assign67(omega = _avzagnartkrarn61)); unassign('_avzagnartkrarn61'), close(omega[2][1])
    catch "file descriptor not in use": null("processing of an open I/O channel")
    end try;
    b := [`if`(omega[2][3] = RAW, open, fopen), omega[2][2],
      op(omega[2][5 .. `if`(omega[2][3] = RAW, 5, 6)])]; null(OpenLN(omega[2][1], op(b)),
      filepos(omega[2][1], omega[1])), `if`(1 < nargs, NULL, WARNING(
      "file <%1> has been restored on I/O channel <%2>", omega[2][2], omega[2][1]))
  end if
end proc
```

Typical examples of the procedure use:

> Close(): open("C:/RANS/lectures.rtf", READ), filepos("C:/RANS/lectures.rtf", 1942), seq(fopen(cat("C:/RANS/",k,".txt"), WRITE),k=1..6), iostatus();

0, 1942, 1, 2, 3, 4, 5, 6, [7, 0, 7, [0, "C:/RANS/lectures.rtf", *RAW*, *FD* = 7, *READ*, *BINARY*],
[1, "C:/RANS/1.txt", *STREAM*, *FP* = 2013506696, *WRITE*, *TEXT*],
[2, "C:/RANS/2.txt", *STREAM*, *FP* = 2013506728, *WRITE*, *TEXT*],
[3, "C:/RANS/3.txt", *STREAM*, *FP* = 2013506760, *WRITE*, *TEXT*],
[4, "C:/RANS/4.txt", *STREAM*, *FP* = 2013506792, *WRITE*, *TEXT*],
[5, "C:/RANS/5.txt", *STREAM*, *FP* = 2013506824, *WRITE*, *TEXT*],
[6, "C:/RANS/6.txt", *STREAM*, *FP* = 2013506856, *WRITE*, *TEXT*]]

> holdof(hold);

0

> open("C:/RANS/ABS.txt", READ): iostatus();

[7, 0, 7, [0, "C:/RANS/ABS.txt", *RAW*, *FD* = 7, *READ*, *BINARY*],
[1, "C:/RANS/1.txt", *STREAM*, *FP* = 2013506696, *WRITE*, *TEXT*],
[2, "C:/RANS/2.txt", *STREAM*, *FP* = 2013506728, *WRITE*, *TEXT*],
[3, "C:/RANS/3.txt", *STREAM*, *FP* = 2013506760, *WRITE*, *TEXT*],
[4, "C:/RANS/4.txt", *STREAM*, *FP* = 2013506792, *WRITE*, *TEXT*],
[5, "C:/RANS/5.txt", *STREAM*, *FP* = 2013506824, *WRITE*, *TEXT*],
[6, "C:/RANS/6.txt", *STREAM*, *FP* = 2013506856, *WRITE*, *TEXT*]]

> holdof(restore); iostatus(), filepos("C:/RANS/lectures.rtf");

Warning, file <C:/RANS/lectures.rtf> has been restored on I/O channel <0>

[7, 0, 7, [0, "C:/RANS/ABS.txt", *RAW*, *FD* = 7, *READ*, *BINARY*],
[1, "C:/RANS/1.txt", *STREAM*, *FP* = 2013506696, *WRITE*, *TEXT*],
[2, "C:/RANS/2.txt", *STREAM*, *FP* = 2013506728, *WRITE*, *TEXT*],
[3, "C:/RANS/3.txt", *STREAM*, *FP* = 2013506760, *WRITE*, *TEXT*],
[4, "C:/RANS/4.txt", *STREAM*, *FP* = 2013506792, *WRITE*, *TEXT*],
[5, "C:/RANS/5.txt", *STREAM*, *FP* = 2013506824, *WRITE*, *TEXT*],
[6, "C:/RANS/6.txt", *STREAM*, *FP* = 2013506856, *WRITE*, *TEXT*]]

Adrive – check of the active disk drives

Call format of the procedure:

Adrive()

Formal arguments of the procedure: **no**

Description of the procedure:

The procedure call **Adrive()** returns the sequence of logical names ascribed to the disk drives which are active at moment. The drives names are returned as the strings. This procedure seems useful enough tool in a number of cases of operation with datafiles of the different types that are located in external computer memory. The **Adrive** procedure allows to identify drivers that can be used as units that are ready for access to datafiles.

```
Adrive:= proc()
local cd, L, N, k, p;
```

```
   assign(cd = currentdir(), L = {seq(convert([65 + p], bytes), p = 0 .. 25)}, N = {});
   for k in L do
      try currentdir(cat(k, ":\\"))
      catch "file or directory does not exist": N := {k, op(N)}
      catch "file I/O error": N := {k, op(N)}
      catch "permission denied": N := {k, op(N)}
      end try
   end do;
   op([L minus N, currentdir(cd)][1])
end proc
```

Typical example of the procedure use:

> Adrive();

"A", "C", "D"

All_Close – closing of all open files or pipes

Call format of the procedure:

All_Close()

Formal arguments of the procedure: **no**

Description of the procedure:

This procedure closes all open files which have been opened either implicitly or by functions ***fopen***, ***`open`***, ***pipe***, or ***popen***. The procedure call **All_Close()** returns the *NULL* value, i.e. nothing. The files closing provide writing of all data actually to a disk. In addition, the **All_Close** procedure does not affect nor the *`default`*, nor *`terminal`* input or output stream. The procedure is rather useful tool at programming of access to datafiles of different types and nature.

```
All_Close:= proc()
local a, k;
   assign(a = iostatus()), close(`if`(a = [0, 0, 7], RETURN(NULL),
   seq(`if`(member(a[k][2], {terminal, default}), NULL, a[k][2]), k = 4 .. nops(a))))
end proc
```

Typical examples of the procedure use:

> [open(`C:/Temp/Grodno.rtf`, WRITE)], [open(`C:/Archive/ABS.rtf`, READ)], iostatus();

[0], [1], [2, 0, 7, [0, "C:/Temp/Grodno.rtf", *RAW, FD* = *7, WRITE, BINARY*], [1, "C:/Archive/ABS.rtf", *RAW, FD* = 8, *READ, BINARY*]]

> All_Close(); iostatus();

[0, 0, 7]

Close – closing of all open files and pipes whose names or paths are equivalent in register-insensitive mode

Call format of the procedure:

Close({F})

Formal arguments of the procedure:

F - (optional) sequence of names, files descriptors or full paths to files which should be closed

Description of the procedure:

This procedure closes all open files which have been opened either implicitly or by functions ***fopen***, ***`open`***, ***pipe***, or ***popen***. The procedure call **Close({F})** returns the *NULL* value, i.e. nothing. The files closing provide writing of all data actually to a disk. In addition, the **Close** procedure does not affect nor the *`default`*, nor *`terminal`* input or output stream. The procedure allows to close all open files and pipes whose names, descriptors or paths are equivalent in register-insensitive mode with files defined by the actual arguments **F.** The procedure call **Close()** closes all pipes and files, opened in a current *Maple* session. The procedure is a rather useful tool at programming of access to files of different types and nature providing register-insensitive mode of files coding. The procedure handles all basic erratic and especial situations with output of appropriate warnings for providing of calculations continuity. In contrast to the standard ***`close`*** procedure, the **Close** procedure is characterized by compatibility with all *Maple* releases since the ***sixth*** release.

```
Close:= proc()
local k, p, h, omega, j, nu, t, psi;
  `if`(nargs = 0, RETURN(All_Close()), assign(h = iostatus(), omega = [], nu = [])),
  `if`(h = [0, 0, 7], RETURN(), seq(assign('omega' = [op(omega), h[k][1 .. 2]]), k = 4 .. nops(h)));
  for k to nargs do
    if not (type(eval(args[k]), {string, symbol}) or type(args[k], {p $ (p = 0 .. 7)})) then
      WARNING("%-1 argument <%3> should be symbol, string or integer from range 0..7 but
        has received <%2>", k, whattype(eval(args[k])), args[k]);
      next
    elif member(cat("", args[k]), {"default", "terminal"}) then next
    else
      for j to nops(omega) do
        try `if`(CF(omega[j][2]) = CF(cat("", args[k])) or args[k] = omega[j][1],
          [close(omega[j][1]), assign('nu' = [op(nu), j])], NULL)
        catch "file descriptor not in use": next
        end try
      end do;
      assign('omega' = subsop(seq(t = NULL, t = nu), omega)),
      `if`(omega = [], RETURN(), assign('nu' = []))
    end if
  end do
end proc
```

Typical examples of the procedure use:

```
> map(fopen, ["C:/RANS/lectures.rtf", "c:/avz", "c:/AVZ", "C:\\Agn", ian,IAN,"c:/agn"],
  WRITE): RANS:=matrix([]): iostatus();

  [7, 0, 7, [0, "C:/RANS/lectures.rtf", STREAM, FP = 2013506696, WRITE, TEXT],
  [1, "c:/avz", STREAM, FP = 2013506728, WRITE, TEXT],
  [2, "c:/AVZ", STREAM, FP = 2013506760, WRITE, TEXT],
  [3, "C:\\Agn", STREAM, FP = 2013506792, WRITE, TEXT],
  [4, "ian", STREAM, FP = 2013506824, WRITE, TEXT],
  [5, "IAN", STREAM, FP = 2013506856, WRITE, TEXT],
  [6, "c:/agn", STREAM, FP = 2013506888, WRITE, TEXT]]
```

> Close(IAN, Grodno, "C:/aVz", RANS, "c:\\rans\\Lectures.rtf", "c:\\agn"); iostatus();

Warning, 4th argument <RANS> should be symbol, string or integer from range 0..7 but has received <array>

[0, 0, 7]

> map(fopen, ["C:/RANS/lectures.rtf", "c:/avz", "c:/AVZ", "C:\\Agn",ian,IAN, "c:/agn"], WRITE): RANS:=matrix([]): Close(); iostatus();

[0, 0, 7]

> map(fopen, ["C:/RANS/lectures", "C:/AVZ.txt"], WRITE), Close(0,1), iostatus();

[0, 1], [0, 0, 7]

Fnull – cleaning of datafiles

Call format of the procedure:

Fnull(F)

Formal arguments of the procedure:

F – filename or full path to a datafile (may be *symbol* or *string*)

Description of the procedure:

In a series of problems dealing with datafiles processing, a datafile cleaning is needed, i.e. the datafile should has null length. The procedure call **Fnull(F)** returns the *NULL* value, cleaning a datafile specified by the actual **F** argument. The procedure call result is a closed *null* datafile **F.**

```
Fnull:= F::file -> null(Close(F), writebytes(F, []), close(F))
```

Typical examples of the procedure use:

> fopen("C:/RANS\\Res5.txt", READ, TEXT), open("C:/RANS/RES5.txt", WRITE): iostatus();

[2, 0, 7, [0, "C:/RANS\\Res5.txt", *STREAM*, *FP* = 2013506696, *WRITE*, *TEXT*],
[1, "C:/RANS/RES5.txt", *RAW*, *FD* = 8, *WRITE*, *BINARY*]]

> Fnull("C:/RANS\\Res5.txt"); iostatus(), IsFempty("C:/RANS\\Res5.txt");

[0, 0, 7], *true*

Users – *Maple* serial number and full path to the USERS directory

Call format of the procedure:

Users(h {, t})

Formal arguments of the procedure:

h - an arbitrary name
t - (optional) an arbitrary valid *Maple* expression

Description of the procedure:

The USERS directory of *Maple* contains at least the *Maple* initialization files of the different level. Information of these files is useful for many appendices. The procedure call **Users(h)** returns full path to the USERS directory in a format allowable for correct use by the standard *Maple* means of access to datafiles, and through actual **h** argument additionally returns the *serial number* of a current *Maple* installation. At absence of serial number, the **h** argument is returned unevaluated. If the optional second argument **t** has been coded, then full path to the USERS directory is returned in a format allowable for correct use by the functions *`system`* and *`ssystem`*. The procedure is a rather useful at programming of access to files of different types and in many other appendices.

```
Users:= proc(h::evaln)
local a, b, c;
  a := fopen(cat(CCM(), "\\MapleSys.ini"), READ, TEXT);
  while not Fend(a) do b := readline(a);
    if b[1 .. 7] = "UserDir" then c := `if`(1 < nargs, Path(b[15 .. -1]), b[15 .. -1])
    elif b[1 .. 6] = "Serial" then unprotect('h'), assign('h' = SD(b[14 .. -1]))
    end if
  end do;
  close(a), c
end proc
```

Typical examples of the procedure use:

> Users(h), h, Users(t, 7), t;

"C:\Program Files\Maple 9\Users", 917864204, "c:\"program files"\"maple 9"\users", 917864204

> DirF(Users(h));

["c:\program files\maple 9\users\maple9.ini", "c:\program files\maple 9\users\links.mws", "c:\program files\maple 9\users\maple.ini"]

---

fpathdf – determination of full path to a datafile or directory

Call format of the procedure:

fpathdf(C, L)

Formal arguments of the procedure:

L - symbol, string, set or list of symbols and/or strings
C - a letter defining logical name of a disk drive

Description of the procedure:

In a series of tasks dealing with datafiles processing, the problem of determination of *full path* to a datafile or directory arises. The problem can be solved by the following procedure **fpathdf.** The procedure call **fpathdf(C, L)** is intended for determination in a file system located on a disk drive with logical name given by the first actual argument **C** of full paths to datafiles and/or directories given by the second actual argument **L.** The datafiles are defined by their full names, for instance, "Maple.ini".

The successful procedure call **fpathdf(C, L)** returns 2-element sequence, the first element of which defines list of full paths to sought directories, whereas the second element defines list of full paths to sought datafiles. If any of the above list is returned as the *empty* list, i.e. [], the appropriate components have not been found on the given disk drive **C.** The procedure handles basic especial and erroneous situations.

```
fpathdf:= proc(C::{letter}, L::{symbol, string, set({symbol, string}), list({symbol, string})})
local a, c, d, h, k, f, t, p;
```

```
 `if`(member(convert(CF2(C), bytes), {[k] $ (k = 97 .. 122)}), `if`(member(CF2(C),
   map(CF2, {Adrive()})), 15, ERROR("drive <%1> is idle", Case(C, upper))), 7);
 assign(a = cat(CDM(), "\\_$ArtKr$_"), d = [], f = [], c = map(CF2, {op(L)}));
 WARNING("Analysis of a file system on <%1> is in progress, please wait!", Case(C, upper));
 system(cat("Dir /A/B/S ", C, ":\\*.* > ", blank(a))), assign(p = fopen(a, READ, TEXT));
 while not Fend(p) do h := CF2(readline(p)); t := Search(h, "\\");
   if member(h[t[-1] + 1 .. -1], c) then
     if type(h, dir) then d := [op(d), h] else f := [op(f), h] end if
   end if
 end do;
 delf1(a, `*`), d, f
end proc
```

Typical examples of the procedure use:

> fpathdf(C, {"Maple.ini", USERS});

Warning, Analysis of a file system on <C> is in progress, please wait!

["c:\program files\maple 6\users", "c:\program files\maple 7\users", "c:\program files\maple 8\users", "c:\program files\maple 9\users"],
["c:\program files\maple 6\users\maple.ini", "c:\program files\maple 7\users\maple.ini", "c:\program files\maple 8\users\maple.ini", "c:\program files\maple 9\users\maple.ini"]

CFF – extracting of components of a path or a datafile name
CCF

Call format of the procedures:

CFF(P)
CCF(P)

Formal arguments of the procedures:

P - path to a file; its type should belong to the set {*symbol, string*}

Description of the procedures:

The procedure call **CFF(P)** returns the components of path to a datafile defined by the **P** argument as the list of strings. The procedure fulfills the full decomposition onto components, and at the same time, the procedure does not require the real existence of path to datafile. The procedure tests a path onto its formal correctness only and can correct some syntax errors.

The procedure call **CCF(P)** returns the components of path to a datafile defined by the **P** argument as the list of strings in the following slits (**1**) chain of subdirectories to the datafile, (**2**) the datafile name, and (**3**) the extension of datafile name. As against the previous procedure, the procedure **CCF** returns the required component-wise decomposition only for real path to datafile; otherwise, an appropriate erroneous situation is identified. The above two procedures seem rather useful tools at operation with file system of a computer.

```
CFF:= proc(F::{string, symbol})
local h;
  try assign(h = CF(F, string)), `if`(h[-1] = "\\" and h[-2] <> ":", assign('h' = h[1 .. -2]), NULL)
  catch "<%1> is invalid path to a directory or a file": ERROR("<%1> is invalid path", F)
  end try;
```

```
  `if`(h[2] = ":" and h[3] <> "\\" or 1 < nops(Search2(h, {":"})) or h[1] = "\\",
     ERROR("<%1> is invalid path", F), NULL), [op(SLD(h, "\\"))]
end proc

CCF:= proc(P::path)
local a, b, k;
  assign(a = CF(P, string)), `if`(a[-1] = "\\" and a[-2] <> ":", assign('a' = a[1 .. -2]), NULL),
  `if`(a[2] <> ":", assign('a' = cat(currentdir(), "\\", a)), NULL), `if`(type(a, dir), RETURN([a]),
  assign(b = Search2(a, {"\\"}))), [a[1 .. b[-1] - 1], op(SLD(a[b[-1] + 1 .. -1], "."))]
end proc
```

Typical examples of the procedures use:

```
> CFF("C:/AVZ_BMA_2001\\Mixture/Demo\\"), CFF("C:Library\\Maple_8/Maple_8.doc");
        ["c:", "avz_bma_2001", "mixture", "demo"], ["c:", "library", "maple_8", "maple_8.doc"]
> CCF("C:/Program Files\\Maple 9/Lib\\UserLib\\Maple.hdb");
                    ["c:\program files\maple 9\lib\userlib", "maple", "hdb"]
> CCF("C:/Program Files\\Maple 9/Lib\\UserLib\\Maple1.hdb");
   Error, invalid input: CCF expects its 1st argument, P, to be of type path, but received
   C:/Program Files\Maple 9/Lib\UserLib\Maple1.hdb
```

CF – standardization of definition of full path to a datafile or directory
CF1
CF2

Call format of the procedures:

CF(F {, T})
CF1(F)
CF2(F {, 1})

Formal arguments of the procedures:

F – a string or symbol defining full path to a datafile or directory
T – (optional) argument defining a type of the returned result (may be *symbol* or *string*)
1 – (optional) argument defining a processing type

Description of the procedures:

The procedure call **CF(F)** returns the so-called standardized *Maple* format of path to a datafile or directory defined by the first actual **F** argument; the resultant format has the lowercase and contains entries of directory separators `\\` instead of directory separators `/`. By default, the standardized path has a type defined by a type of the first actual **F** argument. However, if the second optional **T** argument has been coded, then the standardized path returned by the procedure call **CF(F, T)** will has a type, defined by the actual **T** argument {*symbol, string*}. In addition, the procedure call **CF(F)** tests the path **F** onto its formal correctness only and in a series of cases can correct the detected errors.

In addition, generally speaking, the above-standardized *Maple* format not is acceptable for MS DOS environment at use within the *Maple* environment of access MS DOS to datafiles. For these purposes may be used the **Path** procedure. Use of the standardized *Maple* format allows to eliminate a series of erroneous situations at operations with datafiles, and essentially to simplify programming of problems of access to datafiles.

As it was already marked, at the organization of access to datafiles the method of coding of their names or paths to them is rather essential and is characterized both by the used separators of directories {"/", "\\"}, and the register used at coding of symbols composing a name or path to the file. In particular, by means of two procedure calls **open("C:/Temp\\file.txt", READ)** and **open("C:/Temp\\File.txt", WRITE)** we can open factually the same file "file.txt" on two different logical I/O channels in two different modes (*READ* and *WRITE*).

Naturally, in a lot of cases such opportunity is enough interesting and allows to organize more refined and effective ways of the organization of access to datafiles first of all having the specific organization and demanding non-standard methods of access. However, generally, the same opportunity promotes occurrence of especial and erroneous situations. For example, use of various formats of coding of path to the same datafile can catastrophically increase quantity of used logical I/O channels, cause a mistiming of access to datafile, etc. The given circumstance supposes special attention of the user in work with datafiles.

In this connection, the role of simple **CF** procedure is being represented for us rather essential. This procedure provides standardization of coding of the above components of file system of a computer, namely: the delimiter "\\" and *lower case* register are used in a file descriptor of the *string*-type.

In our opinion, the means similar to the **CF** procedure, it would be desirable to include in the standard *Maple* tools providing basic access to datafiles (***open, fopen, writeline, readdata, readline, writebytes, writedata, readbytes,*** etc.) to provide the above standardization. In particular, similar standardization is supported by the base operational system (MS DOS). In this case, the user can use arbitrary format at coding of files descriptors identifying datafiles within of a current session with the *Maple* package.

The following procedure **IsPtf** illustrates an example of use of an opportunity of the different coding formats of files descriptors that allows to work with the same *txt*-file in different modes.

```
IsPtf:= proc(F::file)
local d, k, f, f1, n, m;
  assign(f = cat("", F)), `if`(IsFtype(f, txt), [assign(d = iolib(6, f, infinity), f1 = cat(`if`(type(f[1],
    upper), Case(f[1]), Case(f[1], upper)), f[2..-1])), iolib(7, f)], ERROR("%1 is not a txt-file", F));
  for k from 0 to floor(1/2*d) - 1 do iolib(6, f, k), iolib(6, f1, d-k-1);
     if iolib(4, f) <> iolib(4,f1) then RETURN(iolib(7, f, f1), false) end if
  end do;
  null(iolib(7, f, f1)), true
end proc
```

> IsPtf("C:/Temp/DEMO\\DemoLib.txt"), IsPtf("C:/Temp/DEMO\\DemoLib1.txt");

true, false

The procedure returns the *true* value if a string which has been written down into a *txt*-file, is *palindrome*; otherwise, the *false* value is returned. The essence of the algorithm implemented by the procedure consists in the following. Reading and comparison of opposite positions (relatively to the center of the same *txt*-file open by the procedure ***filepos*** on two logical I/O channels in view of use of various formats of coding of descriptor of initial **F** file) are done. The reception described above and used in the procedure **IsPtf,** of opening of the same datafile on different logical I/O channels provides an effective algorithm of testing of a *txt*-file for its contents be *palindrome.* Similar approach can be useful enough and in a series of other problems dealing with organization of access to datafiles. In particular, the above approach allows to organize processing the same datafile by a few different procedures, increasing the efficiency of processing in a series of cases. However, it demands synchronization of access processes.

In contrast to the **CF** procedure, the procedure call **CF1(F)** returns the representation of full path to a datafile **F** with symbol **"/"** as the separator of subdirectories, what in a series of cases seems useful enough reception. This procedure is based on the previous procedure and inherits its basic features. Both procedures **CF** and **CF1** do not demand real existence of full path to a datafile or directory defined by their actual argument; however provide definite testing onto correctness of their definitions. In addition, if condition **cat("", F)[2] <> ":"** takes place, the both procedures **CF** and **CF1** define location of the **F** datafile in a current *Maple* directory.

At last, procedure **CF2** is similar to procedures **CF** and **CF1**, however as against them does not support check of the first actual F argument onto correctness from the point of view of the *Maple* language syntax used for a coding of files and paths to them. The procedure call **CF2(F, 1)** is similar to the procedure call **CF1(F)**, whereas the procedure call **CF2(F)** is similar to the procedure call **CF(F, *string*)** with taking into consideration of the made remark.

```
CF:= proc(F::{string, symbol})
local h, omega, u;
  omega, u:= () -> ERROR("<%1> is invalid path to a directory or a file", F), interface(warnlevel);
  `if`(Red_n(cat(" ", F), " ", 2) = " " or member(F, {``, ""}), omega(), interface(warnlevel = 0));
  assign('h' = Red_n(Subs_All("/" = "\\", Case(cat("", F)), 1), "\\", 2)), interface(warnlevel = u);
  assign('u' = Search(h, ":")), `if`(u <> [], `if`(1 < nops(u) or 2 < u[1], omega(), NULL), NULL);
  assign('h' = `if`(h[1] = "\\" and h[-1] = "\\", h[2 .. -2], `if`(h[-1] = "\\", h[1 .. -2],
    `if`(h[1] = "\\", h[2..-1], h)))); `if`(length(h) = 2 and h[2] = ":", RETURN(cat(h, "\\")), NULL);
  `if`(h[2]= ":" and h[3] <> "\\", assign('h' = cat(h[1..2], "\\", h[3..-1])), assign('h' = `if`(h[2] <> ":",
    cat(currentdir(), "\\", h), h))), `if`(nargs = 1, h, convert(h, args[2]))
end proc

CF1:= (F::{string, symbol}) -> convert(Subs_All("\\" = "/", CF(F), 1), whattype(F))

CF2:= (F::{string, symbol}) -> SUB_S(`if`(nargs = 2 and args[2] = 1, ["\\" = "/"], ["/" = "\\"]),
       Case(convert(F, string)));
```

Typical examples of the procedures use:

```
> CF("C:/Temp/RANS\\IAN/DownLoad\\Library/ArtKr.158");
                    "c:\temp\rans\ian\download\library\artkr.158"
> CF("C:MAPLE 5_9/MAPLE_EDU/MAPLE 9 EDU\\AVZ 61/\\\Demo.ZIP");
                "c:\maple 5_9\maple_edu\maple 8 edu\avz 61\demo.zip"
> CF("/MAPLE_EDU/MAPLE 9 EDU\\AVZ 61/DemoLib.ZIP\\");
      "C:\Program Files\Maple 9\bin.win\maple_edu\maple 9 edu\avz 61\demolib.zip"
> CF1(`C:\\Program Files\\Maple 9/Bin.Win/license/license.dat`);
                      c:/program files/maple 9/bin.win/license/license.dat
> CF(`C::abc`, string);
   Error, (in omega) <C::abc> is invalid path to a directory or a file
> CF2("C:RANS/ian/RAC.rea", 1), CF2("C:RANS/ian/RAC.rea");
                          "c:rans/ian/rac.rea", "c:rans\ian\rac.rea"
```

---

cfln – check of a filename given at its creation or last renaming

Call format of the procedure:

cfln(F)

Formal arguments of the procedure:

F – a filename or full path to a datafile (may be *symbol* or *string*)

Description of the procedure:

The *Maple* package uses *context-sensitive* names, therefore in a series of cases the necessity of determination of filename given at its creation or its last renaming arises. The procedure call **cfln(F)** returns the context-sensitive filename given at creation/renaming of a datafile defined by the actual **F** argument. The filename is returned in the *string*-format. The given tool seems a rather useful in light of the concept of context-insensitive filenames of the underlying operating system and a possibility to work in different modes with the same datafile open on different I/O channels.

```
cfln:= proc(F::{string, symbol})
local h, z, p, v, r;
  `if`(not type(F, file), ERROR("<%1> is not a file", F), system(cat("dir ", Path(F), " > __")));
  assign(p = CFF(F)[-1]), assign(z = length(p), v = cat(" ", p, convert([10], bytes)));
  assign(h = readbytes("__", TEXT, infinity)), fremove("__"), assign(r = searchtext(v, h)),
  h[r + 1 .. r + z]
end proc
```

Typical examples of the procedure use:

> cfln("C:\\Academy\\Examples\\HelpBase\\mws_files\\userlib!!!.mws");

"UserLib!!!.mws"

> map(cfln, ["C:/Academy/readme.htm",`rans_ian.61`,"C:/temp/avzagn.txt"]);

["ReadMe.htm", "RANS_IAN.61", "AvzAgn.txt"]

DEL_F – deletion of all temporary datafiles created by `*rtable*` function of *Maple*

Call format of the procedure:

DEL_F()

Formal arguments of the procedure: **no**

Description of the procedure:

As a side effect of the ***rtable*** function call (directly or indirectly, for example, via the ***Array***-function) is a `*clogging*` of system TEMP directory by temporary files "mpnnnnn.TMP" and "mpnnnnn.TMP.m" that are not deleted after terminating a current *Maple* session. These files can be deleted outside of the *Maple* environment by means of the corresponding commands of MS DOS or Windows. However, the problem can be easily solved by means of the **DEL_F** procedure inside the *Maple* environment. Meanwhile, after the procedure call **DEL_F()** restoration of a ***rtable***-objects history is impossible in full measure.

```
DEL_F:= proc()
local a, d;
  try null(assign(a = currentdir()), assign(d = WT()), currentdir(d), system("Del *.TMP"),
    system("Del *.TMP.m"), currentdir(a))
  catch "file or directory does not exist": RETURN(NULL)
  end try
end proc
```

The successful procedure call **DEL_F()** deletes the above temporary datafiles of system TEMP directory and returns the *NULL* value, i.e. nothing.

Typical examples of the procedure use:

> DEL_F();

Con_Mws – search of *mws*-files with the *Maple* documents

Call format of the procedure:

Con_Mws(K, Kr, Q)

Formal arguments of the procedure:

K - a string or symbol defining full path to a directory
Kr - a pattern (*symbol* or *string*)
Q - an assignable name

Description of the procedure:

The procedure call **Con_Mws(K, Kr, Q)** returns the list of *mws*-files, located in a directory defined by the first actual **K** argument, which contain a pattern defined by the second actual **Kr** argument. Through the third actual **Q** argument, the list of all *mws*-files located in the **K** directory is returned.

```
Con_Mws:= proc(K::{string, symbol}, Kr::{string, symbol}, Q::evaln)
local k, Art, R, p, F, N, L, G;
  interface(warnlevel = 0), assign(G = cat([libname][1][1 .. 2], "\\$AGN$"), F = [], L = []),
    system(cat("Dir ", CF(Path(K), string), " > ", G)), interface(warnlevel = 3);
  Art:= proc(X::string, Y::string)
    local n;
      n := 0;
      do if searchtext(Y, X, length(X) - n - 5 .. length(X) - n - 5) <> 0 then break else n := n + 1
        end if
      end do;
      length(X) - n - 4
    end proc;
  `if`(filepos(G, infinity) = 0, RETURN(), [close(G), assign(p = fopen(G, READ, TEXT))]);
  while R <> 0 do R := readline(p);
    if length(R) < 5 then next
    elif not member(substring(R, -4 .. -1), {".MWS", ".mws"}) then next
    elif substring(R, -4 .. -4) = "." and substring(R, -5 .. -5) = "`" then N := Art(R, "`")
    else N := Art(R, " ")
    end if;
    F := [op(F), substring(R, N .. length(R))]
  end do;
  [close(G), fremove(G), assign(Q = [seq(convert(F[k], symbol), k = 1 .. nops(F))])];
  for k to nops(F) do do R := readbytes(cat(K, "\\", F[k]), 60000, TEXT);
      if R = 0 then break
      elif searchtext(Kr, convert(R, symbol)) = 0 then next
      else close(cat(K, "\\", F[k])); L := [op(L), F[k]]; break
      end if
    end do
```

```
  end do;
  [seq(convert(L[k], symbol), k = 1 .. nops(L))]
end proc
```

Typical examples of the procedure use:

```
> Con_Mws("C:/Program Files/Maple 8/MPL9MWS", "AVZ", Kr); nops(%), nops(Kr);
   [Apmv.mws, CCF.mws, CF.mws, Case.mws, Close.mws, Con_Mws.mws, CorMod.mws,
   CrNumMatrix.mws, Currentdir.mws, D_ren.mws, DoF.mws, FBcopy.mws, FLib.mws,
   FO_state.mws, F_T.mws, F_atr.mws, Ftype.mws, Iddn.mws, Insert.mws, IsOpen.mws,
   LibUser.mws, MAM.mws, MPL_txt.mws, MkDir.mws, Mwsin.mws, N_Cont.mws,
   Nstring.mws, ParProc.mws, RTab.mws, Read1.mws, Red_n.mws, Rmf.mws, SDF.mws,
   Save.mws, SaveMP.mws, Search.mws, Search_D.mws, TRm.mws, Type_D.mws, XSN.mws,
   avm_VM.mws, belong.mws, cmtf.mws, conSV.mws, convert,module.mws, deletab.mws,
   extrS.mws, filePM.mws, filename.mws, holdof.mws, howAct.mws, insL.mws, isRead.mws,
   maxl.mws, open2.mws, pathtf.mws, sMf.mws, save2.mws, sstr.mws, swmpat.mws,
   type,Table.mws, type,boolproc.mws, type,digit.mws, type_dir.mws, writedata1.mws]
                                   65, 338
```

Dir – output of contents of a directory

Call format of the procedure:

Dir(d)

Formal arguments of the procedure:

d – a name or full path to a directory; (may be *name, symbol* or *string*)

Description of the procedure:

The procedure call **Dir(d)** outputs the contents of a directory given by the actual **d**-argument. The given procedure in environment {*MS DOS* | *Windows*} is fulfilled stably and is useful in the practical respect in work with datafiles of different nature. The **Dir** procedure usage has a sense in interactive mode only, since result output by it cannot be directly used by active tools of a current *Maple* session. Although usage of mechanism of **`*disks transits*`** allows to do it.

```
Dir:= proc(K::{name, string, symbol})
local a, n, F;
  `if`(not type(K, path), ERROR("<%1> is not a path", K), [assign(F = cat(convert(K,string)[1..2],
    "\\$$"), n = 61, a =CF(Path(K), string)), system(cat("dir ", a, `if`(a[-1] = "\\", "*.* > ",
    "\\*.* > "), Path(F)))]);
  while n <> 0 do n := readline(F); `if`(n = "" or n = 0, NULL, printf("%s
      ", convert(n, symbol))) end do;
  fremove(F)
end proc
```

Typical example of the procedure use:

```
> Dir("C:/Program Files\\Maple 9/Users\\");
    Volume in drive C is ALADJEV_V_Z
    Volume Serial Number is 1626-13F3
    Directory of C:\Program Files\Maple 9\Users
    .        <DIR>      09-06-03  5:48p .
    ..       <DIR>      09-06-03  5:48p ..
```

```
MAPLE9        INI      2,475  09-26-03   7:16p  maple9.ini
PROCLI~1      MWS    323,082  09-25-03  11:39a  ProcLib_6_7_8_9.mws
HELPMAN       MWS        794  09-25-03   4:11p  helpman.mws
MAPLE         INI         58  09-08-03  10:16a  Maple.ini
READ          MWS      1,438  09-26-03   8:50a  read.mws
READDA~1  MWS      1,556  09-22-03  10:30a  Readdata1.mws
        6 file(s)       329,403 bytes
        2 dir(s)      3,261.08 MB free
```

DirE – check of an arbitrary directory upon emptiness

Call format of the procedure:

DirE(F)

Formal arguments of the procedure:

F – a name or full path to a directory (may be *symbol* or *string*)

Description of the procedure:

The procedure call **DirE(F)** returns the *true* value, if **F** is empty directory and the *false* value otherwise. If **F** is not a directory, the *FAIL* value is returned with output of appropriate warning. The given tool seems rather important in light of file concept of the underlying operating system and of key distinctions in handling of subdirectories and datafiles. The given procedure extends standard tools of the *Maple* package, providing an opportunity of testing of an arbitrary directory of file system of a computer upon emptiness. In addition, it is necessary to have in mind; the **DirE** procedure operates with the directories exclusively, ignoring both existing datafiles, and nonexistent elements of file system of a computer.

```
DirE:= proc(F::{name, string, symbol})
local a;
  assign(a=CF(F)), `if`(not type(a,dir),RETURN(FAIL,WARNING("<%1> is not a directory",F)),
    system(cat("dir /A /B", Path(a), " > _$")));
  `if`(iolib(6, "_$", infinity) = 0, op([iolib(8, "_$"), true]), op([iolib(8, "_$"), false]))
end proc
```

Typical examples of the procedure use:

> map(DirE, ["C:\\Temp/DemoLib/", "C:/Temp/demo", "C:/Academy\\Books/"]);

Warning, <C:\Temp/DemoLib/> is not a directory
Warning, <C:/Temp/demo> is not a directory

$$[FAIL, FAIL, false]$$

> map(DirE, ["C:\\Academy/ReadMe.htm", "C:\\Lasnamae", `C:/Temp`, "C:/"]);

Warning, <C:\Academy/ReadMe.htm> is not a directory

$$[FAIL, true, false, false]$$

DirF – receiving of list of all files of a directory

Call format of the procedure:

DirF(F {, r} {, *only*})

Formal arguments of the procedure:

F – a name or full path to a directory (may be *symbol* or *string*)
r – (optional) a case of the returned paths to datafiles (may be *uppercase* or *lowercase*)
only – (optional) key word defining a search level of datafiles

Description of the procedure:

The procedure call **DirF(F)** returns the list of full paths to files located in a directory specified by the actual **F** argument and in all its subdirectories. If the optional **r**-argument is used, it defines the case {*uppercase, lowercase*} of the returned paths to datafiles (by default, the *lowercase* value is used), whereas their type is defined by a type of the **F** argument. The optional key word **`*only*`** bounds the datafiles search in the specified **F** directory only without its subdirectories. At absence of the sought datafiles, the procedure call returns the empty list, i.e. []. Furthermore, if **F** is not a directory, the procedure call **DirF(F)** similarly to the procedure call ***listdir*(F)** generates error. The given procedure in environment {*MS DOS | Windows*} runs stably and is a rather useful in the practical respect at operation with datafiles or at programming in the *Maple* environment of a series of problems dealing with access to datafiles of different types and nature.

```
DirF:= proc(F::dir)
local n, k, p, L, f, t;
  assign(p = 97 .. 122), `if`(member(convert(CF2(F), bytes), {[k] $ k = p, [k, 58] $ k = p,
    [k, 58, 92] $ k = p}), assign(t = 7, L = []), assign(L = [])); assign(f = cat(CDM(),
    "\\_$ArtKr$")), system(cat("dir /A/B", `if`(member(only, {args}), " ","/S "),
    `if`(t = 7, cat(cat("", F)[1], ":"), Path(F)), "\\*.* > ", blank(f)));
  do n := iolib(2, f); if n = 0 then break else L := [op(L), convert(n, whattype(F))]
    end if
  end do;
  if member(only, {args}) then assign(p = 0, 'L' = map2(cat, cat(F, "/"), L));
    do
      if p = nops(L) then break else p := 0 end if;
      for k to nops(L) do if type(L[k], dir) then L:= subsop(k = NULL, L); break else p:= p + 1
        end if
      end do
    end do
  end if;
  iolib(8, f), map(convert, sls(L, "\\", `<`), `if`(member(uppercase, {args}), uppercase, lowercase))
end proc
```

Typical examples of the procedure use:

```
> DirF("C:/RANS\\RAC/REA\\IAN", uppercase);
  ["C:\RANS\RAC\REA\IAN\TRG2.TXT", "C:\RANS\RAC\REA\IAN\TRG1.TXT",
  "C:\RANS\RAC\REA\IAN\TRG.TXT", "C:\RANS\RAC\REA\IAN\RESULT1.RTF",
  "C:\RANS\RAC\REA\IAN\RESULT1.TXT", "C:\RANS\RAC\REA\IAN\LECTURES.RTF",
  "C:\RANS\RAC\REA\IAN\ICCS_2004.TXT", "C:\RANS\RAC\REA\IAN\FMERGE.TXT",
  "C:\RANS\RAC\REA\IAN\FMERGE.RTF"]
> DirF("C:\\Program Files/Maple 9\\Users/");
  ["c:\program files\maple 9\users", "c:\program files\maple 9\users\readdata1.mws",
  "c:\program files\maple 9\users\read.mws", "c:\program files\maple
  9\users\proclib_6_7_8_9.mws", "c:\program files\maple 9\users\maple9.ini",
  "c:\program files\maple 9\users\maple.ini", "c:\program files\maple
  9\users\helpman.mws"]
```

> listdir("C:\\Program Files/Maple 9\\Users/");

["." , "..", "maple9.ini", "ProcLib_6_7_8_9.mws", "helpman.mws", "Maple.ini", "read.mws", "Readdata1.mws"]

> map(DirF, ["C:/RANS\\RAC/REA", "C:/RANS\\RAC/"], only);

[[], []]

> DirF(`C:/RANS`);

[c:\rans\trg2.txt, c:\rans\trg1.txt, c:\rans\rac, c:\rans\trg.txt, c:\rans\result1.rtf, c:\rans\result1.txt, c:\rans\lectures.rtf, c:\rans\iccs_2004.txt, c:\rans\rac\rea, c:\rans\rac\rea\ian, c:\rans\rac\rea\ian\trg2.txt, c:\rans\rac\rea\ian\trg1.txt, c:\rans\rac\rea\ian\trg.txt, c:\rans\rac\rea\ian\result1.rtf, c:\rans\rac\rea\ian\result1.txt, c:\rans\rac\rea\ian\lectures.rtf, c:\rans\rac\rea\ian\iccs_2004.txt, c:\rans\rac\rea\ian\fmerge.txt, c:\rans\rac\rea\ian\fmerge.rtf]

> DirF("C:/RANS", only, uppercase);

["C:/RANS/TRG2.TXT", "C:/RANS/TRG1.TXT", "C:/RANS/TRG.TXT", "C:/RANS/RESULT1.RTF", "C:/RANS/RESULT1.TXT", "C:/RANS/LECTURES.RTF", "C:/RANS/ICCS_2004.TXT"]

> DirF("A");

["a:\setramd.bat", "a:\readme.txt", "a:\ramdrive.sys", "a:\oakcdrom.sys", "a:\msdos.sys", "a:\io.sys", "a:\himem.sys", "a:\flashpt.sys", "a:\findramd.exe", "a:\fdisk.exe", "a:\extract.exe", "a:\ebd.sys", "a:\ebd.cab", "a:\drvspace.bin", "a:\config.sys", "a:\command.com", "a:\btdosm.sys", "a:\btcdrom.sys", "a:\autoexec.bat", "a:\aspicd.sys", "a:\aspi8u2.sys", "a:\aspi8dos.sys", "a:\aspi4dos.sys", "a:\aspi2dos.sys"]

In the eighth *Maple* release, a new iolib function ***listdir*(F)** has been introduced providing the return of contents of a directory **F** as the list. In addition, the directories "**.**" (*current*) and "**..**" (directory that is one level higher) are included in the returned list, as that the last example of the fragment below illustrates. Meanwhile, comparative analysis of the procedure **DirF** and the function ***listdir*** gives preference to the first tool, which relative to the ***listdir*** function gives more full information about the specified directory, namely: full paths to files and subdirectories forming the specified directory. This procedure is used by a number of other procedures dealing with processing of file system of a computer.

DirFT – check of types of all datafiles in an arbitrary directory

Call format of the procedure:

DirFT(F)

Formal arguments of the procedure:

F - a symbol or a string defining full path to a directory

Description of the procedure:

The procedure call **DirFT(F)** returns the set of the files types contained in a directory defined by the actual **F** argument. The actual **F** argument should define *full path* to a directory, otherwise the erroneous situation arises. If the actual **F** argument defines a file, then the *false* value is returned with output of appropriate warnings. This procedure (similarly to the procedure **Ftype**) as the *`type`* understand a type defined by expansion of the file name; for example file "Order.rtf" has "**.**rtf" type. Presence in the returned set additionally of key words *`dir`* and *`undefined`* speaks about presence in the tested directory of subdirectories and files without name expansions accordingly, whereas the *FAIL* value identifies an unidentified element of a PC file system.

```
DirFT:= proc(F::path)
local h, R;
  assign(h = length(DoF(F)), R = {}), `if`(not type(F, dir),
    RETURN(WARNING("<%1> is a file of %2-type", F, Ftype(F)), false), `if`(DirE(F),
    RETURN([]), system(cat("Dir /B ", `if`(search(F, " "), blank(CF(F)), CF(F)), " > _$$$_"))));
  do h := readline(`_$$$_`);
    if h = 0 then RETURN(fremove(`_$$$_`), R) else R := {op(R), Ftype(cat(F, "/", h))} end if
  end do
end proc
```

Typical examples of the procedure use:

```
> map(DirFT, [`C:/My Documents`, "C:/Temp/", `C:/Archive\\`, "A:/ICCS_2004.txt"]);
  Warning, <A:/ICCS_2004.txt> is a file of .txt-type
            [{dir}, {".doc", ".tmp"}, {FAIL, dir, ".hdb", ".pdf", ".rtf", ".mws"}, false]
```

cfdd – copying of a directory into other directory

Call format of the procedure:

cfdd(F1, F2)

Formal arguments of the procedure:

F1 – a string or symbol defining full path to a directory
F2 – a string or symbol defining full path to a target directory

Description of the procedure:

The successful procedure call **cfdd(F1, F2)** returns the *NULL* value, providing a copying of a directory **F1** along with its files but without its subdirectories into a directory, defined by the second actual **F2** argument. However, it is necessary to note, if the target directory contains **F1** subdirectory then the files contained in it are updated; at absence of **F1** subdirectory in **F2** directory, it will be created before a copying. In addition, the procedure copies the files with attributes {*system, archive, read only*) only, assigning to them in the target directory the `*archive*`-attribute. Abnormal terminating of the procedure causes one of the following erroneous situations, namely:

`One or both directories are absent`
`One of actual arguments represents a file but not a directory`

```
cfdd:= proc(G::{string, symbol}, S::{string, symbol})
local a, b, c;
  assign(a = map(DoF1, [G, S]), b = map(CF, [G, S], string)),
    assign(c = b[1][Search2(b[1], {"\\"})[-1] + 1 .. length(b[1])]);
  `if`(belong(0, a), ERROR(`One or both directories are absent`), NULL);
  `if`(belong({op(a)}, {3, 4}), null([MkDir(cat(b[2], "\\", c)), system(cat("Copy ", blank(b[1]),
    "\\*.*", blank(cat(b[2], "\\", c))))]),
    ERROR(`One of actual arguments represents a file but not a directory`))
end proc
```

Typical examples of the procedure use:

```
> cfdd("C:\\Book\\BBS/My Documents", "C:\\Temp\\DemoLib/Tallinn\\Lasnamae");
```

DoF – check of properties of a directory or datafile
DoF1
isFile
isDir

Call format of the procedures:

DoF(S)
DoF1(S)
isFile(S {, 'h'})
isDir(S)

Formal arguments of the procedures:

S - a name or full path to a directory or datafile (may be *symbol* or *string*)
h - (optional) an unevaluated name

Description of the procedures:

The procedure call **DoF(S)** classifies the elements of file system of a computer from the point of view of characteristics such as *`empty directory`*, *`full directory`*, *`file`*, *`file is open`* and *`file or directory does not exist`*. The procedure call **DoF(S)** returns one of these values as the *symbol*, characterizing a datafile or directory defined by the actual **S** argument. The given tool seems rather important in light of the file concept of an underlying operating system and of key distinctions in handling of datafiles and subdirectories.

The procedure call **DoF1(S)** is equivalent to the above procedure call, returning the positive integers, namely:

- **0,** if file or directory **S** does not exist;
- **1,** if **S** is a *closed file;*
- **2,** if **S** is an *open file;*
- **3,** if **S** is an *empty directory;*
- **4,** if **S** is a *full directory.*

The **DoF1** procedure is more acceptable for use in program constructions dealing with access to file system of a computer.

The procedure call **isFile(S)** returns the *true* value, if and only if the actual **S** argument defines an *existent* file; otherwise, the *false* value is returned. In addition, an open **S** file preserves own current state. If the procedure call **isFile(F, 'h')** returns the *true* value, then via the second optional **h** argument the file status is returned, namely: *`close`* or *`open`*.

At last, the procedure call **isDir(S)** returns the *true* value, if and only if the actual **S** argument defines an *existent* directory; otherwise, the *false* value is returned. The procedures **DoF, DoF1, isDir** and **isFile** extend standard tools of the package, providing an opportunity of check of properties of an arbitrary directory or a file of file system of a computer. They allow to comfortably automate the handling of erroneous and especial situations arising at operations with basic elements of file system of a computer.

```
DoF:= proc(F::{string, symbol})
local a, b, k, omega;
  assign(a = cat("", F), b = iostatus(), omega = (n -> interface(warnlevel = n))), omega(0);
  `if`(member(a[-1], {"\\", "/"}), assign('a' = a[1 .. -2]), 1), `if`(nops(b) = 3, 14, seq(`if`(CF(b[k][2],
    string) = CF(a, string), RETURN(omega(3), `file is open`), 6), k = 4 .. nops(b)));
  `if`(DirE(F) = true, RETURN(omega(3), `empty directory`), `if`(DirE(F) = false,
```

```
    RETURN(omega(3), `full directory`), 1));
  try fopen(a, READ)
  catch "file or directory does not exist": RETURN(omega(3), `file or directory does not exist`)
  end try;
  RETURN(omega(3), close(a), file)
end proc

DoF1:= proc(F::{string, symbol})
local a, b, k, omega;
  assign(a = cat("", F), b = iostatus(), omega = (n -> interface(warnlevel = n))), omega(0);
  `if`(member(a[-1], {"\\", "/"}), assign('a' = a[1 .. -2]), 1), `if`(nops(b) = 3, NULL,
    seq(`if`(CF(b[k][2], string) = CF(a, string), RETURN(omega(3), 2), NULL), k = 4..nops(b)));
  `if`(DirE(F) = true, RETURN(omega(3), 3), `if`(DirE(F) = false, RETURN(omega(3), 4), 157));
  try fopen(a, READ)
  catch "file or directory does not exist": RETURN(omega(3), 0) end try;
  RETURN(omega(3), close(a), 1)
end proc

isFile:= proc(F::{string, symbol})
local a, n, k;
  assign(a = iostatus(), n = nops(iostatus())), `if`(n=3, NULL, seq(`if`(CF(a[k][2]) = CF(cat("",F)),
    RETURN(true, `if`(nargs = 2, assign('args'[2] = open), NULL)), NULL), k = 4 .. n));
  try fopen(F, READ)
  catch "file or directory does not exist": RETURN(false)
  catch "file I/O error": RETURN(false)
  end try;
   close(F), true, `if`(nargs = 2, assign('args'[2] = close), NULL)
end proc

isDir:= F::{string, symbol} -> member(DoF1(F), {3, 4})
```

Typical examples of the procedures use:

> map(DoF, ["C:/temp", `C:/Belarus/Grodno`, `C:/temp/vgtu.txt`, "c:/academy/readme.htm", "C:\\Temp/Tallinn/Lasnamae"]);

[*full directory, empty directory, file, file is open, file or directory does not exist*]

> map(DoF1, [`C:/temp`, `c:/Belarus/Grodno`, `c:/temp/vgtu.txt`, "c:/academy/readme.htm", "C:\\Temp/Tallinn/Lasnamae"]);

[4, 3, 1, 2, 0]

> map(isFile, [`c:/rans/readme.htm`,`c:/temp/AVZ`]), [isFile("c:/rans/readme.htm", 'h'), h];

[*true, true*], [*true, open*]

> map(isDir, ["C:/Temp\\", "C:\\Belarus/Grodno/", "C:/Temp/AVZ", "C:\\Academy"]);

[*true, true, false, true*]

> map3([DoF, DoF1, isFile, isDir], ["C:\\Program Files\\Maple 8\\Users"], 0);

[*full directory*, 4, *false, true*]

The above procedures are essentially used in a series of other procedures providing an access to components of file system of a computer. In particular, the procedure **isFile** provides the correct test of *real existence* of an arbitrary file whereas use for these purposes of the built-in ***close***-function is release-dependent: for releases **6** and **7** the procedure call **close(F)** initiates error, whereas for releases **8** and **9/9.5** the same procedure call returns the *NULL* value on non-existent **F** file.

FO_state – check of all files which have been opened in a current *Maple* session

Call format of the procedure:

FO_state()

Formal arguments of the procedure: **no**

Description of the procedure:

At operation with external datafiles often, the necessity of receiving of the state over files open in a current session arises. The certain information of such type is given by the built-in *iolib*-function **iostatus;** however, in a series of cases this information is insufficient.

The procedure call **FO_state()** returns the table of states of all files open in a current session. The returned table has the *numbers of I/O channels* on which files have been opened as *indices*, and 5-element lists as *entries;* elements of each such list are: **(1)** a file qualifier, **(2)** an access mode, **(3)** a type of file, **(4)** a scanned position of file and **(5)** a file volume in bytes accordingly.

For files of a type {*direct, process pipe*}, the number of logical **I/O** channel additionally appears as the first element of the corresponding list. If in a current session the open files do not exist, then the call **FO_state()** returns the *NULL* value with output of the warning `*All files are closed*`.

```
FO_state:= proc()
local k, n, P, S, T;
  [assign(S = iostatus()), assign(n = nops(S))], `if`(S = [0, 0, 7],
      RETURN(WARNING("All files are closed")),
   (proc() try seq(assign(T[S[k][1]], [S[k][2], S[k][5], S[k][6], filepos(S[k][1]), filepos(S[k][1],
             infinity)]), k = 4 .. n)
      catch: `Processing of especial situation of access to files {direct, process, pipe}`
      end try
    end proc)());
  seq((proc() try filepos(S[k][1], T[S[k][1]][4]) catch: T[S[k][1]] := S[k] end try end proc)(),
  k = 4..n); `if`(type(eval(T), symbol), NULL, eval(T))
end proc
```

Typical examples of the procedure use:

> restart: iostatus(), FO_state();

Warning, All files are closed

[0, 0, 7]

> [fopen("c:/temp/AVZ", READ, TEXT), fopen("c:/temp/vgtu.txt", WRITE, BINARY), open("c:/temp/testdata.dat", READ), open("c:/temp/testdata1.dat", WRITE)], FO_state();

[**0, 1, 2, 3**], table([**0** = ["c:/temp/AVZ", *WRITE, TEXT*, 0, 6780], **1** = ["c:/temp/vgtu.txt", *WRITE, BINARY*, 0, 0], **2** = ["c:/temp/testdata.dat", *READ, BINARY*, 0, 10], **3** = ["c:/temp/testdata1.dat", *WRITE, BINARY*, 0, 0]])

SDF – determination of paths to the required files or directories
SSF
FSSF

Call format of the procedures:

SDF(U, S {, 1})
SSF(U, F)
FSSF()

Formal arguments of the procedures:

U - a logical drive name (may be *symbol* or *string*)
S - a name of the required datafile or directory (may be *symbol* or *string*)
1 - (optional) (*one*) - value defining a search mode
F - names of the required datafiles (may be *symbol, string* or *list of symbols* and/or *strings*)

Description of the procedures:

In many problems of data access, an expediency of search of full paths containing the required subdirectories and/or datafiles arises. The **SDF** procedure solves this problem. The procedure call **SDF(U, S)** returns the list of full paths containing the required subdirectory or datafile, defined by the second actual **S** argument, on a drive whose logical name is defined by the first actual **U** argument. In addition, the second actual **S** argument is coded in a view **`/N`** or **"/N"**, where **N** is a name of the required subdirectory or full name of a datafile.

By default, the **SDF** procedure returns the list of all full paths containing the required subdirectory or datafile irrespective of its location in the path. While a coding of the value **1** as the third actual argument defines the return of full paths which contain the required subdirectory or file only at the end of their.

The **SSF(U, F)** procedure in many respects is similar to the **SDF** procedure, however has essential distinctions, namely: **(1)** if the second actual **F** argument defines a *symbol* or *string*, the procedure call returns the list of directories containing a datafile, defined by the second actual **F** argument, on a drive whose logical name is defined by the first actual **U** argument. At absence of the required file, the *empty* list is returned, i.e. []; **(2)** if the second actual **F** argument defines a *list* of *symbols* and/or *strings*, the procedure call returns the table whose indices define dataliles from **F**, whereas its entries define sequences of directories containing the appropriate datafiles, on a drive whose logical name is defined by the first actual **U** argument. At absence of all required datafiles, the *empty* list, i.e. [], is returned with output of the appropriate warning; at absence of several datafiles, the above table is returned with output of appropriate warnings for the missing datafiles.

At last, the **FSSF()** procedure from indeterminate quantity of the actual arguments defining the full names of sought datafiles is intended for search on all *active* (*online*) drives of datafiles given by the actual arguments. The procedure returns the sequence of nested lists; the first element of such lists defines full name of the sought datafile, whereas the second element defines the list of full paths to directories containing the datafile. Type of the first element of such lists corresponds to a type of filename defined at the **FSSF** call, whereas the sublist elements have type *`list(string)`*. At absence of the tested files, the **FSSF** procedure returns for them the empty lists, i.e. []. The procedure call outputs warning about all online-drives. In many procedures dealing with access to datafiles the procedures of this group play a rather important part, providing strict identification of paths to datafiles.

```
SDF:= proc(U::{symbol, string}, S::{symbol, string})
local s, h, t, L;
   system(cat("Dir /S /B ", U, ":\\*.* > $$Art_Kr")), assign(L=[],s=CF2(S), interface(warnlevel=0));
   do assign('h' = readline("$$Art_Kr"), 't' = 't');
      if h = 0 then break
      else `if`(Search1(CF2(h), s, 't') and belong(right, t) or
         `if`(nargs = 3 and args[3] = 1, false, search(CF2(h), cat(s, "\\"))), assign('L' = [op(L), h]), 7)
      end if
   end do;
   map(CF2, L), fremove("$$Art_Kr"), interface(warnlevel = 3)
end proc
```

```
SSF:= proc(C::{string, symbol}, F::{string, symbol, list({string, symbol})})
local k, h, Dr, R, L, t, tab, f, n;
  assign(f=`if`(type(F,list),F,[F])), assign('f'=map(Case,map2(cat, "",f)), h=cat("",C,":\\$Kr_Art"));
  assign(tab = table([seq(f[k] = NULL, k = 1 .. nops(f))]), n = nops(f));
  system(cat("Dir /S ", C, ":\\ > ", h)), close(h);
  while R <> "0" do R := convert(readline(h), string);
    if search(R, "Directory of") then Suffix(R, " ", t, del), assign('Dr' = t[14 .. -1])
    else for k to n do if search(Case(R), cat(" ", f[k])) then tab[f[k]] := tab[f[k]], Dr end if end do
    end if
  end do;
  if not type(F, list) then fremove(h), entries(tab)
  else
    if Empty(tab) then RETURN(fremove(h),[],WARNING("files %1 do not exist on <%2>",F,C))
    elif  add(`if`([entries(tab)][k] = [], 0, 1), k = 1 .. n) = n then NULL
    else for k to n do
        if tab[f[k]] = NULL then WARNING("file <%1> does not exist on <%2>", F[k], C) end if
      end do
    end if;
    fremove(h), eval(tab)
  end if
end proc
```

```
FSSF:= proc()
local a, b, k, p, Lambda;
  op(op([`if`(nargs = 0 or {seq(whattype(eval(args[k])), k = 1 .. nargs)} minus
  {string, symbol} <> {}, ERROR(`Arguments are invalid: %1`, [args]),
  assign('a' = [Adrive()])), assign([seq(unassign(convert(k, symbol)), k = 1 .. nargs)]),
  assign('b' = [seq(convert(k, symbol), k = 1..nargs)]), seq(assign('b'[k] = [args[k]]), k = 1..nargs),
  assign(Lambda = (x -> `if`(1 < nops(x), x, NULL))), WARNING("Online-drives are %1", a),
  seq(seq(assign('b'[p] = [op(b[p]), SSF(a[k], args[p])]), k = 1 .. nops(a)), p = 1 .. nargs),
  map(Lambda, b), unassign('a', seq(convert(k, symbol), k = 0 .. nargs))]))
end proc
```

Typical examples of the procedures use:

```
> map2(SDF, C, ["/ReadMe.htm", "/VGTU.txt", `/AVZ`, "/Grodno"]);
  [["c:\academy\readme.htm", "c:\academy\books\readme.htm",
  "c:\academy\examples\readme.htm",
  "c:\academy\books\mpl9book\readme.htm", "c:\academy\examples\helpbase\readme.htm",
  "c:\academy\userlib9\readme.htm", "c:\program files\common
  files\system\ado\mdacreadme.htm",
  "c:\program files\regcleaner\readme.htm", "c:\windows\readme.htm"], ["c:\temp\vgtu.txt"],
  ["c:\temp\avz"], ["c:\belarus\grodno"]]
> SSF(C, "MapleSys.ini");
        ["C:\Program Files\Maple 8\bin.win", "C:\Program Files\Maple 6\BIN.WNT",
                          "C:\Program Files\Maple 7\BIN.WNT"]
> SSF(C, ["Tallinn.mws", "RAC_REA.txt", "VGTU.uni", `Processing.doc`]);
  Warning, files [Tallinn.mws, RAC_REA.txt, VGTU.uni, Processing.doc] do not exist on <C>
                                        []
```

> SSF(C, ["ProcLib_6_7_8_9.mws", "RANS_IAN.txt", "grsu.txt", `Fultus.doc`]);

Warning, file <grsu.txt> does not exist on <C>

table(["rans_ian.txt" = "C:\TEMP", "grsu.txt" = (), "proclib_6_7_8_9.mws" = ("C:\Program Files\Maple 8\Users", "C:\Program Files\Maple 6\Users", "C:\Program Files\Maple 7\Users"), "fultus.doc" = "C:\BOOK"])

> FSSF("ProcLib_6_7_8_9.mws", `Fultus.doc`);

Warning, Online-drives are [A, C]

["ProcLib_6_7_8_9.mws", [], ["C:\Program Files\Maple 8\Users", "C:\Program Files\Maple 6\Users", "C:\Program Files\Maple 7\Users"]], [*Fultus.doc*, [], ["C:\BOOK"]]

> FSSF("ProcLib_6_7_8_9.mws", "RANS_IAN.txt", "grsu.txt", `Fultus.doc`);

Warning, Online-drives are [C]

["ProcLib_6_7_8_9.mws", ["C:\Program Files\Maple 8\Users", "C:\Program Files\Maple 6\Users", "C:\Program Files\Maple 7\Users"]], ["RANS_IAN.txt", ["C:\TEMP"]], ["grsu.txt", []], [*Fultus.doc*, ["C:\BOOK"]]

The procedures **FSSF, SDF** and **SSF** are useful tools for operation with file system of a computer.

---

sfd – classifying search of sought elements of file system of a computer

Call format of the procedure:

sfd(C, F)

Formal arguments of the procedure:

C - a logical drive name (may be *symbol* or *string*)
F - a name of the required datafile or directory (may be *symbol* or *string*)

Description of the procedure:

In many problems of data access, an expediency of search of full paths containing the required directories and/or datafiles arises. The above **SDF** procedure solves this problem. The following procedure sfd expands facilities of **SDF,** providing the more comprehensive search with classifying of a sought element of file system of a computer from the point of view of directories and datafiles.

The procedure call **sfd(C, F)** returns the sequence of lists whose the first element defines the type of the sought element (*dir* - directories or *file* - datafiles), whereas the others define full paths to the appropriate **F** elements, situated on drive **C.** In addition, the procedure handles main especial and erroneous situations with output of the accurate diagnostic warnings. Moreover, as a drive name the first symbol of **C** is chosen.

```
sfd:= proc(C::{string, symbol}, F::{string, symbol})
local a, b, h, f, fl, dr, k, s, psi, Sigma;
   assign(a = cat("", C)[1], fl = {}, dr = {}, b = Case(cat("\\", F))), assign(f = cat(a, ":\\$ArtKr"));
   `if`(not type(a, letter), ERROR("invalid coding of a drive <%1>", C),
      `if`(member(Case(cat("", a)), map(Case, [Adrive()])), 7,
      ERROR("drive <%1> is off-line", cat(a, ":\\"))));
   system(cat("Dir /S /B /A ", a, ":\\ > ", f)), assign(Sigma = (() -> `if`(nargs = 0, [], args))),
      assign(psi = (() -> seq(`if`(nops(args[k]) = 1, NULL, args[k]), k = 1 .. nargs)));
   while h <> "0" do h := Case(convert(readline(f), string)); s := Search2(h, {b});
```

```
      if s <> [] then for k to nops(s) do
          if h[s[k] + length(F) + 1] = "\\" then dr := {op(dr), h[1 .. s[k] + length(F)]}
          elif s[k] + length(F) = length(h) then
             if type(h, dir) then dr := {h, op(dr)} else fl := {h, op(fl)} end if
          end if
        end do
      end if
   end do;
   delf(f), Sigma(psi([file, op(fl)], [dir, op(dr)]))
end proc
```

Typical examples of the procedure use:

```
> sfd(C, "MapleSys.ini");
    [file, "c:\program files\maple 6\bin.wnt\maplesys.ini",
    "c:\program files\maple 7\bin.wnt\maplesys.ini",
    "c:\program files\maple 8\bin.win\maplesys.ini"]
> sfd(C, "Temp");
    [file, "c:\temp\temp"],
    [dir, "c:\windows\system32\config\systemprofile\local settings\temp",
    "c:\documents and settings\default user\local settings\temp",
    "c:\windows\pchealth\helpctr\temp", "c:\windows\temp", "c:\temp",
    "c:\documents and settings\aladjev\local settings\temp",
    "c:\documents and settings\networkservice\local settings\temp",
    "c:\documents and settings\localservice\local settings\temp"]
> sfd(C, "Cookies");
    [dir, "c:\documents and settings\aladjev\cookies",
    "c:\documents and settings\default user\cookies",
    "c:\windows\system32\config\systemprofile\cookies",
    "c:\documents and settings\localservice\cookies"]
> sfd(a, Lasnamae); sfd(G, "RANS.ian"); sfd("62", "RANS.ian");
                                        []
    Error, (in sfd) drive <G:\> is off-line
    Error, (in sfd) invalid coding of a drive <62>
```

F_T – check of type of a datafile (*TEXT* or *BINARY*)

Call format of the procedure:

F_T(F)

Formal arguments of the procedure:

F – a symbol or string defining a filename or path to a required datafile

Description of the procedure:

The procedure call **F_T(F)** returns the type {*TEXT, BINARY*} of a datafile defined by the actual **F** argument; in addition, the datafile does not vary own status {*open* | *close*}. If the tested datafile is open then it does not vary the *open* status with preservation of a current position of scanning. If the actual **F** argument defines no datafile, then the *FAIL* value is returned with output of appropriate warning. The **F_T** procedure seems useful enough tool for operation with datafiles of any type and nature.

In addition, it is necessary to have in mind, the type returned by the **F_T** procedure does not characterize the essence itself of the text datafiles (*txt*-files); it is linked with logical organization of a file only, namely: a *TEXT* datafile consists of logical records (*lines*) ending by the symbols `hex(0D0A)`, while a *BINARY* datafile presents an unstructured set of bytes. In particular, from a ***TEXT*** datafile we can read information both by records (*lines*) by means of the ***readline*** procedure, and by bytes by means of the ***readbytes*** procedure, whereas from a *BINARY* datafile we can read information by bytes by means of the ***readbytes*** procedure, or the datafile as a whole by means of the ***readline*** procedure if the datafile does not contain the especial symbol `hex(1A)`. Namely, neither the ***readline*** procedure nor the ***readbytes*** procedure can read a datafile in the *TEXT* mode after the first entry into the datafile of symbol `hex(1A)`. This problem can be solved by the procedure ***readbytes*** in the *BINARY* mode only. The last three examples of a fragment below very clearly illustrate the told.

```
F_T:= proc(F::{name, string, symbol})
local a, h;
   assign(h = DoF(F)), `if`(h = file, [filepos(F, filepos(F, infinity) - 2) + 2,
   RETURN(`if`(readbytes(F, 2) = [13, 10], TEXT, BINARY), close(F))],
   `if`(h = `file is open`, [assign(a = filepos(F)), filepos(F, filepos(F, infinity) - 2) + 2,
   RETURN(`if`(readbytes(F, 2) = `if`(Fend(F), [10], [13, 10]), TEXT, BINARY),
   null(filepos(F, a)))], RETURN(FAIL, WARNING("<%1> is not a file", F))))
end proc
```

Typical examples of the procedure use:

```
> map(F_T, ["C:/Temp/avz", "C:/Temp/data5.txt", `C:/Temp/ArtKr.147`, "C:/Academy.ian"]);
   Warning, <C:/Temp/ArtKr.147> is not a file
   Warning, <C:/Academy.ian> is not a file
                          [BINARY, TEXT, FAIL, FAIL]
> F:="C:/AVZ/text": writeline(F,"RANS_IAN"), writeline(F,"RANS_IAN"), close(F),
  F_T(F), readline(F), readline(F), close(F);
                      9, 9, TEXT, "RANS_IAN", "RANS_IAN"
> writebytes(F,"RANS_IAN"), writebytes(F,"RANS_IAN"), close(F), F_T(F), readbytes(F,
  TEXT, infinity), close(F), readline(F), close(F);
            8, 8, BINARY, "RANS_IANRANS_IAN", "RANS_IANRANS_IAN"
> writebytes(F, cat("ArtKr", convert([26], bytes), "ArtKr")), close(F), readline(F), readline(F);
                               11, "ArtKr", 0
> close(F); readbytes(F, TEXT, infinity); close(F);
                                  "ArtKr"
> close(F); readbytes(F, infinity); close(F);
                [65, 114, 116, 75, 114, 26, 65, 114, 116, 75, 114]
```

`type/rlb` – check of a datafile to be *rlb*-datafile

Call format of the procedure:

type(F, *txt*)

Formal arguments of the procedure:

F – symbol or string defining filename or path to a datafile which should be checked

Description of the procedure:

Of description of the above procedure **F_T** and the *Maple* concepts of datafile types *TEXT* and *BINARY* follows, that they do not cover the text datafiles as a whole. In this connection, it is desirable to define the wider filetype embracing all datafiles, which would be correctly read in the *TEXT* mode both by the procedure ***readline*** and by the readbytes {by means of the procedure call ***readbytes*(F, *TEXT*, n)**}.

Of our considerations and experience of use of *Maple* for problem solving dealing with access to text datafiles we have defined a new filetype named as the *rlb*-type. Under the *rlb*-datafile, we understand such datafile that would be correctly read in the *TEXT* mode both by the procedure ***readline*** and by the ***readbytes***. Such datafile should not contain the especial symbols `hex(1A)` and `hex(0)`, which would split sense information defined by symbols from range `hex(32)..hex(255)`. In a sense the symbol `hex(1A)` can be considered as the factual end of a datafile of any of types *TEXT* or *BINARY*, whereas the symbol `hex(0)` is the factual end of a datafile read by the ***readline*** procedure, and defines the end of a record (*line*) for the ***readbytes*** procedure in the *TEXT* mode. An *rlb*-datafile is correctly read in the *TEXT* mode both by the ***readline*** procedure, and by the ***readbyte*** procedure. In many cases, it is very essential ability.

The procedure call **type(F, *rlb*)** returns the *true* value, if a datafile defined by the actual **F** argument is the *rlb*-datafile, otherwise the *false* value is returned. The procedure call closes an open datafile. In addition, it is necessary to have in mind, the procedures **Ftype** and **IsFtype** verifies the type of a datafile according to extension of its name only, what generally does not ensure its belonging to the *rlb*-datafiles. For these purposes the procedure **`type/rlb`** is intended. Thus, any *rlb*-datafile is the *txt*-datafile, whereas the opposite is incorrectly in general, namely. Any datafile can be renamed as *txt*-datafile and recognized by the procedures **Ftype** and **IsFtype** as the *txt*-datafile, being some other type.

```
`type/rlb`:= proc(F::file)
local h, t, k, s, omega;
  `if`(IsFempty(F), RETURN(true), [assign(omega = {seq(k, k = 0 .. 31)}), Close(F)]);
  do assign('h' = readbytes(F, 8192));
    if not (member(0, h, 't') or member(26, h, 't')) then if Fend(F) then break else next end if
    elif not belong({op(h[t + 1 .. -1])}, omega) then Close(F), RETURN(false)
    else s := 7; next
    end if;
    if not belong({op(h)}, omega) and s = 7 then Close(F), RETURN(false) end if
  end do;
   null(close(F)), true
end proc
```

Typical examples of the procedure use:

> map(type, ["c:/windows/setup.bmp","c:/program files/maple 8/bin.win/maplew8.exe"], rlb);

[false, false]

> map(type, ["c:/temp/info.txt", "c:/Academy/IAN/readme.htm", "c:/rans/lectures.rtf"], rlb);

[true, true, true]

`convert/rlb` – converting of an arbitrary datafile into appropriate *rlb*-file

Call format of the procedure:

convert(F, *rlb*)

Formal arguments of the procedure:

F – a datafile to be converted

Description of the procedure:

Of description of the above procedure **`type/rlb`** and the *Maple* concepts of file types *TEXT* and *BINARY* follows, that in a series of cases it is desirable to convert a file having a type different from *rlb*-type into an appropriate *rlb*-file which in the concept attitude is equivalent to the source file. The procedure **`convert/rlb`** solves the problem. The procedure call **convert(F, *rlb*)** provides a converting of a datafile defined by the actual **F** argument into *rlb*-file **cat(F, ".rlb")** with returning the *two*-element sequence whose the first element defines the file containing the converting result and the second element defines the filesize with output of warning about location of the created file. The created file in full volume can be read in the *TEXT* mode both by the ***readline*** procedure, and by the ***readbytes*** procedure.

A converting algorithm is based on substitutions of especial symbols `hex(0)`, `hex(0D)` and `hex(1A)` of a source file onto symbols `hex(7)`, `hex(0A)` and `hex(0A)` accordingly. If datafile **F** has the *rlb*-type the above converting is not done, however the procedure call returns the result in the above view. Thus, in contrast to the standard ASCII-files the files converted by the above manner are characterized by a missing of especial symbols `hex(1A)` and `hex(0)`. Meantime, relative to the ***readline*** procedure the converted file consists of lines of variable length and allows to read information contained in it directly in the *string*-format. In a series of cases, it is very useful property. In particular, at processing of executable program files.

```
`convert/rlb`:= proc(F::file)
local k, m, f, nu;
  `if`(type(F, rlb), RETURN(F, filepos(F, infinity), WARNING("file <%1> has the rlb-type", F),
     Close(F)), 7);
  assign67(m = ceil(1/2048*filepos(F, infinity)), nu = interface(warnlevel), f = cat("", F, ".rlb"),
     close(F)); interface(warnlevel = 0), MkDir(f, 1), interface(warnlevel = nu);
  for k to m do writeline(f, convert(subs(0 = 7, 13 = 10, 26 = 10, readbytes(F, 2048)), bytes))
  end do;
  f, filepos(f, infinity), close(f), WARNING("the converting result is in file <%1>", f)
end proc
```

Typical examples of the procedure use:

1. Converting of the *exe*-file WinZip32.exe into the appropriate *rlb*-file.

```
> convert("C:/Program Files/WinZip/WinZip32.exe", rlb);
   Warning, the converting result is in file <C:/Program Files/WinZip/WinZip32.exe.rlb>
                    "C:/Program Files/WinZip/WinZip32.exe.rlb", 1438148
```

2. Line-by-line read of the *rlb*-file WinZip32.exe.rlb with return of quantity of lines and the last line

```
> T:= 0: F:= "C:/Program Files/WinZip/WinZip32.exe.txt": do assign('h'=readline(F));
  if h=0 then close(F): break else assign('R'=h, 'T'=T+1) end if end do; T, R;
                                        12682,
  "NB10\a\a\a\a§ìý8□\a\a\aG:\\NMC\WZRel80\winzip\src\\NTSHARE\winzip32.pdb\a"
> convert("C:/RANS\\Academy/IAN\\ICCS_2004.txt", rlb);
   Warning, file <C:/RANS\Academy/IAN\ICCS_2004.txt> has the rlb-type
                        "C:/RANS\Academy/IAN\ICCS_2004.txt", 1556
```

```
> convert("C:/Program Files/Maple 8/Bin.W9X/Maplew8.exe", rlb);
    Warning, the converting result is in file <C:/Program Files/Maple
    8/Bin.W9X/Maplew8.exe.rlb>
                  "C:/Program Files/Maple 8/Bin.W9X/Maplew8.exe.rlb", 1984419
```

Ftype – check of type of a datafile (according to extension of its name)
IsFtype
IsFempty

Call format of the procedures:

Ftype(F)
IsFtype(F, T)
IsFempty(F)

Formal arguments of the procedures:

F - symbol or string defining a datafile or directory, or full path to them
T - symbol or string defining a datafile type (for example, `**doc**`, `.**doc**`, "**doc**" or "**.doc**")

Description of the procedures:

The procedure call **Ftype(F)** returns the type of a datafile defined by the actual **F** argument, otherwise a value {*FAIL* | *dir*} is returned. In addition, if the actual **F** argument defines a directory, then the *dir* value is returned; otherwise, the *FAIL* value is returned. The procedure as the `*type*` understands a type defined by extension of a filename; for instance, file "TRG.rtf" has ".rtf" type (or *rtf*-type) and file "RANS.html" has ".html" type (or *html*-type). If a filename has not any extension, the file's type is supposed as `*undefined*`.

The procedure call **IsFtype(F, T)** returns the *true* value, if a datafile defined by the first actual argument **F** has a type, defined by the second actual argument **T,** otherwise the *false* value is returned. The second argument **T** can be coded in one of the following formats `hhh`, `.hhh`, "hhh", ".hhh", `*undefined*` or "*undefined*", where hhh – extension of a filename. Both procedures are rather useful tools for check of types of datafiles of different nature.

In addition, it is necessary to have in mind, the procedures **Ftype** and **IsFtype** verifies the type of a file according to extension of its name only, what generally does not ensure its belonging to the *rlb*-files. For these purposes the procedure **`type/rlb`** is intended. Thus, any *rlb*-file is the *txt*-file, whereas the opposite is incorrectly in general, namely. Any datafile can be renamed as *txt*-file and recognized by the procedures **Ftype** and **IsFtype** as the *txt*-file, being some other type.

At last, the call **IsFempty(F)** returns the *true* value, if a file defined by the first actual argument **F** does exist and is *empty*, i.e. its length equals *zero*. If an existing file **F** is closed and its length is different from zero, then the *false* value is returned. Otherwise, the *FAIL* value is returned (if the **F** file is open) or erratic situation arises (if the **F** file does not exist).

```
Ftype:= proc(F::{string, symbol})
local a, b;
  `if`(type(F, dir), RETURN(dir), `if`(not type(F, file), RETURN(FAIL), assign(a = CFF(F))));
  assign(b = Search2(a[-1], {"."})), `if`(b = [], undefined, a[-1][b[-1] .. -1])
end proc

IsFtype:= proc(F::file, t::{string, symbol})
local h, p;
```

```
  assign(h = Ftype(F), p = cat("", t)), `if`(p[1] = ".", assign('p' = cat(``, p[2 .. -1])),
  assign('p' = cat(``, p))), `if`(h = cat(".", p), true, `if`(h = undefined and p = undefined, true, false))
end proc

IsFempty:= proc(F::{string, symbol})
local a, k, h;
  assign('h' = iolib(13)), assign(a = nops(h)), `if`(3 < a, seq(`if`(cat("", F) = h[k][2],
    RETURN(FAIL), NULL), k = 4 .. a), 7); `if`(type(F, file),
    `if`(iolib(6, F, infinity) = 0, true, false), ERROR("<%1> is not a file", F)), close(F)
end proc
```

Typical examples of the procedures use:

> map(Ftype, ["maple.ini", `ProcLib_6_7_8_9.mws`, "c:/Belarus", "Grodno", "c:/temp/avz", "c:/temp/vgtu.txt", "C:/Academy/Books/ReadMe.htm"]);

[".ini", ".mws", *dir*, *FAIL*, *undefined*, ".txt", ".htm"]

> IsFtype("C:/temp/vgtu.txt", `.txt`), IsFtype("C:/Academy/Books/readme.htm", "htm");

true, *true*

> map(IsFempty, ["C:/temp/testdata.dat", "C:/temp/vgtu.txt", "C:/Temp/AVZ.61"]);

[*false*, *true*, *true*]

F_atr – check-up and/or change of attributes of a datafile or directory
Atr
F_atr1
F_atr2

Call format of the procedures:

F_atr(F)
Atr(D)
F_atr1(F {, *atr*})
F_atr2(D {, *atr1*})

Formal arguments of the procedures:

F – string or symbol defining a filename or full path to a datafile
D – string or symbol defining a filename or full path to a datafile or directory
atr – (optional) new attributes for a datafile or directory **F**
atr1 – (optional) key symbol **`?`**, or new attributes for a datafile or directory **D**

Description of the procedures:

The procedure call **F_atr(F)** returns the *attributes* of a file defined by the actual **F** argument. The returned result is represented by the 2-element sequence whose the first element is list of the file attributes {*Read-only, Archive, Hidden, System*}, while the second element defines list with volume of the **F** file in bytes, with date and time of its creation accordingly. In addition, the first list can contain the message about absence for the file of attributes assigned to it, namely: *`file has not any attributes`*.

In contrast to the previous procedure, the procedure call **Atr(D)** returns the *attributes* of a file or directory defined by the actual **D** argument. The returned result is the list whose elements are {"A" – *archive*, "H" – *hidden*, "S" – *system* or "R" – *read-only*). At absence of any attributes the *empty* list is returned, i.e. []. In addition, the actual **D** argument must define the full path to a datafile or directory; otherwise, the erroneous situation arises.

The procedure call **F_atr1(F)** returns the *attributes* of a file defined by the actual **F** argument. The returned result is presented by the list whose elements define the attributes ascribed to the **F** file; one-symbol strings present the elements, namely: **"R"** - *read-only*, **"A"** - *archive*, **"H"** - *hidden*, and **"S"** - *system* file accordingly. At absence of the attributes for an **F** file, the *empty* list is returned, i.e. [].

If the optional actual ***atr*** argument has been coded, then successful procedure call **F_atr1(F, *atr*)** returns the *NULL* value, i.e. nothing, providing a setting for an **F** file of attributes defined by the actual ***atr*** argument. The second actual ***atr*** argument is coded by a list of values of view {**`+h`** | **"+h"**}, where **h** accepts a value {**R** | **A** | **H** | **S**}, while the sign {**+**} defines a setting of the **h**-attribute for a file defined by the first actual **F** argument. The attributes, which are defined by the second actual argument ***atr*** of the **F_atr1** procedure, are ascribed to the file **F** only. Thus, the second actual ***atr*** argument set a new file attributes, canceling the previous file attributes. For cancel of all file attributes the second actual argument ***atr*** is coded by the *empty* list, i.e. [].

At last, the **F_atr2(D)** procedure is intended for a reinstallation of attributes (**r** - only for reading**, a** - archive, **h** - hidden, and **s** - system) both for the concrete file, and for all files of a subdirectory (including subdirectory, all its files and all subdirectories with files contained in them) defined by the first actual **D** argument.

At absence of the second optional ***atr1*** argument, the canceling of all attributes for all directories, of subdirectories and files defined by the actual **D** argument is done. The second optional ***atr1*** argument is coded as a string or a symbol of view {` | "}{+ | -}**r** {+ | -}**a** {+ | -}**h** {+ | -}**s**{` | "}; in addition, the presence in front of the attribute indicator of a prefix {**+** | **-**} means {installation | cancel} of an attribute accordingly. A set of such settings is arbitrary; they are not doubled and are parted by *blanks*. In addition, the successful call of the **F_atr2** procedure returns the *NULL* value, i.e. nothing.

If the **F_atr2(D, *atr1*)** procedure has been applied to a subdirectory **D,** then all subdirectories which are located in it, and also files contained in them obtain identical settings of attributes, defined by the second actual argument ***atr1*.** At absence of the second argument ***atr1*** the canceling of all attributes for elements of all subtrees of subdirectories and files with a root directory defined by actual **D** argument is done. The settings of attributes are done both for closed files and for open files, excepting files shared at the moment by other *Windows*-applications. In addition, at running of the **F_atr2** procedure the MS DOS queries of the following view can arise:

> ***Sharing violation reading drive* <X>**
> **Abort, Retry, Fail ?**

where **X** - logical name of a drive whose file system contains path **D** to a file or a directory. Reply to this query by **{A | a}!** The given situation arises if the directory **D** or a file(s) located in it, are being shared at the given moment by *Windows*-applications (for example, a file of the directory is open in a current *Maple* session); otherwise, the above query does not arise. If the **`?`**-symbol has been coded as the second actual argument ***atr1*,** the procedure call **F_atr2(D, `?`)** returns the list of attributes ascribed to a directory or a file defined by the first actual **D** argument.

```
F_atr:= proc(F::{string, symbol})
local d, k, h, t, Art, Kr, V, G, w;
  assign(w = interface(warnlevel)), interface(warnlevel = 0);
  assign(Art =[], V =table([A = Archive, R = Read_only, S = System, H = Hidden]), Kr =Path(F)),
     `if`(not isFile(F), [interface(warnlevel = w), ERROR("file <%1> does not exist", F)],
  [assign(d = filepos(F, infinity)), close(F), assign(G = [seq(cat("$dir.$", k), k=[R,A,H,S,"-D"])])]);
  for k in G do system(cat("Dir ", CF(Kr, string), "  /A:", k[-`if`(k[-2] = "-", 2, 1) .. -1], "  >  ", k))
  end do;
  for k in G[1 .. -2] do `if`(searchtext(CFF(F)[-1], readbytes(k, TEXT, infinity)) <> 0,
```

```
      assign('Art' = [op(Art), V[convert(k[-1], symbol)]]), NULL), fremove(k) end do;
   do h := readline(G[-1]);
      if searchtext(CFF(F)[-1], h) <> 0 then t := h[1]; fremove(G[-1]); break
      end if
   end do;
   seq(assign('t' = cat(t, `if`(h[k .. k + 1] = "  ", ``, h[k + 1]))), k = 1 .. length(h) - 1);
   assign('t' = SLD(t, " ")[2 .. -2]), `if`(Art = [], [`file has not any attributes`], Art),
      [d, seq(`if`(Search2(t[k], {".", ":", "-"}) <> [], t[k], NULL), k = 1 .. nops(t))],
      interface(warnlevel = w)
end proc

Atr:= proc(D::{file, dir})
local a, b, c, t, p, k, f;
   assign(p = 97 .. 122, f = `if`(CF2(D)[-1] = "\\", CF2(D)[1 .. -2], CF2(D))),
      `if`(member(convert(f, bytes), {[k, 58, 92] $ (k = p), [k] $ (k = p), [k, 58] $ (k = p)}),
      `if`(member(f[1], map(CF2, {Adrive()})), RETURN([]),
      ERROR("drive <%1> is idle", Case(f[1], upper))), assign(b = "$ArtKr$", c = CFF(f)[-1]));
   system(cat("Attrib ", blank(f), " > ", b)), assign(a = CF2(readline(b))), fremove(b),
      ewsc(a, ["~"], [infinity], t);
   map(Case, convert(Red_n(sub_1([f = NULL, c = NULL], t), " ", 1), list), upper)
end proc

F_atr1:= proc(F::{string, symbol})
local Art, Kr, S, G, V, k, T, U, w;
   assign(w = interface(warnlevel)), interface(warnlevel = 0),
      `if`(isFile(F), 7, ERROR("<%1> is not a file", F));
   Art:= proc(F, r)
      local b, s, g, h, k, D, p, z;
         assign(s = "$Art.$Kr", g = (s -> cat(seq(`if`(s[k] = " ", "", s[k]), k = 1 .. length(s)))));
         assign(D = currentdir(), b = CFF(CF(F, string))), currentdir(`if`(nops(b) = 1, cat(b[1], "\\"),
            cat(seq(cat(b[k], "\\"), k = 1 .. nops(b) - 1))));
         system(cat("Attrib", r, blank(b[-1]), `if`(r = " ", cat(" > ", s), ""))), `if`(r <> " ", goto(E),
            assign(h = readbytes(s, TEXT, infinity))), fremove(s);
         assign(p = SLD(FNS(Sub_st(["  " = " "], h[1 .. searchtext(CF(F, string), h) - 1], z)[1],
            " ", 3), " ")), currentdir(D), interface(warnlevel = w), `if`(nops(p) = 1, RETURN([]),
            RETURN(convert(g(cat(seq(p[k], k = 1 .. nops(p) - 1))), list)));
         E; delf(s), currentdir(D)
      end proc;
   assign67(T = [`-R`, `-A`, `-H`, `-S`, `+R`, `+A`, `+H`, `+S`]),
      `if`(nargs = 0, RETURN(interface(warnlevel = w)), `if`(nargs = 1, RETURN(Art(F, " ")),
      `if`(nargs = 2 and type(args[2], list({string, symbol})) and
      belong({op(map(convert, args[2], symbol))}, T), assign(Kr = cat(seq(cat(" ", args[2][k], " "),
      k = 1 .. nops(args[2]))), U = cat(seq(cat(" ", T[k], " "), k = 1 .. 4))),
      ERROR("invalid values for attributes %1", args[2], interface(warnlevel = w)))));
   Art(F, U), Art(F, Kr), interface(warnlevel = w)
end proc

F_atr2:= proc(F::{name, string, symbol})
local h, k, t, G, L, w, p;
   assign(p = 97 .. 122), `if`(member(convert(CF2(F), bytes), {[k] $ k = p, [k, 58] $ k = p,
      [k, 58, 92] $ k = p}), `if`(member(CF2(F)[1], map(CF2, {Adrive()})), RETURN([]),
```

```
    ERROR("drive <%1> is idle", F)), assign67(w = interface(warnlevel)));
  `if`(not type(F, path), ERROR("<%1> is not a path", F), [assign(h = SUB_S(["/" = "\\"],
    cat("", F))), `if`(h[-1] = "\\", assign('h' = h[1 .. -2]), 7)]);
  `if`(nargs = 2 and args[2] = `?`, RETURN(Atr(h)), interface(warnlevel = 0)),
    `if`(nargs = 2, `if`(type(args[2], {string, symbol}) and belong({op(SLD(args[2], " "))},
    {"-s", "+r", "+a", "-h", "+s", "+h", "-r", "-a"}) and nops(SLD(args[2], " ")) =
    nops({op(SLD(args[2], " "))}), assign(G = args[2]), [interface(warnlevel = w),
    ERROR(`Invalid the 2rd argument: %1`, args[2])]), assign(G = "-r -a -h -s ")),
    `if`(member(DoF(F), {`file is open`, file}), assign(L = [h]), `if`(Search1(DoF(h), directory, 't')
    and t = [right], assign('L' = [h, op(DirF(h))]), RETURN(interface(warnlevel = w),
    `file or directory does not exist`)));
  for k in L do system(cat("Attrib ", G, " ", blank(k))) end do;
    interface(warnlevel = w)
end proc
```

Typical examples of the procedures use:

> map(F_atr, ["C:/AVZ/rans", "C:/AVZ/invoice2.doc", "C:/AVZ/iccs_2004.txt"]);

[[file has not any attributes], [2003, "10-14-03", "3:07p"], [Read_only, Archive], [30720, "05-28-03", "10:02p"], [Archive, Hidden, System], [1556, "09-27-03", "4:13p"]]

> map(F_atr1, ["C:/AVZ/rans", "C:/AVZ/invoice2.doc", "C:/AVZ/iccs_2004.txt"]);

[["A", "H"], ["A", "S", "R"], ["A", "S", "H", "R"]]

> F_atr1("C:/AVZ/rans", [`+S`,`+A`]), F_atr1("C:/AVZ/rans"), F_atr1("C:/AVZ/rans", [`+A`, "+S", "+H", "+R"]), F_atr1("C:/AVZ/rans");

["A", "S"], ["A", "S", "H", "R"]

> F_atr1("C:/AVZ/Invoice.4"), F_atr1("C:/AVZ/Invoice.4", ["+R", "+A", "+S"]), F_atr1("c:/AVZ/invoice.4"), F_atr1("c:/AVZ/invoice.4",[]), F_atr1("c:/AVZ/invoice.4");

["A", "S", "H", "R"], ["A", "S", "R"], []

> F_atr1("c:/ICCS_2004.txt"), F_atr1("c:/ICCS_2004.txt", []), F_atr1("c:/ICCS_2004.txt");

["A", "S", "H", "R"], []

> Atr("c:/Program Files\\Maple 8\\users/maple.ini");

["A"]

> Atr(`C:/Program Files\\aa bb cc.txt`), Atr("C:/aaa bbb");

["A", "S", "H"], []

> map(F_atr2, ["C:/Program Files/folder.htt", "C:/Program Files/", "C:/AVZ/"], `?`);

[["H"], ["R"], ["A", "H"]]

> F_atr2("C:/AVZ/", `?`), F_atr2("C:/AVZ\\", "+a -h"), F_atr2("C:/AVZ/", `?`);

["A", "H"], ["A"]

> F_atr2("C:/A V Z\\", `?`), F_atr2("C:/A V Z", "-s -a -h -r"), F_atr2("C:/A V Z", `?`);

["A", "S", "H", "R"], []

> F_atr2("C:/A V Z\\", `?`), F_atr2("C:/A V Z", "+s +r"), F_atr2("C:/A V Z", `?`);

[], ["S", "R"]

> F_atr2("C:/A V Z\\iccs_2004.txt", `?`), F_atr2("C:/A V Z/iccs_2004.txt", "-s -r +h +a"), F_atr2("C:/A V Z/ICCS_2004.txt", `?`);

["S", "R"], ["A", "H"]

> F_atr2("C:/AVZ/", `?`), F_atr2("C:/AVZ\\", "-a +h +s -r"), F_atr2("C:/AVZ/", `?`);

["A", "R"], ["S", "H"]

> map(F_atr2, ["C:/Program Files/folder.htt", "C:/Program Files/", "C:\\Program Files\\aa bb cc.txt"], `?`);

[["H"], ["R"], ["A", "S", "H"]]

> F_atr2("C:\\Program Files\\aa bb cc.txt"), F_atr2("C:\\Program Files\\aa bb cc.txt",`?`);

[]

> F_atr2("C:\\Program Files\\aa bb cc.txt", "+s +h"), F_atr2("C:\\Program Files\\aa bb cc.txt",`?`);

["S", "H"]

> map(F_atr2, [`C`, "C:", "C:\\", "C:/", `a`, `d`, `D:`, `D:/`], `?`);

[[], [], [], [], [], [], [], []]

> map(Atr, [`C`, "C:", "C:\\", "C:/", `a`, `d`, `D:`, `D:/`]);

[[], [], [], [], [], [], [], []]

> F_atr2("G:\\", `?`);

Error, (in **F_atr2**) drive <G> is idle

> Atr("G");

Error, invalid input: Atr expects its 1st argument, D, to be of type {file, dir}, but received G

The procedures **F_atr, F_atr1, Atr** and **F_atr2** seem useful tools at operation with file system of a computer, extending the standard *Maple* tools intended for providing of access to datafiles.

F_ren – renaming of a datafile with saving its attributes

Call format of the procedure:

F_ren(F1, F2)

Formal arguments of the procedure:

F1 – string or symbol defining a filename or full path to a datafile
F2 – a new name for the initial datafile **F1** (may be *string* or *symbol*)

Description of the procedure:

The procedure call **F_ren(F1, F2)** returns the *NULL* value, providing renaming of a file defined by the first actual **F1** argument, onto a new name defined by the second actual **F2** argument. The file renaming is fulfilled with saving of all attributes assigned to the source datafile. This procedure has more wide possibilities relative to the MS DOS command **ren**, providing renaming of datafiles with attribute *`hidden`*. Attempt to rename a directory causes the appropriate error.

```
F_ren:= proc(G::file, S::{string, symbol})
local b, c, D, T, W, k;
  assign(D = currentdir()), close(G), assign(T = F_atr1(G), b = CFF(G));
  c := `if`(nops(b) = 1, cat(b[1], "\\"), cat(seq(cat(b[k], "\\"), k = 1 .. nops(b) - 1)));
```

```
  currentdir(c), `if`(T = [], 1, assign(W = [seq(cat(`+`, T[k]), k = 1 .. nops(T))]));
  F_atr1(G, [`-R`, `-A`, `-H`, `-S`]), system(cat("Rename ", blank(b[-1]), " ", blank(S)));
  if T <> [] then F_atr1(cat(c, S), W); null(currentdir(D)) end if
end proc
```

Typical examples of the procedure use:

```
> F_atr2("C:/A V Z/rans ian.u", "+r +h +a +s"); F_atr2("C:/A V Z/rans ian.u", `?`),
  F_ren("C:/A V Z/rans ian.u", "A G N.u"), F_atr2("C:/A V Z/A G N.u", `?`);
                    ["A", "S", "H", "R"], ["A", "S", "H", "R"]
> F_ren("C:/Academy", Booklets);
   Error, invalid input: F_ren expects its 1st argument, G, to be of type file, but received
   C:/Academy
> Atr("C:/Program Files\\aa bb cc.txt");
                              ["S", "H"]
> F_ren("C:/Program Files\\aa bb cc.txt", `RANS_IAN RAC`);
> Atr(`C:/Program Files\\RANS_IAN RAC`);
                              ["S", "H"]
```

Dir_ren – renaming of a directory or a datafile

Call format of the procedure:

Dir_ren(F1, F2)

Formal arguments of the procedure:

F1 – a string or symbol defining full path to a directory or a datafile
F2 – a new name for the initial directory or datafile **F1**

Description of the procedure:

The successful procedure call **Dir_ren(F1, F2)** returns the *NULL* value, providing renaming of a directory or datafile defined by the first actual **F1** argument, onto a new name defined by the second actual **F2** argument. The renaming is fulfilled with saving of all attributes assigned to the initial directory or datafile **F1**. Furthermore, all elements of tree of directories and files with a root directory **F1** receive attributes ascribed to directory **F1.** This procedure has more wide possibilities relative to the MS DOS command **ren**, providing renaming of datafiles or directories with attribute `*hidden*`. Attempt to rename a directory containing the open datafiles or an open datafile causes the error with appropriate diagnostics.

However, it is necessary to note, that as a result of renaming of a directory the datafiles located in it receive attributes of the directory or lose their, if the initial directory has no attributes. Abnormal terminating of the procedure **Dir_ren** causes one of the following erroneous situations, namely:

`directory <%1> contains an open file`
`file <%1> is open`
`invalid input: %1 expects its %-2 argument, %3, to be of type %4, but received %5`

```
Dir_ren:= proc(S::path, N::{string, symbol})
local at, atr, d, h, p, k, omega, Sigma;
  assign67(p = iostatus(), interface(warnlevel = 0), omega = (() -> interface(warnlevel = 3)),
```

```
    Sigma = `if`(CF(S)[-1] = "\\", CF(S)[1 .. -2], CF(S))), `if`(type(S, dir) and 3 < nops(p) and
    intersection({seq(CF(p[k][2]), k = 4 .. nops(p))}, {op(map(CF, DirF(Sigma)))}) <> {},
    [omega(), ERROR(`directory <%1> contains an open file`, Sigma)], 7);
  assign(h = length(DoF(Sigma))), `if`(h = 12, [omega(), ERROR(`file <%1> is open`, Sigma)],
    `if`(h = 4, RETURN(null(omega(), F_ren(Sigma, N))), `if`(15 < h, [omega(),
    ERROR(`file or directory <%1> does not exist`, Sigma)], NULL)));
  assign(atr = F_atr2(Sigma, `?`)), F_atr2(Sigma), system(cat("Ren ", blank(CF(Sigma)), " ",
     blank(N)));
  at := `if`(atr = [], NULL, cat(seq(cat("+", Case(atr[k]), " "), k = 1 .. nops(atr))));
  assign(d = CFF(Sigma)), assign('d' = cat(seq(cat(d[k], "\\"), k = 1 .. nops(d) - 1), N)),
    `if`(atr = [], NULL, F_atr2(d, at))
end proc
```

Typical examples of the procedure use:

> Dir_ren("C:/Belarus/Grodno/Academy", RANS_IAN);

> Dir_ren("C:/Book/BBS_rus\\Orders/Invoice_Form", Maple_Books);

Error, (in **Dir_ren**) directory <c:/book/bbs_rus/orders/invoice_form> contains an open file

> Dir_ren("C:/Archive/ABS.rtf", "ABS147.rtf");

> Atr(`C:/Program Files\\Vilnius`); D_ren("C:/Program Files\\Vilnius",`Grodno Tallinn`);

["A", "H", "R"]

> Atr("C:/Program Files\\Grodno Tallinn");

["A", "H", "R"]

D_ren – renaming of a directory with saving its attributes

Call format of the procedure:

D_ren(F1, F2)

Formal arguments of the procedure:

F1 – string or symbol defining full path to a directory
F2 – a new name for the directory **F1** (may be *symbol* or *string*)

Description of the procedure:

The procedure call **D_ren(F1, F2)** returns the *NULL* value, providing renaming of a directory whose full path is defined by the first actual **F1** argument, onto a new name defined by the second actual **F2** argument. The directory renaming is fulfilled with saving of all attributes assigned to the source directory **F1**. Furthermore, all elements of tree of directories and files with a root directory **F1** receive attributes ascribed to directory **F1.** This procedure has more wide possibilities relative to the MS DOS command **ren**, providing renaming of directories with attribute *`hidden`*. Attempt to rename a file causes the appropriate error. It is necessary to note one important circumstance. As contrast to the procedure **Dir_ren**, the procedure **D_ren** allows to rename directories containing the open files, however this operation demands the high discretion.

In addition, at running of the **D_ren** procedure the MS DOS queries of the following view can arise:

***Sharing violation reading drive* <X>**
Abort, Retry, Fail ?

where **X** – logical name of a drive whose file system contains path **D** to a directory. Reply to this query by **{A | a}!** The given situation arises if the directory files are being shared at the moment by other *Windows*-applications (for example, a file of the directory is open in a current *Maple* session); otherwise the above query does not arise.

```
D_ren:= proc(G::dir, S::{string, symbol})
local h, a, omega, k;
  assign(a = F_atr2(G, `?`), h = CF(G)), F_atr2(G);
  `if`(a = [], 1, assign(omega = cat(seq(cat(`+`, Case(a[k], lower), ` `), k = 1 .. nops(a)))));
  system(cat("Rename ", blank(h), " ", blank(S)));
  `if`(a <> [], F_atr2(clsd([op(CFF(G)[1 .. -2]), S], "\\"), cat("", omega)[1 .. -2]), NULL)
end proc
```

Typical examples of the procedure use:

```
> F_atr2("C:/I A N/AVZ/AGN/K r", `?`), D_ren("C:/I A N/AVZ/AGN/K r", `A r t`),
  F_atr2("C:/I A N/AVZ/AGN/A r t", `?`);
                    ["A", "S", "H", "R"], ["A", "S", "H", "R"]
> F_atr2("C:/AVZ", `?`), D_ren("C:/AVZ", `A G N`), F_atr2("C:/A G N", `?`);
                            ["A", "S"], ["A", "S"]
> D_ren("C:/AVZ_AGN", `RANS_IAN`);
  Error, invalid input: D_ren expects its 1st argument, G, to be of type dir, but received
  C:/avz_agn
> Atr("C:/RANS_IAN RAC");
                                  ["A", "R"]
> D_ren("C:/RANS_IAN RAC", `Grodno Tallinn`); Atr("C:/Grodno Tallinn");
                                  ["A", "R"]
```

UbF – a special renaming of datafiles

Call format of the procedure:

UbF(F {, t {, h}})

Formal arguments of the procedure:

F - string or symbol defining a filename or full path to a datafile
t, h - (optional) strings or symbols of length **1**

Description of the procedure:

The procedure call **UbF(F)** returns the *NULL* value, providing renaming of a datafile defined by the first actual **F** argument, onto a new name, which is created by means of replacement in it of all entries of *blank* symbol onto symbol "_" (*underline*). At the procedure call **UbF(F, t, h)** with the optional second and third arguments **t** and **h** a new name of the **F** datafile will be formed by means of replacement of all entries of a symbol defined by the second actual **t** argument, onto a symbol defined by the third argument **h;** in addition, the procedure call **UbF(F, t)** is equivalent to the procedure call **UbF(F, t, " ").**

The given procedure provides a datafiles renaming which is not supported by the function ***rename*** (***ren***) of an underlying operating system. In particular, the procedure provides a renaming of datafiles with names containing *`blank`* symbols with saving of all attributes ascribed to them. The examples represented below illustrate standard appendices of the procedure for files renaming.

```
UbF:= proc(F::file)
local a, b, omega;
  assign(b = CFF(F)[-1]), `if`(nargs = 1 or 1 < nargs and search(b, args[2]), NULL,
    ERROR("<%1>-symbol does not belong to filename", args[2]));
  omega:= proc(s::{string, symbol}, h::{string, symbol}, z::{string, symbol})
    local a, b, c, k, n;
      `if`(search(s, h), assign(a = cat("", s), b = "", c = cat("", h), n = length(s)), RETURN(s));
      convert(cat(seq(`if`(a[k] = c, z, a[k]), k = 1 .. n)), whattype(s))
    end proc;
  assign(a = omega(b, `if`(nargs = 1, op([" ", "_"]), `if`(nargs = 2, op([args[2], " "]), args[2 .. 3])))),
    close(F), F_ren(F, a)
end proc
```

Typical examples of the procedure use:

```
> F_atr1("C:/A V Z/A G N.u"), UbF("C:/A V Z/A G N.u"), F_atr1("C:/A V Z/A_G_N.u");
                    ["A", "S", "H", "R"], ["A", "S", "H", "R"]
> UbF("C:/A V Z/A_G_N.u", "_", "$"), F_atr1("C:/A V Z/A$G$N.u");
                         ["A", "S", "H", "R"]
> UbF("C:/A V Z/A_G_N.u", "_", "$"), F_atr1("C:/A V Z/A$G$N.u");
> UbF("C:/A V Z/A$G$N.u", "$"), F_atr1("C:/A V Z/A G N.u");
                         ["A", "S", "H", "R"]
> F_atr1("C:/Prog\\RANS_IAN RAC"), UbF("C:/Prog\\RANS_IAN RAC"),
  F_atr1("C:/Prog\\RANS_IAN_RAC");
                         ["S", "H"], ["S", "H"]
> UbF("C:/Prog\\RANS_IAN_RAC","_","$"), F_atr1("C:/Prog\\RANS$IAN$RAC");
                              ["S", "H"]
> UbF("C:/A V Z/ICCS_2004.txt", "z", "y");
  Error, (in UbF) <z>-symbol does not belong to filename
```

IsOpen – testing of a datafile to be open
IsOpen1
IsOpen2
IsOpenF

Call format of the procedures:

IsOpen(F)
IsOpen1(F)
IsOpen2(F {, 'h'})
IsOpenF({F})

Formal arguments of the procedures:

F – a datafile assigned to testing
h – an assignable name

Description of the procedures:

During execution of procedures of access to datafiles, the erroneous and special situations are possible; many of them are initiated by absence of the required open datafile. For testing of a datafile to be open three procedures represented below can be rather useful at programming of access to datafiles.

The procedure call **IsOpen(F)** returns the *true* value, if a file defined by the actual **F** argument is open, otherwise the *false* value is returned. In addition, it is necessary to have in mind an important circumstance; an coding format in predetermined ***iostatus*** variable for an open file is completely adequate to coding at its opening. The given circumstance allows to open the same datafile in the different modes, providing some interesting modes of access to datafiles. Similar questions are discussed in our books [8-14,29-33,43,44]. In this connection, the procedure **IsOpen** is used if the above circumstance is essential enough for data processing.

If the above circumstance is inessential for data processing and it is desirable to receive information about a datafile irrelatively of format of coding at opening of the analyzed **F** file, then the **IsOpen1(F)** procedure is used. In all other respects, the **IsOpen1(F)** procedure is analogous to the above procedure **IsOpen(F).**

As against the previous procedures, the procedure **IsOpen2(F {, 'h'})** admits up to two actual arguments, where the first obligatory **F** argument defines a file assigned to testing, while through the second optional argument **h** an useful additional information about tested file is returned. If **F** file does exist and is closed then the *false* value is returned, while through the **h** argument can be received the *status table* whose indices are numbers of logical I/O channels with open **Fk** files for which take place the following relations **CF(F) = CF(Fk),** i.e. definitions of their paths are equivalent irrelatively of format of coding (for example, "c:/Temp/file.txt" and "C:/temp\\File.txt" are equivalent files). The entries of such table are lists whose elements define (**1**) number of logical I/O channel, (**2**) path to the open file, (**3**) a file mode (*READ* or *WRITE*), and (**4**) a filetype (*TEXT* or *BINARY*) accordingly.

If **F** file does exist and is open then the *true* value is returned, while through the **h** argument can be received the sequence whose the first element is number of logical I/O channel with file **F**, and the second element is the above *status table*. If the table is empty, then instead of it the *NULL* value is returned. If **F** datafile does not exist, the erratic situation arises with output of an appropriate diagnostics. The procedure provides as testing of a file to be open and presence of files open with other formats of coding.

The procedure call **IsOpenF(F)** returns the *status table* of an open file, given by a file qualifier that is defined by actual **F** argument. The returned *table* has a logical number of channel (on which has been opened the given file **F**) as an *index*, and 3-element sequence as an *entry;* elements of this sequence are (**1**) a file qualifier, (**2**) an access mode, and (**3**) a filetype accordingly.

In addition, if in a current session the open files do not exist, then the procedure call **IsOpenF(F)** returns the *false* value. If in a current session any open files do exist, except the given file **F,** the procedure call **IsOpenF(F)** returns the *false* value. If in a current session the open files do exist, the procedure call **IsOpenF()** returns the above *status table* for all open files with the difference that instead of 3-element sequences the **3**-element lists are returned as *entries*. In addition, if in a current session the open files do not exist, then the procedure call **IsOpenF()** returns the *false* value.

```
IsOpen:= proc(F::{string, symbol})
local a, k, omega;
  `if`(type(F, file), assign(a = iostatus(), omega = `if`(member(cat("", F)[-1], {"/", "\\"}),
```

```
  cat("", F)[1 .. -2], cat("", F))), ERROR("<%1> is not a file", F)), seq(`if`(omega = a[k][2],
  RETURN(true), NULL), k = 4 .. nops(a)), false
end proc

IsOpen1:= proc(F::{string, symbol})
local a, k, omega;
  `if`(type(F, file), assign(a = iostatus(), omega = `if`(member(cat("", F)[-1], {"/", "\\"}),
  cat("", F)[1 .. -2], cat("", F))), ERROR("<%1> is not a file", F)), seq(`if`(CF(omega) =
  CF(a[k][2]), RETURN(true), NULL), k = 4 .. nops(a)), false
end proc

IsOpen2:= proc(F::{string, symbol})
local a, h, k, T, n, err;
  assign(h = iostatus(), n = -1), assign(a = nops(h), T = TABLE([]),
    err = (() -> ERROR("file <%1> does not exist", F))),
    `if`(a = 3, `if`(type(F, file), RETURN(false), err()), 7);
  for k from 4 to a do
    if h[k][2] = cat("", F) then n := h[k][1]
    elif CF(h[k][2]) = CF(F) then T[h[k][1]] := [h[k][2], h[k][5], h[k][6]]
    end if
  end do;
  if nargs = 2 then assign67(args[2] = `if`(n = -1, NULL, n), `if`(T = TABLE([]), NULL, eval(T)))
  end if;
  unassign('READ', 'WRITE'), `if`(n = -1, `if`(type(F, file), false, err()), true)
end proc

IsOpenF:= proc()
local k, s, T, omega;
  assign(s = iostatus(), omega = `if`(nargs = 0, 1, `if`(member(cat("", args[1])[-1], {"/", "\\"}),
    cat("", args[1])[1 .. -2], cat("", args[1])))), `if`(nops(s) = 3, RETURN(false), `if`(nargs = 0
    [seq(assign('T'[s[k][1]] = [s[k][2], s[k][5], s[k][6]]), k = 4 .. nops(s)), RETURN(eval(T))], 1));
  for k from 4 to nops(s) do
    if searchtext(omega, s[k][2]) = 0 then next
    else T[s[k][1]] := substring(s[k][2], -length(omega) .. -1), s[k][5], s[k][6]
    end if
  end do;
  unassign('READ', 'WRITE'), `if`(type(eval(T), symbol), false, eval(T))
end proc
```

Typical examples of the procedures use:

```
> F:="C:/AVZ\\iccs_2004.txt/": F1:=`C:/temp/helpman.mws`: IsOpen(F1), open(F1,READ),
  IsOpen(F), IsOpen(F1); All_Close();
                              false, 0, false, true

> F:="C:/Temp\\HelpMan.mws": F1:=`C:/temp/helpman.mws`: IsOpen1(F1),
  open(F1, READ), IsOpen1(F), IsOpen1(F1); All_Close();
                              false, 0, true, true

> open(F1, WRITE), fopen(F, READ, TEXT): IsOpenF(F1), IsOpenF(); All_Close();
    table([0 = ("C:/temp/helpman.mws", WRITE, BINARY)]),
    table([0 = ["C:/temp/helpman.mws", WRITE, BINARY], 1 = ["C:/Temp\\HelpMan.mws",
    WRITE, TEXT], 2 = ["C:\\Temp\\HelpMan.mws", WRITE, TEXT]])
```

> fopen("C:/temp\\Helpman.mws", READ), open("c:/Temp/helpman.mws", WRITE):
fopen("C:\\Temp\\HelpMan.mws", READ), open("c:/temp/Helpman.mws", WRITE):
IsOpen2("C:\\Temp\\helpman.mws", 'h'), eval(h);

false, table([0 = ["C:/temp\\Helpman.mws", *WRITE, TEXT*], 1 = ["c:/Temp/helpman.mws", *WRITE, BINARY*], 2 = ["C:\\Temp\\HelpMan.mws", *WRITE, TEXT*], 3 = ["c:/temp/Helpman.mws", *WRITE, BINARY*]])

> IsOpen2("C:/temp\\Helpman.mws", READ), IsOpen2("c:/Temp/helpman.mws", WRITE);

true, true

> close(0,1,3); IsOpen2("c:\\Temp\\HelpMan.mws", 'h'), eval(h); IsOpen2("c:/academy/ian");

true, 2

Error, (in **err**) file <c:/academy/ian> does not exist

LnFile - check of number of I/O channel with open datafile

Call format of the procedure:

LnFile(F)

Formal arguments of the procedure:

F - a datafile assigned to testing (may be *symbol* or *string*)

Description of the procedure:

The procedure call **LnFile(F)** returns the 2-element list of view [*number of a logic channel, qualifier of a datafile*] for an open file defined by a qualifier **F** or a name. If a file has been open by the **open2** procedure considered below, then the sequence of lists of the above view is returned. At absence in a current session of open files the *false* value is returned, while in case of the closed file **F** the *empty* list is returned, i.e. []. The given procedure is especially useful at implicit opening of a file whose name is known in the program. In addition, the procedure returns all open datafiles whose names or qualifiers are identical without consideration of their registers.

```
LnFile:= proc(F::{string, symbol})
local a, b, c, h;
  h := `if`(member(cat("", F)[-1], {"/", "\\"}), cat("", F)[1 .. -2], cat("", F));
  `if`(type(h, file), op([assign(a = iostatus(), b = []), `if`(nops(a) = 3, RETURN(false),
    seq(assign('b' = [op(b), `if`(search(CF(a[k][2]), CF(h), 'c') and
    length(a[k][2]) - c + 1 = length(h), [a[k][1], a[k][2]], NULL)]), k = 4 .. nops(a))),
    `if`(Empty(b), [], op(b))]), ERROR("<%1> is not a file", h))
end proc
```

Typical examples of the procedure use:

> LnFile("c:/a v z/a g n.u"), fopen("c:/a v z/a g n.u", READ), open("C:/A V Z/A G N.U", READ), open("c:/A v z/A g n.u", READ), LnFile("c:/a v z/a g n.u");

false, 0, 1, 2, [0, "c:/a v z/a g n.u"], [1, "C:/A V Z/A G N.U"], [2, "c:/A v z/A g n.u"]

> LnFile("C:/A V Z/ICCS_2004.txt");

[]

> LnFile("C:/Academy/Books\\Examples/IAN.m");

Error, (in **LnFile**) <C:/Academy/Books\Examples/IAN.m> is not a file

Currentdir – an extending of the standard *Maple* procedure *`currentdir`*

Call format of the procedure:

Currentdir({F})

Formal arguments of the procedure:

F – (optional) a datafile, a directory or full path to them (may be *symbol* or *string*)

Description of the procedure:

The procedure call **Currentdir(F)** provides setting of the last subdirectory of a path defined by the actual **F** argument as a current directory with returning the previous current directory. The **F** argument can define a file, a directory or full path to them. If actual **F** argument does not define a path, the procedure creates a new path on the basis of **F** and does it as a current directory. For instance, the procedure call **Currentdir("c:/abc\\user.h")** creates the directory "c:\\abc\\user.h" as a current directory. At last, the procedure call **Currentdir()** restores the initial current package directory, i.e. the current directory which has been set after the *Maple* initialization. The **Currentdir** procedure appreciably extends the standard procedure ***currentdir***, simplifying programming a series of problems of operation with file system of a computer. The examples presented below illustrate the **Currentdir** usage.

```
Currentdir := proc()
local a, b;
  null(Release('a')), `if`(nargs = 0, RETURN(iolib(27, a)), `if`(map(type, {args},
  {string,symbol}) = {true}, assign(b = CF(cat("", args[1]))),
  ERROR("arguments %1 should be strings or symbols", [args]))),
  `if`(1 < nargs, ERROR("%1 quantity of actual arguments more than 1", [args]),
  `if`(type(args, dir), iolib(27, args), `if`(type(args, file), iolib(27, clsd(CFF(args)[1 .. -2], "\\")),
  op([null(MkDir(args)), iolib(27, args)]))))
end proc
```

Typical examples of the procedure use:

```
> Currentdir("C:/A V Z\\V_G_S/"), currentdir(), Currentdir(), currentdir();
   "C:\Program Files\Maple 9\bin.win", "C:\A V Z\V_G_S", "C:\A V Z\V_G_S",
   "C:\Program Files\Maple 9"
> Currentdir(`Kristo.7\\`), currentdir(), Currentdir(), currentdir();
   "C:\Program Files\Maple 9", "C:\Program Files\\Maple 9\Kristo.7",
   "C:\Program Files\Maple 9\Kristo.7", "C:\Program Files\\Maple 9"
> Currentdir(avz, agn);
   Error, (in Currentdir) [avz, agn] quantity of actual arguments more than 1
> Currentdir(2003);
   Error, (in Currentdir) arguments [2003] should be strings or symbols
> Currentdir("C:/AVZ/iccs_2004.txt"), currentdir(), Currentdir(), currentdir();
   "C:\Program Files\Maple 9", "c:\avz", "c:\avz", "C:\Program Files\Maple 9"
```

---

open2 – opening of an existing datafile on two logical I/O channels

Call format of the procedure:

open2(F, L, P, N {, n {, h, p}})

Formal arguments of the procedure:

F - a datafile assigned to opening (may be *symbol* or *string*)
L - an assignable name
P - an assignable name
N - an assignable name
n - (optional) an access mode in logical I/O channel **L** {by default *READ, WRITE*}
h - (optional) an access mode in logical I/O channel **P** {by default *APPEND, READ, WRITE*}
p - (optional) a filetype in logical I/O channel **P** {by default *TEXT, BINARY*}

Description of the procedure:

Of our experience of the work with the package one conclusion, important for the appendices follows. In the package, it is impossible to open a file twice with the same qualifier, whereas by a variation of formats of its qualifier it can be made, what in many cases of operation with external datafiles can be useful enough. In particular, the variation of the qualifier can be obtained by simple replacement of any lowercase letter onto the uppercase letter, and vice versa. The given approach has been implemented in the procedure **open2(F, L, P, N)** intended for a file opening defined by **F** qualifier of an existing file on two logical channels with return via variables **L, P** and **N** of the numbers of logical channels and file size in bytes accordingly. The source file is opened by means of the standard function *`open`* in logical channel **L** and by the *fopen* function in logical channel **P.** The successful procedure call **open2(F, L, P, N)** returns the *NULL* value, i.e. nothing.

If the fifth optional **n** argument is coded, it defines a mode of access in logical **L** channel; at presence of optional arguments **h** and **p** they define the mode of access and a type of file in the logical **P** channel accordingly, otherwise the values by default for the indicated parameters are accepted. In addition, use of the **open2** procedure supposes certain circumspection, caused by the artificial reception described above, for differentiation of the same file at its opening on various logical channels. For example, the described reception of opening of the same file on several logical channels provides an opportunity of a work with the file in different modes; however, its use supposes certain circumspection.

```
open2:= proc(F::{string, symbol}, L::evaln, P::evaln, N::evaln)
local h, n, J;
  `if`(type(F, file), assign(h = CF(F)), ERROR("<%1> is not a file", F));
  assign(n = op(convert(substring(h, 1), bytes)), N = filepos(h, infinity));
  `if`(65 <= n and n <= 90, assign(J = cat(convert([n + 32], bytes), substring(h, 2 .. length(h)))),
     assign(J = cat(convert([n - 32], bytes), substring(h, 2 .. length(h)))));
  op([close(h), assign(L = open(h, `if`(4 < nargs, args[5], READ)),
     P = fopen(J, `if`(5 < nargs, args[6], READ), `if`(nargs = 7, args[7], BINARY)))])
end proc
```

Typical examples of the procedure use:

> open2("C:/AVZ\\Lectures.rtf/", L, P, N, WRITE, APPEND, TEXT): iostatus(), L, P, N;

[2, 0, 7, [0, "c:\avz\lectures.rtf", *RAW*, FD = 7, *WRITE, BINARY*],
[1, "C:\avz\lectures.rtf", *STREAM*, FP = 2013506696, *READ, TEXT*]], 0, 1, 0

> IsOpenF(), writebytes(0, [61]), readline(1);

table([0=["c:\avz\lectures.rtf", *WRITE, BINARY*], 1=["C:\avz\lectures.rtf", *READ, TEXT*]]), 1, "="

> open2("C:/Academy/Books/Maple9.txt", L, P, N, READ, APPEND, TEXT);

Error, (in **open2**) <C:/Academy/Books/Maple9.txt> is not a file

The reception of opening of the same file on several logical channels, which is supported by the **open2** procedure, has been used in a number of applied problems of data processing, providing possibility of use of effective enough algorithms of processing of datafiles of different purpose.

OpenLN – opening of a datafile on the given logical I/O channel

Call format of the procedure:

OpenLN(N, O, F, *parms*)

Formal arguments of the procedure:

N – an integer defining the number of logical I/O channel; **0 <= N <= 6**
O – an opening mode; should be one of the set {*open, fopen*}
F – string or symbol defining a filename or full path to it
parms – a sequence of actual arguments which should be passed to the opened **F** file

Description of the procedure:

Of our experience of operation with the package one conclusion, important for the appendices follows. In the *Maple* environment it is impossible to open a file directly on the given logical **I/O** channel, while in a number of cases of data processing such possibility appears rather useful in the different problems of access to datafiles.

The given problem is solved by the procedure **OpenLN(N, O, F, *parms*)** intended for a file opening defined by the actual **F** argument at an opening mode **O** on logical I/O channel defined by the first actual **N** argument with passing to the opened **F** file of the actual arguments defined by the fourth actual argument ***parms***. Successful procedure call **OpenLN(N, O, F, *parms*)** returns the **2**-element sequence whose the first element defines the number of a logical I/O channel, i.e. **N** value, while the second – a set of channels numbers remaining open. The fourth ***parms*** argument defines the remaining actual arguments depending on a built-in opening function defined by the **O** argument with admitted values {***open, fopen***}. The procedure **OpenLN** provides a program-controlled handling of the basic especial and erratic situations arising at files opening. This procedure appears as a rather useful tool in datafiles handling when the necessity of restoration of status of an open datafile after its closing is demanded. The given procedure has been successfully approved on the examples of restoring a status of datafiles after their abnormal closing.

```
OpenLN:= proc(N::digit, O::symbol, F::{string, symbol})
local a, k, h, p, L, t, K;
  assign(K = cat("", F)), `if`(6 < N, ERROR("the first argument should be less than 6, but has
    received <%1>", N), `if`(member(O, {open, fopen}), assign(h = iostatus(), L = {}, a =
  DoF1(K)), ERROR("mode of opening should be {open, fopen}, but had received <%1>", O)));
  `if`(a = 2, ERROR("file <%1> is already open", F), `if`(member(a, {3, 4}),
    ERROR("<%1> is a directory", K), 1));
  if nargs = 3 then L := {k $ (k = 0 .. 7)}
  else seq(assign('L' = {op(L), h[k][1]}), k = 4 .. nops(h)), assign(p = {k $ (k = 0 .. 7)} minus L)
  end if;
  `if`(member(N, L), ERROR("logical I/O channel <%1> is busy", N), NULL);
  member(N, p, 't'), seq(open(cat("Art_Kr.$", k), WRITE), k = 1 .. t - 1);
  try O(K, seq(args[k], k = 4 .. nargs)), p minus {N}, seq(fremove(cat("Art_Kr.$", k)), k = 1..t-1)
  catch "too many files already open": seq(fremove(cat("Art_Kr.$", k)), k = 1 .. t - 1)
  end try
end proc
```

Typical examples of the procedure use:

> **All_Close(); OpenLN(6, open, "C:/I A N/rans.txt", WRITE), iostatus();**

6, {0, 1, 2, 3, 4, 5, 7}, [1, 0, 7, [6, "C:/I A N/rans.txt", *RAW*, *FD* = 13, *WRITE*, *BINARY*]]

> **OpenLN(3, open, "C:/Academy/helpman.mws", READ), iostatus();**

3, {0, 1, 2, 4, 5, 7}, [2, 0, 7, [**3**, "C:/Academy/helpman.mws", *RAW*, *FD*=10, *READ*, *BINARY*],
[6, "C:/I A N/rans.txt", *RAW*, *FD* = 13, *WRITE*, *BINARY*]]

> **OpenLN(1, fopen, "C:/Academy/Books\\MPL9Book\\ReadMe.htm", WRITE, TEXT);**

1, {0, 2, 4, 5, 7}

> **IsOpenF();**

table([1 = ["C:/Academy/Books\\MPL9Book\\ReadMe.htm", *WRITE*, *TEXT*],
3 = ["C:/Academy/helpman.mws", *READ*, *BINARY*],
6 = ["C:/I A N/rans.txt", *WRITE*, *BINARY*]])

> **OpenLN(6, open, "C:/I A N/rans.txt", WRITE);**

Error, (in **OpenLN**) file <C:/I A N/rans.txt> is already open

---

Fopen – opening of files – an extension of *Maple* procedures *`fopen`* and *`open`*
Open

Call format of the procedures:

Fopen(F, m {, t})
Open(F, p)

Formal arguments of the procedures:

F - a symbol or string defining a datafile or full path to a datafile to be opened
m - an opening mode; may be: *READ*, *WRITE*, or *APPEND*
p - an opening mode; may be: *READ* or *WRITE*
t - (optional) a datafile type; may be: *TEXT* or *BINARY*

Description of the procedures:

The standard *Maple* procedures *`fopen`* and *`open`* serves for *buffered* and *unbuffered* reading or writing of a datafile accordingly, by returning a *file descriptor* - integer from the range **0..7**. However, in a case of a non-existing file in the opening mode *READ* or in the opening mode {*WRITE* | *APPEND*} when *Maple* cannot create path to datafile the erratic situation arises with diagnostics "file or directory does not exist". In many problems, dealing with processing of datafiles it is very unwanted phenomenon. The **Fopen** and **Open** procedures extending the standard procedures *`fopen`* and *`open`* accordingly provide the error-free opening of datafiles.

The call coding format of the procedures **Fopen** and **Open** is fully analogous to the standard procedures *`fopen`* and *`open`* accordingly, likewise as the result returned by them. However, an essential difference takes place, namely: in any case, at appearance of the above erratic situation a datafile that must be opened is created with *null* length (*empty* datafile) and is opened in a specified opening mode with return of a file descriptor. In addition, the calls of procedures can output a warning that informs about full path to an open datafile. The warning appears only if the procedure creates a datafile with full path different from a specified at the procedure call. These procedures appear as rather useful tools when error-free datafiles handling is demanded. Our experience confirms a rather high efficiency of the procedures use at implementation of means dealing with datafiles processing.

```
Fopen:= proc(F::{string, symbol}, t)
  try fopen(args)
  catch "file or directory does not exist": fopen(pathtf(F), args[2 .. -1])
  end try
end proc

Open:= proc(F::{string, symbol}, t)
  try open(args)
  catch "file or directory does not exist": open(pathtf(F), args[2 .. -1])
  end try
end proc
```

Typical examples of the procedures use:

```
> fopen("C:/RANS/IAN/Vgtu\\ABS.txt", READ, BINARY);
  Error, (in fopen) file or directory does not exist
> open("C:/RANS/IAN/Vgtu\\Book.txt", WRITE);
  Error, (in open) file or directory does not exist
> Fopen("C:/RANS/IAN/Vgtu\\ABS.txt", READ, BINARY);
  Warning, target path <c:\rans\_ian\vgtu\abs.txt> has been activated instead of the required
  path <c:\rans\ian\vgtu\abs.txt>
                                          0
> Open("C:/RANS/IAN/Vgtu\\Book.txt", WRITE);
  Warning, target path <c:\rans\_ian\vgtu\book.txt> has been activated instead of the required
  path <c:\rans\ian\vgtu\book.txt>
                                          1
> Fopen("C:/Academy/Examples/Grsu.txt", WRITE, TEXT),
  Open("C:/Academy/Examples/TRU.txt", WRITE);
                                         2, 3
> iostatus(); Close();
  [4, 0, 7, [0, "c:\rans\_ian\vgtu\abs.txt", STREAM, FP = 2013506696, WRITE, TEXT],
  [1, "c:\rans\_ian\vgtu\book.txt", RAW, FD = 8, WRITE, BINARY],
  [2, "C:/Academy/Examples/Grsu.txt", STREAM, FP = 2013506728, WRITE, TEXT],
  [3, "C:/Academy/Examples/TRU.txt", RAW, FD = 10, WRITE, BINARY]]
```

MSDcom – check of availability of the external MS DOS commands
isMSDcom

Call format of the procedures:

MSDcom()
isMSDcom({n})

Formal arguments of the procedures:

n - (optional) a sequence of symbols defining names of external MS DOS commands

Description of the procedures:

The built-in functions ***`system`*** and ***ssystem*** provide an opportunity of execution of commands of an operating system (MS DOS, UNIX, etc.). At the same time, the operating system MS DOS has commands of two levels: internal (for example, ***copy, dir, del*** etc.) and external (for example, ***attrib, deltree*** etc.). The internal commands are located in the MS DOS kernel and are accessible at any moment. While the files with external commands are located in the system subdirectory and the information about their set represents undoubted interest. The given problem is solved by means of two procedures represented below, namely **MSDcom** and **isMSDcom.**

The call **MSDcom()** returns the list of names of the executable files containing external commands of MS DOS. At absence of such files the *empty* list is returned, i.e. []. The procedure **isMSDcom** has the more extended possibilities, namely. The call **isMSDcom()** is equivalent to the call **MSDcom().** While the call **isMSDcom(n)** returns the *true* value, if the system subdirectory contains external MS DOS command with the given name **n;** otherwise the *false* value is returned. At last, the call **isMSDcom(n1,n2,n3, ..., np)** returns generically the 4-element sequence of the following view:

true, **[x1, x2, x3, ...,xa],** ***false***, **[y1, y2, y3, ..., yb],** where **xk, yj** belong to the set
{n1, n2, n3, ..., np} k=1 .. a, j=1 .. b; a + b <= p

Actual arguments should have the *symbol* type, defining names of external MS DOS commands; otherwise, the appropriate warnings are output. A pair {*true*, [...] | *false*, [...]} defines the names list of {existent | missing} external MS DOS commands accordingly; in addition, the *empty* list suppresses return of an appropriate pair. The above procedures seem rather useful at use of the built-in functions ***`system`*** and ***`ssystem`*** of *Maple.*

```
MSDcom:= proc()
local h, n, t, L, Q;
  assign(Q = WS()), system(cat("Dir ", Q, "\\*.* > $Art_Kr")), assign(L = []);
  do assign('h' = readline("$Art_Kr"));
    if h = 0 then break elif h = "" then next end if;
    assign('n' = Search(h, " ")[-1] + 1), `if`(map2(search, h[n .. -1],
      {".COM", ".exe", ".EXE", ".com"}, 't') <> {false},
      assign('L' = [op(L), cat(``, h[n .. -1][1 .. t - 1])]), 1)
  end do;
  fremove("$Art_Kr"), map(Case, sort(L))
end proc

isMSDcom:= proc()
local a, b, c, k;
  assign67(a = {op(MSDcom())}, b = [], c = []), `if`(nargs = 0, RETURN(MSDcom()),
    `if`(nargs = 1, `if`(type(args, symbol), RETURN(member(Case(args), a)),
    ERROR("actual argument has a type different from symbol")), 7));
  for k to nargs do
    if not type(args[k], symbol) then
      WARNING("argument <%1> is different from symbol", args[k])
    elif member(Case(args[k]), a) then b := [op(b), args[k]] else c := [op(c), args[k]] end if
  end do;
  if b = [] then false elif c = [] then true else true, b, false, c end if
end proc
```

Typical examples of the procedures use:

> MSDcom(); **# for *Windows 98 SE***

[arj32, attrib, chkdsk, choice, cvt, debug, deltree, diskcopy, doskey, edit, extract, fc, fdisk, find, format, iextract, keyb, label, mem, mode, more, move, mscdex, nlsfunc, scandisk, scanreg, sort, start, subst, sulfnbk, sys, xcopy, xcopy32, cscript]

> isMSDcom();

[arj32, attrib, chkdsk, choice, cvt, debug, deltree, diskcopy, doskey, edit, extract, fc, fdisk, find, format, iextract, keyb, label, mem, mode, more, move, mscdex, nlsfunc, scandisk, scanreg, sort, start, subst, sulfnbk, sys, xcopy, xcopy32, cscript]

> isMSDcom(Attrib), isMSDcom(Salcombe); isMSDcom("RANS");

true, false

Error, (in **isMSDcom**) actual argument has a type different from symbol

> isMSDcom(Attrib, Find, DelTree, Arj32), isMSDcom(Salcombe, rans, IAN);

true, false

> isMSDcom(Salcombe, agn, "rans", "IAN", Attrib, Find, DelTree, Arj32);

Warning, argument <rans> is different from symbol
Warning, argument <IAN> is different from symbol

true, [Attrib, Find, DelTree, Arj32], false, [Salcombe, agn]

> isMSDcom(Attrib, Find, DelTree, Arj32, Salcombe, rans, Format, Unerase, Scanreg, IAN);

true, [Attrib, Find, DelTree, Arj32, Format, Scanreg], false, [Salcombe, rans, Unerase, IAN]

MkDir – creation of a directory of any level of nesting and/or a datafile

Call format of the procedure:

MkDir(*dirName* {, 1})

Formal arguments of the procedure:

dirName – a name, a chain of directories or a path to create (may be *symbol* or *string*)
1 – (optional) indicator of a datafile creation

Description of the procedure:

The **MkDir** procedure creates a directory, a directories chain and/or a file in file system of the underlying operating system. The ***dirName*** argument that should be the *Maple* string or symbol specifies a directories chain that should be created. The procedure admits a coding of the actual ***dirName*** argument by the string or symbol of the following view:

<disk unit:\\Directory/subdirectory_1\\.../subdirectory_n>

The set of characters that are permitted in the directories names is system-dependent. Likewise, a character, used to separate the components of a directories chain is system-dependent. If the *backslash* character appears in the string, it must be doubled up, because the *Maple* strings use the *backslash* character as *escape* character. For separators the symbols {"\\", "/"} can be used; however they are not equivalent, generally. So, usage of "/"-separator in the procedures ***`system`*** or ***ssystem*** causes the error as a rule.

If the at procedure call **MkDir(*dirName*, 1)** the second actual argument **1** has been coded, then the procedure considers the last element of a chain defined by the first actual ***dirName*** argument as a file which is created as the *empty* closed datafile. The created file is located in the last subdirectory of the created directories chain. For example, the procedure call

MkDir("C:/Temp/Dir/Test147.txt", 1)

creates both the directories chain "C:/Temp/Dir" (*within missing components*) and the empty closed datafile "C:/Temp/Dir/Test147.txt".

It is necessary to note the following circumstance. The file conception of operating MS DOS system does not distinguish a file itself and a directory. Therefore, even the MS DOS command ***mkdir*** cannot create subdirectory in a directory containing file of the same name. For elimination of the given lack, the procedure **MkDir** uses an artificial reception consisting in the following. If at creation of the required subdirectory, the procedure finds out presence of file of the same name in the catalogue of previous level the created subdirectory receives the new name formed of the given name by means of adding to it of prefix "_". Analogous situation takes place if creation of a required file **MkDir** finds out presence of subdirectory of the same name; about that, the appropriate warnings are output. In general case this warning has the following view "*Path element* <%1> *has been found, it has been renamed on* <%2>".

A successful call of the **MkDir(F)** procedure keeps a current directory and returns the full path to the required directory or file given by the actual **F** argument. The procedure call **MkDir({`` | ""})** returns the *NULL* value only. In addition, if the required path is absent, it is being created with output of an appropriate warning of the above view if it is necessary. An unsuccessful call causes an exception. In contrast to the standard procedure ***mkdir*** the successful call **MkDir** always returns the required full path. The returned path has the standard *Maple* notation in lowercase. Such approach allows to simplify programming of many problems dealing with access to datafiles. The **MkDir** procedure essentially extends opportunities of the built-in ***mkdir*** function that allows to create the directories of one level of nesting only. The procedure in a lot cases allows to simplify programming of problems of processing of elements of file system of a computer and problems dealing with access to datafiles of different nature.

```
MkDir:= proc(F::{string, symbol})
local cd, r, k, h, z, K, L, Lambda, t, d, omega, omega1, u, f, s, g, v;
   s := "Path element <%1> has been found, it has been renamed on <%2>";
   `if`(Empty(F), RETURN(), `if`(type(F, dir) and nargs = 1, RETURN(CF(F)), `if`(type(F, file)
      and nargs = 2, RETURN(CF(F)), assign(cd = iolib(27)))));
   omega1, u, K := () -> ERROR("<%1> is invalid path to a directory or a file", F),
      interface(warnlevel), CF(F);
   assign(Lambda = (x -> close(open(x, WRITE)))), assign(L = CFF(K), omega = 0);
   assign(r = cat(L[1], "\\")), `if`(nargs = 2 and args[2] = 1, assign('L' = L[1 .. -2],
      'omega' = L[-1], t = 1), 1);
   if L = [] then assign(g = cat(r, r[1]), v = cat(r, "_", r[1]));
      `if`(t <> 1, RETURN(r), `if`(type(g, dir), RETURN(null(Lambda(v)), v), `if`(type(g, file),
         RETURN(g), RETURN(null(Lambda(g)), g))))
   end if;
   for k from 2 to nops(L) do iolib(27, r), assign('g' = cat(r, L[k]));
      if type(g, dir) then try iolib(29, g) catch "": NULL end try
      elif type(g, file) then assign('L'[k] = cat("_", L[k]), 'd' = 7);
         try iolib(29, cat(r, L[k]))
         catch "file I/O error": d := 7
         catch "directory exists and is not empty": d := 7
         end try;
```

```
      assign('d' = 7), WARNING(s, L[k][2 .. -1], L[k])
    else iolib(29, cat(r, L[k]))
    end if;
    assign('r' = cat(r, L[k], "\\"))
  end do;
  if t = 1 then `if`(type(cat(r, omega), dir), op([Lambda(cat(r, "_", omega)),
    WARNING(s, omega, cat("_", omega))]), RETURN(`if`(d = 7, cat(r, omega), CF(F)),
    Lambda(cat(r, omega)), null(iolib(27, cd)))), cat(r, "_", omega), null(iolib(27, cd))
  else null(iolib(27, cd)), `if`(d = 7, r[1 .. -2], CF(F))
  end if
end proc
```

Typical examples of the procedure use:

```
> MkDir("C:/Temp/rans/ian\\VGTU/AGN\\Art/Kr\\Svet/Arne");
                "c:\temp\rans\ian\vgtu\agn\art\kr\svet\arne"
> MkDir("C:/Temp/rans/ian\\VTU/AGN\\Art/Kr\\Svet/Arne\\RANS.ian/", 1);
                "c:\temp\rans\ian\vtu\agn\art\kr\svet\arne\rans.ian"
> MkDir("C:/Temp/rans/ian\\VTU/AGN\\Art/Kr\\Svet/Arne\\RANS.ian/");
  Warning, Path element <rans.ian> has been found, it has been renamed on _rans.ian
                "c:\temp\rans\ian\vtu\agn\art\kr\svet\arne\_rans.ian"
> MkDir("C:/temp/rans/ian/vtu/agn/art/kr/svet/arne/_rans.ian/Art.Kr", 1);
                "c:\temp\rans\ian\vtu\agn\art\kr\svet\arne\_rans.ian\art.kr"
> type("C:/temp/rans/ian/vtu/agn/art/kr/svet/arne/_rans.ian/Art.Kr", file), iostatus();
                true, [0, 0, 7]
> MkDir("c:\\temp\\avz\\ASV\\art\\kristo\\svet\\aaa"); type(%, dir);
  Warning, Path element <avz> has been found, it has been renamed on <_asv>
                "c:\temp\_avz\asv\art\kristo\svet\aaa"
                true
> MkDir("C:\\temp\\avz\\ASV\\art\\kristo\\svet\\Order.txt",1); type(%, file);
  Warning, Path element <avz> has been found, it has been renamed on <_asv>
                "c:\temp\_avz\asv\art\kristo\svet\order.txt"
                true
> MkDir("C:"), MkDir("C:/"), MkDir("C:/Temp\\"), MkDir("A:/Academy/Rans_IAN", 1);
                "c:\", "c:\", "c:\temp", "a:\academy\rans_ian"
> MkDir("C:/Temp/Temp/Temp\\Temp/Temp"), map(MkDir, [``, ""]);
                "c:\temp\temp\temp\temp\temp", []
> MkDir("c:/Temp/Temp/Temp\\Temp/Temp", 1);
  Warning, Path element <temp> has been found, it has been renamed on <_temp>
                "c:\temp\temp\temp\temp\_temp"
```

However, in spite of the marked advantages of the procedure **MkDir**, we recommend to avoid duplication of names of files and subdirectories in the same catalogue; that will provide more simple algorithms of processing of elements of file system of a computer and handling of erratic situations arising at access to datafiles of different nature. The procedure essentially simplifies programming of tasks dealing with access to datafiles of different purposes and nature.

pathtf – creation of an achievable existing target datafile

Call format of the procedure:

pathtf(F)

Formal arguments of the procedure:

F - symbol or string defining name or path to a target datafile

Description of the procedure:

In many of the problems dealing with access to datafiles, the question of identification of an *achievable* target datafile is topical enough. In a mode of automatic running of a program, it is important to provide existence of a target (accepting) datafile whose absence can be caused by a number of reasons, namely: a mistake of path indication to the datafile, presence in an accepting directory of subdirectory with the same name, etc. Similar situations cause mistakes stopping a process of calculations. For maintenance of automatic processing of similar situations, the procedure **pathtf** can successfully be used.

The procedure call **pathtf(F)** returns the achievable path to an accepting datafile whose path or name is determined by the actual **F** argument. The successful call of the procedure returns the *name* (for a current directory) or *full path* to a required target file. At absence of path to such file, the name or full path to its analog (*achievable empty closed file*) is returned with output of an appropriate warning. The principle of creation of missing path elements to a target file is described in the previous procedure **MkDir.** The procedure **pathtf** is widely used in a series of other procedures dealing with processing of access to datafiles of different types and nature.

```
pathtf:= proc(F::{string, symbol})
local h, t, r, omega;
  `if`(type(eval(F), {string, symbol}), [assign(h = CF(F), omega = interface(warnlevel)),
     interface(warnlevel = 0)], ERROR("argument should be a string or symbol but has received
     <%1>", whattype(eval(F))));
  try  assign67(r = `if`(type(h, file), RETURN(h, interface(warnlevel = omega)), MkDir(h, 1)));
     interface(warnlevel = omega), r, `if`(nops(Search2(h, {"_"})) <> nops(Search2(r, {"_"})),
 WARNING("target path <%1> has been activated instead of the required path <%2>",r,h),NULL)
  catch "invalid input: %1 expects its %-2 argument": procname(args)
  end try
end proc
```

Typical examples of the procedure use:

```
> pathtf("C:\\Temp\\ASV\\AVZ\\Art\\Kristo\\Sveta/Books/Order.txt", 1);
                        "c:\temp\asv\avz\art\kristo\sveta\books\order.txt"
> pathtf("C:\\Archive\\ABS.TXT\\Invoice.pdf", 1);
   Warning, target path <c:\archive\_abs.txt\invoice.pdf> has been activated instead of the
   required path <c:\archive\abs.txt\invoice.pdf>
                                   "c:\archive\_abs.txt\invoice.pdf"
```

```
> map(pathtf, ["C:/Temp/icss.txt", "C:/Art", "C:/Book/Nook.doc", "A:/Kr.7", "C:/Temp"]);
   Warning, target path <c:\_temp> has been activated instead of the required path <c:\temp>
           ["c:\temp\icss.txt", "c:\art", "c:\book\nook.doc", "a:\kr.7", "c:\_temp"]
> writeline(pathtf("C:/Academy"), "International Academy of Noosphere, 1995-2004");
   Warning, target path <c:\_academy> has been activated instead of the required path
   <c:\academy>
                                          46
> All_Close(); readline("C:/_Academy"); fremove("C:/_Academy");
                       "International Academy of Noosphere, 1995-2004"
> pathtf(`convert/rlb`, 1);
   Error, (in pathtf) argument should be a string or symbol but has received <procedure>
```

RmDir – removing of a directory of file system of a computer
rmdir1
_rd

Call format of the procedures:

RmDir(F {, p})
rmdir1(F)
_rd(F)

Formal arguments of the procedures:

F - path to a directory which should be removed
p - (optional) a valid *Maple* expression defining a mode of directory removing

Description of the procedures:

The **RmDir(F)** procedure removes a directory **F** out of file system of an underlying operating system. All file structure for which a subdirectory defined by **F** argument is the root directory, is completely deleted. The **F** argument, which should be a *Maple* string or symbol, specifies path to a directory that should be removed. Successful procedure call returns the *NULL* value, i.e. nothing.

In addition, the procedure call with the second argument **p** saves the *empty* structure (without files) of all subdirectories with **F** directory as the root directory. If the second actual argument has the value **0** (**p** = **0**), the list of all components of the processed directories structure is returned, otherwise the *NULL* value is returned.

By the functionalities, the **RmDir** procedure essentially exceeds the built-in ***rmdir*** function of the *Maple* package and function of the same name of the operating MS DOS system. Furthermore, the **RmDir** extends the built-in function ***fremove*** in case of files with attributes {**r, h, s**}. In environment of some versions of *Windows*-platform the call of the given procedure can cause erroneous situation with diagnostics *"file I/O error"*. In such case, we recommend to fulfill callback of the procedure. The procedure is effective enough with respect to comparatively small systems of subdirectories, but with growth of their size, its effectiveness is reduced.

The procedure call **rmdir1(F)** removes an *empty* directory **F** (*irrelatively of attributes ascribed to it*) out of file system of an underlying operating system. If an **F** directory is not *empty*, the erroneous situation arises. In particular, the procedure does not allow to delete whole file system of a disk volume. In many cases, this procedure is very useful at datafiles processing.

The procedure call **_rd(F)** removes a directory **F** out of file system of an underlying operating system. A file structure for which the subdirectory defined by the **F** argument is root directory is deleted. The procedure **_rd** uses the external ***DelTree*** function of operating MS DOS system; that guesses its presence in an underlying operating system. The successful call **_rd** returns the value **0;** at unsuccessful call, an exception situation will be raised. In addition, if the above ***DelTree*** command is absent in an underlying operating system, the procedure call is returned unevaluated.

```
RmDir:= proc(F::dir)
local a, k, p;
  assign(a = DirF(F)), `if`(a = [], RETURN(iolib(30, F)), NULL);
  a := [seq(`if`(member(DoF1(a[k]), {1, 2}), null([F_atr2(a[k]), iolib(8, a[k])]), a[k]),
    k = 1 .. nops(a))];
  assign('a' = Rlss(a, "\\", `<`)), `if`(nargs = 1, null(seq([F_atr2(p), iolib(30, p)], p = a)),
    `if`(nargs = 2 and args[2] = 0, a, NULL))
end proc

rmdir1:= proc(D::dir)
local a, d, k;
  if not DirE(D) then ERROR("directory is full")
  else assign(a = 97 .. 122, d = `if`(CF2(D)[-1] = "\\", CF2(D)[1 .. -2], CF2(D)));
    `if`(member(convert(d, bytes), {[k] $ (k = a), [k, 58] $ (k = a), [k, 58, 92] $ (k = a)}),
      `if`(member(d[1], map(CF2, {Adrive()})),
      ERROR("deletion of file system of whole disk <%1> is prohibited", Case(d, upper)),
      ERROR("drive <%1> is idle", d)), NULL);
    try rmdir(d)
    catch "file I/O error": F_atr2(D); rmdir(d)
    catch "permission denied": F_atr2(D); rmdir(d)
    end try
  end if
end proc

_rd:= K::dir -> `if`(isMSDcom(DelTree), system(cat("deltree /Y ",Path(CF(K)))),'procname(args)')
```

Typical examples of the procedures use:

```
> DirF("C:\\Temp/Archive");
   [c:\temp\archive, c:\temp\archive\vaganov, c:\temp\archive\proclib_6_7_8_9.mws,
   c:\temp\archive\maple.hdb, c:\temp\archive\lectures.rtf, c:\temp\archive\booklet,
   c:\temp\archive\abs.txt, c:\temp\archive\vaganov\proposal8.htm,
   c:\temp\archive\vaganov\proposal6.rtf, c:\temp\archive\vaganov\proposal5.rtf,
   c:\temp\archive\vaganov\proposal4.rtf, c:\temp\archive\vaganov\proposal3.rtf,
   c:\temp\archive\booklet\b2_3.doc, c:\temp\archive\booklet\b1_4.doc]
> RmDir("C:\\Temp\\Archive", 0);
   [c:\temp\archive\booklet, c:\temp\archive\vaganov, c:\temp\archive]
> RmDir("C:\\Temp/RANS\\IAN/_AVZ"); RmDir("C:\\Temp\\Archive");
> _rd("C:/Temp/rans/ian\\VTU/AGN\\Art/Kr\\Svet/A V Z\\Tallinn/Lasnamae");
                                        0
> null(map(RmDir, ["C:/TRU\\Tartu", "C:\\GRSU/Grodno", "C:\\RANS/Moscow"]));
```

> H:= "C:\\RANS_IAN": type(H, dir), Atr(H); rmdir(H); rmdir1(H), type(H, dir);

true, ["A", "S", "H", "R"]

Error, (in **rmdir**) permission denied

false

The above procedures **MkDir, RmDir** and **rmdir1** are useful at operation with directories and datafiles of file system of a computer.

---

delsc – deletion of a subdirectories tree

Call format of the procedure:

delsc(D {, p})

Formal arguments of the procedure:

D - string or symbol defining full path to a root directory
p - (optional) an arbitrary valid *Maple* expression defining a removal mode

Description of the procedure:

The procedure **delsc** extends the standard procedure *rmdir,* and our procedures **rmdir1** and **RmDir,** providing deletion of the given subdirectories tree. Let us, the first actual **D** argument defines *full path* to a directory and has the following view: **"C:\\D1\\D2\\DX3\\ ... \\Dk".** Successful procedure call **delsc(D)** returns the *NULL* value, i.e. nothing, providing deletion of the subdirectories tree including its root directory "**Dk**". Thus, the root directory "**Dk**" and all its subdirectories with datafiles are deleted irrespective of attributes ascribed to them. In addition, the datafiles are closed if they do exist and are open, and then removed. Deletion of datafiles foregoes to deletion of a directory containing them.

The procedure call **delsc(D, p)** is analogous to the procedure call **delsc(D)** with the sole distinction - the root directory "**Dk**" and its datafiles remain; however all its subdirectories are deleted. The procedure handles basic especial and erroneous situations. In particular, the procedure does not allow to delete whole file system of a disk volume. The procedure has many useful appendices at datafiles processing.

```
delsc:= proc(D::fpath)
local a, b, c, d, k, p, n, t, h, fp;
  assign(d = 97 .. 122, fp = `if`(CF2(D)[-1] = "\\", CF2(D)[1 .. -2], CF2(D))),
    assign(h = nops(CFF(fp)));
  `if`(member(convert(fp, bytes), {[k] $ (k = d), [k, 58] $ (k = d), [k, 58, 92] $ (k = d)}),
    `if`(member(fp[1], map(CF2, {Adrive()})),
    ERROR("deletion of file system of whole disk <%1> is prohibited", Case(fp[1], upper)),
    ERROR("drive <%1> is idle", Case(fp[1], upper))), NULL);
  `if`(type(fp, dir), assign(a = InvL(DirF(fp)), b = [], c = CF2(fp)),
    ERROR("the first argument must define a directory, but has received <%1>", fp));
  if 1 < nargs then
    for k in a do
      if search(k, fp) and nops(CFF(k)) = h + 1 and type(k, file) then
        NULL else b := [op(b), k]
      end if
    end do;
```

```
    assign('a' = b), assign('b' = [])
  end if;
  for k in a do if type(k, file) then delf1(k, `*`) else b := [op(b), k] end if end do;
  for k in [op(b), `if`(1 < nargs, NULL, c)] do rmdir1(k) end do
end proc
```

Typical examples of the procedure use:

```
> delsc("C:/Tallinn\\Vilnius/Grodno");        # subdirectory Grodno will be deleted
> delsc("C:/Tallinn\\Vilnius/Grodno", 7);     # subdirectory Grodno will be kept
> delsc("D:\\");
   Error, (in delsc) deletion of file system of whole disk <D> is prohibited
> delsc("A:\\");
   Error, (in delsc) deletion of file system of whole disk <A> is prohibited
```

Path – standardization of definition of full path to a datafile or directory

Call format of the procedure:

Path(P)

Formal arguments of the procedure:

P – string or symbol defining full path to a datafile or directory

Description of the procedure:

At usage in the *Maple* programs of paths to datafiles or subdirectories is recommended to be guided by agreements accepted in *Maple*, however use of the functions ***`system`*** and ***ssystem*** which provide an access to functional tools of MS DOS, requires an additional tool providing a processing of *blanks* and directory *separators* in names of subdirectories and files to satisfy the coding rules of paths of such type for operating MS DOS system. The simple enough **Path** procedure solves this quite urgent problem.

The procedure call **Path(P)** returns the so-called MS DOS standardized format of path to a file or a directory; the resultant format contains entries of directory separators "\\" instead of directory separators "/", and also necessary processing of *blanks* belonging to the initial format defined by the actual **P** argument is done also. The standardized *path* is returned as a *string*. The procedure call **Path(P)** tests a path **P** onto its formal correctness only and in a series of cases can correct the detected errors. In addition, generally speaking, the above-standardized *format* not is acceptable for the *Maple* environment itself as it illustrates the last example below; however, this format is necessary at use of many MS DOS commands within the *Maple* environment.

```
Path:= proc(K::{name, string, symbol})
local p, k, omega, t, Sigma;
  assign(omega = CF(K), t = [], Sigma = ""), seq(`if`(omega[p] = "\\", assign('t' = [op(t), p]), 0),
    p = 1 .. length(omega));
  omega := [omega[1 .. t[1] - 1], seq(omega[t[k] + 1 .. t[k + 1] - 1], k = 1 .. nops(t) - 1),
    omega[t[-1] + 1 .. length(omega)]];
  [seq(assign('Sigma' = cat(Sigma, `if`(search(omega[k], " "), cat("""", omega[k], """"),
    omega[k]), "\\")), k = 1 .. nops(omega))], RETURN(Sigma[1 .. -2])
end proc
```

Typical examples of the procedure use:

```
> Path("C:/Maple 5 6 7/Maple_edu/Maple 7 edu/LIB/\\\UserLib/");
                  "c:\"maple 5 6 7"\maple_edu\"maple 7 edu"\lib\userlib"
> system(cat("dir ", Path("C:/Program Files/Maple 9/LIB\\UserLib/")," > C:\\Temp\\Dir"));
                                   0
> Path("C:/Program Files/Maple 9/LIB\\UserLib/"); currentdir(%);
                      "c:\"program files"\"maple 9"\lib\userlib"
  Error, (in currentdir) file or directory does not exist
```

Sys_Env – receiving of settings of an underlying operating system

Call format of the procedure:

Sys_Env(U, Z)

Formal arguments of the procedure:

U - logical name of an external memory volume (may be *symbol* or *string*)
Z - an assignable name

Description of the procedure:

The procedure call **getenv(N)** of the standard *Maple* procedure returns the string that contains settings for environment variable of an underlying operating system, whose name is defined by the actual **N** argument. For operating MS DOS system by the environment variables are being understood those, for which values are set by the`*set*` command. However, in a number of cases the opportunities of the **getenv** procedure turns out insufficient. The **Sys_Env** procedure largely closes this question.

The procedure call **Sys_Env(U, Z)** outputs the contents of two system files "Autoexec.bat" and "Config.sys" located on a drive with logical name defined by the actual **U** argument; these files contain the basic settings of an underlying operating system. Whereas via the second actual **Z** argument the nested list with two sublist is returned. The first element of each sublist defines name of one of the above system files, whereas the rest of elements define settings of the variables of the underlying operating system.

```
Sys_Env:= proc(U::{string, symbol}, Z::evaln)
local R, F, G, S;
  `if`(evalb(length(U) = 1) and belong(op(convert(Case(U), bytes)), op(convert(a, bytes)) ..
    op(convert(z, bytes))), assign(G = [], S = []), ERROR("<%1> is not an unit name", U));
  R:= proc(C::{name, string, symbol}, F::{string, symbol})
    local n, L;
      try assign(n = fopen(cat(C, ":/", args[2]), READ, TEXT))
      catch "file or directory does not exist": ERROR("System HDD %1 does not exist", U)
      finally [assign('Z' = args[2], 'G' = [op(G), F]), `if`(search("System HDD", lasterror) <> 0,
        NULL, WARNING("System file: %1", F))]
      end try;
      do L := readline(n);
        if L = 0 then print(`\n`); break  elif L = "" then next
```

```
        elif searchtext("REM", L) <> 0 then next  else lprint(L), assign('G' = [op(G), L])
        end if
      end do;
      fclose(n), assign('S' = [op(S), G]), assign('G' = [])
    end proc;
  R(U, "Autoexec.bat"), R(U, "Config.sys"), assign(Z = S)
end proc
```

Typical examples of the procedure use:

> Sys_Env(C, h): h; **Windows 98SE**

```
"@ECHO OFF"
"SET BLASTER=A220 I5 D1 T4"
"PROMPT $P$G"
"PATH C:\WINDOWS; C:\WINDOWS\COMMAND; C:\; C:\program files"
"SET TEMP=C:\TEMP"
"DOS=HIGH,UMB"
"DEVICEHIGH=C:\WINDOWS\HIMEM.SYS"
"DEVICE=C:\WINDOWS\setver.exe"
"DEVICEHIGH=C:\WINDOWS\EMM386.EXE"
"FILES=99"
"device=C:\WINDOWS\cwdinit.exe /A"
[["Autoexec.bat", "@ECHO OFF", "SET BLASTER=A220 I5 D1 T4", "PROMPT $P$G",
"PATH C:\WINDOWS; C:\WINDOWS\COMMAND; C:\; C:\program files",
"SET TEMP=C:\TEMP"], ["Config.sys", "DOS=HIGH,UMB",
"DEVICEHIGH=C:\WINDOWS\HIMEM.SYS ", "DEVICE=C:\WINDOWS\setver.exe",
"DEVICEHIGH=C:\WINDOWS\EMM386.EXE ", "FILES=99",
"device=C:\WINDOWS\cwdinit.exe /A"]]
```

> Sys_Env(C, h), h; **Windows XP Pro**

```
"mode con codepage prepare=((775) C:\WINDOWS\COMMAND\ega3.cpi)"
"mode con codepage select=775"
"DEVICE=C:\WINDOWS\setver.exe"
"device=C:\WINDOWS\COMMAND\display.sys con=(ega,,1)"
"Country=372,775,C:\WINDOWS\COMMAND\country.sys"
[["Autoexec.bat", "mode con codepage prepare=((775) C:\WINDOWS\COMMAND\ega3.cpi)",
"mode con codepage select=775"], ["Config.sys", "DEVICE=C:\WINDOWS\setver.exe",
"device=C:\WINDOWS\COMMAND\display.sys con=(ega,,1)",
"Country=372,775,C:\WINDOWS\COMMAND\country.sys"]]
```

> Sys_Env(D, t), t;

Error, (in **R**) System HDD D does not exist

Vol – receiving of information about disk volumes

Call format of the procedure:

Vol({G})

Formal arguments of the procedure:

G – (optional) string or symbol defining logical name of a disk drive

Description of the procedure:

The procedure call **Vol()** returns the 2-element sequence whose the first element defines the label of a disk volume containing a current (*active*) directory, whereas the second element defines its serial number. In addition, the procedure calls **Vol(), Vol(``)** and **Vol("")** are identical. At last, the procedure call **Vol(G)** returns the 2-element sequence that is analogous the above, but with respect to a disk volume defined by the actual **v** argument (*logical disk name*). The procedure handles the basic especial and erratic situations.

```
Vol:= proc()
local a, b, d, h, omega, f, c, err, k, R, t, x;
  err:= b -> ERROR("<%1> is idle logical drive name or the drive is not ready for use", b);
  omega, f, c := b -> ERROR("<%1> is not a drive name", b), "#_$ArtKr$#_", [];
  `if`(nargs = 0 or args = `` or args = "", assign(b = CF2(currentdir()[1])),
    assign(b = CF2(cat("", args[1])[1])));
  `if`(belong(op(convert(b, bytes)), op(convert("a", bytes)) .. op(convert("z", bytes))),
    `if`(member(b, {op(map(CF2, {Adrive()}))}), NULL, err(b)), omega(b));
  try assign('b' = Case(b, upper)), system(cat("VOL ", cat(b, ":"), " > ", f)) catch: err(b) end try;
  assign(x = convert([1], bytes)), assign(R = [x, x]), fopen(f, READ, TEXT);
  while not Fend(f) do h := Red_n(FNS(readline(f), " ", 3), " ", 2);
    if member(h, {" ", ""}) then next else c := [op(c), h] end if
  end do;
  fremove(f), assign(a = "has no label", d = "has no serial number");
  for k in c do
    if search(CF2(k), "drive ", 't') and not search(CF2(k), " has no ") then R[1]:= k[t + 11 .. -1]
    elif search(CF2(k), "number is", 't') then R[2] := k[t + 10 .. -1]
    end if
  end do;
  if R[1] = x and R[2] = x then cat("<", b, "> ", a), cat("<", b, "> ", d)
  elif R[1] = x and R[2] <> x then cat("<", b, "> ", a), R[2]
  elif R[1] <> x and R[2] = x then cat("<", b, "> is ", R[1]), cat("<", b, "> ", d)
  else cat("<", b, "> is ", R[1]), R[2]
  end if
end proc
```

Typical examples of the procedure use:

```
> map(Vol, [Dos, Cveta, Aladjev]);
     ["<D> is Z", "280F-FE5A", "<C> is ALADJEV_V_Z", "1751-19F4", "<A> is G", "0F3E-18D9"]
> map(Vol, [A, D]);
     ["<A> is PLA9503_V36", "0000-0000", "<D> is XR2PFRE_EN", "4CC6-4207"]
> map(Vol, [`Create`, "a"]);
     ["<C> is ALADJEV_V_Z", "1751-19F4", "<A> is as no label", "2B73-7D81"]
> Vol();
     "<C> is ALADJEV_V_Z", "1751-19F4"
> Vol(D);
Error, (in err) <d> is idle logical drive name or the drive is not ready for use
```

```
> Vol("###"); map(Vol, [A, D]);
    Error, (in omega) <#> is not a drive name
        ["<A> has no label", "2B73-7D81", "<D> is MsOfficeXP", "D787-3B79"]
```

WD, WT, WS, WDS – determination of full paths to basic *Windows* directories

Call format of the procedures:

WD(), WT(), WS(), WDS()

Formal arguments of the procedures: **no**

Description of the procedures:

The procedure call **WD()** returns the full path to the main directory of the current *Windows* environment. The given procedure has enough great meaning for the *Maple* users dealing with problems that use means of the host operational system. Procedure provides reliable identification of the full path to main *Windows* directory of versions 95/98/98SE/ME/NT/XP/2000/2003 and their clones even under condition of presence on a computer of any set of the specified systems. The essence of the algorithm used by the procedure, is based on the information of system files "BOOT.ini" and "MsDos.sys", and also on the artificial reception that uses inaccessibility in the *Maple* environment to files of the register of a current *Windows* system. Procedure handles the basic especial and erroneous situations.

Procedure **WDS** is based on the previous procedure **WD** and system utilities "Start.exe" and "MsInfp32.exe", providing (*at their presence*) more simple identification of a full path to the main catalogue of the current *Windows* system. The procedure call **WDS()** returns the list whose the first element defines full path to the main directory of the current *Windows* system, whereas the second - full path to its system directory. On the some *Windows* configurations the procedure call can provoke occurrence of an alert condition with diagnostics "*Windows cannot find ...*" to which it is necessary to answer "OK". The same answer it is necessary to give on the alert condition with diagnostics "*System Information: Access is denied*". After that, the call returns correct result. Procedure handles the basic especial and erroneous situations.

The procedure call **WT()** returns the full path to the **TEMP** directory of the current *Windows* environment. The procedure call **WS()** returns the full path to the *Windows* directory containing the external commands of MS DOS. These procedures seem useful enough tools in a number of cases of work with datafiles of the different types and nature, as well as at operation with file system of a computer as a whole.

```
WD:= proc()
local a, b, v, k, t1, t2, f, V, x, w, G, S;
  V:= (x, y) -> op({seq(`if`(search(x, k), RETURN(k), false), k = y)});
  G:= proc(k, w, x)
    local n, h, z, r;
      z := cat(k, ":\\Boot.ini"); n := fopen(z, READ, TEXT);
      while not Fend(n) do
        h := CF2(readline(n));
        if Search2(h, {"signature(", "multi("}) <> [] and search(h, w) and search(h, x) then
          Close(z); r := Case(cat(k, ":", FNS(sextr(h, "\\", ssign)[1], " ", 3)), upper);
          if S(w, r, t2) then RETURN(r) end if  end if
      end do;
```

```
      Close(z), ERROR("system file BOOT.INI on <%1> is improper or corrupted", k)
    end proc;
  S:= proc(t, wd, t2)
    local a;
      `if`(member(t, t2), assign(a = cat(wd, "\\system32\\software")),
        assign(a = cat(wd, "\\system.dat")));
      if type(a, file) then false else true end if
    end proc;
  assign(a = map(CF2, {Adrive()}), x = "windows", t1 = {" me", " mill", " 95", " 98"},
    t2 = {" nt", " 2003", " xp", " 2000"}, f = "_$ArtKr$_");
  system(cat("Ver  >  ", f)), assign(v = CF2(readbytes(f, TEXT, infinity))),
    delf1(f, `*`), assign(w = V(v, t1 union t2));
  if member(w, t2) then
    for k in a do
      try RETURN(G(k, w, x))
      catch "file or directory does not exist":  next
      catch "system file BOOT.INI on <%1> is improper":
        WARNING("system file BOOT.INI on <%1> is improper", Case(k, upper));  next
      end try
    end do;
    ERROR("fatal system error: proper file BOOT.INI has not been found on %1",
        map(Case, a, upper))
  elif member(w, t1) then
    for k in a do
      f := cat(k, ":\\MsDos.sys");  if not type(f, file) then next  end if;
      do  b := readline(f);
        if b = 0 then  WARNING("system file MSDOS.SYS on <%1> is corrupted", k);
          close(f);  break
        end if;
        if search(CF2(b), "windir") then close(f); b:= FNS(b[Search(b, "=")[1] + 1..-1], " ", 3);
          if S(w, b, t2) then RETURN(b) end if;  break
        end if
      end do
    end do;
    ERROR("fatal system error: proper file MSDOS.SYS has not been found on %1",
      map(Case, a, upper))
  else ERROR("System error: improper host operating system <%1>", v)
  end if
end proc

WDS:= proc()
local a, b, c, f, t;
  assign(f = map2(cat, blank(CDM()), ["\\_$ArtKr$_", "\\_$SveGal$_"]), t=interface(warnlevel)),
    assign(a = system(cat("Start MsInfo32 /report ", f[1], " /Categories SystemSummary")));
  if a <> 0 then map(Case, [WD(), WS()], upper)
  else
    try a := subsLS([0 = NULL, 9 = 32, 20 = 32, 254 = NULL, 255 = NULL],
        readbytes(f[1], infinity))
    catch "file or directory does not exist":  RETURN(map(Case, [WD(), WS()], upper))
    end try;
```

```
    Close(f[1]), writebytes(f[2], a), Close(f[2]), fopen(f[2], READ, TEXT), assign(c = ["", ""]);
    while not Fend(f[2]) do interface(warnlevel = 0);
      b := FNS(Red_n(CF2(readline(f[2])), " ", 2), " ", 3);  interface(warnlevel = t);
      if search(b, "windows directory") then  c[1] := b[19 .. -1]
      elif search(b, "system directory") then  c[2] := b[18 .. -1]
      end if
    end do;
    if member("", c) then ERROR("at least one of Windows directories has not been
      defined - %1",c) else delf1(op(f), `*`), map(Case, c, upper) end if;
  end if
end proc

WT:= () -> cat("", WD(), "\\Temp")

WS:= proc()
local a, k;
  a := map2(cat, WD(), ["\\command", "\\system", "\\system32"]);
  for k to nops(a) do
    if type(cat(a[k], "\\format.com"), file) then RETURN(a[k]) end if
  end do;
  ERROR("System violation: external MS DOS commands have not been found in Windows
    directories <Command, System, System32>")
end proc
```

Typical examples of the procedures use:

> WD(), WT(), WS(); # *for Windows 2000*

"C:\WINNT", "C:\WINNT\Temp", "C:\WINNT\system32"

> WD(), WT(), WS(); # *for Windows 98 SE*

"C:\WINDOWS", "C:\WINDOWS\Temp", "C:\WINDOWS\command"

> WD(), WT(), WS(); # *for Windows XP Pro*

"C:\WINDOWS", "C:\WINDOWS\Temp", "C:\WINDOWS\system32"

> WD();

Warning, system file BOOT.INI on <C> is improper
Error, (in WD) fatal system error: proper file BOOT.INI has not been found on {A, C, D}

> WD();

Warning, system file BOOT.INI on <A> is improper
Warning, system file BOOT.INI on <D> is improper
Error, (in WD) fatal system error: proper file BOOT.INI has not been found on {A, C, D}

> WD();

Error, (in WD) System error: improper host operating system <Windows 2004 Home>

> WDS();

["C:\WINDOWS", "C:\WINDOWS\SYSTEM32"]

Group of four procedures **WD, WT, WS** and **WDS** have essential enough value at operation with a file system of both the host operating system, and the package. They are essentially used in a number of the procedures represented in the present monograph.

Vol_Free_Space – determination of free space of disk volumes

Call format of the procedure:

Vol_Free_Space(A)

Formal arguments of the procedure:

A – a string or symbol defining logical name of an external memory volume

Description of the procedure:

The procedure call **Vol_Free_Space(A)** returns free space of an external memory volume (HDD, FD or CD ROM) with logical name given by the actual **A** argument. The **A** argument should define the logical name of a volume, i.e. should be coded as a letter in the *string* or *symbol* format, for example **"C", `C`, AGN, Art, "Demo"** etc The procedure call **Vol_Free_Space(A)** returns the 2-element list whose the first element defines free space and the second element defines its *unity* (*bytes*, *MB*, *GB*). In addition, the procedure calls **Vol_Free_Space()** and **Vol_Free_Space({"" | ``})** are completely identical, defining free space for all disk volumes available at the moment with returning the nested list whose elements are 3-element sublists. The first element of such sublists defines logical name as a string, while the rest elements are analogous to the above case of the procedure call. The procedure handles the basic especial and erratic situations.

```
Vol_Free_Space:= proc(A::{string, symbol})
local a, b, d, fs, omega, nu, t;
  omega, nu := b -> ERROR("<%1> is not a drive name", b),
    b -> ERROR("<%1> is idle logical drive name", b);
  fs:= proc(V::string)
    local h, t;
      system(cat("Dir ", cat(V, ":\\ > "), "_##_"));
      do assign('h' = readline("_##_"), 't' = readline("_##_"));
        if h = 0 and t = 0 then fremove("_##_"); nu(V)
        elif t = 0 then break end if
      end do;
      fremove("_##_"), SLD(Red_n(h, " ", 2), " ")[-3 .. -2];
      [cat("", V, ""), SD(sstr(["," = NULL, "." = NULL], %[1])), cat(``, %[2])]
    end proc;
  assign(d = [Adrive()]), `if`(nargs = 0 or args = "" or args = ``,
    assign(b = Case(currentdir()[1]), t = 7), assign(b = Case(cat("", args[1])[1])));
  `if`(not belong(op(convert(b, bytes)), map(op, map(convert, "a" .. "z", bytes))), omega(b),
    `if`(t = 7, fs(b), fs(b)[2 .. 3]))
end proc
```

Typical examples of the procedure use:

```
> map(Vol_Free_Space, ["C", AGN, ``, ""]);
                 [[96065, MB], [936, bytes], ["c", 95665, MB], ["c", 95665, MB]]
> Vol_Free_Space(D); Vol_Free_Space("$ArtKr");
   Error, (in nu) <d> is idle logical drive name
   Error, (in omega) <$> is not a drive name
```

The procedures represented in the present item have general purpose and are substantially used in a lot of other procedures represented below, which are oriented onto the more special cases of datafiles processing and operation with file system of a computer as a whole.

10.2. Software for operation with *TEXT* datafiles

The next four items of the section contain tools supporting a number of useful functions of a work with datafiles of types *`TEXT`* and *`BINARY`*, with the *Maple* files of other types, and a lot of tools for special datafiles processing. In particular, tools of these items provide procedures such as an effective handling of the situation *"end of BINARY file"*, methods of organization of direct access to the *TEXT* files, an extracting of correct email-addresses out of a *txt*-file, a coding/ decoding of a *TEXT* file, useful restructurings of the *TEXT* files, and many others.

Datafiles of the *TEXT* format are the most used files with which *Maple* and its numerous appendices deal. For datafiles processing of this type *Maple* has appropriate built-in functions of access to the *TEXT* files. However, in many cases at the solution in the *Maple* environment of various problems dealing with datafiles processing, these tools appear insufficient. In the given item, the tools of access to the *TEXT* datafiles oriented to the widest use in problems of access to datafiles of such type are represented. The procedures represented here extend tools of *Maple* at operation with the *TEXT* datafiles, allowing to solve a whole series of problems dealing with the *TEXT* datafiles the more effectively. These procedures have been used in a whole series of the manifold applications dealing with datafiles processing of the *TEXT* type, and in the more general case with so-called *text* datafiles or datafiles of the *rlb*-type. The technique used at programming of these procedures can be useful at practical programming of problems dealing with organization of access to the text datafiles of different purpose. At present, these tools are essentially used at processing of text datafiles of different purpose, confirming own efficiency. A series of them is used in other procedures of datafiles processing.

---

writedata1 - write of data into ASCII datafiles

Call format of the procedure:

writedata1(F, *data*)

Formal arguments of the procedure:

F - a symbol or string defining name or full path to an ASCII datafile
data - list, nested list, vector, matrix or Matrix with data which should be saved
h - (optional) a valid *Maple* expression defining a write mode

Description of the procedure:

The standard function **writedata(F, *data*)** writes numeric ***data*** from a *Maple* vector, matrix, list, or nested list into a text file **F.** If ***data*** is a vector or list of values, each value is printed on a separate line. If the ***data*** is a matrix or a nested list of values, the data is printed one row per line with values parted by *tabs***.** Normally the ***data*** must consist of integers or floating-point numbers. Provision for handling non-numeric data and complex numeric data is provided, however it demands of programming of special procedures. The standard function ***readdata*** is the corresponding function for reading numerical data from a text file created by the ***writedata*** function into a current *Maple* session. However, in many cases the necessity of saving of arbitrary data in ASCII files is needed. In particular, the data can contain the values of *fraction*-type and arbitrary valid *Maple* expressions. This problem can be solved by the **writedata1** procedure.

The procedure call **writedata1(F, *data*)** writes arbitrary ***data*** from a *Maple* vector, matrix, list, nested list or *NAG* Matrix into an ASCII file **F.** If ***data*** is a vector or list of values, each value is printed on a separate line. If the ***data*** is a matrix, a nested list or Matrix of values, the ***data*** is printed one row per line with values parted by *blanks***.** Normally the ***data*** admit arbitrary valid *Maple* expressions, extending sphere of appendices of the procedure.

The procedure call **writedata1(F, *data*)** creates a new file **F** every time it is called (irrespective of the existence of **F**). By the way, the similar remark can be attributed to the standard procedure ***writedata*** also. In addition, the file qualifier **F** can define a chain of directories of arbitrary level of nesting. The procedure creates the chain if it is necessary.

If the procedure call **writedata1(F, *data*, h)** contains the optional third argument **h** (an arbitrary *Maple* expression), the ***data*** is appended to the file **F** if it already exists. For the avoidance of possible data contentions, a filename **F** is given by the *file qualifier* (instead of a *file descriptor*). The successful call of the procedure **writedata1** returns the *NULL* value, i.e. nothing, with output of an appropriate warning about location of the target datafile with the saved ***data.***

The procedure allows to save arbitrary data of the above structure in ASCII files that can be read by the **readdata1** procedure considered below. The procedure **writedata1** has shown oneself as useful tool in many appendices dealing with the access to datafiles of different nature.

```
writedata1:= proc(F::{string, symbol}, data::{Matrix, matrix, vector, list, listlist})
local k, Lambda, h, f, R;
  Lambda := proc()
    local n, m, k, j, mu, mu1, st, L;
      assign(f = `if`(type(F, file), F, pathtf(F))), assign(h = fopen(f, `if`(nargs = 2,
        APPEND, WRITE)));
      assign(mu = convert([255], bytes), mu1 = convert([254], bytes), st = "");
      if type(args[1], {vector, list}) then  R := convert(args[1], list);
        for k to nops(R) do
          if type(R[k], string) then R[k] := cat(R[k], mu)
          elif type(R[k], symbol) then R[k] := cat(R[k], mu1)
          end if;
          null(writeline(h, convert(R[k], string)))
        end do
      else assign(R = convert(args[1], Matrix)), assign(n = Pind(R)[1], m = Pind(R)[2],
        L = convert(R, listlist));
        for k to n do
          for j to m do
            if type(L[k][j], string) then L[k][j] := cat(L[k][j], mu)
            elif type(L[k][j], symbol) then L[k][j] := cat(L[k][j], mu1)
            else L[k][j] := convert(L[k][j], string)
            end if;
            st := cat(st, L[k][j], " ")
          end do;
          null(writeline(h, st[1 .. -2]), assign('st' = ""))
        end do
      end if
    end proc;
  `if`(nargs = 3, Lambda(data, 7), Lambda(data)), close(f),
    WARNING("data have been saved in datafile <%1>", f)
end proc
```

Typical examples of the procedure use:

Creation of the *NAG* matrix **M** and *Maple* vector **V:**

```
> writedata1("C:\\LBZ\\TestData1.dat", [[62, 57, 37], [42, 8, 15], [a, b, c]]);
   Warning, data have been saved in datafile <c:\lbz\testdata1.dat>
```

```
> M:= Matrix(3, 4, [[8,"2",C, sin(7)], [8,`15`,"G",`ln(7)`], ["AVZ",Art,"Kr", 8+15*I]]):
  V:= [8, "C", G, "sin(15)", (a-b)/(c-d), a+b*I]: [M, V];
```

$$\left[\begin{bmatrix} 8 & "2" & C & \sin(7) \\ 8 & \mathit{15} & "G" & \mathit{ln(7)} \\ "AVZ" & Art & "Kr" & 8+15\,I \end{bmatrix}, \left[8, "C", G, "\sin(15)", \frac{a-b}{c-d}, a+b\,I\right]\right]$$

Saving of the above *Maple* objects in the ASCII files in modes of *WRITE* and *APPEND:*

```
> writedata1("C:/Temp/TestData1.dat", M); writedata1("C:/Temp/TestData1.dat", M, 7);
    Warning, data have been saved in datafile <C:/Temp/TestData1.dat>
    Warning, data have been saved in datafile <C:/Temp/TestData1.dat>
> writedata1("C:/Temp/TestData2.dat", V); writedata1("C:/Temp/TestData2.dat", V, 14);
    Warning, data have been saved in datafile <C:/Temp/TestData2.dat>
    Warning, data have been saved in datafile <C:/Temp/TestData2.dat>
```

The results of the saving will be tested below by the procedure **readdata1.**

---

readdata1 – read of data from ASCII datafiles

Call format of the procedure:

readdata1(F, p {, h})

Formal arguments of the procedure:

F - a symbol or string defining name or full path to an ASCII datafile
n - a positive integer, or range or infinity defining the number of columns of readable data
h - (optional) a valid *Maple* expression

Description of the procedure:

The standard ***readdata*** function reads numeric data from a text file into a current *Maple* session. The data in the file must consist of integers or floating-point values arranged in columns, separated by *blanks*. If only one column of the data is read, the output is a list of the data. If more than one column is read, the output is a nested list of data corresponding to the rows of data in the file for the specified columns. However, in many cases the necessity of readout of data from ASCII datafiles containing the data of the *numeric* and *string* types as the valid *Maple* expressions is needed. In particular, the data can contain the values of *fraction*-type and arbitrary valid *Maple* expressions. This problem can be solved by the **readdata1** procedure.

The **readdata1** procedure reads data from an ASCII file into a current *Maple* session. The data in the file must consist of valid *Maple* expressions arranged in columns, separated by *blanks*. The readable data are returned as a rectangular *Matrix;* in addition, if a datafile has been prepared outside of *Maple* (for instance, by means of the text processor Word 2002) with saving of the above format, the procedure call returns the ***data*** saved in the file with saving all their types irrespective of quantity of actual arguments of the call (*see examples below*).

A rather different situation takes place at reading a datafile prepared by means of the procedure **writedata1,** namely: (**1**) the procedure call **readdata1(F, p)** with two actual arguments only returns the ***data*** saved in a datafile **F** with saving all their types, defined at the file creation; (**2**) while the procedure call **readdata1(F, p, h)** with three actual arguments returns the ***data*** received from the saved ***data*** by a converting their into the appropriate *Maple* expressions by means of acceptable cancellation of types {*string, symbol*} for the ***data*** (see examples below). The given possibility allows to use in the *Maple* environment the ASCII datafiles of the above format that have been prepared by other software (*FORTRAN, Basic, Pascal, Word,* etc.).

The procedure call **readdata1(F, p)** returns the *Matrix* with data defined by the **p**th column or of by a range **p** of serial columns. If as the actual **p** argument the *infinity* has been coded, the **F** datafile is entirely read. The file **F** is closed when **readdata1** returns. By the way, the similar remark can be attributed to the standard function ***readdata*** also. For the avoidance of possible data contentions, a file name **F** is given by the *file qualifier* (instead of a *file descriptor*).

The procedure **readdata1** allows to read arbitrary data from ASCII files of the above structure that can be prepared by both the popular programming languages, and the text processors, and the above procedure **writedata1.** However, it is necessary to have in mind, as the *blank* symbol is used as a data delimiter, the data should not contain *blanks*. Otherwise infringement of reliability of data processing is possible. Meanwhile, the given restriction can be removed easily, for example, by replacement of the *blank* symbol of as a delimiter on symbol of *tabulation*. It will require a simple enough updating of the procedure. The procedure **readdata1** has shown oneself as useful tool in many appendices dealing with the access to datafiles of different nature.

```
readdata1:= proc(F::file, p::{infinity, posint, range})
local k, j, L, s, t, n, gs, h, Omega, z, V, Xi, Xi1, v, w;
  `if`(IsFempty(F), ERROR("datafile <%1> is empty", F), `if`(type(F, rlb), assign(w = {}),
    ERROR("datafile <%1> has a type different from TEXT", F)));
  Omega:= proc(L::Matrix)
    local k, j, g;
    global __Art_Kr_, _mu7;
      assign('__Art_Kr_' = Matrix(args[2], args[3]));
      for k to op(1, L)[1] do for j to op(1, L)[2] do
          try g := 14;  __Art_Kr_[k, j] := SN(L[k, j])
          catch "Invalid value <%1>": `if`(L[k, j][-1] = Xi, assign('g' = 7, '_mu7' =L[k, j][1..-2]),
               `if`(L[k, j][-1] = Xi1, assign('g' = 7, '_mu7' = cat(``, L[k, j][1 .. -2])),
               assign('_mu7' = L[k, j])));
          if nargs = 3 and g = 7 then __Art_Kr_[k, j] := _mu7
          else writeline("$AVZ$", cat("__Art_Kr_[", k, ",", j, "]:=", _mu7, ":")), close("$AVZ$");
               (proc(x) read x; fremove(x) end proc)("$AVZ$");
               next
             end if
           end try
         end do
      end do;
      __Art_Kr_, unassign('__Art_Kr_', '_mu7')
    end proc;
  `if`(IsOpen(F), close(F), NULL), assign(s = []);
  assign(gs = interface(warnlevel)), interface(warnlevel = 0);
  do v := SLD(Red_n(FNS(readline(F), " ", 3), " ", 2), " "); w := w union {nops(v)};
    if 1 < nops(w) then interface(warnlevel = gs);
      ERROR("incorrect dimension of data array and/or occurrence of blanks in data")
    else s := [op(s), v]
    end if;
    if Fend(F) then interface(warnlevel = gs); break
    end if
  end do;
  if not type(s, nestlist) then s := [s] end if;
  close(F), assign(t=nops(s[1]), n =nops(s), Xi=convert([255], bytes), Xi1=convert([254], bytes));
```

```
  if p = infinity then RETURN(Omega(convert(s, Matrix), nops(s), t, `if`(nargs = 2, NULL, 147)))
  elif type(p, posint) and p <= t then h := p .. p; z := 1
  elif type(p, range) and belong(lhs(p), 1 .. t) and belong(rhs(p), 1 .. t) then
     h := p; z := rhs(h) - lhs(h) + 1
  else ERROR("the second argument <%1> is invalid, it should be in range 1..%2", p, t)
  end if;
  Omega(convert([seq(s[k][h], k = 1 .. n)], Matrix), n, z, `if`(nargs = 2, NULL, 147))
end proc
```

Typical examples of the procedure use:

Suppose that three ASCII datafiles prepared outside of *Maple* (for instance, by means of the text processor Word 2002) have the above format and contain the following data, namely:

TestData3.dat:

```
57         62         37         -41         8      15
1          "2"        3          4           5      6/7
4.2        -4.7       5/6        4/7         -5/7   6/5
`sin(7)`   sqrt(15)   ln(57)     x*sin(x)    5!     (a+b)/(c-d)
"V"        "G"        "S"        "Ar"               `Art` 7+15*I
```

TestData4.txt:

```
61
4.7
-7/15
"57"
`sin(2)`
7+15*I
sqrt(15)
```

TestData5.txt:

```
61  4.7  -7/14  "56"  `sin(2)`  7+15*I  sqrt(15)
```

Reading of all data from the above three datafiles by the **readdata1** procedure:

> readdata1("C:/Temp/Academy\\Book/TestData3.dat", infinity);

$$\begin{bmatrix} 57 & 62 & 37 & -41 & 8 & 15 \\ 1 & 2 & 3 & 4 & 5 & \frac{6}{7} \\ 4.2 & -4.7 & \frac{5}{6} & \frac{4}{7} & \frac{-5}{7} & \frac{6}{5} \\ sin(7) & \sqrt{15} & \ln(57) & x\sin(x) & 120 & \frac{a+b}{c-d} \\ "V" & "G" & "S" & "Ar" & Art & 7+15I \end{bmatrix}$$

Reading of all columns from the datafile with use of the third optional argument (7):

> readdata1("C:/Temp/Academy\\Book/TestData3.dat", infinity, 7);

$$\begin{bmatrix} 57 & 62 & 37 & -41 & 8 & 15 \\ 1 & 2 & 3 & 4 & 5 & \frac{6}{7} \\ 4.2 & -4.7 & \frac{5}{6} & \frac{4}{7} & \frac{-5}{7} & \frac{6}{5} \\ sin(7) & \sqrt{15} & \ln(57) & x\sin(x) & 120 & \frac{a+b}{c-d} \\ "V" & "G" & "S" & "Ar" & Art & 7+15I \end{bmatrix}$$

> **readdata1("C:/Temp/Academy\\Book/TestData3.dat", 3 .. 5);**

$$\begin{bmatrix} 37 & -41 & 8 \\ 3 & 4 & 5 \\ \frac{5}{6} & \frac{4}{7} & \frac{-5}{7} \\ \ln(57) & x\sin(x) & 120 \\ \text{"S"} & \text{"Ar"} & Art \end{bmatrix}$$

> **readdata1("C:/Temp/Academy\\Book/TestData4.txt", infinity);**

$$\begin{bmatrix} 61 \\ 4.7 \\ \frac{-7}{15} \\ \text{"57"} \\ sin(2) \\ 7 + 15\,I \\ \sqrt{15} \end{bmatrix}$$

> **readdata1("C:/Temp/Academy\\Book/TestData5.txt", infinity, 14);**

$$\begin{bmatrix} 61 & 4.7 & \frac{7}{14} & \text{"56"} & sin(2) & 7 + 15\,I & \sqrt{15} \end{bmatrix}$$

Hence, the data saved in a datafile prepared by the above manner save all own types as a result of reading by the **readdata1** procedure. The given possibility allows to use in *Maple* the ASCII datafiles of the above format that have been prepared by other software (*FORTRAN, Word,* etc.).

Reading of the above datafiles prepared by the procedure **writedata1**. Reading of all ***data*** from the both datafile **TestData1.dat** and **TestData2.dat**:

> **readdata1("C:/Temp/TestData1.dat", infinity);**

$$\begin{bmatrix} 8 & \text{"2"} & C & \sin(7) \\ 8 & 15 & \text{"G"} & ln(7) \\ \text{"AVZ"} & Art & \text{"Kr"} & 8 + 15\,I \end{bmatrix}$$

> **interface(rtablesize=20): readdata1("C:/Temp/TestData2.dat", infinity, 7);**

$$\begin{bmatrix} 8 \\ C \\ G \\ \sin(15) \\ \frac{a-b}{c-d} \\ a + b\,I \end{bmatrix}$$

Reading of all data from the both datafile **TestData1.dat** and **TestData2.dat** with use of the third optional argument at the call:

Reading of the columns **2..4** from the datafile:

```
> readdata1("C:/Temp/TestData1.dat", 2 .. 4, 7);
```

$$\begin{bmatrix} 2 & C & \sin(7) \\ 15 & G & \ln(7) \\ Art & Kr & 8 + 15\,I \end{bmatrix}$$

Reading of a non-existent column from the datafile:

```
> readdata1("C:/Academy/Examples/Data\\Read1.dat", 14);
```

Error, (in **readdata1**) the second argument <14> is invalid, it should be in range 1..6

In many cases, the procedures **writedata1** and **readdata1** are more preferable than the *Maple* functions **writedata** and **readdata** at programming of access to datafiles of different nature.

DAopen – a method of organization of direct access to the *TEXT*-files
DAread
DAclose

Call format of the procedures:

DAopen(F)
DAread(F, n)
DAclose(F)

Formal arguments of the procedures:

F - symbol or string defining full path to an *TEXT* datafile
n - a positive integer defining the number of sought line in the *TEXT* datafile

Description of the procedures:

As an *TEXT* datafile consists of logical records (values of the *string*-type) parted by symbols of line feed and carriage return {hex(0D0A)}, then in a lot of cases at operation with files of such type the knowledge of lines positions forming the file is useful enough. The given problem is solved by two procedures **FTabLine** and **FTabLine1** that will be considered below in the item.

For providing of direct access to lines of a *TEXT*, datafile (in general case of an *rlb*-file) by their numbers the basic three procedures **DAopen, DAread** and **DAclose** can be represented. The first procedure **DAopen(F)** is intended for an opening of an arbitrary *TEXT* datafile, given by full path **F** to it, in a mode of direct access to its records (*lines*) according to their numbers. As a result of successful call of the given procedure, a special file (available in the same subdirectory as the initial **F** file) is created. This file contains table whose *indices* are the numbers of lines of the **F** file, while *entries* - positions which correspond to them in the file. As a result of the procedure call **DAopen(F)** the activation of the relevant table for the **F** file in a current *Maple* session is done.

After that, by means of the procedure call **DAread(F, n)** in a current *Maple* session we obtain an access immediately to **n**th record of the **F** file, returning its value. Simultaneously with such file, it is possible to work and by standard facilities of access, taking into account influence of procedure **DAread(F, n)** onto positioning of pointer of a current record of the **F** file. The closing of the file **F**, opened by means of procedure **DAopen(F)** is fulfilled by the procedure call **DAclose(F)** that provides both naturally closing of the **F** file, and deleting of file with the table corresponding to it, and also deactivation of the table in a current *Maple* session. Examples of a fragment, represented below, give concrete appendices of the above procedures for a reading of the file lines, given by their numbers. In addition, the estimation of the time of access is done. On the basis of the above three procedures an effective enough algorithms of access to *TEXT* datafiles (in general case to *rlb*-files) can be programmed, providing essential reduction of access time to the data.

```
DAopen:= proc(F::rlb)
local _Tab, k, h, Z, p, r;
  `if`(IsFempty(F), ERROR("datafile <%1> is empty", F),
     assign(h = cat("", F, "_da"), _Tab[1] = 0, k = 1));
  do assign('p' = readline(F), 'k' = k + 1);
     if p = 0 then close(F); break else _Tab[k] := _Tab[k - 1] + length(p) + 2 end if
  end do;
  (proc(x) save _Tab, x end proc)(h), assign(Z = readbytes(h, TEXT, infinity)), close(h);
  writeline(h, cat("_Tab_DA_", sub_1([" " = "_", "." = "_"], CFF(F)[-1]),
     Z[5 .. length(Z)])), close(h);
  read h
end proc

DAread:= proc(F::file, n::posint)
local r, h;
  assign(h = cat(_Tab_DA_, sub_1([" " = "_", "." = "_"], CFF(F)[-1])));
  `if`(nops(op(op(h))) <= n, ERROR("the %-1 record is absent", n),
     [filepos(F, h[n]), came(readline(F))][-1])
end proc

DAclose:= proc(F::file)
local h, r, p, s;
  assign(h=cat("", F, "_da")), assign(s=cat(_Tab_DA_, sub_1([" " = "_", "." = "_"], CFF(F)[-1])));
  try Close(F), fremove(h), unassign(s)
  catch "file or directory does not exist": RETURN()
  end try
end proc
```

Typical examples of the procedures use:

```
> F:="C:/Academy/TRG.txt": for k to 10000 do writeline(F,cat("", k)) end do: Close(F);
  t:=time(): for k to 10000 do readline(F) end do: time()-t; All_Close();

                                  0.424

> DAopen(F); t:=time(): DAread(F, 3000); time()-t; t:=time(): DAread(F, 6000); time()-t;

                              "3000", 0.066
                              "6000", 0.066

> DAread(F, 12003);

        Error, (in DAread) the 12003rd record is absent

> t:=time(): DAread(F, 8000); time()-t; t:=time(): DAread(F, 10000); time()-t; DAclose(F);

                              "8000", 0.069
                              "10000", 0.068
```

Of the examples of the fragment follows, the represented approach to organization of direct access to the *TEXT* datafiles allows to receive a very essential saving of time at operation with separate lines of a *TEXT* datafile. The more detailed description of the algorithms used by these procedures can be found in our books [32,33,43,44]. The main essence of these algorithms is clear enough from source codes of the procedures represented above.

XTfile – **operations with *TEXT* files in the *Maple* environment (*group 1*)**
RTfile
FT_subs
Extract
Reduce_T

Call format of the procedures:

XTfile(F {, P})
RTfile(F, K {, P})
FT_subs(F, P, L)
Extract(F, K, P {, n {, m}})
Reduce_T(F, P, R)

Formal arguments of the procedures:

F - symbol or string defining name or full path to an *TEXT* file
P - (optional) symbol or string defining name or full path to the target file
K - set or list of symbols or strings defining a search pattern (*prefix*) of the file lines
L - a set or list of equations (*substitutions*)
n - (optional) argument admits the value **1**
m- (optional) argument admits the value **2**
R - a positive integer accepting two values **1** or **2** only

Description of the procedures:

At operation with *TEXT* datafiles in a series of cases, the inversion problem of such files arises, i.e. update of existing file or creation on its basis of a new file of the same type, but with the inverse order of lines. The given problem is solved by the procedure **XTfile(F {, P})** which at presence of one actual **F** argument updates a file specified by the argument in situ whereas in case of two actual arguments creates a new target file defined by the second argument **P.** The successful termination of the procedure returns the full path to a target datafile with output of the appropriate warning. In many appendices, dealing with text datafiles the procedure has been successfully used.

The procedure **RTfile(F, K {, P})** provides the indefinite deletion out of an *TEXT* file specified by the first actual **F** argument, of all lines (excepting a line which has been added into the file by last) which have as a prefix the patterns specified by the second actual **K** argument (*set* or *list*). At call with two indicated arguments the updating of the initial **F** file is done in situ, otherwise the result of deletion of lines is saved in a file specified by the third actual argument **P.** The successful termination of the procedure returns the full path to a target datafile with output of the appropriate warning.

The procedure **FT_subs(F, P, L)** on the basis of a *TEXT* file specified by the first actual **F** argument creates a file specified by the second actual **P** argument that will contain only those **F** lines on which the replacements of entries of patterns were done according to substitutions given by a list or a set **L.** The substitutions are done up to their full exhaustion in each line of **F.** Arbitrary *Maple* expressions can be as the members of substitutions **L.** If the substitutions have not been made the procedure call returns the *false* value (in this case a target **P** file is not created), otherwise the call returns the sequence *`true,* **tf**`, where **tf** – the full path to a target file **P** with output of the appropriate warning.

The procedure call **Reduce_T(F, P, h)** on the basis of a *TEXT* file, specified by the first actual **F** argument, creates a file specified by the second actual **P** argument, which will contains lines of the first file with multiplicity **1.** In addition, in a **P** file are saved only those lines out of the groups of identical lines that have been written down by the last. The third formal **h** argument of the procedure accepts only two values, namely: **1** – saving of the input file **F** and **2** – its deletion. The successful termination of the procedure returns the full path to a target datafile with output of the appropriate warning.

The procedure call **Extract(F, K, P {, n {, m}})** puts into a **P** file the **F** file lines which contain entries of a pattern specified by **K** argument. The optional arguments **n** and **m** receive values **1** and **2,** defining not case sensitive searching of a **K** pattern and an output of the numbers of the found lines into the target file **P** accordingly. The successful termination of the procedure returns the full path to a target datafile **P** with output of the appropriate warning; in addition, the created datafile **P** will has the following content:

<Empty line>
---------- full path to an initial F *file*
{[*line number*]} *<line containing a pattern* K>
====================================
{[*line number*]} *<line containing a pattern* K>

```
XTfile:= proc(F::{string, symbol})
local L, k, j, h, G, p, nu, d, z, s, t, x, Lambda;
  `if`(2 < nargs, ERROR("quantity of arguments > 2"), `if`(not type(F, file),
    ERROR("<%1> is not a file", F), `if`(not type(F, rlb),
    ERROR("file <%1> has a type different from the rlb-type", F), `if`(not IsFempty(F),
    assign(L = [], p = filepos(F, infinity)), ERROR("<%1> is empty file", F))))),
    `if`(nargs = 1, assign(G = cat(F, ".k")), assign(G = `if`(type(args[2], file),
    args[2], MkDir(args[2], 1)))), assign(x = 128, Close(F)),
    assign(Lambda = (proc(G, h) local a, j; a := InvL(SLD(convert(h, bytes), "\n"));
      for j to nops(a) do writeline(G, a[j]) end do end proc));
  if p <= 16384 then assign('L' = InvL(SLD(convert(subs(13 = NULL, readbytes(F, infinity)),
    bytes), "\n"))), Close(F);
    (proc(x, y) local k; for k to nops(x) do writeline(y, x[k]) end do; Close(y) end proc)(L, G)
  else
    assign(d = ceil(p/x), s = []), assign(z = p - (d - 1)*x);
    for k to d - 1 do
      filepos(F, p - k*x), assign('h' = [op(subs(13 = NULL, readbytes(F, x))), op(s)]);
      if not member(10, h, 't') then s := h; h := []
      elif 1 < t then s := h[1 .. t - 1]; h := h[t + 1 .. -1]
      else s := []; h := h
      end if;
      if h = [] then next else Lambda(G, h) end if
    end do;
    filepos(F, 0), assign('h' = [op(subs(13 = NULL, readbytes(F, z))), op(s)]),
      Lambda(G, h), Close(F, G)
  end if;
```

```
  `if`(nargs = 1, null([writebytes(F, readbytes(G, infinity)), fremove(G), Close(F)]), NULL),
    WARNING("the inversion result is in file <%1>", [F, G][nargs]), [F, G][nargs]
end proc

RTfile:= proc(F::{string, symbol}, K::{set({string, symbol}), list({string, symbol})})
local a, k, h, p, g, omega, nu, Lambda, w;
  omega:= proc(x, y)
    local t;
      if map2(Suffix, x, {op(y)}, t) = {false} then false else true end if
    end proc;
  assign(w = interface(warnlevel)), interface(warnlevel = 0), `if`(3 < nargs,
    ERROR("quantity of arguments > 3"), `if`(not type(F, file), ERROR("<%1> is not a file", F),
    `if`(not type(F, rlb), ERROR("file <%1> has a type different from the rlb-type", F),
    `if`(not IsFempty(F), assign(p = 0, a = map(convert, K, string)),
    ERROR("<%1> is empty file", F))))), `if`(nargs = 2, assign(g = cat(F, ".k")),
    assign(g = `if`(type(args[3], file), args[3], MkDir(args[3], 1)))), XTfile(F),
    assign(Lambda = (u -> WARNING("the deletion outcome is in file <%1>", [F, g][u])));
  do h := readline(F);
    if h = 0 then break else if not omega(h, a) then writeline(g, h) end if end if
  end do;
  null(Close(F, g), `if`(nargs = 2, [writebytes(F, readbytes(g, infinity)), fremove(g)], Close(g)),
    Close(F), XTfile(F), `if`(nargs = 3, XTfile(g), NULL), interface(warnlevel = w),
    Lambda(nargs - 1)), `if`(nargs = 3, CF2(g), CF2(F))
end proc

FT_subs:= proc(G::{string, symbol}, S::{string, symbol}, L::{list(equation), set(equation)})
local k, wl, h, v, p, F, tf, t;
  `if`(not type(G, file), ERROR("<%1> is not a file", G), `if`(not type(G, rlb),
    ERROR("<%1> is not a TEXT-file", G), [Close(G), open(G, READ)]));
  assign(wl = interface(warnlevel)), interface(warnlevel =0), assign(F = 0, p = 0, tf = MkDir(S,1));
  t := [seq(convert(lhs(L[k]), string) = convert(rhs(L[k]), string), k = 1 .. nops(L))];
  while not Fend(G) do
    assign('h' = readline(G)), seq(`if`(Search(h, lhs(t[k])) <> [],
    assign('h' = Subs_all1(t[k], h), 'p' = p + 1), NULL), k = 1 .. nops(t));
    `if`(p = 0, NULL, [writebytes(tf, h), assign('F' = F + 1), assign('p' = 0)])
  end do;
  Close(G), interface(warnlevel = wl); `if`(F = 0, op([false, fremove(tf)]), op([Close(tf), true, tf,
    WARNING("the resultant file is in <%1>", tf)]))
end proc

Reduce_T:= proc(F::{string, symbol}, P::{string, symbol}, R::{1, 2})
local f, nu, k, omega;
description `The procedure reduces the multiple strings in a TEXT file`;
  assign(omega = interface(warnlevel)), `if`(not type(F, file), ERROR("<%1> is not a file", F),
    `if`(not type(F, rlb), ERROR("<%1> is not a TEXT-file", F),
    [interface(warnlevel = 0), assign67(fopen(F, READ, TEXT), f = MkDir(P, 1), nu = [])]));
  while not Fend(F) do nu := [op(nu), readline(F)] end do;
  Close(F), assign('nu' = redL(InvL(nu)), interface(warnlevel = omega)), assign('nu' = InvL(nu));
  for k to nops(nu) do writeline(f, nu[k]) end do;
  Close(f), `if`(R = 1, NULL, fremove(F)), WARNING("the reduced file is in <%1>", f), CF2(f)
end proc
```

```
Extract:= proc(F::{string, symbol}, K::{string, symbol}, P::{string, symbol})
local omega, f;
  assign(omega = interface(warnlevel)), `if`(not type(F, file), ERROR("<%1> is not a file", F),
  `if`(not type(F, rlb), ERROR("<%1> is not a TEXT-file", F), null(interface(warnlevel = 0),
  assign(f = MkDir(P, 1)), system(cat("FIND ", `if`(belong({1, 2}, [args[4 .. -1]]), " /N /I ",
  `if`(args[4]=1, " /I ", `if`(args[4]=2, " /N ", " "))), cat("""", K, """"), " ", Path(F), " > ", Path(f)))))),
  interface(warnlevel = omega), WARNING("the result is in <%1>", f), f
end proc
```

Typical examples of the procedures use:

```
> XTfile("C:/RANS/TRG.txt", "C:/RANS/TRG1.txt"),  XTfile("C:/RANS/TRG1.txt");

    Warning, the inversion result is in file <c:\rans\trg1.txt>
    Warning, the inversion result is in file <C:/RANS/TRG1.txt>

                         "c:\rans\trg1.txt", "C:/RANS/TRG1.txt"

#  The TEXT files "TRG.txt" and "TRG1.txt" are identical.
> RTfile("C:/RANS/TRG.txt", {"10aa", "26aaa", `142a`}, "C:/Temp/RANS/TRG42.txt");
    Warning, the deletion outcome is in file <c:\temp\rans\trg42.txt>
                              "c:\temp\rans\trg42.txt"
> RTfile("C:/RANS/TRG1.txt", {"10aa", "26aaa", `142a`});
    Warning, the deletion outcome is in file <C:/RANS/TRG1.txt>
                                 "c:\rans\trg1.txt"
> FT_subs("c:/temp/trg.txt", "c:/temp/trg1.txt", [`7a`=Kr,`14a`=Art+Kr,"aa"="$Euro"]);
    Warning, the resultant file is in <c:\temp\trg1.txt>
                              true, "c:\temp\trg1.txt"
> Reduce_T("C:/rans/Academy/trg.txt", "C:/rans/Academy/trg1.txt", 1);
    Warning, the reduced file is in <c:\rans\academy\trg1.txt>
                             "c:\rans\academy\trg1.txt"
> Reduce_T("C:/rans/Academy/trg.txt", "C:/rans/Academy", 2);
    Warning, the reduced file is in <c:\rans\_academy>
                                "c:\rans\_academy"
> Extract("C:\\RANS\\TRG.txt", "AVZ", "C:\\Temp\\Artur.14", 1, 2);
    Warning, the result is in <c:\temp\artur.14>
                                 "c:\temp\artur.14"
> do readline("C:/temp/Artur.14"); if Fend("C:/temp/Artur.14") then break end if end do;
                                        ""
                           "---------- c:\temp\artur.14"
                                     "[1]AVZ"
                                     "[2]AVZ"
                                     "[3]AVZ"
                                   "[4]3AVZ00"
                                   "[5]4AVZ00"
                                   "[6]5AVZ00"
```

```
> Extract("C:\\RANS\\TRG42.txt", "Art_Kr_147", "C:\\Academy", 1);
    Warning, the result is in <c:\_academy>
                                   "c:\_academy"
```

The procedures represented above have shown itself as useful enough tools at solution of a whole series of the manifold problems dealing with processing of datafiles of so-called *text* type and datafiles of the *rlb*-type. The providing of return of the full path to a target datafile allows successfully to use the tools of this group both in interactive and in automatic programming mode. The given circumstance is conditioned by the fact; the target file with a processing result can differ from files specified at the procedure call.

GenFT - generating of test *TEXT* datafiles

Call format of the procedure:

GenFT(F, L, N)

Formal arguments of the procedure:

F - a symbol or string defining name or full path to a target *TEXT* datafile
L - a positive integer (*posint*)
N - a positive integer (*posint*)

Description of the procedure:

The procedure **GenFT(F, L, N)** provides creation of a *TEXT* datafile specified by the first actual **F** argument whose lines are generated by a random manner from symbols of the Roman alphabet, decimal digits and special keyboard symbols. The second actual **L** argument defines a maximal length of the file lines, while the third actual **N** argument - a quantity of lines in the required *TEXT* datafile **F**.

The procedure call **GenFT(F, L, N)** returns the integer **3**-element list whose the first and the second elements define minimal and maximal length of lines, written into the file **F** accordingly, whereas the third element defines length of the file **F** in bytes. The procedure return is accompanied by the appropriate warning about real location of the target datafile **F**. The procedure gives useful enough tool for creation of the test *TEXT* files intended for debug of procedures and program modules that solve the different problems dealing with datafiles of such type. The procedure has many interesting appendices linked with *TEXT* datafiles processing. Above all, the procedure is enough widely used in debug purposes, in particular, at debugging of tools dealing with text datafiles.

```
GenFT:= proc(F::{string, symbol}, L::posint, N::posint)
local h, h1, h2, k, A, T, t, min, max, nu, f;
  A := [seq(cat("", k), k = 0 .. 9), op(convert("`-=;',./~!@#$%^&*()_+{}:<>? | ", list)),
    op(map2(cat, "", [a, b, c, d, e, f, g, h, i, j, k, l, m, n, o, p, q, r, s, t, u, v, w, x, y, z]))];
  assign(h = rand(1 .. L), h1 = rand(1 .. 64), h2 = rand(1 .. 2), min = L, max = 1,
    nu = interface(warnlevel)), interface(warnlevel = 0), assign(f = MkDir(F, 1));
  for k to N do T := cat("", op([seq(Case(A[h1()], `if`(h2() = 2, upper, NULL)), k = 1 .. h())]));
    assign('t' = length(T)), `if`(t < min and max < t, assign('min' = t, 'max' = t), `if`(t < min,
      assign('min' = t), `if`(max < t, assign('max' = t), 6))), writeline(f, T)
  end do;
  [min, max, filepos(f, infinity), Close(f)], interface(warnlevel = nu),
    WARNING("the resultant file is in <%1>", f)
end proc
```

Typical examples of the procedure use:

```
> n:=62: GenFT("C:\\Temp\\Orders/ABS\\File_Art.147", 72, n);
    Warning, the resultant file is in <c:\temp\orders\abs\file_art.147>
                              [1, 71, 2627]
> n:=57: GenFT("C:\\Temp\\Orders/ABS", 70, n);
    Warning, the resultant file is in <c:\temp\orders\_abs>
                              [2, 70, 2338]
> for k to 9 do readline("C:\\temp\\orders/_abs") end do; Close("C:\\temp\\orders/_abs");
  "$dY#z&o=&1_*_WF8W~Y?>s:hLqy?O'.!P/:}#&77Gw!v|)=)F<#5oOK#MZ,<?@PG^}kJ<v"
  "=xD7;.ck@>@*#<{7_}'{?f60|}Tx8^0"
  "vuW#H3DS;Z;w9p8u}8cE({(|7S:$VA824k8H~G~#9_|6>-03"
  "<L$6@^z&mH,<'=r=.gk4e+_W{T=4+dY0zo:'Ij"
  "=2mo9w|?I1T@.@D<<8t0c/wCb?"
  "G&?4*^mSqG66*Ow8X1%"
  "9RQ|#==S#$}#1"
  ">mLK{FA5;YL'f/j!oOQ1Z4k|vu#dOoKvA"
  "g^9/;T575|O{myQb3E9<ra+%37Rw4xvjco!3B?D@k2_c*it*95pe!?"
```

Nstring – **operations with *TEXT* datafiles in the *Maple* environment (*group 2*)**
FindFSK
LTfile
FTabLine
FTabLine1
nLine

Call format of the procedures:

```
Nstring(F, S {, F1})
FindFSK(F, Ko {, F1 {, `sensitive`}})
LTfile(F)
FTabLine(F {, FLDT {, 'h'}})
FTabLine1(F, R {, FLDT {, 'h'}})
nLine(F, T, n)
```

Formal arguments of the procedures:

- **F** – symbol or string defining name or full path to a *TEXT* datafile
- **F1** – (optional) symbol or string defining name or full path to a target *TEXT* datafile
- **S** – set or list of positive integers
- **Ko** – set or list of symbols or strings defining the search patterns
- ***sensitive*** – (optional) key word defining a search mode
- **h** – (optional) an assignable name
- **R** – a positive integer range
- **T** – a table
- **n** – a positive integer, symbol, list of positive integers or list of symbols
- ***FLDT*** – (optional) key word defining a saving mode of the *file lines disposition table*

Description of the procedures:

The procedure call **Nstring(F, S, F1)** returns the full path to a datafile specified by the actual argument **F1** into which the lines of a *TEXT* datafile specified by the **F** qualifier, according to their numbers specified by a set or list **S** are put with output of the necessary warnings. At absence of the third optional argument **F1** the procedure call **Nstring(F, S)** returns the list with lines of an **F** file that satisfy the above condition and with output of the necessary warnings.

The procedure call **FindFSK(F, Ko)** provides searching of lines in a file specified by the first actual **F** argument that contain a pattern from a set or list specified by the second actual **Ko** argument. The procedure call **FindFSK(F, Ko)** returns the sequence of two-element lists whose the first element defines the serial number of a **F**-file line which contains any of the given patterns **Ko,** whereas the second defines the line with this number. If the key optional *`sensitive`* argument has been coded, then search of patterns is fulfilled in a case-sensitive mode, otherwise not. If a target datafile **F1** has been coded additionally, then the received lines are put into it. The lines of the *TEXT* datafile **F1** have the following view:

nnnaxxxxxxxxxxxxx

where **nnn** – the serial number of a line **xxxxxxxxxxxxx** in the file **F** and **a:=convert([7], bytes).** Such organization of the target *TEXT* file **F1** allows to easily receive both the extracted lines and their serial numbers in the source file **F.** The procedure call returns the full path to a target file. At absence in the file **F** of lines with the given patterns **Ko** the *NULL* value, i.e. nothing, or empty list (at absence of the target file **F1**), i.e. [], is returned. The returned result is accompanied by the appropriate warnings.

The procedure call **LTfile(F)** returns the list whose the first element defines quantity of lines of a *TEXT* datafile specified by the file qualifier **F,** while the second defines the file size in bytes.

Whereas a *TEXT* file consists of logical records (values of *string*-type) parted by symbols of line feed and carriage return {hex(0D0A)}, in a lot of cases at operation with files of such type the knowledge of lines positions which compose the file is useful enough. The given problem is solved by means of the procedure **FTabLine.**

The procedure call **FTabLine(F)** returns the special *file lines disposition table* (***FLDT***) whose indices are the numbers of lines of a *TEXT* datafile specified by the actual **F** argument, whereas its entries are 2-element lists whose elements define initial and final positions of an appropriate line in the file **F.** Moreover, at coding of the optional second argument **h** via it the 3-element list of view **[d, [k, *Min*], [p, *Max*]]** is returned, where: **d** – a lines quantity in the file **F** and **k, p** – line number with minimal (*Min*) and maximal (*Max*) length accordingly. The knowledge of the given information, in particular, allows to organize a quick access to lines on the basis of their numbers for large enough datafiles of the *TEXT* type.

At coding of the optional key word ***FLDT*** the procedure additionally allows to save the ***FLDT*** for a file **F** in file whose path is defined as **cat(F, ".m").** The subsequent procedure calls **FTabLine(F)** in case of availability of such file activate the global variable with name defined by the output warning. In general case the name has the following view **aaa_bbb_fldt,** where **aaa.bbb** is name of the **F** file. Such possibility seems useful enough if an **F** file is subjected to rare enough updates. The following procedure **FTabLine1** extends possibilities of the above procedure **FTabLine.**

The procedure **FTabLine1** provide an opportunity of organization of access to lines of a ***TEXT*** file not over their *numbers*, but over *keys* that represent a content of fields of lines located in their identically fixed positions. The line keys have the *symbol*-type and after activation of the returned ***FLDT*** **T** (*similarly to a case of the previous* **FTabLine** *procedure*) for a datafile, their list can be obtained by means of construction **with(T).**

The first formal **F** argument of the procedure call **FTabLine1(F, R)** is analogous to the first argument of the above procedure **FTabLine.** The second formal **R** argument sets an integer range defining positions of keys in each line of the **F** file. In addition, in the returned table of the **F** file the information only about those lines of the file is located which have keys; the other lines remain in the **F** file but are not subjected to further processing.

At coding of the third optional **h** argument via it the **3**-element list (similarly to a case of the **FTabLine** procedure) is returned with that only difference, that instead of *numbers* of lines with minimal and maximal length their *positions* in the **F** file are indicated. For a receiving of all keys defined by the procedure for file **F**, can be used the clause **with(T),** where **T** is a ***FLDT*** returned by the procedure **FTabLine1(F, R).** The returned table of the file **F** as the indices uses keys of lines, while the entries of the table are analogous to a case of the previous procedure **FTabLine.**

At coding of the optional key word ***FLDT*** the procedure additionally allows to save the ***FLDT*** for a file **F** in file whose path is defined as **cat(F, ".m")** analogously to the **FTabLine** procedure case described above. The examples of appendices of the **FTabLine1** procedure well enough illustrate the told.

At last, for support of direct access to lines of a *TEXT* file over their numbers or keys created on the basis of two procedures **FTabLine** or **FTabLine1** represented above, the procedure **nLine(F, T, n)** can be used. This procedure in dependence from expediency of evaluation of the ***FLDT*** **T** of a **F** file returns one or several lines by their numbers or keys defined by the actual **n** argument for a file defined by the **F** argument and for which the ***FLDT*** **T** by means of the procedure {**FTabLine | FTabLine1**} has been activated. The procedure returns the required line as a list; at line absence, the false value is returned with output of the appropriate warning. Such approach allows to provide the continuous calculations along with exact identification of absent lines. Last examples of a fragment represented below give concrete appendices of the procedure **nLine** for a reading of the file lines given by keys and their numbers along with estimations of access time to datafile lines.

Before use of the procedure **nLine** for organization of direct access to lines of a *TEXT* datafile over their numbers or keys, beforehand it is necessary by means of the above procedures ***FTabLine*** or **FTabLine1** to activate ***FLDT*** **T** for the used datafile. With a view of rise of reactivity of such access mechanism the procedure **nLine** does not provide check of the passed ***FLDT*** **T** onto correspondence to the processed file. Furthermore, the procedure a priori supposes, the **F** datafile has the *TEXT* type and for the file, the appropriate ***FLDT*** **T** has been activated in a current session. In this case, the responsibility for a correctness of datafile processing rests on the user.

The above procedures **nLine, FtabLine** and **FTabLine1** provide a mechanism of direct access to the *TEXT* datafiles of the different purpose. The given tools provide an effective enough access to the *TEXT* datafiles represented by records (*lines*) of variable or constant length.

The procedures represented above have shown itself as useful enough tools at solution of a whole series of the manifold problems dealing with processing of datafiles of so-called *text* type and datafiles of the *rlb*-type. The providing of return of the full path to a target datafile allows successfully to use the tools of this group both in interactive and in automatic programming mode. The given circumstance is conditioned by the fact; the target file with a processing result can differ from files specified at the procedure call.

```
Nstring:= proc(F::{string, symbol}, S::{set(posint), list(posint)})
local k, p, s, f, omega, omega1, omega2, nu, nu1;
  `if`(type(F, file) and type(F, rlb), `if`(Empty(S), ERROR("the second argument is empty"),
     NULL), ERROR("<%1> is not a file or has a type different from TEXT", F));
  omega1, p, k := () -> WARNING("all required lines are absent in file"), [Close(F)], 0;
```

```
  omega2, nu1 := nu -> WARNING("lines with numbers %1 are absent in file", nu),
     () -> `if`(type(f, list), true, false);
  `if`(nargs = 3 and type(eval(args[3]), {string,symbol}), [assign67(omega=interface(warnlevel)),
     interface(warnlevel = 0), assign(f = MkDir(args[3], 1))], assign(f = []));
  assign('p' = fopen(F,READ,TEXT), nu = {op(S)}), `if`(nu1(f), 1, interface(warnlevel = omega));
  while not Fend(p) do [assign('s' = readline(p), 'k' = k + 1)];
     if member(k, nu) then `if`(nu1(), assign('f' = [op(f), s]), writeline(f, s)); nu := nu minus {k}
     end if
  end do;
  if nu = {op(S)} then RETURN(omega1(), `if`(nu1(), f, fremove(f)), Close(F))
  elif nu <> {} then omega2(nu), `if`(nu1(f), RETURN(f, Close(F)), NULL)
  else
  end if;
  Close(F), `if`(nu1(f), f, op([WARNING("the resultant file is in <%1>", f), Close(f), f]))
end proc

FindFSK:= proc(F::{string, symbol}, Ko::{list({string, symbol}), set({string, symbol})})
local a, b, k, p, s, f, omega, omega1, omega2, nu, nu1;
  `if`(type(F, file) and type(F, rlb), `if`(Empty(Ko), ERROR("the second argument is empty"),
     NULL), ERROR("<%1> is not a file or has a type different from TEXT", F));
  omega1, p := () -> WARNING("lines with patterns %1 are absent in file", Ko), [Close(F)];
  omega2 := proc()
    local a, b, k;
      if nargs = 2 then RETURN([``, []])
      else a := {args[3 .. -1]} minus {sensitive}
      end if;
      if a = {} and member(sensitive, {args}) then RETURN([1, []])
      else for k to nops(a) do
            if type(eval(a[k]), {string, symbol}) then a[k]; break else 7
            end if
         end do
      end if;
      `if`(% = 7, `if`(member(sensitive, {args}), [1, []], [``, []]),
         `if`(member(sensitive, {args}), [1, %], [``, %]))
    end proc;
  nu1, k, a, b := () -> `if`(type(f, list), true, false), 0, convert([7], bytes), 0;
  `if`(omega2(args)[2] = [], assign(f = []), [assign67(omega = interface(warnlevel)),
     interface(warnlevel = 0), assign(f = MkDir(omega2(args)[2], 1))]);
  assign('p' = fopen(F,READ,TEXT), nu={op(Ko)}), `if`(nu1(f), 1, interface(warnlevel = omega));
  while not Fend(p) do [assign('s' = readline(p), 'k' = k + 1)];
    if `if`(omega2(args)[1] = ``, Search2(Case(s), {op(map(Case, nu))}), Search2(s, nu)) <> [] then
       `if`(nu1(), assign('f' = [op(f), [k, s]]), writeline(f, cat("", k, a, s))); b := 147
    end if
  end do;
  if b = 0 then RETURN(omega1(), `if`(nu1(), NULL, fremove(f)), Close(F)) end if;
  Close(F), `if`(nu1(f), op(f), op([WARNING("the resultant file is in <%1>", f), Close(f), f]))
end proc

LTfile:= proc(F::file)
local a, p;
```

```
  `if`(type(F, rlb), [Close(F), assign(a = 0, p = fopen(F, READ, TEXT))],
    ERROR("<%1> has a type different from TEXT", F));
  while not Fend(p) do readline(p), assign('a' = a + 1) end do;
  [a, filepos(p, infinity), Close(F)]
end proc

FTabLine:= proc(F::file)
local d, k, h, m, n, p, T, x, y, t, avz, omega;
  assign(omega = cat(F, ".m"), nu = cat(``, sstr(["." = "_"], CFF(F)[-1], avz), "_fldt")),
    `if`(type(omega, file), (proc() read omega;
         RETURN(WARNING("the FLDT has been activated as <%1>", nu))
    end proc)(), 7);
  `if`(type(F, rlb), assign(k = 1, T[1] = 0, m=0, n = infinity, Close(F), t = fopen(F,READ,TEXT)),
    ERROR("<%1> has a type different from TEXT", F));
  while not Fend(t) do readline(t); k := k + 1; T[k] := filepos(t)  end do;
  for k to nops(op(2, eval(T))) - 1 do T[k] := [T[k], T[k + 1] - 3] end do;
  Close(F), (proc(t) T[t] := 'T[t]'; NULL end proc)(k), table(op(2, eval(T))),
    null(`if`(1 < nargs and args[2] <> FLDT, [assign(d = nops(op(2, eval(T)))),
    add(_1N(0, [assign('p' = T[k][2] - T[k][1] + 1), `if`(m < p, assign('m' = p, 'x' = [k, p]), 1),
    `if`(p < n, assign('n' = p, 'y' = [k, p]), 1)]), k = 1 .. d), assign(args[2] = [d, y, x])], 1));
  if member(FLDT, {args}) then
    save1(cat(``, sstr(["." = "_"], CFF(F)[-1], avz), "_fldt"), T, omega);
    eval(T), WARNING("the FLDT for file <%1> has been saved in file <%2>", F, omega),
    unassign(nu)
  else eval(T)
  end if
end proc

FTabLine1:= proc(F::file, R::range(posint))
local a, d, k, h, m, n, p, T, x, y, z, v, t, omega, nu;
  assign(omega = cat(F, ".m"), nu = cat(``, sstr(["." = "_"], CFF(F)[-1], avz), "_fldt")),
      `if`(type(omega, file), (proc()  read omega;
      RETURN(WARNING("the FLDT has been activated as <%1>", nu)) end proc)(), 7);
  `if`(type(F, rlb), assign(T = table([]), z = 0, v = infinity, a = fopen(F, READ, TEXT)),
    ERROR("<%1> has a type different from TEXT", F));
  while not Fend(a) do assign('m' = filepos(a)), assign('h' = readline(a)), assign('n' = filepos(a));
    if rhs(R) <= length(h) - 1 then T[cat(``, substring(h, R))] := [m, n - 3] else next end if
  end do;
  Close(F), eval(T), assign(d = with(T)), null(`if`(2 < nargs and args[2] <> FLDT,
    [assign(t = nops(d)), add(_1N(0, [assign('p' = T[k][2] - T[k][1] + 1),
    `if`(z < p, assign('z' = p, 'x' = [k, p]), 1), `if`(p < v, assign('v' = p, 'y' = [k, p]), 1)]),
    k = d), assign(args[3] = [t, y, x])], 1)), unassign(op(d));
  if member(FLDT, {args}) then
    save1(cat(``, sstr(["." = "_"], CFF(F)[-1], avz), "_fldt"), T, omega);
    eval(T), WARNING("the FLDT for file <%1> has been saved in file <%2>", F, omega),
      unassign(nu)
  else eval(T)
  end if
end proc

nLine:= proc(F::file, T::table, n::{posint, symbol, list(symbol), list(posint)})
```

```
local a, p, k, omega, L;
  omega, L := a -> [filepos(F, a), readline(F)[1 .. -2]][2], [];
  if type(n, {symbol, list(symbol)}) then if type(n, symbol) then p := [n] else p := n end if;
    for k to nops(p) do a := T[p[k]];
      if type(a, list) then L := [op(L), omega(a[1])]
      else WARNING("line with key <%1> does not exist", p[k]); L := [op(L), false]
      end if
    end do;
    `if`(type(n, {symbol, list(symbol)}), unassign(seq(p[k], k = 1 .. nops(p))), 7),
     RETURN(close(F), L)
  end if;
  if type(n, posint) then p := [n] else p := n end if;
  for k to nops(p) do
    try L := [op(L), omega(T[p[k]][1])]
    catch "file position must be an integer or infinity": L := [op(L), false];
      WARNING("line with %1-th number does not exist", p[k]); next
    end try
  end do;
  close(F), L
end proc
```

Typical examples of the procedures use:

> Nstring("C:/RANS/TRG.txt", [7, 15, 37, 57, 62], "C:/Temp/Order/ABS.txt");

Warning, lines with numbers {57, 62} are absent in file
Warning, the resultant file is in <c:\temp\order\abs.txt>

"c:\temp\order\abs.txt"

> Nstring("C:/RANS/TRG.txt", [22, 15, 28, 57, 62]);

Warning, lines with numbers {57, 62} are absent in file

["7aaaaaaaaaa", "17aaaaaaaa", "14aaaaaaa"]

> Nstring("C:/RANS/TRG.txt", [1995, 2004], "C:/Temp/Order/ABS.txt");

Warning, all required lines are absent in file

> Nstring("C:/RANS/TRG.txt", [1995, 2004]);

Warning, all required lines are absent in file

[]

> Nstring("C:/RANS/TRG.txt", [7, 15, 28], "C:/Temp/Order/ABS.txt");

Warning, the resultant file is in <c:\temp\order\abs.txt>

"c:\temp\order\abs.txt"

> Nstring("C:/RANS/TRG.txt", [7, 15, 28]);

["7aaaaaaaaaa", "17aaaaaaaa", "14aaaaaaa"]

> Nstring("C:/RANS/TRG.txt", [], "C:/Temp/Order/ABS.txt");

Error, (in **Nstring**) the second argument is empty

> FindFSK("C:/RANS/TRG.txt", ["17aaa", `10aaa`], 62, "C:/Temp/ABS", 57);

Warning, the resultant file is in <c:\temp\abs>

"c:\temp\abs"

> FindFSK("C:/RANS/TRG.txt", ["17aaa", `10aaa`]);

[2, "17aaaaaaaa"], [5, "17aaaaaaaa"], [14, "17aaaaaaaa"], [18, "10aaaaaaaaaa6"],
[19, "10aaaaaaaaaa6"], [20, "10aaaaaaaaaa"], [21, "10aaaaaaaaaa6"], [22, "10aaaaaaaaaa"],
[36, "17aaaaaaaa"]

> FindFSK("C:/RANS/TRG.txt", ["17AAA", `10AAA`], 57, sensitive, 62);

Warning, lines with patterns [17AAA, 10AAA] are absent in file

> FindFSK("C:/RANS/TRG.txt", ["17AAA", `10AAA`], 57, 2004, 62);

[2, "17aaaaaaaa"], [5, "17aaaaaaaa"], [14, "17aaaaaaaa"], [18, "10aaaaaaaaaa6"],
[19, "10aaaaaaaaaa6"], [20, "10aaaaaaaaaa"], [21, "10aaaaaaaaaa6"], [22, "10aaaaaaaaaa"],
[36, "17aaaaaaaa"]

> FindFSK("C:/RANS/TRG.txt", [], 57, 2004, 62);

Error, (in **FindFSK**) the second argument is empty

> map(LTfile, ["C:/rans/trg.txt", "C:/temp/147.txt", "C:/temp/test", "C:/temp/info.txt"]);

[[1000, 10893], [96, 34567], [234, 456754], [95, 2004]]

> GenFT("C:/Temp/Veeroja.SV", 62, 20); T:=FTabLine("C:/Temp/Veeroja.SV");

Warning, the resultant file is in <c:\temp\veeroja.sv>

[1, 62, 1064]

T := table([1 = [0, 3], 2 = [6, 44], 3 = [47, 99], 4 = [102, 127], 5 = [130, 130], 6 = [133, 187],
7 = [190, 227], 8 = [230, 275], 9 = [278, 293], 10 = [296, 311], 11 = [314, 354], 12 = [357, 395],
13 = [398, 412], 14 = [415, 474], 15 = [477, 484], 16 = [487, 543], 17 = [546, 550],
18 = [553, 567], 19 = [570, 626], 20 = [629, 654], 21 = [657, 676], 22 = [679, 739],
23 = [742, 784], 24 = [787, 792], 25 = [795, 847], 26 = [850, 904], 27 = [907, 938],
28 = [941, 1002], 29 = [1005, 1045], 30 = [1048, 1061]])

> FTabLine("C:/Temp/Veeroja.SV", 'h'): h;

[30, [5, 1], [28, 62]]

> F:= "C:/RANS/Orders/ABS\\AAA.txt": t:= time(): Tab1:= FTabLine(F): time() – t;

62.942

> nLine(F, Tab1, [100, 200, 500000, 1000]);

Warning, line with 500000th number does not exist

["100", "200", *false*, "1000"]

> T:=FTabLine("C:\\Temp\\Veeroja.SV", FLDT);

Warning, the FLDT has been activated as <veeroja_sv_fldt>
Warning, the saving result is in datafile <c:/temp/veeroja.sv.m>
Warning, the FLDT for file <C:\Temp\Veeroja.SV> has been saved in file
<C:\Temp\Veeroja.SV.m>

T := table([1 = [0, 21], 2 = [24, 68], 3 = [71, 86], 4 = [89, 108], 5 = [111, 159], 6 = [162, 211], 7 = [214, 265], 8 = [268, 310], 9 = [313, 358], 10 = [361, 413], 11 = [416, 422], 12 = [425, 464], 13 = [467, 498], 14 = [501, 503], 15 = [506, 519], 16 = [522, 577], 17 = [580, 632], 18 = [635, 656], 19 = [659, 698], 20 = [701, 700], 21 = [703, 762], 22 = [765, 824], 23 = [827, 885]])

> eval(veeroja_sv_fldt);

veeroja_sv_fldt

> FTabLine("C:/Temp/Veeroja.sv", 'h'): h;

Warning, the FLDT has been activated as <veeroja_sv_fldt>

[23, [20, 0], [21, 60]]

> eval(veeroja_sv_fldt);

table([1 = [0, 21], 2 = [24, 68], 3 = [71, 86], 4 = [89, 108], 5 = [111, 159], 6 = [162, 211], 7 = [214, 265], 8 = [268, 310], 9 = [313, 358], 10 = [361, 413], 11 = [416, 422], 12 = [425, 464], 13 = [467, 498], 14 = [501, 503], 15 = [506, 519], 16 = [522, 577], 17 = [580, 632], 18 = [635, 656], 19 = [659, 698], 20 = [701, 700], 21 = [703, 762], 22 = [765, 824], 23 = [827, 885]])

> Tab:= FTabLine1("C:/RANS\\Family.6", 1 .. 4, 'U');

Tab := table([*1996* = [89, 104], *1989* = [72, 86], *1947* = [18, 33], *1962* = [56, 69], *1967* = [36, 53], *1942* = [0, 15]]), [6, [*1962*, 14], [*1967*, 18]]

> U, [U[2][1], U[3][1]];

[6, [*1962*, 14], [*1967*, 18]], [*1962*, *1967*]

> Tab1:= FTabLine1("C:/RANS\\Paskula.147", 1 .. 4, 'G', FLDT), G;

Warning, the FLDT for file <c:/rans\Paskula.147> has been saved in file <c:/rans\Paskula.147.m>

Tab1 := table([*1967* = [50, 72], *1947* = [25, 47], *1989* = [100, 122], *1942* = [0, 22], *1996* = [125, 148], *1962* = [75, 97]]), [6, [*1942*, 23], [*1996*, 24]]

> FTabLine1("C:/RANS\\Paskula.147", 1 .. 4);

Warning, the FLDT has been activated as <paskula_147_fldt>

table([*1967* = [50, 72], *1947* = [25, 47], *1989* = [100, 122], *1942* = [0, 22], *1996* = [125, 148], *1962* = [75, 97]])

> eval(paskula_147_fldt);

table([*1967* = [50, 72], *1947* = [25, 47], *1989* = [100, 122], *1942* = [0, 22], *1996* = [125, 148], *1962* = [75, 97]])

> nLine("C:/RANS\\Family.6", Tab, [`1989`, `1995`, `1996`, `1947`, `2003`]);

Warning, line with key <1995> does not exist
Warning, line with key <2003> does not exist

["1989-Artur-2003", *false*, "1996-Kristo-2003", "1947-Galina-2003", *false*]

> mapN(nLine, 3, "C:/RANS\\Family.6", Tab, n, {`1989`, `1996`, `1947`});

{["1989-Artur-2003"], ["1996-Kristo-2003"], ["1947-Galina-2003"]}

Creating of a simple *TEXT* datafile containing 100.000 lines (for *Maple* of release **9**):

A test datafile is created by means of simple cyclic construction and procedure ***writeline.***

```
> F:= "C:/Temp/AAA.txt": for k to 10^5 do writeline(F, cat("", k)) end do: Close(F):
```

For the created file by means of the procedure **FTabLine** the *FLDT* T is evaluated, requiring for this purpose about 160 secs on PC **TARGA II** (Pentium II, *Windows* 98 SE) with 266 MHz.

```
> t:= time(): T:= FTabLine(F): time() - t;
                                   161.942
```

Next, by means of the procedure **nLine** an access to any line of the file over its number does not exceed 0.01 sec on the **PC TARGA II** with **64** Mb RAM in the environment *Maple* **9.** Thus, the combined use of the above procedures **FTabLine, FTabLine1** and **nLine** allows to organize for the *TEXT* datafiles to some extent an indexed access method both by numbers of their lines and by keys which are in the identical fixed positions of lines, what in a lot of cases of operation with large files allows to gain an considerable enough temp scoring.

```
> t:= time(): nLine(F,T,2003), time()-t, assign('t' = time()), nLine(F,T,10^3), time()-t,
  assign('t'=time()), nLine(F,T,50000), time()-t, assign('t'=time()), nLine(F,T,90000), time()-t;
           ["2003"], 0.009, ["10000"], 0.008, ["50000"], 0.008, ["90000"], 0.008
```

Comparative timing analysis of sequential access to lines of the datafile and by means of the procedures *FTabLine* and *nLine:*

```
> n:=90000: t:= time(): for k to n do h:=readline(F) end do: h,time()-t; Close(F);
                                 "90000", 6.594
> t:= time(): nLine(F, T, 90000), time()-t;
                                ["90000"], 0.007
> t:=time(): for k to 3000 do nLine(F, T, k) end do: time()-t;
                                   35.189
> t:=time(): for k to 3000 do close(F): for j to k do readline(F) end do: end do: time()-t;
                                   256.995
```

In a case of large volume (the amount of lines of a *TEXT* datafile can exceed 100.000), a need in organization of access to *lines* of such datafile by their numbers and/or keys arises. The above two mechanisms of such access are based on the procedures **DAopen, DAread, DAclose** and **nLine, FTabLine, FTabLine1** accordingly. The given tools provide an effective enough access to the *TEXT* datafiles represented by records (*lines*) of variable or constant length. Practical use of these mechanisms had confirmed their high enough effectiveness relative to the sequential access to lines of the *TEXT* datafiles of different purpose.

---

mtf – merging/partition and copying of *TEXT* datafiles
FT_part
FTcopy
FTmerge

Call format of the procedures:

mtf(F, P, r, n)
FT_part(F, R, d, p)
FTcopy(F, R)
FTmerge(F, R {, F1})

Formal arguments of the procedures:

F, R, F1 - symbol or string defining a name or full path to datafile or directory
P - symbol or string defining a filename
d, r, n - symbols or strings
p - a positive integer (*posint*)

Description of the procedures:

As that was noted in our books [8-14,29-33,43,44] the interpretations of *TEXT* datafiles in the *Maple* environment and MS DOS essentially differ. So, in the MS DOS environment any file of ASCII format is considered as a text file, while in the *Maple* environment not each ASCII file is considered as a text file, i.e. file of the *TEXT* type. Meantime, in many cases a problem of merging of several files of ASCII format into alone file arises. For solution of such problems, the **mtf** procedure is intended. The procedure has many useful appendices at operation with datafiles of the above type.

The procedure call **mtf(F, P, r, n)** provides a merging of text files of a subdirectory specified by the first actual **F** argument, into a file specified by the second actual **P** argument (only basic name of the file without its extension should be coded), which will be located in the same subdirectory **F.** In addition, the 3rd actual **r** argument specified a files delimiter in an output datafile, and the fourth actual **n** argument defines a name extension of the merged files of the **F** subdirectory; at coding for argument **n** of the value "*.*" all text datafiles of the indicated subdirectory **F** are subjected to merging into a datafile with filename extension **".txt"** (*txt*-file). If **n <> "*.*"**, only datafiles with extension **n** will be merged, in addition the resultant filename receives an extension **n.** The successful procedure call returns the full path to a datafile with the merge result and with output of an appropriate warning. The procedure **mtf** handles the basic erroneous situations.

The procedure **FT_part** is inverse in relation to the previous **mtf** procedure. The procedure call **FT_part(F, R, d, p)** provides partition of a text datafile specified by the first actual **F** argument onto subfiles whose names will be generated on the basis of a path to the resultant files (**R** argument) with the indication in it only of basic name of the file (i.e. filename without its extension). The partition of an **F** file is done with respect to a delimiter specified by the actual **d** argument, with extracting of parts containing **p** delimiters in each part (except, may be, of the last part). In a result of procedure call is created a set of *txt*-subfiles in the subdirectory specified by the second actual **R** argument and is returned the 2-element sequence, whose the first element defines size of the last subfile, while the second defines a total of delimiters **d** in the source file **F.** The partition result is accompanied by an appropriate warning. If the procedure determines an impossibility to fulfill a partition of the source file **F,** the appropriate warning is output. The procedure handles the basic erroneous situations.

The procedure call **FTcopy(F, R)** provides copying of a text file specified by the first actual **F** argument into a file specified by the second actual **R** argument. The successful termination of the procedure returns the full path to a datafile with the copy result and with output of an appropriate warning. If a target file **R** already exists, it is being updated, if the directory path to a target file is absent, it is created with output of appropriate warning. The procedure allows to copy into a new file being in a new directory of any nesting level. The procedure handles the basic erratic situations.

The procedure call **FTmerge(F, R)** provides merging of two text files specified by own file qualifiers **F** and **R** (by means of adding of the second file into end of the first file) with return of the full path to a datafile with the merge result and with output of an appropriate warning in case of successful terminating. If the actual arguments define the same file in the different coding, the first file will be doubled; otherwise, the second file is added in the end of the first file. The procedure call **FTmerge(F, R, F1)** provides merging of two text files specified by own file qualifiers **F** and **R** with putting of the merge result into a datafile specified by the third optional **F1** argument. The procedure allows to merge into a new datafile being in a new directory of any nesting level. The procedure handles the basic erratic situations. The procedure algorithm uses the above artificial method consisting in opening of the same datafile on two different logical I/O channels by means of coding of filenames or paths to them in different transcription.

The providing of return of the full path to a target datafile allows successfully to use the tools of this group both in interactive and in automatic programming mode. The given circumstance is conditioned by the fact; the target file with a processing result can differ from files specified at the procedure call. The tools of the given group are effective enough for *TEXT* datafiles processing.

```
mtf:= proc(F::{string, symbol}, P::{string, symbol}, r::{string, symbol}, n::{name, string})
local a, g, cd, nu, k, R, h;
  if not type(F, dir) then ERROR("<%1> is not a directory", c)
  elif evalb(assign(a = DirF(F, only)), a = []) then ERROR("directory <%1> is empty", F)
  else
    assign(g = cat("", n), cd = currentdir()), assign(nu = [seq(`if`(type(a[k], dir), NULL,
      `if`(g = "*.*" and type(a[k], rlb), a[k], `if`(g <> "*.*" and IsFtype(a[k], n), a[k], NULL))),
      k = 1 .. nops(a))]), `if`(nu = [],
      ERROR("files for merging do not exist in directory <%1>", F), 7);
    currentdir(F), assign(R = cat(P, `if`(g = "*.*", ".txt", `if`(g[1] = ".", g, cat(".", g)))))
  end if;
  for h in nu do writebytes(R, cat(readbytes(h, TEXT, infinity), r)); close(h) end do;
  close(R), currentdir(cd), WARNING("the resultant file is in <%1>", cat(F,"\\", R)), cat(F,"\\",R)
end proc

FT_part:= proc(F::{string, symbol}, R::{string, symbol}, d::{string, symbol}, n::posint)
local b, k, p, S, L, nu, omega, h;
  assign(omega = interface(warnlevel)), `if`(type(F, file), `if`(type(F, rlb), `if`(length(d) <> 1,
     ERROR("the 3rd argument should has length 1 but had received %1", length(d)),
     [assign(S = readbytes(F, TEXT, infinity)), close(F), assign(L = Search(S, d))]),
    ERROR("file <%1> has a type different from TEXT", F)), ERROR("<%1> is not a file", F));
  `if`(L <> [] and n <= nops(L), assign(nu = [L[n*k] $ (k = 1 .. nops(L)/n)]),
    RETURN(WARNING("<file <%1> cannot be exposed to a partition", F)));
  interface(warnlevel = 0), assign(h = MkDir(R, 1)), interface(warnlevel = omega);
  writebytes(cat(h, "0.txt"), S[1 .. nu[1] - 1]), close(cat(h, "0.txt")), assign(p = nops(L));
  for k to nops(nu) - 1 do b:= cat(h, k, ".txt"); writebytes(b, S[nu[k] + 1 .. nu[k + 1] - 1]), close(b)
  end do;
  writebytes(cat(h, nops(nu), ".txt"), S[nu[-1]+1..-1]), close(cat(h, nops(nu), ".txt")), p, fremove(h),
    WARNING("the partition result is in files <%1> - <%2>", cat(h, "0.txt"), cat(h, k, ".txt"))
end proc

FTcopy:= proc(F::{string, symbol}, R::{string, symbol})
local k, p, h, omega;
  `if`(type(F, file), `if`(type(F, rlb), assign(k = fopen(F, READ, TEXT)),
    ERROR("file <%1> has a type different from TEXT", F)), ERROR("<%1> is not a file", F)),
    assign(omega = interface(warnlevel)), interface(warnlevel = 0), assign(h = MkDir(R, 1)),
    interface(warnlevel = omega), assign(p = fopen(h, WRITE, TEXT));
  while not Fend(k) do writeline(p, readline(k)) end do;
  close(k, p), WARNING("the resultant file is in <%1>", h), h
end proc
```

```
FTmerge:= proc(F::{string, symbol}, R::{string, symbol})
local b, m, n, p, h, omega, nu, t;
  `if`(map(type, {F, R}, file) <> {true}, ERROR("<%1> or/and <%2> are not files", F, R),
    `if`(map(type, {F, R}, rlb) <> {true},
    ERROR("<%1> or/and <%2> have a type different from TEXT", F, R),
    assign(omega = (() -> `if`(nargs = 3 and type(args[3], {string, symbol}),
    [true, MkDir(args[3], 1)], [false])))), assign(t = interface(warnlevel)),
    interface(warnlevel = 0), assign(b = omega(args)), interface(warnlevel = t);
  assign(n = fopen(F, `if`(b[1], READ, APPEND), TEXT), p = fopen(R, READ, TEXT));
  nu := (x, y) -> writebytes(x, readbytes(y, TEXT, infinity));
  `if`(b[1], assign(m = fopen(b[2], WRITE, TEXT)), NULL);
  if b[1] then nu(m, n), nu(m, p), close(n, m, p) else nu(n, p), close(n, p) end if;
  assign('t' = `if`(b[1], b[2], F)), WARNING("the merge result is in file <%1>", t), t
end proc
```

Typical examples of the procedures use:

Merging of datafiles of the indicated subdirectory with partitioning of their by symbols of line feed and carriage return. At absence in the subdirectory of sought datafiles, the error arises.

```
> mtf("C:\\RANS/RAC\\REA\\IAN", Fmerge, "\n", "*.*");
  Warning, the resultant file is in <C:\RANS/RAC\REA\IAN\Fmerge.txt>
                    "C:\RANS/RAC\REA\IAN\Fmerge.txt"
> mtf("C:\\RANS/RAC\\REA\\IAN", Fmerge, "\n", "rtf");
  Warning, the resultant file is in <C:\RANS/RAC\REA\IAN\Fmerge.rtf>
                    "C:\RANS/RAC\REA\IAN\Fmerge.rtf"
> mtf("C:\\RANS/", Fmerge, "\n", "*.*"), mtf("C:\\RANS/", Fmerge, "\n", "rtf");
  Warning, the resultant file is in <C:\RANS\Fmerge.txt>
  Warning, the resultant file is in <C:\RANS\Fmerge.rtf>
                "C:\RANS\Fmerge.txt", "C:\RANS\Fmerge.rtf"
```

Partition of a *txt*-file containing symbols "," onto the subfiles with names "Emailp.txt" (p=0, 1,2,..., 61). Subfiles should be located in the specified subdirectory and each contains 250 delimiters (except, may be, of the last subfile). The last subfile has length 5545, whereas the source datafile contains 2738 delimiters.

```
> FT_part("C:/Temp\\Result.txt", "C:/RANS/Orders\\Email", ",", 250);
  Warning, the partition result is in files <c:\rans\orders\email0.txt> -
  <c:\rans\orders\email61.txt>
                          5545, 2738
> FT_part("C:/aaa/bbb/ccc/result1.txt", "C:/aaa/bbb/ddd/res1", ",", 7);
  Warning, file <C:/aaa/bbb/ccc/result1.txt> cannot be exposed to a partition
```

Copying of a text datafile with creating a new directories path for a target datafile.

```
> FTcopy("C:/Temp/AVZ.txt", "C:/Temp\\Vilnius/Grodno\\Moscow/Tallinn\\AGN.txt");
  Warning, the resultant file is in <c:\temp\vilnius\grodno\moscow\tallinn\agn.txt>
              "c:\temp\vilnius\grodno\moscow\tallinn\agn.txt"
```

Merging of two text datafiles (different and identical ones).

> FTmerge("C:/rans/result1.txt", "C:/rans/trg.txt", "C:/Temp/rans/ian/Merge.txt");

Warning, the merging result is in file <c:\temp_rans\ian\merge.txt>

"c:\temp_rans\ian\merge.txt"

> FTmerge("C:/RANS/TRG.txt", "C:/rans/copy.mws");

Warning, the merging result is in file <C:/RANS/TRG.txt>

"C:/RANS/TRG.txt"

> FTmerge("C:/RANS/TRG.txt", "C:/rans/trg.txt");

Warning, the merging result is in file <C:/RANS/TRG.txt>

"C:/RANS/TRG.txt"

`convert/TEXT` – converting of a datafile into *TEXT* datafile

Call format of the procedure:

convert(F, TEXT)

Formal arguments of the procedure:

F – a symbol or string defining filename or full path to a datafile to be converted
TEXT – a key word

Description of the procedure:

The procedure call **convert(F, TEXT)** provides a converting of an arbitrary file defined by the actual **F** argument into file of the *TEXT* type. The successful procedure call **convert(F, TEXT)** returns the *true* value without change of the source file defined by the first actual **F** argument if it has the *TEXT* type; otherwise, the procedure converts it into datafile of the *TEXT* type with return of the *NULL* value, i.e. nothing, and with output of the appropriate warning.

```
`convert/TEXT`:= proc(F::{string, symbol})
local h;
  `if`(not type(F, file), ERROR("<%1> is not a file", F), `if`(F_T(F)=TEXT, RETURN(true), 7));
  close(F), assign(h=readbytes(F, infinity)), close(F), writebytes(F, subs([0=7, 26=10, 13=10], h));
  null(writebytes(F, [13, 10]), close(F)),
     WARNING("file <%1> has been converted into TEXT-type", F)
end proc
```

Typical examples of the procedure use:

1. Converting of the *exe*-file `RegClean.exe` into the appropriate *TEXT* file.

> F:="c:/program files/maple 7/bin.wnt/cmaple.exe": F_T(F), convert(F, TEXT), F_T(F);

Warning, file <c:/program files/maple 7/bin.wnt/cmaple.exe> has been converted into TEXT-type

BINARY, TEXT

2. Line-by-line read of the *TEXT* file `Cmaple.exe` with return of quantity of lines and the last line.

> T:= 0: do assign('h' = readline(F)); if h = 0 then close(F): break else assign('R' = h, 'T' = T + 1) end if end do; T, R;

636, "\a\a\asystem calls are disabled\a\a\asystem and ! are not available under Windows ..."

The useful **`convert/TEXT`** procedure concerns the general tools of providing of access to the *TEXT* files; it provides (in contrast to the similar **`convert/txt`** procedure) a converting in situ of an arbitrary file into file of the *TEXT* type. The procedure has a series of useful appendices.

FT_part1 – operations with *TEXT* datafiles in the *Maple* environment (*group 3*)
FT_restr
QFline
FmF
sorttf

Call format of the procedures:

FT_part1(F, R, d, n)
FT_restr(F, d1, P)
QFline(F)
FmF(F, S, M {, G})
sorttf(F, {, S} {, sf} {, 1})

Formal arguments of the procedures:

F - symbol or string defining filename or full path to a text datafile
R - symbol or string defining a prefix of path to the target datafiles
d - a delimiter (may be *symbol* or *string*)
n - a positive integer (*posint*)
P - symbol or string defining filename or full path to a target datafile
d1 - a marker (may be *string, symbol, set*({*symbol, string*}) or *list*({*symbol, string*}))
G - symbol or string defining filename or full path to a target datafile
S - symbol or string defining filename or full path to a target datafile
M - a mode of files manipulation {**`union`** | **`intersect`** | **`minus`**}
1 - (optional) a key symbol defining a type of the sorted datafile
sf - (optional) an ordering function (*boolean function of two arguments*)

Description of the procedures:

Procedure **FT_part1(F, R, d, n)** is intended for creation on the basis of an input *TEXT* file **F** of a set of the files composed of tuples of symbols of length **n** of the first file provided that they contain full entrances of successively going tuples of the input file which are limited by the given marker **d.** Full paths to target files have view "Rppp.txt", where <ppp> is an integer and **R** is the prefix of path to the target files, i.e. a directories path with a main filename without its extension. The procedure **FT_part1** handles the basic erratic and especial situations.

Successful termination of the procedure returns the *NULL* value with output of warning about quantity of target files and full paths to them. In particular, the procedure **FT_part1** seems useful at necessity to split big files containing email addresses lists parted by *comma* or *blank* onto sublists that can be completely located in the receiving buffer of addresses of a post server.

Procedure **FT_restr(F, d1, P)** is intended for a restructuring of an input file **F** into a text file consisting of the lines received by replacement of all entrances of **d1**-marker of the file **F** onto the control symbol "**\n**" (*newline* character). Markers can be specified by a list, a set, separate symbol or by a line of appropriate length. The third actual argument **P** specifies a target file. Successful termination of the procedure returns the full path to the target file (which can differ from **P**) with output of warning about quantity of lines of the restructured file or about impossibility of such restructuring. In the last case, the procedure call returns the *NULL* value and deletes the target file.

In addition, if the third actual **P** argument specifies a non-existent directories path to a target file, it will be created. The procedure has useful appendices at operating with text datafiles when their line-by-line processing is needed.

Simple procedure **QFline(F)** returns the quantity of lines contained in a *text* file specified by its specifier **F** (*filename* or *full path* to a datafile); in addition, the **F** file may has a ***text*** type different from *TEXT*, i.e. the actual **F** argument can generically specify a datafile of the *rlb*-type.

The procedure call **FmF(F, S, M {, G})** provides a set-theoretic manipulation of two text files specified by their qualifiers **F** and **S** on a level of lines composing them. For the third argument **M** a value from set {**`union`**, **`minus`**, **`intersect`**} is valid only. In addition, the result is put into a datafile specified by the fourth optional argument **G,** or at its absence replaces an initial **F** datafile. Successful termination of procedure returns the full path to a target file **G**. In addition, the target file lines are not being duplicated and sorted in lexicographical order. If the fourth actual **G** argument has been coded and it defines a non-existent directories path to a target file, the directories path to the **G** datafile will be created. The given procedure has useful enough appendices at operating with text datafiles.

The procedure call **sorttf(F {, S} {, sf} {, 1})** provides a quick line-by-line sorting of a text file specified by the first actual **F** argument according to a specified ordering *sf*-function. If the optional actual **S** argument has been coded and it specifies a non-existent directories path to a target datafile, the directories path to the **S** file will be created. At absence of the target file **S** an initial file **F** will be sorted in situ; at absence of an ordering *sf*-function the sorting is being made according to ***lexorder***-function. At last, at coding of optional key symbol **`1`** in the sorted datafile the *multiple* entries of lines are absent. Successful termination of the procedure returns the full path to a target datafile with output of the appropriate warning.

```
FT_part1:= proc(F::{string, symbol}, R::{string, symbol}, d::{string, symbol}, n::posint)
local a, k, S, L, p, h, nu, omega;
  `if`(not type(F, file), ERROR("<%1> is not a file", F), `if`(not type(F, rlb),
    ERROR("file <%1> has a type different from TEXT", F), `if`(length(d) <> 1,
    ERROR("the 3rd argument should has length 1"), assign(nu = interface(warnlevel)))));
  interface(warnlevel = 0), assign(S = Red_n(readbytes(F, TEXT, infinity), " ",
    `if`(d <> " " and d <> ` `, 1, 2)), p = 0), close(F), assign(a = CFF(R)),
    assign(h = cat(MkDir(clsd(a[1 .. -2], "\\")), "\\", a[-1])), interface(warnlevel = nu),
    assign('S' = redss(S, d, n), omega = (x -> `if`(x = 1, x, x - 1)));
  `if`(S = NULL, ERROR("conditions of partition are incompatible", 'procname(args)'), NULL);
  do
    if length(S) <= n then null(writebytes(cat(h, p + 1, ".txt"), S)), close(cat(h, p + 1, ".txt"));
      break
    else assign('p' = p + 1, 'L' = Search2(S, {d})) end if;
    for k to nops(L) do
      if n <= L[k] then null(writebytes(cat(h, p, ".txt"), S[1 .. L[omega(k)] - 1])),
          close(cat(h, p, ".txt")), assign('S' = S[L[omega(k)] + 1 .. -1]);
        break
      end if
    end do
  end do;
  WARNING("%1 files <%2> have bee n generated where X belongs to diapason [1..%1]",
    p + 1, cat(h, "X.txt"))
end proc
```

```
FT_restr:= proc(F::file, d1::{string, symbol, set({string, symbol}), list({string, symbol})},
                    P::{string, symbol})
local k, S, p, n, G, h, w;
  assign(w = interface(warnlevel)), interface(warnlevel = 0), assign(h = MkDir(P, 1));
  writebytes(h, readbytes(F, infinity)), Close(F, h), `if`(type(h, rlb), NULL, convert(h, rlb));
  assign(S = readbytes(h, TEXT, infinity)), close(h), assign(n = nops({op(d1)})),
    assign(G = [Sub_st([seq({op(d1)}[k] = "\n", k = 1 .. n)], S, p)]), interface(warnlevel = w),
    `if`(p = true, assign('S' = Red_n(G[1], "\n", 2)),
    RETURN(WARNING("file <%1> cannot be restructured", F), fremove(h)));
  WARNING("quantity of lines in file <%2> is equal %1", nops(Search2(S, {"\n"})), h),
    null(writebytes(h, S)), Close(h), h
end proc

QFline:= F::file -> `if`(type(F, rlb), op([Qsubstr(readbytes(F, TEXT, infinity), "\n"), close(F)]),
          ERROR("file <%1> has a type different from rlb-type", F))

FmF:= proc(F::file, S::file, M::symbol)
local Art, Kr, h, p;
  `if`(member(M, {`intersect`, `minus`, `union`}), assign(Art = {}, Kr = {}),
    ERROR(`3rd argument is wrong`));
  do h := readline(F); if h = 0 then close(F); break else Art := {h, op(Art)} end if end do;
  do h := readline(S); if h = 0 then close(S); break else Kr := {h, op(Kr)} end if end do;
  assign('Art' = M(Art, Kr)), assign('h' = `if`(nargs = 3, F, pathtf(args[4])));
  seq(null(writeline(h, p)), p = sort([op(Art)], lexorder)), Close(h), h
end proc

sorttf:= proc(F::file)
local nu, h, k, p, f, omega, sf, t;
  `if`(IsFempty(F), ERROR("file <%1> is empty", F), `if`(type(F, rlb),
    assign(nu = [], p = 0, f = F, sf = lexorder), ERROR("<%1> is not a text file", F)));
  do h := readline(F); if h = 0 then close(F); break else nu := [op(nu), h] end if end do;
  if 1 < nargs then for k from 2 to nargs do `if`(type(eval(args[k]), {string, symbol}),
      assign('f' = args[k], 'p' = 7), `if`(type(args[k], boolproc), assign('sf' = args[k]),
      `if`(args[k] = 1, assign('nu' = redL(nu)), NULL)))
    end do
  end if;
  if p = 7 then t := interface(warnlevel); interface(warnlevel = 0); f := MkDir(f, 1);
    interface(warnlevel = t)
  end if;
  null(seq(writeline(f, h), h = sort(nu, sf))), close(f), f,
    WARNING("the sorting result is in file <%1>", f)
end proc
```

Typical examples of the procedures use:

> FT_part1("C:/Temp/Orders\\Result.txt", "C:/Academy/Result", ",", 250);

Warning, 323 files <c:\temp\resultX.txt> have been generated where X belongs to diapason [1..323]

> FT_part1("C:/Windows/command/Edit.com", "C:/Academy/Result", ",", 250);

Error, (in **FT_part1**) file <C:/Windows/command/Edit.com> has a type different from TEXT

> FT_part1("C:/rans/merge", "C:/temp/R", ",", 7);

Error, (in **FT_part1**) conditions of partition are incompatible, FT_part1(c:/rans/merge,c:/temp/r,,,7)

> FT_part1("C:/rans/E5_mail.txt", "C:/temp/Res", ",", 250);

Warning, 148 files <c:\temp\resX.txt> have bee n generated where X belongs to diapason [1..148]

> FT_restr("C:/RANS/E5_mail.txt", {",", ` `}, "C:/Temp/Galina\\Sveta/Artur\\AGN.56");

Warning, quantity of lines in file <c:\temp\galina\sveta\artur\agn.56> is equal 389

"c:\temp\galina\sveta\artur\agn.56"

> FT_restr("C:/RANS/E5_mail.txt", "\t", "C:/Temp/Galina\\Sveta/Artur\\AGN.56");

Warning, file <C:/RANS/E5_mail.txt> cannot be restructured

> FT_restr("C:/RANS/E5_mail.txt", {",", ` `}, "C:/Temp/Galina");

Warning, quantity of lines in file <c:\temp_galina> is equal 389

"c:\temp_galina"

> map(QFline,["C:/RANS\\Fmerge1.rtf","C:/Temp_Galina.56"]);

[1995, 2003]

> QFline("C:/RANS\\Format.com");

Error, (in **QFline**) file <C:/RANS\Format.com> has a type different from rlb-type

> FmF("c:/RANS/Res1.txt","c:/RANS/Res2.txt",`*intersect*`,"c:/Temp/Galina/Sveta/Artur");

"c:\temp\galina\sveta\artur"

> FmF("C:/RANS/Res1.txt","C:/RANS/Res5.txt", `*union*`, "C:/Temp/Galina/Sveta/");

"c:\temp\galina_sveta"

> FmF("C:/RANS/Res1.txt","C:/RANS/Res1.txt", `minus`, "C:/Temp/Galina/");

"c:\temp_galina"

> FmF("C:/RANS/Res1.txt","C:/RANS/Res5.txt", `*union*`);

"C:/RANS/Res1.txt"

> sorttf("C:/RANS/Res2.txt", "C:/Temp/RANS/IAN/Sort.txt");

Warning, the sort result is in file <c:\temp\rans\ian\sort.txt>

"c:\temp\rans\ian\sort.txt"

> sorttf("C:/RANS/Res2.txt", "C:/Temp/RANS/IAN/Sort1.txt", 1);

Warning, the sort result is in file <c:\temp\rans\ian\sort1.txt>

"c:\temp\rans\ian\sort1.txt"

An ordering function specification:

> psi:= (a::string, b::string) -> `if`(op(convert(a[-1], bytes)) < op(convert(b[-1], bytes)), true, false): sorttf("C:/RANS/Res3.txt", "C:/Temp/RANS/IAN/Sort2.txt", 1, psi);

Warning, the sort result is in file <c:\temp\rans\ian\sort2.txt>

"c:\temp\rans\ian\sort2.txt"

> sorttf("C:/RANS/Res3.txt", 1, psi);

Warning, the sort result is in file <C:/RANS/Res3.txt>

"C:/RANS/Res3.txt"

The providing of return of the full path to a target datafile allows successfully to use the tools of this group both in interactive and in automatic programming mode. The given circumstance is conditioned by the fact; the target file with a processing result can differ from files specified at the procedure call. The tools of the given group are effective enough for *TEXT* datafiles processing, and in general case for so-called *text* datafiles and datafiles of the *rlb*-type.

10.3. Software for operation with *BINARY* datafiles

As against the software of the previous item providing access to external files of *Maple* at a level of logical records (lines, sequences of *Maple* expressions, the whole *Maple* clauses), tools represented below, provide access to datafiles at a level of separate symbols (*bytes*). For datafiles processing of this type, the package *Maple* has appropriate built-in functions of access to *BINARY* datafiles. The procedures represented below extend tools of the package at processing of *BINARY* datafiles by allowing to solve a number of problems dealing with *BINARY* datafiles the more effectively. Some of them have made a good showing from the viewpoint of practical programming.

Fend – handling of especial situation *"the end of a datafile"*

Call format of the procedure:

Fend(F {, 'h'})

Formal arguments of the procedure:

F - string or symbol defining a *file qualifier* (filename or full path), or an integer from range **0..7**
h - (optional) an assignable name

Description of the procedure:

Essential distinctions between datafiles of types *TEXT* and *BINARY* are observed relative to handling of an especial situation ***"the end of a datafile"***. For *BINARY* datafiles, *Maple* has not tools of handling of the specified situation and in each concrete case this problem of handling lays on the user, what not always is desirable. The given question is considered in our books [8-14,29-33,43,44] enough in detail. With the purpose of solution of the above problem, the **Fend** procedure has been introduced.

Extremely useful procedure **Fend** provides handling of the situation ***"the end of a datafile"*** for datafiles of any type. The procedure call **Fend(F {, 'h'})** returns the *true* value in case of detection of situation ***"the end of a datafile"*** and the *false* value otherwise for a datafile specified by the first actual **F** argument (qualifier or number of a logical I/O channel). In case of coding of the second optional **h** argument (in case of the *false* value) via it the list of view [*scanned position of a file, length of the rest of a file*] is returned. The **Fend** procedure makes sense only for an open datafile and does handling of the basic especial and erroneous situations. The durable experience of usage of the given procedure has shown its quite satisfactory operating characteristics. The procedure generates erratic situations "*file descriptor <%1> is invalid*", "*<%1> is not a file*", "*all datafiles are closed*" and "*<%1>: file is closed or file descriptor is not used*" whose sense is quite clear.

```
Fend:= proc(F::{integer, string, symbol})
local k, a, b, p, h;
  `if`(type(F, integer), `if`(member(F, {k $ (k = 0 .. 7)}), 1,
```

```
    ERROR("file descriptor <%1> is invalid", F)), `if`(type(F, file), 1,
    ERROR("<%1> is not a file", F))), assign(a = iolib(13));
  `if`(a = [0, 0, 7], ERROR("all datafiles are closed"), seq(`if`(a[k][1] = F or a[k][2] = cat("", F),
    assign('h' = 6), NULL), k = 4 .. nops(a))), `if`(h <> 6,
    ERROR("<%1>: file is closed or file descriptor is not used", F), NULL);
  try assign(p = iolib(6, F)), `if`(iolib(6, F, infinity) <= p,
    RETURN(assign(b = iolib(6, F, p)), true), RETURN(false, `if`(nargs = 1,
    op([NULL, assign(b = iolib(6, F, p))]), op([assign([args][2] = [p, iolib(6, F, infinity) - p]),
    assign(b = iolib(6, F, p))]))))
  catch: null("Processing of an especial situation with datafiles {direct, process, pipe}")
  end try
end proc
```

Typical examples of the procedure use:

> filepos("C:/RANS/ABS.txt", infinity), Fend("C:/RANS/ABS.txt", 'h'), h;

917, *true*, [10]

> filepos("C:/RANS/ABS.txt", 61), Fend("C:/RANS/ABS.txt", 'h'), h;

61, *false*, [61, 856]

> F:=fopen("C:/rans/Simona.txt", READ, BINARY): while not Fend(F) do h:=readbytes(F): writebytes("C:/Temp/Simona.txt", h) end do: Close(); Fequal("C:/RANS/Simona.txt", "C:/Temp/Simona.txt");

true

> Fend("C:/Temp/Demo/Demo/DemoLib/Maple.hdb");

Error, (in **Fend**) <C:/Program Files/Maple 9/Intro.mws>: file is closed or file descriptor is not used

---

FBcopy – certain useful operations with *BINARY* datafiles
FBBcopy
Fequal
Fappend
ModFile

Call format of the procedures:

FBcopy(F, P)
FBBcopy(F, P, {, N})
Fequal(F, P {, 'h'})
Fappend(F, P {, N})
ModFile(T, T1, F {, t})

Formal arguments of the procedures:

F – symbol or string defining filename or full path to a datafile
P – symbol or string defining filename or full path to a datafile
h – (optional) an assignable name
T, T1 – symbol or string defining a pattern for replacements
N – (optional) a positive integer (*posint*)
t – (optional) an arbitrary valid *Maple* expression

Description of the procedures:

Two procedures **FBcopy(F, P)** and **FBBcope(F, P)** have the same purpose, providing a copying of a datafile specified by the first actual **F** argument, into a datafile specified by the second actual **P** argument. If the first procedure has a high reactivity, but on source files of very large size can cause a system error (`System error, `, "out of memory"), then the second procedure operates correctly on files of any type and size, having a smaller reactivity. If path to a target file **P** is absent, then it will be created with output of appropriate warning; otherwise, the existing target datafile will be updated with output of an appropriate warning. The successful procedure call returns the full path to a target datafile as a *string*.

As against the **FBcopy**, the procedure call **FBBcopy(F, P, {N})** admits the third optional actual **N** argument allowing to define size in **KBs** (*by default* N = 30 KB) of program I/O buffer for a file copying. A maximal value of this argument depends on the available main memory for a current session; at increasing of this value, the time of copying essentially decreases for large enough source files that should be copied. If path to a target file **P** is absent, then it will be created with output of appropriate warning; otherwise, the existing target file will be updated with output of an appropriate warning. The successful procedure call returns the full path to a target datafile as a *string*.

The procedure call **Fequal(F, P {, 'h'})** tests the identity of two datafiles specified by actual arguments **F** and **P**, and at detection of the given fact returns the *true* value; otherwise the *false* value is returned. At coding of the optional third argument **h** through it, the list is returned whose elements define the datafiles positions in that they differ. This procedure has a sense for small datafiles (at most 3 - 5 KB).

The procedure call **Fappend(F, P)** provides a merging of two files specified by own qualifiers **F** and **P** (by means of adding of the second file into the end of the first file) with return of the *NULL* value in case of successful terminating. In addition, it is necessary to note, that the semantically correct merging is done only for *text* datafiles. At coding both actual arguments of the procedure by identical ones (in semantic instead of syntactical sense) a source file is doubled. The procedure call **Fappend(F, P {, N})** admits the third optional actual **N** argument allowing to define size in **KB** s (*by default* N = 30 KB) of program I/O buffer for a datafiles merging. A maximal value of this argument depends on the available main memory for a current session; at increasing of this value, the merging time essentially decreases for large source datafiles.

The procedure call **ModFile(T, T1, F {, t})** provides replacement in a file specified by the 3rd actual **F** argument, of each entry of a **T** pattern onto a **T1** pattern with return of the list whose the first element defines new length of the modified file, whereas the second element defines a quantity of done replacements. By default, at patterns replacement the case sensitive mode is used, whereas coding of the fourth optional argument **t** allows to use the insensitive mode. The given procedure provides modification of a file whose size does not exceed 260 KB. In addition, it is necessary to have in view; the procedures of the given item (*excepting the* **Fend** *procedure*) admit both filenames and qualifiers (*filenames* or *full paths* to datafiles) as actual arguments, but not the numbers of logical I/O channels, ascribed to them.

```
FBcopy:= proc(F::{string, symbol}, P::{string, symbol})
local a, b, c, t;
  assign(t = interface(warnlevel)), `if`(not type(F, file), ERROR("<%1> is not a file", F),
    `if`(type(P, file), [WARNING("target file <%1> has been updated", P),
    assign(b = P, c = 7), Close(P)], `if`(not type(eval(P), {string, symbol}),
    ERROR("<%1> is not a file path", P), NULL)));
```

```
  `if`(c <> 7, [interface(warnlevel = 0), assign('b' = MkDir(P, 1)), interface(warnlevel = t)], 7);
  null(iolib(5, b, iolib(4, F, infinity))), Close(F, b), WARNING("the target file is <%1>", b), b
end proc

FBBcopy:= proc(F::{string, symbol}, P::{string, symbol})
local b, c, p, t, n;
  assign(t = interface(warnlevel)), `if`(not type(F, file), ERROR("<%1> is not a file", F),
    `if`(type(P, file), [WARNING("target file <%1> has been updated", P),
    assign(b = P, c = 7), Close(P)], `if`(not type(eval(P), {string, symbol}),
    ERROR("<%1> is not a file path", P), NULL))), Close(F);
  `if`(c <> 7, [interface(warnlevel = 0), assign('b' = MkDir(P, 1)), interface(warnlevel = t)], 7);
  assign(p = fopen(F, READ, BINARY), n = `if`(2 < nargs and
    type(args[3], posint), 1024*args[3], 30720));
  while not Fend(p) do writebytes(b, readbytes(F, n)) end do;
  Close(F, b), WARNING("the target file is <%1>", b), b
end proc

Fequal:= proc(F::file, P::file)
local k, n, p, t, L;
  `if`(CF(F) = CF(P), RETURN(true), [assign(n = map(filepos, [F, P], infinity)), Close(F, P)]);
  `if`(n[1] <> n[2], RETURN(false, WARNING("files have different length")),
    assign(p = map(fopen, [F, P], READ, BINARY), t = 0, L = []));
  while not Fend(p[1]) do t := t + 1;
    if iolib(4, p[1]) <> iolib(4, p[2]) then L := [op(L), t] end if
  end do;
  `if`(L = [close(op(p))], true, op([false, `if`(nargs = 3 and type(args[3], assignable),
    assign(args[3] = L), NULL)]))
end proc

Fappend:= proc(F::file, P::file)
local n, p, f;
  assign(n = `if`(2 < nargs and type(args[3], posint), 1024*args[3], 30720)), close(F, P);
  assign(f = Case(F)), filepos(f, filepos(f, infinity)),
    assign(p = fopen(Case(P, upper), READ, BINARY));
  (proc() while not Fend(p) do writebytes(f, readbytes(p, n)) end do; Close(F, P) end proc)()
end proc

ModFile:= proc(T::{string, symbol}, T1::{string, symbol}, F::{string, symbol})
local R, n, p;
  `if`(type(F, file), assign67(R = readbytes(F, infinity, TEXT), p = 0, Close(F)),
    ERROR("<%1> is not a file", F));
  `if`(map(type, map(eval, {T, T1}), {string, symbol}) = {true}, NULL,
    ERROR("the first two arguments should be symbols or strings but have received %3", T, T1,
    map(whattype, map(eval, [T, T1]))));
  `if`(search(`if`(3 < nargs, Case(T1), T1), `if`(3 < nargs, Case(T), T)),
    ERROR("substitution  <%1 = %2> initiates the infinite process", T, T1), NULL);
  do
    if search(`if`(3 < nargs, Case(R), R), `if`(3 < nargs, Case(T), T), 'n') then
      assign('R' = cat(R[1 .. n - 1], T1, R[n + length(T) .. length(R)]), 'p' = p + 1)
    else break
    end if
```

```
   end do;
   [close(F), null(writebytes(F, R)), filepos(F, infinity), p, close(F)]
end proc
```

Typical examples of the procedures use:

> FBcopy("C:/RANS/ABS.txt", "C:/Program Files/RANS_IAN.txt");

Warning, target file <C:/Program Files/RANS_IAN.txt> has been updated
Warning, the target file is <C:/Program Files/RANS_IAN.txt>

"C:/Program Files/RANS_IAN.txt"

> FBcopy("C:/Program Files/Maple 6/Bin.Wnt/Wmaple.exe", "C:/RANS");

Warning, the target file is <c:_rans>

"c:_rans"

> FBcopy("C:/Program Files/Maple 8/LIB/Maple.hdb", "C:/RANS/Maple.hdb");

System error, , "out of memory"

> FBBcopy("C:/Program Files/Maple 8/LIB/Maple.hdb", "C:/RANS/Maple.hdb");

Warning, the target file is <c:\rans\maple.hdb>

"c:\rans\maple.hdb"

> FBBcopy("C:/Program Files/Maple 8/LIB/Maple.hdb", "C:/RANS/Maple.hdb", 60);

Warning, target file <C:/RANS/Maple.hdb> has been updated
Warning, the target file is <C:/RANS/Maple.hdb>

"C:/RANS/Maple.hdb"

> Fequal("C:/RANS/Simona.txt", "c:\\rans\\Simona.txt");

true

> Fequal("C:/RANS/Simona.txt", "c:\\RANS\\ABS.txt");

Warning, files have different length

false

> Fequal("C:/RANS/Simona.txt", "c:\\RANS\\Order.rlb", 'h'), h;

false, [85, 494, 556, 705, 801, 802, 803, 804, 805, 806]

> Fappend("C:/Temp/Orders\\ICSS1.txt", "C:/Temp/Orders\\ICSS1.txt");

> Fappend("C:/RANS/ABS.txt", "C:/RANS/Order.rlb", 40);

> ModFile(`PP`, `RANS`, "C:/RANS/Lectures.rtf");

Error, (in **ModFile**) the first two arguments should be symbols or strings but have received [procedure, symbol]

> ModFile(`Books`, `Monographs`, "C:/RANS/ABS.txt");

[2003, 14]

> ModFile(`Books`, `Monographs`, "C:/RANS/ABS.txt", 157);

[2053, 24]

> ModFile(`IAN`, `RANS_IAN`, "C:/Temp/Orders\\Books\\RANS.ian");

Error, (in **ModFile**) substitution <IAN = RANS_IAN> initiates the infinite process

ExtrF – extracting of bytes out of a datafile

Call format of the procedure:

ExtrF(F, a, b {, P})

Formal arguments of the procedure:

F – symbol or string defining a filename or full path to it
a, b – integers
P – (optional) filename or full path to a target datafile for extracted information

Description of the procedure:

In many problems, dealing with datafiles processing arises the necessity of *extraction* from a file of some quantity of successively going bytes of information without infringement of the file status (a scanned position and mode of opening). The given problem is solved by means of the procedure **ExtrF** represented below.

The procedure call **ExtrF(F, a, b)** returns the string containing contents of successively going bytes of information of a datafile specified by the first **F** argument. In addition, the second argument (**a**) and the third argument (**b**) define an initial position of the read-out bytes and their quantity accordingly.

The fourth optional **P** argument specifies a filename or full path to a target datafile intended for saving of extracted information; in addition, the procedure processes the **P** argument in concordance with the above procedure **pathtf.** At any rate, the procedure call returns the string containing the sought information out of a datafile specified by the first actual **F** argument. For an open datafile, the procedure keeps a file status, namely: number of logical I/O channel, a scanned position, and other datafile characteristics. The procedure handles basic especial and erratic situations arising at datafiles processing and is quite useful at programming of access to datafiles of different purpose.

```
ExtrF:= proc(F::{string, symbol}, a::integer, b::integer)
local k, h, p, t, d, Er, Res, Rd, io, v, G, Sv, f;
  assign(f = cat("", F), Rd = ((x, y, z) -> [iolib(6, x, y), iolib(4, x, z, TEXT), iolib(7, x)][2]),
    Sv = ((x, y) -> [iolib(5, x, y), iolib(7, x), WARNING("the result is in file <%1>", x)]),
    Er = ((x, y, z) -> `if`(z < x + y - 1, [iolib(7, f),
    ERROR("at least one of arguments %1 is invalid", [x, y])], 6))), `if`(nargs < 4, NULL,
    `if`(type(eval(args[4]), {string, symbol}), assign(d = pathtf(args[4]), t = 7),
    WARNING("<%1> is not a path", args[4])));
  `if`(isFile(f, 'h'), `if`(h = close, goto(A), goto(B)), ERROR("<%1> is not a file", F));
  A;
  assign(p = iolib(6, f, infinity)), Er(a, b, p), iolib(7, f), RETURN(Rd(f, a, b),
    null(`if`(t = 7, Sv(d, Rd(f, a, b)), NULL)));
  B;
  io := iolib(13);
  for k from 4 to nops(io) do
    if io[k][2] = f then G := io[k][2]; v := [io[k][1], iolib(6, G), io[k][3], io[k][5], io[k][6]];
      break
    else next
    end if
  end do;
  assign(p = iolib(6, f, infinity)), Er(a, b, p), iolib(7, G), assign(Res = Rd(G, a, b)),
```

```
    `if`(iolib(13) <> [0, 0, 7], iolib(7, v[1]), NULL), OpenLN(v[1],
    `if`(v[3] = RAW, open, fopen), G, v[4], v[5]), iolib(6, G, v[2]),
    RETURN(Res, null(`if`(t = 7, Sv(d, Res), NULL)))
end proc
```

Typical examples of the procedure use:

> readbytes("C:/RANS/ABS.txt", 100);

[118, 105, 100, 101, 111, 64, 97, 98, 115, 98, 111, 111, 107, 46, 99, 111, 109, 13, 10, 13, 10, 68, 101, 97, 114, 32, 68, 114, 46, 32, 74, 97, 121, 32, 82, 111, 122, 101, 110, 58, 13, 10, 13, 10, 78, 111, 118, 101, 109, 98, 101, 114, 32, 49, 115, 116, 44, 32, 50, 48, 48, 51, 32, 105, 110, 116, 111, 32, 121, 111, 117, 114, 32, 97, 100, 100, 114, 101, 115, 115, 58, 13, 10, 13, 10, 65, 77, 66, 65, 83, 83, 65, 68, 79, 82, 32, 66, 79, 79, 75]

> iostatus(), filepos("C:/RANS/ABS.txt");

[1, 0, 7, [0, "C:/RANS/ABS.txt", *STREAM*, *FP* = 2013506696, *READ*, *BINARY*]], 100

> ExtrF("C:/RANS/ABS.txt", 1519, 110, "C:/Temp/Sveta.txt");

Warning, the result is in file <c:\temp\sveta.txt>

"Tentatively I inform you, in January 2004 should be prepared our new electronic book being the second extended"

> iostatus(), filepos("C:/RANS/ABS.txt");

[1, 0, 7, [0, "C:/RANS/ABS.txt", *STREAM*, *FP* = 2013506696, *READ*, *BINARY*]], 100

---

reconf – substitutions into a datafile

Call format of the procedure:

reconf(L, F, G {, `*sensitive*`})

Formal arguments of the procedure:

L – a set or list whose elements are equations (*substitutions rules*)
F – symbol or string defining a filename or full path to a datafile to be processed
G – symbol or string defining a filename or full path to a target datafile
`*sensitive*` – (optional) key word defining a search mode at substitutions

Description of the procedure:

The procedure call **reconf(L, F, G)** provides replacement in a file specified by the second actual **F** argument, of each entry of the left member of an element from **L** onto its right member. In addition, arbitrary valid expressions can be as members of **L** equations. At patterns replacement the case insensitive mode is used, however use of the optional fourth argument **`*sensitive*`** supposes the case sensitive mode. The substitutions result is saved in a datafile specified by the third argument **G** that will be created if it does not exist. The procedure call returns the *NULL* value with output of the appropriate warnings. However, if the first actual argument **L** is *empty*, the procedure call is returned unevaluated. The given procedure provides modification of a datafile containing non-printable symbols, what, for example, is useful enough at executable files processing.

```
reconf:= proc(L::{list(equation), set(equation)}, F::file, G::{string, symbol})
local a, b, h, f, p, k, v, t;
  `if`(Empty(L), RETURN('procname(args)'), assign(f = G,
    v = mapLS(convert, [op(L)], 2, string)));
```

```
  if not type(G, file) then b := interface(warnlevel); interface(warnlevel = 0);
     f := MkDir(G, 1); interface(warnlevel = b)
  end if;
  assign(p = (() -> fopen(F, READ)), k = 512, t = mapLS(LSL, mapLS(convert, v, 2, bytes), 2)),
     assign67(a = `if`(nargs = 4 and args[4] = sensitive, sensitive, NULL));
  WARNING("Processing of datafile <%1> with result saving in datafile <%2>,
     Please wait!", F, f);
  if filepos(F, infinity) <= 10*k then close(F),
     writebytes(f, LSL(SUB_S(t, LSL(readbytes(F, infinity)), a)))
  else close(F); p();
     while not Fend(F) do
        writebytes(f, LSL(SUB_S(t, LSL(readbytes(F, k)), a)))
     end do
  end if;
  close(F, f), WARNING("the processing result has been saved in datafile <%1>", f)
end proc
```

Typical examples of the procedure use:

> reconf(["At that"="In addition", "at that"="in addition"], "C:/RANS/Maple1.hdb", "C:/RANS/Maple2.hdb", `sensitive`);

Warning, Processing of datafile <C:/RANS/Maple1.hdb> with result saving in datafile <C:/RANS/Maple2.hdb>, Please wait!

Warning, the processing result has been saved in datafile <C:/RANS/Maple2.hdb>

> reconf([], "C:/RANS/ACRI_2004.txt", "C:/RANS/62.txt", `sensitive`);

reconf([], "C:/RANS/ACRI_2004.txt", "C:/RANS/62.txt", sensitive)

---

statf – check of open datafiles relative to their type `*TEXT*` and `*BINARY*`

Call format of the procedure:

statf()

Formal arguments of the procedure: **no**

Description of the procedure:

In a number of cases along with the standard function ***iostatus***, the **statf** procedure can be useful enough. The procedure call **statf()** returns the *NULL* value, if in a current session the open files are absent. Otherwise the call **statf()** returns the table whose indices define two types of files {*TEXT, BINARY*}, whereas its entries define sequences of 5-element lists whose elements define **(1)** number of logical I/O channel, **(2)** a filename or path to a file, **(3)** kind of a file {*STREAM, RAW, PIPE, PROCESS, DIRECT*}, **(4)** a scanned file position, and **(5)** access mode {*APPEND, READ, WRITE*}. The procedure **statf** has a lot useful appendices at programming of access to datafiles.

```
statf:= proc()
local a, t, k;
  assign67(a = iolib(13), t[TEXT] = NULL, t[BINARY] = NULL), `if`(nops(a) = 3, NULL,
  [seq(assign67('t[a[k][-1]]' = t[a[k][-1]], [op(a[k][1 .. 3]), iolib(6, a[k][1]), a[k][-2]]),
  k = 4 .. nops(a)), RETURN(eval(t))])
end proc
```

Typical examples of the procedure use:

> Close(); filepos("C:\\RANS/ABS.txt", 1942): fopen("C:\\temp\\Lib.doc", WRITE, TEXT): filepos("C:/rans\\Simona.txt", infinity): open("C:/rans/Lectures.rtf", WRITE): iostatus();

[**4**, 0, 7, [0, "C:\RANS/ABS.txt", *STREAM*, *FP* = 2013506696, *READ*, *BINARY*],
[**1**, "C:\Temp\Lib.doc", *STREAM*, *FP* = 2013506728, *WRITE*, *TEXT*],
[**2**, "C:/RANS\Simona.txt", *STREAM*, *FP* = 2013506760, *READ*, *BINARY*],
[**3**, "C:/RANS/Lectures.rtf", *RAW*, *FD* = 10, *WRITE*, *BINARY*]]

> statf();

table([*TEXT* = [**1**, "C:\Temp\Lib.doc", *STREAM*, 0, *WRITE*],
BINARY = ([**0**, "C:\RANS/ABS.txt", *STREAM*, 1942, *READ*],
[**2**, "C:/RANS\Simona.txt", *STREAM*, 829, *READ*],
[**3**, "C:/RANS/Lectures.rtf", *RAW*, 0, *WRITE*])])

The above procedures extend the *Maple* tools at operation with *BINARY* datafiles, allowing to solve a series of problems dealing with ***BINARY*** files the more effectively. Above all, that concerns the **Fend** procedure providing a reliable method of situation handling ***"the end of a datafile"***.

10.4. Software for operation with *Maple* files

The given item represents a number of procedures providing a few additional possibilities at operation with *Maple* files, i.e. files whose names have expansions {*mws, m*} or can be successfully read by means of the built-in `*read*` function. Files of the given kinds occupy an especial place in the *Maple* file system. The *mws*-files document a job of the user in the *Maple* environment, whereas the *m*-files form elements of both the *Maple* libraries and the user libraries.

---

porf – a restructuring of *TEXT* files created by the standard `*save*` function

Call format of the procedure:

porf(F)

Formal arguments of the procedure:

F - string or symbol defining a filename or full path to it

Description of the procedure:

The standard function **save(n1, n2, ..., nk, *filename*)** writes the specified variables **n1, n2, ..., nk** into ***filename*** as a sequence of assignment statements. If ***filename*** ends with the characters **".m"** then **np** and the assigned values are written into the file in the internal *Maple* format. Otherwise, the input *Maple* language format is used.

The **read(F)** statement is used to read *Maple* language files saved by the standard `***save***` function as a rule into a current session. If the file is in the input *Maple* language format, the statements in the file are read and executed as if they were being typed in. However, the statements are not echoed unless **interface(*echo*)** is set to two or higher, or the call `***read***` does not end by colon (**":"**). The following simple enough procedure **porf(F)** by the switch principle reorganizes a **F** file of the input *Maple* language format thus, what at any conditions the call **read(F)** does not screen the statements saved in the **F** file. The successful call **porf(F)** returns the **F** value with the above reorganization of the **F** file; otherwise, an appropriate erroneous situation arises. The repeated procedure call **porf(F)** restores an initial file organization, i.e. formally the relation **F= porf(porf(F))** takes place. In a series of cases, the procedure seems useful enough, above all inside *Maple* procedures.

```
porf:= proc(F::file)
local a, k, p, h, w;
  `if`(cat("", F[-2 .. -1]) = ".m", ERROR("file <%1> has internal Maple format", F),
    `if`(not type(F, rlb), ERROR("file <%1> has a type different from TEXT", F), NULL));
  try assign(w = interface(warnlevel)), interface(warnlevel = 0),
    parse(readbytes(F, TEXT, infinity), statement)
  catch "incorrect syntax": close(F); interface(warnlevel = w);
    ERROR("file <%1> is not intended for read", F)
  end try;
  close(F), assign(p = fopen(F, READ, TEXT), a = "_$$$_");
  while not Fend(p) do h := readline(p); writeline(a, `if`(h[-1] = ";", cat(h[1 .. -2], ":"),
      `if`(h[-1] = ":", cat(h[1 .. -2], ";"), h)))
  end do;
  null(close(F, a), writebytes(F, readbytes(a, infinity)),
    close(F), fremove(a), interface(warnlevel = w)), F
end proc
```

Typical examples of the procedure use:

```
> V:=1942: G:=1947: Art:=(a+b)/(c-d): Kr:= () -> `+`(args)/nargs: Arn:=(a+b)^2:
> save(V, G, Art, Kr, Arn, "C:/Academy\\test.txt"); read("C:/Academy\\test.txt"):
> read("C:/Academy\\test.txt");
```

$$V := 1942$$
$$G := 1947$$
$$Art := \frac{a+b}{c-d}$$
$$Kr := (\ ) \rightarrow \frac{`+`(\text{args})}{\text{nargs}}$$
$$Arn := (a+b)^2$$

```
> read(porf("C:/Academy\\test.txt"));
> read(porf("C:/Academy\\test.txt"));
```

$$V := 1942$$
$$G := 1947$$
$$Art := \frac{a+b}{c-d}$$
$$Kr := (\ ) \rightarrow \frac{`+`(\text{args})}{\text{nargs}}$$
$$Arn := (a+b)^2$$

```
> porf("C:/Academy/RANS/REA/RAC\\test.m");
    Error, (in porf) file <C:/Academy/rans/rea/rac\test.m> has internal Maple format
> porf("C:/Academy/RANS/REA/RAC\\helpman.mws");
    Error, (in porf) file <C:/Academy/rans/rea/rac\helpman.mws> is not intended for read
```

> porf("C:/Academy\\Maplew9.exe");

Error, (in **porf**) file <C:/Academy\Maplew9.exe> has a type different from TEXT

> porf(porf("C:/Academy\\test.txt")) = "C:/Academy\\test.txt";

"C:/Academy\test.txt" = "C:/Academy\test.txt"

TRm – check of a file of the internal or input *Maple* format (*m*-file or *mws*-file)

Call format of the procedure:

TRm(F)

Formal arguments of the procedure:

F – string or symbol defining a filename or full path to it

Description of the procedure:

The procedure call **TRm(F)** returns the *true* value, if a file specified by the actual **F** argument is *m*-file compatible with a current *Maple* release, or the **F** argument defines a *mws*-file, or a file readable by the `**read**` statement; otherwise the *false* value is returned. The procedure has a series of rather useful appendices at a processing of the *Maple* files.

```
TRm:= proc(F::file)
local r, h, t, w, Lambda;
  assign(Lambda = (t -> interface(warnlevel = t)), h = CFF(F)[-1], w = interface(warnlevel));
  `if`(5 <= length(h) and h[-4 .. -1] = ".mws", RETURN(true), `if`(2 <= length(h) and
     h[-2 .. -1] = ".m", [assign67('h' = iolib(4, F, TEXT, 4), r = Release(), iolib(7, F))], NULL));
  `if`(h[1 .. 2] = "M7" and member(r, {7, 8, 9}) or h[1 .. 2] = "M6" and
     r = 6, RETURN(true), NULL);
  try Lambda(2); parse(readbytes(F, TEXT, infinity), statement)
  catch "incorrect syntax in parse": close(F); Lambda(w); RETURN(false)
  end try;
  close(F), Lambda(w), true
end proc
```

Typical examples of the procedure use:

> map(TRm, ["C:/AVZ/pp.m", "C:/AVZ/newmail.txt", "C:/AVZ/aaa.m", "C:/AVZ/ccc", "C:/AVZ/acc.mws", "C:/AVZ/h.txt", "C:/AVZ/xxx.m", "C:/Academy/ReadMe.htm"]), type(TRm, boolproc);

[*true*, *false*, *true*, *true*, *true*, *true*, *true*, *false*], true

---

transmf7_6 – converting of a *m*-datafile for the current *Maple* release

Call format of the procedure:

transmf7_6(F, G)

Formal arguments of the procedure:

F – symbol or string defining name or full path to a source *m*-file
G – symbol or string defining name or full path to a target *m*-file

Description of the procedure:

A file of the input *Maple* language format contains statements conforming to the syntax of the *Maple* language. These are the same as statements that can be entered interactively. *Maple* language files can be created by using a text editor or via the *`save`* function. Usually, the *`save`* function is used to save results or procedures that were entered into *Maple* interactively. *Maple* language files are read using the *`read`* function. The statements within the file are read as if they were being entered into a current *Maple* session interactively. Because the *Maple* language may change from release to release, a *Maple* language file that was saved by one release may not be readable by the next release. However, each new release is shipped with a conversion utility that translates *Maple* language files from the previous release to the current one. Having internal text format, a file of the input *Maple* language format is easily analyzed and manually adapted to the current *Maple* release. Within releases **6 – 9 /9.5,** the *Maple* language files are fully compatible. Entirely other situation takes place for *Maple* files of the internal format.

A file of the *Maple* internal format is used to store procedures and objects in a more compact, easy-for-*Maple*-to-read format. Objects stored in internal format can be read by *Maple* faster than objects stored in *Maple* language format. *Maple* internal format files are identified by filenames ending with the two characters "**.m**" (*m-files*). *Maple* internal format files are created using the *`save`* function, they are read using the *`read`* function. The fact that the specified filename ends in "**.m**" tells *Maple* that the file is in internal format. Generally, the format of internal files changes with each new release of *Maple* in order to accommodate new features; it stipulates an incompatibility of *m*-files created by different releases. However, *Maple* language files created by the *Maple* of releases **6 – 9/9.5** are acceptable for all last four releases of *Maple*, whereas a *m*-file created in a release **{7 | 8 | 9 | 9.5}** is not compatible with the 6th *Maple* release. The main ground of such situation lays in different representations of internal data structures of the above *Maple* releases – the representations in releases **6** and **{7, 8, 9/9.5}** are different. In a lot cases the procedure **transmf7_6** solves this problem.

The procedure call **transmf7_6(F, G)** return the *NULL* value, trying to convert a *m*-file specified by the first actual **F** argument and created in *Maple* of releases **7 – 9/9.5** into the *m*-file compatible with the 6th release. The result is put into an *m*-file specified by the second actual argument **G.** A procedure call outputs an appropriate warning. Our experience reveals that the procedure provides the successful converting for many *m*-files created in *Maple* of releases **7 – 9/9.5**. The procedure has a number of useful appendices in problems dealing with *Maple* files processing.

```
transmf7_6:= proc(F::file, G::{string, symbol})
local a, b, r, nu, f, h;
  `if`(Ftype(F) = ".m", assign(r = Release()), ERROR("<%1> is not a m-datafile", F));
  nu:= () -> WARNING("datafile <%1> is compatible with the current Maple release", F);
  assign(a = readbytes(F, TEXT, 4)), close(F);
  `if`(belong(r, 7 .. 9) or r = 6 and a = "M6R0", RETURN(nu()), assign(f = G));
  if not type(G, file) then
    if cat(" ", G)[-2 .. -1] <> ".m" then h := cat(G, ".m") else h := G end if;
    assign(b = interface(warnlevel)), interface(warnlevel = 0), assign('f' = MkDir(h, 1));
    interface(warnlevel = b)
  elif Ftype(G) <> ".m" then assign('f' = cat(G, ".m")) end if;
  writebytes(f, SUB_S(["M7R0"="M6R0", "f*" = "R"], readbytes(F, TEXT, infinity))), close(F,f);
  try read f
  catch: WARNING("file <%1> cannot be converted for Maple 6", F); RETURN(delf(f))
  end try;
```

```
  WARNING("the converted file <%1> has been read into a current Maple6 session", f)
end proc
```

Typical examples of the procedure use:

> P:=()->`+`(args)/nargs: M:=module() export a; a:=61 end module: agn:=56: P1:=proc() proc() [args] end proc(args) end proc: save(P,`agn`,M,P1, "C:/rans/test9.m"); *# Maple 9*

> transmf7_6("C:/rans/test6.m"", "C:/rans/Res1.m");

Warning, datafile <C:/rans/test6.m> is compatible with the current Maple release

> transmf7_6("C:/RANS/Test9.m", "C:/rans/Res2.m"); *# Maple 6*

Warning, the converted file <C:/rans/Res2.m> has been read into a current Maple6 session

> transmf7_6("C:/RANS/Test9.m", "C:/rans/Res3.m");

Warning, file <C:/RANS/Test9.m> cannot be converted for release 6

Rmf – check of a file to be the *Maple* file

Call format of the procedure:

Rmf(F)

Formal arguments of the procedure:

F – any valid *Maple* expression

Description of the procedure:

Files with software saved in the *input Maple* language format by means of the built-in function *`save`* are transportable relative to all releases of *Maple* (though they can be syntactically release-dependent) whereas files of the *internal Maple* format (*m*-files) always are release-dependent. Attempt of reading of a *m*-file which is not corresponding to a current release, by means of the built-in function *`read`* always causes an erroneous situation whereas reading of a file in the *input Maple* language format runs correctly in any release if definitions of procedures and modules located in it do not contain any release-dependent syntactic features. We will call files that are correctly read by means of the function *`read`* by the *Maple* files.

The procedure call **Rmf(F)** returns the *true* value, if a file defined by the actual **F** argument is the *Maple* file and the *false* value otherwise. The procedure has a series of rather useful appendices at a processing of the *Maple* files.

```
Rmf:= F::file -> TRm(F)
```

Typical examples of the procedure use:

> map(Rmf, ["C:/AVZ/pp.m", "C:/AVZ/newmail.txt", "C:/AVZ/aaa.m", "C:/AVZ/ccc", "C:/AVZ/acc.mws", "C:/AVZ/h.txt", "C:/AVZ/xxx.m", "C:/Academy/ReadMe.htm"]);

[*true, false, true, true, true, true, true, false*]

---

isRead – check of correct read of *Maple* objects from files by the *`read`* function

Call format of the procedure:

isRead(F)

Formal arguments of the procedure:

F – a datafile destined to testing

Description of the procedure:

At usage of the built-in *`read`* function for read of files containing syntactically accurate *Maple* expressions, in a lot of cases a problem of testing of files, which can be read by the above function without an initiation of special or erroneous situations, arises. The **isRead** procedure solves the given problem in a great extent.

The procedure call **isRead(F)** returns the *true* value, if a datafile specified by actual **F** argument will be read by function *`read`* correctly; otherwise the *false* value is returned. In addition, it is necessary to have in mind; the *true* value does not guarantee a correctness of running of program objects contained in a datafile **F.** The *true* value guarantees only with rather high probability a correctness of reading of the **F** datafile into a current session of the package.

At the same time, the given procedure seems useful tool in problems of automatic processing of datafiles containing definitions of variables, procedures and program modules, by fulfilling in some sense a special *"filter"* at reading of datafiles in auto mode by means of the built-in *`read`* function.

```
isRead:= proc(F::file)
local k, t, p, nu, omega;
  seq(`if`(Search1(cat("", F), cat(".", k), 't') and t = [right], RETURN(false), NULL),
    k = [abt, acg, acl, acq, acs, adx, al2, ani, aot, api, apl, arc, arj, as, aun, avc, aw, awe, ax, bak,
    bar, bat, bbr, bdb, bif, bin, bis, bm_, bmp, bpf, btl, cab, cag, cb, cbk, cdr, cfg, cgm, chi, chm,
    chs, chw, cif, clf, cls, clx, cmp, cnt, cnv, com, cpe, cpi, cpl, cpt, cpx, cr, crl, csf, css, dap, daz,
    db, ddd, def, des, deu, dhs, dib, dic, dict, did, dir, diz, dl_, dll, dls, dlx, dmv, doc, dos, dot, dpc,
    dpf, drv, drw, dsn, dun, dxf, e, eif, ell, elm, emf, eng, env, epi, eps, er, esp, exe, fcs, fed, fin, flt,
    fnt, fon, fpr, fra, fsd, fts, gid, gif, gra, hdb, hlp, htc, htf, htm, html, htt, htx, htz, hun, hyp, i_s,
    i_sf, iad, icc, icm, ico, icw, idf, idx, in_, inc, ind, inf, ini, ins, iqy, isp, isu, ita, ja, jar, jpeg, jpg,
    jpn, jrt, js, jsp, kbd, klb, kor, lcd, lex, lgc, lha, lib, lic, lmd, lng, lnk, log, lsd, lst, lzh, map, mbx,
    mdb, mds, mfd, mfm, mfs, mif, mml, mmm, mmw, modp, mof, msc, msg, mws, ndx, nls, obe,
    ock, ocr, ocx, odc, olb, ops, pce, pci, pct, pcx, pdd, pdf, pdr, pdx, pfb, pfm, pif, pip, pl, plc,
    png, ppa, prc, pre, prefs, prf, prn, prv, prx, ps, pub, pwl, q3x, qtc, rar, rat, rcd, re, ref, reg, rep,
    rll, rsc, rsd, rsp, rst, rtf, sam, scf, scp, scr, sdb, sdc, sequ, set, sfc, sid, sll, spd, spl, sql, srg, stc,
    std, stf, stp, swi, syd, sys, syx, tar, tdf, tga, thd, tif, tip, tlb, tlx, tmp, toc, tpl, trn, tsk, ttf, tvl,
    twd, tx_, typ, udc, url, vbs, vbx, ver, vgw, vm, vnd, vwp, vxd, wab, wav, wbm, wdf, wiz, wkg,
    wld, wll, wmf, wpc, wpg, wpx, x32, xla, xml, xsl, zip]),
   assign(nu = interface(warnlevel), p = Release(), omega = (x -> interface(warnlevel = x)));
  omega(1), assign('t' = convert(iolib(4, F, infinity), 'bytes')), iolib(7, F),
    `if`(cat(" ", F)[-2 .. -1] = ".m", `if`(t[2] = "6" and p = 6 or member(p, {7, 8, 9}),
    RETURN(omega(nu), true), RETURN(omega(nu), false)), NULL);
  try parse(t, 'statement')
  catch "incorrect syntax in parse:": RETURN(omega(nu), false)
  end try;
  true, omega(nu)
end proc
```

Typical examples of the procedure use:

> map(isRead, ["c:/temp/UserLib.avz", "c:/Academy/ReadMe.htm","c:/Archive\\ABS.txt", "C:/Academy/helpman.mws", "C:/Archive\\lectures.rtf", "C:/Archive\\Maple.agn"]);

$$[true, false, false, false, false, true]$$

The above procedures **isRead, TRm** and **Rmf** seem rather useful tools at programming of numerous problems dealing with different datafiles, when out of them is being fulfilled a load of definitions of program elements such as variables, procedures, program modules, etc. These tools provide some elements of check of such load.

---

ftpd – check of types of all datafiles in an arbitrary directory

Call format of the procedure:

ftpd(F, T {, *only*})

Formal arguments of the procedure:

F - symbol or string defining name or full path to a directory
T - an assignable name
only - (optional) a files analysis mode

Description of the procedure:

The procedure **ftpd** is a modification of the above procedure **DirFT** that helps to obtain more specific information about types of files composing the file system of a computer. The information provides more exact settings of the first part of the above procedure **isRead.** The procedure call **ftpd(F, T, *only*)** returns the list of types of datafiles contained in a directory specified by the actual **F** argument, excluding its subdirectories. Whereas through the second actual **T** argument the **(n, 2)-**matrix is returned whose the first column defines the types of datafiles and the second – the quantities corresponding to them. The actual **F** argument should define name or full path to a real directory; otherwise, an erroneous situation arises.

The procedure call **ftpd(F, T)** returns the list of types of datafiles contained in a directory specified by the actual **F** argument, including all its subdirectories. This procedure (*similarly to the procedure* **Ftype**) as a type understands a type defined by extension of a filename; for example, file "Orders.rtf" has *`rtf`* type. Presence in the returned list additionally of key words *`dir`* and *`undefined`* speaks about presence in the tested directory of subdirectories and files without name extension accordingly. At impossibility of the classification, the *FAIL* value is returned. The procedure has a number of useful appendices in problems dealing with datafiles processing.

```
ftpd:= proc(F::dir, T::evaln)
local a, Tab, omega;
  assign(Tab = table([]), omega = (t -> `if`(type(t, symbol), t, cat(``, t[2 .. -1])))),
    system(cat("dir /A/B", `if`(member(only, {args}), " ", "/S "), Path(CF(F)), "\\*.* > $ArtKr$"));
  do a := Ftype(`if`(member(only, {args}), cat(F, "\\", iolib(2, "$ArtKr$")), iolib(2, "$ArtKr$")));
    `if`(member(a, map(op, [indices(Tab)])), assign('Tab[a]' = Tab[a] + 1), assign('Tab[a]' = 1));
    if Fend("$ArtKr$") then break end if
  end do;
  iolib(8, "$ArtKr$"), assign(T = tabar(mapTab(omega, Tab, 0), type, quantity)),
    map(omega, map(op, [indices(Tab)]))
end proc
```

Typical examples of the procedure use:

> ftpd("C:/Program Files\\Maple 6", G, only), eval(G);

$$[dir], \begin{bmatrix} type & quantity \\ dir & 4 \end{bmatrix}$$

> ftpd("C:/Program Files\\Maple 6", G), eval(G);

$$[undefined, ini, dll, doc, js, htm, rtf, txt, jpg, mws, gif, dir, css], \left[\begin{array}{cc} type & quantity \\ gif & 68 \\ jpg & 24 \\ doc & 17 \\ txt & 15 \\ dir & 15 \\ htm & 14 \\ js & 5 \\ dll & 4 \\ mws & 4 \\ css & 3 \\ rtf & 3 \\ ini & 2 \\ undefined & 1 \end{array}\right]$$

Mwsin – deletion of *Output*-paragraphs out of *mws*-files

Call format of the procedure:

Mwsin(F, G)

Formal arguments of the procedure:

F – symbol or string defining a filename of a *mws*-file or directory with them
G – symbol or string defining a target directory for processed *mws*-files

Description of the procedure:

The procedure call **Mwsin(F, G)** is intended for deletion out of an *mws*-file specified by the actual **F** argument, or out of all *mws*-files located in a directory specified by the first **F** argument of all *Output*-paragraphs, i.e. results of calculations of the corresponding *Maple* document saved in a current session or previous ones. The files search is done in an **F** directory, excluding its subdirectories. Resultant *mws*-files are put into a directory specified by the second actual **G** argument; if such directory does not exist then it will be created. In any case, the procedure outputs warning about path to a target directory in which the processed *mws*-files have been saved. The resultant file receives name of a source *mws*-file with addition of "$"-prefix. The procedure call returns the full path to a target directory with the processed *mws*-files with output of an appropriate warning. This procedure handles basic erratic and especial situations and can be useful for automation of the above processing of *mws*-files without loading them into a current *Maple* session.

In particular, this procedure has arisen in connection with a problem of effective packing of a directory with many *mws*-files that have contained *Output*-paragraphs. Use of the procedure with posterior application of archiver *Arj32* to a directory with *mws*-files – result of the procedure call **Mwsin** – has allowed to essentially decrease size of the packed *mws*-files. The procedure has and other rather useful appendices.

```
Mwsin:= proc(F::{file, dir}, G::{string, symbol})
local a, c, h, k, b, d, s, t, omega, x, v, p, z;
  if not type(eval(G), {string, symbol}) then
    ERROR("<%1> should be string or symbol but has received <%2>", G, whattype(eval(G)))
  elif type(F, file) and Ftype(F) = ".mws" then x := [F]
  elif type(F, dir) then v := DirF(F, only); x := [seq(`if`(Ftype(v[k]) = ".mws", v[k], NULL),
      k = 1 .. nops(v))]
  else ERROR("<%1> is not a mws-file and not a directory", F)
  end if;
  if evalb(x = []) then ERROR("<%1> does not contain mws-files", F)
  elif type(G, dir) then z := G
  else t := interface(warnlevel); interface(warnlevel = 0); z := MkDir(G); interface(warnlevel = t)
  end if;
  assign(b = "{EXCHG {PARA 0 "> " 0 "" {MPLTEXT"), assign(d = length(b));
  omega := proc(n::posint)
    local a, t;
      `if`(search(h[n + d .. -1], "" }}", t), cat(h[n .. n + d + t], "" }}}"), false)
    end proc;
  for p in x do assign('h' = Subs_all1("\n" = NULL, iolib(4, p, TEXT, infinity))), iolib(7, p);
    assign('a' = Search(h, b)), assign('s' = h[1 .. a[1] - 1]);
    for k to nops(a) do c := omega(a[k]); if c <> false then s := cat(s, c) end if end do;
    assign('h' = cat(z, "/", cat("$", CFF(p)[-1]))), null(iolib(5, h, cat(s, "}")), iolib(7, h))
  end do;
  WARNING("the processed mws-files are in directory <%1>", z), z
end proc
```

Typical examples of the procedure use:

```
> Mwsin("C:/Program Files/Maple 8/Users/Ex.mws", "C:/Temp/AGN");
    Warning, the saved mws-files are in directory <C:\Temp\AGN>
                              "C:/Temp/AGN"
> Mwsin("C:/RANS/came.mws", "C:/Temp/AVZ");
    Warning, the processed mws-files are in directory <c:\temp\avz>
                              "C:/Temp/AGN"
> Mwsin("C:/Book", "C:/Temp/AGN");
    Error, (in Mwsin) <C:/Book> does not contain mws-files
> Mwsin("C:/RANS/", "C:/RANS/Lectures.rtf");
    Warning, the processed mws-files are in directory <c:\rans\_lectures.rtf>
                              "c:\rans\_lectures.rtf"
> RANS_IAN_RAC:= matrix([]): Mwsin("C:/RANS/", RANS_IAN_RAC);
    Error, (in Mwsin) <RANS_IAN_RAC> should be string or symbol but has received <array>
> Mwsin("C:\\LBZ", "C:\\Temp\\Test.mws");
    Warning, the processed mws-files are in directory <c:\temp\test.mws>
                              "c:\temp\test.mws"
```

Iddn – an attribute identifying length of software name
Iddn1

Call format of the procedures:

Iddn(n)
Iddn1(S {, 'h' {, p}})

Formal arguments of the procedures:

n - a positive integer or symbol; an expression length or expression name
S - a string or a symbol
h - (optional) an assignable name
p - (optional) any valid *Maple* expression

Description of the procedures:

A record containing any valid named *Maple* expression saved in a file of the internal *Maple* format (*m*-file) begins with the construction of the following simple view:

I<*Atr*><*Name*>... **(a)**

where ***Atr*** is an attribute identifying length of an expression name, and ***Name*** – its name itself. In a series of problems dealing with processing of *m*-files, the question of definition of the attribute for the given length of a name represents undoubted interest. The given problem is solved by means of the **Iddn** procedure represented below.

The procedure call **Iddn(n)** returns the attribute identifying an expression name or its length specified by the actual **n** argument. The required ***Atr*** attribute is returned as a *string*. In a certain sense, the **Iddn1** procedure is inverse to the previous procedure, namely: the procedure call **Iddn1(S)** returns *Name* as a symbol if and only if the actual **S** argument has structure of the above view (**a**). In addition, the procedure call **Iddn1(S, 'h')** with the second optional **h** argument additionally through **h** returns the table whose indices are the above attributes ***Atr*** and entries are lengths appropriate to them. If in the procedure call **Iddn1(S, 'h', p)** has been coded the third optional **p** argument (**p** is any valid *Maple* expression) then it defines a mode of interchange of indices and entries of the above table being returned through **h** argument. For change of view of the returned table the **Iddn1** procedure uses the **InvT** procedure useful for work with the *Maple* objects of type `*table*`.

```
Iddn:= proc(n::{posint, symbol})
local a, k, h, m, F, S, t;
global `0`;
  `if`(type(n, symbol) and length(n) = 1 or n = 1, RETURN("""), [assign(a = `0`), unassign('`0`')]);
  `if`(type(n, posint), assign(m = n), `if`(not type(n, symbol) or length(n) = 0,
     ERROR("<%1> is invalid actual argument", n), assign(m = length(n)))),
     assign(cat(seq(`0`, k = 1 .. m)) = "AVZAGNVSVArtKr", F = "_$$$_.m", S = "");
  (proc() save args, F end proc)(cat(seq(`0`, k = 1 .. m))), iolib(2, F);
  do h := iolib(2, F); if h = 0 then iolib(8, F); break else S := cat(S, h) end if end do;
  unassign('`0`'), null(search(S, `0`, 't')), S[2 .. t - 1], assign(`0` = a)
end proc

Iddn1:= proc(G::{string, symbol})
local k, s, t, d;
option remember;
  `if`(type(eval(G), {string, symbol}), assign(s = cat("", G), d = length(G)),
```

```
    ERROR("<%1> should be symbol or string but had received <%2>", G, whattype(eval(G)))),
    seq(assign('t'[Iddn(k)] = k), k = 1 .. 255), `if`(nargs = 2, assign('args'[2] = eval(t)),
    `if`(nargs = 3, assign('args'[2] = InvT(t)), NULL));
  `if`(s[1] = "I", `if`(whattype(t[s[2]]) = integer and t[s[2]] + 2 <= d, cat(``, s[3 .. t[s[2]] + 2]),
    `if`(whattype(t[s[2 .. 3]]) = integer and t[s[2 .. 3]] + 3 <= d,
    cat(``, s[4 .. t[s[2 .. 3]] + 3]), false)), false)
end proc
```

Typical examples of the procedures use:

> map(Iddn, [7,14,36,40,56,61,450]), map(Iddn, [Tallinn,Grodno,Vilnius,Moscow,Art,Kr]);

["(", "/", "E", "I", "Y", "hn", "]gl"], ["(", "'", "(", "'", "$", "#"]

> Iddn(MkDir);

Error, (in **Iddn**) <**MkDir**> is invalid actual argument

> Iddn1("I)RANS_IAN*AVZ-AGN"), Iddn1("I'Art_Kr2003", 'h', 147), eval(h);

RANS_IAN, *Art_Kr*, table([120 = "cr", 241 = "\z", 121 = "dr", 1 = "\"", 242 = "]z", 122 = "er",

==

116 = "_r", 237 = "hy", 117 = "`r", 238 = "iy", 118 = "ar", 239 = "jy", 119 = "br", 240 = "[z"])

---

mmp – procedure heading and module heading in the internal *Maple* format

Call format of the procedure:

mmp(S {, 'h'})

Formal arguments of the procedure:

S - a string
h - (optional) an assignable name

Description of the procedure:

A record containing any procedure or module saved in a datafile of the internal **Maple** format (*m*-file) or in a library structurally similar to the main *Maple* library begins with the construction (*heading*) of the following simple view:

I<*Atr*><*Name*><Z>... **(b)**

where ***Atr*** is an attribute identifying length of an expression name, ***Name*** – its name itself, and **Z** has a value depending on an expression type and the *Maple* release, namely: **"`6"** a module for releases 6 – 9/9.5, **"R6"** – a procedure for release **6** and **"f*"** – a procedure for releases 7 – 9/9.5. In a series of problems dealing with processing of *m*-files or *Maple* libraries, the question of check of a procedure heading and module heading represents undoubted interest. The given problem is solved by means of the **mmp** procedure represented below.

The procedure call **mmp(S)** returns the *true* value if and only if a string specified by the first actual argument **S** has structure of the above view (**b**); otherwise, the *false* value is returned. If at the procedure call **mmp(S, h)** has been coded the second optional **h** argument then through it a name of a procedure or module with heading part **S** in the internal *Maple* format is returned, the procedure call returns the *true* value. The procedure successfully handles *m*-datafiles created in any of releases 6 – 9/9.5. In particular, the procedure is a rather useful tool for work with *m*-files and *Maple* libraries. In many appendices the procedure has made a good showing.

```
mmp:= proc(S::string)
local a;
  a := Search2(S, {"`6", "f*", "R6"}); if S[1] <> "I" or a = [] then false
  elif a[1] <= 59 and S[2] = Iddn(a[1] - 3) then true,
    `if`(nargs = 2, assign(args[2] = convert(S[3 .. a[1] - 1], symbol)), NULL)
  elif a[1] <= 257 and S[2 .. 3] = Iddn(a[1] - 4) then true,
    `if`(nargs = 2, assign(args[2] = convert(S[4 .. a[1] - 1], symbol)), NULL)
  elif a[1] <= 4097 and S[2 .. 4] = Iddn(a[1] - 5) then true,
    `if`(nargs = 2, assign(args[2] = convert(S[5 .. a[1] - 1], symbol)), NULL)
  else false
  end if
end proc
```

Typical examples of the procedure use:

> [mmp("I\"0R6""F$",'p'), p], [mmp("I%mmp2`6#%aGb6#%+thismoduleG6F)",'p'), p];

[true, 0], *[true, mmp2]*

> mmp("IOabcdfrtdertsgrtyhjnfghjkyutresdfrtyukngfdsweqa`6kkkkk", 'p'), p;

true, abcdfrtdertsgrtyhjnfghjkyutresdfrtyukngfdsweqa

---

isLnkMws – check of a *mws*-file onto presence of links in it

Call format of the procedure:

isLnkMws(F {, 'h'})

Formal arguments of the procedure:

F - a filename or full path to a file
h - (optional) an unevaluated name

Description of the procedure:

The following procedure **IsLnkMws** is intended for a testing of files of *mws*-type with *Maple* documents for the purpose of presence in them of various type of links (*hyperlinks, URLs, bookmarks, worksheets*). The procedure call **isLnkMws(F {, 'h'})** returns the *true* value, if a file specified by the actual **F** argument is a *mws*-file and contains *hyperlinks* or/and *URLlinks;* otherwise the *false* value is returned. If a file specified by the actual **F** argument is not a *mws*-file or does not exist, the appropriate erroneous situation arises.

If as the second optional **h** argument an unevaluated name has been coded, then through it the 3-element sequence of lists is returned. In addition, elements of the first list define the names of *hyperlinks* to the package Help database, including help databases of all libraries logically linked with the main *Maple* library; elements of the second list define the names of other hyperlinks {*bookmarks, worksheets*}, and elements of the third list define *URLlinks* contained in the **F** file accordingly. This procedure is rather useful tool for operation with *Maple* documents.

```
isLnkMws:= proc(F::file)
local a, b, c, d, k, omega, h, A, G, S;
  omega:= proc(s::string, n::posint)
    local p;
      p := searchtext("}{", s, n .. length(s)); SLD(s[n + 1 .. n + p - 2], " ")
    end proc;
```

```
  `if`(Ftype(F) <> ".mws", ERROR("<%1> is not a worksheet file", F),
    [assign(a = readbytes(F, TEXT, infinity), b = "{HYPERLNK 17",
    c = "{URLLINK 17"), close(F)]), assign('a' = cat(op(SLD(a, "\\n")))),
    assign(d = map2(Search, a, [b, c]));
  `if`(d = [[], []], RETURN(false), assign(A = [], G = [], S = []));
  for k to nops(d[1]) do assign('h' = omega(a, d[1][k])),
    `if`(h[4] = "2", assign('A' = [op(A), cat(``, h[3][2 .. -2])]),
    assign('G' = [op(G), cat(``, h[3][2 .. -2])]))
  end do;
  for k to nops(d[2]) do h := omega(a, d[2][k]); S := [op(S), cat(``, h[5][2 .. -2])] end do;
  RETURN(true, `if`(nargs = 2, assign(args[2] = op([A, G, S])), NULL))
end proc
```

Typical examples of the procedure use:

> isLnkMws("c:/Academy/Examples\\HelpBase\\mws_files/UserLib!!!.mws", 'g'), ***nops*****(g[1]);**

true, 433

> isLnkMws(`C:/Academy/Examples\\HelpBase\\MWS_files/isLnkMws.mws`, 'g'), g;

true, [*file, hyperlinks, hyperlinks, file, file_types, bookmark, hyperlinks*], [], []

The represented procedures, above all **isRead, Rmf** and **TRm** seem rather useful tools at operation with *Maple* files. They provide testing of the datafiles for belonging of them to the group of so-called *Maple* files.

10.5. Some special tools for operation with datafiles

The procedures represented in the given item, not only illustrate use of the package tools for an access to datafiles of the both types {*TEXT, BINARY*}, including a number of useful and non-standard receptions of programming, but also are of interest themselves as additional tools of operation with external datafiles of the package. In particular, the procedures **E_mail** and **email** are effective enough at organization of **e**-commerce and are widely used in this direction. Other tools of the item have rather useful appendices also.

E_mail – choice of correct *email*-addresses from datafiles
email

Call format of the procedures:

E_mail(F, H {, h})
email(F, G)

Formal arguments of the procedures:

F – a name or full path to a datafile
H, G – a filename or full path to a target file
h – (optional) an arbitrary valid *Maple* expression

Description of the procedures:

The procedure call **E_mail(F, H)** provides choice of correct *email*-addresses out of a text datafile specified by the first actual **F** argument, and puts their into a target file specified by the second actual **H** argument. In addition, in case of absence of path to such file it will be created with output of appropriate warning. Result of execution of the given procedure is not only creation of a text file with required *email*-addresses, suitable to the immediate use by means of *e-mail,* but also a providing of output of the basic statistics whose purpose easily can be seen of the fragment represented below.

If the procedure call **E_mail(F, H, h)** uses the optional third argument **h** (**h** - any valid *Maple* expression), then each correct *email*-address is put into a target file **H** as separate line, otherwise *email*-addresses are parted by symbols **", "** {hex(44), hex(32)} and form the sole line. All suspicious addresses (*if they do exist*) are put into a separate text **$$$**-file whose main name is identical to a name of the source file **F.** Afterwards on the basis of review of the given file very much often it is possible to correct some part of the \`@\`-constructions located in it into correct *email*-addresses. The procedure also outputs other useful statistics about the source file processing.

The procedure **email(F,G)** is based on the previous procedure, inherits the same two first formal arguments, extending the first procedure onto datafiles of types {*doc, rtf, htm*}, and in general case onto files of *rlb*-type. In case of impossibility to extract *email*-addresses the **email** procedure returns the *NULL* value, i.e. nothing with output of appropriate warning. The procedures **E_mail** and **email** are rather useful tools for extract of *email*-addresses from datafiles of wide enough set of types.

```
E_mail:= proc(InF::file, OutF::{string, symbol})
local In, Out, Art, S, G, k, p, T, V, Kr, R, F, Y, Z, E, L, t, r, omega, nu;
  `if`(not type(InF, rlb), ERROR("<%1> is not a text file", InF),
    `if`(not type(eval(OutF), {string, symbol}), ERROR("<%1> is not a file path", OutF),
    `if`(type(OutF, file), assign(omega = OutF, L = [], t = 0), [assign(nu = interface(warnlevel),
    L = [], t = 0), interface(warnlevel = 0), assign(omega = MkDir(OutF, 1)),
    interface(warnlevel = nu)])));
  T := {45, 46, (48 + k) $ (k = 0 .. 9), (64 + k) $ (k = 0 .. 58)} minus {64, 91, 92, 93, 94, 96};
  [assign(Art = time(), S = iolib(4, InF, TEXT, infinity), R = 0), iolib(7, InF),
    assign(G = cat(" ", S, " "))], `if`(search(S, "@"), NULL,
    ERROR("file <%1> does not contain email addresses", InF)),
   assign(V = [seq(`if`(G[k] = "@", k, NULL), k = 1 .. length(G))]);
  for k in V do assign('Kr' = "@", 'p' = k);
    do p := p - 1;
      if member(op(convert(G[p], bytes)), T) then Kr := cat(G[p], Kr) else p := k; break end if
    end do;
    do p := p + 1;
      if member(op(convert(G[p], bytes)), T) then Kr := cat(Kr, G[p]) else break end if
    end do;
    `if`(Kr[1] = "@" or searchtext(".", Kr, searchtext("@", Kr) .. length(Kr)) = 0,
      assign('R' = R + 1, 'F' = 61, 'Y' = 4), NULL);
    if F = 61 or Kr[-1] = "." or  not search(Kr[-4 .. -2], ".") then
      iolib(5, cat(substring(InF, 1 .. -4), "$$$"), Kr), iolib(5, cat(substring(InF, 1 .. -4), "$$$"),
      `if`(2 < nargs, [10], [44, 32])), assign('E' = 3), unassign('F')
    else assign('L' = [op(L), Kr]), assign('Z' = 12)
    end if
  end do;
  [assign('t' = nops(L)), assign('L' = map(Case, {op(L)})), assign('r' = nops(L))];
  for k to nops(L) do iolib(5, omega, L[k]), iolib(5, omega, `if`(2 < nargs, [10], [44, 32]))
  end do;
  iolib(7, `if`(Z = 12, omega, NULL), `if`(Y = 4, cat(substring(InF, 1 .. -4), "$$$"), NULL));
```

```
  WARNING("%1 - total of constructions `@` in file <%2>
    %3 - quantity of correct email addresses in file <%4>
    %5 - quantity of multiple email addresses
    %6 - quantity of suspicious email addresses in file <%7>
    run time of the procedure = %8 minute(s)", nops(V), InF, r, omega, t - r, R,
    cat(substring(InF, 1 .. -4), "$$$"), round(1/60*time() - 1/60*Art))
end proc

email:= proc(F::file, G::{string, symbol})
local h, k, t, omega, nu;
  `if`(not type(eval(G), {string, symbol}), ERROR("<%1> is not a file path", G),
    `if`(type(G, file), assign(omega = G), [assign(nu = interface(warnlevel)),
    interface(warnlevel = 0), assign(omega = MkDir(G, 1)), interface(warnlevel = nu)])),
    assign('nu' = cat(F, ".$$$"));
  assign(h = subs([seq(t = NULL, t = [k $ (k = 0 .. 31)])], iolib(4, F, infinity))), iolib(5, nu, h),
    iolib(7, F, nu);
  try E_mail(nu, omega, 147), iolib(7, nu)
  catch "file <%1> does not contain email addresses": WARNING("email addresses have not
    been found, however they can be encoded in file <%1>", F)
  end try
end proc
```

Typical examples of the procedures use:

> E_mail("C:/Temp\\Record.txt", "C:/Temp\\AVZ\\Orders\\Result.txt", 7);

Warning, **384** - total of constructions `@` in file <C:/Temp\Record.txt>

26 - quantity of correct email addresses in file <c:\temp_avz\orders\result.txt>
42 - quantity of multiple email addresses
15 - quantity of suspicious email addresses in file <C:/Temp\Record.$$$>
Run time of the procedure = 2 minute(s)

> email("C:/RANS/My_email.rtf", "C:/Temp/Email/Academy.txt");

Warning, **20** - total of constructions `@` in file <C:/RANS/My_email.txt>

20 - quantity of correct email addresses in file <C:/RANS/Email/Email.txt>
0 - quantity of multiple email addresses
0 - quantity of suspicious email addresses in file <C:/RANS/My_email.$$$>
run time of the procedure = 0 minute(s)

> email("C:/RANS/Invoice.doc", "C:/Temp/Email/Invoice.txt");

Warning, **190** - total of constructions `@` in file <C:/RANS/Invoice.doc.$$$>

61 - quantity of correct email addresses in file <c:\temp\email\invoice.txt>
7 - quantity of multiple email addresses
14 - quantity of suspicious email addresses in file <C:/RANS/Invoice.doc.$$$>
run time of the procedure = 0 minute(s)

> email("C:/Archive/Booklet/Booklet1.rtf", "C:/Temp/Email/Booklet1.txt");

Warning, email addresses have not been found, however they can be coded in file

<C:/Archive/Booklet/Booklet1.rtf>

> email("C:/RANS/My_email.htm", "C:/Temp/My Email/My_email.txt");

Warning, **38** – total of constructions `@` in file <C:/RANS/My_email.htm.$$$>

25 – quantity of correct email addresses in file <c:\temp\my email\my_email.txt>
13 – quantity of multiple email addresses
0 – quantity of suspicious email addresses in file <C:/RANS/My_email.htm.$$$>
run time of the procedure = 0 minute(s)

IAN_REA – extracting of symbols from a datafile

Call format of the procedure:

IAN_REA(X)

Formal arguments of the procedure:

X – symbol or string defining a filename or full path to a datafile

Description of the procedure:

The procedure **IAN_REA(X)** is intended for extraction from an arbitrary file specified by the actual **X** argument of symbols with decimal codes from range [32 .. 126], i.e. *Latin letters* and *special symbols*. A target file will have the *txt*-format with length of line no more than 76; the file will be located in the same subdirectory as a source file, and its name will be formed of name of the source file by addition of *txt*-extension. The successful termination of the procedure returns the full path to the target file with output of appropriate warning about its location. At the existence of target file, the procedure call returns the *NULL* value, i.e. nothing, with output of an appropriate warning.

The **IAN_REA** procedure seems rather useful tool in problems of analysis of text components of the executable program files and in other appendices, for example, in case of necessity of the preliminary analysis of unknown program file, providing extraction of text information out of them. The procedure can be used and as a separate tool at operation with *txt*-datafiles. The given procedure has been successfully used at analysis of many executable program files and has confirmed own efficiency.

```
IAN_REA:= proc(f::file)
local k, p, nu, g, h, L;
  `if`(nargs <> 1, ERROR("quantity <%1> of actual arguments is invalid", nargs),
    assign(nu, cat(f, ".txt")));
  `if`(type(nu, file), RETURN(WARNING("target file <%1> exists - work has been completed",
    nu)), 7);
  assign(g = open(f, READ), h = open(nu, WRITE));
  while not Fend(g) do L := subs((k = NULL) $ (k = 0 .. 31), (k = NULL) $ (k = 127 .. 255),
      readbytes(g, 10240));
    assign('L' = insL(L, {nops(L), (76*k) $ (k = 1 .. trunc(1/76*nops(L)))}, 10)), writebytes(h, L)
  end do;
  close(g, h), WARNING("target file <%1> has been created", nu), nu
end proc
```

Typical examples of the procedure use:

> IAN_REA("C:/RANS/RegClean.exe");

Warning, target file <C:/RANS/RegClean.exe.txt> has been created

"C:/RANS/RegClean.exe.txt"

> IAN_REA("C:/RANS/RegClean.exe");

Warning, target file <C:/RANS/RegClean.exe.txt> exists - work has been completed

The procedure has been repeatedly used for analysis of text information of *exe*-files and *com*-files.

---

conwf – a special *Windows* datafiles converting

Call format of the procedure:

conwf(F, n)

Formal arguments of the procedure:

F - symbol or string defining full path to a datafile
n - a positive integer defining maximal length of string of the target text datafile

Description of the procedure:

The *Windows* system has a series of *ASCII* datafiles that represent a certain interest owing to presence in them of useful information (for example, datafiles of types {".sav", ".log", ".tlb", ".cfg", etc.}). Having the *ASCII* format, these datafiles have a rather inconvenient format both for viewing and for program use by the users having a slight acquaintance with *Windows*. The successful procedure call **conwf(F, n)** returns the *full path* to a target file created on the basis of a source datafile **F.** The target file will have the *txt*-format with length of line no more than **n;** the file will be located in the same subdirectory as a source file **F,** and its name will be formed of name of the source file by addition of *txt*-extension. The successful termination of the procedure returns the full path to the target datafile.

```
conwf:= proc(F::file, n::posint)
local a, b, f, k, j, t;
  assign(f = cat(F, ".txt"), t = interface(warnlevel), b = "");
  fopen(F, READ, BINARY), interface(warnlevel = 0);
  while not Fend(F) do  a := subsLS([0 = NULL, seq(k = 32, k = 1 .. 31),
        seq(j = 32, j = 127 .. 255)], [op(convert(b, bytes)), op(readbytes(F, 1024))]);
    a := frss(Red_n(cat(op(map(convert, map(convert, a, list), bytes))), " ", 2), n);
    if a = [] then next
    elif nops(a) = 1 then b := ""; a := map(cat, a, "\r")
    elif length(a[-1]) < 1/2*n then b := a[-1]; a := map(cat, a[1 .. -2], "\r")
    else b := ""; a := map(cat, a, "\r")
    end if;
    writebytes(f, FNS(cat(op(a)), " ", 3))
  end do;
  `if`(b <> "", null(writebytes(f, FNS(b, " ", 3))), NULL), interface(warnlevel = t),
     Close(F, f), CF2(f)
end proc
```

Typical examples of the procedure use:

> conwf("C:\\Windows\\System32\\Config\\Default.sav", 72);

"c:\windows\system32\config\default.sav.txt"

The procedure has been repeatedly used for analysis of text information of the *Windows* datafiles both in interactive and in program mode. These files have specific internal organization, uncomfortable for processing by means oriented onto standard text datafiles.

Cookies – cleaning the Cookies directories from unwanted datafiles

Call format of the procedure:

Cookies(C, L)

Formal arguments of the procedure:

L - symbol, string, set or list of symbols and/or strings
C - a letter defining logical name of a disk drive

Description of the procedure:

A "cookie" is a small text datafile that a web server can store temporarily with your web browser. This is useful for having your browser remember some specific information that the web server can later retrieve. One of the primary purposes of cookies is to provide a convenience feature that you can use to save time. When you visit the same Web site, the information you previously provided can be retrieved, so you can easily use the Web site features that you previously chose.

However, in a series of cases you desire to selectively delete the unwanted cookies basing on certain information containing in them. The procedure call **Cookies(C, L)** is intended for deletion from all Coookies directories located on a disk drive with logical name given by the first actual argument C of *txt*-files containing patterns given by the second actual argument **L.**

The successful procedure call returns the *NULL* value, i.e. nothing with deletion of the above-mentioned datafiles and with output of the appropriate warnings. The procedure handles basic especial and erroneous situations. The procedure has been repeatedly used for cleaning of Cookies directories from the unwanted datafiles. Above all, the procedure has been used for deleting the extremely unwanted datafiles relating to viruses, spyware, spam etc.

```
Cookies:= proc(C::{letter}, L::{string, symbol, set({string, symbol}), list({string, symbol})})
local a, b, c, d, h, k, j, t, s;
  `if`(member(convert(CF2(C), bytes), {[k] $ (k = 97 .. 122)}), `if`(member(CF2(C),
    map(CF2, {Adrive()})), 15, ERROR("drive <%1> is idle", Case(C, upper))), 7);
  assign(a = cat(CDM(), "\\_$ArtKr$_"), b = [], c = map(CF2, {op(L)}), s = 0);
  WARNING("Analysis of Cookies on <%1> is in progress, please wait!", Case(C, upper));
  system(cat("Dir /A/B/S ", C, ":\\*.* > ", blank(a)));
  do h := readline(a);
    if h = 0 then break else
      if Search1(CF2(h), "\\cookies", 't') and t = [right] and type(h, dir) then b:= [op(b), h] end if
    end if
  end do;
  delf(a), `if`(b = [], ERROR("Cookies directories do not exist on drive <%1>",
    Case(C, upper)), NULL);
  for k in b do  d := DirF(k);
     for j in d do
        if IsFtype(j, ".txt") and Search2(CF2(readbytes(j, TEXT, infinity)), c) <> [] then
            delf1(j, `*`); s := s + 1
        else close(j)
        end if
     end do
  end do;
  WARNING("Work has been done; %1 files have been deleted", s)
end proc
```

Typical examples of the procedure use:

> Cookies("G", {MarcoMedia, `RANS_IAN`, "REA_RAC"});

Error, (in **Cookies**) drive <G> is idle

> Cookies("$", {MarcoMedia, `RANS_IAN`, "REA_RAC"});

Error, invalid input: Cookies expects its 1st argument, C, to be of type {letter}, but received $

> Cookies("A", {MarcoMedia, `RANS_IAN`, "REA_RAC"});

Warning, Analysis of Cookies on <A> is in progress, please wait!
Error, (in **Cookies**) Cookies directories do not exist on drive <A>

> Cookies("C", {MarcoMedia, `RANS_IAN`, "REA_RAC"});

Warning, Analysis of Cookies on <C> is in progress, please wait!
Warning, Work has been done; 62 files have been deleted

DeCoder – coding / decoding of a datafile
DeCod

Call format of the procedures:

DeCoder(F, T)
DeCod(F {, P})

Formal arguments of the procedures:

F - filename or full path to a source datafile
T - filename or full path to a target file (result of coding/decoding of the source datafile)
P - (optional) a permutation procedure of bits in the bytes

Description of the procedures:

The procedure call **DeCoder(F, T)** (on the basis of the *switch* principle) {*codes* | *decodes*} the contents of a datafile specified by the first actual **F** argument with putting of the result into a file specified by the second **T** argument. The appendices of the given procedure is most expedient for datafiles of types {*txt, rtf*}, and in more general case for datafiles of *rlb*-type. If successful, the procedure call **DeCoder** returns the full path to a target file with output of the appropriate warning.

The procedure call **DeCod(F {, P})** (on the basis of the *switch* principle) {*codes* | *decodes*} the contents of a file specified by the first **F** argument in situ. The coding/decoding is carried out on the basis of a procedure specified by the optional second actual **P** argument. The **P** procedure returns the result of permutation of elements of an integer 8-element list of view **[1,2,3,4,5,6,7,8]** on the basis of the *switch* principle. The procedure call **DeCod(F)** without the second optional **P** argument assumes that by default will be used the **P** procedure which *inverts* the above list. The appendices of the given procedure is the most expedient for files types {*txt, doc, rtf*}, and in more general case for datafiles of any type. If successful, the procedure call **DeCod** returns the *NULL* value, i.e. nothing.

```
DeCoder:= proc(F::file, T::{string, symbol})
local a, b, c, d, k, f, t, nu;
  `if`(not type(F, rlb), ERROR("<%1> is not a text file", F),
    `if`(not type(eval(T), {string, symbol}), ERROR("<%1> is not a file path", T),
    `if`(not type(T, file), [assign(nu = interface(warnlevel)), interface(warnlevel = 0),
    assign(f = MkDir(T, 1)), interface(warnlevel = nu)], assign(f = T)))),
    assign(a = (n -> 10*n - 99*floor(1/10*n)), b = iolib(4, F, infinity, TEXT), c = "");
```

```
  mul([assign('d' = op(convert(b[k], bytes))), assign('c' = cat(c, `if`(charfcn[0, 62](d) = 1, b[k],
     `if`(charfcn[1, 2](length(d)) = 1, convert([a(d)], bytes), `if`(member(d, [seq(seq(10*h + t,
     h = 0 .. 4), t = [206, 207, 208, 209])]), b[k], convert([100*floor(1/100*d) +
     a(d - 100*floor(1/100*d))], bytes))))))], k = 1 .. length(b));
  null(iolib(5, f, c)), iolib(7, F, f), WARNING("the result is in file <%1>", f), f
end proc

DeCod:= proc(F::file)
local a, b, c, k;
  `if`(1 < nargs and type(args[2], procedure), assign(c = args[2]),
     assign('c' = (L -> [L[8 - k] $ (k = 0 .. 7)])));
  null([assign(a = iolib(4, F, infinity)), iolib(7, F), iolib(5, F, [seq(Bit1(a[b], c), b = 1 .. nops(a))]),
     iolib(7, F)])
end proc
```

Typical examples of the procedures use:

```
> DeCoder("C:/RANS/RegClean.exe", "C:/Academy/IAN/RAE/decod.txt");
   Error, (in DeCoder) <C:/RANS/RegClean.exe> is not a text file
> DeCoder("C:/RANS/Lectures.rtf", "C:/Academy/IAN/RAE/decod.txt");   # coding
   Warning, the result is in file <c:\academy\ian\rae\decod.txt>
                                  "c:\academy\ian\rae\decod.txt"
> DeCoder("C:/Academy/IAN/RAE/decod.txt", "C:/RANS/Lectures1.rtf"); # decoding
   Warning, the result is in file <c:\rans\lectures1.rtf>
                                      "c:\rans\lectures1.rtf"
```

Prm – a permutation procedure of bits in the bytes

```
> Prm:=(L::list(integer)) -> [L[8], L[2], L[6], L[4]+1 mod 2, L[5]+1 mod 2, L[3], L[7], L[1]]:
  DeCod("C:/RANS/Lectures.rtf", Prm);                                    # coding
> DeCod("C:/RANS/Lectures.rtf", Prm);                                    # decoding
> DeCod("C:/RANS/Calc.exe", Prm);                                        # coding
> DeCod("C:/RANS/Calc.exe", Prm);                                        # decoding
```

FBfile – a structuring of a datafile

Call format of the procedure:

FBfile(F, H, p)

Formal arguments of the procedure:

F – symbol or string defining a filename or full path to a datafile
H – symbol or string defining a target filename
p – a positive integer (*posint*)

Description of the procedure:

The procedure call **FBfile(F, H, p)** provides structuring of a datafile specified by the first actual **F** argument onto lines of length **p** by inserting into the corresponding positions the control symbol `hex(0A)` {*carriage return*} and removing the necessary control symbols; the result of such structuring is put into a file specified by its **H** name. Target file **H** is located in the same directory as the source file **F.** In addition, if target file **H** exists, the procedure allows to update it or to define a new filename for it. The procedure call returns the full path to a target file with output of an appropriate warning. Meanwhile, the given procedure has some useful practical appendices, for example, for a viewing of large datafiles, a data printing, etc.

```
FBfile:= proc(F::file, H::{string, symbol}, p::posint)
local cd, h, x, y, psi, omega, nu;
  `if`(not type(eval(H), {string, symbol}), ERROR("<%1> is not a file path", H),
    [Close(F), assign(cd = currentdir(), y = open(F, READ), nu = H), Currentdir(F)]);
  do
    if not type(nu, file) then x := open(nu, WRITE); omega := cat(currentdir(), "\\", nu); break
    else WARNING("file %1 exists - update (y) / create a new file (n)?", nu),
        assign('psi' = readline(terminal));
      if Case(psi) = "y" then close(nu); x := open(nu, WRITE);
        omega := cat(currentdir(), "\\", nu);  break
      end if;
      WARNING("Introduce a new filename: "), assign('psi' = readline(terminal)),
      assign('nu' = psi)
    end if
  end do;
  while not Fend(y) do h := [op(subs(0 = NULL, 10 = NULL, 13 = NULL,
      26 = NULL, readbytes(y, p))), 10];
    if h = [10] then next else writebytes(x, h) end if end do;
  close(x, y), WARNING("the result is in file <%1>", omega), currentdir(cd), omega
end proc
```

Typical examples of the procedure use:

> FBfile("C:/RANS/calc.exe", "Tallinn", 62);

Warning, file Tallinn exists - update (y) / create a new file (n)?

> n

Warning, Introduce a new filename:

> Grodno

Warning, file Grodno exists - update (y) / create a new file (n)?

> n

Warning, Introduce a new filename:

> Vilnius

Warning, file Vilnius exists - update (y) / create a new file (n)?

> y

Warning, the result is in file <c:\rans\Vilnius>

"c:\rans\Vilnius"

The given procedure has been repeatedly used for analysis of large datafiles of different types. Above all, the procedure is rather useful at analysis of contents of executable program files and datafiles containing control symbols.

Fword – choice of words out of a datafile according to the given conditions

Call format of the procedure:

Fword(F, K, D, P {, v})

Formal arguments of the procedure:

F – filename or full path to a datafile
K – a set or a list of contexts of type *`string`* or *`symbol`*
D – words delimiters in the file; may be *string, symbol*, list({*string,symbol*}), set({*string,symbol*})
P – a location mode of context in sought words; may be [*left*], [*inside*], [*right*], [*left,inside*], [*left, right*], [*inside,right*], [*left, inside, right*]
v – (optional) a valid *Maple* expression

Description of the procedure:

The procedure call **Fword(F, K, D, P)** considers a text file specified by the first actual **F** argument as a *string* and forms a list of words formed by means of fragmentation of the string on the basis of delimiters specified by the actual argument **D.** Then, out of the above list of words, the **Fword** procedure forms a new set of words of *symbol*-type that contain at least one context from a set or a list **K** (its elements are *symbols* or *strings*) with a location mode specified by the 4th **P** argument. If at procedure call the fifth argument **v** has been coded then the procedure simultaneously saves the chosen words in a target file in line-by-line format with return of full path to the datafile. The procedure call returns the sequence with one or two elements (depending on presence of the 5th argument **v**) whose the first element defines the set of chosen words and the second argument (*at its presence*) defines the full path to a text target file with chosen words as lines.

Let's suppose, the basic location modes of a context **w** in a word **X** are presented as **[***left***]**, **[***inside***]** or **[***right***]**, if word **X** can be presented as **X=wX1, X2wX3** or **X4w** accordingly. Along with the basic location modes for a context the following location modes are entirely admissible **[***left, inside***]**, **[***left, right***]**, **[***inside, right***]** and **[***left, inside, right***]**. All these location modes can be specified as the fourth **P** argument. Joint use of arguments **K** and **P** defines a *`filter`* through which the **Fword** chooses from the above list, received as a result of fragmentation of a text datafile **F** by delimiters **D** the sought words.

```
Fword:= proc(F::file, K::{set({string, symbol}), list({string, symbol})}, D::{string, symbol,
                    set({string, symbol}), list({string, symbol})}, P::list(symbol))
local h, k, p, j, t, a, nu;
  `if`(not type(F, rlb), ERROR("<%1> is not a text file", F), [assign(nu = {op(D)},
     p = cat(F, ".ian")), assign(h = SLS(readbytes(F, TEXT, infinity), nu), a = {})]);
  for k to nops(h) do for j to nops(K) do
       if Search1(h[k], K[j], 't') and  t = P then a := {op(a), cat(``, h[k])};  next end if
     end do
  end do;
  a, close(F), `if`(4 < nargs and a <> {}, [map2(writeline, p, a), close(p), p][-1], NULL)
end proc
```

Typical examples of the procedure use:

Choice out of text files of words delimited by the given symbols according to the specified contexts and their location modes with saving their in a target file in line-by-line format or without saving.

> Fword("C:/Book/MBL.txt", {`.com`,`.uk`,`.ee`,`.lt`,`.by`,`.ru`, `.lv`}, {",", " "}, [right], 15);

{nik@grsu.unibel.by, ian@uni.ac.uk, hunt@evr.ee, vaganov@lenta.ru, marius@ti.vtu.lt, aladjev@hotmail.com, noosphere_academy@yahoo.com}, "C:/Book/MBL.txt.ian"

```
> Fword("C:/RANS/My_email.htm", {`.com`,`.uk`,`.ee`,`.lt`,`.by`,`.ru`}, {"<", ">"}, [right]);
    {Viktor.Aladjev@mail.ee, J.J.Klimovich@rambler.ru, Julvagfam@aol.com, info@exponenta.ru,
    noosphere_academy@yahoo.com, V.Vaganov@rambler.ru, aladjev@lenta.ru,
    veetousme@stat.ee, valadjev@yahoo.com, academy60@hotmail.com,
    olegas@ti.vtu.lt,hunt@evr.ee,vaganov@lenta.ru, aladjev@hotmail.com,
    aladjev.victor@rambler.ru, marius@ti.vtu.lt}
```

MPL_txt – a support of mechanism of the *"disk transits"*

Call format of the procedure:

MPL_txt(F, T {, N})

Formal arguments of the procedure:

F - symbol or string defining a filename or full path to a *mws*-file
T - symbol or string defining a filename or full path to a target datafile
N - (optional) a valid *Maple* expression

Description of the procedure:

In a series of our works [1-14,29-33] the method of so-called ***"disk transits"*** developed by us has been used in many appendices enough effectively. In particular, the given method can be used the most advantageously in many problems, needing in a generating of on the-fly executable fragments of the *Maple* programs. The essence of the given mechanism in conditions of the *Maple* environment is well illustrated in our last books [29-33,43,44]. For automation of the mechanism of *"disk transits"*, the procedure **MPL_txt** can be useful enough, returning the *NULL* value or full path to a target text datafile in case of successful termination. The essence of an algorithm implemented by the procedure can be described in a general way as follows.

The first actual **F** argument of the procedure **MPL_txt(F, T {, N})** specifies qualifier of an initial *mws*-file (filename or full path to it), whereas the second actual **T** argument specifies qualifier of a target datafile with a *Maple* document of a text format, suitable for subsequent use by the built-in ***`read`*** function. At last, the optional third argument **N** of the procedure specifies a mode of interactive execution of the *mws*-file indicated by the first actual **F** argument, in a current session on the basis of reading by ***`read`*** function of the **T** datafile appropriate to it.

As a result of successful running of the procedure all tools and results whose definitions are in the initial file **F**, become accessible in a current session. The procedure call **MPL_txt(F, T, N)** with three actual arguments provides creating of a text equivalent of a required *mws*-file **F** with the opportunity of its subsequent running by means of the built-in ***`read`*** function with return of the *NULL* value and with output of warning about full path to a target text datafile **T.** Along with that, all tools of the **F** file become available in a current session. While the procedure call with two actual arguments provides only creating of a text equivalent **T** with return of full path to it. The subsequent call **read(T)** provides availability in a current session of the above tools of the **F** file.

The given procedure provides a definite level of automation of the mechanism of *`disk transits`* and is founded on the structural analysis of source *mws*-files, what in case of their large sizes requires essential time expenditures. Therefore, use of the procedure can be effective enough relative to *mws*-files of the size no more than 20 - 30 KB; however, this estimation depends on characteristics of main resources of a computer. Whereas for cases of large *mws*-files we recommend to use the reception described in our books [11,12].

```
MPL_txt:= proc(M::file, T::{string, symbol})
local a, b, k, h, G, S, c, A, R, d, f, p, Q, r, t, Z, nu;
```

```
  `if`(Ftype(M) = ".mws", assign(Z = "MPLTEXT", Q = "", R = "", A = "", S = "",
    nu = interface(warnlevel), f = T), ERROR("<%1> should has mws-type but had received %2",
    Ftype(M))), `if`(not type(eval(T), {string, symbol}), ERROR("<%1> is not a file path", T),
    `if`(type(T, file), 7, [interface(warnlevel = 0), assign('f' = MkDir(T, 1)),
    interface(warnlevel = nu)]));
  assign67(a = fopen(M, READ, BINARY), close(f), b = fopen(f, WRITE, BINARY),
    c = "", t = 1, h = 1), assign(G = convert(iolib(4, a, infinity), bytes));
  for k to length(G) do `if`(G[k] = "\n" or G[k] = "", NULL, assign('S' = cat(S, G[k]))) end do;
  for k to length(S) do `if`(S[k .. k + 3] = "\\\\", assign('c' = cat(c, "/"), 'k' = k + 3),
    assign('c' = cat(c, S[k]))) end do;
  for k to length(c) do `if`(c[k] = "\\", NULL, assign('A' = cat(A, c[k]))) end do;
  do assign('r' = length(A)), assign('d' = searchtext(Z, A, h .. r) + h + 11);
    if d = h + 11 then
      while t < length(R) do `if`(search(R[t .. t + 2], `" "`), assign('t' = t + 3),
        `if`(search(R[t .. t + 2], "+if"), assign('Q' = cat(Q, "if"), 't' = t + 3),
        `if`(search(R[t .. t + 3], "+end"), assign('Q' = cat(Q, "end"), 't' = t + 4),
        assign('Q' = cat(Q, R[t]), 't' = t + 1))))
      end do;
      RETURN(null(iolib(5, b, convert(Q[2 .. -2], bytes)), iolib(7, a, b)), `if`(2 < nargs,
          (proc() WARNING("the result is in file <%1>", f); read f end proc)(f), f))
    else NULL
    end if;
    assign('d' = d + searchtext(" ", A, d .. r)), assign('p' = d + searchtext("}{", A, d .. r));
    while A[p] <> " " do p := p - 1 end do;
    assign('R' = cat(R, A[d .. p]), 'h' = p + 2)
  end do
end proc
```

Typical examples of the procedure use:

```
> MPL_txt("C:/RANS/helpman.mws", "C:/Temp\\AVZ\\Orders/Svetlana.txt", 7);
  Warning, the result is in file <C:/Temp\AVZ\Orders/Svetlana.txt>
> map(evalf, [SR(62,57,41,37,15,7), DS(62,57,41,37,15,7)]);
                    [36.50000000, 22.01590334]
> restart: MPL_txt("C:/RANS/helpman.mws", "C:/Temp\\AVZ\\Orders");
                    "c:\temp\avz\_orders"
> map(evalf, [SR(62,57,41,37,15,7), DS(62,57,41,37,15,7)]);
                    [SR(62,57,41,37,15,7), DS(62,57,41,37,15,7)]
> read("c:\\temp\\avz\\_orders"); map(evalf, [SR(62,57,41,37,15,7), DS(62,57,41,37,15,7)]);
```

$$SR := (\,) \to \frac{\text{`+`(args)}}{\text{nargs}}$$

$$DS := (\,) \to \sqrt{\frac{\sum_{k=1}^{\text{nargs}} (\text{SR(args)} - \text{args}_k)^2}{\text{nargs} - 1}}$$

[36.50000000, 22.01590334]

In particular, this procedure can successfully be applied at operation with large enough *mws*-files, whose structure and purpose are not known or they are enough poorly documented. The procedure **MPL_txt** chooses from such files all definitions of the procedures contained in them.

Images – a documents making in HTML format

Call format of the procedure:

Images(D, F, P {, 'h'})

Formal arguments of the procedure:

D - string defining name of a directory
F - symbol or string defining a filename
P - symbol or string defining a subdirectory name
h - (optional) an unevaluated name

Description of the procedure:

The *Maple* package has effective enough tools for creation of documents in *HTML* format what is being provided by any of two functional chains of the *Graphical User Interface* (**GUI**), namely: (**1**) *File – > Export As – > HTML* or (**2**) *File – > Save As – > HTML Source*. In addition, in both cases a *name* of subdirectory is inquired (by default `*images*`), in which *gif*-files with graphical objects created on the basis of a source document will be located.

However, in the publishing respect the *Maple* package essentially yields to such widespread word processor as *Ms Word 97*, for example. Therefore in such case into documents prepared in the environment *Ms Word 97* are being inserted any objects from a *Maple* document via *Clipboard*. Next a *Word*-document created thus, is saved in *HTML* format. However, in contrast to *Maple*, here there is not any opportunity of setting of a directory, in which the files of formats {*gif, jpg*} with graphical objects created on the basis of a source document will be located. These files are put into the same directory with the saved file of *HTML* format, what creates certain inconveniences at operation with files created thus. The **Images** procedure represented below eliminates the above-mentioned shortcoming and in many cases can be by rather useful tool.

The procedure call **Images(D, F, P)** has three formal arguments, namely: **D** – a directory with a processed file of *HTML* format; **F** – a full file name, and **P** – a subdirectory name of the **D** directory intended for accompanied graphical files of formats {*gif, jpg*}. Before procedure call, an **F** file with a *Word* document it is necessary to copy in a **D** directory, then load it into the *Word* environment with the subsequent saving in the *HTML* format. After that, we have all necessary information for the procedure **Images** call.

The procedure creates a specified **P** subdirectory for files of formats {*gif, jpg*}, translocates into it all files {*gif, jpg*} from a **D** directory and then does necessary processing of a source file of the *HTML* format. The result of successful procedure call is creation of the required file organization for a source *HTML*-file with output of appropriate warning. At attempt of repeated processing of the earlier processed ***HTML*** file by means of the **Images** procedure arises erratic situation with diagnostics "*invalid attempt of repeated processing of* <%1>". If at procedure call, the fourth optional **h** argument has been coded then through it, the size of a processed **F** file is returned in bytes. The procedure **Images** has been repeatedly used for creation of electronic books on *Maple*.

```
Images:= proc(D::dir, F::{symbol, string}, P::{symbol, string})
local a, k, p, h, L, S, G, Q, t, nu;
  `if`(not type(cat(D, "\\", F), file), ERROR("<%1> is not a datafile", F), NULL);
  `if`(type(eval(P), {symbol, string}), assign(p = 61, L = [], h = 1, assign(nu = cat(D, "\\", P))),
     ERROR("<%1> should be symbol or string but had received <%2>", P, whattype(eval(P))));
```

```
  if not type(nu, dir) then t := interface(warnlevel); interface(warnlevel = 0);
    nu := MkDir(nu); interface(warnlevel = t)
  end if;
  assign(Q = cat(CFF(nu)[-1], "\\")), ssystem(cat("copy ", cat(Path(D), "\\*.jpg "),
    Path(nu), " /Y")), ssystem(cat("copy ", cat(Path(D), "\\*.gif "), Path(nu), " /Y"));
  ssystem(cat("Del ", cat(Path(D), "\\*.jpg"))), ssystem(cat("Del ", cat(Path(D), "\\*.gif")));
  assign(a = cat(D, "\\", F)), assign(S = iolib(4, a, TEXT, infinity)), iolib(7, a);
  try
    while p <> 0 do p:= searchtext("="Image", S, h .. length(S)); h:= p + h + 1; L:= [op(L), h]
    end do;
    assign('L' = L[1 .. nops(L) - 1]), assign('G' = S[1 .. L[1] - 1]),
      seq(assign('G' = cat(G, Q, S[L[k] .. L[k + 1] - 1])), k = 1 .. nops(L) - 1),
      assign('G' = cat(G, Q, S[L[nops(L)] .. length(S)])), null(iolib(5, a, G), iolib(7, a),
      `if`(nargs = 4, [assign([args][4] = iolib(6, a, infinity)), iolib(7, a)], 0),
      WARNING("file <%3> is located in <%1>, while its accompanied files {jpg, gif}
         are located in <%2>", D, nu, F))
  catch "invalid subscript selector": iolib(7, a);
        ERROR("invalid attempt of repeated processing of <%1>", F)
  end try
end proc
```

Typical examples of the procedure use:

```
> Images("C:/RANS/Academy", "Invoice.htm", GIF_JPG, 'G');  G;
   Warning, file <Invoice.htm> is located in <C:/RANS/Academy>, while its accompanied files
   {jpg, gif} are located in <c:\rans\academy\_gif_jpg>
                                    552144
> Images("C:/RANS\\academy", "Invoice.htm", GIF_JPG, 'G');
   Error, (in Images) invalid attempt of repeated processing of <Invoice.htm>
> RACREA:= table([a=b, c=d]):  Images("C:/RANS", "Invoice.htm", RACREA, 'G');
   Error, (in Images) <RACREA> should be symbol or string but had received <table>
```

The procedure has been used many times for preparing of our electronic books in computer algebra and computer science.

AnalT – an useful statistics on a text datafile or string

Call format of the procedure:

AnalT(F {, 'h'} {, 'p'})

Formal arguments of the procedure:

F - a string or a symbol, or text datafile
h - an assignable name
p - an assignable name

Description of the procedure:

Procedure **AnalT(F {, 'h'} {, 'p'})** is intended for a gathering of an useful statistics concerning the symbols composing a text datafile or string specified by the first actual **F** argument. Through the second optional **h** argument, the table determining quantities of entries of the Latin letters (irrespectively from their register) is returned. The indices of the table define the Latin letters in string notation and its entries define quantities of their entries into an analyzed text datafile or string. Whereas through the third optional **p** argument the plot of the above table is returned. The plot is accompanied by the equations list that defines the one-to-one correspondence between the table entries and points of X-axis. The procedure returns the 5-element list whose elements are: **(1)** quantity of the register switching of a keyboard at creation of the analyzed text, **(2)** quantity of entries of symbols of upper-register of US keyboard, **(3)** quantity of entries of symbols of lower-register of US keyboard, **(4)** quantity of entries of the Latin letters and **(5)** quantity of entries of other symbols of US keyboard.

```
AnalT:= proc(F::{string, symbol})
local A, K, H, U, L, N, P, B, Z, Tab;
  `if`(not type(eval(F), {string, symbol}),
     ERROR("<%1> is not a symbol, not a string, and not a file path", F), 7);
  `if`(type(F, file), `if`(type(F, rlb), [assign(H = readbytes(F, TEXT, infinity)), close(F)],
     ERROR("<%1> is not a text file", F)), assign(H = cat("", F))), assign(P = length(H), U = 0,
     L = 0, N = 0, Z = 0, B = 0);
  Tab:= table(["q" = 0, "w" = 0, "e" = 0, "r" = 0, "t" = 0, "y" = 0, "u" = 0, "i" = 0, "o" = 0, "p" = 0,
     "a" = 0, "s" = 0, "d" = 0, "f" = 0, "g" = 0, "h" = 0, "j" = 0, "k" = 0, "l" = 0, "z" = 0, "x" = 0,
     "c" = 0, "v" = 0, "b" = 0, "n" = 0, "m" = 0]);
  for K to P do
     if type(H[K], letter) then B := B + 1; A := cat(Case(H[K])); Tab[A] := Tab[A] + 1  end if;
     if type(H[K], Lower) then L := L + 1; if Z = 1 then N := N + 1 end if;  Z := 2
     else  U := U + 1; if Z = 2 then N := N + 1 end if;  Z := 1
     end if
  end do;
  [N, U, L, B, P - B], `if`(1 < nargs, [assign(args[2] = eval(Tab)), `if`(2 < nargs,
     assign(args[3] = plotTab(Tab, thickness = 2, color = magenta)), NULL)], NULL)
end proc
```

Typical examples of the procedure use:

```
> AnalT("C:/RANS\\LabTalk.txt", 'omega', 'nu'); nu; eval(omega);
```

[1 = "q", 2 = "w", 3 = "e", 4 = "r", 5 = "t", 6 = "y", 7 = "u", 8 = "i", 9 = "o", 10 = "p", 11 = "a", 12 = "s", 13 = "d", 14 = "f", 15 = "g", 16 = "h", 17 = "j", 18 = "k", 19 = "l", 20 = "z", 21 = "x", 22 = "c", 23 = "v", 24 = "b", 25 = "n", 26 = "m"]

[22731, 18463, 44602, 38965, 24100], []

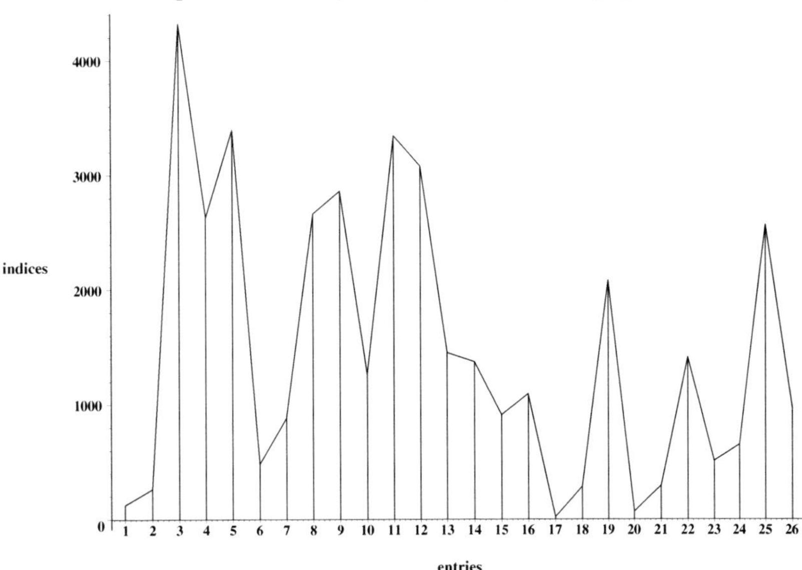

table(["m" = 967, "f" = 1371, "w" = 264, "g" = 908, "e" = 4322, "h" = 1090, "r" = 2632, "j" = 20, "t" = 3393, "k" = 284, "y" = 481, "l" = 2080, "u" = 880, "z" = 66, "i" = 2664, "x" = 291, "o" = 2860, "c" =1412, "p" = 1262, "v" = 508, "a" = 3340, "b" = 649, "s" = 3078, "n" = 2567, "d" = 1451, "q" = 125])

> RANS_IAN:= "Aladjev V.Z., Bogdevicius M.A., Vaganov V.A. Systems of Computer Algebra: New Software ToolBox for Maple, Tallinn, 2003, 420 p. + CD-ROM, ISBN 9985-9277-8-8": AnalT(RANS_IAN, 'h'), eval(h);

[51, 51, 105, 98, 58], [], table(["i" = 4, "e" = 8, "t" = 5, "a" = 11, "k" = 0, "d" = 3, "g" = 3, "n" = 5, "q" = 0, "u" = 2, "j" = 1, "z" = 1, "c" = 3, "b" = 4, "s" = 6, "p" = 3, "r" = 5, "f" = 3, "h" = 0, "l" = 6, "x" = 1, "m" = 5, "w" = 2, "o" = 10, "y" = 1, "v" = 6])

Information being given by the *AnalT* procedure can seem rather useful at analyses of various sort of text information (an used quantity of register switching of a keyboard, quantities of entries of Latin letters and special symbols, etc.).

Kernels – identification in *Windows* environment of active *Maple* sessions

Call format of the procedure:

Kernel()

Formal arguments of the procedure: **no**

Description of the procedure:

At presence in the current *Windows* session of a few active *Maple* sessions, the problem of their identification arises. Successful procedure call **Kernels()** returns the sequence of 2-element lists whose the first element defines Process **ID** of an active *Maple* session, while the second – the filename with its kernel. Order of the sequence elements corresponds to order of *Maple* sessions activation. The procedure handles basic especial and erroneous situations.

```
Kernels:= proc()
local a, b, c, d, h, t, k;
  assign('c' = "$ArtKr$", 't' = interface(warnlevel)), assign('a' = system(cat("tlist.exe > ", c)));
  interface(warnlevel = 0), com_exe2({`tlist.exe`}), interface(warnlevel = t);
  assign67(b = fopen(c, READ), d = NULL, h = " Maple ");
  while not Fend(c) do a := readline(c);
    if search(a, h) then k:= SLD(Red_n(a, " ", 2), " "); d:= d, `if`(search(k[2], "maple"),
        [came(k[1]), k[2]], NULL)
    end if
  end do;
  d, delf(c)
end proc
```

Typical example of the procedure use:

> Kernels();

[284, "maplew8.exe"], [1348, "wmaple.exe"], [1848, "maplew8.exe"], [1880, "wmaple.exe"]

The procedure is rather useful at work in *Windows* with a few *Maple* sessions simultaneously. This procedure allows to analyze in program mode the presence in the current *Windows* session of active *Maple* sessions different from a current *Maple* session.

Chapter 11. Software for solving problems of mathematical analysis

In the given section the tools extending the possibilities of *Maple* of releases **6 - 9/9.5** for problem solving of mathematical analysis is represented. In this connection, the descriptions of these tools are represented below in the above aspects.

&Shift - an expanding of the shift operator

Call format of the procedure:

&Shift(G, [h1, h2, ... ,hn], x1, x2, ... ,xn)

Formal arguments of the procedure:

G - a function name
hj - a list of shifts on leading variables of the function
xj - a sequence of leading variables of the function

Description of the procedure:

In the *Maple* environment the user (*neutral*) `&` operator is represented by call of an appropriate procedure. In addition, the *infix* notation is allowable only for a case of two operands, while the *prefix* notation - for any quantity of operands. A definition method of the user operators described in our books [10-14,29-33,43,44], in a view of the *Maple* procedures mechanism is transparent enough and has the important applied meaning, by allowing to introduce own `&` operators for special operations. As one of appendices of the given method an expansion of the **shift** operator is given. The procedure call **&Shift (G, [h1,h2,...,hn], x1,x2,...,xn)** returns the *shift* result of a function specified by a name **G** on its leading variables **xj**, where **hj** specify values of *shifts* on the appropriate leading variables, i.e.

&Shift (G, [h1, h2, ... ,hn], x1, x2, ... ,xn) -> G(x1 + h1, ... ,xn + hn)

In addition, between elements of the second and third operands of the operator the one-to-one correspondence is supposed.

```
`&Shift`:= proc()
local k;
  `if`(nargs < 3, ERROR("incorrect quantity <%1> of actual arguments", nargs),
    `if`(not (type(args[1], symbol) and type(args[2], list)),
    ERROR("incorrect type of the first argument <%1> and/or the second argument <%2>",
    args[1], args[2]), `if`(nops(args[2]) <> nargs - 2, ERROR("incorrect quantity of shifts <%1>
    and/or leading variables of function <%2>", nops(args[2]), nargs - 2), `if`(member(false,
```

```
    {op(map(type, [args['k'] $ ('k' = 3 .. nargs)], name))}) = true,
    ERROR("incorrect types of actual arguments %1", 3 .. nargs), NULL))));
  args[1]((args['k' + 2] + args[2]['k']) $ ('k' = 1 .. nops(args[2])))
end proc
```

Typical examples of the procedure use:

> &Shift (Art, [a,b,c,d,g], x,y,z,t,u), &Shift (Kr, [42,47,67,62,89,96], x, y, z, t, h, p);

Art(x + a, y + b, z + c, t + d, u + g), Kr(x + 42, y + 47, z + 67, t + 62, h + 89, p + 96)

> &Shift(Art, [42,47,67,62,89,95,a,b,c], x1,x2,x3,x4,x5,x6,x7,x8,x9);

Art(x1 + 42, x2 + 47, x3 + 67, x4 + 62, x5 + 89, x6 + 95, x7 + a, x8 + b, x9 + c)

CoefTaylor – evaluation of Taylor series coefficients

Call format of the procedure:

CoefTaylor(T, V, N)

Formal arguments of the procedure:

T – an algebraic expression
V – list or set of names or equations
N – non-negative integer (*nonnegative*) – the `*truncation order*` of the series

Description of the procedure:

The simple **CoefTaylor** procedure has the coding format, similar to the ***mtaylor*** function of *Maple*, and returns the 2-element sequence of lists of *Taylor* series coefficients and the terms appropriate to them. The first sublist defines the *Taylor* series coefficients, whereas the second defines the terms appropriate to them. If **V[k]** is an equation, then the left member of **V[k]** is a variable, and the right member is a point of expansion. If **V[k]** is a name, then **V[k] = 0** is assumed as the expansion point. The used concept `*truncation order*` specifies total degree in the variables. Analogously to the ***mtaylor***, the **CoefTaylor** restricts its domain to `*pure*` *Taylor* series, those series with non-negative powers in the variables.

```
CoefTaylor:= proc(T::algebraic, V::{list({equation, symbol}), set({equation, symbol})},
                  N::nonnegative)
local h, nu, omega, psi;
  assign(nu = (x -> `if`(type(x, equation), x, x = 0)), psi = (x -> `if`(type(x, equation), lhs(x), x)));
  assign(omega = map(nu, V)), [coeffs(sort(expand(mtaylor(T, omega, N)), lexorder),
    map(psi, V), 'h')], [h]
end proc
```

Typical examples of the procedure use:

> CoefTaylor((8*x^2+18)*exp(x)+x*sin(y)*cos(z)-x^2+z^2, [x=0, y=Pi, z=Pi], 7);

$$\left[18, 18+\frac{2}{3}\pi^3-\pi-\frac{2}{15}\pi^5, \frac{\pi^2}{2}-\frac{1}{2}, \frac{1}{12}, \frac{1}{24}, \pi-\frac{2}{3}\pi^3, \frac{\pi^2}{6}-\frac{1}{6}, 16, 11, \frac{19}{4}, \frac{89}{60}, \frac{1}{3}\pi^4-\pi^2+1, 1, \frac{43}{120}, \frac{1}{120}, \frac{1}{2}\pi-\frac{1}{3}\pi^3, \frac{1}{3}\pi^4-\pi^2, \frac{\pi^2}{2}, -\frac{\pi}{6}, -\frac{\pi}{4}, \frac{1}{2}\pi-\frac{1}{3}\pi^3, \frac{\pi^2}{6}, -\frac{\pi}{24}, -\frac{\pi}{24}, -\frac{\pi}{6}\right], [1, x, x\,y\,z^2, x\,y^3\,z^2, x\,y\,z^4, x\,y\,z, x\,y^3, x^2, x^3, x^4, x^5, x\,y, z^2, x^6, x\,y^5, z^2\,x, z\,x, z\,y^2\,x, z^3\,y\,x, z^2\,y^2\,x, y^2\,x, z^3\,x, z^4\,x, y^4\,x, z\,y^3\,x]$$

> CoefTaylor(14*(3*x^2+10)*exp(x) + 7*x*sin(y)*cos(z), [x=a, y, z=b], 5);

$$\Big[140\,e^{a} - 140\,a\,e^{a} + 7\,e^{a}\,a^{5} + \frac{7}{4}\,e^{a}\,a^{6} + 70\,e^{a}\,a^{2} - \frac{70}{3}\,e^{a}\,a^{3} + \frac{35}{6}\,e^{a}\,a^{4},$$
$$-7\,e^{a}\,a^{5} + 70\,e^{a}\,a^{2} - \frac{70}{3}\,e^{a}\,a^{3} + 140\,e^{a} - 35\,e^{a}\,a^{4} - 140\,a\,e^{a}, -\frac{7}{6}\,a\sin(b)\,b^{3}, 14\,a\,e^{a} + \frac{161}{6}\,e^{a} + \frac{7}{4}\,e^{a}\,a^{2},$$
$$-7\,e^{a}\,a^{3} - 49\,e^{a}\,a^{2} - \frac{196}{3}\,a\,e^{a} + \frac{196}{3}\,e^{a}, -\frac{7}{6}\,a\sin(b)\,b, -\frac{7}{2}\cos(b)\,b^{2} + 7\sin(b)\,b + 7\cos(b), -\frac{7}{6}\cos(b), \frac{7}{6}\,a\sin(b),$$
$$\frac{7}{6}\,a\sin(b), -\frac{7}{2}\cos(b), -\frac{7}{2}\,a\sin(b)\,b, 7\cos(b)\,b - 7\sin(b), \frac{7}{2}\,a\sin(b)\,b^{2},$$
$$\frac{21}{2}\,e^{a}\,a^{4} - 112\,a\,e^{a} + 56\,e^{a}\,a^{2} + 112\,e^{a} + 63\,e^{a}\,a^{3}\Big],$$
$$[1, x, y, x^{4}, x^{3}, y^{3}, x\,y, x\,y^{3}, z\,y^{3}, z^{3}\,y, z^{2}\,y\,x, z^{2}\,y, z\,y\,x, z\,y, x^{2}]$$

> CoefTaylor(12*(3*x^2+10)*exp(x) + 6*x*sin(y)*cos(z), [x, y, z], 7);

$$\Big[120, 120, 23, 56, 6, -1, -3, \frac{5}{3}, 7, \frac{1}{20}, \frac{1}{2}, \frac{1}{4}, 96\Big], [1, x, x^{4}, x^{3}, x\,y, x\,y^{3}, z^{2}\,y\,x, x^{6}, x^{5}, x\,y^{5}, x\,y^{3}\,z^{2}, x\,y\,z^{4}, x^{2}]$$

---

etf – evaluation of values of tabular functions

Call format of the procedure:

etf(x, F {, h})

Formal arguments of the procedure:

x - a value of *numeric* type
F - a tabular function {*array* | *Array* | *matrix* | *Matrix* | *file*}
h - (optional) a valid *Maple* expression

Description of the procedure:

The procedure call **etf(x, F)** returns the value of a function specified by an *array, Array, matrix, Matrix* or *text file* **F** of dimensionality **(n, 2)**, in a point **x.** The first column defines tabular values of a function argument, whereas the second column defines the function values appropriate to them. If the second actual argument **F** defines a text datafile, the data in the file must consist of integers, floating-point values or fractional values arranged in two columns parted by *blanks*, i.e. a tabular function admits the values of *numeric* type. If the procedure call uses only two actual arguments, a tabular function is approximated by linear function by the least squares method between the tabulated points, otherwise the approximation is done by square function.

```
etf:= proc(x::numeric, F::{Matrix, matrix, Array, array, file})
local n, k, T, z, f, nu;
option remember;
  assign(f = a*z + b, nu = z^2 + b*z + c), `if`(type(F, file), `if`(type(F, rlb),
    assign(T = readdata1(F, infinity)), ERROR("<%1> is not a text datafile", F)),
    `if`(type(F, Matrix), assign(T = F), assign(T = convert(F, Matrix))));
  for k to op(1, T)[1] - 1 do
    if belong(x, T[k, 1] .. T[k + 1, 1]) then
      if x = T[k, 1] then RETURN(T[k, 2])
      elif x = T[k + 1, 1] then RETURN(T[k + 1, 2])
```

```
        else RETURN(subs(z = x, CurveFitting[LeastSquares]([[T[k, 1], T[k, 2]], [T[k + 1, 1],
          T[k + 1, 2]]], z, curve = `if`(nargs = 2, f, nu))))
        end if
      else next
      end if
   end do;
   ERROR("value <%1> is outside of diapason of tabular function", x)
end proc
```

Typical examples of the procedure use:

> evalf(etf(85, "C:/Academy/Examples/Family.dat", 7));

-74.14814815

> evalf(etf(85, "C:/Academy/Examples/Family.dat"));

17.85185185

> evalf(etf(127/3, "C:/Academy/Examples/Family.dat"));

60.66666667

> etf(460, "C:/Academy/Examples/Family.dat");

Error, (in **etf**) value <460> is outside of diapason of tabular function

> etf(61.56, "C:/Academy/Examples/Family.dat", 14);

-37.7664

---

diffP – evaluation of partial derivatives in the given points

Call format of the procedure:

diffP(V, L)

Formal arguments of the procedure:

V - an algebraic expression
L - a list of initial conditions

Description of the procedure:

The procedure call **diffP(V, L)** returns the partial derivatives of an algebraic expression specified by the first actual **V** argument with respect to the *initial conditions* specified by the second actual **L** argument. By elements of an **L** list can be both the separate variables and the equations of view ***`variable = pnt`***, where ***pnt*** - a point in which the value of appropriate derivative should be evaluated. The procedure allows to calculate derivatives of algebraic expressions from many leading variables in especial points, for example, at condition ***`variable= ±infinity`***.

```
diffP:= proc(V::algebraic, L::list)
local W, h, k, R, x;
   h, W, R := indets(V, name), x -> WARNING("condition <%1> is invalid", x), V;
   try for k to nops(L) do
        if type(L[k], `=`) and member(lhs(L[k]), h) then
          try R := eval(subs(L[k], diff(R, lhs(L[k]))))
          catch "numeric exception:": R := limit(diff(R, lhs(L[k])), L[k])
          end try;
          next
```

```
      elif not type(L[k], `=`) and member(L[k], h) then R := diff(R, L[k]); next
      else W(L[k]); next
      end if
    end do
  catch "numeric exception": WARNING("situation - %1 in point %2", lasterror, L[k])
  end try;
  R
end proc
```

Typical examples of the procedure use:

```
> diffP(36*x^3+6*x^2*sin(x)+14*cos(x)+3, [x=Pi, c=56]), diffP(a/(a+b), [a=7, b=14]),
  diffP(56*x^3+61*x^2*sin(y)+14*cos(y)+6, [x=a+b,y=a^b,z=61]), diffP(x/(a+y),[x=0,y=0]),
  diffP(cos(x)/x, [x=0]), diffP(x*ln(y)/y^2, [x=1, y=0]), diffP(tan(x)/x^2, [x=0]);
    Warning, condition <c = 56> is invalid
    Warning, condition <z = 61> is invalid
```

$$102\,\pi^2,\ \frac{-1}{1323},\ 122\,(a+b)\cos(a^b),\ -\frac{1}{a^2},\ -\infty,\ \mathit{undefined},\ -\infty$$

INT – full integration of algebraic expressions

Call format of the procedure:

INT(expr {, 't'})

Formal arguments of the procedure:

expr - an algebraic expression
t - (optional) an unevaluated name

Description of the procedure:

The procedure call **INT(*expr*)** returns the integration result of an algebraic expression specified by the actual ***expr*** argument over all independent variables, being in it. The procedure call **INT(*expr*, 't')** with the second optional actual **t** argument provides return through it of a list of independent variables additionally. The order of the variables of this list defines order of integration of an expression specified by the first actual ***exrp*** argument. If a name **t** belongs to an integration variables list then it will be renamed as **cat(`_`, t, 1)** with output of an appropriate warning.

```
INT:= proc(V::algebraic)
local k, L, S, W;
global `&*`;
  S, L, W, `&*` := sort([op(indets(evalf(V), name) minus {constants})]), [], V, '`&*`';
  for k to nops(S) do L := [op(L), `if`(depends(W, S[k]), S[k], NULL)] end do;
  for k in L do W := simplify(int(W, k)) end do;
  if nargs = 1 then W
  elif nargs = 2 and member(args[2], S) then assign(cat(_, args[2], 1) = S), W,
  WARNING("the integration variables list has been assigned to variable <%1>",cat(_, args[2],1))
  else assign(args[2] = S), W
  end if
end proc
```

Typical examples of the procedure use:

> INT(4*(a+b)/(c+d)*h/p, 'z'), z;

b*a*h^2*ln(p)*(a*ln(c+d)*c+a*ln(c+d)*d-a*c-a*d+b*ln(c+d)*c+b*ln(c+d)*d-b*c-b*d), [a, b, c, d, h, p]

> INT(4*(a+b)/(c+d)*h/p, 'a'), _a1;

Warning, the integration variables list has been assigned to variable <_a1>

b*a*h^2*ln(p)*(a*ln(c+d)*c+a*ln(c+d)*d-a*c-a*d+b*ln(c+d)*c+b*ln(c+d)*d-b*c-b*d), [a, b, c, d, h, p]

Nint – numeric integrating of algebraic expressions

Call format of the procedure:

Nint(*args*)

Formal arguments of the procedure:

args – a sequence of actual arguments

Description of the procedure:

The procedure call **Nint(*args*)** returns the result of numeric integrating of an algebraic expression specified by the first actual argument, whereas the other actual arguments correspond to the standard procedure **int.** The procedure returns the result as the 2-element list whose elements represent accordingly result itself and time of evaluation of the definite integral of an algebraic expression specified by the first actual argument.

```
Nint:= proc()
local Pr, Op, R, T, n;
   [assign(n, nargs), assign(Op, _CCquad), assign(Pr, Digits)];
   if n < 4 or 6 < n then ERROR("quantity <%1> of arguments is invalid", n) end if;
      Pr := `if`(n = 5 or n = 6, `if`(whattype(args[5]) = integer, args[5], Pr), Pr);
      Op := `if`(n = 5 or n = 6, `if`(whattype(args[5]) = string, args[5], Op), Op);
   [assign(T, time()), assign(R, evalf(Int(args[1], args[2] = args[3] .. args[4], Pr, Op)))];
   RETURN([R, time() - T])
end proc
```

Typical examples of the procedure use:

> Nint(cos(x)^3*sin(x)^3, x, Pi/4, Pi/3, 7, `_Dexp`), Nint((x^3+60)/(x^4+sqrt(x^2+6)), x, 7, 14, 7, `_CCquad`);

[0.02864583, 0.009], [0.7431976, 0.005]

IP – a polynomial data interpolation

Call format of the procedure:

IP(x, p, b)

Formal arguments of the procedure:

x – a value or name {may be integer, float, rational, name, symbol}
p – a list of values of ordinates
b – a list of values of abscisses

Description of the procedure:

The procedure call **IP(x, p, b)** returns the value **P(x)** of an interpolation polynomial **P** obtained on the basis of lists **p** and **b** of values of axes **X** and **Y** accordingly. If the first actual **x** argument specifies an indefinite variable, the procedure call **IP(x, p, b)** returns an *interpolation polynomial* itself. In addition, if lists **p** and **b** have different length, the procedure instead of them uses the lists **p[1..n]** and **b[1..n]**, where **n = min(nops(p), nops(b)).**

```
IP:= proc(x::{rational, float, integer, name, symbol}, p::list, b::list)
local c, n;
  assign(n = min(nops(p), nops(b))), subs(c = x, interp(p[1 .. n], b[1 .. n], c))
end proc
```

Typical examples of the procedure use:

```
> Digits:=7: G:= [1,0,29.1,70.8,91.6,95.8,97.9]: S:= [0,10.1,47.1,78.3,86.8,92.7,100]:
  IP(14,G,S), IP(h, G, S);
        -2.9824, 0.2094562e-7*h^6 – 0.6656078e-5*h^5 + 10.1 + 0.8093605e-3*h^4 + 1.252231*h^2
                                   -11.30642*h – 0.4661084e-1*h^3
> Digits:=6: G:= [1,0,29.1,70.8,91.6,95.8,97.9]: S:= [0,10.1,47.1,78.3,86.8, 92.7]: IP(14,G,S), IP(t, G, S);
     -23.736, -0.617580e-6*t^5 + 0.184282e-3*t^4 +10.1 – 0.195441e-1*t^3 + 0.847119*t^2 – 10.9278*t
```

For convenience of definition of a set of *linear constraints,* it is possible to offer two useful enough procedures **LO** and **Linear_Const,** which provide a self-acting generating of constraints on the basis of the given conditions. The procedures seem a rather effective at problems solving of linear optimization by the *simplex*-method by means of the package module **simplex.** Above all, the procedures are rather effective at making a various sort of procedures of linear optimization on the basis of the *simplex*-method.

Linear_Const – generating of linear constraints
LO

Call format of the procedures:

Linear_Const(A, X, B)
LO(A1, B1, y)

Formal arguments of the procedures:

A – a matrix or array of coefficients
X – a list or vector of independent variables
B – a vector of values of right members of inequalities for linear constraints
A1 – a *NAG*-matrix of coefficients
B1 – a *NAG*-vector of values of right members of inequalities for linear constraints
X – a name

Description of the procedures:

The procedure call **Linear_Const(A, X, B)** returns a set of linear constraints on the basis of a coefficients matrix specified by actual argument **A**, of a vector or list **X** of independent variables and a vector or list **B** of values of the right members of inequalities for *linear constraints.*

The **LO(A1,B1,x)** procedure has been implemented on the basis of tools of the **LinearAlgebra** module and provides the self-acting generating of linear constraints on the basis of the given conditions, namely: a *NAG*-matrix **A1** of coefficients, a *NAG*-vector **B1** of the absolute terms and an independent variable **x.**

```
Linear_Const:= proc(A::{matrix, array}, X::{vector, list}, B::{vector, list})
local n, m, k, p;
  assign(n = linalg[coldim](A), m = linalg[rowdim](A)),
  `if`(n <> nops(X), ERROR("columns quantity of %1 and length of list %2 are different", A, X),
  `if`(m <> nops(B), ERROR("rows quantity of %1 and length of list %2 are different", A, B),
  {seq(sum(A[k, p]*X[p], p = 1 .. n) <= B[k], k = 1 .. m)}))
end proc

LO:= proc(A::Matrix, B::Vector, x::{name, symbol})
local a, b, k;
  assign(a = [op(1, A)][2]), `if`(a <> nops(op(2, B)),
  ERROR("mismatch of dimensions of Matrix and Vector"), NULL),
  assign('b' = (A . (Vector(a, [seq(x[k], k = 1 .. a)])) – B), {seq(b[k] <= 0, k = 1 .. a)}
end proc
```

Typical examples of the procedures use:

> A:= matrix(3,2, [7,15,37,41,57,62]): X:=[x,y]: B:= [1995, 300, 2004]: Linear_Const(A,X,B);

{7*x+15*y <= 1995, 37*x+41*y <= 300, 57*x+62*y <= 2004}

> M:= <<a,b,c,r> | <d,e,f,t> | <g,h,p,a> | <r,t,v,h>>: B1:=Vector(4, [seq(b[k], k=1..4)]): LO(M,B1,z);

$$\{a z_1 + d z_2 + g z_3 + r z_4 - b_1 \le 0, b z_1 + e z_2 + h z_3 + t z_4 - b_2 \le 0, c z_1 + f z_2 + p z_3 + v z_4 - b_3 \le 0, r z_1 + t z_2 + a z_3 + h z_4 - b_4 \le 0\}$$

Indets – evaluation of independent variables of an algebraic expression

Call format of the procedure:

Indets(*expr*)

Formal arguments of the procedure:

expr – any valid *Maple* expression

Description of the procedure:

A call **indets(*expr, name*)** of the built-in *Maple* function returns a set containing all independent variables of an algebraic expression ***expr,*** including such constants as {***false, gamma, infinity, true, Catalan, FAIL,* Pi**}. Whereas in a lot cases it is necessary to find all independent variables of an algebraic expression, i.e. the **X** variables that satisfy the determinative relation **type(evalf(X), name) -> *true.*** The **Indets** procedure solves this problem. The procedure call **Indets(*expr*)** returns a set containing all independent variables of an algebraic expression ***expr,*** which satisfy the above determinative relation.

```
Indets:= A::anything -> indets(A, name) minus {constants}
```

Typical examples of the procedure use:

> Indets(Pi*cos(x)+Catalan*sin(y)-2*exp(1)*gamma*z-sqrt(6)*ln(14)/(t+h)+G*S*V*Art*Kr);

{*x, y, Kr, t, V, h, z, Art*}

The researching problem of behavior of functional dependences consists in determination for them of the special and singular points on the leading variables, first, linked with examination of the continuity of functions and their asymptotical behavior. A series of functional tools of the *Maple* package (procedures ***discount, ascent, and discount, singular***) allow solving these problems for a case of algebraic expressions as a whole. The following rather useful procedure **F_Analys** in certain respects generalizes these tools.

F_Analys – analysis of an algebraic expression or function from one variable

Call format of the procedure:

F_Analys(G, a, b)

Formal arguments of the procedure:

G - a function or algebraic expression from one variable
a, b - ends of an interval of investigation of **G** function (should be *real constants*)

Description of the procedure:

The procedure call **F_Analys(G,a,b)** returns the table whose indices are the calculated singular points of a **G** function (*algebraic expression*) on the interval **[a, b]**, and its entries are 2-element sequences whose elements define accordingly a type of a point of discontinuity or singularity and the limit values of the expression in the point **[**at the left of point, to the right of point**]**. Besides, the returned *table* contains indices defining the interval minimum and interval maximum of the tested algebraic expression **G**, and also the index defining a set of the points of inflection (*if such points exist*) of the **G** expression on the given interval **[a, b]**.

```
F_Analys:= proc(G::algebraic, a::realcons, b::realcons)
local k, h, p, x, C, S, r, TOC, L, R, omega;
  omega:= proc(C, a, b)
    local c, x, y, h, p, k, o, t, n, u, s, z, v;
      assign(x = {}, n = -1, u = 6, s = 14);
      for k to nops(C) do
        assign('p' = rhs(C[k])), assign('h' = [op(p)]), assign('y' = convert(h[-1], string),
            'c' = convert(h[1], string));
        if y[1] = "_" or c[1] = "_" then
          assign('t' = op(0, p)), `if`(t = symbol, assign('r' = 1, 'o' = `*`),
            assign('r' = `if`(y[1] = "_", t(op(p)[1 .. -2]), t(op(p)[2 .. -1])), 'o' = t));
          do  assign('n' = n + 1), assign('z' = o(r, n), 'v' = o(r, -n));
            if a <= evalf(z) and evalf(z) <= b then x := {op(x), lhs(C[k]) = z}; u := 1
            else u := 0 end if;
            if a <= evalf(v) and evalf(v) <= b then x := {op(x), lhs(C[k]) = v}; s := 1
            else s := 0 end if;
            if u + s <> 0 then u, s := 0, 0; next
            else u, s, n := 0, 0, -1; break end if
          end do;
          next
        elif belong(evalf(rhs(C[k])), a .. b) then x := {C[k], op(x)}; next end if
      end do;
      [op(x)]
    end proc;
```

```
assign(x = op(indets(evalf(args[1]), name)), r = {}), `if`(nops([x]) = 1, NULL,
   ERROR("quantity of variables is more than 1, namely - %1", [x]));
if iscont(args[1], x = a .. b, closed) then WARNING("function is continuous"); goto(toc)
else S := singular(args[1], x); C := discont(args[1], x) end if;
C := omega(map(op, [op({S} union {seq({x = C[k]}, k = 1 .. nops(C))})]), a, b);
for k to nops(C) do
  if abs(rhs(C[k])) = infinity then assign('L' = limit(args[1], C[k]), 'R' = limit(args[1], C[k]));
    TOC[C[k]] := `Asymptotic behaviour on infinity`, [R]; next
  else assign('L' = limit(args[1], C[k], left), 'R' = limit(args[1], C[k], right)) end if;
  if L = R and abs(R) <> infinity then
    TOC[C[k]] := `Discontinuity point of the eliminated type`, [L, R]
  else
    if L <> R and abs(L) <> infinity and
    abs(R) <> infinity then TOC[C[k]] := `Discontinuity point of the first type`, [L, R]
    else TOC[C[k]] := `Discontinuity point of the second type`, [L, R]
    end if
  end if
end do;
toc;
TOC[{x = -infinity}] := `Asymptotic behaviour on -infinity`, [limit(G, x = -infinity)];
TOC[{x = infinity}] := `Asymptotic behaviour on +infinity`, [limit(G, x = infinity)];
h := {fsolve(diff(G, x) = 0, x = args[2] .. args[3])};
assign(p = evalf([subs(x = args[2], G), subs(x = args[3], G), `if`(h = {}, NULL,
   op([seq(subs(x = h[k], G), k = 1 .. nops(h))]))]));
r := {seq(`if`(subs(x = k, diff(G, x, x)) = 0, k, NULL), k = [op(h), args[2], args[3]])};
TOC[`Interval minimum`] := sort([op(p)])[1];
`if`(r = {}, NULL, assign(TOC[`inflection points`] = evalf(r)));
assign(TOC[`Interval maximum`] = sort([op(p)])[-1]), eval(TOC)
end proc
```

Typical examples of the procedure use:

> h:=evalf(2*Pi): F_Analys(sin(x)*cos(x)/(x - 3) + x*cos(x)/(x + 2), -h, h);

table([$(x = 3) = (\textit{Discontinuity point of the second type}, [\infty, -\infty])$,
$(x = -2) = (\textit{Discontinuity point of the second type}, [-\infty, \infty])$, $\textit{Interval minimum} = 0.02050693317$,
$\{x = \infty\} = (\textit{Asymptotic behaviour on +infinity}, [-1 .. 1])$, $\textit{Interval maximum} = 1.466942207$,
$\{x = -\infty\} = (\textit{Asymptotic behaviour on -infinity}, [-1 .. 1])$
])

> G:= x -> piecewise(x<=0, 2-x^2, x<=1, 9*x, 12): F_Analys(G(x)*(x + 5)/(x – 12), -25, 25);

table([$(x = 1) = \left(\textit{Discontinuity point of the first type}, \left[\frac{-54}{11}, \frac{-72}{11}\right]\right)$,
$(x = 12) = (\textit{Discontinuity point of the second type}, [-\infty, \infty])$, $\textit{Interval minimum} = -336.7567567$,
$\{x = -\infty\} = (\textit{Asymptotic behaviour on -infinity}, [-\infty])$, $(x = 0) = \left(\textit{Discontinuity point of the first type}, \left[\frac{-5}{6}, 0\right]\right)$,
$\{x = \infty\} = (\textit{Asymptotic behaviour on +infinity}, [12])$,
$\textit{Interval maximum} = 27.69230770$
])

> F_Analys(a/(x-b)+sin(b*x)/c, -2*Pi, 2*Pi);

Error, (in **F_Analys**) quantity of variables is more than 1, namely - [a, x, b, c]

> F_Analys(7/((x-7)^3*(x-14)^2*(x+7)^3), -20, 15);

table([(*x* = 7) = (*Discontinuity point of the second type* , [$-\infty$, ∞]),
(*x* = -7) = (*Discontinuity point of the second type* , [∞, $-\infty$]),
(*x* = 14) = (*Discontinuity point of the second type* , [∞, ∞]), *Interval minimum* = -0.2812442381 10^{-6},
{*x* = $-\infty$ } = (*Asymptotic behaviour on -infinity* , [0]), {*x* = ∞ } = (*Asymptotic behaviour on +infinity* , [0]),
Interval maximum = 0.1283985255 10^{-5}
])

> F_Analys(cos(x)/sin(Pi*x), -3, 3);

table([(*x* = -1) = (*Discontinuity point of the second type* , [∞, $-\infty$]),
(*x* = -2) = (*Discontinuity point of the second type* , [∞, $-\infty$]),
(*x* = 1) = (*Discontinuity point of the second type* , [∞, $-\infty$]),
(*x* = 3) = (*Discontinuity point of the second type* , [$-\infty$, ∞]), *Interval minimum* = -0.8044662594 10^{9},
(*x* = -3) = (*Discontinuity point of the second type* , [$-\infty$, ∞]),
{*x* = $-\infty$ } = (*Asymptotic behaviour on -infinity* , [*undefined*]),
(*x* = 0) = (*Discontinuity point of the second type* , [$-\infty$, ∞]),
{*x* = ∞ } = (*Asymptotic behaviour on +infinity* , [*undefined*]), *Interval maximum* = 0.8044662594 10^{9},
(*x* = 2) = (*Discontinuity point of the second type* , [∞, $-\infty$])
])

Problems dealing with searching of a minimum and/or maximum (*minimax*) of an expression have not only theoretical significance for qualitative aspects of a lot of mathematical fields, but in the greater measure in numerous appendices of the analysis. For the problem, solving of the given kind the *Maple* package gives a number of tools of a various level. Here we shall present the **MiniMax** procedure resumptive to a certain extent the standard *Maple* tools for search of *minimax* of algebraic expressions.

MiniMax – searching of minimax of algebraic expressions

Call format of the procedure:

MiniMax(Kw, ex {, *adarg*})

Formal arguments of the procedure:

Kw – a key word defining a minimax type (may be ***`maxi`***, ***`mini`***)
ex – an expression for which minimax should be found
adarg – (optional) options permissible by standard *minimax* functions which define an area of *minimax* searching

Description of the procedure:

The **MiniMax(Kw, *ex*)** procedure has at least two arguments and returns a result of maximizing or minimizing (depending on the value of the first actual **Kw** argument) of an algebraic expression specified by the second actual ***ex*** argument. Through the rest optional actual arguments ***adarg,*** the options permissible by standard *minimax* functions for defining of an area of *minimax* searching are specified.

For a case of one leading variable the **MiniMax** procedure returns a sequence of the following view: a *minimax* type, its value and a nested list, whose the first element defines a list of points with the same value of the found *minimax*, whereas the second element defines a list of values of the *discrepancies* appropriate to them; for a multivariate case a *minimax type* and its *value* are returned only. For 3D-case the procedure **minmax3d** represented below solves this problem.

```
MiniMax:= proc()
local k, R, p, L;
  `if`(args[1] <> mini and args[1] <> maxi, ERROR("the first argument is wrong: %1", args[1]),
    `if`(nargs < 2, ERROR("quantity of actual arguments is wrong: %1", nargs),
    assign(R = evalf(cat(args[1], mize)(seq(args[k], k = 2 .. nargs))))));
  cat(args[1], mize), R, `if`(nops(Indets(args[2])) = 1, [`in points`, assign(L=[fsolve(args[2] = R)])
    L, [seq(subs(op(Indets(args[2])) = L[p], args[2] - R),p = 1 .. nops(L))]], NULL)
end proc
```

Typical examples of the procedure use:

> MiniMax(`maxi`, 7*sin(x)+14*cos(y)+61), MiniMax(`mini`, 7*sin(x)+14*cos(y)+61);

maximize, 82.0, *minimize*, 40.0

> MiniMax(`mini`, 7*x^3+14*x^2-36*x-40, x=-3..3);

minimize, -56.25620769, [*in points*, [-3.605190214], [-.1e-7]]

The package has a series of rather effective tools, which can be united into a common **Solve** group of solvers of the equations, their systems and inequalities both in the numerical, and in the algebraic form. The procedure ***`solve`*** constitutes the basis of the given group. The **V_Solve** procedure represented below extends expressive means of the basic procedure ***`solve`*** of the last five releases of the *Maple* package.

V_Solve – solving of an equation or system of equations with return of discrepancies

Call format of the procedure:

V_Solve(U, P, n, V)

Formal arguments of the procedure:

U - an equation or system of equations
P - a name or a set of names (unknowns to solve for)
n - an accuracy of solution (should be a positive integer – *posint*)
V - an assignable name

Description of the procedure:

The procedure call **V_Solve(U, P, n, V)** returns solution of an equation or a system of equations specified by the first actual **U** argument with respect to the leading variables specified by the second actual **P** argument, with an accuracy specified by the third actual **n** argument; whereas through the fourth actual **V** argument the discrepancies of the obtained solution are returned, i.e. values of sought equations at substitution into them of the obtained roots. The discrepancies are returned as *equations*, whose left members are the *obtained values* and right members are *exact values;* if sought equations have been given not in form **a = b,** then the discrepancies are returned as *values***.** The procedure has found useful usage in many numerical calculations.

```
V_Solve:= proc(U, P, n::integer, V::evaln)
local a, k;
  [assign(a = [evalf(solve(U, P), n)]), assign(V = `if`(type(P, 'set'), seq(eval(subs(a[k], U)),
  k = 1 .. nops(a)), seq(eval(subs(P = a[k], U)), k = 1 .. nops(a)))), RETURN(op(a))]
end proc
```

Typical examples of the procedure use:

```
> Digits:=6: V_Solve({6.1*v+5.6*g-3.6*s=9.8, 4.2*v-4.7*g+6.7*s=6.2, 2*v+ 14*g-35*s = 61},
  {v, s, g}, 6, H), H;
        {s = -2.74474, g = -2.88059, v = 2.63119}, {61.0000 = 61, 9.80006 = 9.8, 6.2000 = 6.2}
```

Residue – computation of algebraic residues of an algebraic expression

Call format of the procedure:

Residue(G, L)

Formal arguments of the procedure:

G - an arbitrary algebraic expression
L - an equation **x=p** or list of equations **[x=p1, ..., x=pk]** defining points at which algebraic residues should be evaluated

Description of the procedure:

The procedure call **Residue(G, L)** returns the *algebraic residue* or list of *algebraic residues* of an algebraic expression **G** for a variable **x** around a point **p** or a set of **pk** points specified by actual **L** argument defining an equation **x=p** or list of equations **[x=p1,...,x=pk].** The implemented algorithm is based on structure analysis of the generalized series.

```
Residue:= proc(V::algebraic, Y::{equation, list(equation)})
local a, k, R, omega;
  assign(R = []), `if`(type(Y, list), assign(omega = Y), assign(omega = [Y]));
  for k to nops(omega) do assign('a' = numapprox[laurent](V, omega[k])),
    add(`if`(op(a)[2*k] = -1, assign('R' = [op(R), op(a)[2*k - 1]]), 0), k = 1 .. 1/2*nops([op(a)]))
  end do;
  `if`(type(Y, list), R, op(R))
end proc
```

Typical examples of the procedure use:

```
> Residue(b*x^3/(x-a)^3, x=a), Residue(64*(61*x^3+56*x-2003)/((x-7)^2*(x-14)*(x-36)),
  [x=7, x=14, x=36]);
                    3*b*a, [161721664/41209, -5317280/539, 91072928/9251]
```

Root – computation of roots of a polynomial from one independent variable

Call format of the procedure:

Root(P, x {, h})

Formal arguments of the procedure:

P - a polynomial from one independent variable
x - a name (unknown to solve for)
h - (optional) key word defining a mode of return of roots (`*complex*`)

Description of the procedure:

The procedure call **Root(P, x)** returns all *real roots* of a polynomial specified by the first actual **P** argument with respect to the leading variable specified by the second actual **x** argument. If the optional third `*complex*` argument has been coded, the procedure call returns all *complex roots* of the polynomial **P** with respect to leading variable **x.**

```
Root:= (P, x) -> `if`(2 < nargs and member(complex, {args}), {fsolve(P, x, complex, fulldigits)}
    minus {fsolve(P, x, fulldigits, maxsols = degree(P, x))}, {fsolve(P, x, maxsols = degree(P, x),
    fulldigits)})
```

Typical examples of the procedure use:

> Digits:=14: map(Root, [h^2+3*h-9, h^3+9*h+3, Pi*h^3-gamma*h+6], h, r);

[{-4.8541019662497, 1.8541019662497}, {-0.32936339745546}, {-1.2900387776967}]

> Root(61*x^6-56*x^5+36*x^4-40*x^3+14*x^2-7*x+2003, x, `complex`);

{1.69423 + 0.884493*I, 1.69423 – 0.884493*I, -1.36622 + 0.925364*I, 0.131010 + 1.81226*I, -1.36622 – 0.925364*I, 0.131010 – 1.81226*I}

SS – calculation of coefficients of a series expansion

Call format of the procedure:

SS(S)

Formal arguments of the procedure:

S - a series

Description of the procedure:

If **S** is a structure of the series type, then for it takes place a practically useful relation of the following view: **[op(S)[-2], op(S)[-1]] -> [O(1),n],** where **O(1)** – an expansion remainder and **n** – an exponent at expansion variable in it. The procedure call **SS(S)** returns a sequence of **2**-element lists **[k, p],** where **k** – coefficient at term **(x – a)^p** of expansion in the structure **S.** The procedure **SS** has a number of useful appendices in the analysis.

```
SS:= proc(S::series)
local k, p;
  seq([op(S)[k], op(S)[k + 1]], k = {seq(`if`(type(p, odd), p, NULL), p = 1 .. nops(S) – 2)})
end proc
```

Typical examples of the procedure use:

> ASV:=series(61+56*y+36*y^2+40*y^3+7*y^4+14*y^5+14*y^6+O(y^13),y,12): SS(ASV);

[61, 0], [56, 1], [36, 2], [40, 3], [7, 4], [14, 5], [14, 6]

---

minmax3d - approximation of *minimax* points for algebraic expressions

Call format of the procedure:

minmax3d(F, x, y, n {, t})

Formal arguments of the procedure:

F - valid algebraic expression from two independent variables **(x, y)**
x - range for **x**-variable
y - range for **y**-variable
n - demanded calculation accuracy
t - (optional) an assignable name

Description of the procedure:

The procedure **minmax3d** is intended for approximation of *minimax* points for an algebraic expression from two independent variables with given accuracy. An exploration array of the algebraic expression is limited by boundary conditions given by ranges of the form **x=a..b** and **y=c..d** where **a, b, c** and **d** are *realnum*-values. Demanded calculation accuracy is defined as **10^(-n);** it is being taken into consideration at approximation of *minimax* points.

The procedure call **minmax3d(F, x=a..b, y=c..d, n),** in general, returns the 2-element nesting list whose the first element represents the list of *minimal* points and the second element represents the list of *maximal* points of an algebraic **F** expression, which is given over an *realnum*-array defined by the second and the third actual argument; whereas the fourth actual argument **n** defines the demanded calculation accuracy. Each *minimax* point is 3-element *realnum* list whose the first element defines the value itself of *minimax* (*minimum* or *maximum*) whereas the second and the third element define **(x,y)**-coordinates of the *minimax*. If the optional fourth argument **t** is given, through it the surface of an **F** expression is returned. In case of impossibility to evaluate a *minimax*, the procedure call returns the *NULL* value, i.e. nothing, with output of appropriate warning. The procedure handles basic especial and erroneous situations.

```
minmax3d:= proc(F::algebraic, x::equation, y::equation, n::posint)
local `1`, `2`, `3`, `4`, a, b, dx, dy, k, x1, y1, xd, yd, xn,yn, min, max, min1, max1,
      min2, max2, tol, r, Z, h;
  if map(type, map(rhs, {x, y}), range(realnum)) = {true} then
     assign(x1 = map(evalf, x), y1 = map(evalf, y), tol = 10^(-n));
     assign(`1` = lhs(rhs(x1)), `2` = rhs(rhs(x1)), `3` = lhs(rhs(y1)), `4` = rhs(rhs(y1))),
       varsort('`1`', '`2`'), varsort('`3`', '`4`')
  else ERROR("2nd or/and 3rd argument is invalid - %1", [args[2], args[3]])
  end if;
  assign(xd = abs(`2` - `1`), yd = abs(`4` - `3`), xn = 30); yn := yd*xn/xd;
  assign(dx = xd/xn, dy = yd/yn); h:= p -> `if`(1 < nops(p), p, op(p));
  Z:= (a, b) -> subs([lhs(x) = a, lhs(y) = b], F);
  assign(min1 = [infinity, infinity, infinity], max1 = [-infinity, -infinity, -infinity],
     min2 = [infinity, infinity, infinity], max2 = [-infinity, -infinity, -infinity]);
  Cycle;
  for a from `1` by dx to `2` do  for b from `3` by dy to `4` do
        r := evalf(Z(a, b));  if  type(r, realnum)  then
           if  evalb(r <= min1[1]) then min2 := [r, a, b]  end if;
           if  evalb(max1[1] <= r) then max2 := [r, a, b]  end if
        end if
     end do
  end do;
  if tol < abs(min1[1] - min2[1]) or tol < abs(max1[1] - max2[1]) then
     min1 := min2;  max1 := max2;  dx := 1/2*dx;  dy := 1/2*dy;  goto(Cycle)
  end if;
  min2, max2, assign(min = [min2], max = [max2]);
  for a from `1` by dx to `2` do  for b from `3` by dy to `4` do
        r := evalf(Z(a, b));
        if  type(r, realnum)  then
           if  abs(r - min2[1]) <= tol  then  min := [op(min), [r, a, b]]  end if;
           if  abs(r - max2[1]) <= tol  then  max := [op(max), [r, a, b]]  end if
        end if
```

```
    end do
  end do;
  a := h(redL(min)), h(redL(max));
  if [a] = [[infinity, infinity, infinity], [-infinity, -infinity, -infinity]] then
    WARNING("Expression <%1> has not a realnum-minimax over the given array", F)
  else
    if 4 < nargs and type(args[5], symbol) then assign(args[5] = plot3d(args[1..3])) end if; a
  end if
end proc
```

Typical examples of the procedure use:

> minmax3d(y*sin(y)+x*cos(x), x=2*Pi..-2*Pi, y=-2*Pi..2*Pi, 6, h); h;

[[2.959337959, 6.283185296, 5.654866766], [2.959338008, 6.283185296, -5.654866778]], [[6.283185226, 6.283185296, 6.283185296], [6.283185301, 6.283185296, -6.283185308], [6.283185282, 6.283185296, -3.141592658], [6.283185296, 6.283185296, -0.60e-8], [6.283185320, 6.283185296, 3.141592646]]

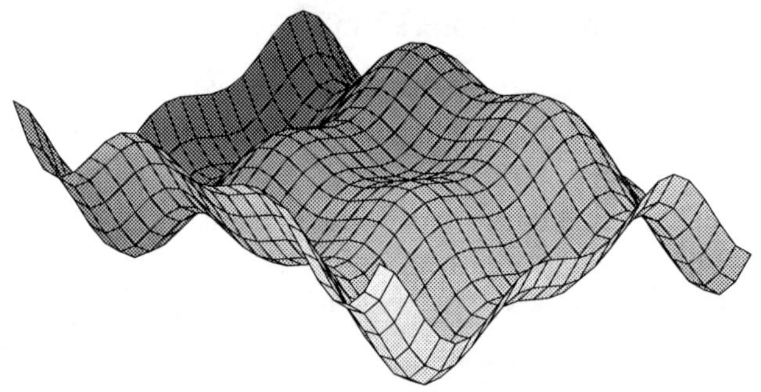

> minmax3d(sqrt(y*sin(y)+x*cos(x)), x=2*Pi..-2*Pi, y=-2*Pi..2*Pi, 6, t); t;

[[1.720272641, 6.283185296, 5.654866766], [1.720272655, 6.283185296, -5.654866778]], [[2.506628258, 6.283185296, 6.283185296], [2.506628273, 6.283185296, -6.283185308], [2.506628270, 6.283185296, -3.141592658], [2.506628272, 6.283185296, -0.60e-8], [2.506628277, 6.283185296, 3.141592646]]

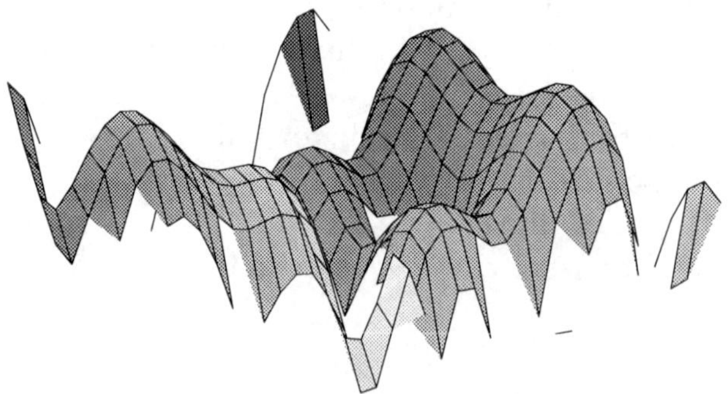

> minmax3d(a*sin(y)+b*cos(x), x=2*Pi..-2*Pi, y=-2*Pi..2*Pi, 6, z);

Warning, Expression <a*sin(y)+b*cos(x)> has not a realnum-minimax over the given array

The procedure **minmax3d** can be rather useful at numeric analysis of algebraic expressions.

The procedures represented in the present section provide a series of simple enough tools simplifying solution of a series of problems of the analysis with elimination of a few shortcomings of the standard tools. The procedures usage in a series of appendices has confirmed their rather high efficiency. The technique used in these procedures can be useful for implementation of other similar tools.

Chapter 12.
Software for solving problems of linear algebra

In the given section, the tools extending possibilities of *Maple* of releases **6 - 9/9.5** at problem solving of linear algebra are represented. These procedures are classified concerning tools of two basic package modules that are focused first on problems of *linear algebra,* namely: the traditional *Maple* module **linalg** and the implanted package module **LinearAlgebra** of the *NAG* ltd.

12.1. General purpose software

The given item represents additional tools providing a solution of some mass problems of linear algebra. First of all, our experience of use of *Maple* of releases **4 - 9/9.5** allows to draw an quite unequivocal conclusion about its insufficiently developed basic I/O system, what in a lot of cases creates essential difficulties at programming of the applied problems which deal with a data of great volume in external memory. In section 10 a series of useful enough procedures for providing of access to the datafiles has been represented. The **CrNumMatrix** procedure relates to the similar tools, providing as a reading of numerical data of matrix structure out of datafiles of text type, and interactive creation of a new datafile with data in case of absence of the first. The given procedure is intended for wide use with a data of the matrix organization, providing input into a current session of an initial numeric data of various kinds.

CrNumMatrix – interactive creating of numeric rectangular matrices

Call format of the procedure:

CrNumMatrix(F, h)

Formal arguments of the procedure:

F - string or symbol defining a name or full path to a text datafile
h - a mode of matrices processing (may be **1**, **2** or **3** only)

Description of the procedure:

The procedure call **CrNumMatrix(F, h)** in the dialog mode line by line forms a text datafile specified by the **F** qualifier, returning the three-element list. The first element of the list defines a required rectangular numerical matrix, whereas the second and the third element define a quantity of its rows and columns accordingly. A mode of datafile processing is specified by the second actual argument **h** (**1** - *update,* **2** - *appending,* **3** - *return of the saved matrix*). In addition, if path to a target file **F** does not exist, then it is created with output of appropriate warning, if it is necessary. At creation of a new datafile the procedure requests the data input which are parted by *blanks* and/or *horizontal tabs* for the next row of the formed numerical matrix; input of a row is completed by *Enter*-key, whereas a data input as a whole is completed by the symbol `@`. The value **3** for **h** argument has a sense only for an existing **F** datafile; otherwise, the procedure call **CrNumMatrix(F, 3)** is equivalent to the procedure call **CrNumMatrix(F, 2).**

At presence of a datafile specified by the actual **F** argument, the procedure returns 3-element list of the structure, mentioned above. However, before that the procedure requests input of a next row of data, providing an opportunity of updating of the file by a new data. The principle of data addition is similar to a case of creation of a new datafile described above. In case of absence of the specified datafile the opportunity of interactive creating of a numerical matrix with the subsequent saving of it in a datafile **F** is given. It is supposed, that numerical data are in a text datafile the each line of which contains values of columns parted by *blanks* and/or *horizontal tabs* of the corresponding row of the matrix.

In inputted strings blanks and/or horizontal tabs part the numeric values. The values of elements of a rectangular matrix are formed line by line; inputs of rows are terminated by *Enter*-key, while input in a whole by the symbol `@`. The example represented below, clearly illustrates the principle of data input for matrixes creating. The procedure **CrNumMatrix** allows effectively enough to form the numerical values of appropriate rows of a numerical matrix with elements of type {*integer, float, fraction*}. The **CrNumMatrix** procedure provides as reading of numerical data of matrix structure out of datafiles of text type, and interactive creating of a new datafile with the data in case of its absence. The procedure is oriented onto wide use with data of the matrix structure, providing input into a current *Maple* session of an initial data of various purposes.

```
CrNumMatrix:= proc(F::{string, symbol}, h::{1, 2, 3})
local a, k, p, m, n, n1, L, Ln, M, Z, omega, x, nu, t;
  nu := L -> map(came, SLS(Red_n(L, [" ", "    "], 2), {" ", " "}));
  assign(n = 0, m = 0), `if`(h = 3 and type(F, file), [assign(omega = F, x = 61), goto(Exit)],
  [assign(omega= `if`(type(F,file), F, pathtf(F))), close(omega), fopen(omega, APPEND,TEXT)]);
  do [lprint("Introduce a next row: @ – end of input"), assign('L', iolib(2, terminal))];
    if L = "@" then break end if; writeline(omega, L)
  end do;
  Exit;
  t := () -> interface(`if`(nargs = 0, warnlevel, warnlevel = args));
  assign(a = t()), t(0), close(omega), assign('Ln', fopen(omega, READ, TEXT));
  do L := iolib(2, Ln);
    if member(L, {0, ""}) then break end if;
    m := m + 1
  end do;
  try [close(Ln), assign('L', iolib(2, omega)), assign('n1', nops(nu(L))),
    close(omega), assign('M', convert(convert(array(1 .. m, 1 .. n1), Matrix), matrix))]
  catch "invalid input:": t(a); ERROR("datafile <%1> is empty", omega)
  end try;
  Ln := fopen(omega, READ, TEXT);
  try for k to m do [assign('L', readline(Ln)), assign('Z', nu(L))];
      if nops(Z) <> n1 then  t(a);  ERROR("column definition %1th is invalid", n1)
      end if;
      for p to n1 do M[k, p] := Z[p] end do
    end do
  catch "invalid input:":  t(a); ERROR("datafile <%1> is empty", omega)
  end try;
  t(a), `if`(x <> 61, WARNING("the created matrix has been saved in datafile <%1>",
```

```
    omega), NULL);
  RETURN(iolib(7, Ln, omega), `if`(x = 61, evalm(M), [evalm(M), m, n1]))
end proc
```

Typical examples of the procedure use:

> CrNumMatrix("C:/RANS\\AVZ\\Tallinn\\Vilnius\\Grodno.ian", 1);

"Introduce a next row: @ – end of input"

> 42 47 67 62 89 96

"Introduce a next row: @ – end of input"

> 61 56 36 40 14 7

"Introduce a next row: @ – end of input"

> 1995 300 50 25 36 2003

"Introduce a next row: @ - end of input"

> 1/7 2 3 14.8 5 6

"Introduce a next row: @ – end of input"
> @

Warning, the created matrix has been saved in datafile
<c:\rans\avz\tallinn\vilnius\grodno.ian>

$$\left[\begin{bmatrix} 42 & 47 & 67 & 62 & 89 & 96 \\ 61 & 56 & 36 & 40 & 14 & 7 \\ 1995 & 300 & 50 & 25 & 36 & 2003 \\ \frac{1}{7} & 2 & 3 & 14.8 & 5 & 6 \end{bmatrix}, 4, 6\right]$$

> CrNumMatrix("C:/RANS\\AVZ\\Tallinn\\Vilnius\\Grodno.ian", 2);

"Introduce a next row: @ – end of input"

> 62 72 82 92 95 7

"Introduce a next row: @ – end of input"
> @

Warning, the created matrix has been saved in datafile
<c:\rans\avz\tallinn\vilnius\grodno.ian>

$$\left[\begin{bmatrix} 42 & 47 & 67 & 62 & 89 & 96 \\ 61 & 56 & 36 & 40 & 14 & 7 \\ 1995 & 300 & 50 & 25 & 36 & 2003 \\ \frac{1}{7} & 2 & 3 & 14.8 & 5 & 6 \\ 62 & 72 & 82 & 92 & 95 & 7 \end{bmatrix}, 5, 6\right]$$

> M:= CrNumMatrix("C:/RANS\\AVZ\\Tallinn\\Vilnius\\Grodno.ian", 3);

$$M := \begin{bmatrix} 42 & 47 & 67 & 62 & 89 & 96 \\ 61 & 56 & 36 & 40 & 14 & 7 \\ 1995 & 300 & 50 & 25 & 36 & 2003 \\ \frac{1}{7} & 2 & 3 & 14.8 & 5 & 6 \\ 62 & 72 & 82 & 92 & 95 & 7 \end{bmatrix}$$

The given procedure allows to work effectively enough with matrix data structures formed and saved in the disk text datafiles. Use of the procedure **CrNumMatrix** in engineering problems, in physical and statistical problems has shown its high enough operational characteristics at data processing [29,33,43,44].

At the polynomial algebra provided by *Maple* often enough the modular function ***recipoly*** is used whose definition is in the **polytools** module of the package. The function has the coding format **recipoly(P, x {, 'p'})** and defines whether a **P** polynomial is a *self-reciprocal* (a *symmetric*) with respect to the leading variable **x.** A polynomial **P** has the given property if and only if the following determinative relation takes place:

coeff(**P, x, k**) = *coeff*(**P, x, h-k**) *for all* **k = 0..h,** *where* **h** = *degree*(**P, x**) **(2)**

In addition, if **h** is an even value and the procedure call uses the optional third argument **p**, then through **p** a polynomial **R** of degree **h/2** such, that **x^(h/2)*R(x+1/x) = P** is returned. In the Help is incorrectly indicated the value returned through the third actual **p** argument in case of an odd degree of a tested polynomial. In our books [8-14] the valid value has been represented which is tested by the simple **Test** procedure with two formal arguments.

Test – a testing of the recipoly function of *Maple*

Call format of the procedure:

Test(P, x)

Formal arguments of the procedure:

P - a polynomial
x - symbol defining name of a leading variable

Description of the procedure:

The first actual **P** argument of the **Test** procedure specifies a *symmetric polynomial* of an odd degree, whereas the second actual **x** argument specifies a leading variable; if **P** specifies a symmetric polynomial of an odd degree, the procedure returns an initial polynomial, forming it on the basis of a value passed through the third actual **p** argument of the ***recipoly*** procedure and the above determinative relation (**2**). If **P** argument specifies a non-symmetric polynomial, the *false* value is returned. The procedure **Test** bears the auxiliary character.

```
Test:= proc(P::polynom, x::symbol)
local h;
  `if`(type(degree(P, x), even), RETURN(`Polynomial has even degree`),
     polytools[recipoly](P, x, h));
  `if`(type(h, name), false, expand(subs(x = x + 1/x, h)*(x + 1)*x^(1/2*degree(P) - 1/2)))
end proc
```

Typical examples of the procedure use:

> Test(61*x^8+14*x^7+56*x^6+36*x^5+1*x^4+35*x+14, x)

Polynomial has even degree

> Test(61*x^9+14*x^8+56*x^7+36*x^6+1*x^5+35*x^4+7*x^3+14*x^2+6*x+1, x);

false

> Test(62*x^7+15*x^6+37*x^5+37*x^4+8*x^3+37*x^2+15*x, x);

false

> Test(61*x^9+14*x^8+56*x^7+36*x^6+40*x^5+40*x^4+36*x^3+56*x^2+14*x+61, x);

61*x^9+14*x^8+56*x^7+36*x^6+40*x^5+40*x^4+36*x^3+56*x^2+14*x+61

FFP – full factorizing of a polynomial from one leading variable

Call format of the procedure:

FFP(P, x)

Formal arguments of the procedure:

P - a polynomial with coefficients of the complexcons type
x - a leading variable of the polynomial

Description of the procedure:

The procedure call **FFP(P, x)** returns the result of full factorizing of a **P** polynomial from one leading variable **x** with coefficients of the *complexcons* type. The procedure is useful in a series of problems of polynomial algebra.

```
FFP:= proc(P::polynom(complexcons), x)
local a, k;
  assign(a = [coeffs(P, x)]), `if`({seq(type(a[k], complexcons), k = 1 .. nops(a))} = {true}, NULL,
    ERROR("polynomial <%1> has non-numeric coefficients %2", P, a));
  product(op(indets(P)) - [evalf(solve(P, op(indets(P))))][k], k = 1 .. degree(P, op(indets(P))))
end proc
```

Typical examples of the procedure use:

> Digits:= 7: P:= x^6 + 61*x^5 - 56*x^4 + 36*x^3 - 40*x^2 + 14*x - 7: FFP(P, x);

(x - 0.9103154) (x - 0.2058206 - 0.4455678 **I**) (x + 0.2039580 - 0.6884604 **I**) (x + 61.91404)
(x + 0.2039580 + 0.6884604 **I**) (x - 0.2058206 + 0.4455678 **I**)

MatrSort – a special sorting of a matrix (array) or Matrix

Call format of the procedure:

MatrSort(M, t, F {, *sf*})

Formal arguments of the procedure:

M - a matrix (array) or Matrix with elements of *realnum*-type
t - a symbol {***row* | *column***}
F - a procedure name
sf - an ordering function (boolean function from two arguments)

Description of the procedure:

Procedure **MatrSort(M,t,F {, *sf*})** has three obligatory arguments and one optional argument and provides sorting of a matrix (array) or Matrix **M** with elements of ***realnum***-type {*row-by-row* | *column-by-column*} depending on a value {***row*** | ***column***} of the second argument **t**. The third actual argument **F** specifies a name of a procedure calculating weight of a row/column, relative to which sorting should be made; quantity of formal arguments of the weight function **F** should be equal to quantity of elements of a row/column of the sorted **M** matrix. The sorting is made concerning weights of rows/columns and specified by the fourth optional ordering *sf*-function for which **'<'** function is supposed by default. The given procedure is of interest in a series of data processing problems that have a matrix organization.

```
MatrSort:= proc(M::{matrix(realnum), Matrix(realnum), array(realnum)}, t::symbol, F::symbol)
local k, n, L, G, v, A;
  `if`(type(M, Matrix), assign(A = convert(M, matrix), v = [op(3, M)]), assign(A = M));
  assign(n = nops([op(1, eval(F))]), L = convert(`if`(t = column, linalg[transpose](A), A), listlist));
  `if`(type(F, procedure) and member(n, {0, nops(L[1])}), assign(G = []),
    ERROR("3-rd argument is wrong"));
  for k to nops(L) do G := [op(G), [op(L[k]), evalf(F(op(L[k])))]] end do;
  assign('L' = SLj(G, nops(G[1]), `if`(nargs = 4, args[4], `<`))), assign('G' = []);
  for k to nops(L) do G := [op(G), L[k][1 .. -2]] end do;
  A := `if`(t = column, linalg[transpose](convert(G, matrix)), convert(G, matrix));
  `if`(type(M, Matrix), Matrix(op(2, eval(A)), {op(op(3, eval(A)))}, op(subseqn([shape = NULL,
    storage = NULL, outputoptions = NULL], [op(v)], `=`))), eval(A))
end proc
```

Typical examples of the procedure use:

```
> F:= (x,y,z,t) -> x^2-6*y+14*z^2-t: F1:= (x,y,z) -> x^2-6*y+14*z^2: assign('M'=matrix([[1,-3, sqrt(3),
  7], [5,6,7,11], [0,1, sqrt(2), 14]]), 'M1' =matrix([[1,-3,3,7], [5,6,7,11], [0,1,2,14]])), map(evalm, [M, 1])
```

$$\left[\begin{bmatrix} 1 & -3 & \sqrt{3} & 7 \\ 5 & 6 & 7 & 11 \\ 0 & 1 & \sqrt{2} & 14 \end{bmatrix}, \begin{bmatrix} 1 & -3 & 3 & 7 \\ 5 & 6 & 7 & 11 \\ 0 & 1 & 2 & 14 \end{bmatrix}\right]$$

```
> H:= () -> `+`(args): MatrSort(M1, column, H, `>`), MatrSort(M1, row, H, `>`),
  MatrSort(M, column, F1, `>`);
```

$$\begin{bmatrix} 7 & 3 & 1 & -3 \\ 11 & 7 & 5 & 6 \\ 14 & 2 & 0 & 1 \end{bmatrix}, \begin{bmatrix} 5 & 6 & 7 & 11 \\ 0 & 1 & 2 & 14 \\ 1 & -3 & 3 & 7 \end{bmatrix}, \begin{bmatrix} 7 & \sqrt{3} & -3 & 1 \\ 11 & 7 & 6 & 5 \\ 14 & \sqrt{2} & 1 & 0 \end{bmatrix}$$

```
> assign(A = array(1..4, 1..3, [[1,ln(3),6], [1/2,sqrt(5),7/42], [1/3, -tan(5),7], [1/2, exp(0.5),-7]])),
  evalm(A), MatrSort(A, row, F1, `<`);
```

$$\begin{bmatrix} 1 & \ln(3) & 6 \\ \frac{1}{2} & \sqrt{5} & \frac{1}{6} \\ \frac{1}{3} & -\tan(5) & 7 \\ \frac{1}{2} & 1.648721271 & -7 \end{bmatrix}, \begin{bmatrix} \frac{1}{2} & \sqrt{5} & \frac{1}{6} \\ 1 & \ln(3) & 6 \\ \frac{1}{3} & -\tan(5) & 7 \\ \frac{1}{2} & 1.648721271 & -7 \end{bmatrix}$$

```
> with(LinearAlgebra): assign('RM' = RandomMatrix(5, 5, generator=1..15, outputoptions=
  [shape=triangular[lower]])), RM, MatrSort(RM, row, H, `>`);
```

$$\begin{bmatrix} 10 & 0 & 0 & 0 & 0 \\ 8 & 7 & 0 & 0 & 0 \\ 8 & 2 & 4 & 0 & 0 \\ 13 & 2 & 2 & 8 & 0 \\ 3 & 15 & 3 & 1 & 6 \end{bmatrix}, \begin{bmatrix} 3 & 15 & 3 & 1 & 6 \\ 13 & 2 & 2 & 8 & 0 \\ 8 & 7 & 0 & 0 & 0 \\ 8 & 2 & 4 & 0 & 0 \\ 10 & 0 & 0 & 0 & 0 \end{bmatrix}$$

```
> assign('RM' = RandomMatrix(9, 6, generator=7 .. 15, outputoptions= [shape=
  triangular[lower]])), RM, MatrSort(RM, column, H, `<`), MatrSort(RM, row, H, `>`);
```

$$\begin{bmatrix} 7 & 0 & 0 & 0 & 0 & 0 \\ 12 & 11 & 0 & 0 & 0 & 0 \\ 7 & 12 & 7 & 0 & 0 & 0 \\ 9 & 7 & 13 & 10 & 0 & 0 \\ 8 & 9 & 15 & 12 & 14 & 0 \\ 11 & 15 & 15 & 14 & 7 & 14 \\ 10 & 11 & 13 & 8 & 9 & 8 \\ 13 & 8 & 7 & 14 & 9 & 12 \\ 15 & 10 & 13 & 14 & 13 & 10 \end{bmatrix}, \begin{bmatrix} 0 & 0 & 0 & 0 & 0 & 7 \\ 0 & 0 & 0 & 11 & 0 & 12 \\ 0 & 0 & 0 & 12 & 7 & 7 \\ 0 & 0 & 10 & 7 & 13 & 9 \\ 0 & 14 & 12 & 9 & 15 & 8 \\ 14 & 7 & 14 & 15 & 15 & 11 \\ 8 & 9 & 8 & 11 & 13 & 10 \\ 12 & 9 & 14 & 8 & 7 & 13 \\ 10 & 13 & 14 & 10 & 13 & 15 \end{bmatrix}, \begin{bmatrix} 11 & 15 & 15 & 14 & 7 & 14 \\ 15 & 10 & 13 & 14 & 13 & 10 \\ 13 & 8 & 7 & 14 & 9 & 12 \\ 10 & 11 & 13 & 8 & 9 & 8 \\ 8 & 9 & 15 & 12 & 14 & 0 \\ 9 & 7 & 13 & 10 & 0 & 0 \\ 7 & 12 & 7 & 0 & 0 & 0 \\ 12 & 11 & 0 & 0 & 0 & 0 \\ 7 & 0 & 0 & 0 & 0 & 0 \end{bmatrix}$$

FOR_DO – dynamically generated nested cyclic *"for_do"* constructions

Call format of the procedure:

FOR_DO(*args*)

Formal arguments of the procedure:

args – a sequence of actual arguments

Description of the procedure:

In a series of cases of operating with data structures of types {*array, Array*} the necessity of dynamic generation of nested cyclic constructions arises, when level of cyclic nesting is not known beforehand. In this case, the procedure **FOR_DO** and similar ones to it can be useful enough. The procedure **FOR_DO** provides dynamic generation of *"for_do"* constructions of the following types that are used enough widely, namely:

(1) *for* k1 *to* M *do* Expr[k1] *end do;* [k1, M]
(2) *for* k2 *by* step *to* M *do* Expr[k2] *end do;* [k2, *step*, M]
(3) *for* k3 *from* a *by* step *to* M *do* Expr[k3] *end do;* [k3, a, *step*, M]
(4) *for* k4 *in* M *do* Expr[k4] *end do;* [k4, `*in*`, M]

On any nesting level, the above constructions may be used only. The procedure call is coded as follows:

FOR_DO([k1,M], [k2,`*in*`,M], [k3,*step*,M], [k4,a,*step*,M], ..., [kp,...], "Expr[k1,k2,k3, ...,kp]")

Quantity of actual arguments should be more than **1** and the first **nargs – 1** arguments specify a sequence of lists describing the corresponding types of cyclic `*for_do*` constructions as represented above. Whereas the last actual argument in *string*-format represents a clauses sequence **Expr[k1, k2, ..., kp]** that should be done repeatedly accordingly to the given nesting level and types of used *`for_do`* constructions. Successful procedure call returns the *NULL* value, i.e. nothing, providing executing a dynamically generated cyclic *`for_do`* construction with a cycle ***body*** specified by the last actual argument. The procedure handles the basic erroneous and especial situations.

The essence of such method of *dynamic* generation briefly consists in the following. The required cyclic construction is generated as a *string* that next is saved in temporary datafile. The datafile is read into a current session by means of ***`read`*** function, activating execution of the cyclic construction whereupon the datafile is deleted. The result of execution of a cyclic ***body*** is available in a current *Maple* session as the examples below demonstrate.

```
FOR_DO:= proc()
local k, T, E, N, n, p, R;
  `if`(nargs < 2, ERROR("quantity of actual arguments should be more than 1"),
    `if`({true} <> {seq(type(args[k], list), k = 1 .. nargs - 1)} or not type(args[-1], string),
    ERROR("actual arguments are invalid"), seq(`if`(type(args[k], list) and
    type(args[k][1], symbol) and member(nops(args[k]), {2, 3, 4}), NULL,
    ERROR("a cycle parameters %1 are invalid", args[k])), k = 1 .. nargs - 1)));
  assign(N = "", n = 0, R = [], E = cat(seq("end do; ", k = 1 .. nargs - 1)));
  T := table([2 = [`for`, `to`], 3 = [`for`, `by`, `to`], 4 = [`for`, `from`, `by`, `to`]]);
  for k to nargs - 1 do assign('R' = [op(R), cat("", args[k][1])]);
  if member(`in`, args[k]) then N:=cat(N, " for ", args[k][1], " in ", convert(args[k][3], string), " ")
    else for p in T[nops(args[k])] do n := n + 1; N := cat(N, p, " ", args[k][n], " ") end do end if;
    assign('n' = 0, 'N' = cat(N, "do "))
  end do;
  iolib(5, "$ArtKr$", cat(N, " ", args[nargs], " ", E)), iolib(7, "$ArtKr$");
  read "$ArtKr$";
  iolib(8, "$ArtKr$"), unassign(op(map(convert, R, symbol)))
end proc
```

Typical examples of the procedure use:

> ArtKr:=Array(1..2, 1..4, 1..5, 1..6, 1..7): FOR_DO([k,2], [j,2,4], [h,2,1,3], [p,2,3,6], [z,1,3,7], "ArtKr[k,j,h,p,z]:= k*j*h*p*z;");

> ArrayElems(ArtKr); FOR_DO();

{(1, 1, 2, 2, 1) = 4, (1, 1, 2, 2, 4) = 16, (1, 1, 2, 2, 7) = 28, (1, 1, 2, 5, 1) = 10, (1, 1, 2, 5, 4) = 40, (1, 1, 2, 5, 7) = 70, (1, 1, 3, 2, 1) = 6, (1, 1, 3, 2, 4) = 24, (1, 1, 3, 2, 7) = 42, (1, 1, 3, 5, 1) = 15, (1, 1, 3, 5, 4) = 60, (1, 1, 3, 5, 7) = 105, (1, 3, 2, 2, 1) = 12, (1, 3, 2, 2, 4) = 48, (1, 3, 2, 2, 7) = 84, (1, 3, 2, 5, 1) = 30, (1, 3, 2, 5, 4) = 120, (1, 3, 2, 5, 7) = 210, (1, 3, 3, 2, 1) = 18, (1, 3, 3, 2, 4) = 72, (1, 3, 3, 2, 7) = 126, (1, 3, 3, 5, 1) = 45, (1, 3, 3, 5, 4) = 180, (1, 3, 3, 5, 7) = 315, (2, 1, 2, 2, 1) = 8, (2, 1, 2, 2, 4) = 32, (2, 1, 2, 2, 7) = 56, (2, 1, 2, 5, 1) = 20, (2, 1, 2, 5, 4) = 80, (2, 1, 2, 5, 7) = 140, (2, 1, 3, 2, 1) = 12, (2, 1, 3, 2, 4) = 48, (2, 1, 3, 2, 7) = 84, (2, 1, 3, 5, 1) = 30, (2, 1, 3, 5, 4) = 120, (2, 1, 3, 5, 7) = 210, (2, 3, 2, 2, 1) = 24, (2, 3, 2, 2, 4) = 96, (2, 3, 2, 2, 7) = 168, (2, 3, 2, 5, 1) = 60, (2, 3, 2, 5, 4) = 240, (2, 3, 2, 5, 7) = 420, (2, 3, 3, 2, 1) = 36, (2, 3, 3, 2, 4) = 144, (2, 3, 3, 2, 7) = 252, (2, 3, 3, 5, 1) = 90, (2, 3, 3, 5, 4) = 360, (2, 3, 3, 5, 7) = 630}

Error, (in **FOR_DO**) quantity of actual arguments should be more than **1**

> FOR_DO([k,2], [j,2,4], [h,2,1,3], [p,2,3,6], [z,1,3,7,14], "ArtKr[k,j,h,p,z]:= k*j*h*p*z;");

Error, (in **FOR_DO**) a cycle parameters [z, 1, 3, 7, 14] are invalid

```
> ArtKr:=Array(1..2, 1..4, 1..5, 1..6, 1..7): FOR_DO([k,2], [j,`in`,{2,3,4}], [h,2,1,3], [p,2,3,6],
  [z, `in`, [1,3,7]], "ArtKr[k,j,h,p,z]:= k*j*h*p*z;"); ArrayElems(ArtKr);
```

{(1, 2, 2, 2, 1) = 8, (1, 2, 2, 2, 3) = 24, (1, 2, 2, 2, 7) = 56, (1, 2, 2, 5, 1) = 20, (1, 2, 2, 5, 3) = 60, (1, 2, 2, 5, 7) = 140, (1, 2, 3, 2, 1) = 12, (1, 2, 3, 2, 3) = 36, (1, 2, 3, 2, 7) = 84, (1, 2, 3, 5, 1) = 30, (1, 2, 3, 5, 3) = 90, (1, 2, 3, 5, 7) = 210, (1, 3, 2, 2, 1) = 12, (1, 3, 2, 2, 3) = 36, (1, 3, 2, 2, 7) = 84, (1, 3, 2, 5, 1) = 30, (1, 3, 2, 5, 3) = 90, (1, 3, 2, 5, 7) = 210, (1, 3, 3, 2, 1) = 18, (1, 3, 3, 2, 3) = 54, (1, 3, 3, 2, 7) = 126, (1, 3, 3, 5, 1) = 45, (1, 3, 3, 5, 3) = 135, (1, 3, 3, 5, 7) = 315, (1, 4, 2, 2, 1) = 16, (1, 4, 2, 2, 3) = 48, (1, 4, 2, 2, 7) = 112, (1, 4, 2, 5, 1) = 40, (1, 4, 2, 5, 3) = 120, (1, 4, 2, 5, 7) = 280, (1, 4, 3, 2, 1) = 24, (1, 4, 3, 2, 3) = 72, (1, 4, 3, 2, 7) = 168, (1, 4, 3, 5, 1) = 60, (1, 4, 3, 5, 3) = 180, (1, 4, 3, 5, 7) = 420, (2, 2, 2, 2, 1) = 16, (2, 2, 2, 2, 3) = 48, (2, 2, 2, 2, 7) = 112, (2, 2, 2, 5, 1) = 40, (2, 2, 2, 5, 3) = 120, (2, 2, 2, 5, 7) = 280, (2, 2, 3, 2, 1) = 24, (2, 2, 3, 2, 3) = 72, (2, 2, 3, 2, 7) = 168, (2, 2, 3, 5, 1) = 60, (2, 2, 3, 5, 3) = 180, (2, 2, 3, 5, 7) = 420, (2, 3, 2, 2, 1) = 24, (2, 3, 2, 2, 3) = 72, (2, 3, 2, 2, 7) = 168, (2, 3, 2, 5, 1) = 60, (2, 3, 2, 5, 3) = 180, (2, 3, 2, 5, 7) = 420, (2, 3, 3, 2, 1) = 36, (2, 3, 3, 2, 3) = 108, (2, 3, 3, 2, 7) =252, (2, 3, 3, 5, 1) = 90, (2, 3, 3, 5, 3) =270, (2, 3, 3, 5, 7) = 630, (2, 4, 2, 2, 1) = 32, (2, 4, 2, 2, 3) = 96, (2, 4, 2, 2, 7) = 224, (2, 4, 2, 5, 1) = 80, (2, 4, 2, 5, 3) = 240, (2, 4, 2, 5, 7) = 560, (2, 4, 3, 2, 1) =48, (2, 4, 3, 2, 3) = 144, (2, 4, 3, 2, 7) =336, (2, 4, 3, 5, 1) = 120, (2, 4, 3, 5, 3) = 360, (2, 4, 3, 5, 7) = 840}

Algorithm of the procedure uses the so-called method of *"disk transits"* considered above, and which can be useful at programming of dynamically generated cyclic constructions of other valid types (*seq, mul, add,* etc.; see the **SEQ** procedure below).

SEQ - dynamically generated nested cyclic *"seq"* constructions

Call format of the procedure:

SEQ(*args*)

Formal arguments of the procedure:

args - a sequence of actual arguments

Description of the procedure:

In a series of cases of operating with data structures of indexed types {*array, Array,* etc.} the necessity of dynamic generation of nested cyclic constructions arises, when level of cyclic nesting is not known beforehand. In this case, the procedure **SEQ** and similar ones to it can be useful enough. The procedure **SEQ** provides dynamic generation of *"seq"* constructions of the following types that are used enough widely, namely:

(1)	**seq(Expr[k], k=a .. b);**	**[k, a..b]**
(2)	**seq(Expr[k], k=V);**	**[k, V]**

On any nesting level, the above constructions may be used only. The procedure call is coded as follows:

SEQ([k1, a1..b1], [k2, a2..b2], [k3, V], ..., [kp, ap..bp], "Expr[k1, k2, k3, ..., kp]")

Quantity of actual arguments should be more than **1** and the first **nargs - 1** arguments specify a sequence of lists describing the corresponding types of cyclic *"seq"* constructions as represented above. Whereas the last actual argument in *string*-format specifies a clauses sequence **Expr[k1, k2, ..., kp]** that should be done repeatedly accordingly to the given nesting level and type of used *"seq"* construction. Successful procedure call **SEQ** returns the *NULL* value, i.e. nothing, providing executing a dynamically generated cyclic *"seq"* construction with cycle *body* specified by the last actual argument. The result of execution of a cyclic ***body*** is available in a current *Maple* session as the examples below demonstrate. The procedure handles the basic erroneous and especial situations. The procedure usage in a lot of appendices whose implementations have used the dynamically generated cyclic constructions of ***seq***-type has confirmed its rather high efficiency.

```
SEQ:= proc()
local k, E, h, R;
  `if`(nargs < 2, ERROR("quantity of actual arguments should be more than 1"),
    `if`(not type(args[nargs], string), ERROR("the last actual argument should be a string but
    has received [%1]", whattype(args[nargs])), seq(`if`(type(args[k], list) and
    type(args[k][1], symbol), NULL, ERROR("actual argument %1 is invalid", args[k])),
    k = 1 .. nargs - 1))), assign(R = []), assign(E = cat(seq("seq(", k = 1 .. nargs - 1), args[-1]));
  assign('R' = [op(R), cat("", args[k][1])]);
  for k to nargs - 1 do E := cat(E, ",", args[k][1], "=", convert(args[k][2], string), ")") end do;
  iolib(5, "$ArtKr$", cat(E, ";")), iolib(7, "$ArtKr$");
  read "$ArtKr$";
  iolib(8, "$ArtKr$"), unassign(op(map(convert, R, symbol)))
end proc
```

Typical examples of the procedure use:

```
> ArtKr:= Array(1..2, 1..3, 1..4, 1..5, 1..7): SEQ([k,1..2], [j,2..3], [h,2..4], [p,3..5], [z,[1,5,7]],
  "assign('ArtKr'[k,j,h,p,z]=k*j*h*p*z)"); ArrayElems(ArtKr); SEQ();

  {(1, 2, 4, 3, 1) = 24, (1, 2, 4, 4, 1) = 32, (1, 2, 4, 5, 1) = 40, (1, 3, 2, 3, 1) = 18, (1, 2, 2, 5, 1) =
  20, (1, 2, 3, 3, 1) = 18, (1, 2, 3, 4, 1) = 24, (1, 2, 3, 5, 1) = 30, (1, 3, 2, 5, 1) = 30, (1, 3, 3, 3, 1) =
  27, (1, 3, 3, 4, 1) = 36, (1, 3, 3, 5, 1) = 45, (1, 3, 2, 4, 1) = 24, (1, 2, 2, 4, 7) = 112, (1, 2, 2, 3, 5) =
  60, (1, 2, 2, 3, 1) = 12, (1, 2, 2, 4, 1) = 16, (2, 2, 2, 4, 1) = 32, (2, 2, 2, 5, 1) = 40, (2, 2, 3, 3, 1) =
  36, (2, 2, 3, 4, 1) = 48, (2, 2, 3, 5, 1) = 60, (1, 2, 2, 3, 7) = 84, (1, 2, 2, 4, 5) = 80, (1, 2, 2, 5, 5) =
  100, (1, 2, 2, 5, 7) = 140, (1, 2, 3, 3, 5) = 90, (1, 2, 3, 3, 7) = 126, (1, 2, 3, 4, 5) = 120, (1, 2, 3, 4, 7) =
  168, (1, 2, 3, 5, 5) = 150, (1, 2, 3, 5, 7) = 210, (1, 2, 4, 3, 5) = 120, (1, 2, 4, 3, 7) = 168, (1, 2, 4, 4, 5) =
  160, (1, 2, 4, 4, 7) = 224, (1, 2, 4, 5, 5) = 200, (1, 2, 4, 5, 7) = 280, (1, 3, 2, 3, 5) = 90, (1, 3, 2, 3, 7) =
  126, (1, 3, 2, 4, 5) = 120, (1, 3, 2, 4, 7) = 168, (1, 3, 2, 5, 5) = 150, (1, 3, 2, 5, 7) = 210, (1, 3, 3, 3, 5) =
  135, (1, 3, 3, 3, 7) = 189, (1, 3, 3, 4, 5) = 180, (1, 3, 3, 4, 7) = 252, (1, 3, 3, 5, 5) = 225, (1, 3, 3, 5, 7) =
  315, (1, 3, 4, 3, 5) = 180, (1, 3, 4, 3, 7) = 252, (1, 3, 4, 4, 5) = 240, (1, 3, 4, 4, 7) = 336, (1, 3, 4, 5, 5) =
  300, (1, 3, 4, 5, 7) = 420, (2, 2, 2, 3, 5) = 120, (2, 2, 2, 3, 7) = 168, (2, 2, 2, 4, 5) = 160, (2, 2, 2, 4, 7) =
  224, (2, 2, 2, 5, 5) = 200, (2, 2, 2, 5, 7) = 280, (2, 2, 3, 3, 5) = 180, (2, 2, 3, 3, 7) = 252, (2, 2, 3, 4, 5) =
  240, (2, 2, 3, 4, 7) = 336, (2, 2, 3, 5, 5) = 300, (2, 2, 3, 5, 7) = 420, (2, 2, 4, 3, 5) = 240, (2, 2, 4, 3, 7) =
  336, (2, 2, 4, 4, 5) = 320, (2, 2, 4, 4, 7) = 448, (2, 2, 4, 5, 5) = 400, (2, 2, 4, 5, 7) = 560, (2, 3, 2, 3, 5) =
  180, (2, 3, 2, 3, 7) = 252, (2, 3, 2, 4, 5) = 240, (2, 3, 2, 4, 7) = 336, (2, 3, 2, 5, 5) = 300, (2, 3, 2, 5, 7) =
  420, (2, 3, 3, 3, 5) = 270, (2, 3, 3, 3, 7) = 378, (2, 3, 3, 4, 5) = 360, (2, 3, 3, 4, 7) = 504, (2, 3, 3, 5, 5) =
  450, (2, 3, 3, 5, 7) = 630, (2, 3, 4, 3, 5) = 360, (2, 3, 4, 3, 7) = 504, (2, 3, 4, 4, 5) = 480, (2, 3, 4, 4, 7) =
  672, (2, 3, 4, 5, 5) = 600, (2, 3, 4, 5, 7) = 840, (2, 2, 4, 3, 1) = 48, (2, 2, 4, 4, 1) = 64, (2, 2, 4, 5, 1) =
  80, (2, 3, 2, 3, 1) = 36, (2, 3, 2, 4, 1) = 48, (2, 3, 2, 5, 1) = 60, (2, 3, 3, 3, 1) = 54, (2, 3, 3, 4, 1) =
  72, (2, 3, 3, 5, 1) = 90, (2, 3, 4, 3, 1) = 72, (2, 3, 4, 4, 1) = 96, (2, 3, 4, 5, 1) = 120, (1, 3, 4, 3, 1) =
  36, (1, 3, 4, 4, 1) = 48, (1, 3, 4, 5, 1) = 60, (2, 2, 2, 3, 1) = 24}

  Error, (in SEQ) quantity of actual arguments should be more than 1

> ArtKr:= Array(1..2, 1..3, 1..4, 1..5, 1..7): SEQ([7,1..2], [j,2..3], [h,2..4], [p,3..5], [z, [1,5,7]],
  "assign('ArtKr'[k,j,h,p,z]=k*j*h*p*z)");

  Error, (in SEQ) actual argument [7, 1 .. 2] is invalid
```

```
> SG:=Array(1..2,1..3,1..4): SEQ([k,1..2], [j,{1,2,3}], [h,2..3], "assign('SG'[k,j,h]=k+j+h)");
> ArrayElems(SG); SEQ([k,1..2], [j,{2,3}], [h,{1,2,3,4}], "assign('SG'[k,j,h]=k+j+h)");
   {(1, 2, 2) = 5, (1, 2, 3) = 6, (1, 3, 2) = 6, (1, 3, 3) = 7, (2, 2, 2) = 6, (2, 2, 3) = 7, (2, 3, 2) = 7,
   (2, 3, 3) = 8, (1, 1, 2) = 4, (1, 1, 3) = 5, (2, 1, 2) = 5, (2, 1, 3) = 6}
> ArrayElems(SG);
   {(1, 2, 2) = 5, (1, 2, 3) = 6, (1, 3, 2) = 6, (1, 3, 3) = 7, (2, 2, 2) = 6, (2, 2, 3) = 7, (2, 3, 2) = 7,
   (2, 3, 3) = 8, (1, 1, 2) = 4, (1, 1, 3) = 5, (2, 1, 2) = 5, (2, 1, 3) = 6, (1, 2, 1) = 4, (1, 2, 4) = 7,
   (1, 3, 1) = 5, (1, 3, 4) = 8, (2, 2, 1) = 5, (2, 2, 4) = 8, (2, 3, 1) = 6, (2, 3, 4) = 9}
```

MEM – check of entry of the given element into a *Maple* object
Mem
Mem1

Call format of the procedures:

MEM(M, h)
Mem(A, h)
Mem1(B, h)

Formal arguments of the procedures:

A – a table, Matrix, Vector, Array
B – an array, matrix, vector
M – an array, Array, matrix, Matrix, Table, vector, Vector
h – an arbitrary valid *Maple* expression

Description of the procedures:

The procedure call **Mem(A, h)** returns the sequence whose the first element defines quantity of entries into an object **A** {*Table, Array, Matrix, Vector*} of an element specified by the second actual **h** argument, whereas the rest of elements define the lists – coordinates of the found elements. If the first actual **A** argument is a *Table*, then the first element of the returned sequence represents quantity of entries of **h** into the *Table* **A**, while the other elements define the *indices* corresponding to the *entries* with the given value **h.** The procedure call **Mem1(B, h)** returns the sequence whose the first element defines quantity of entries into an object **B** {*array, matrix, vector*} of an element specified by the second actual **h** argument, whereas the rest of elements define the lists – coordinates of the found elements. At last, the **MEM(M, h)** procedure generalizes the above two procedures, allowing the *Maple* objects of a type {*array, Array, matrix, Matrix, Table, vector, Vector*} as its first actual **M** argument.

```
MEM:= proc(M::{Matrix, Vector, matrix, vector, Array, array, Table}, h::anything)
local a, b, k, p, t, n, m, L, omega, S;
  assign(t = numboccur(eval(M), h)), `if`(t <> 0, assign(omega = Pind(M)),
    ERROR("object does not contain <%1>", h)), `if`(type(M, Table), [assign(a = op(op(1, M))),
    assign(b = [seq(`if`(rhs(a[k]) = h, lhs(a[k]), NULL), k = 1 .. nops(a))]),
    RETURN(nops(b), op(b))], NULL), `if`(type(M, Vector), [assign(L = [], n = 0, m = omega[1]),
    seq(`if`(M[k] = h, assign('n' = n+1, 'L' = [op(L), k]), NULL), k = 1..m), RETURN(op([n, L]))],
  assign(L = [(proc(omega, M, h)
    local k, _k, a, b, p, t;
    global __v, __S;
      assign67(a = "", b = "", '__v' = NULL, '__S' = M, p =[seq(cat(_k, k), k = 1..nops(omega))]);
      map(unassign, p), assign(t = cat("", seq(op([p[k], ","]), k = 1 .. nops(omega)))[1 .. -2]);
```

```
    for k to nops(omega) do a := cat(a, "for ", p[k], " to ", omega[k], " do "); b := cat(b, " end do;")
      end do;
      t := cat("", a, " ", cat(" if __S[", t, "]=", h, " then __v:=__v,[", t, "] end if "), " ", b);
      null(writeline("_$_", t), iolib(7, "_$_")),
        (proc(z) read z end proc)("_$_"), null(map(unassign, p)), __v
    end proc)(omega, M, h)]));
  t, op(L), iolib(8, "_$_"), unassign('__v', '__S')
end proc

Mem:= proc(M::{Matrix, Vector, Array, table}, h::anything)
local a, b, k, p, n, m, L, Q;
  `if`(numboccur(eval(M), h) = 0, ERROR("%1 does not contain element: %2", M, h),
    assign(Q = Pind(M)));
  `if`(type(M, table), [assign(a = op(op(1, M))), assign(b = [seq(`if`(rhs(a[k])=h, lhs(a[k]), NULL),
    k = 1 .. nops(a))]), RETURN(nops(b), op(b))], NULL);
  `if`(type(M, Vector), [assign(L = [], n = 0, m = Q[1]), seq(`if`(M[k] = h,
    assign('n' = n + 1, 'L' = [op(L), k]), NULL), k = 1 .. m), RETURN(op([n, L]))], NULL);
  for k to Q[1] do for p to Q[2] do if M[k, p] = h then L := L, [k, p] end if end do end do;
  assign('L' = [seq([L][k], k = 2 .. nops([L]))]), RETURN(nops(L), op(L))
end proc

Mem1:= proc(M::{matrix, vector, array}, h::anything)
local k, p, n, m, L;
  `if`(numboccur(eval(M), h) = 0, ERROR("%1 does not contain element: %2", M, h), NULL);
  `if`(type(M, vector), [assign(L = [], n = 0, m = linalg[vectdim](M)), seq(`if`(M[k] = h,
    assign('n' = n + 1, 'L' = [op(L), k]), NULL), k = 1 .. m), RETURN(op([n, L]))], NULL);
  for k to linalg[rowdim](M) do for p to linalg[coldim](M)
    do if M[k, p] = h then L := L, [k, p] end if
    end do
  end do;
  assign('L' = [seq([L][k], k = 2 .. nops([L]))]), RETURN(nops(L), op(L))
end proc
```

Typical examples of the procedures use:

```
> Fn:=(k,j) -> k^2+2*k*j: M:=matrix(3,9,Fn): V1:=Vector([a,b,a,c,d,a]): M1:=Matrix(4,4):
  M1[2,3]:=61: M1[4,4]:=61: A:=Array(1..2,1..3, [[1,2,2], [3,4,4]]): V2:=vector(12, [a1,v,v,h,
  g,d,l,k,h,j,d, h]): T:=table([Vt=60,Gl=55,Sv=35,Vh=60]): x:=array(1..2,1..3,[[61,56,36],
  [40,6,14]]): v:=vector([42,47,42,67]):
> map3([Mem, Mem, Mem, Mem], [[T,60], [M1,61], [A,4], [V1,a]]);
                    [2, Vt, Vh, 2, [2, 3], [4, 4], 2, [2, 2], [2, 3], 3, [1, 3, 6]]
> map3([Mem1, Mem1, Mem1], [[M,63], [v,42], [x,6]]);
                              [1, [3, 9], 2, [1, 3], 1, [2, 2]]
> restart; M:=Matrix(3,5,[]): M[2,3]:=61: M[1,3]:=61: M[1,1]:=61: M[3,3]:=61: m:=matrix(3,5,[]):
  m[2,3]:=61: m[1,4]:=61: m[3,5]:=61: m[3,3]:=61: V:=Vector(4,[61,56,61, 56]): v:=vector(6,
  [36,61,56,61,56,61]): a:=array(1..2,1..3,1..5,1..6,[]): a[1,2,1,4]:=61: a[2,1,1,5]:=61: a[2,3,5,6]:=61:
  A:=Array(1..2,1..3,1..5,1..6,[]): A[1,2,1,4]:=61: A[2,3,3,5]:=61: A[2,3,5,6]:=61: T:=table([V1=61,G=56,
  S=36, V2=61,V3=61,Art=14,Kr=6, Ar=40,V4=61, V5=61]): map(MEM, [a,A,m,M,T,v,V], 61);
  [3, [1, 2, 1, 4], [2, 1, 1, 5], [2, 3, 5, 6], 3, [1, 2, 1, 4], [2, 3, 3, 5], [2, 3, 5, 6], 4, [1, 4], [2, 3],
  [3, 3], [3, 5], 4, [1, 1], [1, 3], [2, 3], [3, 3], 5, V2, V3, V4, V5, V1, 3, [2], [4], [6], 2, [1, 3]]
```

In many contexts, it is not necessary to know the exact value of an expression; it suffices to know that the expression belongs to some broad class, or group, of expressions that share some common properties. These classes or groups are known as ***"types"***. If **T** represents a type, then we say that an expression is of a type **T** if it belongs to the class that **T** represents. For example, an expression is said has the type ***'table'*** if it belongs to the class of expressions denoted by the type name ***`table`***, which is the set of objects created by the built-in function ***`table`***. These objects are widely used in the *Maple* environment, however, starting with the *sixth* release they are incorrectly being tested by means of the standard procedure ***`type/table`***. Namely, the call **type(T, *table*)** returns the *true* value on a **T** object of a type {*array, table, matrix, vector*}, what in a whole series of cases is inadmissible. Omitting details, we note only, the implantation of the *NAG*-module **LinearAlgebra** into the *Maple* environment is the most plausible reason of such misunderstanding. The simple procedure **'type/Table'** allows to remedy the given deficiency, providing correct testing of the *Maple* objects for their belonging to the type ***'table'***.

`type/Table` – check of a *Maple* object to be a table with indices and entries of the specified type

Call format of the procedure:

type(T, *Table*)
type(T, *Table*(t {, t1}))

Formal arguments of the procedure:

T - a valid *Maple* expression
t - a valid *Maple* type or set of valid *Maple* types
t1 - (optional) a valid *Maple* type or set of valid *Maple* types

Description of the procedure:

The procedure call **type(T, Table)** returns the *true* value, if an expression specified by the actual **T** argument is a table; otherwise, the *false* value is returned. Whereas the procedure call **type(T, Table(t))** returns the *true* value, if **T** is a table with *entries* of type **t**; otherwise, the *false* value is returned. At last, the procedure call **type(T, Table(t, t1))** returns the *true* value, if **T** is a table with *indices* of type **t** and with *entries* of type **t1**; otherwise, the *false* value is returned. Obviously, the procedure call **type(T, Table)** is equivalent to the procedure call **type(T, Table(*anything*, *anything*))**. The procedure allows differentially to check types of indices and entries of a table. The procedure provides strict identification of the tablular *Maple* objects that play important part in numerous appendices. The procedure handles basic especial and erratic situations.

```
`type/Table`:= proc(T::anything)
local a, omega, c, t;
  if not type(T, table) or `or`(op(map2(type, T, [array, rtable]))) then RETURN(false)
  else assign(a = convert('procname(args)', string)),
    `if`(search(a, ",", t), assign67(c = came(a[t + 1 .. -2])), assign(c = 61));
    omega:= () -> ERROR("a type specification error - %1", [args]);
    if 2 < nops([c]) then omega(c) end if
  end if;
  if c = 61 then true
  elif nops([c]) = 1 then
    if type(c, type) then
      andN(op(map(type, map(rhs, op(2, eval(T))), c)))
    else omega(c)
```

```
    end if
  elif nops([c]) = 2 then
    if type(c[1], type) and type(c[2],type) then
      andN(op(map(type, map(rhs, op(2, eval(T))), c[2])))
      and andN(op(map(type, map(lhs, op(2, eval(T))), c[1])))
    else omega(c)
    end if
  end if
end proc
```

Typical examples of the procedure use:

> M:=Matrix(3,5,[]): M[2,3]:=61: M[1,3]:=61: M[1,1]:=61: M[3,3]:=61: m:=matrix(3,5,[]): m[2,3]:=61: m[1,4]:=61: m[3,5]:=61: m[3,3]:=61: V:=Vector(4,[61,56,61,56]): v:=vector(6, [36,61,56,61,56,61]): a:=array(1..2,1..3,1..5,1..6,[]): a[1,2,1,4]:=61: a[2,1,1,5]:=61: a[2,3,5,6]:=61: A:=Array(1..2,1..3,1..5,1..6,[]): A[1,2,1,4]:=61: A[2,3,3,5]:=61: A[2,3,5,6]:= 61: T:=table([V1=61,G=56,S=36,V2=61,V3=61,Art=14,Kr=6, Ar=40]): map(type, [a,A,m, M,v,V, Vr,Vc,T], table), map(type, [a,A,m,M,v,V,Vr,Vc,T], Table(integer));

[*true, false, true, false, true, false, false, false, true*], [*false, false, false, false, false, false, false, false, true*]

> T:=table([61=avz, agn=56, [Kr,Art]={7,15}, `1995_2004`=60]);

T := table([[*Kr, Art*] = {7, 15}, 61 = *avz, agn* = 56, 1995_2004 = 60])

> type(T, Table({integer, symbol}, symbol)), type(T, Table(anything, anything));

false, true

> type(T, Table(anything, {set(integer), symbol, posint}));

true

> type(T, Table), type(T, Table({list(symbol), symbol, posint}, anything));

true, true

> type(T, Table(symbol, string, integer));

Error, (in **omega**) a type specification error – [symbol, string, integer]

> type(T, Table(anything, {set(integer), symbol, posint, rans}));

Error, (in **omega**) a type specification error – [anything, {posint, symbol, set(integer), rans}]

Pind – dimensionality of an array, Array, table, Matrix, matrix, vector or Vector

Call format of the procedure:

Pind(b)

Formal arguments of the procedure:

b – an array, matrix, vector, table, Matrix, Vector, Array

Description of the procedure:

The procedure call **Pind(b)** returns dimensionality of an object specified by the actual **b** argument in form of a {*one* | *two*}-element list depending on a type of **b** argument {*table, vector, Vector* | *array, Array, matrix, Matrix*} accordingly. For a vector and table **b,** the *one*-element list defining accordingly the quantity of vector elements and indices of a table is returned. The procedure unifies evaluation of dimensionality for the *Maple* objects of the above types.

```
Pind:= proc(K::{Matrix, Vector, matrix, vector, Array, array, table})
local a, b, k, p;
  `if`(type(K, Matrix) or type(K, Vector), RETURN([op(1, K)]), `if`(type(K, Array) or
     type(K, array), RETURN(map(rhs, [op(2, eval(K))])), NULL));
  `if`(whattype(eval(K)) = table, [nops([indices(K)])], op([assign(a = [], b = [op(eval(K))]),
     seq(assign('a' = [op(a), `if`(type(b[k], range), b[k], NULL)]), k = 1 .. nops(b)),
     [seq(rhs(a[k]), k = 1 .. nops(a))]]))
end proc
```

Typical examples of the procedure use:

```
> Fn:= (k,j) -> k^2+2*k*j: M:= matrix(3,9, Fn): V2:= Vector(6): A:= array(1..2,1..2, [[1,2],
  [3,4]]): V1:=Vector([p,b,c,d]): M1:= Matrix([[1,2,3], [4,5,6]]): V:= vector(4,[v,g,s,k]):
  M2:= Matrix(14,15): T:=table([V=61,G=56,S=36]): Art_Kr:= Array(1..6,1..14,1..2,[]):
  art:= array(1..2,1..6,1..14,[]): map(Pind, [M, V2, A, V1, M1, V, T, M2, Art_Kr, art]);
```

[[3, 9], [6], [2, 2], [4], [2, 3], [4], [3], [14, 15], [6, 14, 2], [2, 6, 14]]

PNorm – norm calculation of a polynomial

Call format of the procedure:

PNorm(P, n {, x})

Formal arguments of the procedure:

P - a polynomial
x - (optional) a variable or set/list of the leading polynomial variables
n - norm of the polynomial (may be *infinity* or *realcons*)

Description of the procedure:

The **PNorm(P,n,x)** procedure extends the standard function **norm(P,n,x)** which calculates the **n**th *norm* of a polynomial **P** from the leading variables **x.** The **PNorm(P,n,x)** procedure admits up to three actual arguments, namely: **P** - a polynomial from one or several variables, **n** - a required norm of the polynomial, **x** - a variable or set/list of leading polynomial variables. If **P** polynomial has coefficients of type *realcons*, then the third actual **x** argument is optional. If **P** polynomial has more than **1** leading variables and the third argument **x** is absent, then the procedure call returns the *norm* of **P** relative to its all indeterminates. The procedure **PNorm** is a rather useful tool in a series of problems of polynomial algebra.

```
PNorm:= proc(P::polynom, n::{infinity, realcons})
local a, k;
  op([assign(a = [coeffs(P, `if`(nargs = 3, `if`(type(args[3], {symbol, list(symbol), set(symbol)}),
  args[3], ERROR("the third argument %1 is invalid", args[3])), `if`(1 < nops(indets(P)),
 indets(P), NULL)))]), `if`(n = infinity, max(op(a)), sum(abs(a[k])^n, k = 1 .. nops(a))^(1/n))])
end proc
```

Typical examples of the procedure use:

```
> P:= a*x^7+62*x^6-57*y^5+b*x^4-c*y^3+15*x^2-37*y^3+41*x*y^2-7*x^2*y^2+2004:
  map2(PNorm, P, [7, infinity], {x, y});
```

$$\left[(1298027879174375114165584 + |a|^7 + |b|^7 + |c+37|^7)^{\left(\frac{1}{7}\right)}, \max(2004, b, a, -37-c)\right]$$

> map2(PNorm, P, [7, *infinity*]);

$$\left[1298027879175324432937 20^{\left(\frac{1}{7}\right)}, 2004 \right]$$

RSLU – solution of systems of linear equations in matrix form

Call format of the procedure:

RSLU(M, V)

Formal arguments of the procedure:

M - a square matrix of coefficients
V - a vector of absolute terms

Description of the procedure:

The procedure call **RSLU(M, V)** returns **Y**-vector of the solution of a system of linear equations specified in matrix form **M*Y = V**, where: **M** – a square matrix of coefficients, **V** – a vector of absolute terms, and **Y** – a vector of sought solution. The procedure **RSLU** handles all basic especial and erroneous situations.

```
RSLU:= proc(M::matrix(square), V::vector)
local a, b, c, d, LA;
   assign(a = convert(M, Matrix), b = convert(V, Vector), LA = LinearAlgebra);
   try
      if LA:- RowDimension(a) <> LA:- Dimension(b) then ERROR("Dimensionality")
      elif LA:- Determinant(a) = 0 then ERROR("Singularity")
      else c := LA:- LinearSolve(a, b)
      end if
   catch "Dimensionality": WARNING("dimensions of %1 and %2 are not in match", M, V); d:= 7
   catch "Singularity": WARNING("matrix %1 is singular, its rank is %2", M, LA:-Rank(a)); d:= 7
   finally `if`(d <> 7, convert(c, vector), NULL)
   end try
end proc
```

Typical examples of the procedure use:

> M1:=matrix(6, 6, [42,47,67,62,89,95,61,56,36,95,67,3,7,1,5,7,14,45,23,89,67,12,78,55,33,15, 1995,300,2003,30,7,7,14,14,61,56]): V1:=vector(6, [62,56,36,40,7,14]): M:=matrix(3,3,[a,b, c,d,e,f,g,h,j]): V:= vector([n,m,p]): RSLU(M,V); RSLU(M1,V1);

$$\left[\frac{-b\,m\,j+b\,f\,p+c\,h\,m-c\,e\,p+n\,e\,j-n\,f\,h}{a\,e\,j-a\,f\,h+d\,h\,c-d\,b\,j+g\,b\,f-g\,e\,c}, \frac{a\,m\,j-a\,f\,p+d\,c\,p+f\,g\,n-m\,g\,c-d\,n\,j}{a\,e\,j-a\,f\,h+d\,h\,c-d\,b\,j+g\,b\,f-g\,e\,c}, \frac{h\,d\,n-a\,h\,m+a\,e\,p-e\,g\,n-d\,b\,p+g\,b\,m}{a\,e\,j-a\,f\,h+d\,h\,c-d\,b\,j+g\,b\,f-g\,e\,c} \right]$$

$$\left[\frac{1179507827627}{129532549565}, \frac{-226348961072}{129532549565}, \frac{18219643041}{129532549565}, \frac{-592007394182}{129532549565}, \frac{53256121723}{129532549565}, \frac{-1325915734}{129532549565} \right]$$

> M:= matrix(3, 3, [a,b,c,d,e,f,g,h,j]): V:= vector([n,m,p,k,x,y]): RSLU(M,V);

Warning, dimensions of M and V are not in match

The procedures represented in the present item give a series of simple enough tools simplifying solution of a series of mass problems, arising in the linear algebra and related topics.

12.2. Software for operation with *rtable*-objects

The present item represents many useful procedures of work with so-called *rtable*-objects making a base of algebra that is supported by the **LinearAlgebra** package module. The procedures of the **LinearAlgebra** module are working with four principal data structures, namely *Arrays, Matrices, Vectors* and *scalars*. The implementation of *Arrays, Matrices* and *Vectors* is based on the *rtable* data structure. As a result, *table*-based arrays, matrixes and vectors are not interchangeable with *rtable*-objects that are supported by the **LinearAlgebra** module. The ***rtable*** function is the low-level routine that is used by the package *Maple* to build *Arrays, Matrices* or *Vectors*. In further we shall name *Maple*-objects, generated by the ***rtable*** function, as *rtable*-objects.

Let us term an object as *active*, if its definition has been evaluated in a current session, or the access to it in a current session is provided by its name. Here we shall present more detailed explanation of type of an active *rtable*-object, than it has been reflected in special literature, in technical documentation and Help of the package *Maple*. As that has been set [29-33,35,43,44], the evaluations of definitions of *rtable*-objects variedly are linked with the predetermined ***_rtable***-variable of the package, namely: if evaluation of an *rtable*-object has been accompanied by output of its result into the *Output*-paragraph, the given object is being reflected in the ***_rtable***-variable, otherwise no.

Let us term active *rtable*-objects reflected in the ***_rtable***-variable, as active objects of the *first type* (***Type*_1**), otherwise by active objects of the *second type* (***Type*_2**). On the other hand, the call ***anames*(*rtable*)** returns a sequence of names of all *rtable*-objects, active in a current *Maple* session irrespectively from their type. In this connection, in general, the sets of *rtable*-objects represented by the ***_rtable***-variable, and of all *rtable*-objects, active in a current *Maple* session, are different.

Each active *rtable*-object of the *first* type is identified by an appropriate record in *mws*-file with a *Maple* document, in which it has been evaluated with the subsequent saving of the document in a datafile. For records of such type in the end of an *mws*-file, the appropriate section identified by title of view {RTABLE_HANDLES <identification numbers of *rtable*-objects>} is placed. At loading of such *mws*-document the access to *rtable*-objects saved in it, is provided or immediately by their reevaluation, or through the ***_rtable***-variable (*by means of their identification numbers*). The saving of active *rtable*-objects of the *first* type is done or in a document generating them, or in the special *m*-file (of internal *Maple* format). On these and other facts are founded the processing algorithms put into a foundation of procedures extending the tools of operation with *rtable*-objects in the *Maple* environment. The given procedures appear an useful enough at advanced programming of manifold problems dealing with *rtable*-objects, providing a converting of *Maple*-objects into *NAG*-objects, and vice versa.

---

avm_VM – converting of *Maple* objects into *NAG* objects, and vice versa
Aconv
trConv

Call format of the procedures:

avm_VM(G)
Aconv(S)
trConv(t, r)

Formal arguments of the procedures:

G - an array, vector, matrix, Vector, Matrix
S - an array, matrix, vector, set, Matrix, Vector, list
t - an array, matrix, vector
r - an *rtable*-object

Description of the procedures:

Between *Maple*-objects {*array, vector, matrix*} and *NAG*-objects {*Array, Vector, Matrix*} key distinctions exist which have been considered in our books [29-33] enough in detail. In particular, a classification of the second objects with respect to the test functions **`*type*`** and ***whattype*** favorably differs by uniqueness, whereas the first objects are recognized by the ***whattype*** function as objects of the *array*-type. At the same time, *Maple*-objects are much easier than *NAG*-objects and simply are converted into the second ones. The procedure **avm_VM** provides conversion of *Maple*-objects of types {*array, vector, matrix*} into *NAG*-objects of types {*Vector, Matrix*}, and vice versa - of objects of a type {*Vector, Matrix*} into vectors and matrixes of the *Maple* type. The procedure call **avm_VM(G)** returns the result of converting of an object specified by the actual **G** argument of a type {*array, vector, matrix*} into object of a type {*Vector, Matrix*}, and vice versa - an object of a type {*Vector, Matrix*} into vectors or matrixes of the *Maple* type. The example represented below, is quite clear and of special explanations does not require.

The procedure call **Aconv(S)** provides the more wide functions, providing converting of an object specified by the actual **S** argument, of a type {*array, matrix, vector, set, Matrix, Vector, list*} into the corresponding object of the *Array*-type; in addition, an orientation {*row, column*} of the *NAG*-vectors is supported. An example represented below, of a procedure use is quite clear and of special explanations does not require.

At last, the procedure call **trConv(t, r)** provides conversion of an object specified by the first actual **t** argument, of type {*array, matrix, vector*} into object of an appropriate *NAG* type {*Array, Matrix, Vector*} with passing to the first object of properties of an *rtable*-object specified by the second actual **r** argument. An example represented below, of a procedure use is quite clear and of special explanations does not require. The above procedures are useful enough tools for conversion of *NAG*-objects into *Maple*-objects, and vice versa.

```
avm_VM:= proc(G::{Matrix, Vector, matrix, vector, array})
local k, h, S, H, M;
  S:= H -> rtable(op(2, eval(G)), {op(op(3, eval(G)))}, subtype = H);
  `if`(type(G, array) = true and nops([op(2, eval(G))]) = 1, S(Vector[row]), `if`(type(G, array) =
     true and nops([op(2, eval(G))]) = 2 and nops([seq(k, k = op(2, eval(G))[1])]) = 1,
    rtable(op([assign(h = op(3, eval(G))), [seq(rhs(h[k]), k = 1 .. nops(h))]]),
    subtype = Vector[row]), `if`(type(G, array) = true and nops([op(2, eval(G))]) = 2  and
    nops([seq(k, k = op(2, eval(G))[2])]) = 1, rtable(op([assign(h = op(3, eval(G))),
    [seq(rhs(h[k]), k = 1 .. nops(h))]]), subtype = Vector[column]), `if`(whattype(G) = Matrix,
    op([assign(M = Matrix(op(1, G), op(2, G), subs(readonly = NULL, MatrixOptions(G)))),
    matrix(op(1, G), [op(op(2, G))])]), `if`(whattype(G) = Vector[row],
    vector(`if`(whattype(op(1, G)) = `..`, rhs(op(1, G)), op(1, G)), op([assign(h = op(2, eval(G))),
    [seq(rhs(h[k]), k = 1 .. nops(h))]])), `if`(whattype(G) = Vector[column],
    matrix(`if`(whattype(op(1, G)) = `..`, rhs(op(1, G)), op(1, G)), 1, op([assign(h=op(2, eval(G))),
    [seq(rhs(h[k]), k = 1 .. nops(h))]])), S(Matrix)))))))
end proc

Aconv:= proc(Kr::{set, Matrix, Vector, matrix, vector, array, list})
local k;
  `if`(type(Kr, {list, 'set'}), Array(1 .. nops(Kr), convert(Kr, list)), `if`(type(eval(Kr),
  {matrix, vector, array}), Array(op(k, eval(Kr)) $ (k = 1 .. 2), {op(op(3, eval(Kr)))}),
  `if`(type(Kr, Matrix), Array(1 .. op(1, Kr)[1],1 .. op(1, Kr)[2], op(2, Kr),
  `if`(search(convert([op(3, Kr)], string), shape), op(3, Kr)[1 .. nops([op(3, Kr)]) - 1], op(3, Kr))),
  Array(`if`(op(1, op(0, Kr)) = column, op([op([1 .. op(1, Kr), 1 .. 1]), {seq((lhs(op(2, Kr)[k]), 1) =
```

```
  rhs(op(2, Kr)[k]), k = 1 .. nops(op(2, Kr)))}]), op([op([1 .. 1, 1 .. op(1, Kr)]),
  {seq((1, lhs(op(2, Kr)[k])) = rhs(op(2, Kr)[k]), k = 1 .. nops(op(2, Kr)))}])),
  `if`(search(convert([op(3, Kr)], string), shape), op(3, Kr)[1..nops([op(3, Kr)]) - 1], op(3, Kr))))))
end proc
```

```
trConv:= (t::{matrix, vector, array}, r::{Matrix, Vector, Array}) -> `if`(whattype(r) = Matrix,
    whattype(r)(op(0, r), op(1, r), {op(op(convert([op(op(2, convert(t, whattype(r))))], table)))},
    op(3, r)), `if`(op(0, whattype(r)) = Vector, whattype(r)(op(0, r), op(1, r),
    {op(op(convert([op(op(2, convert(t, op(0, whattype(r)))))], table)))}, op(3, r)),
    whattype(r)(op(1, eval(t)), op(2, eval(t)), {op(op(2, convert(eval(t), table)))}, op(4, r))))
```

Typical examples of the procedures use:

> m:=matrix(1,3, [Sv,Art,Kr]): m1:=matrix(3,1,[Sv,Art,Kr]): v:=vector([Sv,Art,Kr]): a:=array(1..2,1..3, [[Sv,Art,Kr],[37,15,7]]): M:=Matrix(3,3,[[36,2,3],[5,15,7]], *shape=symmetric, storage=rectangular*): V1:=Vector[row]([36,7,14]): V2:= Vector[column]([62,57, 37]): avm_VM(m), avm_VM(m1), avm_VM(v), avm_VM(a), avm_VM(M), avm_VM(V1), avm_VM(V2);

$$[Kr, Art, Sv], \begin{bmatrix} Kr \\ Art \\ Sv \end{bmatrix}, [Sv, Art, Kr], \begin{bmatrix} Sv & Art & Kr \\ 37 & 15 & 7 \end{bmatrix}, \begin{bmatrix} 36 & 5 & 0 \\ 5 & 15 & 0 \\ 0 & 0 & 0 \end{bmatrix}, [36, 7, 14], \begin{bmatrix} 62 \\ 57 \\ 37 \end{bmatrix}$$

> restart; a:=array(*symmetric*, 1..2, 1..2, [[1995,300], [300,2004]]): L:=[42,47,67]: m:=linalg[*matrix*](2,2,[Kr,Art,7,14]): v:=vector(3,[*RANS,IAN,VTU*]): V:=Vector(3,[Sv,Kr,Art]): V1:=Vector[*row*](3,[A,G,W]): S:={41,57,62}: M:=evalf(rtable(1..2,1..2, frandom(7..15,0.5), subtype=Matrix, readonly), 3): map(Aconv, [V,V1,M,a,v,m,L,S]);

$$\left[\begin{bmatrix} Sv \\ Kr \\ Art \end{bmatrix}, [A \quad G \quad W], \begin{bmatrix} 7.55 & 0 \\ 0 & 0 \end{bmatrix}, \begin{bmatrix} 1995 & 300 \\ 300 & 2004 \end{bmatrix}, [RANS, IAN, VTU], \begin{bmatrix} Kr & Art \\ 7 & 14 \end{bmatrix}, [42, 47, 67], [41, 57, 62]\right]$$

> restart; m:=matrix(3,3, [[Sv,Art,Kr],[36,14,7],[67,89,96]]): v:=vector(6,[Sv,Kr,Art,Arn,G, AVZ]): V:=convert(v,*Vector*): M:=convert(m,*Matrix*): m:=matrix(3,3,[[42,47,67],[37,8,15], [61,56,40]]): a:=array(*symmetric*, 1..3, 1..3, [[1,x,y],[x,y,z],[y,z,h]]): A:= Array(*symmetric*, 1..3,1..3, [[1,x,y], [x,y,z],[y,z,h]]): trConv(m,M), trConv(v,V), trConv(a,A);

$$\begin{bmatrix} 42 & 47 & 67 \\ 37 & 8 & 15 \\ 61 & 56 & 40 \end{bmatrix}, [Sv, Kr, Art, Arn, G, AVZ], \begin{bmatrix} 1 & x & y \\ x & y & z \\ y & z & h \end{bmatrix}$$

IdR – checking of active *rtable*-objects in a current *Maple* session
Rtable
Arobj

Call format of the procedures:

IdR()
Rtable()
Arobj({`?`})

Formal arguments of the procedures:

`?` - (optional) key symbol defining print of a brief procedure description

Description of the procedures:

The procedure call **IdR()** provides testing of presence in a current session of active *rtable*-objects and their types. The procedure call **IdR()** returns the 2-element sequence whose the first element defines a names set of active *rtable*-objects of the *first type*, and the second element defines a names set of active *rtable*-objects of the *second type*. Definition of the types of *rtable*-objects can be found above. If a current session does not contain any *rtable*-objects, the procedure call **IdR()** returns the *NULL* value with output of the appropriate warning. In addition, the first set can contain names of a view **`RTABLE_SAVE/nnnnn`** which define *rtable*-objects with identification **nnnnn**-numbers saved in an *mws*-datafile with a current *Maple* document. This procedure has important enough significance at operation with *rtable*-objects of different types and purpose.

The call of the simple procedure **Rtable()** returns the *NULL* value and simultaneously does all active *rtable*-objects of a current session by objects of the *first type*, what in a lot of cases simplifies a work with such objects, because allows to unify a programming with use of them. If a current session does not contain *rtable*-objects, the procedure call **Rtable()** returns the *NULL* value with output of an appropriate warning. The examples represented below of appendices of these procedures is clear enough and of special explanations do not require. Furthermore, these examples toppingly illustrate a principal difference between both types of *rtable*-objects.

At last, the procedure call **Arobj()** returns the *classification table* of *rtable*-objects, active in a current session, in the context of their types {*Type_1, Type_2, Array, Matrix, Vector*}; in addition, the result is returned in the form of an *Array*-object, providing simple access to its elements. Names of *rtable*-objects in this table are represented by their string equivalents and/or integer identification numbers, therefore for their subsequent use as the names of *rtable*-objects preliminary converting of their into symbol equivalents should be done; for example by means of expressions of simple view **cat("", X)** or **convert(X, *symbol*),** where **X** - a string equivalent. In addition, objects of the *first type* are identified by their names, while objects of the *second type* by their identification numbers (*indices* of the ***_rtable***-table). Whereas the procedure call **Arobj(`?`)** returns the *NULL* value with output of brief information about the procedure. If a current *Maple* session does not contain *rtable*-objects, the procedure call **Arobj()** returns the *NULL* value with output of an appropriate warning. Procedures of this group have confirmed own availability at advanced processing of *rtable*-objects.

```
IdR:= proc()
local a, b, c, k;
  [assign(a = [anames(rtable)], c = [], 'b' = `if`(eval(_rtable) = _rtable, [],
  [seq(op([entries(_rtable)][k]), k = 1 .. nops([entries(_rtable)]))])), `if`(a = [],
  RETURN(WARNING("a current session does not contain any active rtable-objects")),
  seq(`if`(member(eval(a[k]), b), assign('c' = [op(c), k]), NULL), k = 1 .. nops(a))),
  RETURN({seq([anames(rtable)][k], k = c)}, {anames(rtable)} minus
  {seq([anames(rtable)][k], k = c)})]
end proc

Rtable:= () -> null(eval(anames(rtable)))

Arobj:= proc()
local a, b, k, n, Tab, omega;
  omega:= S -> `if`(search(S, "RTABLE_SAVE/"), SN(cat("", S)[13 .. length(S)]), S);
  `if`(args = `?`, RETURN(lprint("Classification table of rtable-objects of a current session")),
    [assign67(b = IdR()), `if`(b = NULL, RETURN(), assign(a[1] = [b][1], a[2] = [b][2],
    a[3] = {anames(Array)}, a[4] = {anames(Matrix)}, a[5] = {anames(Vector)}))]);
  assign(n = {seq(nops(a[k]), k = 1 .. 5)}[-1]), assign(Tab = Array(1 .. n + 1, 1 .. 5));
  assign('Tab'[1, 1] = Type_1, 'Tab'[1, 2] = Type_2, 'Tab'[1, 3] = Array, 'Tab'[1, 4] = Matrix,
    'Tab'[1, 5] = Vector), null(seq(seq(assign('Tab'[j + 1, k] = omega(cat("", a[k][j]))),
    j = 1 .. nops(a[k])), k = 1 .. 5)), Tab
end proc
```

Typical examples of the procedures use:

> IdR();

{*RTABLE_SAVE/18484820, RTABLE_SAVE/18322972, RTABLE_SAVE/18581344*}, {*A, V, M*}

> Art:=Array(); Kr:=Vector[row]([7,96,14,1986,17.1203], readonly=true); Kr1:=Vector([96, 7, 300,1995,2003], readonly=true): M:=evalf(rtable(1..2,1..2, frandom(5..12,0.5), subtype= Matrix, readonly), 3): IdR();

Art := Array({}, *datatype* = *anything*, *storage* = *rectangular*, *order* = *Fortran_order*)
{*Art, Kr, RTABLE_SAVE/18484820, RTABLE_SAVE/18322972, `RTABLE_SAVE/18581344*},
{*A, V, M, Kr1*}

> map(whattype, [op(%[1])]), map(whattype, [op(%[2])]);

[*Vector*[*row*], *Vector*[*row*], *Array*, *Matrix*, *Array*], [*Vector*[*row*], *Array*, *Vector*[*column*], *Matrix*]

> Rtable(); IdR();

{*RTABLE_SAVE*/5474160, *Art*, *RTABLE_SAVE*/27194600, *Kr*, *Kr1*, *M*,
RTABLE_SAVE/28428484, *RTABLE_SAVE*/28444628}, {}

> Arobj(); Arobj(`?`);

Type_1	*Type_2*	*Array*	*Matrix*	*Vector*
5474160	0	5474160	"M"	27194600
"Art"	0	"Art"	0	"Kr"
27194600	0	28444628	0	"Kr1"
"Kr"	0	28767648	0	28428484
"Kr1"	0	0	0	0
"M"	0	0	0	0
28428484	0	0	0	0
28444628	0	0	0	0
28767648	0	0	0	0

"Classification table of rtable-objects of a current session"

Here it is necessary to note, that the mechanism used by algorithm of the **IdR** procedure, in case of advanced programming may be rather effective for organization of work with copies of the same *rtable*-object when data of different copies can be processed independently, or with the same objects but of different type as that has been shown on a lot of examples in our previous books [29-33,43,44].

---

rID – checking of identification numbers of active *rtable*-objects
rName

Call format of the procedures:

rID(R)
rName(n)

Formal arguments of the procedures:

R - an arbitrary valid *Maple* expression
n - positive integer defining the identification number (*rtable-index*) of an *rtable*-object

Description of the procedures:

The procedure call **rID(R)** returns the *identification number* (*rtable-index*) of an active *rtable*-object **R** of the *first type* and outputs an appropriate warning for an ***rtable***-object of the *second type*. In addition, as a result of the procedure call **rID(R)** the conversion of an *rtable*-object **R** of the *second type* into *rtable*-object of the same name of the *first type* is done. If a current session does not contain any *rtable*-objects, the procedure call **rID(R)** returns the *NULL* value with output of appropriate warning. If a current *Maple* session contains *rtable*-objects, but actual **R** argument does not define an active *rtable*-object, the erratic situation arises.

On the other hand, the procedure call **rName(n)** returns name of an active *rtable*-object of the *first* or *second* type defined by its identification number (*rtable-index*) **n,** or the *NULL* value with output of an appropriate warning. Thus, for an arbitrary active *rtable*-object of any of the above two types take place the following determinative relations:

rID(eval(rName(n))) = n and **rName(rID(R) = R**

where **n** and **R** is an identification number (*rtable-index*) and name of the same active *rtable*-object accordingly. In addition, as a result of the procedure call **rName(n)** the conversion of an active *rtable*-object of the *second* type with identification number **n** into *rtable*-object of the same name of the *first* type is done. If a current *Maple* session does not contain any active *rtable*-objects, the erratic situation arises.

```
rID:= proc(R::anything)
local a, b, c, k;
  assign67(b=IdR()), `if`(b=NULL, RETURN(), `if`(type(R, rtable), `if`(whattype(eval(_rtable))=
  symbol, RETURN(lprint("a current session does not contains rtable-objects of the first type")),1),
  seq(`if`(rhs(op(2, eval(_rtable))[k]) = R, assign(a = lhs(op(2, eval(_rtable))[k])), NULL),
  k = 1 .. nops([indices(_rtable)])), RETURN(`if`(type(a, symbol),
  WARNING("%1 is rtable-object of the second type", R), a))], [assign(c = whattype(eval(R))),
  ERROR(`if`(c=symbol, "%1 is not an active rtable-object", "argument has type <%2>"), R, c)]))
end proc

rName:= proc(n::posint)
local a, b, c, k;
  if type(eval(_rtable), assignable) then
       ERROR("active rtable-objects do not exist in a current session")
  else [assign('a' = [], 'c' = [seq(convert([anames(rtable)][k], string), k=1..nops([anames(rtable)]))],
       'b' = {indices(eval(_rtable))})], Rtable()
  end if;
  seq(assign('a' = [op(a), `if`(search(c[k], "RTABLE_"), NULL, c[k])]), k = 1 .. nops(c)),
    `if`(not member([n], b), RETURN(WARNING("<%1> is not an rtable-index", n)),
    mul(`if`(_rtable[n] = eval(convert(a[k], name)), RETURN(convert(a[k], name)), 1),
    k = 1 .. nops(a)))
end proc
```

Typical examples of the procedures use:

```
> Art:= Vector[row](6, [42,47,67,62,89,96]); Svet:=Array(): Kr:=Vector[row]([96,7,17.12003,
  Paskula],readonly=true); Kr1:=Vector[row]([61,56,36,40,7,14,2003]); Raadiku:=Matrix(3):
  M:=rtable(1..3,1..6, frandom(1..6,0.6), subtype=Matrix, readonly): map(rID, [Art, Svet, Kr,
  Kr1, Raadiku]);
```

$$Art := [42, 47, 67, 62, 89, 96]$$

$$Kr := [96, 7, 17.12003, Paskula]$$
$$Kr1 := [61, 56, 36, 40, 7, 14, 2003]$$

Warning, RTABLE(19220104, Array({}, datatype = anything, storage = rectangular, order = Fortran_order), Array) is rtable-object of the second type

Warning, RTABLE(19267896, MATRIX([[0, 0, 0], [0, 0, 0], [0, 0, 0]]), Matrix) is rtable-object of the second type

[19077276, 19077436, 19232932]

> map(rID, [Art, Svet, Kr, Kr1, Raadiku]);

[19077276, 19220104, 19077436, 19232932, 19267896]

> rans:=[42, 62, 300, RANS, IAN]: rID(rans); rID(RANS_IAN);

Error, (in **rID**) argument has type <list>
Error, (in **rID**) RANS_IAN is not an active rtable-object

> map(rName, [19422004, 19077276, 19220104, 19077436, 19232932, 19267896]);

Warning, 19422004 is not an rtable-index

[*Art*, *Svet*, *Kr*, *Kr1*, *Raadiku*]

> restart; rName(19422004300);

Error, (in **rName**) active rtable-objects do not exist in a current session

---

Rt_s – checking of saving of active *rtable*-objects in a current *Maple* session
RTsave

Call format of the procedures:

Rt_s()
RTsave()

Formal arguments of the procedures: **no**

Description of the procedures:

The procedure call **Rt_s()** returns a set of the *identification numbers* (*rtable-indexes*) of *rtable*-objects reflected only in the predetermined ***_rtable***-variable which saves any call to an *rtable*-object of the *first* type. Otherwise, the *empty* set is returned with output of an appropriate warning.

In turn, the procedure call **RTsave()** returns a set of the *identification numbers* (*rtable-indexes*) of *rtable*-objects processed in a current *Maple* session and saved in *mws*-file of a current document, otherwise the *empty* set is returned with output of an appropriate diagnostic warnings. Procedures of this group have confirmed own availability at advanced processing of *rtable*-objects.

```
Rt_s:= proc()
local k, Z;
  try Z := [indices(eval(_rtable))]
  catch "wrong number (or type) of parameters in function":
    RETURN({WARNING("rtable-objects do not exist or they have the second type only")})
  end try;
  {seq(op(Z[k]), k = 1 .. nops(Z))}
end proc
```

```
RTsave:= proc()
local a, b, p, k;
  `if`({anames(rtable)} <> {}, [null([assign('a' = []), seq([`if`(search([anames(rtable)][k],
  RTABLE_SAVE), assign('a' = [op(a), convert([anames(rtable)][k], string)[13 .. length([anames(
  rtable)][k])]]), NULL)], k = 1 .. nops([anames(rtable)])), RETURN(assign(b = nops(a)),
  `if`(b = 0, {WARNING("active rtable-objects have not been saved in Maple work area")},
  {`if`(b <> 0, seq(add((convert(a[p], bytes)[k] - 48)*10^(length(a[p]) - k), k = 1 .. length(a[p])),
  p = 1 .. b), NULL)}))])],
  {WARNING("a current Maple session does not contain active rtable-objects")})
end proc
```

Typical examples of the procedures use:

> restart: Rt_s(); RTsave();

Warning, rtable-objects do not exist or they have the second type only

{}

Warning, a current Maple session does not contain active rtable-objects

{}

> Art:= Array(1..4, [36, 40, 7, 14]); Svetla:=Array(1..6, [42,47,67,62,89,96]); Kr:=Vector[row] ([96, 7, 26.403], readonly=true); Tallinn:= Vector(): Kr1:= Vector([96,7]): Raadiku:= Matrix(6): M:= evalf(rtable(1..2, 1..2, frandom(7..14, 0.5), subtype=Matrix, readonly), 3): Rt_s();

Art := [36, 40, 7, 14]
Svetla := [42, 47, 67, 62, 89, 96]
Kr := [96, 7, 26.403]]
{19676932, 19677092, 19678292}

[This document has been saved in a *mws*-datafile

> RTsave(); map(rName, %);

{19676932, 19677092, 19678292}
{*Svetla, Art, Kr*}

In certain sense, the procedures **Rt_s** and **RTsave** provide a mode monitoring of saving of *rtable*-objects of a current *Maple* session in *mws*-files, by allowing to create the extended tools of processing of such *Maple* objects.

MwsRtb – checking of presence of *rtable*-objects in *mws*-datafiles

Call format of the procedure:

MwsRtb(F {, 'h'})

Formal arguments of the procedure:

F - symbol or string defining name of a *mws*-file or full path to it
h - (optional) an unevaluated symbol

Description of the procedure:

The procedure **MwsRtb(F)** is intended for analysis of an *mws*-file whose qualifier is specified by the actual **F** argument, for the purpose of presence in it of the saved *rtable*-objects. In case of detection of such objects a list of string equivalents of their calls is returned, otherwise the call returns the *NULL* value with output of an appropriate warning. If the second optional **h** argument has been coded, then through it a list of *identification numbers* (*rtable-indexes*) of *rtable*-objects located in an **F** datafile is returned. Between the main return list and a list returned through h one-to-one correspondence takes place.

In addition, the procedure call creates in the main *Maple* directory the subdirectory "RTBobj" into which datafiles of the internal *Maple* format with *rtable*-objects of a **F** file are saved. The appropriate warning contains a rather detail information about it. The *rtable*-objects from datafile **F** become available in a current *Maple* session. If the actual **F** argument defines non-existing datafile or datafile has a type different from ".mws", the erratic situation arises. The examples represented below of the procedure appendices is quite clear and of special explanations do not require, illustrating the told above.

```
MwsRtb:= proc(F::{name, string, symbol})
local a, b, c, d, k, p, h, t;
  if not type(F, file) or Ftype(F) <> ".mws" then
    ERROR("%1 is not a mws-datafile or does not exist", F)
  else  assign(a = RS([`\n` = ``], readbytes(F, TEXT, infinity))), close(F);
    if not search(a, "{RTABLE_HANDLES", 't') then
      RETURN(WARNING("rtable-objects do not exist in datafile <%1>", F))
    end if;
    assign('a' = a[t + 1 .. -1], c = "{RTABLE "), search(a, "}{", 't'),
      assign(b = map(SD, SLD(a[15 .. t - 2], " ")),
      'a' = a[t + 1 .. -1]), `if`(1 < nargs and type(args[2], symbol), assign(args[2] = b), NULL)
  end if;
  assign(p=SLD(a, c), d = map2(cat, "RTABLE_SAVE/", b), h = MkDir(cat(CDM(), "\\rtbobj")));
  for k to nops(p) do assign('a' = cat(h, "\\_", b[k], ".m")), writeline(a, cat(p[k][1 .. 4], "\n",
       p[k][5 .. -2])), close(a);
    read a
  end do;
  d, WARNING("rtable-objects of datafile <%1> are active in a current session and have been
    saved in directory <%2>; their names are %3", F, h, map2(cat, "_", b, ".m"))
end proc
```

Typical examples of the procedure use:

```
> MwsRtb("C:/Academy\\Examples\\MPL9MWS\\Rt_s.mws", 'h');
```

Warning, rtable-objects do not exist in datafile <C:/Academy\Examples\mpl9mws\Rt_s.mws>

```
> MwsRtb("C:/Academy\\Examples\\MPL9MWS\\IdR.mws", 'h'); h;
```

Warning, rtable-objects of datafile <C:/Academy\Examples\MPL9MWS\IdR.mws> are active in a current session and have been saved in directory

<c:\program files\maple 9\rtbobj>; their names are [_5290620.m, _5603484.m, _5188712.m]
["RTABLE_SAVE/5290620", "RTABLE_SAVE/5603484", "RTABLE_SAVE/5188712"]
[5290620, 5603484, 5188712]

```
> for k in %% do eval(cat(``, k)) end do;
```

Array({}, *datatype* = *anything*, *storage* = *rectangular*, *order* = *Fortran_order*)

[7, 96, 14, 1986, 17.1203]

Type_1	*Type_2*	*Array*	*Matrix*	*Vector*
5474160	0	5474160	"M"	27194600
"Art"	0	"Art"	0	"Kr"
27194600	0	28444628	0	"Kr1"
"Kr"	0	28767648	0	28428484
"Kr1"	0	0	0	0
"M"	0	0	0	0
28428484	0	0	0	0
28444628	0	0	0	0
28767648	0	0	0	0

mws789_6 – converting of *mws*-files containing *rtable*-objects

Call format of the procedure:

mws789_6(F)

Formal arguments of the procedure:

F – symbol or string defining name of a *mws*-file or full path to it

Description of the procedure:

If an *mws*-file does not contain saved *rtable*-objects, then it can be successfully loaded into a current session of *Maple* of any of releases **6 – 9/9.5**. In addition, at attempt to load an *mws*-file prepared in *Maple* of release **7 – 9/9.5** into *Maple* **6** can arise especial situation with diagnostics

*"**Maple 6:** There were problems during the loading process. Your worksheet may be incomplete"*

As a rule, this situation does not influence upon correctness of a worksheet load. Radically different situation takes place at attempt to load into *Maple* **6** a *mws*-file prepared in *Maple* of releases **7 – 9/9.5** which contains *rtable*-objects. In this case the crucial situation with diagnostics

*"**Wmaple:** This program has performed an illegal operation and will be shut down"*

invoking the abnormal ending of a current *Maple* session. Reason of that consists in the following.

Maple saves *rtable*-objects in an *mws*-file in the internal *Maple* format (*m*-format), which is release-dependent, namely: an *m*-file prepared in *Maple* of releases **6 – 9/9.5** is correctly read by any of the above releases, whereas *Maple* **6** cannot read *m*-files prepared in releases **7 – 9/9.5.** At loading of an *mws*-file into *Maple* **6,** all *rtable*-objects contained in it are evaluated invoking the above abnormal ending of a current session. The procedure **mws789_6** allows to bypass this situation.

The procedure **mws789_6(F)** is intended for conversion of a *mws*-file whose qualifier is specified by the actual **F** argument, and which has been prepared in *Maple* of releases **7 – 9/9.5** and contains *rtable*-objects into *mws*-file available for successful load into *Maple* **6.** If an *mws*-file with name "aaa.mws" has been prepared in *Maple* **6** or does not contain *rtable*-objects, the procedure call returns the *NULL* value with output of the appropriate warning – the file is compatible with *Maple* **6**. Otherwise, the procedure call returns the *NULL* value with creation of file with name "aaa6.mws" compatible with *Maple* **6.** This file is located in the same directory as an initial *mws*-file.

```
mws789_6:= proc(F::{name, string, symbol})
local a, h, t, omega;
  if not type(F, file) or Ftype(F) <> ".mws" then
```

```
    ERROR("%1 is not a mws-file or does not exist", F)
  else assign(a = RS([`\n` = ""], readbytes(F, TEXT, infinity))),
    search(a, "{RTABLE_HANDLES", t);
    assign(omega = (x -> WARNING("file <%1> is compatible with Maple 6", x))), close(F)
  end if;
  if type(t, symbol) then RETURN(omega(F))
  elif not search(a[t .. -1], "M7R0") then RETURN(omega(F))
  else
    a := a[1 .. Search2(a, {"{PAGENUMBERS"})[-1] - 1];
    assign(h = cat(cat("", F)[1 .. -5], 6, ".mws")), null(writebytes(h, a), close(h))
  end if;
  WARNING("datafile <%1> is now compatible with Maple 6", h)
end proc
```

Typical examples of the procedure use:

> mws789_6("C:/Academy\\Examples\\MPL9MWS\\Rt_s.mws");

Warning, file <C:/Academy\Examples\MPL9MWS\Rt_s.mws> is compatible with Maple 6

> mws789_6("C:/Academy\\Examples\\MPL9MWS\\IdR.mws");

Warning, datafile <c:/academy\examples\mpl9mws\IdR6.mws> is now compatible with Maple 6

> mws789_6("C:/Academy\\Examples\\MPL9MWS\\IdR7.mws");

Error, (in mws789_6) c:/academy\examples\mpl9mws\idr7.mws is not a mws-file or does not exist

At last, we shall present the important enough **ResRtb** procedure that provides a work with rtable-objects processed in the previous sessions of *Maple* of releases **6 - 9/9.5.** The procedure allows to trace a history of calculations in the environment of module **LinearAlgebra** of the package. In contrast to *Maple*-objects for *rtable*-objects, it is possible to watch a history of operation with them during the previous sessions with *Maple.* In particular, the given circumstance gives an opportunity to process *rtable*-objects from a history of work with **LinearAlgebra** module. This possibility seems a useful in many appendices dealing with linear algebra and related topics.

ResRtb - a history restoring of *rtable*-objects of the previous *Maple* sessions

Call format of the procedure:

ResRtb(R {, K} {, `?`})

Formal arguments of the procedure:

R - an unevaluated name
K - (optional) a directory containing datafiles with *rtable*-objects
`?` - (optional) key symbol defining an output mode of warnings

Description of the procedure:

The procedure **ResRtb(R {, K} {, `?`})** admits from one up to three actual arguments. Through the first **R** argument is returned the nested list of *rtable*-objects which have been saved in the system directory ***"C:\\{Windows | WinNT}\\Temp"***; the 2-elements sublists define a saved *rtable*-object and its multiplicity accordingly; the sublists are sorted lexicographically regarding to their first element. The optional **K** argument, if it is different from symbol `**?**`, specifies a directory containing *m*-files associated with the required *rtable*-objects. The *rtable*-objects restored in a current *Maple* session belong to the *first type* according to the above classification. Absence of the saved *rtable*-objects causes an erratic situation.

Simultaneously with that, the procedure call outputs the warnings about identifiers which have been assigned to the restored *rtable*-objects, what allows to use them explicitly in a current session. In these warnings, the saved *rtable*-objects are identified by a procedure of saving and by their *identification numbers* (*rtable-indexes*), and by names over which is provided access to them in a current session. In addition, at output of the procedure the appearance of identical *rtable*-objects under different names is possible, what in a number of cases of advanced programming seems as a rather useful tool. The examples represented below of the procedure use illustrate the told.

The procedure **ResRtb** provides access to earlier saved *rtable*-objects through names ascribed to them that have view **G[k]**, where **G** – the first actual argument and **k** is number of an appropriate *rtable*-object in a current *Maple* session. At last, at coding of the optional key symbol `**?**` as the second or third argument the output of informational warnings of the following view:

***Warning, rtable-object* <RTABLE_SAVE/nn> *saved `data` at `time` has been activated as* G[k]**

are suppressed, excluding the general information warning

***Warning, rtable-objects have been activated under names* G[k], *where* k=1 .. p**

about names of restored *rtable*-objects and their quantity. The indexed names represented by the given warnings provide a simple method of access to earlier saved *rtable*-objects created in the previous sessions. At procedure call **ResRtb** the user has the possibility to define the indexed names under which the earlier saved *rtable*-objects will be returned. The datafiles with *rtable*-objects saved in the system TEMP directory can be deleted in the *Maple* environment by means of the procedure call **DEL_F()**.

```
ResRtb:= proc(R::evaln)
local a, b, c, omega, L, Z, St, h, k, S, N, Q, f, nu, epsilon;
global _agavsvartkr;
  St:= proc(g::string)
    local a, k;
      a := length(g);
      for k do if g[a - k] = " " then RETURN([a - k + 1 .. a]) end if end do
    end proc;
  assign(L = "1995-2003 * RANSIAN = Tallinn", epsilon = {args} minus {R, `?`}, Z = [], a = [],
    f = "_$$_");
  assign(omega = `if`(epsilon = {}, WT(), `if`(type(epsilon[1], dir), Path(epsilon[1]), WT()))),
   system(cat("Dir ", omega, "\\MPL*.TMP.m > _$$_")),
    assign(b = fopen(f, READ, TEXT));
  while not Fend(b) do L := iolib(2, b);
    if not search(Case(cat("", L)), ".tmp.m") then next
    else assign('Z' = [op(Z), L], 'a' = [op(a), SLD(L, " ")[-3 .. -2]])
    end if
  end do;
  `if`(Z = [], [iolib(8, b), ERROR("<%1> does not contain rtable-files", omega)],
    assign(c = nops(Z)));
  assign('Z' = array(1 .. c, Z)), seq(assign('Z[k]' = Z[k][op(St(Z[k]))]), k = 1 .. c);
  assign(N = array(1 .. c, [seq(convert(k, symbol), k = 1 .. c)]),
    nu = interface(warnlevel));
```

```
  for k to c do S := [iolib(2, cat(omega, "\\", Z[k])), iolib(2, cat(omega, "\\", Z[k]))][2];
    if S = 0 then break else NULL end if;
    iolib(7, cat(omega, "\\", Z[k])), (proc(x) read x end proc)(cat(omega, "\\", Z[k]));
    assign(Q = convert(S[3 .. searchtext("X", S) - 1], symbol), 'N[k]' = eval(Q));
    `if`(member(`?`, {args}), interface(warnlevel = 0), NULL);
    WARNING("rtable-object <%1> saved %3 at %4 has been activated as <%2>", Q,
      cat(convert(R, name), convert([k], name)), a[k][1], a[k][2]), iolib(7, cat(omega, "\\", Z[k]))
  end do;
  assign(_agavsvartkr = convert(N, list)), assign(R = _agavsvartkr), iolib(8, b),
    interface(warnlevel = nu);
  WARNING("rtable-objects have been activated under name s %1, where k=1..%2",
    cat(R, "[k]"), k - 1);
  `if`(member(`?`, {args}), NULL, gelist(_agavsvartkr)), unassign('_agavsvartkr')
end proc
```

Typical examples of the procedure use:

```
> ResRtb(G);
  Warning, rtable-object <RTABLE_SAVE/28418016> saved 12-20-03 at 10:42a
  has been activated as <G[1]>
  ==============================================================
  Warning, rtable-object <RTABLE_SAVE/28418016> saved 12-20-03 at 10:42a
  has been activated as <G[14]>

  Warning, rtable-objects have been activated under names G[k], where k=1..14
```

Type_1	*Type_2*	*Array*	*Matrix*	*Vector*
5474160	0	5474160	"M"	27194600
"Art"	0	"Art"	0	"Kr"
27194600	0	28444628	0	"Kr1"
"Kr"	0	28767648	0	28428484
"Kr1"	0	0	0	0
"M"	0	0	0	0
28428484	0	0	0	0
28444628	0	0	0	0
28767648	0	0	0	0

, 4],

[Array({ }, *datatype* = *anything*, *storage* = *rectangular*, *order* = *Fortran_order*), 4], [[7, 96, 14, 1986, 17.1203], 5],

[[96, 6, 100, 1995, 2003], 1]

> ResRtb(G, `?`);

Warning, rtable-objects have been activated under names G[k], where k=1..14

> G[1], G[8], G[15];

[7, 96, 14, 1986, 17.1203],

Type_1	*Type_2*	*Array*	*Matrix*	*Vector*
5474160	0	5474160	"M"	27194600
"Art"	0	"Art"	0	"Kr"
27194600	0	28444628	0	"Kr1"
"Kr"	0	28767648	0	28428484
"Kr1"	0	0	0	0
"M"	0	0	0	0
28428484	0	0	0	0
28444628	0	0	0	0
28767648	0	0	0	0

,

Array({ }, *datatype* = *anything*, *storage* = *rectangular*, *order* = *Fortran_order*)

> ResRtb(S, "C:/Windows/Temp", `?`);

Warning, rtable-objects have been activated under names S[k], where k=1..14

> S[1], S[8], S[15];

[7, 96, 14, 1986, 17.1203],

Type_1	*Type_2*	*Array*	*Matrix*	*Vector*
5474160	0	5474160	"M"	27194600
"Art"	0	"Art"	0	"Kr"
27194600	0	28444628	0	"Kr1"
"Kr"	0	28767648	0	28428484
"Kr1"	0	0	0	0
"M"	0	0	0	0
28428484	0	0	0	0
28444628	0	0	0	0
28767648	0	0	0	0

,

Array({ }, *datatype* = *anything*, *storage* = *rectangular*, *order* = *Fortran_order*)

> DEL_F(): ResRtb(Art_Kr_Sv);

Error, (in **ResRtb**) <C:\Windows\Temp> does not contain rtable-files

It is necessary to note close connection of results of running of the procedure **ResRtb** with saving of *rtable*-objects of a current session as a result of the *ABEND* of the package, or the basic operational system. The given opportunity of the procedure is actual enough depending on our operational experience, namely with growth of numbers of *Maple* releases from **6** until **9.5** the stability of their running is reduced, unfortunately. In this relation, the procedure **ResRtb** allows to restore results of a work with *rtable*-objects after the above erratic situations. The above procedure possibility is provided by the circumstance, that *Maple* for a number of *rtable*-objects of the *first type* creates temporary files in system TEMP directory that are saved after a package *ABEND* or a system *ABEND*. That is one more reason in favor of *rtable*-objects use of the *first type*.

Since the *sixth* release, the *Maple* package contains the **LinearAlgebra** module of *NAG* ltd. that is an efficient and robust suite of routines for doing computational linear algebra. Full rectangular and sparse matrices are fully supported at the data structure level, as well as upper and lower triangular matrices, unit triangular matrices, banded matrices, and a variety of others. Further, symmetric, skew-symmetric, hermitian, and skew-hermitian are known qualifiers that are used appropriately to reduce storage and select amongst algorithms.

The following data types are handled efficiently: hardware floating-point numbers (both *real* and *complex*), hardware integers of various sizes, arbitrary-precision floating-point numbers (both *real* and *complex*), and general symbolic expressions. For increased compatibility with external routines, matrices can be stored in either ***C*** or ***FORTRAN*** order.

The **LinearAlgebra** module has been designed to accommodate different sets of usage scenarios: casual use and programming use. Correspondingly, there are functions and notations designed for easy casual use (*sometimes at the cost of some efficiency*), and some functions designed for maximal efficiency (*sometimes at the cost of ease-of-use*). In this way, the **LinearAlgebra** facilities scale easily from first year classroom use to heavy industrial usage, emphasizing the different qualities that each type of use needs.

Implementation of the **LinearAlgebra** module is based on *rtable* data structures that can be used outside the module context, for example *rtable*-objects can be successfully used for data processing and creation of tools dealing with access to datafiles. In view of the told, the role of *rtable*-objects at their usage both for a programming of applied tasks in the package environment, and in interactive mode is well apparent. Therefore, group of the above tools providing the advanced processing of *rtable*-objects represents undoubted interest in many respects. With questions of handling of *rtable*-objects of *Maple* of *five* last releases based on procedures represented above, it is possible to familiarize oneself in our books [29-33,39,43,44] in more details.

Chapter 13.
Software to support simple statistics problems

For support of the statistical data analysis the package *Maple* has a series of tools provided by the procedures ***rand, randomize,*** and by modules **stats** and **RandomTools** which support definitions of a number of procedures useful to solution of probabilistic and statistical problems. In the given section, some useful additional tools that extend the package tools orientated to the statistical data analysis are represented. Representation of these means is carried out according to four basic aspects of simple statistics: descriptive statistics, elements of regression analysis, testing of statistical hypotheses and elements of dynamic (*time*) series. Software is represented both by the procedures, and by the program modules. In many respects, they well supplement standard package means for the problems solving of the given type.

13.1. Software for solving problems of descriptive statistics

This item represents a number of additional means of problem solving of *descriptive statistics.* The *Maple* package has software for realization of statistical analysis at two levels: *built-in* and *modular.* On the first level the built-in ***rand*** procedure implementing a generator of pseudorandom numbers appears. Whereas the modular means are in two modules **RandomTools** (*by starting with the seventh release*) and **stats.** The sphere of the appendices of the **rand** generator is rather extensive, including a generating of the debug data and test sets of the statistical data. However, essential imperfection of the given functional facility is that the **rand** procedure generates only so-called normal-similar distributions, which should additionally be analyzed. Above all, it concerns an estimation of a degree of deviation of distributions, generated by the ***rand*** procedure, from the *standard normal distribution.* The given problem to the considerable extent is being solved by the **rand_Histo** procedure. The standard submodule **stats[random, ...]** provides a series of tools for generating of pseudo-random numbers with a given distribution. The technique used in the **rand_Histo** can be used for estimation of real distributions of such pseudo-random numbers.

---

rand_Histo – testing of built-in Generator of pseudo-random numbers

Call format of the procedure:

rand_Histo(a, h, n, t)

Formal arguments of the procedure:

a - an integer
h - an integer step of division of an interval
n - quantity of such steps
t - quantity of experiments

Description of the procedure:

With the purpose of providing of the general preliminary analysis of the above problem, the **rand_Histo** procedure that displays histogram of distribution generated by the built-in ***rand*** procedure on the given interval has been created. The procedure call **rand_Histo(a, h, n, t)** outputs the histogram of a distribution generated by the ***rand*** procedure, on interval **[a, a+h*m],** where: **a** - integer; **h** - integer step of division of the interval; **n** - quantity of such steps; **t** - quantity of experiments (amount of values generated by the ***rand*** procedure); **m=n,** if **n** is odd and **m=n+1** otherwise.

Really, the obtained histogram of distribution generated by the ***rand*** procedure on the given interval, allows to do the certain conclusions concerning a degree and character of deviation of explored distribution from the normal distribution. In this connection, the procedure **rand_Histo** can be considered as a rather useful additional tool at use of the built-in ***rand*** generator. The received histogram of distribution generated by the ***rand*** procedure on interval [1942 .. 2005], allows to do quite defined conclusions concerning a degree and character of deviation of the researched distribution from the normal distribution. It is necessary to note, to a certain extent the procedure **rand_Histo** is a program analog of the Galton's mechanical device illustrating regularity of the *binomial distribution* whose extreme case is the *normal distribution*.

```
rand_Histo:= proc(a::integer, h::integer, n::integer, t::integer)
local omega, b, Kr, k, p, m, v;
  assign(m = `if`(type(n, odd), n, n + 1)), assign(omega = rand(a .. a + h*m),
    b = [seq(0, v = 1 .. m)]);
  for k to t do for p to m do
      if omega() <= a + p*h and a + (p - 1)*h < omega() then b[p]:= b[p] + 1; break
      else
      end if
    end do
  end do;
  Kr:= stats[statplots, histogram]([seq(Weight(a + (p - 1)*h .. a + p*h, b[p]), p = 1 .. m)]);
  plots[display](Kr, axes = boxed, axesfont = [TIMES, BOLD, 10],
    title = `Histogram of distribution generated by rand-procedure`,
    titlefont = [TIMES, BOLD, 12])
end proc
```

Typical examples of the procedure use:

> rand_Histo(1942, 7, 9, 500000);

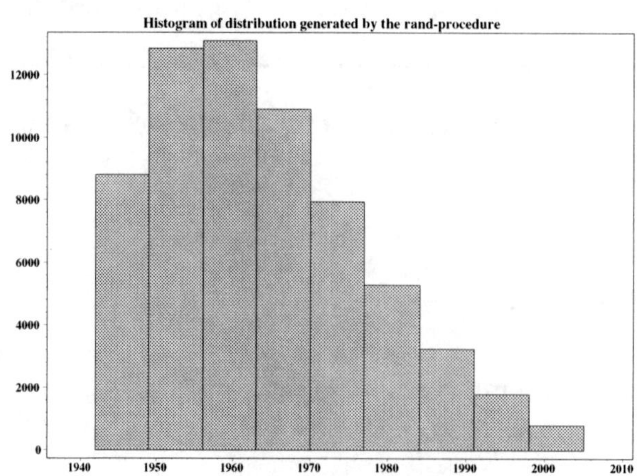

Dist_rand_n_2 – analysis of distribution of median decimal digits of pseudo-random integers

Call format of the procedure:

Dist_rand_n_2(A, H, T)

Formal arguments of the procedure:

A - a range
H - a positive integer
T - an assignable name

Description of the procedure:

For analysis of distribution of median decimal digits of numbers generated by the built-in ***rand*** generator, the **Dist_rand_n_2(A, H, T)** procedure serves. The first actual **A** argument of the procedure defines a range of values of integers generated by means of the ***rand*** generator; the second actual **H** argument defines length of a generated sequence of pseudorandom numbers, while via the third actual **T** argument a matrix of *Matrix*-type is returned, whose the second row defines decimal digits and the first row defines frequencies of appearance their in the middle of the generated numbers. The procedure returns the graph of frequency distribution of median decimal digits of the numbers generated by the built-in ***rand*** generator according to the given conditions.

Of the below-mentioned example of an application of the **Dist_rand_n_2** procedure follows what at concrete initial conditions takes place the following rather interesting result: the maximum frequency falls to digit **0**, and minimum frequency falls to digit **9**, while frequencies of remaining digits are enough close among themselves and difference between them diminishes with increase of length of a generated sequence of pseudorandom numbers. In addition, the given behavior of frequency distribution had place for all ranges **A** of generated numbers tested by us. The procedure admits a simple modification allowing to investigate behaviour of any digit of numbers generated by the standard **`rand`** generator of pseudo-random numbers. The approach can be extended onto other similar generators also.

```
Dist_rand_n_2:= proc(A::range, H::posint, T::evaln)
local a, b, k, p, h, t, omega;
  omega:= n::posint -> [assign('b' = (proc() `if`(args[1] <= length(args[2]),
      op(convert(convert(args[2], string)[args[1]], bytes)) - 48, 1) end proc)),
      assign('a' = length(n)), `if`(a = 1, RETURN(n), RETURN(b(round(1/2*a), n)))];
  [assign(h = rand(A)), assign(t = Matrix(2, 10)), seq(assign('t[2, k]' = k - 1), k = 1 .. 10)];
  for k to H do  p := omega(h());
    if p = 0 then t[1, 1] := t[1, 1] + 1 else t[1, p] := t[1, p] + 1 end if
  end do;
  PLOT(CURVES([seq([k - 1, 1/10*t[1, k]], k = 1 .. 10)], COLOR(RGB, 0, 0, 0)),
    THICKNESS(2), AXESTICKS([seq(k = k, k = 0..9)], 10), FONT(TIMES, BOLD,10),
    AXESLABELS("Digits", "Frequence")), assign(T = t)
end proc
```

Typical examples of the procedure use:

> Dist_rand_n_2(1..5000, 1000000, h): h;

$$\begin{bmatrix} 198948 & 100135 & 99607 & 100354 & 100170 & 100071 & 100034 & 100047 & 100634 & 0 \\ 0 & 1 & 2 & 3 & 4 & 5 & 6 & 7 & 8 & 9 \end{bmatrix}$$

> Dist_rand_n_2(1 .. 2004, 1000000, T); T;

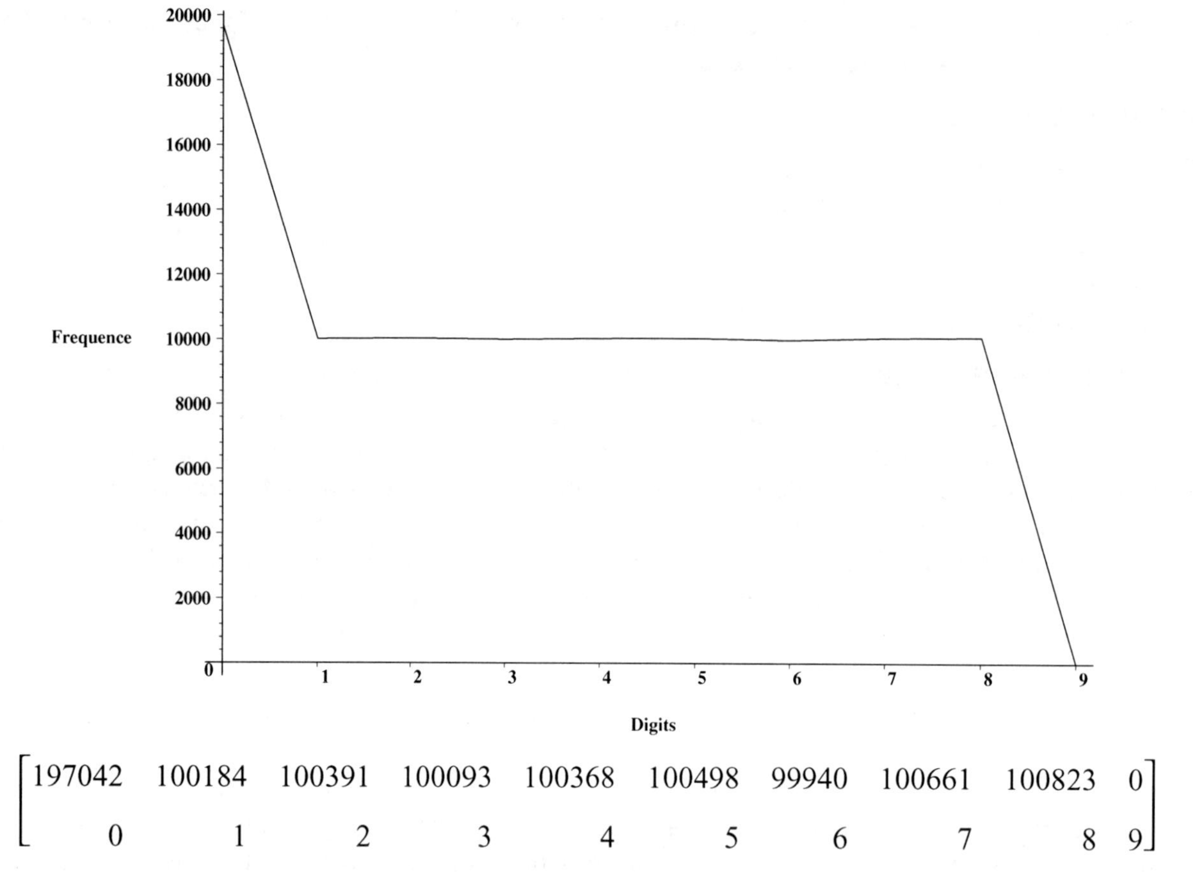

SA_text – a statistical analysis of text datafiles

Call format of the procedure:

SA_text(SF, L, S)

Formal arguments of the procedure:

SF - a datafile descriptor or filename
L - a symbol
S - an assignable name

Description of the procedure:

The procedure call **SA_text(SF, L, S)** returns the frequency distribution of letters composing a text contained in text file defined by the first actual **SF** argument. The second actual **L** argument of the procedure defines the language characteristic of the text (**R** – Russian, **E** – English). While through third optional **S** argument a matrix of quantity of entries into the text of letters of the corresponding alphabet is returned. Of the result of experiments with the procedure which were done by us a rather interesting corollary follows, namely: the frequency distributions of letters of the Russian and the English texts of the same contents structurally are similar enough, their obtained graphics representations testify about that. By default, if dimensionality of the returned *Matrix* more than 10x10, then it is returned via **S** argument as a placeholder which is made up of a set of square brackets and noneditable text fields describing its structure. To view the output, the built-in *Structured Data Browser* is used, or the appropriate setting for ***rtablesize***-parameter of the ***`interface`*** function can be used. Inasmuch as the returned matrix has (**2xn**)-dimensionality, where **n** - number of letters of the used alphabet, the setting ***interface***(***rtablesize***=**32)** is quite sufficient.

```
SA_text:= proc(SF::file, L::symbol, S::evaln)
local n, d, k, t, h, v, z, F, p, A, B, P, Q, omega;
  `if`(member(L, {R, E}), assign(z=0), ERROR("the second argument <%1> is invalid", L));
  A := convert([seq(k, k = 224 .. 255), seq(k, k = 192 .. 223)], bytes);
  B := convert([seq(k, k = 97 .. 122), seq(k, k = 65 .. 90)], bytes);
  assign(n = iolib(6, SF, infinity), d = round(1/4*kernelopts(bytesalloc))), iolib(7, SF),
  `if`(L = R, [assign(p=Matrix(1..2, 1..32)), seq(assign('p[2, k]' = convert(A[k], symbol)),
  k = 1..32)], [assign(p = Matrix(1..2,1..26)), seq(assign('p[2, k]' = convert(B[k], symbol)),
    k = 1 .. 26)]);
  `if`(n <= d, assign(omega = iolib(4, SF, TEXT, infinity)), assign(omega = 6));
  for k to n do t := `if`(omega = 6, iolib(4, SF, 1), convert(omega[k], bytes));
    h := `if`(t = [32], "", convert(t, bytes));
    if L = R and search(A, h, 'v') then
        p[1, `if`(v <= 32, v, v - 32)]:= p[1, `if`(v <= 32, v, v - 32)] + 1; z := z + 1
    elif L = E and search(B, h, 'v') then
        p[1, `if`(v <= 26, v, v - 26)]:= p[1, `if`(v <= 26, v, v - 26)] + 1; z := z + 1
    end if
  end do;
  [assign(S = p, iolib(7, SF)), RETURN(PLOT(CURVES([seq([k, 100*p[1, k]/z],
    k = 1 .. `if`(L = R, 32, 26))]), COLOR(RGB, 0, 0, 0), AXESTICKS([seq(k = p[2, k],
    k = 1 .. `if`(L = R, 32, 26))], DEFAULT), AXESLABELS("", `Frequence in %`),
    FONT(TIMES, BOLD, 10), THICKNESS(2)))]
end proc
```

Typical examples of the procedure use:

> SA_text("C:\\Academy\\Books\\Maple6789.txt", E, S), interface(rtablesize=32); S;

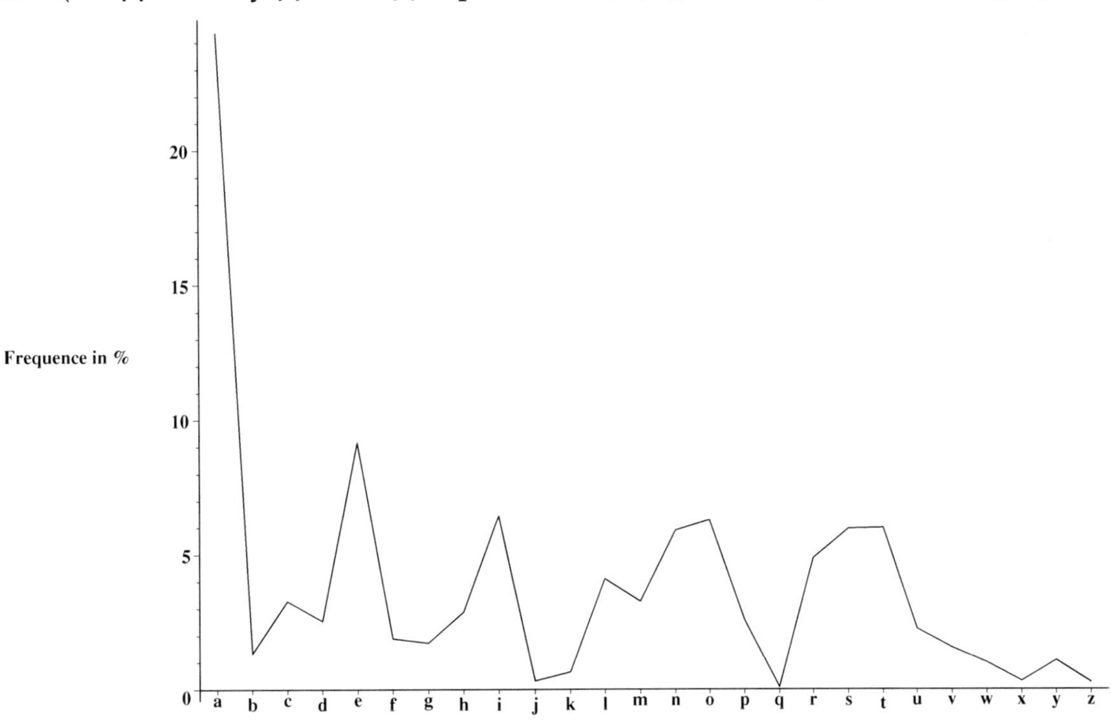

[4202 , 230 , 566 , 439 , 1581 , 326 , 298 , 496 , 1113 , 54 , 111 , 708 , 563 , 1017 , 1084 , 444 , 13 , 840 , 1028 , 1033 , 387 , 268 , 171 , 52 , 185 , 44]
[$a , b , c , d , e , f , g , h , i , j , k , l , m , n , o , p , q , r , s , t , u , v , w , x , y , z$]

DDT – an analysis of distribution of decimal digits of irrational numbers

Call format of the procedure:

DDT(C::*realnum*, n::*posint* {, 'P'})

Formal arguments of the procedure:

C - an irrational number
n - a positive integer
P - an assignable name

Description of the procedure:

For analysis of distribution of decimal digits of the irrational numbers, the **DDT(C, n {, 'P'})** procedure serves. The first actual **C** argument of the procedure defines an ***irrational*** number (for instance, **Pi, sqrt(2), ln(2), e,** etc.), the second actual **n** argument defines length of an analyzed sequence of decimal digits of irrational number after decimal point, while through third optional **P** argument a histogram of decimal digits of irrational number is returned. The procedure returns the table whose indices are decimal digits **0 .. 9** and entries are quantities of the corresponding digits composing fractional part of the analyzed *irrational* number.

```
DDT:= proc(C::realnum, n::posint)
local a, d, k, h, p, Tab;
  assign(Tab = table([seq(k = 0, k = 0 .. 9)])); Digits:= n + floor(C); h:= eval(C, n + floor(C));
  for k to n do a := NDP(h, k); Tab[a] := Tab[a] + 1 end do;
  eval(Tab), `if`(nargs = 3, assign(args[3] = stats[statplots, histogram]
    ([seq(Weight(k .. k + 1, Tab[k]), k = 0 .. 8)], xtickmarks = [p $ (p = 0 .. 9)],
    axesfont = [TIMES, BOLD, 10])), NULL)
end proc
```

Typical examples of the procedure use:

> Digits, DDT(Pi, 2004, 'P'), Digits; P;

10, table([0 = 182, 1 = 212, 2 = 207, 3 = 188, 4 = 196, 5 = 206, 6 = 201, 7 = 197, 8 = 202, 9 = 213]), 10

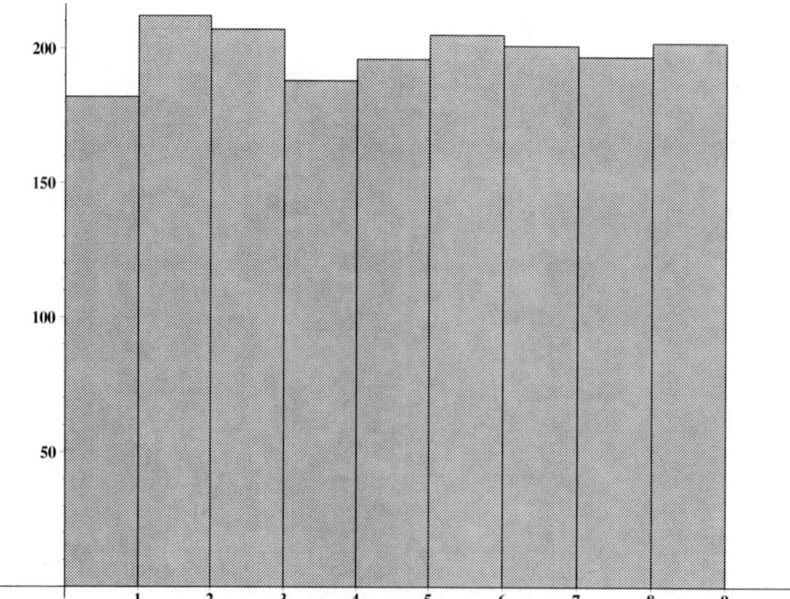

> Digits, Order, proc() Digits:=7: Order:=15: [Digits,Order] end proc(), Digits, Order;

10, 6, [7, 15], 10, 6

Incidentally, here quite pertinently to make one essential enough remark concerning redefinition of predetermined variables ***Digits*** and ***Order*** in a procedures body. As a result of such operation the area of its action is limited only to area of definition of procedure, i.e. up to procedure call and after an exit out of it the values of the given variables are identical, whereas in area of procedure definition of their value are being defined by the made redefinitions. The second example above very well illustrates the told.

Weights – tools of operation with the weighted data
Weights_L
Weight_LF

Call format of the procedures:

Weights(L)
Weights_L(L)
Weight_LF(F, R, M, Z)

Formal arguments of the procedures:

L - a list
F - path to a datafile with statistical data (may be *symbol* or *string*)
R - a range
M - an assignable name
Z - an assignable name

Description of the procedures:

The procedure call **Weights(L)** returns the 2-dimensional array whose first row contains the sorted set of elements of the given statistical list **L**, while its second row contains *weights* or *frequencies* corresponding to these elements. The procedure call **Weights_L(L)** returns the 2-element sequence for the given list **L** of statistical data. The first element of the sequence is similar to result of the procedure call **Weights(L)**, while the second element represents a statistical list corresponding to definition of the statistical data of the **\`stats\`** module of *Maple*.

The procedure call **Weight_LF(F, R, M, Z)** returns the statistical list in the terms of the *weighed* data for the given text file **F** of values of some observation and **R** interval of permissible values. The data in the **F** file must consist of integers or floating-point values, parted by symbols of carriage return and line feed (*Enter* key; hex code \`0D0A\`). It is necessary to note that at presence among the given data values of *rational*-type their numerators are chosen only. In addition, through third argument **M** the special matrix is returned, while through the fourth **Z** argument a list of the initial statistical data is returned. The first row of the special matrix of dimensionality (2xn) contains values of data elements, whereas the second row contains the *weights* corresponding to them.

```
Weights:= proc(L::list)
local k, p, G, R;
  assign(G = [op(sort({op(L)}))]), assign(R = array(sparse, 1 .. 2, 1 .. nops(G)));
  for k to nops(G) do R[1, k] := G[k];
    for p to nops(L) do  if G[k] = L[p] then R[2, k] := R[2, k] + 1 end if
    end do
  end do;
  evalm(R)
end proc

Weights_L:= proc(L::list)
```

```
local k, p, G, R;
  assign(G = [op(sort({op(L)}))]), assign(R = array(sparse, 1 .. 2, 1 .. nops(G)));
  for k to nops(G) do R[1, k] := G[k];
    for p to nops(L) do if G[k] = L[p] then R[2, k] := R[2, k] + 1 end if
    end do
  end do;
  evalm(R), [seq('Weights'(R[1, k], R[2, k]), k = 1 .. nops(G))]
end proc

Weight_LF:= proc(F::file, R::range({float, integer}), M::evaln, Z::evaln)
local k, p, G, T, L, Q;
  [assign(Q = readdata(F, float)), iolib(7, F), assign(Z = Q)];
  L := [seq(`if`(Q[k] <= rhs(R) and lhs(R) <= Q[k], Q[k], NULL), k = 1 .. nops(Q))];
  [assign('G' = [op(sort({op(L)}))]), assign(T = array(sparse, 1 .. 2, 1 .. nops(G)))];
  for k to nops(G) do  T[1, k] := G[k];
    for p to nops(L) do  if G[k] = L[p] then T[2, k] := T[2, k] + 1 end if
    end do
  end do;
  assign(M = convert(T, Matrix)), [seq(Weight(T[1, k], T[2, k]), k = 1 .. nops(G))]
end proc
```

Typical examples of the procedures use:

> H:= [3, 8, 3,2, 6, 3, 5, 1,7, 1, 4, 8,4, 6, 3, 1,3,2, 14, 1, 1, 1,0, 2, 6, 6, 8, 8, 14, 14, 8, 8, 15, 7, 7, 7, 8, 8, 8, 15, 15, 15, 7, 7, 7, 8, 8, 15, 15, 8, 7, 7, 7, 7, 8, 15, 15, 15, 7]: Weights(H);

$$\begin{bmatrix} 0 & 1 & 2 & 3 & 4 & 5 & 6 & 7 & 8 & 14 & 15 \\ 1 & 6 & 3 & 5 & 2 & 1 & 4 & 12 & 13 & 3 & 9 \end{bmatrix}$$

> Weights_L(H);

$$\begin{bmatrix} 0 & 1 & 2 & 3 & 4 & 5 & 6 & 7 & 8 & 14 & 15 \\ 1 & 6 & 3 & 5 & 2 & 1 & 4 & 12 & 13 & 3 & 9 \end{bmatrix}, [\text{Weights}(0, 1), \text{Weights}(1, 6), \text{Weights}(2, 3), \text{Weights}(3, 5),$$
$$\text{Weights}(4, 2), \text{Weights}(5, 1), \text{Weights}(6, 4), \text{Weights}(7, 12), \text{Weights}(8, 13), \text{Weights}(14, 3), \text{Weights}(15, 9)]$$

> G:=rand(1..62): F:="C:/Academy/Examples/RANS_IAN.dat": MkDir(F,1): for k to 5000 do writeline(F,convert(G(),string)) end do: close(F); Weight_LF(F,7..15,M,Z), M;

$$[\text{Weight}(10., 68), \text{Weight}(7., 90), \text{Weight}(14., 79), \text{Weight}(15., 75), \text{Weight}(8., 75), \text{Weight}(12., 94), \text{Weight}(9., 107),$$
$$\text{Weight}(11., 82), \text{Weight}(13., 72)], \begin{bmatrix} 10. & 7. & 14. & 15. & 8. & 12. & 9. & 11. & 13. \\ 68 & 90 & 79 & 75 & 75 & 94 & 107 & 82 & 72 \end{bmatrix}$$

To the certain extent, the **Weights** procedure generalizes two procedures ***statsort*** and ***frequency*** from submodule **transform** of the module **stats.** At last, at operation with statistical data (*in terms of the* **stats** *module*) the procedures **Weights_L** and **Weight_LF** can be useful also. They have useful appendices in a simple statistic analysis.

MA – calculation of the moving averages
MAM

Call format of the procedures:

MA(n, L)
MAM(L, n {, p})

Formal arguments of the procedures:

n - an integer
L - a list or a text datafile of statistical data
p - (optional) a positive integer (*posint*)

Description of the procedures:

The procedure call **MA(n, L)** returns the list of all permissible moving averages with length **n** of a sliding segment for the given list **L** of statistical data. Whereas the procedure call **MAM(L, n)** returns a list of all permissible *moving averages* with length **n** of a sliding segment for the given list **L** of statistical data. If the third optional argument **p** has been coded, the result is being returned with **p** significant digits.

The data in the **L** file should consist of integers or floating-point values, parted by symbols of carriage return and line feed (*Enter* key; hex code `0D0A`). It is necessary to note that at presence among the given data values of *rational* type their numerators are chosen only. It is conditioned by usage in procedure **MAM** of standard procedure ***readdata***. The **MAM1(L, n)** procedure basing on our **readata1** procedure avoids this disadvantage.

```
MA:= proc(n::integer, L::{list, file})
local a, k, R, Tb_MA;
  `if`(type(L, list), assign(a = L), assign(a = readdata(L, 1))),
    assign(R = a, k = 0, Tb_MA = table());
  while k*(n - 1) <= nops(a) - n do k := k + 1;
    R := stats[transform, moving[n, mean]](R);  Tb_MA[k] := R
  end do;
  Tb_MA['k'] $ ('k' = 1 .. k)
end proc

MAM:= proc(L::{list, file}, n::posint)
local a, b, k, p;
  `if`(type(L, list), assign(b = L), assign(b = readdata(L, 1)));
  assign(a = []), [seq(assign('a' = [op(a), sum(b[p], p = k..k+n-1)/n]), k = 1..nops(b) - n+1),
    `if`(nargs = 3 and type(args[3], posint) and 2 <= args[3], evalf(op(a), args[3]), op(a))]
end proc
```

Typical examples of the procedures use:

> A:=[3,8,3,2,6,3,5, 1,7,1, 4, 8,4, 6, 3,1,3,2,14,1, 1,1, 0,2,6,6,8,8,14,14,15,15,14,7,8,7,8,15]:
MA(14, A); MAM(A, 3);

$$\left[\frac{61}{14}, \frac{61}{14}, \frac{27}{7}, \frac{27}{7}, \frac{27}{7}, \frac{31}{7}, \frac{30}{7}, 4, 4, \frac{7}{2}, \frac{25}{7}, \frac{26}{7}, \frac{25}{7}, \frac{27}{7}, 4, \frac{67}{14}, \frac{40}{7}, \frac{46}{7}, \frac{15}{2}, \frac{15}{2}, \frac{111}{14}, \frac{59}{7}, \frac{62}{7}, \frac{66}{7}, \frac{145}{14}\right],$$
$$\left[\frac{773}{196}, \frac{192}{49}, \frac{387}{98}, \frac{200}{49}, \frac{419}{98}, \frac{127}{28}, \frac{233}{49}, \frac{983}{196}, \frac{1045}{196}, \frac{159}{28}, \frac{299}{49}, \frac{1291}{196}\right]$$

$$\left[\frac{14}{3}, \frac{13}{3}, \frac{11}{3}, \frac{11}{3}, \frac{14}{3}, 3, \frac{13}{3}, 3, 4, \frac{13}{3}, \frac{16}{3}, 6, \frac{13}{3}, \frac{10}{3}, \frac{7}{3}, 2, \frac{19}{3}, \frac{17}{3}, \frac{16}{3}, 1, \frac{2}{3}, 1, \frac{8}{3}, \frac{14}{3}, \frac{20}{3}, \frac{22}{3}, 10, 12, \frac{43}{3}, \frac{44}{3}, \frac{44}{3}, 12, \frac{29}{3}, \frac{22}{3}, \frac{23}{3}, 10\right]$$

> AVZ:= [42, 61, 47, 56, 67, 36, 62, 40, 89, 14, 96, 7]: MAM(AVZ, 4, 6);

[51.5000, 57.7500, 51.5000, 55.2500, 51.2500, 56.7500, 51.2500, 59.7500, 51.5000]

```
> MA(3, "C:\\Academy\\Examples\\Sample2.dat");
```

[73.6666667, 74.6666667, 81.3333333, 89.3333333, 90.6666667, 87.3333333, 87.0000000, 82.3333333, 77.3333333], [76.5555555, 81.7777778, 87.1111111, 89.1111111, 88.3333333, 85.5555555, 82.2222222], [81.8148148, 86.0000000, 88.185185, 87.6666666, 85.3703704], [85.3333333, 87.2839506, 87.0740741], [86.5637860]

```
> MAM("C:\\Academy\\Examples\\Sample2.dat", 3);
```

[73.66666667, 74.66666667, 81.33333333, 89.33333333, 90.66666667, 87.33333333, 87.00000000, 82.33333333, 77.33333333]

```
> MAM1("C:\\Temp\\Common\\Mws6789\\Data.txt", 3);
```

[46.20000000, 28.80000000, 3.091666667, 1.836111111, 1.169444444, 8.544444443]

CDiag – visualization of statistical pie charts

Call format of the procedure:

CDiag(Pr, Cl, Obj)

Formal arguments of the procedure:

Pr – a sequence of percents
Cl – a sequence of colors
Obj – legends of objects

Description of the procedure:

The **CDiag** procedure outputs the *pie chart* on the basis of the given statistical data. As formal arguments of the **CDiag(Pr, Cl, Obj)** procedure the percents (**Pr**), colourings (**Cl**) and legends (**Obj**) of the represented objects appear. The total sum of percents should not exceed 100 and the order of coding of actual arguments should correspond to the mentioned above; in addition, each percent value should match a coloring and one legend of an object. As the values for a coloring, the names of colors, which are admitted by color-option of the standard plot-function, should be used. An erratic coding of the actual arguments at procedure call causes an appropriate erratic situation. The procedure appears as a rather useful tool both for quick output of pie chart and for use in combination with other procedures dealing with statistical data processing.

```
CDiag:= proc()
local a, n, k, L, E;
  assign(L = [], E = (x -> ERROR("arguments are wrong %1", x)), a = [args]),
    `if`(nargs = 0, E(a), assign(n = 1/3*nargs)), `if`(type(n, integer) <> true, E(a),
    `if`(100 < `+`(args[k] $ (k = 1 .. n)), E(a), [with(plots, display, textplot),
    with(plottools, disk, pieslice)]));
  for k to n do L := [op(L), pieslice([0, 0], 4, `if`(k = 1, 0, sum(1/50*Pi*args[p],
     p = 1 .. k - 1)) .. sum(1/50*Pi*args[p], p = 1 .. k), color = args[n + k]),
    disk([4.9, 0.5*(-1)^k*k], 0.3, color = args[n + k]), textplot([5.5, 0.5*(-1)^k*k,
    cat(``, ` - `, args[k], `% - `, args[2*n + k])], align = RIGHT, color = args[n + k])]
  end do;
  display(L, scaling=constrained, axes=none, titlefont=[HELVETICA, BOLDOBLIQUE,
    14], font = [TIMES, BOLD, 11], title = "Pie Chart", thickness = 2)
end proc
```

Typical examples of the procedure use:

> CDiag(35, 20, 10, 10, 10, 15, khaki, red, magenta, green, blue, brown, Tallinn, Moscow, Minsk, Vilnius, Gomel, Grodno);

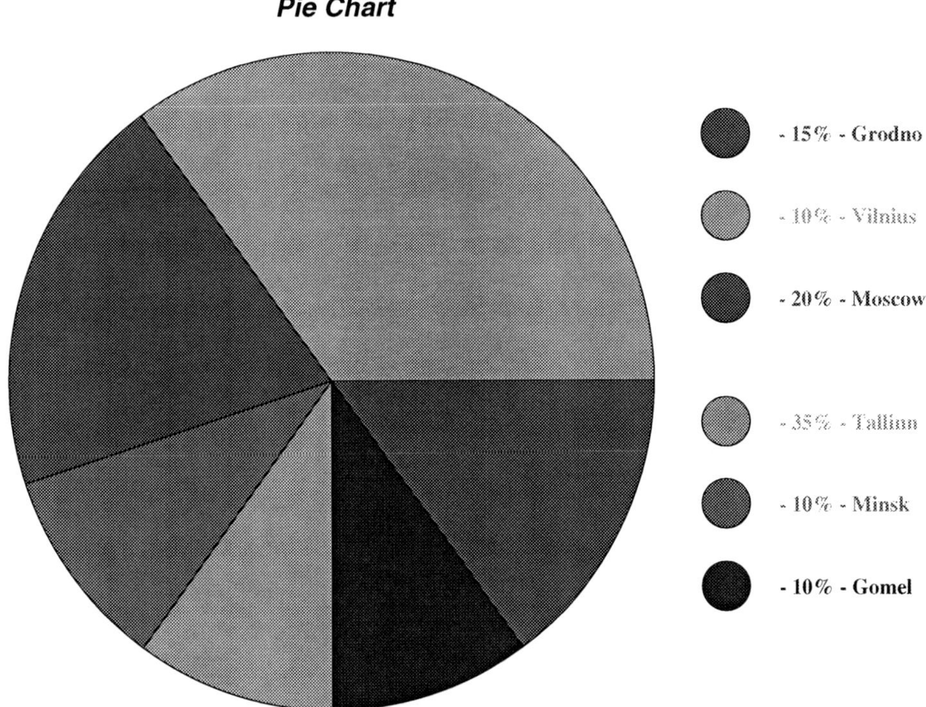

The procedures represented above, can be useful enough at solving problems dealing with simple processing of statistical data of the different nature and purpose.

13.2. Software for solving problems of regression analysis

For smoothing of statistical data and creation of regression models of two types (*linear* - **Y=a*X+b** and *nonlinear* - **Y=a*X^2+b*X+c**) in statistics is widely used the well-known *least squares method.* With the purpose of some generalization of solution of task of creation of linear and nonlinear (*quadratic*) models of regression for the given resultant and factor variables, and also for calculation of *coefficients of correlation* (**CC**) and *correlation ratio* (**CR**) with graphical representation in an united coordinate system of initial data and model of regression in environment of the *Maple* package of releases **6 - 9/9.5** the special **LRM_NRM** procedure has been created. The procedure appears as rather useful tool for quick analysis of statistical data correlation.

LRM_NRM - one-factor regression models

Call format of the procedure:

LRM_NRM(A, B, T, P, CR)

Formal arguments of the procedure:

A - a list, vector or text datafile (a *resultant* variable)
B - a list, vector or text datafile (a *factor* variable)
T - a type of sought model of regression (may be ***LRM*** or ***NRM***)
CR - an assignable name
P - an assignable name

Description of the procedure:

For smoothing of statistical data and building of the above regression models, the special procedure **LRM_NRM(A,B,T,P,CR)** with five formal arguments has been created. The formal arguments of the **LRM_NRM** procedure have the following sense. The first and second actual **A** and **B** arguments define lists, vectors or text files of data of *resultant variable* and a *factor variable* accordingly. The third actual **T** argument defines a type of sought regression model (*LRM* – **y=a*x+b**, *NRM* – **y=a*x^2+b*x+c**). At last, through fourth and fifth actual arguments **P** and **CR** a graph of distribution of the initial statistical data and *regression curve* (i.e. the *regression model*), and *correlation ratio* are returned accordingly.

The data in files **A** and **B** should consist of integers or floating-point values, parted by symbols of carriage return and line feed (*Enter* key; hex code `0D0A`). It is necessary to note that at presence among the given data values of the *rational* type their numerators are chosen only. In addition, if length of one of the samples is less than 3, the erratic situation arises, whereas at detection of samples with different lengths their leveling by the minimal length is done with output of appropriate warning.

The procedure **LRM_NRM** directly returns the sought model of one-factor regression. The following fragment represents an example of procedure use for solution of a task of creation of linear and nonlinear models of regression for resultant **U**-variable and factor **A**-variable whose values have been chosen in accordance with annual domestic and foreign citing of our scientific works in the *Cellular Automata* theory and computer science [1-3].

```
LRM_NRM:=proc(A::{vector,file,list},B::{vector,file,list},T::name, P::evaln, CR::evaln)
local a, b, k, m, n, p, L, M, N, r, omega, o, G, X, Y, P1, P2, F, sr, ds, v;
  `if`(type(A, list), assign(a = A), `if`(type(A, vector), assign(a = convert(A, list)),
    assign(a = readdata(A, 1))));
  `if`(type(B, list), assign(b = B), `if`(type(B, vector), assign(b = convert(B, list)),
    assign(b = readdata(B, 1))));
  assign(o = {nops(b), nops(a)}), `if`(o[1] < 3, ERROR("One of samples is too small"),
    [assign('a' = a[1 .. o[1]], 'b' = b[1 .. o[1]]), `if`(1 < nops(o), WARNING("Lengths of
    samples is different, a levelling by minimal length has been done"), NULL)]);
  `if`(nops(a) <> nops(b), ERROR("different dimensionalities of lists of source statistical
    data"), assign(omega = [seq(1, p = 1 .. nops(a))], G = [], F = [TIMES, BOLD, 9]));
  assign(v = (() -> op([assign('L' = []), seq(assign('L' = [op(L), product(args[m][p],
    m = 1 .. nargs)]), p = 1 .. nops(args[1])), L])));
  assign(sr = (() -> `+`(args)/nargs), ds = (() -> `+`(seq((args[k] - sr(args))^2,
    k = 1 .. nargs))/nargs)), `if`(T = LRM, assign('M' = Matrix(2, 2, [[sr(op(v(b, b))),
    sr(op(b))], [sr(op(b)), 1]]), 'N' = Vector(2, [sr(op(v(a, b))), sr(op(a))])), `if`(T = NRM,
    assign('M' = Matrix(3, 3, [[sr(op(v(b, b, b, b))), sr(op(v(b, b, b))), sr(op(v(b, b)))],
    [sr(op(v(b, b, b))), sr(op(v(b, b))), sr(op(b))], [sr(op(v(b, b))), sr(op(b)), 1]]),
    'N' = Vector(3, [sr(op(v(a, b, b))), sr(op(v(a, b))), sr(op(a))])),
    ERROR("regression model type <%1> is inadmissible", T)));
  assign('r' = convert(evalf(LinearAlgebra:-LinearSolve(M, N)), list));
  assign(Y = `if`(nops(r) <> 1,
    proc(X) local n; sum(r[n]*X^(nops(r) - n), n = 1..nops(r)) end proc,
    proc(X) sign(op(r))*X end proc)), `if`(nops(r) = 1, assign('CR' = sign(op(r))), NULL);
  `if`(nops(r) = 2, assign('CR' = evalf(sqrt((ds(op(a)) - sr(op(v(a - r[1]*b - r[2]*omega,
    a - r[1]*b - r[2]*omega))))/ds(op(a))))), `if`(nops(r) = 3, assign('CR' =
    evalf(sqrt((ds(op(a)) - sr(op(v(a - r[1]*v(b, b) - r[2]*b - r[3]*omega, a - r[1]*v(b, b) -
    r[2]*b - r[3]*omega))))/ds(op(a))))), NULL));
```

```
  seq(assign('G' = [op(G), [b[n], a[n]]]), n = 1 .. nops(a));
  P1 := plots[pointplot](G, color = blue, thickness = 3, symbol = CIRCLE);
  P2 := plot(Y(X), X = b[1] .. b[-1], thickness = 2);
  assign(P = plots[display]([P1, P2], axesfont = F, scaling = UNCONSTRAINED, labels =
  [convert("B,X", name), convert("A,Y", name)], labelfont = F)), RETURN(Y(X))
end proc
```

Typical examples of the procedure use:

> U:= [0,0,5,7,12,13,14,14,15,18,27,24,22,29,42,48,56,64,68,68,70,65,58,55,60,63,67,75,95]: A:= [33,38,42,76,94,96,98,102,125,145,174,186,205,220,222,236,240,266,284,292,304,314, 316, 320,336,345,384,395,405]: LRM_NRM(U, A, NRM, *Figure*, CR), CR; *Figure;*

-8.714233847 + 0.00003166933948*X^2 + 0.2148197209*X, 0.9582099422

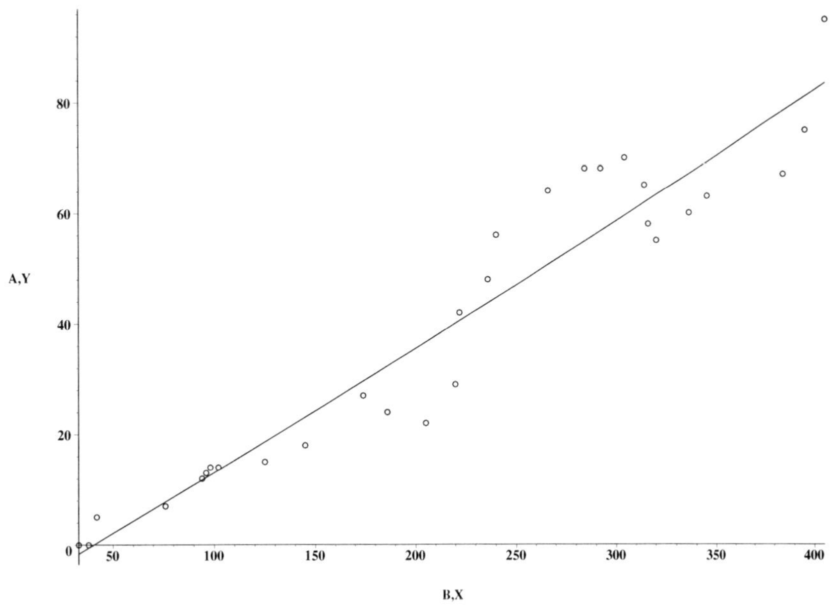

> LRM_NRM("c:/academy/examples/members.dat", "c:/academy/examples/age.dat", NRM, Figure, CR), CR; Figure;

5.554988480 - 0.001080042656*X^2 - 0.004403992540*X, 0.9425931978

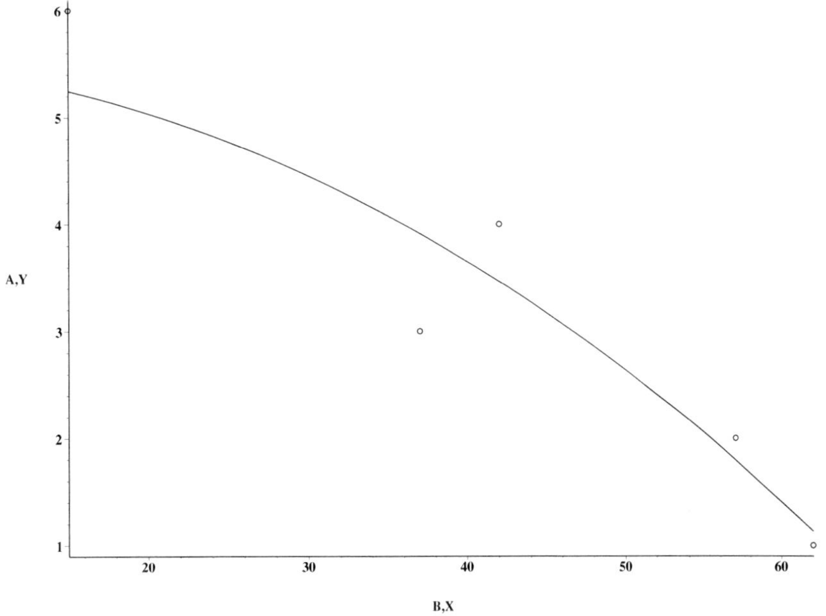

As a result of execution of the **LRM_NRM** procedure with given values for its formal arguments the value CR(U,A)=0.9582099422 for the correlation ratio has been obtained. Through actual *Figure* argument of the procedure the united graph of sought quadratic model of regression and distribution of points of the initial statistical data (U, A) is returned. The obtained values of correlation ratio on the basis of both *LRM*, and *NRM* speak about the presence of a close enough connection between the above variables U and A of the observation.

LSF – polynomial data smoothing by means of the least squares method

Call format of the procedure:

LSF(A, x, n, t, G {, H})

Formal arguments of the procedure:

A - an (2xn) *Array* with numeric elements (the *smoothed data*)
x - a name (a *leading variable* of sought polynomial)
n - a positive integer **> 2**
t - precision of calculations (a positive integer - *posint*)
G - an assignable name
H - (optional) an assignable name

Description of the procedure:

For polynomial data smoothing by means of the method of least squares the **LSF(A, x, n, t, G)** procedure can be useful, which outputs the joint graph of a sought smoothing polynomial and allocation of points of the source data given by numerical Array **A** of dimensionality (2xn). While via **G** argument the sought smoothing polynomial of degree (**n-1**), calculated with **t** precision, from **x** variable is returned. The first and second row of the **A** Array define values of factor variable and resultant variable accordingly. The data with any semantic value can appear as elements of the array **A.** The **LSF** procedure allows to obtain the *smoothing polynomials* of dimensionality not above than **d = rhs([ArrayDims(A)][2] - 1.**

The returned *smoothing polynomial* **G** does not support direct calculation of its values in the required points. For these purposes, for example, the construction **subs(x=b, G)** can be used, where **b** - a required point. However, coding the optional sixth **H** argument, through it is being returned a procedure that allows to directly calculate values of the smoothing polynomial **G.** The **LSF** procedure has a series of useful appendices in data analysis, above all, for a smoothing of data, providing their subsequent analysis by algebraic methods.

```
LSF:= proc(A::Array(numeric), x::symbol, n::posint, t::posint, G::evaln)
local a, p, k, j, v, h, b, z, r;
  assign(h = rhs([ArrayDims(A)][2]), a = [seq(cat(a, k), k = 1 .. n)]),
    `if`(2 <= n and n <= h, assign(z = (x -> sum(a[k]*x^(n - k), k = 1 .. n))),
    ERROR("3rd argument %1 is invalid", n));
  b := plots[pointplot]([seq([A[1, v], A[2, v]], v = 1..h)], symbol =circle, symbolsize =14);
  G := evalf(subs(fsolve({seq(diff(expand(add((A[2, j] - z(A[1, j]))^2, j = 1 .. h)),
    a[p]) = 0, p = 1 .. n)}), z(x)), t);
  if nargs = 6 and type(args[6], symbol) then assign(args[6] = (t -> subs(x = t, eval(G))))
  end if;
  plots[display](b, plot(G(x), x = A[1, 1] .. A[1, h]), thickness = 2, axesfont = [TIMES,
    BOLD, 9])
end proc
```

Typical examples of the procedure use:

> Art:=Array(1..2,1..6, [[1, 2, 3, 4, 5, 6], [62, 57, 37, 41, 8, 15]]): LSF(Art, y, 6, 8, Kr, H); Kr;

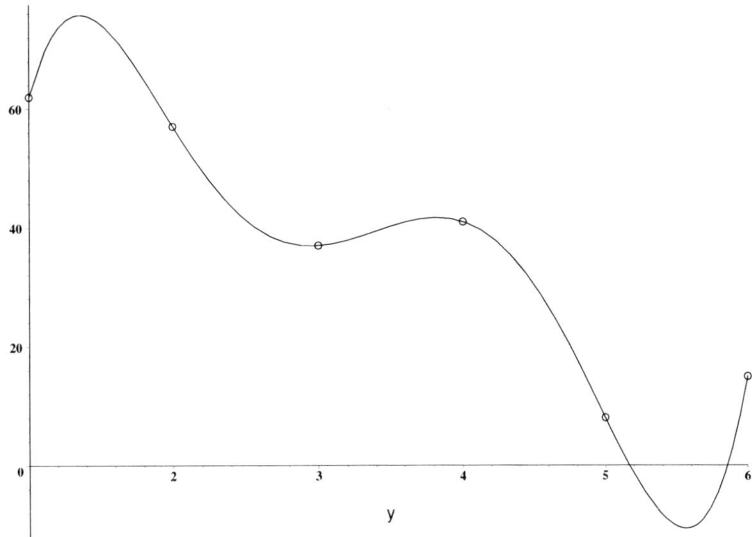

1.98333*y^5 - 33.91667*y^4 + 216.75000*y^3 - 638.58333*y^2 + 840.76667*y - 325.

> evalf(seq(H(p), p = [1, 2, 3, 4, 5, 6, 7, 8, 9, 10]), 7);

62.0001, 57.000, 37.000, 41.002, 8.004, 15.000, 514.997, 2574.964, 8113.950, 20140.94

The procedures of this item can be useful at problem solving of regression analysis and data smoothing.

13.3. Software for testing of statistical hypotheses

Testing of reliability of hypotheses plays an important part in problems of the statistical analysis. Here a series of useful procedures that solve the given problem of check of so-called null hypothesis (**Ho**-*hypothesis*) on the basis of both *parametrical* tests, and *nonparametric* ones is represented.

F_test_Ds – the Fisher F-test of check of the Ho-hypothesis

Call format of the procedure:

F_test_Ds(L1, L2, n)

Formal arguments of the procedure:

L1 - a list or text datafile of statistical data (*small sample*)
L2 - a list or text datafile of statistical data (*small sample*)
n - a positive integer (*accuracy of calculations*)

Description of the procedure:

With the purpose of simplification of estimation of *reliability* of the **Ho**-*hypothesis* concerning *equality of variances* of populations on the basis of the *Fisher* **F**-test a rather simple procedure **F_test_Ds(L1, L2, n)** which returns the list of kind [k1, k2, Ff], has been created. As formal arguments of the procedure the lists or text datafiles **L1** and **L2** of values of elements of researched samples and the given accuracy **n** of calculations appear. The given procedure returns the list with three elements: **(1)** quantity of degrees of freedom **k1** for the *greater* variance, **(2)** quantity of degrees of freedom **k2** for the *smaller* variance and **(3)** **Ff**-value calculated with **n**-accuracy.

The data in the text files should consist of integers or floating-point values, parted by symbols of carriage return and line feed (*Enter* key; hex code `0D0A`). It is necessary to note that at presence among the given data values of the *rational* type their numerators are chosen only. In addition, if length of one of the samples is less than 3, an appropriate erratic situation arises.

For illustration of use of the **F_test_Ds(L1,L2,n)** procedure created on the basis of the *Fisher* **F**-test by means of the built-in ***rand***-generator of pseudorandom numbers, two small samples **L1** and **L2** of sizes **n=20** and **n=22** accordingly were created. The procedure call **F_test_Ds(L1,L2,n),** used with respect to the given samples, returns the list [k1=22, k2=20, Ff=1.082], defining the quantities of degrees of freedom k1=22, k2=20 and a value Ff=1.082. Then, on the basis of the table (*see statistical tables, for example* [49,50]) of critical points of the **F**-criterion for a confidence level **a=5%** (**a=1%**) and for quantities of degrees of freedom k1=22 and k2=20 we easily obtain values Fst=2.07 (Fst=2.83) of the critical points accordingly. Since for both significance levels takes place the relation **Ff << Fst,** the **Ho**-*hypothesis* can be accepted on a high significance level and distinction between variances of both researched samples can be considered random.

The procedure **F_test_Ds** can be enough easily updated by means of implantation into its body of appropriate reference to the table (preliminary saved in a file, for example, of *m*-format) of critical points of the **F**-criterion; that will allow to obtain the answers in terms of the *true* value (**Ho**-*hypothesis* is accepted) and the *false* value (**Ho**-*hypothesis* is rejected).

```
F_test_Ds:= proc(L1::{list, file}, L2::{list, file}, n::integer)
local a, b, d, c, p, k1, k2, R, L, Sr, Ds, Ff, x, y, o;
  `if`(type(L1, list), assign(x = L1), [assign(x = readdata(L1, 1))]);
  `if`(type(L2, list), assign(y = L2), [assign(y = readdata(L2, 1))]);
  assign(o = {nops(y), nops(x)}), `if`(o[1] < 3,
    ERROR("One of samples is too small"), NULL);
  assign67(Sr = (() -> `+`(args)/nargs), a = op(x), b = op(y), d = nops(x), c = nops(y));
  Ds:= () -> sum((args[p] - Sr(args))^2, p = 1 .. nargs)/(nargs - 1);
  R := `if`(Ds(b) <= Ds(a), [k1 = d, k2 = c, Ff = Ds(a)/Ds(b)],
    [k1 = c, k2 = d, Ff = Ds(b)/Ds(a)]); RETURN([R[1], R[2], evalf(R[3], n)])
end proc
```

Typical examples of the procedure use:

```
> L1, L2:= [], []: Kr:=rand(42..99): Art:=rand(47..99): seq(assign('L1'=[op(L1),Kr()]),
  j=1..20); seq(assign('L2'= [op(L2), Art()]), j=1..22); L1, L2; F_test_Ds(L1, L2, 4);
    [58, 78, 75, 70, 83, 96, 71, 85, 59, 75, 51, 49, 70, 79, 46, 49, 87, 84, 67, 95], [86, 62, 73, 89, 82, 97, 93,
    72, 96, 79, 57, 70, 85, 92, 72, 48, 95, 50, 89, 53, 57, 79]
                              [k1 = 22, k2 = 20, Ff = 1.082]
> F_test_Ds(L1, "C:\\Academy\\Examples/Sample2.dat", 4);
                              [k1 = 20, k2 = 22, Ff = 1.066]
```

T_test_AV – the Student t-criterion of check of the Ho-hypothesis

Call format of the procedure:

T_test_AV(L1, L2, n)

Formal arguments of the procedure:

L1 – a list or text datafile of statistical data (*small sample*)
L2 – a list or text datafile of statistical data (*small sample*)
n – a positive integer (*accuracy of calculations*)

Description of the procedure:

With the purpose of simplification of estimation of reliability of the **Ho**-*hypothesis*, concerning the *equality* of *averages* of populations a rather simple **T_test_AV(L1, L2, n)** procedure that returns the list of kind [k, tf], has been created. As formal arguments of the procedure the lists or text datafiles **L1** and **L2** of values of elements of researched samples and the given accuracy **n** of calculations appear. The given procedure returns the list with only two elements, namely: (**1**) quantity of degrees of freedom **k** and (**2**) **tf**-value calculated with **n**-accuracy.

The data in the text files should consist of integers or floating-point values, parted by symbols of carriage return and line feed (*Enter* key; hex code `0D0A`). It is necessary to note that at presence among the given data values of the *rational* type their numerators are chosen only. In addition, if length of one of the samples is less than **3**, an appropriate erratic situation arises.

For illustration of use of the **T_test_AV(L1,L2,n)** procedure created on the basis of the *Student* **t**-criterion, by means of the built-in ***rand***-generator of pseudorandom numbers two small samples **L1** and **L2** of sizes **n=20** and **n=22** accordingly were created. The procedure call **T_test_AV(L1, L2, n),** used with respect to the given samples, returns the list [k=40, tf=0.9147], defining quantity of degrees of freedom k=40 and a value **tf**=0.9147. Then, on the basis of the table (*see statistical tables, for example* [49,50]) of critical points of the *Student* **t**-criterion for a confidence level **a=0.1%** and for quantity of degrees of freedom k=40 we easily obtain a value **tst=3.55** of a critical (*standard*) point. Inasmuch as for both significance levels takes place the relation **tf << tst,** the **Ho**-*hypothesis* can be accepted on high significance level (*confidence level* P=0.999) and distinction between averages of both researched samples can be considered random.

The procedure **T_test_AV** can be enough easily updated by means of implanting into its body of appropriate reference to the table (preliminary saved in a datafile, for example, of *m*-format) of critical points of the *Student* **t**-criterion; that will allow to obtain the answers in terms of the *true* value (**Ho**-*hypothesis* is accepted) and the *false* value (**Ho**-*hypothesis* is rejected). The above statistical table can be implemented as *Maple* table or array, or *NAG* matrix.

```
T_test_AV:= proc(L1::{list, file}, L2::{list, file}, n::integer)
local a, b, p, h, k, L, H, Sr, Ds, Sd, tf, x, y, o;
  `if`(type(L1, list), assign(x = L1), [assign(x = readdata(L1, 1))]);
  `if`(type(L2, list), assign(y = L2), [assign(y = readdata(L2, 1))]);
  assign(o = {nops(y), nops(x)}), `if`(o[1] < 3,
     ERROR("One of samples is too small"), NULL);
  assign(Sr = (a -> `+`(op(a))/nops(a)), a = nops(x), b = nops(y));
  Ds:= (c, d) -> sum((c[p] - Sr(c))^2, p =1 .. nops(c)) + sum((d[h] - Sr(d))^2, h =1 .. nops(d));
  if a = b then Sd := sqrt(Ds(x, y)*a/(a - 1))
  else Sd := sqrt(Ds(x, y)*(a + b)/((a + b - 2)*a*b))
  end if;
  [k = a + b - 2, tf = abs(evalf((Sr(x) - Sr(y))/Sd, n))]
end proc
```

Typical examples of the procedure use:

```
> L1, L2:=[], []: Kr:=rand(42..99): Art:=rand(47..99): seq(assign('L1'=[op(L1),Kr()]),
  j=1..20): seq(assign('L2'=[op(L2), Art()]),j=1..22): T_test_AV(L1, L2, 4);
                              [k = 40, tf = 0.9147]
> T_test_AV(L1, "C:\\Academy\\Examples/Sample2.dat", 4);
                              [k = 40, tf = 1.784]
```

Two procedures represented above, were based on so-called *parametric* tests of check of the **Ho**-*hypothesis*, while the next two procedures are based on the *nonparametric* criteria of *Van der Waerden* and *Mann-Whitney* [13,14,34,35,42].

U_test_MW – the Mann-Whitney U-criterion of check of the Ho-hypothesis

Call format of the procedure:

U_test_MW(L1, L2, n)

Formal arguments of the procedure:

L1 – a list or text datafile of statistical data (*small sample*)
L2 – a list or text datafile of statistical data (*small sample*)
n – a positive integer (*accuracy of calculations*)

Description of the procedure:

With the purpose of simplification of estimation of reliability of the **Ho**-*hypothesis* by means of the nonparametric criterion of *Mann-Whitney* a rather simple procedure **U_test_MW(L1, L2, n),** which returns the three-element list of kind [n1, n2, Uf], has been created. As formal arguments of the procedure the lists or text datafiles **L1** and **L2** of values of elements of researched samples and the given accuracy **n** of calculations appear. The given procedure returns a list with three elements, namely: **(1)** quantity of elements of the first sample **L1, (2)** quantity of elements of the second sample **L2,** and **(3)** actual **Uf**-value of the **U**-criterion calculated with **n**-accuracy.

The data in the text files should consist of integers or floating-point values, parted by symbols of carriage return and line feed (*Enter* key; hex code `0D0A`). It is necessary to note that at presence among the given data values of the *rational* type their numerators are chosen only. In addition, if length of one of the samples is less than 3, an appropriate erratic situation arises.

For illustration of use of the **U_test_MW(L1, L2, n)** procedure, created on the basis of the *Mann-Whitney* **U**-criterion, two small samples **L1** and **L2** of sizes n1=9 and n2=11 accordingly are chosen. The procedure call **U_test_MW(L1, L2, 5),** used with respect to the given samples, returns list [n1=9, n2=11, Uf=29.], defining quantities of elements of the samples and actual **Uf**-value for the **U**-criterion accordingly. Then, on the basis of the table (*see statistical tables, for example* [49,50]) of critical points of the **U**-criterion we easily obtain a value **Ust=19** for numbers n1=9, n2=11 and for confidence level **1%.** However, since the relation **Uf=29 >> Ust=19** takes place, the **Ho**-*hypothesis* can be accepted with the above significance level – **1%**, i.e. distinction between the compared samples can be considered random.

The procedure **U_test_MW** can be enough easily updated by means of implanting into its body of appropriate reference to the table (preliminary saved in a datafile, for example, of *m*-format) of critical points of the **U**-criterion; that will allow to obtain the answers in terms of the *true* value (**Ho**-*hypothesis* is accepted) and the *false* value (**Ho**-*hypothesis* is rejected).

```
U_test_MW:= proc(L1::{list, file}, L2::{list, file}, n)
local S, n1, n2, G, Srt, k, Z, r1, r2, z, x, y, N, L, a, b, o;
  `if`(type(L1, list), assign(a = L1), [assign(a = readdata(L1, 1))]);
  `if`(type(L2, list), assign(b = L2), [assign(b = readdata(L2, 1))]);
  assign(o = {nops(b), nops(a)}), `if`(o[1] < 3,
    ERROR("One of samples is too small"), NULL);
  assign(G = (z -> `if`(whattype(z) = `*`, op(1, z), z)), L = [seq(1, k = 1 .. nops(a))]);
  Srt:= (x, y) -> `if`(G(x) <= G(y), true, false); seq(assign('L'[k] = a[k]*S), k =1..nops(a));
```

```
  assign(r1 = 0, r2 = 0, Z = sort([op(L), op(b)], Srt));
  for k to nops(Z) do
    if whattype(Z[k]) = `*` then r1 := r1 + k else r2 := r2 + k end if
  end do;
  assign(n1 = 1/2*nops(a)*(nops(a) + 1), n2 = 1/2*nops(b)*(nops(b) + 1));
  RETURN([nops(a), nops(b), evalf(min(r1 - n1, r2 - n2), n)])
end proc
```

Typical examples of the procedure use:

> L1:= [64,68,70,72,75,76,79,80,83]: L2:= [60,60,62,66,68,69,70,71,73,78, 80]: U_test_MW(L1, L2, 5);

[9, 11, **29.**]

> U_test_MW(L1, "C:\\Academy\\Examples/Sample2.dat", 4);

[9, 11, **31.**]

X_test_VW – the Van der Waerden X-criterion of check of the Ho-hypothesis

Call format of the procedure:

X_test_VW(L1, L2, n)

Formal arguments of the procedure:

L1 - a list or text datafile of statistical data (*small sample*)
L2 - a list or text datafile of statistical data (*small sample*)
n - a positive integer (*accuracy of calculations*)

Description of the procedure:

With the purpose of simplification of estimation of reliability of the **Ho**-*hypothesis* by means of the nonparametric **X**-criterion of *Van der Waerden* a rather simple **X_test_VW(L1,L2,n)** procedure which returns the two-element sequence of kind **N, [R/(N+1)]**, has been created. As formal arguments of the procedure the lists or text datafiles **L1** and **L2** of values of elements of researched samples and the given accuracy **n** of calculations appear. The given procedure returns two-element sequence, namely: **(1)** total quantity of elements of the compared samples, and **(2)** a list of values **R/(N+1)** calculated with **n**-accuracy.

The data in the text files should consist of integers or floating-point values, parted by symbols of carriage return and line feed (*Enter* key; hex code `0D0A`). It is necessary to note that at presence among the given data values of the *rational* type their numerators are chosen only. In addition, if length of one of the samples is less than 3, an appropriate erratic situation arises. Indeed, statistical samples of less length do not need to be considered.

For illustration of use of the **X_test_VW(L1, L2, n)** procedure created on the basis of the **X**-criterion of *Van der Waerden*, two small samples **L1** and **L2** are used. The procedure call **X_test_VW(L1, L2,3)**, used with respect to the given samples, returns a sequence defining total quantity of elements of the compared samples and also a list of values **R/(N+1)**, namely:

N = 20, [0.190, 0.286, 0.429, 0.571, 0.667, 0.714, 0.810, 0.857, 0.952]

Then, on the basis of the table of values of ψ-function we compute the values **ψ[R/(N+1)]** for each element of list returned by procedure **X_test_VW** as the second element of the sequence, namely:

ψ(0.190)=-0.88, ψ(0.286)=-0.57, ψ(0.429)=-0.18, ψ(0.571)=+0.18, ψ(0.667)=+0.43, ψ(0.714)=+0.57, ψ(0.810)=+0.88, ψ(0.857)=+1.07, ψ(0.952)=1.66

By summarizing the obtained results (*with taking into consideration of their signs*), we obtain a value **Xf = Σ $_{\psi}$[R/(N+1)]=3.16.** Then, for number **N=n1+n2=9+11=20** and for confidence level **5%** with taking into consideration of difference **n1-n2=11-9=2** in the special table (*for example* [49,50]) of critical points **Xst** for the **X**-criterion we easily obtain a value **Xst=3.84.** However, since the relation **Xf=3.16 < Xst= 3.84** takes place, the **Ho**-*hypothesis* can be accepted with the above significance level **- 5%.**

The procedure **X_test_VW** can be enough easily updated by means of implanting into its body of appropriate reference to the table (preliminary saved in a datafile, for example, of *m*-format) of critical points **Xst** for **X**-criterion; that will allow to obtain the answers in terms of the *true* value (**Ho**-*hypothesis* is accepted) and the *false* value (**Ho**-*hypothesis* is rejected).

```
X_test_VW:= proc(L1::{list, file}, L2::{list, file}, n)
local S, L, Lx, Gal, Sr, k, h, omega, z, x, y, N, a, b, o;
  `if`(type(L1, list), assign(a = L1), [assign(a = readdata(L1, 1))]);
  `if`(type(L2, list), assign(b = L2), [assign(b = readdata(L2, 1))]);
  assign(o = {nops(a), nops(b)}), `if`(o[1] < 3,
    ERROR("One of samples is too small"), NULL);
  Gal:= z -> `if`(whattype(z) = `*`, op(1, z), z);
  Sr:= (x, y) -> `if`(Gal(x) <= Gal(y), true, false);
  if nops(a) <= nops(b) then L := a; Lx := b else L := b; Lx := a  end if;
  seq(assign('L'[k] = L[k]*S), k = 1 .. nops(L)),
    assign(omega = [], h = sort([op(L), op(Lx)], Sr));
  for k to nops(h) do
    if whattype(h[k]) = `*` then omega := [op(omega), k/(nops(h) + 1)] end if
  end do;
  RETURN(N = nops(h), evalf(omega, n))
end proc
```

Typical examples of the procedure use:

> L:= [64,68,70,72,75,76,79,80,83]: P:= [60,60,62,66,68,69,70,71,73,78,80]: X_test_VW(L,P, 3);

N = **20**, [0.190, 0.286, 0.429, 0.571, 0.667, 0.714, 0.810, 0.857, 0.952]

> X_test_VW(L, "C:\\Academy\\Examples/Sample2.dat", 4);

N = **20**, [0.1429, 0.1905, 0.2381, 0.2857, 0.4286, 0.4762, 0.5238, 0.6190, 0.7143]

The above four procedures have useful enough appendices in statistic analysis. In more details with the procedures intended for testing of statistical hypotheses the reader can familiarize oneself in our previous books [11-13,32,34,35,43].

13.4. Elements for simple analysis of time (dynamic) and variation series

In the given item we shall present the program module **SimpleStat,** which exports a number of useful functional tools, intended for providing of simple statistical analysis, first of all, for analysis of variation and time (dynamic) series. The tools of this program module are intended for support of *simple statistical analysis.*

SimpleStat – some maintenance functions of simple statistics

Call format of the variables exported by the module:

SimpleStat:- X(*args*) or **with(SimpleStat): X(*args*)**

Formal arguments of the variables calls exported by the module:

X   - a name of function exported by the **SimpleStat** module
args - formal arguments corresponding to the exported function

Description of the module:

The simple module **SimpleStat** exports a series of useful tools intended for providing of simple statistical analysis, above all, for the analysis of variation and time (dynamic) series. The module **SimpleStat** exports thirteen tools for support of simple statistical analysis, namely:

ACC, MCC, PCC, LT, FD, MAM, CR, CC, Sko, Ds, SR, LRM_NRM, Weights

SR(L) - *mean* of statistical data (**SD**) from a list, vector or text datafile **L;**

Ds(L) - *dispersion* (variance) of **SD** from a list, vector or text datafile **L;**

Sko(L) - *standard deviation* of **SD** from a list, vector or text datafile **L;**

CC(L1, L2 {, n}) - *correlation coefficient* between **SD** from lists, vectors or text datafiles **L1** and **L2;** (*optional*) positive integer argument **n** defines accuracy of the returned result; otherwise, **n=Digits** is supposed;

CR(L1, L2 {, n}) - *correlation relation* between **SD** from lists, vectors or text datafiles **L1** and **L2;** (*optional*) positive integer argument **n** defines accuracy of the returned result; otherwise, **n=Digits** is supposed;

MAM(L, n) - *moving averages* of levels of a time series which are defined by a list, vector or text datafile **L,** where **n** is length of a moving interval;

FD(L) - sequence of lists of *differences* of all orders of levels of a time series that are defined by a list, vector or text datafile **L;**

LT(L, F, X, t) - serves for an evaluating of *trend* of a dynamic (*time*) series defined by a list, vector or text datafile **L** of values of its levels. The others formal arguments have the following sense: **F** - an identifier of the returned function defining a sought *linear trend* of the time series; **X** - an identifier of independent variable of function **F** of *linear trend* of the time series; **t** - required accuracy of the returned result. The **LT** procedure not only return the equation describing the function of trend, but also evaluates definition of the given linear function, what provides access to it in a current session at a level of the main library *Maple* functions;

ACC(L) - *autocorrelation coefficient* of **SD** from a list, vector or text datafile **L;**

Weights(L) - *weights* of values of **SD** from a list, vector or text datafile **L;** the result is returned as (2xn) array whose first row represents *values* of **SD** from **L,** whereas the second row represents *weights* corresponding to them;

PCC(X,Y,Z) - *partial correlation coefficient* between **SD** from lists, vectors or text datafiles **X, Y, Z;**

MCC(X,Y,Z) - *multiple correlation coefficient* between **SD** from lists, vectors or text datafiles **X,Y,Z;**

LRM_NRM(A,B,*args*) - *linear/nonlinear regression model* of **SD** defined by lists, vectors or text datafiles **A** and **B;** description of other formal ***args*** arguments can be found in item **13.2** above.

The **SD** in the text datafiles should consist of integers or floating-point values, parted by symbols of carriage return and line feed (*Enter* key; hex code `0D0A`). It is necessary to note that at presence among the given data values of the *rational* type their numerators are chosen only. In addition, the appropriate handling of the situations of too small samples and disparity of samples lengths is being done.

```
> SimpleStat:= module ()
local n, V, L, G, SUM, Gdnt;
export SR, # mean of statistical data (SD) from a list, vector or text datafile
   Ds, # dispersion (variance) of SD from a list, vector or text datafile
   Sko, # standard deviation of SD from a list, vector or text datafile
   CC, # correlation coefficient between SD from lists, vectors or text datafiles
   CR, # correlation relation between SD from lists, vectors or text datafiles
   MAM, # moving average of levels of a time series which are defined by a list, vector
           or text datafile
   FD, # sequence of lists of differences of all orders of levels of a time series which are
        defined by a list, vector or text datafile
   LT, # linear trend of levels of a time series which are defined by a list, vector or text
        datafile
   PCC, # partial correlation coefficient between SD from lists, vectors or text datafiles
   MCC, # multiple correlation coefficient between SD from lists, vectors or text datafiles
   ACC, # autocorrelation coefficient of SD from a list, vector or text datafile
   Weights, # weights of values of SD from a list, vector or text datafile
   LRM_NRM; # linear/nonlinear regression model of statistical data
description "Simple statistical data analysis with Maple of releases 6, 7, 8 and 9/9.5";
options `CopyRight (c) RANS_IAN = Tallinn-Grodno-Vilnius-Moscow; 21.09.2003`,
          load = [V, SUM], package;
   assign(SR = proc(L::{list, vector, file}) local a; `if`(type(L, list), assign(a = L),
     `if`(type(L, vector), assign(a = convert(L, list)) ,assign(a = readdata(L, 1))));
     `+`(op(a))/nops(a) end proc, Ds = proc(L::{list, vector, file}) local a,k; `if`(type(L, list),
     assign(a=L), `if`(type(L ,vector), assign(a =convert(L,list)), assign(a=readdata(L, 1))));
      sum((a[k]-SR(a))^2, k=1..nops(a))/nops(a) end proc), assign(Sko = proc(L::{list, file})
      local a; `if`(type(L,list), assign(a=L), assign(a=readdata(L, 1))); sqrt(Ds(a)) end proc);

CC:= proc(A::{list, vector, file}, B::{list, vector, file})
local a, b, o;
   `if`(type(A, list),assign(a = A), `if`(type(A, vector), assign(a = convert(A, list)),
     assign(a = readdata(A,1)))); `if`(type(B, list), assign(b = B),`if`(type(B, vector),
     assign(b = convert(B, list)), assign(b = readdata(B, 1))));
   assign(o = {nops(a), nops(b)}), `if`(o[1] < 3, ERROR("One of samples is too small"),
     [assign('a' = a[1 .. o[1]], 'b' = b[1 .. o[1]]), `if`(nops(o) > 1, WARNING("Lengths of
     samples is different, a levelling by minimal length has been done"), NULL)]);
   evalf((SR(V(a, b)) - SR(a)*SR(b))/sqrt(Ds(a)*Ds(b)), `if`(nargs = 3 and type(args[3],
     posint), args[3], Digits))
end proc;

MAM:= proc(L::{list, vector, file}, n::posint)
local a, k, p, G;
   `if`(type(L, list), assign(a = L), `if`(type(L, vector), assign(a = convert(L, list)),
     assign(a = readdata(L, 1)))); [assign('G' = []), seq(assign('G' = [op(G), sum(a[p],
     p = k .. k + n - 1)/n]), k=1 .. nops(a) - n + 1), op(G)]
end proc;

LT:= proc(L::{list, vector, file}, F::symbol, X::symbol, t::posint)
local a, k, n, b;
   `if`(type(L,list), assign(b = L), `if`(type(L, vector), assign(b = convert(L, list)),
     assign(b = readdata(L, 1)))); op([assign(a = [assign(n = nops(b)), assign(('%s') =
```

```
    6*sum((2*k - n - 1)*b[k], k = 1 .. n)/(n^3 - n)), assign(('%g') = sum(b[k],k = 1.. n)/n),
    %s, %g - 1/2*(n + 1)*%s, unassign('%s', '%g')]), RETURN(F(X) = evalf(a(b), t)[1]*
    X + evalf(a(b), t)[2], assign(F = proc(X) options operator, arrow; evalf(a(b), t)[1]*X +
    evalf(a(b), t)[2] end proc))])
end proc;

FD:= proc(L::{list, vector, file})
local k, p, FD1, A, K, b;
  `if`(type(L, list), assign(b = L), `if`(type(L, vector), assign(b = convert(L, list)),
    assign(b = readdata(L, 1))));
  FD1:= K -> [assign('A' = []), seq(assign('A' = [op(A), K[p + 1] - K[p]]), p=1 .. nops(K) - 1),
   op(A)]; b, FD1(b), seq(FD1(A), k=2 .. nops(b) - 1)
end proc;

LRM_NRM:= proc(A::{vector, list, file}, B::{vector, list, file},T::name, P::evaln, CR::evaln)
local a, b, k, m, n, p, L, M, N, r, omega, G, X, Y, P1, P2, F, sr, ds, v, o;
  `if`(type(A, list), assign(a = A), `if`(type(A, vector), assign(a = convert(A, list)),
    assign(a = readdata(A, 1)))); `if`(type(B, list), assign(b = B), `if`(type(B, vector),
    assign(b = convert(B, list)), assign(b = readdata(B, 1))));
  assign(o = {nops(a), nops(b)}), `if`(o[1] < 3, ERROR("One of samples is too small"),
    [assign('a' = a[1 .. o[1]], 'b' = b[1 .. o[1]]), `if`(nops(o) > 1, WARNING("Lengths of
    samples is different, a levelling by minimal length has been done"), NULL)]);
  assign(omega = [seq(1, p = 1 .. nops(a))], G = [], F = [TIMES, BOLD, 9]);
  assign(v = proc() options operator, arrow; op([assign('L' = []), seq(assign('L' = [op(L),
    product(args[m][p], m = 1 .. nargs)]), p = 1 .. nops(args[1])), L]) end proc);
  assign(sr = proc () options operator, arrow; `+`(args)/nargs end proc, ds = proc ()
  options operator, arrow; `+`(seq((args[k] - sr(args))^2, k = 1 .. nargs))/nargs end proc),
  `if`(T = LRM, assign('M' = Matrix(2, 2, [[sr(op(v(b, b))), sr(op(b))], [sr(op(b)), 1]]),
    'N' = Vector(2, [sr(op(v(a, b))), sr(op(a))])), `if`(T = NRM, assign('M' = Matrix(3, 3,
    [[sr(op(v(b, b, b, b))), sr(op(v(b, b, b))), sr(op(v(b, b)))], [sr(op(v(b, b ,b))),
    sr(op(v(b, b))), sr(op(b))], [sr(op(v(b, b))), sr(op(b)), 1]]),
    'N' = Vector(3, [sr(op(v(a, b, b))), sr(op(v(a, b))), sr(op(a))])), ERROR("regression
    model type <%1> is inadmissible",T)));
  assign('r' = convert(evalf(LinearAlgebra:- LinearSolve(M, N)), list));
  assign(Y = `if`(nops(r) <> 1,
    proc(X) local n; sum(r[n]*X^(nops(r) - n), n = 1 .. nops(r)) end proc,
    proc(X) sign(op(r))*X end proc)), `if`(nops(r) = 1, assign('CR' = sign(op(r))), NULL);
  `if`(nops(r) = 2, assign('CR' = evalf(sqrt((ds(op(a)) - sr(op(v(a - r[1]*b - r[2]*omega,
    a - r[1]*b - r[2]*omega))))/ds(op(a))))), `if`(nops(r) = 3, assign('CR' = evalf(sqrt((ds(op(a)) -
    sr(op(v(a - r[1]*v(b, b) - r[2]*b - r[3]*omega, a - r[1]*v(b, b) - r[2]*b -
    r[3]*omega))))/ds(op(a))))), NULL));
  seq(assign('G' = [op(G), [b[n], a[n]]]), n=1 .. nops(a));
  P1:= plots[pointplot](G, color = blue, thickness = 3, symbol = CIRCLE);
  P2:= plot(Y(X), X = b[1] .. b[-1], thickness = 2);
  assign(P = plots[display]([P1, P2], axesfont = F, scaling = UNCONSTRAINED, labels =
    [convert("B, X", name), convert("A, Y", name)], labelfont = F)), RETURN(Y(X))
end proc;

CR:= proc(A::{list,vector, file}, B::{list, vector, file})
local a, b, p, cr;
  `if`(type(A, list), assign(a = A), `if`(type(A, vector), assign(a = convert(A, list)),
```

```
    assign(a = readdata(A, 1))));
  `if`(type(B, list), assign(b = B), `if`(type(B, vector) ,assign(b = convert(B, list)),
    assign(b = readdata(B, 1))));
  LRM_NRM(a, b, LRM, p, cr);
  evalf(cr, `if`(nargs = 3 and type(args[3], posint), args[3], Digits))
end proc;

PCC:= proc(Z::{list, vector, file}, X::{list, vector, file}, Y::{list, vector, file})
local a, b, c, o;
  `if`(type(Z, list), assign(a = Z), `if`(type(Z, vector), assign(a = convert(Z, list)),
    assign(a = readdata(Z, 1))));
  `if`(type(X, list), assign(b = X), `if`(type(X, vector), assign(b = convert(X, list)),
    assign(b = readdata(X, 1))));
  `if`(type(Y, list), assign( c= Y), `if`(type(Y, vector), assign(c = convert(Y, list)),
    assign(c = readdata(Y, 1))));
  assign(o = {nops(a), nops(b), nops(c)}), `if`(o[1] < 3,
    ERROR("One of samples is too small"), [assign('a' = a[1 .. o[1]], 'b' = b[1 .. o[1]],
    'c' = c[1 .. o[1]]), `if`(nops(o) > 1, WARNING("Lengths of samples is different,
    a levelling by minimal length has been done"), NULL)]);
  evalf((CC(a, b) - CC(a, c)*CC(b, c))/sqrt((1 - CC(a ,c)^2*(1 - CC(b ,c)^2))))
end proc;

MCC:= proc(Z::{list, vector, file}, X::{list, vector, file}, Y::{list, vector, file})
local a, b, c, k, delta, E, B, R, o;
  `if`(type(Z, list), assign(a = Z), `if`(type(Z, vector), assign(a = convert(Z, list)),
    assign(a= readdata(Z, 1))));
  `if`(type(X, list), assign(b = X), `if`(type(X, vector), assign(b = convert(X, list)),
    assign(b = readdata(X, 1))));
  `if`(type(Y, list), assign( c= Y), `if`(type(Y, vector), assign(c = convert(Y, list)),
    assign(c = readdata(Y, 1))));
  assign(o = {nops(a), nops(b), nops(c)}), `if`(o[1] < 3, ERROR("One of samples is too
   small"), [assign('a' = a[1 .. o[1]], 'b' = b[1 .. o[1]], 'c' = c[1 .. o[1]]), `if`(nops(o) > 1,
   WARNING("Lengths of samples is different, a levelling by minimal length has been
   done"), NULL)]);
  E:= []: for k to nops(a) do E:= [op(E), 1] end do: k:= 'k': delta:= matrix(3, 3, [SUM(b,b),
   SUM(b, c), SUM(b, E), SUM(b, c), SUM(c, c), SUM(c, E), SUM(b, E), SUM(c ,E),
   nops(a)]);  B:= vector(3, [SUM(a, b), SUM(a, c), SUM(a, E)]);
  R:= evalf(linalg[linsolve](delta, B));
  evalf(sqrt((R[1]*PCC(a, b, c)*sqrt(Ds(b)) + R[2]*PCC(a, c, b)*sqrt(Ds(c)))/sqrt(Ds(a))))
end proc;

V:= proc()
local k, p, L;
  op([assign(L = []), seq(assign('L' = [op(L), product(args[k][p], k=1 .. nargs)]),
    p=1 .. nops(args[1])), L])
end proc;

SUM:= proc(A::list, B::list) local k; sum(A[k]*B[k], k=1 .. nops(A)) end proc;
module Gdnt ()
local p, h, n, Sr, Ds, S;
export ACC;
```

```
description "Calculation of autocorrelation coefficient";
  Sr:= (S, p) -> sum(S[k], k = p .. `if`(p = 1, nops(S) - 1, nops(S)))/(nops(S) - 1);
  Ds:= (S, h) -> sum(S[k]^2, k = h .. `if`(h = 1, nops(S) - 1, nops(S)))/(nops(S) -1) - Sr(S,h)^2;
  ACC:= S -> evalf(sum((S[n + 1] - Sr(S, 2))*(S[n] - Sr(S, 2)),
      n=1 .. (nops(S)-1))/(sqrt(Ds(S, 1)*Ds(S, 2))*(nops(S) - 1)))
end module;

ACC:= proc(L::{list, vector, file})
local a;
  `if`(type(L, list), assign(a = L), `if`(type(L, vector), assign(a = convert(L, list)),
      assign(a = readdata(L, 1)))); Gdnt:- ACC(a)
end proc;

Weights:= proc(L::{list, vector, file})
local a, k, p, G, R;
  `if`(type(L, list), assign(a = L), `if`(type(L, vector), assign(a = convert(L, list)),
     assign(a = readdata(L, 1)))); G:=[op(sort({op(a)}))]: R:= array(sparse, 1 .. 2,1 .. nops(G)):
  for k to nops(G) do R[1,k]:= G[k]:
  for p to nops(a) do
  if G[k] = a[p] then R[2,k]:= R[2,k] + 1 else next end if
    end do
   end do;
   evalm(R)
  end proc
end module
```

Typical examples of the module use:

> with(SimpleStat); F:="C:/academy/examples/sample2.dat":

$[ACC, CC, CR, Ds, FD, LRM\_NRM, LT, MAM, MCC, PCC, SR, Sko, Weights]$

> SR(F), Ds(F), Sko(F);

81.30769231, 137.2899410, 11.71707903

> CC("C:/Academy/Examples/Members.dat", "C:/Academy/Examples/Age.dat");

-0.9830255251

> MAM(F, 3); LT(F, H, X, 5); FD(F);

[73.66666667, 74.66666667, 81.33333333, 89.33333333, 88.33333333, 89.66666667, 86.00000000, 78.00000000, 70.00000000, 62.00000000, 62.00000000, 65.66666667, 69.33333333, 73.00000000, 78.33333333, 83.66666667, 89.00000000, 89.00000000, 89.00000000, 86.66666667, 84.33333333, 87.00000000, 92.00000000, 97.00000000]

H(X) = 0.58735 X + 73.378

[86., 62., 73., 89., 82., 97., 86., 86., 86., 62., 62., 62., 62., 73., 73., 73., 89., 89., 89., 89., 89., 82., 82., 97., 97., 97.], [-24., 11., 16., -7., 15., -11., 0., 0., -24., 0., 0., 0., 11., 0., 0., 16., 0., 0., 0., 0., -7., 0., 15., 0., 0.], [35., 5., -23., 22., -26., 11., 0., -24., 24., 0., 0., 11., -11., 0., 16., -16., 0., 0., 0., -7., 7., 15., -15., 0.], [-30., -28., 45., -48., 37., -11., -24., 48., -24., 0., 11., -22., 11., 16., -32., 16., 0., 0., -7., 14., 8., -30., 15.], [2., 73., -93., 85., -48., -13., 72., -72., 24., 11., -33., 33., 5., -48., 48., -16., 0., -7., 21., -6., -38., 45.], 71., -166., 178., -133., 35., 85., -144., 96., -13., -44., 66., -28., -53., 96., -64., 16., -7., 28., -27., -32., 83.], ...]

> LRM_NRM("c:/academy/examples/Members.dat", "c:/academy/examples/Age.dat", NRM, a, b), b; a;

5.554988480 - 0.001080042656*X^2 - 0.004403992540*X, 0.9425931978

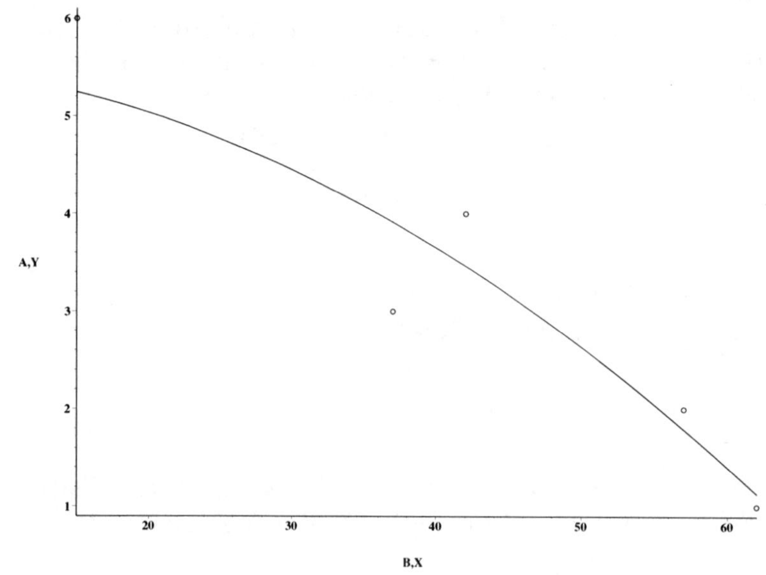

> CR("C:/Academy/Examples/Members.dat", "C:/Academy/Examples/Age.dat");

0.9830255261

> PCC("C:/Academy/Examples/Members.dat", "C:/Academy/Examples/Age.dat", F);

-0.7446177489

> MCC("C:/Academy/Examples/Members.dat","C:/Academy/Examples/Age.dat", F);

0.8548484302

> ACC(\`c:/academy/examples/age.dat\`), Weights(\`c:/academy/examples/sample2.dat\`);

$$0.9194529096, \begin{bmatrix} 86. & 62. & 73. & 89. & 82. & 97. & 61. & 56. & 7. & 87. \\ 6 & 7 & 11 & 6 & 1 & 3 & 6 & 1 & 1 & 1 \end{bmatrix}$$

The software represented in the present section can serve as addition to statistical tools of the *Maple* package, providing the solution of a wide enough range of the problems dealing with simple processing of statistical data. By a series of possibilities and characteristics, the represented tools well enough supplement already available means of the package and are successfully used in tasks of simple statistical analysis of the data of various nature and purpose.

Chapter 14.
Software for operation with the user libraries

The *Maple* package of releases **6 – 9/9.5** has a series of tools for creation of mechanisms for operation with the user libraries structurally analogous to the main *Maple* library; these libraries allow to use them in the *Maple* environment at a level of access, analogous for built-in tools of the package. In the present section, we shall present three most effective organization levels of the user libraries of procedures, modules and functions. The procedures represented below allow to substantially simplify and to extend operation with the user libraries. As is shown by our experience these procedures extend capabilities of the user at libraries organization of own software.

`type/mlib` – check of an object to be a library similar to the main *Maple* library

Call format of the procedure:

type(L, *mlib*)

Formal arguments of the procedure:

L - symbol or string defining a library name or full path to it

Description of the procedure:

The procedure call **type(L, *mlib*)** returns the *true* value if and only if the actual argument **L** specifies a name or full path to a library structurally analogous to the main *Maple* library; otherwise, the *false* value is returned. Generally, the **L** argument should specify full path to a library, otherwise it will be supposed as name of a library located in the main *Maple* directory or subdirectory with the main *Maple* library. The procedure is a rather useful at operation with the user libraries. Above all, the procedure introduces a new type allowing to test libraries to be the *Maple* libraries.

```
`type/mlib`:= proc(L::{string, symbol})
local a, omega, nu;
  Release(a), assign(omega = ((a, L) -> map2(cat, cat(a, "\\"),
    map2(cat, L, {"\\maple.lib", "\\maple.ind"})))));
  nu:= x -> map(type, omega(a, L), file) = {true} or
    map(type, omega(cat(a, "\\lib"), L), file) = {true};
  try
    if not type(L, dir) then if cat("", L)[2] = ":" then false elif nu(L) then true else false end if
    else if map(type, omega(L, ""), file) = {true} or nu(L) then true else false end if
    end if
  catch "<%1> is invalid path to a directory or a file": false
  end try
end proc
```

Typical examples of the procedure use:

```
> type("C:/Program Files/Maple 6/Lib", mlib);
                                   true
> type("C:/Program Files/Maple 7/Lib", mlib);
                                   true
> type("C:/Program Files/Maple 8/Lib", mlib);
                                   true
> type("C:/Program Files/Maple 9/Lib", mlib);
                                   true
> type("C:/Academy/UserLib9/UserLib", mlib);
                                   true
> map(type, [LIB, Library, UserLib, RANS], mlib),  map(type, [sin,``,"",`    `,"    "], mlib);
                  [true, false, true, false], [false, false, false, false, false]
```

exeP – selective calls of library procedures, program modules, etc.

Call format of the procedure:

exeP(P, L {, *pr*})

Formal arguments of the procedure:

P - symbol defining name of a procedure, a program module, etc.
L - symbol or string defining path to a required library
pr - (optional) actual arguments passed to the **P** object

Description of the procedure:

As a rule, along with the main *Maple* library, the user deals with other libraries of *procedures* and *modules*, including own libraries of tools. In a series of cases, the necessity of a procedure call being in a concrete library arises. This problem is the more urgent in two basic cases, namely: **(1)** if a few libraries are logically linked with the main *Maple* library and contain procedures of the *same name*, and **(2)** a required procedure is in a library that logically is not linked with the main *Maple* library. The useful **exeP** procedure solves this problem.

The procedure call **exeP(P, L {, *pr*})** returns the result of call of a **P** procedure, program module or other *Maple* object (for which takes place the following relation **eval(P) <> P**) being in a library specified by a path **L** to it on actual arguments specified by the optional ***pr*** argument. For a library **L** the full path should be specified. If the third optional ***pr*** argument (a sequence of valid *Maple* expressions) is absent, the procedure call **P()** is supposed. If the first actual argument specifies a **P** program module, then the third argument ***pr*** should specify its an export name and actual arguments passed to it. In other cases, the procedure call returns the evaluating result of an object being in a library **L**, ignoring all actual arguments since the third argument if they exist, of course. An unsuccessful **exeP** call causes an appropriate erratic situation. After any procedure call, the previous *Maple* state is fully restored.

Usage of the given procedure is the most effective if necessary to call the tools of the same name that are located in various libraries and logically not connected with the main *Maple* library, i.e.

which are not represented in the predetermined *libname*-variable of the package. In particular, it well meets the conditions of tools debugging.

```
exeP:= proc(P::{name, symbol}, L::mlib)
local h, psi, Res, f, omega, epsilon;
global libname, `_$$`, `_$`;
  (proc(P) f := "_$$_"; if eval(P) <> P then save P, f end if  end proc)(P);
  assign(omega = (y -> `if`(1 < nops(y), op(y[2 .. -1]), NULL)),
    epsilon = (() -> op([unprotect(args[1]), unassign('args'[1])])));
  psi:= proc()
    local a, b, k, g;
      try assign67(a = march('list', args[2]), b = 0, '`_$$`' = libname)
      catch "there is no existing archive in "%1"":  delf(f);
        ERROR("for library the full path should be specified - <drive>:/<dir1>/ .../%1", args[2])
      end try;
      for k to nops(a) do
        if a[k][1] = cat("", args[1], ".m") then b := 1;  break  else  NULL  end if
      end do;
      `if`(b = 0, [unassign('`_$$`', delf(f)), ERROR("<%1> does not exist in <%2>", args[1],
        args[2])], [epsilon(args), assign67('libname' = args[2], libname),
        WARNING("name <%1> has been unprotected and redefined", args[1])]);
      if type(args[1], procedure) then
        try 61, args[1](`if`(nargs = 2, NULL, args[3 .. nargs])), epsilon(args),
          assign67('libname' = `_$$`)
        catch:  epsilon(args), assign67('libname' = `_$$`);
          ERROR("invalid arguments for procedure <%1>", args[1])
        end try
      elif type(args[1], `module`) then
        try assign67(g = "_$E_", '`_$`' = NULL), writeline(g, cat("`_$`:=", args[1],
          "[", args[3], "](", `if`(3 < nargs,  seqstr(args[4 .. nargs]), ``), "):")),
          (proc() close(g);  read g; fremove(g), omega([56, `_$`]), epsilon(args),
              assign67('libname' = `_$$`)
          end proc)()
        catch:  epsilon(args), assign67('libname' = `_$$`), ERROR("invalid module export call")
        end try
      else 36, eval(args[1]), epsilon(args), assign67('libname' = `_$$`)
      end if
    end proc;
  try Res := psi(args)
  catch "invalid arguments for procedure <%1>": h := 7
  catch "invalid module exports call": h := 14
  end try;
  (proc(f) if type(f, file) then  read f; delf(f)  end if;  unassign('`_$$`', '`_$`')  end proc)(f);
  `if`(h = 7, ERROR("invalid arguments for procedure <%1>", args[1]),
    `if`(h = 14, ERROR("invalid module export call"), omega([Res])))
end proc
```

Typical examples of the procedure use:

```
> libname;
          "c:/program files/maple 9/lib/userlib", "C:\Program Files\Maple 9/lib"
```

> exeP(PP, "c:/Tallinn/Grodno/SveGal", 61, 56, 36, 40, 7, 14);

Warning, name <PP> has been unprotected and redefined

table([kr = 7, avz = 61, agn = 56, sv = 36, art = 14])

> libname, eval(PP);

"c:/program files/maple 9/lib/userlib", "C:\Program Files\Maple 9/lib", true

> exeP(PP1, "c:/Tallinn/Grodno/SveGal", 61, 56, 36, 40, 7, 14);

Error, (in **psi**) <PP1> does not exist in <c:/Tallinn/Grodno/SveGal>

> libname;

"c:/program files/maple 9/lib/userlib", "C:\Program Files\Maple 9/lib"

> exeP(Vol, "c:/Tallinn/Grodno/SveGal", 61, 56, 36, 40, 7, 14);

Warning, name <Vol> has been unprotected and redefined

[61, 56, 36, 40, 7, 14, 214]

> exeP(RANS, "c:/Tallinn/Grodno/SveGal", Art_Kr, 61, 56, 36, 40, 7, 14);

Warning, name <RANS> has been unprotected and redefined
Error, (in **unknown**) module does not export `Art_Kr`

> libname;

"c:/program files/maple 9/lib/userlib", "C:\Program Files\Maple 9/lib"

> exeP(RANS, "c:/Tallinn/Grodno/SveGal", SR, 61, 56, 36, 40, 7, 14);

Warning, name <RANS> has been unprotected and redefined

35.66666667

> libname;

"c:/program files/maple 9/lib/userlib", "C:\Program Files\Maple 9/lib"

> exeP(SimpleStat, "UserLib", Sr, 61, 56, 36, 40, 7, 14);

Error, (in **psi**) for library the full path should be specified - <drive>:/<dir1>/ .../UserLib

---

permlib – an elements permutation of the libname variable

Call format of the procedure:

permlib(L)

Formal arguments of the procedure:

L - a list of non-negative integers

Description of the procedure:

The **permlib** procedure is intended for permutation of libraries paths being in the *Maple* variable ***libname*** with possibility of deletion of their. The procedure call **permlib(L)** returns the result of elements permutation of the ***libname*** variable in conformity with an integer list specified by the actual argument **L.** The **L** elements are integers from range **0..nops([*libname*])** which define the positions of permutation and/or deletion of the ***libname*** elements, namely: if **L[k] = 0** then out of the returned result an element **[*libname*][L[k]]** is deleted, otherwise element **[*libname*][k]** in the returned result will has position **[*libname*][L[k]]** with correction for length reduction of the returned result because of elements deletion. The procedure has many appendices in the problem solving which deal with the user libraries, providing dynamic change of access priority to the tools contained in libraries logically linked with the main *Maple* library.

```
permlib:= L::list(nonnegative) -> op([assign('libname' = op(perml([libname], L))), eval(libname)])
```

Typical examples of the procedure use:

> libname;

libname:="C:\Program Files\Maple 8/lib", "c:/program files/maple 8/lib/userlib", "c:/program files/maple 8/lib/sveta"

> permlib([2, 1, 3]);

"c:/program files/maple 8/lib/userlib","C:\Program Files\Maple 8/lib", "c:/program files/maple 8/lib/sveta"

> permlib([1, 2, 0]);

"c:/program files/maple 8/lib/userlib", "C:\Program Files\Maple 8/lib"

FLib – **creation of the simple user libraries (*the first level*)**
FLib1
SaveProc
ReadProc

Call format of the procedures:

FLib(L, N {, V})
FLib1(L {, N} {, V})
SaveProc(L, p)
ReadProc(L {, N2})

Formal arguments of the procedures:

L - symbol or string defining name of a library datafile or full path to it
N - a non-negative integer defining a mode of work with library **L; N={0|1|2}**
V - (optional) list or set defining names of procedures and/or program modules, and/or strings with their definitions
p - a set or list of strings with definitions of procedures or modules, and/or symbols defining names of procedures and/or modules
N2 - (optional) list or set of symbols defining names of procedures or program modules

Description of the procedures:

Here we shall present the simplest way of creation of the user libraries and organization of access to them. The mechanism of such access consists in the following. Into a text datafile by means of any permissible way, the strings of the following view (*definitions of procedures* or *program modules*) are put, namely:

"Name := {*Procedure definition* | *Module definition*} {:|;}"

where *Name* is name of a procedure or program module. Three procedures **SaveProc, ReadProc, FLib** and **FLib1** represented below, support an operation with the user libraries of such simple kind. Our experience of usage of these tools has shown, that in a series of cases such approach is more preferable than usage of standard tools.

The procedure call **SaveProc(L, p)** returns the *NULL* value, creating or updating a text datafile specified by the first actual **L** argument; processing of the datafile is done by means of putting into it of procedures and/or program modules whose names or definitions are specified by the second actual **p** argument. The definitions of procedures or modules should be **(1)** evaluated in a current *Maple* session or should be located in a library logically linked with the main *Maple* library, or **(2)** be represented by strings of the above view. An access to all procedures or program modules saved by such manner is supported by means of the call **read(L),** where **L** is the first actual argument of the **SaveProc** procedure. In addition, if path to the **L** datafile is absent, it is created with output of an appropriate warning, if it is necessary. In this case, afterwards a path to the library datafile that is defined in the warning should be used. If **L** defines a datafile not being a library of the above type, then it is fully updated, otherwise the datafile is appended-only updating. The procedure handles the basic especial situations.

The procedure **ReadProc(L {, N2})** is reverse to the **SaveProc** procedure; its call **ReadProc(L)** with one actual **L** argument returns the set of names of all tools contained in a text datafile specified by the **L** argument, ensuring access to them in a current *Maple* session, while its call **ReadProc(L, N2)** with two actual arguments also returns the set of names, ensuring access in a current *Maple* session to those tools contained in **L** whose names are specified by the second argument **N2.** In both cases, the procedure call returns the set of names with output of warnings about tools activated by **ReadProc.** If a library **L** contains tools with the same names, then the procedure call will activate the tools that have been saved by the last. The procedure handles the basic especial and erroneous situations.

The procedure **FLib** basing in a great extent on the above two procedures provides a number of useful functions of handling of a library created by the method described above. The procedure admits up to three actual arguments of which two are obligatory. The first formal **L** argument specifies a name or full path to a datafile destined for the user library organized by the described manner; an actual value for it should have a type {*string, symbol*}. The procedure call **FLib(L)** is equivalent to the procedure call **ReadProc(L).**

The second formal **N** argument has the greater semantic load, namely. The procedure call **FLib(L,0)** returns a set of names of tools being in a library **L,** whereas the procedure call **FLib(L, 1)** provides interactive update of a library **L** in the appending mode. At last, the procedure call **FLib(L, 2)** provides a compression of a library **L,** when in it only the latest versions of tools with the same names remain.

In interactive mode in reply to the procedure inquiry

"Introduce a name or string with definition of a procedure or module: "

the user should introduce name of an active procedure or program module (available in a current *Maple* session) or definition of a procedure or module of the format represented above; an input is being completed according to the procedure ***readstat.*** The next inquiry **`Continue?(y/n): `** provides possibility to continue (**y**) update of an **L** library or to terminate it (**n**). The successful procedure call **FLib(L, 1)** returns the set of names of tools being in a library **L.**

If the 3rd actual **V** argument has been coded and the second actual argument **N=1,** then the procedure automatically updates a library **L** on the basis of a set or list **V** defining names of procedures and/or program modules, and/or strings with their definitions. These tools should be available in a current session. The strings with definitions of procedures and program modules must have the above format, when their names, operator ":=" and bodies are parted by one *blank*. In addition, the procedure call **FLib(L, 1, V)** returns the set of names of tools being in a library **L.**

If the third actual **V** argument has been coded and the second actual argument **N=2,** then the procedure call **FLib(L, 2, V)** is analogous to the procedure call **ReadProc(L, V).** On the basis of a series of procedures dealing with text datafiles, the above procedure **FLib** can be enough simply modified, providing some additional basic operations at work with the user libraries, similar to those with which the procedure **FLib** operates.

At last, the procedure call **FLib1(L {, N} {, V})** has the same set of formal arguments as the above **FLib** procedure. Its implementation is based on the above three procedures **SaveProc, ReadProc** and **FLib.** In contrast to the **FLib,** the **FLib1** procedure works with both existent datafiles and non-existent datafiles. If a datafile **L** is the library in the above-mentioned sense, then the procedure call **FLib1(L, *args*)** is equivalent to the procedure call **FLib(L, *args*).**

If a datafile **L** does not exist then the procedure call **FLib1(L, *args*)** allows to confirm creation of the empty library **L** with the subsequent call **FLib(L, *args*)** or to terminate the procedure. If a datafile **L** does exist but is not the library in the above-mentioned sense then the procedure call **FLib1(L, *args*)** allows to terminate the procedure or to confirm update of the **L** datafile with interactive creation of the library on the basis of an introduced set of names of procedures and/or modules, and/or their definitions with the subsequent procedure call **FLib(L, *args*).** Work with the above procedures is clear enough and does not demand any additional clarifications. The examples represented below well illustrate the told.

```
SaveProc:= proc(L::{string, symbol}, p::{list({string, symbol}), set({string, symbol})})
local a, k, nu, r;
  `if`(Empty(p), ERROR("2nd argument must be non-empty set or list"), assign(nu = {}));
  for k to nops(p) do
    if type(p[k], string) then
      try parse(p[k], statement)
      catch "incorrect syntax in parse: ":  nu := {op(nu), p[k]}
      end try
    elif type(p[k], {procedure, `module`}) then next
    else nu := {op(nu), p[k]}
    end if
  end do;
  if nops(p) = nops(nu) then ERROR("2nd argument contains neither procedures nor modules")
  elif nu <> {} then r := {op(p)} minus nu;
     WARNING("%1 are neither procedures nor modules", nu)
  else r := p
  end if;
  `if`(type(L, file), `if`(type(L, rlb) and readbytes(L, TEXT, 5) = "av61:", assign(a = L),
    assign67(Close(L), writebytes(L, "av61:"), a = L)),
    [assign(a = pathtf(L)), writebytes(a, "av61:")]), Close(a), fopen(a, APPEND);
  for k in r do writeline(a, `if`(type(k, procedure), cat(cat("", k), " := ", convert(eval(k),string),":"),
    `if`(type(k, `module`), cat(null(with(k)), "", k, " := ", convert(`if`(type(k, mod1),
    NULL, mod21(k)), eval(k), string), ":"), k)))
  end do;
  Close(a)
end proc

ReadProc:= proc(L::file)
local a, b, c, h, N, omega;
  `if`(type(L, rlb) and readbytes(L, TEXT, 5) = "av61:", assign(c = "_$$$_", N = {}, h = 7),
     ERROR("<%1> is not a library datafile", L));
  do  h := readline(L);
    if h = 0 then Close(L); break  elif  search(h, " := ", 't')  then
```

```
      N := {op(N), cat(``, h[1 .. t - 1])}
    end if
  end do;
  if nargs = 1 then unprotect(op(N));
    read L;
    RETURN(N, WARNING("the names %1 have been redefined and unprotected", N))
  elif type(args[2], {list(symbol), set(symbol)}) then  omega := N intersect {op(args[2])};
    if omega = {} then ERROR("the names %1 are absent in library <%2>", args[2], L)
    else do  h := readline(L);
        if h = 0 then  Close(L, c), unprotect(op(omega));
          read c;
          fremove(c), RETURN(omega, WARNING(cat("the names %1 have been redefined
            and unprotected", `if`(omega = {op(args[2])}, "", "; the names %2 are absent in
            library <%3>")), omega, {op(args[2])} minus omega, L))
        elif search(h, " := ", 't') and member(cat(``, h[1 .. t - 1]), omega) then  writeline(c, h)
        end if
      end do
    end if
  else ERROR("2nd argument must be a set or list of symbols but has received <%1>", args[2])
  end if
end proc

FLib:= proc(L::file, N::{0, 1, 2})
local a, b, c, h, P, omega;
  Close(L), `if`(type(L, rlb) and readbytes(L, TEXT, 5) = "av61:", assign(c = "_$$$_", P={}, h=7),
    ERROR("<%1> is not a library datafile", L));
  do h := readline(L);
    if h = 0 then Close(L); break
    elif search(h, " := ", 't') then P := {op(P), cat(``, h[1 .. t - 1])}
    end if
  end do;
  if nargs = 1 then RETURN(ReadProc(args))
  elif nargs = 2 and N = 0 then RETURN(P)
  elif nargs = 2 and N = 1 then
    do  h := readstat("Introduce a name or string with definition of a procedure or module: ");
      SaveProc(L, {`if`(type(cat(``, h), {`module`, procedure}), cat(``, h), h)});
      if Case(readstat("Continue?(y/n): ")) <> y then  break  end if
    end do;
    Close(L), RETURN(procname(L, 0))
  elif nargs = 2 and N = 2 then  assign(omega = interface(warnlevel)),
      interface(warnlevel = 0), XTfile(L, c), readline(L),
      writebytes(L, readbytes(c, TEXT, infinity)), Close(L, c);
    interface(warnlevel = omega), writeline(c, readline(L));
    do  h := readline(L);
      if h = "av61:" then Close(L, c), writeline(L, readbytes(c, TEXT, infinity)), Close(L),
          fremove(c);
        break
      elif search(h, " := ", 't') and member(cat(``, h[1 .. t - 1]), P) then writeline(c, h);
        P := P minus {cat(``, h[1 .. t - 1])}
      end if
```

```
    end do;
    RETURN(WARNING("library <%1> has been compressed with names %2",
    L, procname(L, 0)))
  elif nargs = 3 and N = 1 and type(args[3], {set({string, symbol}), list({string, symbol})}) then
    SaveProc(L, args[3]), procname(L, 0)
  elif nargs = 3 and N = 2 and type(args[3], {list(symbol), set(symbol)}) then
    ReadProc(L, args[3])
  end if
end proc

FLib1:= proc(L::{string, symbol}, N::{0, 1, 2})
local a, h;
  assign(a = Path(L));
  if not type(a, file) then h:= readstat("Datafile does not exist, do you desire to create it?(y/n): ");
    if Case(h) <> y then  RETURN(WARNING("Work has been completed!"))
    else  a := pathtf(a);  writeline(a, "av61:");  Close(a), procname(a, args[2 .. -1])
    end if
  end if;
  Close(a), `if`(type(a, rlb) and readbytes(a, TEXT, 5) = "av61:", RETURN(FLib(args)), Close(a));
  if type(a, rlb) and readbytes(a, TEXT, 5) <> "av61:" then  Close(a),
      assign('h' = readstat("File is not a library, do you desire to update it?(y/n): "));
    if Case(h) <> y then  RETURN(WARNING("Work has been completed!"))
    else do
        h := readstat("Introduce set of names of procedures and/or modules, and/or their
          definitions: ");
        if not type(h, set) or h = {} or map(type, h, {string, symbol}) <> {true} then
          h := readstat("Invalid input. Continue(y/n)?: ");
          if h <> y then RETURN(WARNING("Work has been completed!"))
          end if
        else  a := pathtf(a);  SaveProc(a, h);  FLib(args);
          break
        end if
      end do
    end if
  else ERROR("<%1> is not a text datafile", L)
  end if
end proc
```

Typical examples of the procedures use:

> Art:= () -> `+`(args)/nargs: Kr:= proc() local k; `if`(nargs=0, NULL, sqrt(sum((args[k] - Art(args))^2,k=1..nargs)/nargs)) end proc: M1:= module () export sr; sr:= ()->`+`(args)/ nargs end module: module M2 () export sr1; sr1:= () -> `+`(args)/nargs end module:

> SaveProc("C:/Temp/AVZ\\AGN/Svetla.txt", [Art, Kr, M1, M2]);

Warning, target path <c:\temp\avz_agn\svetla.txt> has been activated instead of the required path <c:\temp\avz\agn\svetla.txt>

> restart; ReadProc("C:/Temp/AVZ_AGN/Svetla.txt");

Warning, the names {Art, Kr, M1, M2} have been redefined and unprotected

{Art, Kr, M1, M2}

```
> 6*map2([Art, Kr, M1:-sr, M2:-sr1], 62, 57, 37, 41, 7, 15);
                              [219, 14541^(1/2), 219, 219]
> SaveProc("C:/Temp/AVZ\\_AGN/Svetla.txt", [Art, Kr, M1, M2, RANS_IAN, RAC, REA]);
   Warning, {RANS_IAN, RAC, REA} are neither procedures nor modules
> SaveProc("C:/Temp/AVZ\\_AGN/Svetla.txt", [AV,AG,VS,"ArtKr := () -> /`+`(args)"]);
   Error, (in SaveProc) 2nd argument contains neither procedures nor modules
> ReadProc("C:/Temp/AVZ\\_AGN/Svetla.txt", RANS_IAN);
   Error, (in ReadProc) 2nd argument must be a set or list of symbols but has received
   <RANS_IAN>
> ReadProc("C:/Temp/AVZ\\_AGN/Svetla.txt", {RANS, IAN});
   Error, (in ReadProc) the names {IAN, RANS} are absent in library
   <C:/Temp/AVZ\_AGN/Svetla.txt>
> ReadProc("C:/Temp/AVZ\\_AGN/Svetla.txt", {Art, Kr, V, G});
   Warning, the names {Art, Kr} have been redefined and unprotected; the names {G, V} are absent
   in library <C:/Temp/AVZ\_AGN/Svetla.txt>
                                        {Art, Kr}
> FLib("C:/Temp/AVZ\\_AGN/Svetla.txt", 1);
[ Introduce a name or string with definition of a procedure or module: "AGN1 := () ->
  sqrt(`+`(args)):";
[ Continue?(y/n):  y;
[ Introduce a name or string with definition of a procedure or module: "AGN2 := ()->
  `+`(args)/nargs:";
[ Continue?(y/n):  n;
                          {M2, M1, Kr, Art, AGN2, AGN1, AGN}
> RANS:=() -> [args, nargs, `+`(args)/nargs]:
> FLib("C:/Temp/AVZ\\_AGN/Svetla.txt", 1);
[ Introduce a name or string with definition of a procedure or module: RANS;
[ Continue?(y/n):  n;
                       {M2, M1, Kr, Art, RANS, AGN2, AGN1, AGN}
> FLib("C:/Temp/AVZ\\_AGN/Svetla.txt", 2);
   Warning, library <C:/Temp/AVZ\_AGN/Svetla.txt> has been compressed with names {M2,
   M1, Kr, Art, RANS, AGN2, AGN1, AGN}
> restart;  FLib("C:/Temp/AVZ\\_AGN/Svetla.txt", 2, {Art, Kr, RANS, SveGal});
   Warning, the names {Kr, Art, RANS} have been redefined and unprotected; the names {SveGal}
   are absent in library <C:/Temp/AVZ\_AGN/Svetla.txt>
                                     {Kr, Art, RANS}
> FLib("C:/Temp/AVZ\\_AGN/Svetla.txt", 1, {Art,Kr,"IAN := () -> [args, nargs]:"});
                    {M2, M1, Kr, Art, RANS, AGN2, AGN1, AGN, IAN}
```

> FLib("C:/Temp/AVZ_AGN/Svetla.txt", 2);

Warning, library <C:/Temp/AVZ_AGN/Svetla.txt> has been compressed with names {M2, M1, Kr, Art, RANS, AGN2, AGN1, AGN, IAN}

> FLib1("C:/Tallinn/Grodno/Vilnius/ArtKr.147", 1, [PP, ACC, RANS]);

Warning, {RANS} are neither procedures nor modules

{PP, ACC}

> FLib1("C:/Tallinn/Grodno/Vilnius/ArtKr.147", 2);

Warning, library <C:/Tallinn/Grodno/Vilnius/ArtKr.147> has been compressed with names {PP, ACC}

> FLib1("C:/Tallinn/Grodno/Vilnius/ArtKr.147", 2, [PP, ACC, RANS]);

Warning, the names {PP, ACC} have been redefined and unprotected; the names {RANS} are absent in library <C:/Tallinn/Grodno/Vilnius/ArtKr.147>

{PP, ACC}

> FLib1("C:/Tallinn/Grodno/Vilnius/ArtKr.txt", 1, [PP, ACC, REA]);

[File is not a library, do you desire to update it?(y/n): y;

[Introduce set of names of procedures and/or modules, and/or their definitions: [SaveProc, ReadProc, "AGN := () -> `+`(args)/nargs:"];

[Invalid input. Continue(y/n)?: y;

[Introduce set of names of procedures and/or modules, and/or their definitions: {Release,PP, ParProc, RAC, RANS, REA};

Warning, {RANS, RAC, REA} are neither procedures nor modules
Warning, {REA} are neither procedures nor modules

{ParProc, Release, PP, ACC}

> FLib1("C:/Tallinn/Grodno/Vilnius/ArtKr.txt");

Warning, the names {ParProc, Release, PP, ACC} have been redefined and unprotected

{ParProc, Release, PP, ACC}

> FLib1("C:/Tallinn/Grodno/Vilnius/RANS.300", 1, [PP, ACC, REA]);

[Datafile does not exist, do you desire to create it?(y/n): y;

[Introduce set of names of procedures and/or modules, and/or their definitions: {Release, PP, ParProc, RAC, RANS, REA};

Warning, {RANS, REA, RAC} are neither procedures nor modules
Warning, {REA} are neither procedures nor modules

{Release, PP, ACC, ParProc}

> FLib1("C:/Tallinn/Grodno/Vilnius/RANS.300", 2);

Warning, library <C:/Tallinn/Grodno/Vilnius/RANS.300> has been compressed with names {Release, ParProc, PP, ACC}

> FLib1("C:/Tallinn/Grodno/Vilnius/RANS.300");

Warning, the names {ParProc, Release, ACC, PP} have been redefined and unprotected

{ParProc, Release, ACC, PP}

Thus, the above four procedures **SaveProc, ReadProc, FLib** and **FLib1** support a rather simple and user-friendly mechanism of work with the user libraries, created by means of the above manner.

For providing of the ***first level*** of connection of the user libraries of procedures and modules, it is possible to offer the following rather convenient mechanism. On the first step, the user library is being created as a datafile of the internal *Maple* format (*m*-file) containing definitions of the user procedures. With this purpose in a current document, the definitions of all procedures and modules, which are supposed to be included into a created text file or *m*-file, should be evaluated. Thereafter, by means of function call **save(P, CF)** the necessary saving of definitions of the software is done, where the first actual **P** argument defines a *sequence of names* of the saved software, whereas the second **CF** argument specifies a target file qualifier. The datafile created by such manner contains definitions of software in the input *Maple* format or in the internal *Maple* format and can be located in any subdirectory of the file system of a computer. For automation of a *logical connection* of a file created by the described manner, with the main *Maple* library the special procedure **LibUser** has been created. This method defines another approach to creation of the simple user libraries that is useful enough in the applied respects. Whereas the procedure **mPM** and **cmtf** provide a viewing of libraries created thus.

LibUser – logical linking of the user libraries with the main *Maple* library with their viewing
mPM
cmtf

Call format of the procedures:

LibUser(L {, h})
mPM(L)
cmtf(F)

Formal arguments of the procedures:

L - full path to an user library
h - (optional) an arbitrary valid *Maple* expression
F - symbol or string defining a datafile with software

Description of the procedures:

The procedure call **LibUser(L {, h})** admits use up to two actual arguments. The first obligatory actual **L** argument specifies a qualifier pointing full path to a required file with the user software. The successful procedure call **LibUser(L)** returns the *NULL* value with output of an appropriate warning, by providing access to the software located in the **L** file. By the procedure call **LibUser(L, h)** with two actual arguments the deletion out of the initialization file "Maple.ini" of the call of the user library specified by the first actual **L** argument is done; in addition, if the file is factually empty, it is being deleted. Thus, in general case a datafile **L** remains, while direct access to a software contained in it becomes inaccessible after command ***`restart`*** as the warning "Warning, software of datafile <%1> will be inaccessible after restart!" informs. For restoring of access to software located in the **L** file, the repeated procedure call **LibUser(L)** is needed.

As the information about logical link of the user library **L** (datafile with tools definitions) is in the initialization file "Maple.ini", then {*activation* | *deactivation*} of the link will be done only after call in a current session of the ***`restart`*** clause. If the library file has been registered in "Maple.ini", then each new ***`restart`*** clause or package loading will begin with a *Maple* document with warning "Software of datafile <%1> has become accessible now!".

It is necessary to note the essential circumstance. Depending on the *Maple* release (**6 - 9**), the initialization file "Maple.ini" is used by *Maple* at its start or restart if it is located in one or in a few subdirectories of the main *Maple* directory, namely: **BIN** (directory **BIN.WNT** for releases **6** and **7, BIN.W9X** (*Windows 98SE*) and **BIN.WIN** (*Windows XP Pro*) for release **8,** and **BIN.WIN** for release **9**), **LIB** and **USERS.**

There and then quite pertinently to note inadmissibility of a similar deviation from standardization of file structure of *Maple* when one of its basic subdirectories is renamed into dependences already from basic operating system. From the reasonable point of view, it is rather difficult to motivate it.

In addition, the **USERS** subdirectory is supposed by default, but it can be redefined at the *Maple* installation. The following useful table defines expedient locating of the "Maple.ini" file in the above-mentioned directories, where: **B** – **BIN**, **L** – **LIB** and **U** – **USERS**, **{+ | +1 | +2}** – *Maple* uses this file with {absolute | the first | the second} priority, "-" – *Maple* ignores the file.

Release	*B*	*L*	*U*	*B, L*	*B, U*	*L, U*	*B, L, U*
6	+	-	+	B	+1, +2	U	B, U
7	-	+	+	L	U	+1, +2	L, U
8	-	+	+	L	U	+1, +2	L, U
9	+	+	+	+2, +1	+, -	+1, +2	+2, +1

Thus, the shaded **U** column defines the directory **USERS** whose the "Maple.ini" file is used by *Maple* of all last five releases (**6 - 9/9.5**), however for the **9**th release this file is ignored, if the directory **BIN.WIN** also contains the similar file. In columns **<B, L>**, **<B, U>** and **<L, U>** are shaded cells which define the above directories expedient for locating of the "Maple.ini" files for their priority use by *Maple*. These files can contain any useful initialization information of a specific and/or common character, including definitions of procedures, modules or their calls. In our concrete case, this file is located in the **U** directory and is used for organization of logical link with the user libraries. On the basis of this file, the effective and simple mechanisms of library links are supported.

The procedure handles the basic especial and erratic situations, including situations conditioned by absence of *complete compatibility* of releases **6** and **7 - 9/9.5** of the package *Maple*. The described approach is convenient in the respect, that it admits use of a set of library files in various directories, and simplicity of their making and deleting in case of absence of the further necessity in them.

For contents viewing of the user *m*-file created by the manner described above, the procedure **mPM** can be useful. The procedure call **mPM(L)** returns the 2-element sequence, the first element of which represents the list of names of software of procedure type, while the second - the list of names of software of modular type which are contained in a *m*-file, given by the actual **L** argument. In addition, the first element of these lists defines the corresponding type, namely: **"proc"** - procedure and **"mod"** - module, whereas the other elements represent names of the corresponding *Maple* objects contained in the *m*-file. At absence of procedures or modules, the appropriate element of the sequence will be absent. At absence of *Maple* objects of the above types, the *NULL* value is returned with output of an appropriate warning. The procedure handles erratic situation, conditioned by incompatibility of *Maple* of releases **6** and **7 - 9/9.5**.

At last, the **cmtf(F)** procedure essentially extends the above procedure **mPM** on condition that a datafile **F** has been created by means of the procedure **save2.** However, with very high certainty the procedure analyzes datafiles created by the standard procedure **`save`** also (*m*-files are analyzed reliably, whereas files in the input *Maple* format with very high reliability). It is necessary to have in mind, the program modules in *m*-files by both procedures *`save`* and **save2** are saved incorrectly.

The procedure call **cmtf(F)** returns a sequence of lists whose the first element is an object identifier (**"proc"** - procedure, **"mod"** - module and **"others"** - other types), whereas the other elements represent names of the corresponding *Maple* objects contained in a datafile specified by the actual **F** argument. At absence of any of the above types, an appropriate element of the sequence will be absent. At absence of *Maple* objects of all types, the *NULL* value is returned with output of an appropriate warning. The procedure handles erratic situations, conditioned by incompatibility of *Maple* of releases **6** and **7 - 9/9.5,** and by non-readability of a datafile **F** by the standard function *`read`*.

```
LibUser:= proc(G::file)
local a, b, c, s, cd, d, h, H, S, F, Art_Kr, nu, p;
  Release(h), assign(cd = h, F = CF1(cat("", G))), assign(p = cat("try read("", F, ""):
    WARNING(")), RegMW();
  nu:= proc(f, s)
    local h;
      do h := readline(f);
        if h = 0 then close(f), close("_$_"); break
        elif not search(h, s) then writeline("_$_", h)
        end if
      end do;
      null(writebytes(f, readbytes("_$_", TEXT, infinity)), close(f), fremove("_$_"))
    end proc;
  while a <> 0 do [assign('a' = iolib(2, cat(cd, "\\", KCM(), "\\Maplesys.ini"))),
    `if`(search(convert(a, string), "UserDir"), [assign('b' = searchtext("=", a), 'c' = a),
    assign('d' = cat(c[b + 1 .. length(c)], "\\Maple.ini"))], 1)]
  end do;
  `if`(1 < nargs, RETURN(nu(d, p), WARNING("software of datafile <%1> will be inaccessible
    after restart!", F)), NULL);
  S := cat("try read("", F, ""): WARNING("software of datafile <", F, "> has be come accessible
    now!"): catch "could not open": WARNING("file <", F, "> does not exist") end try;");
  `if`(type(d, file), NULL, `if`(nargs = 1, [writeline(d, S), iolib(7, d), assign(H = 61)],
    [WARNING("initialization file <%1> does not exist", d), assign(H = 56)]));
  try assign(s(`if`(H = 56, NULL, `if`(nargs = 1, `if`(H = 61, 0,
    `if`(search(iolib(4, d, TEXT, infinity), S), NULL, [iolib(6, d, iolib(6, d, infinity) - 2),
    writeline(d, cat("\n", S)), iolib(7, d)])), [assign('H' = iolib(4, d, TEXT, infinity)),
    iolib(7, d), `if`(search(H, S), [assign('H' = cat(H[1 .. searchtext(S, H) - 1],
    H[searchtext(S, H) + length(S) .. length(H)])), iolib(5, d, H), iolib(7, d),
    `if`(Sts(iolib(4, d, TEXT, infinity), "\n") or  iolib(6, d, infinity) = 0, iolib(8, d),
    iolib(7, d))], NULL)]))), `if`(nargs = 1, (proc(d) read d end proc)(d), NULL))
  catch: null("Case caused by absence of full compatibility of Maple of releases 6 and {7, 8, 9}")
  end try
end proc
```

```
mPM:= proc(F::file)
local a, b, c, d, e, f, g, h, j, k, n, p, omega;
  iolib(7, F), assign(omega = (x -> `if`(nops(x) = 1, NULL, x)),
    j = (proc(x) local h, p;
      assign(h = ""), seq(assign('h' = cat(h, `if`(x[p] = " ", "~", x[p]))), p = 1 .. length(x)), h
```

```
    end proc));
  assign(e = ["proc"], f = ["mod"]), `if`(Release1() = 6 and iolib(4, F, TEXT, 2)[2] = "7",
    [iolib(7, F), ERROR("< %1> is m-file of Maple of release {7|8|9}", F)],
    iolib(7, F)), assign67('a' = (proc(x) read x; {anames(procedure), anames(`module`)} minus
        {anames(builtin)} end proc)(F)),
    assign('b' = iolib(4, F, TEXT, infinity)), assign('c' = Search(b, "I")),
    assign('d' = [op({seq(seq(`if`(j(convert(a[k], string)) = b[c[n] + 3 .. c[n] + length(a[k]) + 2]
    and (b[c[n] + length(a[k]) + 3 .. c[n] + length(a[k]) + 4] = "f*" or
    b[c[n] + length(a[k]) + 3 .. c[n] + length(a[k]) + 4] = "`6"), a[k], NULL), n = 1 .. nops(c)),
    k = 1 .. nops(a))})]), iolib(7, F), seq(`if`(type(d[k], procedure), assign('e' = [op(e), d[k]]),
    assign('f' = [op(f), d[k]])), k = 1 .. nops(d)), RETURN(assign67('a' = op(map(omega, [e, f]))),
    `if`(a = NULL, WARNING("file <%1> does not contain procedures and modules", F), a))
end proc

cmtf:= proc(F::file)
local a, c, k, omega, p, m, o, h, x, psi, nu;
  psi:= () -> seq(`if`(nops(args[k]) = 1, NULL, args[k]), k = 1 .. nargs);
  assign(p = ["proc"], m = ["mod"], o = ["others"], h = 7, c = Release());
  if not type(F, rlb) then ERROR("<%1> is not a text datafile", F)
  elif cat(" ", F)[-2 .. -1] <> ".m" then assign(a = interface(warnlevel)), interface(warnlevel = 0);
    try close(F), parse(readbytes(F, TEXT, infinity), statement), close(F)
    catch "incorrect syntax in parse: ": close(F), interface(warnlevel = a);
      ERROR("datafile <%1> is not read by function `read`", F)
    catch "invalid left hand side of assignment": close(F), interface(warnlevel = a);
      ERROR("datafile <%1> is not read by function `read`", F)
    end try;
    omega:= proc(x, h)
      local t;
        if search(h[x + 3 .. -1], "proc ", 't') and t = 1 then p := [op(p), cat(``, h[1 .. x - 2])]
        elif search(h[x + 3 .. -1], "module ", 't') and t = 1 then m := [op(m), cat(``, h[1 .. x - 2])]
        else o := [op(o), cat(``, h[1 .. x - 2])]
        end if
      end proc;
    do h := readline(F);
      if h = 0 then close(F); break else search(h, ":=", 'x'); omega(x, h) end if
    end do;
    interface(warnlevel = a), assign67('a' = psi(p, m, o)), `if`(a = NULL,
      WARNING("file <%1> does not contain any Maple objects", F), a)
  else
    nu:= proc(h)
      local a, b;
        assign(a = Iddn1(h)), assign(b = length(a));
        if c = 6 and h[b + 3] = "R" or 6 < c and h[b + 3 .. b + 4] = "f*" then p := [op(p), a]
        elif h[b + 3 .. b + 4] = "`6" then m := [op(m), a]
        else o := [op(o), a]
        end if
      end proc;
    do h := readline(F);
      if h = 0 then close(F); break elif h[1] <> "I" then next else nu(h) end if
    end do;
```

```
    assign67('a' = psi(p, m, o)), `if`(a = NULL,
       WARNING("file <%1> does not contain any Maple objects", F), a)
  end if
end proc
```

Typical examples of the procedures use:

```
> Art:= () -> `+`(args)/nargs:  Kr:= () -> [nargs, `+`(args)/nargs]:  protect(Kr);  AGN:= 57:
  protect('AGN');  Svet:= 37:  RANS:= table([x=y]):  IAN= matrix([]):  M1:= module ()
  export x; option package; x:= () -> `+`(args) end module:  module M2 () export x;
  option package;  x:= () -> `+`(args)  end module:
> save2(Art,Kr,M1,M2,'AGN','Svet',RANS,MkDir,IAN, "C:/rans/AGN/AVZ/Art_Kr.txt");
   Warning, the saving result is in datafile <C:/rans/AGN/AVZ/Art_Kr.txt>
> save2(Art,Kr,M1,M2,'AGN','Svet',RANS,MkDir,IAN, "C:/rans/AGN/AVZ/Art_Kr.m");
   Warning, the saving result is in datafile <C:/rans/AGN/AVZ/Art_Kr.m>
> cmtf("C:/rans/AGN/AVZ/Art_Kr.txt");
         ["proc", Art, Kr, MkDir], ["mod", M1, M2], ["others", AGN, Svet, RANS, IAN]
> cmtf("C:/RANS/rans_ian.txt");
   Error, (in cmtf) datafile <C:/RANS/rans_ian.txt> is not read by function `read`
> cmtf("C:/RANS/rans_ian.m");
   Warning, file <C:/RANS/rans_ian.m> does not contain any Maple objects
> mPM("C:/temp/Academy/Ztest.m");
   Warning, file <C:/temp/Academy/Ztest.m> does not contain procedures and modules
> mPM("C:/rans/AGN/AVZ/Art_Kr.m");
                     ["proc", MkDir, Art, Kr], ["mod", M1, M2]
> LibUser("C:/rans/AGN/AVZ/Art_Kr.txt");  restart;
    Warning, software of datafile <c:/rans/agn/avz/art_kr.txt> has become accessible now!
    Warning, software of datafile <c:/rans/agn/avz/art_kr.txt> has become accessible now!
> 3*map2([Art, Kr],62,57,37,41,7,15), AGN, Svet, eval(RANS), type(MkDir, procedure);
                     [219, 14541^(1/2)], AGN, Svet, RANS, true
> LibUser("C:/rans/AGN/AVZ/Art_Kr.txt", 7);
   Warning, software of datafile <c:/rans/agn/avz/art_kr.txt> will be inaccessible after restart!
> restart;  6*map2([Art, Kr], 62, 57, 37, 41, 7, 15), AGN, Svet, eval(RANS);
         [6*Art(62, 57, 37, 41, 7, 15), 6*Kr(62, 57, 37, 41, 7, 15)], AGN, Svet, RANS
```

The essence of algorithms implemented by the above procedures **LibUser, mPM** and **cmtf** are described in [29-33,39,43,44] in detail enough. They seem a rather useful at operation with the user libraries of non-standard organization and with datafiles containing definitions of the *Maple* objects above all of procedures and modules.

The main *Maple* library contains the most frequently used *Maple* procedures and modules (other than those in the kernel). This library is located in **Lib** directory of the package and contains a files set represented on fig. 1; the library contains three main files "Maple.hdb"`, "Maple.ind" and "Maple.lib", whereas some other files depend on the current *Maple* release. In particular, it concerns the obsolete file "Maple.rep" that is in releases **6** and **7** of *Maple* and which for subsequent releases is absent.

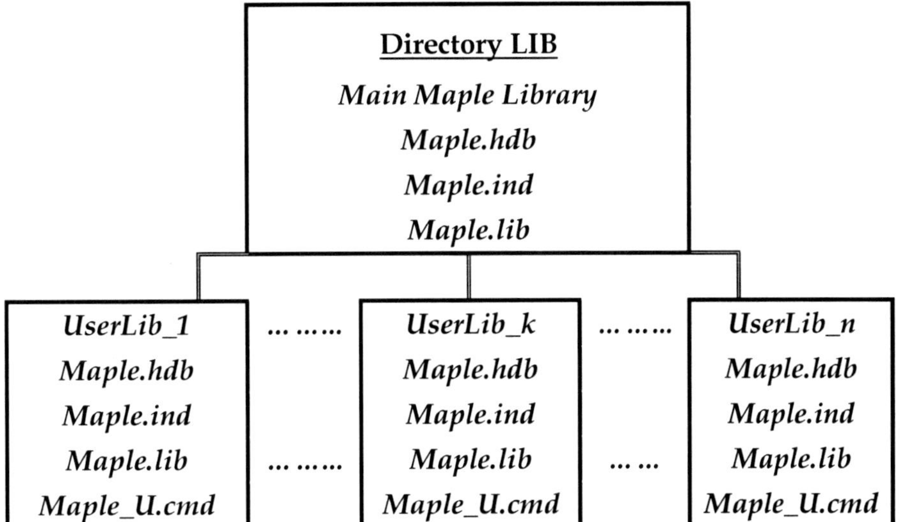

The principle files organization of the main *Maple* library and the user libraries

At the *second* level of the library organization the creation of the user libraries in subdirectories of the **LIB**-directory containing the main *Maple* library, standardly delivered with the package is provided. In this case, the files organization of the user libraries assumes the following simple view, inheriting the structural organization of the main *Maple* library (fig. 1). Each user library is located in separate subdirectory of the **LIB**-directory under a name **UserLib_k (k=1 .. n)**. The first three files of the user library are fully analogous to the appropriate files of the main *Maple* library whereas file "Maple_U.cmd" contains list of the procedures names located in the library and history of operation with the library. Furthermore, depending on a current *Maple* release in a process of operation with an user library in it can appear three extra files "Maple.rep", "elpam.ind" and "elpam.lib" whose description can be found in our previous books [29-33,43,44].

The user's libraries *organization* represented above, is convenient enough and localizes their disposition, providing their rather convenient program processing. Along with it, the given organization allows to use for support of the user libraries the *Maple* tools, namely its built-in *`march`* utility. The procedures represented below, provides a creation, update of the user libraries according to the above file organization, and also their logical linking with the main *Maple* library, what provides access to tools contained in them at a level of the standard *Maple* tools. Use of the library organization supported by *Maple* essentially simplifies work with the user libraries already by the circumstance, what for support of such libraries much software exists.

User_pfl **– creating, updating and viewing of the user libraries (*the second level*)**

User_pflM
User_pflMH
NonaLP
NLP

Call format of the procedures:

User_pfl(F, L, {S {, R}})
User_pflM(F, L, {S {, R}})
User_pflMH(F, L, {S {, R}})
NonaLP(F)
NLP(F {, h})

Formal arguments of the procedures:

F - symbol or string defining name or full path to a *Maple* library
L - set or list of names of procedures and/or program modules which should be saved in a library; this software should be preliminary evaluated in a current session
S - (optional) a size (posint) of a created user library
R - (optional) a mode (*string*) of a library link with the main *Maple* library
h - (optional) an assignable name

Description of the procedures:

The **User_pfl** procedure provides a *creating* or *updating* of the user libraries in conformity with the *Maple* library organization, represented above (fig. 1), and logical linking them with the main library of the package, what provides access to functional means contained in them, at a level of the package tools.

The first actual **F** argument specifies location of an user library. In addition, it is necessary to have in mind, manner of coding of the first argument determines two methods of a library making, namely. If *subdirectory name* with a library has been coded, the library is created in *subdirectory* with the indicated name in the **Lib** directory of the package. If *full path* to a library has been coded, the library is created just according to the indicated path to it. The full path guesses the indication of a logic device name with the library. It allows with a sufficient degree of arbitrariness to create user libraries. In this connection we recommend to locate libraries of the most often used means in the **Lib** subdirectory of the package, whereas all remaining ones - outside of file structure of *Maple*, including portable memory volumes.

The second actual **L** argument specifies a set or list of names of procedures and/or program modules whose definitions have been evaluated in a current session and which should be saved in a created or existing user library. For a saving in the library of a program module before evaluation of its definition, it is necessary to provide presence in its body of the ***`package`*** option of *Maple.*

The third optional integer **S** argument specifies the size of a created library with respect to amount of tools being placed in it (in addition, allowable maximum amount of tools approximately equals to the double value of the third argument). By default, the procedure supposes value 500 for the third argument. If the third actual argument accepts a value {**"U"**, **"M"**}, the procedure provides the logical linking of an user library with a chain of libraries defined by the predetermined ***libname***-variable of *Maple* with the subsequent priority of tools searching, namely: <*user library, libname*> (**"U"**) or <*libname, user library*> (**"M"**). By default, value **"U"** is supposed.

The fourth optional **R** argument has the same sense as the third argument **S** with value {**"U"**, **"M"**}. By default for the given argument is supposed value **"U"** for a new-created library and value **"M"** for an updated library. The procedure uses tools of the MS DOS; therefore, at the end of its running it is necessary to close the dark window "**MS DOS Prompt**" in **2 - 3** sec after its visualization. The successful procedure call returns the current state of the predetermined ***libname***-variable of *Maple* and outputs the warning about a creating/updating of an user library. After that, the user can use tools of the given library analogously to the standard package tools.

The persistent access to the user library created by means of the **User_pfl** procedure is provided on the basis of the created or updated file "Maple.ini", into which a line of view ***`libname:= F, library:`*** or ***`libname:=library, F:`*** is put. The line defines chain of logically linked libraries; location of an user **F** library in the chain defines a priority of access to its tools. File "Maple.ini" is located in a subdirectory defined by ***UserDirectoty***-variable of initialization file "Maplesys.ini". In addition, in any case for a current *Maple* session the procedure sets maximal priority of access for a new-created user library.

The **User_pfl** procedure presumes a work with *case-insensitive* names of the saved tools, what for avoidance of occurrence of possible misunderstandings (erratic, special and unpredictable situations) presumes availability at creation / update of an user library by means of the given procedure of empty crossing of sets of names of tools which are located in the library and/or should be put into it under condition of their reducing to one register (upper case or lower case). At noncompliance of the given condition into the index file "Maple.ind" of the library all names of procedures and modules will be put, whereas for procedures into the basic body "Maple.lib" of the library the definitions only of chronologically last ones with case-insensitive names will be put, i.e. if into a library a procedure with **n**-name is put, then the given name will be put into the index file "Maple.ind", and its definition replaces definition of procedure with the same name which matches to the first, under condition of the *case-insensitivity*. However, it is necessary to have in mind, that the logical but not physical replacement is done, i.e. new and old copies of the tool are retained in file "Maple.lib".

Let us term two symbols **x** and **y** as *case-equivalent* ones, if the following relation **convert(x, *uppercase*) = convert(y, *uppercase*)** takes place. In conditions of the principle of case-sensitivity of names in the *Maple* environment in many cases, it is desirable to have information on presence in a library of *case-equivalent* names. In particular, the given information is rather useful to testing presence in libraries of discrepancies of their index files “Maple.ind” with files “Maple.lib”. This problem is solved by means of the **NonaLP** procedure.

Above all, the procedure supposes that a tested library is the *Maple* library, i.e. it is similar to the main *Maple* library. The procedure call **NonaLP(F)** tests a library specified by the actual **F** argument for the purpose of presence in it of procedures with *case-equivalent* names. If such names do not exist, the *NULL* value is returned. On the other hand, if such names have been found and **F** library is logically linked with the main *Maple* library, and a *discrepancy* of its index file "Maple.ind" with file "Maple.lib" does not exist, then the *NULL* value is returned also, but the procedure additionally outputs an appropriate warning containing the set of such procedures names.

At detection of a *discrepancy* between files “Maple.ind” and “Maple.lib” of a **F** library which is logically linked with the main *Maple* library, the procedure returns the list of view **[***Active* = {P1, …, Pn}, *Symbols* ={P11,…,Pn1}**]**, in which the first set *Active* defines names of available procedures and the second set *Symbols* defines *case-equivalent* names whose indexes are located in index file "Maple.ind" but whose definitions do not exist in the main library file "Maple.lib". In case of absence of logical linking of an analyzed library **F** with the main *Maple* library the *Active* list is returned *empty*, whereas the second *Symbols* list contains names of all procedures and program modules, contained in the **L** library that is structurally analogous to the main *Maple* library.

The procedure **User_pflM** (modification of the above procedure **User_pfl**) with the same formal arguments and rules of their coding at a procedure call automates creation of the user libraries deprived of the above restrictions. The result of its call also is analogous to a case of the **User_pfl** procedure. However essential difference of the **User_pflM** procedure from the **User_pfl** procedure is that it allows to save in an user library the software (procedures and program modules) in case of the *case-sensitivity;* i.e. the procedure correctly saves, for example, procedures with names \`AAA\`, \`AaA\` and \`aAa\` as three different objects. In addition, it is necessary to have in mind; both procedures represented above create an user library with empty Help database, i.e. a *dummy* Help database "Maple.hdb".

At last, the procedure **User_pflMH** with respect to formal arguments, of a rule of their coding at a procedure call, and the returned results is analogous to the above procedures **User_pfl** and **User_pflM** excepting two essential circumstances, namely. First, the procedures **User_pfl** and **User_pflM** provide creation of an user library with a *dummy* Help database in arbitrary **F** directory of file system of a computer, whereas the procedure **User_pflMH** serves for creation of an user library with Help database specified by the user. Therefore, for creation of a library Help database the procedure inquires the user about the Help database location, namely:

Define a setup mode of a Library help database {*dummy, default, path*}:

Next, depending on the received answer the procedure **User_pflMH** fulfills one of three operations:

Dummy – *a dummy Help* database for an user library will be created;
default – *Help* database located in file "C:/maple_6-9/books/libraries/userlib9/maple.hdb" will be selected as a library *Help* database;
path – *Help* database located in a subdirectory or in datafile specified by the given *path* will be selected as a library *Help* database.

In addition, it is supposed that a specified subdirectory contains beforehand prepared Help database as a file "Maple.hdb". Furthermore, this procedure is intended for creation of the user libraries that should have structural organization similar to the main *Maple* library. Such decision does not bear any restrictive character because, in our opinion, the offered organization of the important user libraries in the best way corresponds to essence of the given aspect of file organization of the *Maple* environment.

Thus, the **User_pflMH** procedure relative to the procedures **User_pfl** and **User_pflM** can be considered as the most general and universal tool of creation of the user libraries structurally similar to the main *Maple* library which are logically linked with it. A series of examples represented below illustrate use of the above procedures for creation of the user libraries of modules and procedures.

In general, the above three procedures **User_pfl, User_pflM** and **User_pflMH** execute the following three basic operations:

* *registration* in the system file "Win.ini" of a current release of the package if earlier such registration has not been made;

* *creation* or *updating* of the initialization file "Maple.ini" with the purpose of a providing of logical link of a library with the main *Maple* library;

* *creation* or *updating* of a library (in addition, the last procedure provides an opportunity of creation of a library help database according to the user suggestion).

Procedures output the appropriate warnings about the done work. Thus, at creation or update of a library the user should specify only a library name (if library will be located in the **LIB** directory of *Maple*) or full path to it, and also a set or list of tools names which should be saved in the library and whose definitions have been evaluated in a current *Maple* session. Additionally, the user can define size for a new-created library, its help database and a mode of logical link of the library with the main *Maple* library.

For operational viewing of contents of a library structurally similar to the main *Maple* library the procedure **NLP(F)** can be a rather useful tool which returns the 2-element sequence whose the first element represents the sorted list of names of procedures and calls of the exported variables of program modules contained in a library specified by its name or full path to it in actual **F** argument, while the second element – their quantity. If the procedure call **NLP(F, h)** uses the second optional argument **h,** then through it the list of packages contained in a library **F** is returned. In addition, if a library **F** is logically linked with the main *Maple* library or is located in the main *Maple* directory, then it can be specified by library name; otherwise full path to the library should be specified. A series of the examples represented below illustrate use of the above procedures.

```
User_pfl:= proc(R::{name, string, symbol}, U::{list(symbol), set(symbol)})
local a, b, k, G, Art, W, U_cmd, M, Q, Z, P, V, h, BS, F, MR, L, T, TU, TM, `0`, omega, lib;
```

```
  RegMW(), assign(MR = Release('h'), L = [op(U)],
      BS = (S::{string, symbol} -> sstr(["\\" = "/"], S)));
  try
    `if`(cat("", R)[2] = ":", assign(lib = BS(CF(R))), assign(lib = BS(CF(cat(h, "/Lib/", R)))));
    assign(omega = interface(warnlevel)), interface(warnlevel = 0), assign('lib' = BS(MkDir(lib)));
    interface(warnlevel = omega), march('create', lib,
      `if`(3 <= nargs and type(args[3], posint), args[3], 500))
  catch "directory exists and is not empty": NULL("Especial situation 1")
  catch "there is already an archive in": NULL("Especial situation 2")
  end try;
  U_cmd := proc(S::string, L::{set, list})
    local k, Kr;
      Kr := fopen(cat(S, "/maple_U.cmd"), APPEND, TEXT);
      writeline(Kr, cat("$ New user procedures and modules = ", ssystem("date")[2][17 .. 30])),
        seq(writeline(Kr, L[k]), k = 1 .. nops(L)), iolib(7, Kr)
    end proc;
  writeline("$Help", "Empty database for your Help system"), iolib(7, "$Help");
  makehelp(Empty, `$Help`, convert(lib, symbol)), iolib(8, "$Help");
  [assign('savelibname' = lib, Art = (proc(x) if search(x, "/") or search(x, "*") or
      type(x, `module`) then savelib(x) else save x, cat(h, "/$", x, `.m`) end if
    end proc)), seq(Art(L[k]), k = 1 .. nops(L))];
  V := subsop(seq(`if`(search(L[k], `/`) or search(L[k], `*`) or type(L[k], `module`),
    k = NULL, NULL), k = 1 .. nops(L)), L);
  [assign('W' = march('list', lib)), assign('M' = [seq(convert(W[k][1], symbol), k = 1 .. nops(W))])];
  seq(march(`if`(member(cat(V[k], `.m`), M), 'update', 'add'), lib, cat(h, "/$", V[k], `.m`), cat(V[k],
    `.m`)), k = 1 .. nops(V)), U_cmd(lib, L), delf(seq(cat(h, "/$", V[k], `.m`), k = 1 .. nops(V))),
    map(march, ['pack', 'reindex'], lib), assign(Q = cat(CCM(), "\\Maplesys.ini"));
  assign('Z' = iolib(4, Q, TEXT, infinity)), iolib(7, Q), assign('Z' = cat(Z[searchtext("UserD", Z) +
    14 .. searchtext("\n", Z, searchtext("UserD", Z) .. length(Z)) + searchtext("UserD", Z) - 2],
    "/Maple.ini"));
  assign(TM = cat("libname:=libname,"", lib, "":"), TU = cat("libname:="", lib, "",libname:"));
  try
    assign(P = iolib(4, Z, TEXT, infinity)), iolib(7, Z);
    `if`(search(P, lib), NULL, [fopen(Z, APPEND, TEXT), writeline(Z, `if`(3 <= nargs and
      member(args[3], {"M", "U"}), `if`(args[3] = "U", TU, TM), `if`(nargs = 4 and
      member(args[4], {"M", "U"}), `if`(args[4] = "U", TU, TM), TU)))])
  catch "file or directory does not exist": writeline(Z, `if`(3 <= nargs and
    member(args[3], {"M", "U"}), `if`(args[3] = "U", TU, TM), `if`(nargs = 4 and
    member(args[4], {"M", "U"}), `if`(args[4] = "U", TU, TM), TU)))
  end try;
  (proc() global libname; libname := lib, libname; libname end proc)(),
    WARNING("the user library <%1> has been created/updated", lib), iolib(7, Z)
end proc

NonaLP:= proc(L::{string, symbol})
local active, a, h, k, j, G, omega, s, symbols;
  `if`(type(L, mlib), assign(h = [], a = NLP(L)[1], G = [], s = {}),
    ERROR("<%1> is not a Maple library", L));
  for k in a do
    try `if`(type(eval(k), `module`) or search(cat("", k), ":-"), NULL, assign('h' = [op(h), k]))
```

```
    catch "": null("Exception handling with program modules")
    end try
  end do;
  for k to nops(h) - 1 do
    for j from k + 1 to nops(h) do
      if Case(h[k], lower) = Case(h[j], lower) then G := [op(G), h[k], h[j]]  else next  end if
    end do;
    `if`(G = [], NULL,  [assign('s' = {op(s), op(G)}), assign('G' = [])])
  end do;
  assign(omega = ((x, y) -> `if`(lexorder(Case(x), Case(y)), true, false))), `if`(s = {},
    WARNING("case sensitive names have not been found in library <%1>", L),
    op([seq(`if`(type(eval(s[k]), procedure), NULL, assign('G' = [op(G), s[k]])), k = 1 .. nops(s)),
    `if`(G <> [], [Active = s minus {op(G)}, Symbols = {op(G)}], NULL),
    WARNING("library <%1> contains available case sensitive names %2",
    L, sort([op(s)], omega))]))
end proc

User_pflM:= proc(R::{symbol, name, string}, U::{list(symbol), set(symbol)})
local a, b, c, k, n, G, Art, W, U_cmd, M, Q, Z, P, V, h, BS, F, MR, L, T, TU, TM, `0`, omega;
  RegMW(), assign(MR = Release('h'), L = [op(U)], n = [], BS = (S -> sstr(["\\" = "/"], S)));
  try
    `if`(cat("", R)[2] = ":", assign('G' = BS(cat("", R))), assign('G' = BS(CF(cat(h, "/Lib/", R)))));
    assign(omega = interface(warnlevel)), interface(warnlevel = 0), assign('G' = BS(MkDir(G)));
    interface(warnlevel = omega),
      march('create', G, `if`(3 <= nargs and type(args[3], posint), args[3], 610))
  catch "directory exists and is not empty":  NULL("Especial situation 1")
  catch "there is already an archive in":  NULL("Especial situation 2")
  end try;
  U_cmd:= proc(S::string, L::{set, list})
    local k, Kr;
      Kr := fopen(cat(S, "/maple_U.cmd"), APPEND, TEXT);
      writeline(Kr, cat("$ New user procedures and modules = ", ssystem("date")[2][17 .. 30])),
        seq(writeline(Kr, L[k]), k = 1 .. nops(L)), iolib(7, Kr)
    end proc;
  writeline("$Help", "Help database of the Aladjev's Library * version 2.157 * International
     Academy of Noosphere, The Baltic Branch = Tallinn = May 3, 2003"), iolib(7, "$Help");
  makehelp(Empty, `$Help`, convert(G, symbol)),
    iolib(8, "$Help");
  [assign('savelibname' = G, Art = (proc(x, k)
    local a;
      if search(x, `/`) or search(x, `*`) or type(x, `module`) then savelib(x)
      else  a := cat("/$", x, cat(``, k), `.m`);  save x, cat(h, a);  a
      end if
    end proc)), seq(assign('n' = [op(n), Art(L[k], k)]),  k = 1 .. nops(L))];
  assign('V' = subsop(seq(`if`(search(L[k], `/`) or search(L[k], `*`) or  type(L[k], `module`),
    k = NULL, NULL), k = 1 .. nops(L)), L)), assign(c = nops(V));
  [assign('W' = march('list', G)), assign('M' = [seq(convert(W[k][1], symbol), k = 1 .. nops(W))])];
  seq(march(`if`(member(cat(V[k], `.m`), M), 'update', 'add'), G, cat(h, n[k]), cat(V[k], `.m`)),
    k = 1 .. c),  U_cmd(G, L),  delf(seq(cat(h, n[k]), k = 1 .. nops(n))),
    map(march, ['pack', 'reindex'], G), assign(Q = cat(CCM(), "\\Maplesys.ini")),
```

```
    assign('Z' = iolib(4, Q, TEXT, infinity)), iolib(7, Q);
  assign('Z' = cat(Z[searchtext("UserD", Z) + 14 .. searchtext("\n", Z, searchtext("UserD", Z) ..
     length(Z)) + searchtext("UserD", Z) - 2], "/Maple.ini"));
  assign(TM = cat("libname:=libname,"", G, "":"),  TU = cat("libname:="", G, "",libname:"));
  try
    assign(P = iolib(4, Z, TEXT, infinity)), iolib(7, Z);
    `if`(search(P, G), NULL, [fopen(Z, APPEND, TEXT),
      writeline(Z, `if`(3 <= nargs and member(args[3], {"U", "M"}), `if`(args[3] = "U", TU, TM),
       `if`(nargs = 4 and member(args[4], {"U", "M"}), `if`(args[4] = "U", TU, TM), TU))))])
  catch "file or directory does not exist":  writeline(Z, `if`(3 <= nargs and member(args[3],
    {"U", "M"}), `if`(args[3] = "U", TU, TM), `if`(nargs = 4 and member(args[4], {"U", "M"}),
    `if`(args[4] = "U", TU, TM), TU)))
  end try;
  (proc() global libname;  libname := G, libname; libname  end proc)(),
    WARNING("the User library <%1> has been created/updated", G), iolib(7, Z)
end proc

User_pflMH:= proc(R::{name, string, symbol}, U::{list(symbol), set(symbol)})
local a, b, c, k, Kr, n, G, Art, W, U_cmd, M, Q, Z, P, r, V, h, BS, HDB, F, MR, L, T, TU, TM,
     omega, Sigma, nu, psi;
  RegMW(), assign(r = {"U", "M"}, Kr = (a -> `if`(a = "U", TU, TM))),
    assign(MR = Release('h'), L = [op(U)], n = [], BS = (S -> sstr(["\\" = "/"], S)),
    psi= (x -> WARNING("Help database <%1> does not exist, library has been created
      without Help database", x)));
  try
    `if`(cat("", R)[2] = ":", assign('G' = BS(cat("", R))), assign('G' = BS(cat(h, "/Lib/", R))));
    assign(nu = interface(warnlevel)), interface(warnlevel = 0), assign('G' = MkDir(G));
    interface(warnlevel = nu), march('create', G, `if`(3 <= nargs and
      type(args[3], posint), args[3], 500))
  catch "directory exists and is not empty":  NULL("Especial situation 1")
  catch "there is already an archive in":  NULL("Especial situation 2")
  end try;
  U_cmd:= proc(S::string, L::{set, list})
    local k, Kr;
      Kr := fopen(cat(S, "/maple_U.cmd"), APPEND, TEXT);
      writeline(Kr, cat("$ New user procedures and program modules = ",
        ssystem("date")[2][17 .. 30])), seq(writeline(Kr, L[k]), k = 1 .. nops(L)), iolib(7, Kr)
    end proc;
  HDB:= proc(x)
    local a;
      assign(a = `$Help_database`), writeline(a, "Help database for Maple 6/7/8/9 *
      version 2.157 * International Academy of Noosphere, The Baltic Branch *
     RANS_IAN_RAC_REA; Tallinn - Grodno - Vilnius - Moscow = January, 2004"),
     iolib(7, a), makehelp(Dummy, a, convert(x, symbol)), iolib(8, a)
    end proc;
  omega := cat("", readstat("Define a setup mode of a Library help database {dummy, default,
    path}: "));
  if omega = dummy then HDB(G); Sigma := "dummy" elif omega = "default" then
    Sigma := "C:\\Maple_6-9\\Books\\Libraries\\UserLib9\\Maple.hdb";
    if type(Sigma, file) then system(cat("Copy ", Path(Sigma), " ", Path(G)))
```

```
    else HDB(G); Sigma := "dummy"
    end if
  else
    `if`(type(omega, file) and Ftype(omega) = ".hdb", assign(Sigma = omega),
      [assign('omega' = cat(omega, "\\Maple.hdb")), `if`(type(omega, file),
      assign(Sigma = omega), assign(Sigma = "dummy"))]);
    if Sigma = "dummy" then HDB(G) else system(cat("Copy ", Path(Sigma), " ", Path(G)))
    end if
  end if;
  [assign('savelibname' = G, Art = (proc(x, k)
    local a;
      if search(x, "/") or search(x, "*") or type(x, `module`) then savelib(x)
      else a := cat("/$", x, cat(``, k), `.m`); save x, cat(h, a); a
      end if
    end proc)), seq(assign('n' = [op(n), Art(L[k], k)]), k = 1 .. nops(L))];
  V := subsop(seq(`if`(search(L[k], "/") or search(L[k], "*") or type(L[k], `module`),
    k = NULL, NULL), k = 1 .. nops(L)), L), assign(c = nops(V));
  [assign('W' = march('list', G)), assign('M' = [seq(convert(W[k][1], symbol), k = 1 .. nops(W))])];
  seq(march(`if`(member(cat(V[k], `.m`), M), 'update', 'add'), G, cat(h, n[k]), cat(V[k], `.m`)),
    k = 1 .. c), U_cmd(G, L), delf(seq(cat(h, n[k]), k = 1 .. nops(n))),
    map(march, ['pack', 'reindex'], G), assign(Q = cat(CCM(), "\\Maplesys.ini"));
  assign('Z' = readbytes(Q, TEXT, infinity)), iolib(7, Q), assign('Z' =
    cat(Z[searchtext("UserD", Z) + 14 .. searchtext("\n", Z, searchtext("UserD", Z) ..
    length(Z)) + searchtext("UserD", Z) - 2], "/Maple.ini"));
  assign(TM = cat("libname:=libname,"", G, "":"), TU = cat("libname:="", G, "",libname:"));
  try assign(P = iolib(4, Z, TEXT, infinity)), iolib(7, Z);
    `if`(search(P, G), NULL, [fopen(Z, APPEND, TEXT), writeline(Z, `if`(3 <= nargs and
      member(args[3], r), Kr(args[3]), `if`(nargs = 4 and member(args[4], r), Kr(args[4]), TU)))])
  catch "file or directory does not exist": writeline(Z, `if`(3 <= nargs and member(args[3], r),
    Kr(args[3]), `if`(nargs = 4 and member(args[4], r), Kr(args[4]), TU)))
  end try;
  (proc() global libname; libname := G, libname; libname end proc)(),
    WARNING("the user library <%1> has been created/updated", G), iolib(7, Z),
    WARNING("Help database for the Library has been defined as <%1>", Sigma)
end proc

NLP:= proc(L::{string, symbol})
local a, b, c, d, k, t, g, pack, pm, omega, sf;
  sf, pack, pm, c := (x, y) -> `if`(lexorder(Case(x), Case(y)), true, false), [], [], [libname];
  omega:= () -> `if`(1 < nargs and type(args[2], symbol), args[2], NULL);
  if cat("", L)[2] = ":" then `if`(type(L, mlib), assign(a = march(list, L)),
    ERROR("<%1> is not a Maple library", L))
  else
    for k to nops(c) do Search1(CF2(c[k]), cat("\\", Case(L)), 't');
      if t = [right] then a := march(list, c[k]); break end if
    end do;
    if not type(a, list) then g := cat(CDM(), "\\LIB\\", L); `if`(not type(g, mlib),
      ERROR("<%1> is not a Maple library", L), assign('a' = march(list, g))) end if
  end if;
  for k to nops(a) do
```

```
    if a[k][1][1 .. 2] = ":-" then next else d := a[k][1][1 .. -3]  end if;
    try  b:= [exports(cat(``, d))];  pack:= [op(pack), d]; pm:= [op(pm), seq(cat(d, ":-", k), k = b)]
    catch "wrong number (or type) of parameters in function":  pm := [op(pm), d]
    end try
  end do;
  if 1 < nargs and type(args[2],symbol) then assign(args[2] = map(convert, sort(pack,sf), symbol))
  end if;
  Repeat;
  try map(convert, sort(pm, sf), symbol), nops(pm)
  catch "bad id": goto(Repeat)
  end try
end proc
```

Typical examples of the procedures use:

> Art:=() ->`+`(args)/nargs: Kr:=module () option package; export sr; sr:=() ->sqrt(`+`(args)/ nargs) end module: Sv:= () -> [nargs, {min(args), max(args)}]: module Vict () export gr; gr:=() -> WARNING("actual arguments %1",[args]) end module: Gal:=proc() Close() end proc: User_pfl("C:/RANS/Tallinn\\Grodno\\SVEGAL", [Art,Kr,Sv,Vict,Gal],7,"M");

Warning, the user library <c:\rans\tallinn\grodno\svegal> has been created/updated

"c:\rans\tallinn\grodno\svegal", "c:/program files/maple 9/lib/userlib", "C:\Program Files\Maple 9/lib"

> restart; libname; 6*Art(62, 57, 37, 41, 7, 15), 6*Kr:- sr(62, 57, 37, 41, 7, 15), Sv(62, 57, 37, 41, 7, 15), Vict:- gr(1995, 2004), Gal();

"c:/program files/maple 9/lib/userlib", "C:\Program Files\Maple 9/lib", "c:/rans/tallinn/grodno/svegal"

Warning, actual arguments [1995, 2004]

219, 3*146^(1/2), [6, {7, 62}]

> NonaLP("C:/RANS\\Grodno\\SVEGAL"); NonaLP(UserLib);

Warning, library <C:/RANS\Grodno\SVEGAL> contains available case sensitive names [Art, art, kR, kr, Kr, Sv, sv]

Warning, library <UserLib> contains available case sensitive names [MEM, Mem, mmf, MmF, modproc, ModProc, save1, Save1, save2, Save2, sls, SLS, SSF, ssf, `type/Lower,` `type/lower`, `type/Upper`, `type/upper`]

> NonaLP("C:/RANS\\Grodno\\Svetlana");

Warning, library <C:/RANS\Grodno\Svetlana> contains available case sensitive names [art, Art, kr, kR, sv, Sv]

[Active = {Art}, Symbols = {art, sv, kr, Sv, kR}]

> Art:=() ->`+`(args)/nargs: Kr:=module () option package; export sr; sr:=() ->sqrt(`+`(args)/ nargs) end module: Sv:= () -> [nargs, {min(args), max(args)}]: module Vict () export gr; gr:=() -> WARNING("actual arguments %1", [args]) end module: sv:=proc() Close() end proc: User_pflM("C:/RANS/Tallinn\\Grodno\\VGTU", [Art, Kr, Sv, Vict, sv], 7, "M");

Warning, the User library <c:/rans/tallinn/grodno/vgtu> has been created/updated

"c:/rans/tallinn/grodno/vgtu", "c:/program files/maple 9/lib/userlib", "C:\Program Files\Maple 9/lib", "c:/rans/tallinn/grodno/svegal"

```
> restart; NLP("C:/RANS/Tallinn/Grodno/VGTU"), libname;
```

[Art, `Kr:- sr`, sv, Sv, `Vict:- gr`], 5, "c:/program files/maple 9/lib/userlib", "C:\Program Files\Maple 9/lib", "c:/rans/tallinn/grodno/svegal", "c:/rans/tallinn/grodno/vgtu"

```
> Art:=() ->`+`(args)/nargs: Kr:=module () option package; export sr; sr:=() -> sqrt(`+`(args)/
  nargs) end module: Sv:= () -> [nargs, {min(args), max(args)}]: module Vict () export gr;
  gr:= () -> WARNING("actual arguments %1", [args]) end module: sv:=proc() Close() end
  proc: User_pflMH("C:/RANS/Tallinn\\Vilnius\\VGTU", [Art, Kr, Sv, Vict, sv], 7, "M");
[ Define a setup mode of a Library help database {dummy, default, path}:
  "C:\\Academy\\UserLib9\\UserLib";
```

Warning, the user library <c:\rans\tallinn\vilnius\vgtu> has been created/updated

Warning, Help database for the Library has been defined as <C:\Academy\UserLib9\UserLib\Maple.hdb>

"c:\rans\tallinn\vilnius\vgtu", "c:/program files/maple 9/lib/userlib", "C:\Program Files\Maple 9/lib"

```
> User_pflMH("C:/RANS/Tallinn\\Grodno\\GRSU", [Art,Kr,Sv,Vict,sv], 7, "M");
[ Define a setup mode of a Library help database {dummy, default, path}: default;
```

Warning, the user library <c:\rans\tallinn\grodno\grsu> has been created/updated

Warning, Help database for the Library has been defined as <C:\Maple_6-9\Books\Libraries\UserLib9\Maple.hdb>

"c:\rans\tallinn\grodno\grsu", "c:/program files/maple 9/lib/userlib", "C:\Program Files\Maple 9/lib"

```
> User_pflMH("C:/RANS/Tallinn\\Tartu\\TRU", [Art,Kr,Sv,Vict,sv], 7, "M");
[ Define a setup mode of a Library help database {dummy, default, path}: dummy;
```

Warning, the user library <c:\rans\tallinn\tartu\tru> has been created/updated
Warning, Help database for the Library has been defined as <dummy>

"c:\rans\tallinn\tartu\tru", "c:\rans\tallinn\grodno\grsu", "c:/program files/maple 9/lib/userlib", "C:\Program Files\Maple 9/lib"

```
> NLP(UserLib, h);
```

[&Shift, _0N, _1N, _mMfile, _nnn, _rd, _SL, A_Color, ACC:-acc, ACC:-Ds, ACC:-Sr, Aconv, ACP, Adrive, AFdes, AlgLists:-&*, AlgLists:-&+, AlgLists:-&-, AlgLists:-&/, AlgLists:-&^, All_Close, AnalT, andN, Animate2D, Animate3D, Aobj, Apmv, Aproc, Arobj, ArtMod, Assign, assign6, assign67, assign7, Atr, AtrRW, avm_VM, belong, Bit, Bit1, blank, Builtin, came, Case,

==

Type_D, typeseq, U_test_MW, UbF, Ulib, Uninstall, UpLib, User_pfl, User_pflM, User_pflMH, UserC, Users, V_Solve, VisM, Vol, Vol_Free_Space, VTest, WD, Weight_LF, Weights, Weights_L, winver, writedata1, WS, WT, X_test_VW, xbyte, xbyte1, xNB, xorN, xpack, XSN, XTfile], 527

```
> h;
```

[ACC, AlgLists, DIRAX, SimpleStat, SoLists]

```
> NonaLP("C:\\Program Files\\Maple 6\\LIB\\Svetlana");
```

Warning, library <C:\Program Files\Maple 6\LIB\Svetlana> contains available case sensitive names [KR, Kr, kr, save1, Save1, save2, Save2, sls, SLS, ssf, SSF]

Ulib **– basic operations with libraries structurally similar to the *Maple* library**

DUlib
helpman
InstUlib
LibLink
LibLink1
Plib

Call format of the procedures:

Ulib(L, N)
DUlib(L)
InstUlib(F, P {, 0})
helpman(R, L, U {, Z})
LibLink(D {, `$`} {, `1`})
LibLink1(D {, `$`} {, `1`})
Plib(F)

Formal arguments of the procedures:

L - a library name, or symbol or string defining full path to a *Maple* library
N - an integer defining a mode of operation with a *Maple* library; **N={0|1|2}**
F - symbol or string defining name or full path to a *Maple* library
D - symbol or string defining full path to a *Maple* library
R - symbol defining a mode of operation with a Help database; **R={*insert*|*display*|*delete*}**
U - a library directory, symbol, string, or list of strings or symbols
Z - (optional) a list of strings or symbols
P - symbol or string defining full path to a *Maple* library
0 - (optional) zero value defining a mode of logical libraries linking
`$` - (optional) a key symbol defining a mode of link registration
`1` - (optional) a key symbol defining a priority of library linking

Description of the procedures:

The above procedures **User_pfl, User_pflM** and **User_pflMH** create the user libraries on the basis of the specified tools and automatically provide their logical linking with the main *Maple* library. In this connection, it is expedient to have any tools of automatic *viewing* of contents of any user library created by means of the above procedures and of deletion or restore of its logical linking with the main *Maple* library.

The procedure call **Ulib(L, N)** in accordance with a value of the second actual **N** argument provides the following basic operations with a library **L** with structural organization analogous to the main *Maple* library, namely:

N = 0 - return of the full list of names of procedures and program modules (*m*-files) of a library specified by the first actual **L** argument with determination of their quantity - the second element of the returned sequence;

N = 1 - canceling of logical linking of a library **L** with the main library of the *Maple* package;

N = 2 - creation/restoration of logical linking of a library **L** with the main library of the package; in addition, the restoration of logical link of the library **L** is done with the lowest priority, i.e. in a chain of the predefined *`libname`* variable the path to library **L** will be the latest. By change of order of paths to libraries represented in the ***libname***-variable, the user can control order of the means call of the same names that are located in these libraries.

The procedure call **Ulib(L, N)** returns the *NULL* value, i.e. nothing. The procedure handles the basic erratic and especial situations, and outputs the appropriate warnings about the done work. The told concerning the coding of the first actual argument of the above procedure **User_pflM** concerns the procedure **Ulib** in the full measure. The procedure does not delete the user libraries, allowing if necessary to use afterwards them again.

The immediate deletion of the user libraries can be done outside of the *Maple* environment by commands of the MS DOS or Windows, or for these purposes can be used the procedure **DUlib(L).** The procedure call **DUlib(L)** deletes out of file system of a computer the subdirectory with an user library specified by the actual **L** argument with output of the appropriate warning. A library name (a name of its directory), or full path to a library can be used as a value for the actual **L** argument. Therefore, the procedure provides deletion of any user library irrespective of its location in file system of a computer. At impossibility to remove the library, the procedure displays the appropriate diagnostic warning about files which it could not remove.

However, before deletion of a library **L** by means of the previous procedure call **Ulib(L, 1)** its logical link with the main *Maple* library should be canceled and the ***`restart`*** clause should be fulfilled (otherwise erratic situation "Error, (in **rmdir**) file I/O error" arises, because *Maple* saves information about the library up to the following reboot). After that, the procedure call **DUlib(L)** deletes the **L** directory out of file system of a computer with output of an appropriate warning.

The procedure call **InstUlib(F, P)** provides copying of a beforehand prepared library whose full path is specified by the second actual **P** argument into **LIB** directory with the main *Maple* library. The library receives a name specified by the first actual **F** argument. The procedure provides the logical linking of the specified library, by creating or updating the initialization file "Maple.ini" of the package which is located in **Users** subdirectory of the main *Maple* directory; in addition, if the optional third argument **0** (*zero*) has been coded then a library **F** is located as the first element of the predefined ***libname***-variable, otherwise it will be located at the end of the ***libname***-variable, defining the lowest priority of access to the library tools **F.** The procedure dynamically outputs the protocol of a library creation.

The procedure call **LibLink(D)** returns a current value of the predetermined ***libname***-variable, providing logical linking of an user library whose full path is specified by the first actual **D** argument with the main *Maple* library. If the procedure call does not use the optional argument **`1`** then logical linking of libraries with *higher* priority for the main *Maple* library is set, otherwise higher priority for the user library **D** is set. Simultaneously the procedure cleans the main ***Maple*** directory from temporary files "$*.m" which could remain at creation of an user library by means of any of the above procedures **User_pfl, User_pflM** or **User_pflMH.**

The procedure provides logical linking of the specified libraries, by creating or updating the initialization file "Maple.ini" of the package which is located in subdirectory **USERS** whose full path is defined by variable ***`UserDirectory`*** of the initialization file "MapleSys.ini" located in the **BIN** directory of *Maple* (for releases **6** and **7 – BIN.WNT,** for releases **8** and **9 –** {**BIN.W9X** (*Windows 98SE*), **BIN.WIN** (*Windows XP Pro*)} and **BIN.WIN** accordingly).

At last, at coding of the optional key symbol **`$`** the procedure **LibLink** does the above logical linking of the libraries, but does not create / update the initialization file "Maple.ini". Such approach allows to do active linking of the libraries available *only* during a current session or up to execution of the clause ***`restart`***. The procedure **LibLink1** is analogue to the procedure **LibLink,** having the same formal arguments and providing all same functions on linking of libraries, but with additional putting / updating of initialization file "Maple.ini" into the **LIB** directory of *Maple;* in a series of cases it can be expedient for *Maple* of releases **7 – 9/9.5.**

For interactive making/updating of Help database of the user library the *Maple* gives the tools of Help-group of the *Graphical User Interface* (**GUI**), namely: two its functions ***`Save to Database`*** and ***`Remove Topic`***. However, in view of a series of differences of releases **6 - 9/9.5** the **helpman** procedure providing automatic update of Help database on the basis of the beforehand prepared *mws*-files has been created.

The procedure **helpman(R, L, U {, Z})** has three obligatory and one optional formal argument. The first argument **R** defines a mode of operation with Help database of a library whose full path (*but not a name*) is defined by the second actual **L** argument, namely:

R = ***insert*** - *insertion* into a library of a new *help*-pages (*topics*), whose *names* are defined by the third actual **U** argument of a type {*list, dir*}; if type of **U** argument is *list*, the fourth argument **Z** defines a list of paths to *mws*-files with *help*-topics and the observance of one-to-one equivalence between lists **U** and **Z** is supposed; if type of **U** argument is *dir*, the procedure call should has three actual arguments only and **U** argument defines a directory with *mws*-files with *help*-topics. In the second case, the procedure chooses all *mws*-files from **U** directory (if they do exist) and on their base inserts *help*-pages into the library. In addition, the *help*-pages receive names defined by main names of the corresponding *mws*-files of the directory. For example, *help*-page for a tool `XyZ` is contained in file "XyZ.mws". The procedure supposes that names of *mws*-files for *help*-pages for tools names of a view **`a/b`** should be coded as **"a,b.mws"** at their creation. In addition, the procedure supports the sensitive mode for *help*-pages names, and the names should not contain *blanks*.

R = ***delete*** - *deletion* of a topic with name given by the third **U** argument of a type {*symbol, string*} out of Help database of a library defined by the second argument **L;**

R = ***display*** - *output* onto the screen of a *help* topic whose name is defined by the third argument **U** of a type {*symbol, string*}.

In addition, it is necessary to have in mind the following circumstances: (**1**) access to the main *Maple* library (except mode ***`display`***) is prohibited, and (**2**) modes ***`display`*** and ***`delete`*** suppose one *help*-topic only for the first mode such approach is natural, whereas for the second one it provides the larger level of safety. The procedure processes main erratic and especial situations, causing errors or an output of the corresponding warnings.

Thus, the **helpman** procedure represents a rather convenient tool for automatic update of Help database of the user libraries on the basis of the beforehand prepared *mws*-files, providing an opportunity of making by one call of several Help topics, whereas their viewing and deletion are done one at a time for each call of the given procedure. In addition, it is necessary to note an important circumstance. In some cases by a chain of functions **"Help -> Save to Database"** of the *Maple's* **GUI,** a putting into the user Help database of a worksheet is not guaranteed (at registering it in the corresponding index file "Maple.ind"), whereas the procedure **helpman** is free of this defect.

The procedures **User_pflM, Ulib, DUlib** and **helpman** provide a rather effective automation of support of the user libraries that have organization similar to organization of the main *Maple* library. In this respect, they in a series of cases support the functions of automation of operation with libraries similarly to the utility ***`sed`*** of the operating system *UNIX (Linux).*

At last, the procedure call **Plib(F)** returns full path to an user library specified by the actual **F** argument and which is logically linked with the main *Maple* library. At missing of such library or its logical link with the main *Maple* library the *false* value is returned.

```
Ulib:= proc(V::mlib, S::{0, 1, 2})
local k, a, b, N, G, Q, Z, P, L, L1, Cn, h, R, g, l, r, F;
```

```
global libname;
  N:= W::string -> op([assign('b' = march('list', W)), assign('b' = sort([seq(`if`(b[k][1][1..2] = ":-",
    NULL, convert(b[k][1], name)), k = 1 .. nops(b))])), b, nops(b)]);
  Cn:= (s, h, t) -> [assign('b' = ""), seq(assign('b' = cat(b, `if`(s[k] = h, t, s[k]))),
    k = 1 .. length(s)), b][-1];
  Release('h'), assign(F = convert(V, string)), `if`(F[2] = ":", assign('G' = CF(F, string)),
    assign('G' = CF(cat(h, "/Lib/", F), string))), `if`(DoF(G) <> `full directory`,
    ERROR("the user library <%1> does not exist", G), `if`(S = 0, RETURN(N(G)), 0));
  assign(Q = cat(h, "/", KCM(), "/Maplesys.ini"));
  assign('Z' = iolib(4, Q, TEXT, infinity)), iolib(7, Q),
    assign('Z' = cat(Z[searchtext("UserD", Z) + 14 .. searchtext("\n", Z, searchtext("UserD", Z) ..
    length(Z)) + searchtext("UserD", Z) - 2], "/Maple.ini"));
  assign(R = iolib(4, Z, TEXT, infinity)), iolib(7, Z);
  if S=2 and not search(Cn(R, "\\", "/"), Cn(G, "\\", "/")) then libname:= libname, Cn(G, "\\", "/");
    try fopen(Z, APPEND, TEXT), assign(g = 54), writeline(Z, cat("libname:=libname,"",
      Cn(G, "\\", "/"), "":")), iolib(7, Z)
    catch "file or directory":  writeline(Z, cat("libname:=libname, "", Cn(G, "\\", "/"), "":")),
           iolib(7, Z)
    finally WARNING("logical link of the user library <%1> with Maple has been restored", G)
    end try
  elif S = 1 then  assign(l = [libname], r = nops([libname])), (proc()
      libname := seq(`if`(search(Cn(l[k], "\\", "/"), Cn(G, "\\", "/")), NULL, l[k]), k = 1 .. r)
      end proc)();
    try  eval(libname), iolib(7, Z), assign(P = iolib(4, Z, TEXT, infinity)),
        assign(L = cat("libname:=libname,"", Cn(G, "\\", "/"), "":"),
        L1 = cat("libname:="", Cn(G, "\\", "/"), "",libname:")), iolib(7, Z);
      `if`(search(P, L, 'a'), [iolib(5, Z, cat(P[1 .. a - 1], P[a + length(L) + 1 .. length(P)])),
        iolib(7, Z)], `if`(search(P, L1, 'a'), [iolib(5, Z, cat(P[1 .. a - 1],
        P[a + length(L) + 1 .. length(P)])), iolib(7, Z)], NULL))
    catch "file or directory does not exist":  NULL
    finally `if`(iolib(6, Z, infinity) = 0, [iolib(7, Z), delf(Z)], iolib(7, Z))
    end try;
    WARNING("logical link of library <%1> with Maple has been cancelled", G)
  end if
end proc

DUlib:= proc(L::{string, symbol})
local k, g, h, f;
global libname;
  if cat("", L)[2] = ":" then
    if type(L, mlib) then g := L else ERROR("<%1> is not a Maple library", L)  end if
  else  g := cat(CDM(), "\\LIB\\", L);
    if not type(g, mlib) then  ERROR("<%1> is not a Maple library", L)  end if
  end if;
  f := {};  libname := op(subsLS([CF2(g) = NULL], map(CF2, [libname])));
  WARNING("linkage of <%1> with main Maple library has been canceled", L);
  h := (map2(cat, g, map2(cat, "\\maple.", {ind, lib, hdb, rep}))  union
    map2(cat, g, map2(cat, "\\elpam.", {ind, lib})))  union  {cat(g, "\\maple_u.cmd")};
  for k in h do
    if type(k, file) then
```

```
        try delf1(k, `*`)
        catch "": f := {op(f), CFF(k)[-1]}; next
        end try
      end if
    end do;
    if DirE(g) and Empty(f) then rmdir(g); WARNING("the user library <%1> has been deleted", L)
    else WARNING("DUlib could not remove all files of <%1> - files %2 have not been removed;
      use other tools", L, f)
    end if
end proc

InstUlib:= proc(N::{string, symbol}, P::{string, symbol})
local a, b, k, L, h, seqstr, t, f, z, n;
global libname;
    `if`(type(P, mlib), assign(L = [libname], a = DirF(P, only)),
      ERROR("<%1> is not a Maple library", P));
    seqstr := () -> convert([args], string)[2 .. -2];
    WARNING("Library <%1> is being copied into LIB directory of Maple. Please, wait!", P);
    try b := cat(CDM(), "\\Lib\\", N); mkdir(b)
    catch "directory exists and is not empty":  NULL
    end try;
    for k in a do  assign('n' = fopen(k, READ), 'f' = cat(b, "\", CFF(k)[-1]));
      FBBcopy(k, f), WARNING("file <%1> has been successfully copied.", k)
    end do;
    WARNING("Library <%1> has been created in LIB directory of Maple", N);
    if nargs = 3 and args[3] = 0 then libname := b, libname else libname := libname, b  end if;
    assign(t = cat(Users(h), "\\Maple.ini"), `if`(nargs = 3 and args[3] = 0, assign(z = higher),
      assign(z = lower)));
    null(fopen(t, APPEND), writeline(t, cat("libname:=", seqstr(libname)), ":"), close(t));
    WARNING("Library <%1> has been logically linked with the main Maple library with %2
      priority", N, z)
end proc

helpman:= proc(R::symbol, L::{string, symbol}, N::{string, symbol, dir, list({string, symbol})})
local k, a, b, f, h, p, n, m;
    `if`(not member(R, {delete, insert, display}),
      ERROR("the 1st argument should be {insert,display,delete}, but has received <%1>", R),
      `if`(Path(L) = Path(cat(CDM(), "\\LIB")) and R <> display,
      ERROR("access prohibition to the main Maple library"), `if`(not type(L, dir),
      ERROR("path to library <%1> does not exist", L), `if`(not type(cat(L, "\\", "maple.hdb"), file),
      ERROR("for library <%1> Help database does not exist", L), NULL))));
    `if`(not member(R, {delete, display}), assign(b = 7, f = "$ArtKr$", p = []),
      `if`(type(N, {string, symbol}), RETURN(INTERFACE_HELP(R, topic = N, library = L)),
      ERROR("the 3rd argument should has type {symbol, string}, but has received type <%1>",
       whattype(N))));
    `if`(R = insert and nargs = 4 and type(N, list({string, symbol})) and
      type(args[4], list({string, symbol})), `if`(nops(N) = nops(args[4]), assign('b' = 14),
      ERROR("mismatching of quantities of names and mws-files: %1<>%2", nops(N),
      nops(args[4]))), NULL);
    if b = 14 then
```

```
    for k to nops(N) do
      try makehelp(convert(N[k], symbol), convert(args[4][k], symbol), convert(L, symbol))
      catch "file or directory does not exist":  WARNING("file <%1> does not exist,
            the operation with it has been ignored", args[4][k]);
        next
      end try
    end do;
    RETURN(WARNING("Work has been done!"))
  elif R = insert and type(N, dir) then [assign(n=[], m=[]), writeto(f), Dir(N), writeto(terminal)];
    do h := readline(f);
      if h = 0 then fremove(f); break
      else h := SLD(h, " ");
        `if`(cat("aaa", h[-1])[-4 .. -1] = ".mws", assign('p' = [op(p), h[-1]]), NULL)
      end if
    end do
  end if;
  if p <> [] then
    for k to nops(p) do n := [op(n), p[k][1 .. -5]];  m := [op(m), cat(Path(N), "\", p[k])] end do;
    RETURN(procname(R, L, n, m))
  else RETURN(WARNING("mws-files have not been saved in Help database"))
  end if;
  WARNING("Work has not been done, mistakes at the call encoding: %1", 'procname(args)')
end proc

LibLink:= proc(D::{string, symbol})
local `0`, h, k, t, omega;
global libname;
  omega:= x -> WARNING("library <%1> has been linked with the main Maple library with %2
            priority", D, x);
  `if`(not type(D, mlib), ERROR("1st argument should be  a full path but has received <%1>", D),
    assign(`0` = cat(Users(t), "\\Maple.ini"))), `if`(not belong(`1`, {args}),
    assign(h = cat("libname:=libname", ",", """", CF1(D), """", ":")), assign(h = cat("libname:=", """",
    CF1(D), """", ",", "libname:"))), system(cat("del ", Path(CDM()), `\\$*.m`));
  if member(`$`, {args}) then
    if member(`1`, {args}) then libname := D, libname; omega(higher); RETURN(libname)
    else libname := libname, D;  omega(lower); RETURN(libname)
    end if
  else fopen(`0`, APPEND), writeline(`0`, h), close(`0`),
      RETURN((proc()  read `0`; `if`(member(`1`, {args}), omega(higher), omega(lower));
      libname  end proc)(args))
  end if
end proc

LibLink1:= proc(D::{string, symbol})
local `0`, `2`, h, k, t, omega;
global libname;
  omega:= x -> WARNING("library <%1> has been linked with the main Maple library with %2
            priority", D, x);
  `if`(not type(D, mlib), ERROR("1st argument should be a full path but has received <%1>", D),
    assign(`0` = cat(Users(t), "\\Maple.ini"), `2` = cat(CDM(), "\\Lib\\Maple.ini"))),
```

```
    `if`(not belong(`1`, {args}), assign(h = cat("libname:=libname", ",", """, CF1(D), """, ":")),
    assign(h = cat("libname:=", """, CF1(D), """, ",", "libname:"))),
    system(cat("del ", Path(CDM()), `\\$*.m`));
  if member(`$`, {args}) then
    if member(`1`, {args}) then  libname := D, libname;  omega(higher);  RETURN(libname)
    else  libname := libname, D;  omega(lower);  RETURN(libname)
    end if
  else fopen(`0`, APPEND), writeline(`0`, h), close(`0`), fopen(`2`, APPEND), writeline(`2`, h),
    close(`2`), RETURN((proc()  read `0`;  `if`(member(`1`, {args}), omega(higher),
    omega(lower)); libname end proc)(args))
  end if
end proc

Plib:= proc(L::{string, symbol})
local a, b, c, k, h, p, omega;
  assign(a = [libname], b = cat("/", Case(L, lower)), omega=(() -> Case(sub_1("\\" = "/", args))),
  c = cat("\\", Case(L, lower))), seq(`if`(omega(a[k]) = omega(L) or
  Search1(Case(a[k], lower), b, 'h') and h = [right] or Search1(Case(a[k], lower), c, 'p') and
  p = [right], RETURN(a[k]), NULL), k = 1 .. nops(a)), false
end proc
```

Typical examples of the procedures use:

1. Creation of the user library *Paskula* with viewing its contents by means of the procedure **Ulib:**

> Art:=() -> `+`(args)/nargs: Kr:=module() option package; export sr; sr:=()->sqrt(`+`(args)/ nargs) end module: Sv:=() -> [nargs, {args}]: User_pflM(Paskula, [Art, Kr, Sv], 7);

Warning, the User library <c:\program files\maple 9\lib\paskula> has been created/updated

"c:\program files\maple 9\lib\paskula", "c:/program files/maple 9/lib/userlib", "C:\Program Files\Maple 9/lib"

> restart: Ulib(Paskula, 0), libname;

[*Art.m, Kr.m, Sv.m*], **3**, "c:/program files/maple 9/lib/paskula", "c:/program files/maple 9/lib/userlib", "C:\Program Files\Maple 9/lib"

2. Cancelling of logical link of the *Paskula* library with the main *Maple* library:

> Ulib(Paskula, 1), libname;

Warning, logical link of library <c:\program files\maple 9\lib\paskula> with Maple has been cancelled

"c:/program files/maple 9/lib/userlib", "C:\Program Files\Maple 9/lib"

3. Restoring of logical link of the *Paskula* library with the main *Maple* library:

> Ulib(Paskula, 2), libname;

Warning, logical link of the user library <c:\program files\maple 9\lib\paskula> with Maple has been restored

"c:/program files/maple 9/lib/userlib", "C:\Program Files\Maple 9/lib", "c:/program files/maple 9/lib/paskula"

4. Canceling of logical link of the *Paskula* library with the main *Maple* library with subsequent deleting of the user library out of file system of a computer:

> Ulib(Paskula, 1); restart; DUlib(Paskula);

Warning, logical link of library <c:\program files\maple 9\lib\paskula> with Maple has been cancelled

Warning, linkage of <C:\Program Files\Maple 9\Lib\Paskula> with main Maple library has been canceled

Warning, user library <C:\Program Files\Maple 9\Lib\Paskula> has been deleted

> DUlib(Svetlana);

Warning, linkage of <Svetlana> with main Maple library has been canceled

Warning, DUlib could not remove all files of <Svetlana> – files {maple.ind} have not been removed; use other tools

5. Operations with Help database of the user library:

> helpman(insert, "c:/program files/maple 8/LIB/userlib", ["DoF1"], [`c:/temp/dof1.mws`]); # insertion of help-topic about procedure DoF1 into the user library UserLib

Warning, Work has been done!

> helpman(insert, "C:/Program Files/maple 8/LIB/UserLib", ["AFdes", `a,b`], [`C:/Temp/AFdes.mws`, "C:/Temp/RANS\\IAN/a,b.mws"]);

Warning, file <C:/Temp/RANS\IAN/a,b.mws> does not exist, the operation with it has been ignored

Warning, Work has been done!

> helpman(insert, "C:/program files/maple 9/LIB1", "C:/Academy\\mws_files");

Error, (in **helpman**) for library <c:/program files/maple 9/LIB1> Help database does not exist

> helpman(insert, "c:/program files/maple 8/LIB/userlib", `c:/temp\\`); # insertion of help-topics into the user library **UserLib** on the basis of all *mws*-files of directory **`c:/temp`**

Warning, Work has been done!

> helpman(display, "c:/program files/maple 8/LIB/userlib", "AFdes1");

Error, Could not find any help on "**AFdes1**"

> helpman(display, "c:/program files/maple 8/LIB/userlib", "AFdes"); # output of help-topic about procedure **AFdes**

> helpman(delete, libname[1], Art_Kr); # deleting of help topic *Art_Kr* out of Helpbase of a library, which is located at the beginning of a chain of libraries reflected in the ***libname*** variable

6. Copying of a beforehand prepared library into directory of the main *Maple* library:

> InstUlib(Svetlana, "C:/RANS/Academy/ModProcLib");

Warning, Library <C:/Academy/UserLib9/UserLib> is being copied into LIB directory of Maple. Please, wait!

Warning, the target file is <c:\program files\maple 9\lib\svetlana\maple_u.cmd>
Warning, file <c:/academy/userlib9/userlib/maple_u.cmd> has been successfully copied.
Warning, the target file is <c:\program files\maple 9\lib\svetlana\maple.lib>
Warning, file <c:/academy/userlib9/userlib/maple.lib> has been successfully copied.
Warning, the target file is <c:\program files\maple 9\lib\svetlana\maple.ind>

Warning, file <c:/academy/userlib9/userlib/maple.ind> has been successfully copied.
Warning, the target file is <c:\program files\maple 9\lib\svetlana\maple.hdb>
Warning, file <c:/academy/userlib9/userlib/maple.hdb> has been successfully copied.
Warning, Library <Svetlana> has been created in LIB directory of Maple

Warning, Library <Svetlana> has been logically linked with the main Maple library with lower priority

7. Two methods of logical linking with the main *Maple* library of the user library **ModProcLib:**

> LibLink("C:/RANS/Academy/ModProcLib", \`$\`);

Warning, library <C:/RANS/Academy/ModProcLib> has been linked with the main Maple library with lower priority

"c:/program files/maple 9/lib/userlib", "C:\Program Files\Maple 9/lib", "C:/RANS/Academy/ModProcLib"

> LibLink1("C:/RANS/Academy/ModProcLib", \`$\`, \`1\`);

Warning, library <C:/RANS/Academy/ModProcLib> has been linked with the main Maple library with higher priority

"C:/RANS/Academy/ModProcLib", "c:/program files/maple 9/lib/userlib", "C:\Program Files\Maple 9/lib"

8. Evaluation of full path to an user library logically linked with the main *Maple* library:

> Plib(userlib), Plib(UserLib), Plib("C:/RANS/Academy/ModProcLib"), Plib(Svetlana);

"c:/program files/maple 9/lib/userlib", "c:/rans/academy/modproclib", *false*

---

ulibpack – packing of a libraries structurally similar to the main *Maple* library

Call format of the procedure:

ulibpack(L {, N})

Formal arguments of the procedure:

L - symbol or string defining full path to a *Maple* library
N - (optional) a positive integer defining a packing mode of the *Maple* library

Description of the procedure:

As a result of a library updating by new versions of the tools existing in it, in the library appears a lot of tools with several versions. In an usual mode, the last versions of tools are accessible to the user only. Meanwhile, obsolete versions increase volume of the library. The following procedure **ulibpack** allows to solve the problem of a library actualization structurally similar to the main *Maple* library. Our experience of the procedure usage has shown, that it allows to essentially improve the library functionality along with reduction of disk space.

The procedure call **ulibpack(L)** returns the *NULL* value, providing actualization of a *Maple* library, the full path to directory of which is defined by the first actual **L** argument. The actualization includes two stages, namely: (**1**) garbage collection of the library ***ind***-file to remove index entries referring to inaccessible obsolete tools, and (**2**) removing of obsolete tools from the library ***lib***-file with its packing. In addition, a source library remains without alteration, whereas the actualization result (*if it has been received*) forms a new library in the main *Maple* directory. Protocol of the procedure call defines full path to the result. Actualization is done if and only if a source library contains at least 2 versions (a *new* and *obsolete*) of a library tool. In addition, if the source library **L** has the `*READONLY*`-attribute, the actualization result will receive the `*WRITABLE*`-attribute. Such approach allows to have both the old version of a *Maple* library, and the packed version. If necessary, files {***ind, lib***} of the old version you can update easily by the corresponding files of actualization result, updating old library as a whole.

Unique difference of the procedure call **ulibpack(L, N)** from the previous will be that it makes actualization of a *Maple* library **L** only if the total of obsolete versions of the library tools will be not less than the value specified by the second optional **N** argument. If type of the second argument is different from '*posint*', the procedure call is equivalent to the call **ulibpack(L).** In a series of cases, the procedure allows to essentially increase the library performance.

```
ulibpack:= proc(L::fpath)
local a, b, c, d, k, p, t, w, n;
global _mulvertools;
  assign(n = 1), `if`(nargs = 1, NULL, `if`(type(args[2], posint), assign('n' = args[2]), NULL));
  assign(t = interface(warnlevel), w = (x -> interface(warnlevel = x))), interface(warnlevel = 0);
  `if`(type(L, mlib), assign(a = cat("", L, "\\maple.lib"), b = cat("", L, "\\maple.ind")),
    ERROR("<%1> is not a Maple library", L));
  assign(c = cat(CDM(), "\\_$", a[4 .. -1]), d = cat(CDM(), "\\_$", b[4 .. -1]));
  p := cat(CDM(), "\\_$", cat("", L)[4 .. -1]);
  MkDir(c, 1); FBBcopy(a, c); w(t);
  WARNING("Analysis of library <%1> is in progress, please wait!", L);
  w(0); CureLib(c); w(t);
  if type(_mulvertools, list) and n <= add(_mulvertools[k][2] - 1, k = 1 .. nops(_mulvertools))
  then system(cat("copy ", blank(CF2(b)), " ", blank(CF2(d))));
    WARNING("files {ind, lib} of the packed library are in directory <%1>", p);
    try march('gc', p); march('pack', p)
    catch: AtrRW(p); march('gc', p); march('pack', p)
    end try
  else delf1(c, `*`); rmdir(p); WARNING("library <%1> does not need a packing", L)
  end if;
  unassign('_mulvertools')
end proc
```

Typical examples of the procedure use:

> ulibpack("C:/Temp\\Lasnamae/RANS\\IAN", 15);

Warning, Analysis of library <C:/Temp\Lasnamae/RANS\IAN> is in progress, please wait!
Warning, library <C:/Temp\Lasnamae/RANS\IAN> does not need a packing

> ulibpack("C:/Temp\\Tallinn/Vilnius", 7);

Warning, Analysis of library <C:/Temp\Tallinn/Vilnius> is in progress, please wait!

Warning, files {ind, lib} of the packed library are in directory
<C:\Program Files\Maple 8_$Temp\Tallinn/Vilnius>

Warning, library <C:\Program Files\Maple 8_$Temp\Tallinn/Vilnius> has received the `WRITABLE`-attribute

> ulibpack("C:/Test\\rans/ian\\rac\\rea", 62);

Warning, Analysis of library <C:/Test\rans/ian\rac\rea> is in progress, please wait!

Warning, files {ind, lib} of the packed library are in directory
<C:\Program Files\Maple 8_$Test\rans/ian\rac\rea>

The above procedures **User_pfl, User_pflM, User_pflMH, Ulib, DUlib, InstUlib, LibLink, ulibpack** and **helpman** provide a rather effective automation of support of the user libraries having the structural organization, similar to the organization of the main *Maple* library. In this respect, they in some cases support the advanced functionality of automation of work with libraries similarly to a case of the ***sed*** utility of operating system *UNIX* (*Linux*).

For support of conversion from the *first* level of the user libraries organization (the above procedures **FLib1, FLib, SaveProc, ReadProc, LibUser** and **mPM**) to the *second* level (the above procedures **Ulib, DUlib, User_pfl, User_pflM, User_pflMH, NonaLP, NLP, helpman, InstUlib, ulibpack** and **LibLink**) the following simple reception can be used. As it follows of the told, the result of the first level is a text file or an *m*-file (let us presume the full path to it has been specified by a string or symbol **F**) which contains definitions of procedures or program modules of the user. In view of that, in a current session the following simple fragment is evaluated:

restart: read(F): L:= [anames()]: User_pflM(G, L, nops(L) + 5); Ulib(G, 0);

The result of the given operation is creation of the user library with name **G** containing as sections the *m*-files with definitions of the tools loaded by means of the ***`read`*** function out of the specified **F** file, with the subsequent return of list of the *m*-files saved in the library. In addition, it is presumed the procedures **Ulib** and **User_pflM** are located or in the main *Maple* library, or in an user library logically linked with the main library. It is necessary to remind, that these and other tools represented in the monograph are in the attached Library that can be easily linked with the main *Maple* library.

As it was noted in our books [29-33,43,44], four last releases of *Maple* do not ensure correct saving of program modules (which have been named by means of any of two permissible manners) by means of ***`save`*** function from the standpoint of their subsequent activation in a current session by means of reading by ***`read`*** function of a datafile containing them as in the internal *Maple* format (*m*-file), and in the input *Maple* format. With the purpose of elimination of the given defect, the procedure **SaveMP** was created which ensures for *Maple* of releases **6, 7, 8** and **9/9.5** (*and of the subsequent ones*) the correct execution of the given operation along with others.

SaveMP – a saving method of program modules and procedures

Call format of the procedure:

SaveMP(F {, M | , "list"} {, "load"})

Formal arguments of the procedure:

F – string defining a name or full path to a datafile with software
M – (optional) string, set or list of program modules and/or procedures
"load", "list" – (optional) key words defining a mode

Description of the procedure:

The procedure **SaveMP(F, {M | "list"})** has one optional and two obligatory arguments. The first actual argument **F** specifies full path to a file with program modules and/or procedures. While the second actual argument of the procedure specifies one thing out of a three, namely:

(1) a list or set **M** of names of program modules and/or procedures whose definitions have been evaluated in a current *Maple* session and which are subjected to saving in a datafile specified by the first **F** argument; let's suppose, that definitions of the saved program modules contain the ***`package`*** option;

(**2**) a set or list of names of program modules and/or procedures which are located in the specified **F** file and which should be activated in a current *Maple* session;

(**3**) the option ***"list"*** specifies an output mode of a list of names of program modules and procedures contained in the indicated **F** file; the output names are classed according to a type of tools identified by them.

At last, the third optional argument admits only one ***"load"*** option defining activation in a current session of the program modules and/or procedures located in the indicated **F** file whose names are defined by a set or list from the second actual argument **M.**

The successful procedure call **SaveMP** in the first case provides a saving in the specified **F** file of software specified by the second actual **M** argument, in the *APPEND*-mode, by providing an opportunity of the subsequent appending of the **F** file by a new program modules and procedures. The procedure outputs the warning about path to a file with the saved tools. After that the correct activation of tools contained in the file, by means of reading of the file by the ***`read`*** function will be ensured in a current *Maple* session.

The given procedure can be successfully used as a tool of creation of the user libraries. It is necessary to note, at saving of a program module in a library which is structurally equivalent to the main *Maple* library (see above the procedure **User_pflMH**) the situations similar to described above do not arise. Therefore is recommended to create the user libraries similar to the main *Maple* library.

```
SaveMP:= proc(F::string, M::{set, list, string, set(`module`), list(`module`)})
local d, k, n, m, h, p, t, v, G, z, u, S, Kr, Art, q, r, f, omega;
  `if`(not type(F, file) and args[2] = "list" or not type(F, file) and nargs = 3 and args[3] = "load",
    ERROR("file <%1> does not exist", F), `if`(type(F, file), assign(f = F), assign(f = pathtf(F))));
  if args[2] = "list" or nargs = 3 and args[3] = "load" then assign(u = [], z = [], S = "");
    do assign('h' = iolib(2, f));
      if h = 0 then break
      else `if`(h = "#AVZ_AGN_2003;", [assign('h' = iolib(2,f)),
        assign('n' = searchtext(" := ",h)), `if`(h[n + 4 .. n + 9] = "proc (",
        assign('u' = [op(u), convert(h[1 .. n - 1], symbol)]), `if`(h[n + 4 .. n + 12] = "module ()",
        assign('z' = [op(z), convert(h[1 .. n - 1], symbol)]), NULL))], 1)
      end if
    end do;
    if args[2] = "list" then RETURN(iolib(7, f), `if`(z = [] and u = [],
      WARNING("file <%1> does not contain a software", f), `if`(u = [], op([modules, z]),
      `if`(z = [], op([procedures, u]), op([procedures, u, modules, z])))))
    else
      assign('q' = {op(M)} minus ({op(u), op(z)} intersect {op(M)})), `if`(q = {}, NULL,
        WARNING("tools %1 do not exist in file <%2>", q, f));
      assign('z' = {op(u), op(z)} minus {op(M)});
      seq(assign('S' = cat(S, z[k], ":='", z[k], "':")), k = 1 .. nops(z));
      (proc(a, b) read f; iolib(5, a, b); iolib(7, a); read a; delf(a) end proc)("$", S),
        assign('r' = {op(M)} minus q), RETURN(WARNING("tools %1 have been activated", r))
    end if
  else Art:= proc(S)
    local a, b, c, p;
      [assign('a' = S, 'b' = Search(S, ":= module")), `if`(nops(b) = 1, RETURN(S),
      assign('b' = b[2 .. -1])), seq([assign('c' = searchtext("()", a, p .. length(a)) + p),
      `if`(a[p + 8 .. c - 1] = " ", ``, assign('a' = cat(a[1 .. p + 8], a[c - 1 .. length(a)])))], p=b), a][2]
```

```
    end proc
  end if;
  `if`(f[-2 .. -1] = ".m", assign67(omega = f[1 .. -3], delf(f)), assign(omega = f)),
    fopen(omega, APPEND);
  for v to nops(M) do  assign('n' = length(cat("unprotect('", M[v], "'); ")),
      't' = length(M[v]), 'm' = length(cat("", M[v], ":-")),
      'h' = length(cat(" protect('", M[v], "');")) + 2);
    (proc(x)  save x, "$"  end proc)(M[v]), assign('G' = iolib(4, "$", TEXT, infinity));
    `if`(type(M[v], `module`), [`if`(search(G, cat("unprotect('", M[v], "'); ")),
      assign('G' = G[n + 1 .. -h]), G), assign('z' = Search(G, cat("", M[v], ":-"))),
      `if`(z <> [], [assign('Kr' = G[1 .. z[1] - 1]), seq(assign('Kr' = cat(Kr,
      G[z[k] + m .. z[k + 1] - 1])), k = 1 .. nops(z) - 1), assign('Kr' = cat(Kr,
      G[z[-1] + m .. length(G)]))], assign('Kr' = G)), assign('d' = Search(Kr, cat(" ", M[v], " "))),
      `if`(d = [], assign('d' = "a"), assign('d' = op(d)))], assign('Kr' = G)), delf("$"),
      writeline(omega, "#AVZ_AGN_2003;", `if`(whattype(eval(M[v])) <> `module`, Kr,
      `if`(d <> "a", Art(cat(Kr[1 .. d - 1], Kr[d + t + 1 .. -1])), Kr))), `if`(not (v = nops(M)), 0,
      RETURN(WARNING("tools %1 have been saved in file <%2>", M, omega),
      close(omega)))
  end do
end proc
```

Typical examples of the procedure use:

```
> f1:=()->`+`(args)/nargs: f2:=()->add(args[k]^2,k=1..nargs)/nargs: f3:=()->[`+`(args),nargs]:
f4:=()->`*`(args)/nargs^2: M1:=module () export Sr; option package; Sr:=()->`+`(args)/nargs end
module: module M2 () export Kr; option package;  Kr:= ()-> `*`(args)/nargs end module:
> SaveMP("C:\\Archive/Tallinn\\Book/RANS_IAN.m", [f1, f2, M1, f3, M2, f4]);
    Warning, tools [f1, f2, M1, f3, M2, f4] have been saved in file
    <c:\archive\tallinn\book\rans_ian>
> SaveMP("C:\\Archive/Tallinn\\Book/RANS_IAN", "list");
                        procedures, [f1, f2, f3, f4], modules, [M1, M2]
> restart;  SaveMP("C:\\Archive/Tallinn\\Book/RANS_IAN", [f1, Art_Kr, M1], "load");
    Warning, tools {Art_Kr} do not exist in file <C:\Archive/Tallinn\Book/RANS_IAN>
    Warning, tools {f1, M1} have been activated
> SaveMP("C:\\Archive/Tallinn\\Book/RANS_IAN.m", [f1, f2, M1, f3, M2, f4], "load");
    Error, (in SaveMP) file <C:\Archive/Tallinn\Book/RANS_IAN.m> does not exist
> SaveMP("C:\\Lasnamae\\Test.m", [RANS_IAN_RAC_REA]);
    Warning, unassigned variable `RANS_IAN_RAC_REA` in save statement
    Warning, tools [RANS_IAN_RAC_REA] have been saved in file <c:\lasnamae\test>
```

In view of the explanations made above, the examples of the given fragment are transparent enough and of any additional explanations do not demand.

In the *Maple* environment, the package modules are standardly supported by structures of a type {*table, procedure, module*}. While in a series of cases for these purposes, the other data structures can be useful enough. Here we shall present an approach illustrating a mechanism of use for these purposes of data structures of type {*array, matrix, list*}. In this case, the procedures definitions are plunged into data structure of one of the above types; in addition, the **ArtMod** procedure represented below provides a handling of such package modules.

ArtMod – an useful type of package modules

Call format of the procedure:

ArtMod(L {, f | , `?`} {, *args*})

Formal arguments of the procedure:

L – symbol defining name of a package module
f – (optional) symbol defining name of a procedure contained in the module
args – (optional) an expression or sequence of *Maple* expressions
`?` – (optional) key symbol defining a mode

Description of the procedure:

The following simple example illustrates functionality of the **ArtMod** procedure. Let us suppose that definitions of procedures **{f1, f2, f3, f4}** are plunged into a structure of *`array`* type which next by means of the procedure **User_pflM** has been saved in the ***Svetlana*** library having an organization similar to the main *Maple* library.

For providing of access to non-standard package modules of such type, a simple enough procedure **ArtMod(L, f {, *args*})** has been created. The first formal **L** argument of the procedure specifies name of a package module, the second argument **f** – name of a procedure contained in it, while the other arguments ***args*** are optional and specify actual arguments passed to the **f** procedure. At the call **ArtMod** search of a module specified by its first actual argument **L** is made above all among active tools of a current *Maple* session and only after that in chain of the libraries defined by predetermined ***libname*** variable. If as the second actual argument the value**`?`** has been coded, the list of names of procedures contained in a module **L** is returned; in addition, all they become active (*accessible*) in a current *Maple* session. Thus, the procedure **ArtMod** can be considered as a certain analog of clause "**with**" of *Maple.*

In a series of cases, the similar non-standard approaches to organization of structures of modules appear for us much more effective than standard ones, allowing to fulfill with the structures of such modules necessary symbolic transformations and calculations. Our experience of testing and heavy use of *Maple* of releases from the *fourth* up to the *ninth* inclusive the quite definitely speaks in favor of the given assertion.

```
ArtMod:= proc(L::symbol, f::symbol)
local a, k;
  [`if`(member(whattype(eval(L)), {array, list}), assign('a' = convert(L, list)),
  ERROR("structure type of module <%1> is invalid", L)), `if`(convert(args[2], string) = "?",
  RETURN(map(lhs, a), seq(assign(a[k]), k = 1 .. nops(a))), [add(`if`(lhs(a[k]) = f,
  RETURN(rhs(a[k])(args[3 .. nargs])), 1), k = 1 .. nops(a)),
  ERROR("procedure <%1> does not exist in package module <%2>", f, L)])]
end proc
```

Typical examples of the procedure use:

> Kr:=array(1..2,1..2, [[f1=(() -> `+`(args)/nargs), f2=(() -> add(*args*[k]^2,k=1..nargs)/nargs)], [f3=(() -> [`+`(args), nargs]), f4=(() -> 143*`*`(args)/2004)]]);

$$Kr := \begin{bmatrix} f1 = \left((\,) \to \frac{\text{`+`(args)}}{\text{nargs}} \right) & f2 = \left((\,) \to \frac{\text{add}(\text{args}_k{}^2,\, k = 1 \,..\, \text{nargs})}{\text{nargs}} \right) \\ f3 = ((\,) \to [\text{`+`(args), nargs}]) & f4 = \left((\,) \to \frac{143}{2004}\, \text{`*`(args)} \right) \end{bmatrix}$$

> User_pflM("Svetlana", [Kr], 7);

Warning, the User library <c:\program files\maple 9\lib\svetlana> had been created/updated

"c:\program files\maple 9\lib\svetlana", "c:/program files/maple 9/lib/userlib",
"C:\Program Files\Maple 9/lib"

> restart; 2004*ArtMod(Kr, f4, 42, 47, 67, 89, 96), ArtMod(Kr, f3);

161591766336, [0, 0]

> restart; ArtMod(Kr, \`?\`), 2004*f4(62, 57, 37, 41, 7, 15);

[*f3*, *f4*, *f1*, *f2*], 80496586170

> 670340004*seq(k(62, 57, 37, 41, 7, 15), k=[f1, f2, f3, f4]);

24467410146, 1163821970278, [146804460876, 4022040024], 26926188570451170

> 2007006*map3([f1, f2, f3, f4], [42, 47, 67, 62, 89, 96], 0);

[134803903, 9848712943, [808823418, 12042036], 10033717547101248]

> ArtMod(Kr, RANS_IAN, 1942, 1947, 1995, 2004, 7, 15, 37, 41);

Error, (in **ArtMod**) procedure <RANS_IAN> is absent in package module <Kr>

The above fragment represents creation of the package module **Kr**, using the *\`array\`* structure with subsequent saving of it in the user **Svetlana** library. The remaining examples of the fragment illustrate use of the **ArtMod** procedure for providing of access to the given package module.

At last, the ***third*** level is supported by the built-in ***savelib*** function that provides a putting of tables, program modules and procedures, and in the more broad sense of definitions of many other evaluated expressions, into shared main *Maple* library (*repository*). It allows to use subsequently the saved objects at the logical level of access just as built-in *Maple* tools. Realization of the given mechanism is provided with performance in a current session of a few steps considered in our books [14,29-33,39,43,44] enough in detail. For providing of update by tables, procedures and program modules of the main *Maple* library (***lib***-file and ***ind***-file) the useful enough procedures **MapleLib** and **UpLib** are intended. These procedures also update the user libraries logically similar to the main *Maple* library.

MapleLib – operation with the main *Maple* library and the user libraries having the same structural organization (*the third level*)
UpLib

Call format of the procedures:

MapleLib(P {, L})
UpLib(F, N)

Formal arguments of the procedures:

P - set or list of symbols defining names of procedures, tables and/or program modules
L - (optional) symbol or string defining full path to the user library
F - symbol or string defining library name or full path to the user library
N - list of symbols defining names of procedures, tables and/or program modules

Description of the procedure:

The procedure **MapleLib** is intended for update of both the main *Maple* library and the user libraries having structural organization, similar to the main library, by procedures, tables and program modules whose definitions have been evaluated in a current session. In addition, use of the given procedure presumes existence for a modified library of ***lib***-file and ***ind***-file. The next object (procedure, table or program module) is put into the end of library ***lib***-file without removal of an object of the same name already existing in the library, however in the library index ***ind***-file the pointer is set on the latest object. It is supposed, that the evaluated definitions of program modules contain ***`package`*** option of *Maple.*

The successful procedure call **MapleLib(P)** updates the main *Maple* library by procedures, tables and/or program modules whose names are specified by the first actual **P** argument (set or list of symbols). If the second optional **L** argument has been coded, it specifies full path to the user library having structural organization, similar to the main library. In any case, the procedure outputs an appropriate warning about done work. The procedure handles basic erratic situations linked with absence or damage of the target library, and with absence of the saved tools. At arising of such situations an appropriate erratic warnings are output.

The procedure provides the above functions for libraries of the specified type for releases **6 – 9.** In addition, libraries in the *Maple* environment of releases **6** and **7** receive the *readonly*-attribute (in the MS DOS conception) after update, whereas libraries in the *Maple* environment of releases **8** and **9/9.5** receive the *READONLY*-attribute (in the *Maple* conception) after update. The examples represented below illustrate updating by the procedure ***Art***, table ***Art_Kr*** and module ***Kr*** of both main *Maple* library and the user library with testing of the update result.

The successful procedure call **UpLib(F, N)** updates a library specified by the first actual argument **F** by procedures, tables and/or package modules whose names are specified by the second actual **N** argument (list of symbols). The procedure supposes that the **F** library has structural organization, similar to the main *Maple* library. Successful procedure call returns the *NULL* value, i.e. nothing, with output of the appropriate warnings. The procedure handles basic erratic situations linked with absence of a library, and with absence of the saved tools. At arising of such situations, the appropriate errors arise.

```
MapleLib:= proc(P::{list(symbol), set(symbol)})
local a, b, c, h, r, p, k, l, i;
  assign(r = Release('h'), p = {}, b = {}), `if`(nargs = 1, assign(l = cat(h, "\\lib\\maple.lib"),
    i = cat(h, "\\lib\\maple.ind")), `if`(type(args[2], mlib), assign(l = cat(args[2], "\\maple.lib"),
    i = cat(args[2], "\\maple.ind")), ERROR("library <%1> does not exist", args[2]))),
    assign(a = `if`(nargs = 1, cat(h, "\\lib"), args[2]));
  `if`(map(type, {l, i}, file) = {true}, NULL,
    ERROR("library <%1> does not exist or is damaged", a));
  seq(assign('p' = {`if`(type(P[k], {procedure, table, `module`}), P[k],
    assign('b' = {k, op(b)})), op(p)}), k = 1 .. nops(P)), `if`(p = {},
    ERROR("1st argument does not contain procedures, tables or modules"),
    `if`(nops(p) < nops(P), WARNING("arguments with numbers %1 are invalid", b), NULL));
  map(F_atr1, [l, i], []), assign(c=interface(warnlevel), 'savelibname' = a), interface(warnlevel=0);
  try savelib(op(p))
  catch "unable to save %1 in %2": AtrRW(savelibname); savelib(op(p))
  finally interface(warnlevel = c), AtrRW(savelibname)
  end try;
  WARNING("tools <%1> have been saved/updated in <%2>", p, savelibname),
    unassign('savelibname')
end proc

UpLib:= proc(L::{symbol, string}, N::list(symbol))
```

```
local a, b, c, d, h, k, p, t, n;
  assign(n = nops(N), a = [], p = [libname], t = cat(CDM(), "/lib/", L));
  if member(cat("", L)[2 .. 3], {":/", ":\\"}) and type(L, mlib) then h := cat("", L)
  elif type(t, mlib) then h := t
  else for k to nops(p) do
      if Search1(Case(cat("\\", L)), CF(p[k]), 'd') and d = [right] then h := p[k]; break end if
    end do
  end if;
  `if`(type(h, symbol), ERROR("<%1> is not a Maple library", L),
    seq(`if`(type(N[k], {procedure, table, `module`}), assign('a' = [op(a), N[k]]),
    WARNING("<%1> is not a procedure and not a module, and not a table", N[k])), k = 1 .. n));
  `if`(nops(a) = 0, ERROR("procedures, modules or tables do not exist for saving"),
    assign(b = NLP(L)[1]));
  for k to nops(a) do
    if member(a[k], b) or Search1(cat(a[k], `:-`), a[k], 'd') and d = [left]
    then WARNING("<%1> does exist and will be updated", a[k])
    else WARNING("<%1> does not exist and will be added", a[k])
    end if
  end do;
  assign(c = savelibname), assign('savelibname' = h), savelib(op(a));
  unassign('savelibname'), assign(savelibname = c), WARNING("Library update has been done!")
end proc
```

Typical examples of the procedures use:

```
> Art:= proc() `+`(args)/nargs end proc:                                    # Release 8 - 9
  Kr:= module () export mean; option package; mean:= () -> `+`(args)/nargs end module:
  MapleLib([Art, Kr]);
    Warning, library <C:\Program Files\Maple 9\lib> has received the `READONLY`-attribute
    Warning, tools <{Art, Kr}> have been saved/updated in <C:\Program Files\Maple 9\lib>
> restart; 6*Art(62, 57, 37, 41, 7, 15), 6*Kr:- mean(62, 57, 37, 41, 7, 15);
                                   219, 219
> Art_Kr:=table([V=62,G=57,S=37,A=41,Art=15,Kr=7]): MapleLib([RANS, IAN, Art_Kr]);
    Warning, arguments with numbers {1, 2} are invalid
    Warning, library <C:\Program Files\Maple 9\lib> has received the `READONLY`-attribute
    Warning, tools <{Art_Kr}> have been saved/updated in <C:\Program Files\Maple 9\lib>
> restart; eval(Art_Kr);
               table([A = 41, V = 62, G = 57, Kr = 7, S = 37, Art = 15])
> MapleLib([Art, Kr]);                                                      # Release 7
    Warning, library <C:\PROGRAM FILES\MAPLE 7\lib> has received the `readonly`-attribute
    Warning, tools <{Art, Kr}> have been saved/updated in <C:\PROGRAM FILES\MAPLE
    7\lib>
> restart; 6*Art(62, 57, 37, 41, 7, 15), 6*Kr:- mean(62, 57, 37, 41, 7, 15);
                                   219, 219
> RANS:= () -> [[`+`(args)], min(args), max(args)]: IAN:= module() export z; z:= 2004 end
  module: MapleLib([Art14, Kr7, RANS, IAN], "C:/program files/maple 9/lib/Svetlana");
```

Warning, arguments with numbers {1, 2} are invalid

Warning, library <C:/program files/maple 9/lib/Svetlana> has received the `READONLY`-attribute

Warning, tools <{RANS, IAN}> have been saved/updated in <C:/program files/maple 9/lib/Svetlana>

> restart; RANS(62, 57, 37, 41, 7, 15), IAN:- z;

[[219], 7, 62], 2004

> MapleLib([RANS, IAN, REA, RAC], "C:/program files/maple 9/lib/UserLib");

Error, (in **MapleLib**) 1st argument does not contain procedures, tables or modules

> aaa:=proc() `+`(args)/nargs end proc: bbb:=module() export sr, ds; option package; sr:=() -> `+`(args)/nargs; ds:=() -> sqrt(sum(args[k]^2, k=1..nargs)) end module:

> UpLib("C:/Program Files/Maple 8/Lib/UserLib", [aaa, bbb]);

Warning, <aaa> does exist and will be updated
Warning, <bbb> does exist and will be updated
Warning, Library update has been done!

> UpLib(UserLib, [aaa, bbb]);

Warning, <aaa> does exist and will be updated
Warning, <bbb> does exist and will be updated
Warning, Library update has been done!

> restart; aaa(7,61,56,15,37,40), bbb:- sr(7,61,56,15,37,40), bbb:- ds(7,60,56,14,36,40,18);

36, 36, 101

> UpLib("c:/program files/maple 8/lib/userlib", []);

Error, (in **UpLib**) procedures and modules do not exist for saving

> UpLib("Art_Kr_Svet_Arn", [aaa, bbb]);

Error, (in **UpLib**) <Art_Kr_Svet_Arn> is not a Maple library

---

dslib – check of existence in a *Maple* library of multiple versions of procedures and modules

Call format of the procedure:

dslib(L {, 't '} {, n})

Formal arguments of the procedure:

L - symbol or string defining a library name or full path to a *Maple* library
t - (optional) an assignable name
n - (optional) a positive integer (*posint*)

Description of the procedure:

As a result of a library updating by new versions of an existing procedure they are added into the end of library with an appropriate correcting in an index file "Maple.ind" of the pointer onto a new location in a file "Maple.lib" of the updated procedure. In addition, all old versions of the procedure are kept in library. The procedure **dslib** is intended for check of existence of multiple versions of procedures and modules in a library specified by the first actual argument **L;** the library should be structurally similar to the main *Maple* library.

In general case, the successful procedure call **dslib(L)** returns the 2-element sequence whose the first element is (nx2)-matrix whose the first column defines names of multiple procedures and the second - quantity of their versions; whereas the second element is also (nx2)-matrix whose the first column defines names of multiple modules and the second - quantity of their versions contained in a library **L.** If the library **L** does not contain multiple versions of procedures and/or modules, the *NULL* value is returned instead of the corresponding element of the sequence.

If the procedure call **dslib(L {, 't'})** uses the optional argument **t,** then through it the **2**-element sequence of lists is returned. The first list has **"proc"** as the first element, whereas its second element defines set of procedures names having multiple versions; the second list has **"mod"** as the first element, whereas its second element defines set of modules names having multiple versions in an analyzed library **L.**

At last, use of the optional argument **n** (*by default* **n = 2**) allows to receive the information only on those means of a library which have at least **n** versions. The procedure handles basic erratic situations linked with absence of the library; at arising of such situation an appropriate error arises. The procedure is a rather useful tool for work with the user libraries structurally similar to the main *Maple* library. The procedure usage allows to determine presence of old copies of tools in a library and receive useful statistics that concerns the updating of the library tools.

```
dslib:= proc(L::{string, symbol})
local a, b, d, t, h, p, k, Tp, Tm, omega, u, g;
  assign(p = [libname], g = 2, t = cat(CDM(), "/lib/", L), Tp = table([]), Tm = table([]));
  if member(cat("", L)[2 .. 3], {":/", ":\\"}) and type(L, mlib) then h := L
  elif type(t, mlib) then h := t
  else for k to nops(p) do
      if Search1(Case(cat("\\", L)), CF(p[k]), 'd') and d = [right] then h := p[k]; break end if
    end do
  end if;
  `if`(type(h, symbol), ERROR("<%1> is not a Maple library", L),
    assign('p' = open(cat(h, "/maple.lib"), READ))),
    assign(omega = (() -> `if`(args = lack, NULL, args)), b = "quantity of its versions");
  while not Fend(p) do h := readline(p);
    if mmp(cat("", h), 't') and not search(t, ":-") then a := 2 + length(cat(Iddn(t), t));
      if h[a .. a + 1] = "`6" then Tm[t] := `if`(type(eval(Tm[t]), indexed), 1, Tm[t] + 1)
      else Tp[t] := `if`(type(eval(Tp[t]), indexed), 1, Tp[t] + 1)
      end if
    end if
  end do;
  if 1 < nargs then for k from 2 to nargs do
      if type(args[k], symbol) then u := k elif type(args[k], posint) then g := args[k] end if
    end do
  end if;
  close(p), `if`(type(u, posint), assign67(args[u] = ["proc", sort(map(op, {indices(Tp)}))],
    ["mod", sort(map(op, {indices(Tm)}))]), NULL), omega(tabar(Tp, "procedure name", b, g)),
    omega(tabar(Tm, "module name", b, g))
end proc
```

Typical examples of the procedure use:

> dslib(UserLib_V, 't'), dslib(UserLib_G, 3);

procedure name	*quantity of its versions*
MkDir	5
dslib	4
Iddn	3

,

module name	*quantity of its versions*
SveGal	5
ArtKr	3
RANS_IAN	3

> t;

["**proc**", {*convert/list, F_atr2, Empty, convert/rlb, mtf, SN_7, FT_part, com6_9, Dist_rand_n_2, Polyhedra, DirE, FT_part1, LibElem2, type/dirax, Rlss, Animate2D, type/package, sls, Test, MkDir,*

==

sfd, isDir, SL, Fremove, Search3, winver, VisM, rand_Histo, ExtrF, Fword, Catch_Vir, Sub_list, exeP, map4, StatLib, CureLib, open2, IsOpen2, CrNumMatrix, Pul_Bal_Cor, ChkPnt, seq2, Rt_s, LibLink1, Vol, mod_proc, mPM, Arobj, Chess, LibUser, PNorm, type/rlb, S_D}],
["**mod**", {*SveGal, DIRAX, SimpleStat, ACC, AlgLists, RANS_IAN, SoLists, ArtKr*}]

> dslib(UserLib, 'z', 3);

procedure name	*quantity of its versions*
F_atr2	4
SSF	3
ewsc	3
conSA	3
Suffix	3
type/file	3
FSSF	3
gelist	3
SDF	3
sorttf	3

> dslib("C:/Academy/UserLib9", 't');

Error, (in **dslib**) <C:/Academy/UserLib9> is not a Maple library

---

reprolib – extracting of all versions of a procedure without their deletion

Call format of the procedure:

reprolib(L, P, C)

Formal arguments of the procedure:

P - symbol defining a procedure name
L - symbol or string defining a library name or full path to a *Maple* library
C - symbol or string defining a target directory for datafiles with procedure versions

Description of the procedure:

As a result of a library updating by new versions of an existing procedure they are added into the end of library with an appropriate correcting in an index file "Maple.ind" of the pointer onto a new location in a file "Maple.lib" of the updated procedure. In addition, all old versions of the procedure are kept in library. The procedure **reprolib** is intended for extracting of all versions of a procedure specified by the second actual argument **P** from a library specified by the first actual argument **L;** the library should be structurally similar to the main *Maple* library. In addition, these versions physically do not leave out of library **L**.

The successful procedure call **reprolib(L, P ,C)** extracts all versions of a procedure **P** from a library **L** without their physical deletion out of the library and in a directory specified by the third actual argument **C** creates the set of datafiles which will contain definitions of appropriate versions of the procedure **P** in the input *Maple* language format. The resultant datafiles with definitions of procedure **P** receive names of view **cat(P, n),** where **n** - sequence datafile number. If directory **C** does not exist, then it will be created. The procedure outputs the appropriate warning about full path to a target directory.

The procedure handles basic erratic situations linked with absence of the library, and with absence in it of the sought procedure. At arising of such situations, the appropriate errors arise. The procedure is a rather useful tool for work with the user libraries structurally similar to the main *Maple* library.

```
reprolib:= proc(L::{string, symbol}, P::symbol, C::{string, symbol})
local a, b, c, d, r, k, n, p, t, h, f, z, x, y, v;
  assign(r = Release(), p = [libname], t = cat(CDM(), "/lib/", L), b = interface(warnlevel), v = 0);
  if member(cat("", L)[2 .. 3], {":/", ":\\"}) and
  type(L, mlib) then h := L
  elif type(t, mlib) then h := t
  else for k to nops(p) do
        if Search1(Case(cat("\\", L)), CF(p[k]), 'd') and d = [right] then h := p[k]; break end if
    end do
  end if;
  `if`(type(h, symbol), ERROR("<%1> is not a Maple library", L), assign(a = NLP(h)[1]));
  `if`(member(P, a), assign('h' = cat(h, "/maple.lib")),
    ERROR("<%1> does not exist in <%2>", P, h));
  if r = 6 and readbytes(h, TEXT, 5)[2 .. 5] <> "M6R0" then close(h);
    ERROR("library <%1> is not compatible with current Maple release 6", L)
  else close(h); `if`(r = 6, assign('t' = "M6R0", z = "`6"), assign('t' = "M7R0", z = "f*"))
  end if;
  if type(C, dir) then c := C else interface(warnlevel = 0); c := MkDir(C);
    interface(warnlevel = b)
  end if;
  assign(f = "_$$$7.m", n = 3 + length(Iddn(length(P))) + length(P),
    'z' = cat("I", Iddn(length(P)), P, z));
  do x := readline(h);
    if x = 0 then break
    elif length(x) < n or x[1 .. n] <> z then next
    else writeline(f, t), writeline(f, x);
      do  x := readline(h);
        if x = 0 then y := 7; break elif x = t then break else writeline(f, x) end if
```

```
        end do;
        (proc() close(f); read f; v := v + 1; save P, cat(c, "/", P, v); unassign(P) end proc)();
        if y = 7 then break else next end if
      end if
    end do;
    delf(f), WARNING("extraction result is in directory <%1>", c)
end proc
```

Typical examples of the procedure use:

> reprolib("C:/CORRECT/Svetlana", Art_Kr, "C:/Temp/ReprocLib\\Test/Vgtu");

Warning, extraction result is in directory <c:\temp\reproclib\test\vgtu>

> reprolib("C:/CORRECT/Svetlana", Art_Kr, "C:/Temp/ReprocLib\\Test/Vgtu");

Error, (in **reprolib**) library <C:/correct/Svetlana> is not compatible with current Maple release 6

AtrRW – change of a library attributes such as `*READONLY*` or `*WRITABLE*`

Call format of the procedure:

AtrRW(L)

Formal arguments of the procedure:

L - symbol or string defining full path to a library

Description of the procedure:

For *Maple* of releases **8** and **9** the procedure call **AtrRW(L)** by a switch principle changes the attributes `*READONLY*` or `*WRITABLE*` (in the *Maple* conception) upon opposite ones for a library whose full path is specified by the actual **L** argument. Whereas for *Maple* of releases **6** and **7** the procedure call **AtrRW(L)** by a switch principle changes the attributes `*readonly*` or `*writable*` (in the MS DOS conception) upon opposite ones. Successful procedure call **AtrRW** returns the *NULL* value, i.e. nothing, with output of the appropriate warning about done work. The procedure is a rather useful tool in case of impossibility of use of the standard utility `***march***` or other *Maple* tools of an access to *readonly* library datafiles. The procedure handles the basic erratic and especial situations.

```
AtrRW:= proc(L::path)
local a, r, h, p, l, i, omega;
  assign(r = Release(), l = cat(L, "\\maple.lib"), i = cat(L, "\\maple.ind"));
  `if`(map(type, {l, i}, file) = {true}, NULL,
     ERROR("library <%1> does not exist or is damaged", L));
  omega:= x -> WARNING("library <%1> has received the %2-attribute", L, x);
  if member(r, {6, 7}) then assign(a = map(F_atr1, [l, i])), map(F_atr1, [l, i], []);
     if not member(map2(member, "R", {op(a)}), {{true}, {false, true}}) then
        null(map(F_atr1, [l, i], ["+R"])), omega("`readonly`")
     else omega("`writable`")
     end if
  else
     try filepos(l, 265), assign(p = (l -> filepos(l, 265))), p(l)
     catch "file I/O error": map(F_atr1, [l, i], []), filepos(l, 265),
          assign(p = (l -> filepos(l, 265))), p(l)
```

```
    catch "permission denied": map(F_atr1, [l, i], []); filepos(l, 265),
        assign(p = (l -> filepos(l, 265))), p(l)
    end try;
    if iolib(4, l) = [2] then null(p(l), iolib(5, l, [1])), close(l), omega("`READONLY`")
    else null(p(l), iolib(5, l, [2])), close(l), omega("`WRITABLE`")
    end if
  end if
end proc
```

Typical examples of the procedure use:

> AtrRW("C:/program files/maple 7/lib");

Warning, library <C:/program files/maple 7/lib> has received the `READONLY`-attribute

> AtrRW("C:/program files/maple 7/lib");

Warning, library <C:/program files/maple 7/lib> has received the `WRITABLE`-attribute

> AtrRW("C:/program files/maple 9/lib");

Warning, library <C:/program files/maple 9/lib> has received the `WRITABLE`-attribute

> AtrRW("C:/program files/maple 9/lib");

Warning, library <C:/program files/maple 9/lib> has received the `READONLY`-attribute

Tools of the ***third*** level give an opportunity of extension of the main *Maple* library by a new software. However, such decision should be considered as exclusive, because it reduces a level of mobility of the tools placed in the main library. In this respect, tools of the *third* level are expedient for an applying only relative to the user libraries structurally similar to the main *Maple* library.

Sometimes, a library listing represents a set of members (so-called *`local`* members) with slightly curiously looking names that have view "**:-n.m**", where **n** is a positive integer. These are generated when closures are saved in a library, and are used to maintain the library in an internally consistent state. In this case, members whose names are **`:-n.m`** are created, in addition to some *global* member **`G.m`** for a **G** module itself. These members (as well as the module member **`G.m`**) are useless by itself, and generally cannot be used separately from the entire library. Although the built-in ***`march`*** function and our tools represented below allow you to extract the individual library members, they are not usable (besides procedures) separately outside a library in which they reside. Manipulating library members outside an intact library is not supported, and does not generally yield useful results. Meantime, in a series of cases the extracted *local* library members can be enough successfully used outside a library. Tools for such manipulating become topical, in particular, in a case of a damaged library. Procedure **LocalnL** allows to discover in the given library structurally analogous to the main *Maple* library a presence of the *global* members having the *local* members, accompanied to them. Whereas the procedure **_nnn** provides receiving of minimal and maximal numbers of *local* members of all libraries defined by the predetermined ***libname*** variable of a current *Maple* session.

LocalnL – check of presence of local members in the *Maple* libraries _nnn

Call format of the procedures:

LocalnL(L {, N})
_nnn({'t'})

Formal arguments of the procedures:

L - symbol or string defining full path to a library
N - (optional) symbol defining main name of a global member
t - (optional) an assignable name

Description of the procedures:

The procedure call **LocalnL(L)** returns the nested list whose elements represent sublists describing all *`global`* members of a library specified by the first actual **L** argument. The first element of each sublist defines name of a global member, whereas the others define names **":-n.m"** of *local* members accompanied to it. If a library **L** contains the *global* members only, the procedure call returns the *empty* list, i.e. **[]**, with output of an appropriate warning.

The procedure call **LocalnL(L, N)** returns the list whose the first element defines name of a *global* module **"N.m"** while the others define names of view **":-n.m"** of *local* members accompanied to it. The examples represented below well illustrate the told. If the returned list will contain a name **"N.m"** only, the given *global* member has not *local* members accompanied to it. In process of the procedure execution, the appropriate warnings are output.

A procedure call **_nnn()** returns the **2**-element integer list whose the first element defines the minimal number whereas the second element defines the maximal number **nnn** of *local* members **":-nnn.m"** which are contained in libraries defined by the predetermined ***libname*** variable in a current *Maple* session. If the procedure call **_nnn('t')** uses the optional argument t then through it the set of positive integers lesser than the maximal number of *local* members which have not been used by them is returned. The procedure call result depends on the current *Maple* release and the user libraries logically linked with the main *Maple* library in a current session. The procedure has a series of useful appendices in work with the *Maple* libraries.

```
LocalnL:= proc(L::path)
local k, G, a, Res, S, h, f, Test;
  `if`(not type(L, mlib), ERROR("library <%1> does not exist or is damaged", L)
    , assign(a = 1, G = march('list', L), Res = [], f = cat(L, "\\maple.lib")));
  `if`(nargs = 1, NULL, `if`(type(args[2], symbol), NULL,
    ERROR("2nd argument should has `symbol`-type but has received <%1>", args[2])));
  Test:= proc(f, a, b)
    local h;
      iolib(6, f, a), assign('h' = iolib(4, f, b, TEXT));
      {op(sextr1(h, ":-", {digit}, {letter}))} minus {":-"}
    end proc;
  for k to nops(G) do
    if G[k][1][1 .. 2] = ":-" then next
    elif nargs = 1 then S := Test(f, G[k][3], G[k][4]);
      if S = {} then next
      else Res := [op(Res), [G[k][1], op(map(cat, S, ".m"))]]
      end if
    else
      if G[k][1] = cat("", args[2], ".m") then Res := [G[k][1],
          op(map(cat, Test(f, G[k][3], G[k][4]), ".m"))]; break
      end if
    end if
  end do;
  iolib(7, f), Res, `if`(Res = [], `if`(nargs = 2,
```

```
    WARNING("<%1> does not exist in library <%2>", args[2], L),
    WARNING("local names do not exist in library <%1>", L)), NULL)
end proc

_nnn:= proc()
local a, b, c, d, k, p, t, h;
  assign(a = {}, b = [libname]), `if`(nargs <> 0, `if`(type(eval(args[1]), symbol), assign(h = 7),
    ERROR("the 1st argument must be a symbol but has received <%1>",
    whattype(eval(args[1])))), NULL);
  for k in b do c := march(list, k);
    for p to nops(c) do d := c[p][1];
      if d[1 .. 2] = ":-" then a := {SD(d[3 .. -3]), op(a)} end if
    end do
  end do;
  [a[1], a[-1]], `if`(h = 7, assign(args[1] = {t $ (t = 1 .. a[-1])} minus a), NULL)
end proc
```

Typical examples of the procedures use:

```
> map2(LocalnL,"C:/Program files/Maple 9/Lib/UserLib", [CureLib, MkDir, ParProc,
  Red_n, Close, delf1, ModProc]); LocalnL("C:/RANS/Academy/Libraries/Svetlana");
  [["CureLib.m"],["MkDir.m"],["ParProc.m"],["Red_n.m"],["Close.m"],["delf1.m"],["ModProc.m"]]
   Warning, local names do not exist in library <C:/RANS/Academy/Libraries/Svetlana>
                                  []
> LocalnL("C:/Program files/Maple 9/Lib/UserLib", DIRAX);
  ["DIRAX.m", ":-37.m", ":-28.m", ":-32.m", ":-36.m", ":-35.m", ":-30.m", ":-27.m", ":-34.m", ":-29.m",
                          ":-33.m", ":-31.m"]
> LocalnL("C:/Program files/Maple 9/Lib/UserLib");
  [["ACC.m", ":-39.m", ":-43.m", ":-42.m", ":-46.m", ":-40.m", ":-45.m", ":-44.m", ":-41.m"],
  ["SimpleStat.m", ":-5.m", ":-14.m", ":-4.m", ":-13.m", ":-7.m", ":-16.m", ":-17.m", ":-19.m",
  ":-12.m", ":-3.m", ":-9.m", ":-18.m", ":-11.m", ":-2.m", ":-15.m", ":-6.m", ":-1.m", ":-10.m"],
  ["SoLists.m", ":-52.m", ":-54.m", ":-55.m", ":-53.m"], ["DIRAX.m", ":-37.m", ":-28.m",
  ":-32.m", ":-36.m", ":-35.m", ":-30.m", ":-27.m", ":-34.m", ":-29.m", ":-33.m", ":-31.m"],
  ["AlgLists.m", ":-50.m", ":-48.m", ":-47.m", ":-49.m", ":-51.m"]]
> _nnn('t'), t;                                                         # Maple 6
  [1, 10818], {56, 57, 58, 59, 60, 61, 62, 63, 64, 65, 66, 67, 68, 69, 70, 71, 72, 73, 74, 75, 76, 77,
  ==========================================================================
  10782, 10783, 10784, 10785, 10786, 10787, 10788, 10789, 10790, 10791, 10792, 10793, 10794}
> _nnn('t'), t;                                                         # Maple 7
                              [1, 10738], {}
> _nnn('t'), t;                                                         # Maple 8
                          [1, 10045], {1301, 8570}
> _nnn('t'), t;                                                         # Maple 9
                              [1, 46552], {}
```

At operation with the user libraries having structural organization, similar to the main *Maple* library, in a series of cases the necessity of restoration of libraries arises, when the *`march`* utility appears powerless (for example, in a case of damage or absence of index file "Maple.ind" of a *Maple* library). In a great extent, the given problem is solved by means of rather useful procedures **DestLm, LibElem, LibElem1, LibElem2, CureLib** and **ContLib.**

DestLm – operations with the damaged user libraries
LibElem
LibElem1
LibElem2

Call format of the procedures:

DestLm(L, P {, Z})
LibElem(F, U {, K})
LibElem1(F, U, K)
LibElem2(F, U, K)

Formal arguments of the procedures:

L – symbol or string defining full path to the user library
F – string or symbol defining full path to a library *lib*-file
P – symbol or string defining name of a procedure or program module
U – symbol, set or list of symbols
Z – (optional) string defining main name of a library *lib*-file
K – symbol or string defining full path to a target directory

Description of the procedures:

The successful procedure call **DestLm(L, P {, Z})** creates an *m*-file with a procedure or program module whose name is specified by the second actual **P** argument; it is located in a library whose full path is specified by the first actual **L** argument. If the third optional **Z** argument has been coded, it defines a main name of two library files, namely **"Z.ind"** and **"Z.lib"**. However, processing of a procedure and program module is done variously, namely. If the second actual **P** argument specifies a procedure, then it is saved in datafile with a name **`P.m`** located in **L** directory with its subsequent activation providing direct access to it in a current *Maple* session. The appropriate warning says about that. In addition, it is necessary to have in mind, the procedure processes only the first versions (*according to time of their saving*) of the sought means.

If a library **L** contains both basic files {**".lib", ".ind"**}, the procedure call returns the *NULL* value with output of an appropriate warning. Whereas at existence of library *ind*-datafile only, the procedure call returns contents of a library **L** in format analogous to format of call **march('list', L).**

If the second actual **P** argument specifies a program module, then its head part is saved in the datafile with name **`P.m`** located in **L** directory without its subsequent activation in a current *Maple* session, but with output of a warning about submodules appropriate to its head part, which have names of view **`:-NNN.m`** (NNN – a positive integer) and remain in the source *lib*-file. The procedure returns the list with full names of such submodules. For full extraction of a **P** module, the procedure recommends execute the **LibElem1** procedure being considered below. The examples represented below well illustrate the told.

Furthermore, the **DestLm** provides a handling of an erroneous situations provoked by abnormal execution of the built-in procedure ***`march`*** in case of absence of library file "Maple.ind" or a library of the other release. For release **9,** these cases provoke erroneous system situations with diagnostics "**Cwmaple9:** *This program has performed an illegal operation and will be shut down.*" or "**Maple9:** *The kernel has been shut down. Further computations cannot be performed*" provoking a crash of a current *Maple* session. Whereas for release **8** these cases provoke an erroneous system situation with diagnostics "**Mserver:** *This program has performed an illegal operation and will be shut down*" provoking a crash of a current *Maple* session, whereas for releases **6** and **7** they provoke an erratic package situation *"Error, (in **march**) there is no existing archive in* <%1>" only. The **DestLm** procedure provides a program handling of the above erratic system situations without of the above *Maple* crash.

The procedure **LibElem** from two or three actual arguments directly adjoins to the previous procedure. The procedure **LibElem(F, U)** supports operation both with the main *Maple* library, and with other libraries, organized analogously. The first actual **F** argument of the procedure specifies full path to *lib*-file of a required library, whereas the second actual **U** argument specifies a set or list of names of procedures or program modules, or simply a separate name. In addition, processing of procedures and program modules specified by **U** argument is done by a different manner, namely.

At detection in list or set **U** of a program module, the appropriate warning is output and processing of the given element of **U** argument terminates. Whereas for case of a procedure as an element of **U** argument, the nested 3-element list is returned. The structure of the list is similar to structure of the information returned by the procedure call `**march('list', ...)**`, namely. The first element of such list defines name of an *m*-file with the found procedure, the second - displacement of beginning of *m*-file from beginning of a *lib*-file, and the third element defines *m*-file size in bytes.

If at procedure call, the third optional **K** argument of a type {*symbol, string*} has been coded, then **K** is considered as a chain of subdirectories (at its absence the chain will be created) and the procedure puts into the last subdirectory of the chain the *m*-files (*with names out of the second argument* **U**) which are extracted out of the indicated **F** file without their deleting out of the library. The saved *m*-files afterwards can be read by the built-in function `***read***`, providing access to software contained in them. The **LibElem** procedure provides a rather simple access to the separate elements (or their set) of a library directly by their basic names.

If a file **F** belongs to a library containing both basic files {**".lib", ".ind"**}, the procedure call returns the *NULL* value with output of an appropriate warning. Whereas at existence of library ***ind***-datafile only, the procedure call causes the erratic situation with appropriate diagnostics.

If as the second actual **U** argument a name has been coded, then the procedure returns the nested list whose elements are 3-element sublists of the above structure, which describe all *m*-files located in an **F** file. In addition, irrespective of quantity of actual arguments at the procedure call through global variable ***_ProcMod*** the **3**-element sequence is returned, namely: its the first element defines a list of procedures names, the second - a list of modules names, and the third element - a list of names of *Maple* expressions of types different from types {`*procedure*`, `*module*`} contained in the analyzed **F** file; names are represented by strings. Along with the said, the procedure outputs a series of useful messages. The examples represented below well illustrate the told.

In contrast to the previous procedure, the **LibElem1(F, U, K)** procedure from three formal arguments provides the more full extracting of the required members specified by the second **U** argument (procedures and program modules) out of a library specified by the first **F** argument. For each extracted element (without its deletion out of the library) in a directory specified by the third actual **K** argument one or more *m*-files are created. Analogously to the **LibElem** the **LibElem1** returns through global variable ***_ProcMod*** the 3-element sequence of the above structure, however if a file **F** belongs to a library containing both basic files {**".lib", ".ind"**}, the procedure call returns the global variable ***_ProcMod*** unevaluated.

For an extracted procedure an *m*-file with its name is created, which can be loaded into a current session by the built-in `***read***` function. In this case, the procedure outputs an appropriate warning. While for an extracted module a few *m*-files are created, namely: a file with module name contains main body of the module, whereas submodules accompanied to it are put into *m*-files with names of view `**_NameNNN.m**`, where: ***Name*** - name of the module, and ***NNN*** - number of its submodule in the library. In the library such submodule has name of view `**:-NNN.m**`. The created *m*-files are loaded into a current session by means of the *`read`* function, however the variables exported by a module extracted in such a way, are inaccessible. However, a set of *m*-files created by an extracted module subsequently can be loaded into other library, for example by means of the **User_pflM** procedure considered above. The procedure outputs a series of information messages and warnings.

At last, the procedure call **LibElem2(F, U, K)** extends the functional possibilities of the previous **LibElem1** procedure, having the same set of formal arguments. However, in addition to the second it provides extracting out of program modules specified by the second actual **U** argument, of the exported objects with saving of them in text files with names **`NNN.txt`**, where **NNN** - names of the extracted modules. Messages, output by the procedure, explain a course of procedure running. The examples represented below well illustrate the told.

It is necessary to note, as a result of the call of any of the four procedures **DestLm, LibElem, LibElem1** and **LibElem2** through global variable ***_ProcMod*** the 3-element sequence is returned, namely: its first element defines a list of procedures names, the second - a list of program modules names, and the third element - a list of names of *Maple* expressions of types different from types {*`procedure`*, *`module`*} which are contained in an analyzed library **F;** all names are represented as *strings*. However if a file **F** belongs to a library containing both basic files {**".lib", ".ind"**}, the procedure call returns the global variable ***_ProcMod*** unevaluated.

In addition, it is necessary to have in mind, in contrast to the **DestLm**, the procedures **LibElem, LibElem1** and **LibElem2** return the results relative to the last versions (*according to time of their saving*) of the analyzed means. The above procedures **DestLm, LibElem, LibElem1** and **LibElem2** can be considered as rather useful tools of a restoring of the damaged user libraries structurally analogous to the main *Maple* library.

```
DestLm:= proc(L::path, P::{symbol, string})
local F, S, v, k, h, Kr, g, Ind;
  `if`(2 < nargs and type(eval(args[3]), {symbol, string}), assign(v = cat(L, "/", args[3], ".lib"),
    Ind = cat(L, "/", args[3], ".ind")), assign(v = cat(L, "/maple.lib"), Ind = cat(L, "/maple.ind")));
  if map(type, {v, Ind}, file) = {false} then
    ERROR("library <%1> does not exist or is crucially damaged!", L)
  elif map(type, {v, Ind}, file) = {true} then
    RETURN(WARNING("library <%1> is complete!", L))
  elif not type(v, file) then RETURN(redL(CureLib(L)))
  else
    `if`(not type(Ind, file), WARNING("library index file <%1> does not exist", Ind), NULL);
    WARNING("extraction from file <%1> is being undertaken. Please, wait!", v);
    try S := CureLib(v, 'F')
    catch "CureLib cannot identify library contents":
      ERROR("DestLm cannot identify library contents")
    end try
  end if;
  `if`(belong(cat("", P, ".m"), S), assign(g = `if`(cat("", P)[1 .. 2] = ":-",
    cat(L, "/", "mod_", cat("", P)[3 .. -1], ".m"),
    cat(L, "/", P, ".m"))), ERROR("procedure or module <%1> are not in library <%2>", P, L));
  for k to nops(F) do if F[k][1] = cat("", P, ".m") then Kr := F[k]; break  end if  end do;
  iolib(6, v, Kr[2]), assign('h' = iolib(4, v, Kr[3], TEXT)), iolib(5, g, h), close(v, g);
  h := {op(sextr1(h, ":-", {digit}, {letter}))} minus {":-"};
  if h = {} then  read g;
```

```
    WARNING("<%1> has been activated in a current session and saved in datafile <%2>", P, g)
  else WARNING("head part of module <%1> has been saved in datafile <%2>.
    In a source file the accompanied submodules %3 have remained. In the sequel, You can use
    the LibElem1 procedure!", P, g, h), {seq(cat(h[k], ".m"), k = 1 .. nops(h))}
  end if
end proc

LibElem := proc(L::file, P::{set, list, symbol})
local a, b, c, d, e, k, p, R, f, t, z;
global _ProcMod;
  if type(clsd(CFF(L)[1 .. -2], "\\"), mlib) then
    RETURN(WARNING("library is complete. The `march` utility can be used."))
  elif not type(cat(clsd(CFF(L)[1 .. -2], "\\"), "\\maple.lib"), file)
  then ERROR("library members cannot be extracted or library does not exist")
  end if;
  `if`(2 < nargs, `if`(type(args[3], dir), assign(f = args[3]), `if`(member(whattype(args[3]),
    {string, symbol}), [assign(t = interface(warnlevel)), interface(warnlevel = 0),
    assign(f = MkDir(args[3])), interface(warnlevel = t)],
    ERROR("3rd argument must be a string or symbol but has received <%1>", args[3]))),
    assign(b = 7)), assign(R = {}), unassign('_ProcMod');
  `if`(type(P, {list(symbol), set(symbol)}) and 0 < nops(P) or type(P, symbol), CureLib(L, 'a'),
    ERROR("the 2nd actual argument <%1> is invalid", P)),
    `if`(nargs = 2 and type(eval(args[2]), symbol), RETURN(a), NULL);
  z := {op(P)}  minus  {seq(cat(``, a[k][1][1 .. -3]), k = 1 .. nops(a))};
  `if`(z = {}, NULL, WARNING("names %1 have not been identified in library %2", z, L));
  seq(seq(`if`(a[k][1] = cat("", P[p], ".m"), `if`(member(cat("", P[p]), _ProcMod[1]), `if`(b=7, a[k],
    op([a[k], assign('d' = cat(f, "/", P[p], ".m"), 'R' = {P[p], op(R)}),
    assign('e' = [iolib(6, L, a[k][2]), iolib(5, d, iolib(4, L, a[k][3], TEXT)), iolib(7, L, d)])])),
    WARNING("<%1> is not a procedure", P[p])), NULL), k = 1 .. nops(a)), p = 1 .. nops(P)),
    `if`(b <> 7 and nops(R) <> 0, WARNING("procedures %1 have been saved in directory
    <%2> in m-files with the same main names", R, f),
    WARNING("saving of procedures in m-files has been not done"))
end proc

LibElem1:= proc(F::{name, string, symbol}, P::{list, set}, K::{string, symbol})
local a, b, f, k, p, h, R, Z, Test, M, R1, Q, S;
global _ProcMod;
  `if`(not IsFtype(F, ".lib"), ERROR("library lib-file <%1> does not exist", F),
    `if`(type(K, dir), assign(f = K), [assign(b = interface(warnlevel)), interface(warnlevel = 0),
    assign(f = MkDir(K)), interface(warnlevel = b)])), unassign('_ProcMod'),
    assign(S = CureLib(F, 'a'));
  `if`(type(a, symbol), assign('a' = [seq(subsop(2 = NULL, S[k]), k = 1 .. nops(S))]), NULL);
  Test:= proc(X)
    local h, k, R, f;
      for k to nops(a) do
        if a[k][1] = cat("", X, ".m") then assign('f' = 61), assign('h' = ExtrF(F, a[k][2], a[k][3]));
          R := {op(sextr1(h, ":-", {digit}, {letter}))}  minus  {":-"}
        end if
      end do;
      `if`(f = 61, [h, {op(R)}], op([false, WARNING("object <%1> does not exist in <%2>",
```

```
        cat(X, ".m"), F)]))
    end proc;
  for k to nops(P) do  R := Test(P[k]);
    if R = false then next
    elif R[2] = {} then assign('Z' = cat(f, "/", P[k], ".m")), iolib(5, Z, R[1]), iolib(7, Z);
      WARNING("<%1> can be loaded by `read` function", cat(P[k], ".m"));  next
    else assign('Z' = cat(f, "/", P[k], ".m")), iolib(5, Z, R[1]), iolib(7, Z),
      assign('R1' = R[2], 'Q' = {});
      WARNING("program module <%1> has been saved in files <%2> and <%3>",
        P[k], cat(P[k], ".m"), cat("_", P[k], "nnn.m"));
      A;
      for p to nops(R1) do  M := Test(R1[p]);
        if M = false then next
        else assign('Z' = cat(f, "/_", P[k], R1[p][3 .. -1], ".m")), iolib(5, Z, M[1]), iolib(7, Z),
          assign('Q' = {op(Q), op(M[2])})
        end if
      end do;
      R1 := Q  minus  R1;
      if R1 <> {} then Q := {};  goto(A)  else  iolib(7, Z)  end if
    end if
  end do;
  null(), WARNING("the created m-files have been located in directory <%1>", f)
end proc

LibElem2:= proc(F::{name, string, symbol}, P::{list, set}, K::{string, symbol})
local a, b, f, k, m, p, h, R, Z, Test, M, R1, Q, S, Y, Ex, x, y;
global _ProcMod;
  `if`(not IsFtype(F, ".lib"), ERROR("library lib-file <%1> does not exist", F),
    `if`(type(K, dir), assign(f = K), [assign(b = interface(warnlevel)), interface(warnlevel = 0),
    assign(f = MkDir(K)), interface(warnlevel = b)])), unassign('_ProcMod'),
    assign(S = CureLib(F, 'a'));
  `if`(type(a, symbol), assign('a' = [seq(subsop(2 = NULL, S[k]), k = 1 .. nops(S))]), NULL);
  Test:= proc(X)
    local h, k, R, f;
      for k to nops(a) do
        if a[k][1] = cat("", X, ".m") then assign('f' = 61), assign('h' = ExtrF(F, a[k][2], a[k][3]));
          R := {op(sextr1(h, ":-", {digit}, {letter}))}  minus  {":-"} end if
      end do;
      `if`(f = 61, [h, {op(R)}], op([false, WARNING("<%1> is absent in library <%2>",
        cat(X, ".m"), F)]))
    end proc;
  for k to nops(P) do  R := Test(P[k]);
    if R = false then next
    elif R[2] = {} then assign('Z' = cat(f, "/", P[k], ".m")), iolib(5, Z, R[1]), iolib(7, Z);
      WARNING("procedure <%1> can be loaded by `read` function", cat(P[k], ".m"));  next
    else assign('Z' = cat(f, "/", P[k], ".m")), iolib(5, Z, R[1]), iolib(7, Z),
      assign('R1' = R[2], 'Q' = {});
      read Z;
      WARNING("program module <%1> has been saved in files <%2> and <%3>", P[k],
        cat(P[k], ".m"), cat("_", P[k], "nnn.m"));
```

```
`Files creation for module and its submodules`;
        for p to nops(R1) do  M := Test(R1[p]);
          if M = false then next
          else assign('Z' = cat(f, "/_", P[k], R1[p][3 .. -1], ".m"), 'Y' = {SN(R1[p][3 .. -1]), op(Y)}),
          iolib(5, Z, M[1]), close(Z), (proc() read Z end proc)(), assign('Q' = {op(M[2]), op(Q)})
          end if
        end do;
        R1 := Q  minus  R1;
        if R1 <> {} then  Q := {};  goto(`Files creation for module and its submodules`)
        else  close(Z), assign('Ex' = [exports(P[k])]),  unprotect(op(Ex));
          for m to nops(Ex) do  assign(Ex[m] = eval(cat(`:-`, Y[m]))) end do;
          (proc()  save args, cat("", f, "/", P[k], ".txt")  end proc)(op(Ex)), assign('Y' = {}),
            WARNING("exports of module <%1> have been saved in file <%2>", P[k],
            cat("", f, "/", P[k], ".txt"))
        end if
      end if
    end do;
    null(), WARNING("the created m-files have been located in directory <%1>", f)
end proc
```

Typical examples of the procedures use:

1. An example of use of the **DestLm** procedure for extracting of procedure **MkDir** out of the damaged user library **Svetlana** that contains only library *lib*-file:

> DestLm("C:/rans/academy/libraries/Svetlana", MkDir);

Warning, library index file <C:/rans/academy/libraries/Svetlana/maple.ind> does not exist

Warning, extraction from file <C:/rans/academy/libraries/Svetlana/maple.lib> is being undertaken. Please, wait!

Warning, Analysis of library contents is being done. Please, wait!

Warning, <**MkDir**> has been activated in a current session and saved in datafile <C:/rans/academy/libraries/Svetlana/MkDir.m>

2. An example of use of the **DestLm** procedure for extracting of program module **DIRAX** from the damaged user library **Svetlana** which contains only library *lib*-file:

> DestLm("C:/rans/academy/libraries/Svetlana", DIRAX);

Warning, library index file <C:/rans/academy/libraries/Svetlana/maple.ind> does not exist

Warning, extraction from file <C:/rans/academy/libraries/Svetlana/maple.lib> is being undertaken. Please, wait!

Warning, Analysis of library contents is being done. Please, wait!

Warning, head part of module <**DIRAX**> has been saved in datafile <C:/rans/academy/libraries/Svetlana/DIRAX.m>.

In a source file the accompanied submodules {:-32, :-34, :-36, :-37, :-30, :-28, :-31, :-33, :-29, :-35, :-27} have remained.

In the sequel, You can use the **LibElem1** procedure!

{":-32.m", ":-36.m", ":-37.m", ":-28.m", ":-33.m", ":-30.m", ":-31.m", ":-29.m",":-35.m", ":-27.m", ":-34.m"}

3. Examples of use of the **LibElem** procedure for extracting of procedures out of the damaged user library **Svetlana,** which contains only library *lib*-file:

> LibElem("C:/rans/academy/libraries/Svetlana/maple.lib", [ParProc, MkDir, AlgLists, CureLib]);

Warning, Analysis of library contents is being done. Please, wait!
Warning, **<AlgLists>** is not a procedure
Warning, saving of procedures in m-files has been not done

["ParProc.m", 72687, 756], ["MkDir.m", 291121, 1125], ["CureLib.m", 226578, 1787]

> LibElem("C:/rans/academy/libraries/Svetlana/maple.lib", [ParProc, MkDir, CureLib], "C:/Temp/ABS");

Warning, Analysis of library contents is being done. Please, wait!

Warning, procedures {**MkDir**, **CureLib**, **ParProc**} have been saved in directory <c:\temp_abs> in m-files with the same main names

["ParProc.m", 72687, 756], ["MkDir.m", 291121, 1125], ["CureLib.m", 226578, 1787]

> LibElem("C:/temp/Galina/maple.lib", [MkDir, DIRAX, AlgLists, CureLib]);

Warning, library file <c:\temp\galina\maple.ind> does not exist
Warning, Analysis of contents of library lib-file is being done. Please, wait!
Warning, Analysis of contents of library lib-file has been completed!

Warning, names {CureLib, AlgLists} have not been identified in library C:/temp/Galina/maple.lib

Warning, <DIRAX> is not a procedure
Warning, saving of procedures in m-files has been not done

["MkDir.m", 3020, 1312], ["MkDir.m", 6332, 1942]

4. An example of use of the **LibElem1** procedure for extracting of procedures and program modules out of the damaged user library **Svetlana** that contains only library *lib*-file:

> LibElem1("C:/rans/academy/libraries/Svetlana/maple.lib", [Vilnius, DIRAX, User_pflM, Grodno, SimpleStat, CureLib], "C:/RANS/IAN/RAC/REA/Art_Kr");

Warning, Analysis of library contents is being done. Please, wait!

Warning, object **<Vilnius.m>** does not exist in <C:/rans/academy/libraries/Svetlana/maple.lib>

Warning, program module **<DIRAX>** has been saved in files <DIRAX.m> and <_DIRAXnnn.m>

Warning, **<User_pflM.m>** can be loaded by \`read\` function

Warning, object **<Grodno.m>** does not exist in <C:/rans/academy/libraries/Svetlana/maple.lib>

Warning, program module **<SimpleStat>** has been saved in files <SimpleStat.m> and <_SimpleStatnnn.m>

Warning, **<CureLib.m>** can be loaded by \`read\` function
Warning, the created m-files have been located in directory <c:\rans\ian\rac_rea\art_kr>

> LibElem1("C:\\Tallinn\\Lasnamae\\Paskula");

Error, (in **LibElem1**) library lib-file <C:\Tallinn\Lasnamae\Paskula> does not exist

> _ProcMod;

{"type/package", "Rssl", "Empty", "FTmerge", "type/sequent", "NonaLP", "Test", "pathtf", "redss", "InvT", "readdata1", "nLine", "Sub_all", "mtf", "Fequal", "Mem1", "Gener", "cfdd", "DirE", "sof","Aconv", "PNorm", "type/dirax", "RS", "stype", "F_Type", "Mulel", "seqstr", "winver", "Plib", ":-11", "Sub_st", "ArtMod", "Cls", ":-5", "convert/uppercase", "WS", "assign67", "InvL", "diffP",

==

"Rev", "type/ssign", "type/mod1", "Read1", "N_Cont", "IsOpenF", "FLib", "IP", "type/lower", "F_atr1", "Smart_Plot", "NDT", "CFF", "Weights_L", "ttable", "Release", "LibLink", ":-7", "type/file", "came", ":-35", "StatLib", "Etest"}

5. An example of use of the **LibElem2** procedure for extracting of procedures and program modules out of the damaged user library **Svetlana** that contains only library ***lib***-file:

> LibElem2("C:/rans/academy/libraries/Svetlana/maple.lib", [Vilnius, DIRAX, User_pflM, Grodno, SimpleStat, CureLib], "C:/RANS/IAN/RAC/REA/Galina");

Warning, Analysis of library contents is being done. Please, wait!
Warning, <**Vilnius.m**> is absent in library <C:/rans/academy/libraries/Svetlana/maple.lib>

Warning, program module <**DIRAX**> has been saved in files <DIRAX.m> and <_DIRAXnnn.m>

Warning, exports of module <**DIRAX**> have been saved in file <c:\rans\ian\rac_rea\galina/DIRAX.txt>

Warning, procedure <**User_pflM.m**> can be loaded by `read` function
Warning, <**Grodno.m**> is absent in library <C:/rans/academy/libraries/Svetlana/maple.lib>

Warning, program module <**SimpleStat**> has been saved in files <SimpleStat.m> and <_SimpleStatnnn.m>

Warning, exports of module <**SimpleStat**> have been saved in file <c:\rans\ian\rac_rea\galina/SimpleStat.txt>

Warning, procedure <**CureLib.m**> can be loaded by `read` function
Warning, the created m-files have been located in directory <c:\rans\ian\rac_rea\galina>

CureLib – an operation with the damaged user libraries

Call format of the procedure:

CureLib(F {, 'h' {, 't'}})

Formal arguments of the procedure:

F - symbol or string defining full path to a library or its datafiles *".lib"* or *".ind"*
h - (optional) an assignable name
t - (optional) an assignable name

Description of the procedure:

At operation with the user libraries having structural organization, similar to the main *Maple* library, in a series of cases the necessity of restoration of libraries arises, when the *`march`* utility appears powerless (for example, at a damage or absence of index file "Maple.ind" of a library). In addition, for release **9** these cases provoke erroneous system situations with diagnostics "**Cwmaple9:** *This program has performed an illegal operation and will be shut down.*" or "**Maple9:** *The kernel has been shut down. Further computations cannot be performed*" provoking a crash of a current

Maple session. For release **8** these cases provoke an erroneous system situation with diagnostics "**Mserver:** *This program has performed an illegal operation and will be shut down*" provoking a crash of a current *Maple* session, whereas for releases **6** and **7** they provoke an erratic package situation "*Error,* (in ***march***) *there is no existing archive in* <%1>" only. In these cases the **CureLib** procedure by a series of important functionalities successfully replaces the standard utility ***`march`***. The procedure **CureLib** provides contents analysis of a required library.

The procedure call **CureLib(F {, 'h' {, 't'}})** is equivalent to the call **march('list', F),** where an actual **F** argument specifies full path to a library structurally similar to the main *Maple* library or its datafiles ***".lib"*** or ***".ind"***, if and only if:

1. **F** is a complete library, i.e. its both datafiles ***".lib"*** and ***".ind"*** exist and are correct;
2. for **F** only a correct index datafile ***".ind"*** exists.

In this case, the optional arguments **h** and **t** are returned unevaluated. The following refers to a case when a library has only correct library ***lib***-datafile. In addition, the main name of library files can be arbitrary, however if an actual argument **F** specifies a library directory, then for library files the main name "Maple" is supposed. If an analyzed ***lib***-file contains multiple entries of means, the procedure outputs the appropriate warning with display of such means. In addition, at presence in the analyzed ***lib***-file multiple entries of means, the procedure call through the predefined variable ***_mulvertools*** returns the sorted nested list of their names with multiplicities appropriate to them. See an example below.

The procedure call **CureLib(F)** returns the list of ***m***-files that are located in a library ***lib***-file specified by the first actual **F** argument. The argument defines full path to a "Maple.lib" file or to its analog. Each element of the returned list has *string*-type. In addition, irrespective of quantity of actual arguments at the procedure call through the global variable ***_ProcMod*** a 3-element sequence is returned, namely: its first element defines a list of procedures names, the second - a list of modules names, and the third element - a list of names of *Maple* expressions whose types are different from types {*`procedure`*, *`module`*} contained in an analyzed library; names are represented as *strings*.

The procedure call **CureLib(F, 'h')** returns the list of ***m***-files which are located in a ***lib***-datafile specified by the first actual **F** argument, with return of the nested list through the second optional argument **h.** Elements of the returned list are 3-element sublists whose elements are: (**1**) name of a ***m***-file of *string*-type, (**2**) initial relative position of the ***m***-file in the **F** file, and (**3**) length of ***m***-file in bytes.

At last, the procedure call **CureLib(F,'h','t')** additionally to result of the previous case through the third optional argument **t** returns the nested list similar to a list returned through **h** argument with one distinction, namely: the first elements of the sublists are names of *symbol*-type of program modules and procedures located in a **F** file. In particular, for releases **7, 8** and **9** the given possibility allows to receive information about belonging of ***m***-files of the **F** file (a ***lib***-datafile) with names of view **":-NNN.m"** to modules contained in **F** file (NNN - a positive integer). For these submodules instead of names **":-NNN.m"** of *string*-type take place names ***`Mod:-:-NNN`*** of *symbol*-type, where ***Mod*** - a program module being main to a **":-NNN.m"** submodule. This information is important enough for a restoring of the user library. However, it is necessary to note, such modules connection takes place in a current *Maple* session only and is not saved in an ***mws***-file with a current *Maple* document.

In addition, it is necessary to have in mind, the procedure **CureLib** returns the result relative to all tools versions of the analyzed ***lib***-file, therefore the above procedures **LibElem, LibElem1** and **LibElem2** return the results relative to the last versions (*according to time of their saving*) of the analyzed means.

A minimal level of such library analysis (*at a level of its separate components*) is existence at least of one of two correct library files *".lib"* or *".ind"*. Basic especial or erratic situations provoke the appropriate warnings or errors. Our experience of the procedure use confirms its enough high efficiency at restoring of the user libraries structurally analogous to the main *Maple* library.

```
CureLib:= proc(F::{file, dir})
local d, f, omega;
global _ProcMod, _mulvertools;
  omega:= proc(F)
    local a, g, k, l, i, h, n, s, L, c, Res, b, P, P1, r, r1, t, N, R, p, Mod, Proc, Var, z, sf;
      `if`(type(f, string), NULL, assign(f = "Maple"));
      if type(F, dir) then unassign('_ProcMod'), assign(l = cat(F, "\\", f, ".lib"),
          i = cat(F, "\\", f, ".ind"));
        if map(type, {l, i}, file) = {false} then
          ERROR("library <%1> does not exist or is damaged", F)
        elif map(type, {l, i}, file) = {true} then WARNING("library <%1> is complete!", F);
          RETURN(march('list', F))
        elif not type(l, file) then WARNING("library file <%1> does not exist", l);
          assign(g = cat(convert([1, 1], bytes), "AvAgVsVaArtKr")), writebytes(l, g), close(l);
          RETURN(march('list', F), fremove(l))
        elif not type(i, file) then WARNING("library file <%1> does not exist", i);
          WARNING("Analysis of contents of library lib-file is being done. Please, wait!");
          assign(h = Release1()), assign(r1 = cat("M", h, "R0", convert([10], bytes)),
            L = [], P = [], b = [], Mod = {}, Proc = {}, Var = {},
             r = cat(convert([1], bytes), "M", h, "R0")), seq(assign('c'[Iddn(k)] = k), k = 1 .. 255);
          do assign('s' = iolib(2, l));
            if s = 0 then break
            else
              if s[1 .. 4] = cat("M", h, "R0") or s[1 .. 5] = cat(convert([1], bytes), "M", h, "R0")
               or s[1] = convert([1], bytes)
              then assign('s' = readline(l));
                if s = 0 then break
                else assign('p' = length(s))
                end if;
                Res := `if`(s[1] = "I", `if`(whattype(c[s[2]]) = integer and c[s[2]] + 2 <= p,
                  cat(``, s[3 .. c[s[2]] + 2]), `if`(whattype(c[s[2 .. 3]]) = integer and
                  c[s[2 .. 3]] + 3 <= p, cat(``, s[4 .. c[s[2 .. 3]] + 3]), false)), false);
                if Res <> false then L := [op(L), cat("", Res, ".m")];
                  assign('a' = length(Iddn(t)) + length(Res) + 2);
                  `if`(s[a .. a + 1] = "`6", assign('Mod' = {cat("", Res), op(Mod)}),
                    `if`(member(s[a .. a + 1], {"f*", "R6"}),
                    assign('Proc' = {cat("", Res), op(Proc)},
                    assign('Var' = {cat("", Res), op(Var)})))
                end if
              end if
            end if
          end do;
          if L = [] then ERROR("CureLib cannot identify library contents")
          else z:= gelist(L);
            if nops(L) > nops(z) then z := [seq(`if`(z[k][2] > 1, z[k], NULL), k=1 .. nops(z))];
```

```
                WARNING("library contains multiple entries of the following means %1;
                the sorted nested list of their names with multiplicities appropriate to them is in
                    predefined variable `_mulvertools`", [seq(z[k][1][1 .. -3], k=1 .. nops(z))]);
                sf:= (x, y) -> `if`(x > y, true, false); _mulvertools :=
                    SLj([seq([cat(``, z[k][1][1 .. -3]), z[k][2]], k=1 .. nops(z))], 2, sf);
              end if;
        assign('N' = iolib(6, l, infinity), n = 0), close(l), assign67(_ProcMod = Proc, Mod, Var)
          end if;
          do
            if N - 6 < n then close(l); break
            else  iolib(6, l, n), assign('R' = iolib(4, l, 5, TEXT)[1 .. 5]);
              if R = r or R = r1 then  b := [op(b), n]; n := n + 5; next  else n := n + 1 end if
            end if
          end do;
          if nargs = 1 then L
          else
            assign('P1' = [seq([L[k], b[k], b[k + 1] - b[k]], k = 1 .. nops(b) - 1),
              [L[nops(L)], b[nops(b)], N - b[nops(b)]]]), assign(args[2] = P1);
            if nargs = 3 then assign('P' = [seq([cat(``, L[k][1 .. -3]), b[k],
              b[k + 1] - b[k]], k = 1 .. nops(b) - 1), [cat(``, L[nops(L)][1 .. -3]),
              b[nops(b)], N - b[nops(b)]]]), assign(args[3] = P)
            end if
          end if;
          close(l), L, WARNING("Analysis of contents of library lib-file has been completed!")
        end if
      end if
    end proc;
  try
    if type(F, dir) then  omega(F, `if`(1 < nargs, args[2 .. -1], NULL))
    else
      `if`(member(Ftype(F), {".ind", ".lib"}), NULL, ERROR("<%1> is not a library file", F));
      assign(d = clsd(CFF(F)[1 .. -2], "\\"), f = CCF(F)[-2]),
      omega(d, `if`(1 < nargs, args[2 .. -1], NULL))
    end if
  catch:  Close(l), ERROR("one or both library files {".lib", ".ind"} are crucially damaged")
  end try
end proc
```

Typical examples of the procedure use:

> CureLib("C:/rans/academy/libraries/svetlana", 'z', 'x');

Warning, library <C:/rans/academy/libraries/svetlana> is complete!

[["holdof.m", [2004, 1, 5, 0, 9, 9], 301636, 1030], ["DoF1.m", [2004, 1, 5, 0, 9, 9], 299450, 558], ["SSet.m", [2004, 1, 5, 0, 9, 8], 294829, 218], ["MAM.m", [2004, 1, 5, 0, 9, 7], 291563, 331],

===

["frame_n.m", [2004, 1, 5, 0, 9, 9], 39302,942], ["Kernels.m", [2004, 1, 5, 0, 9, 8], 26999, 2034], ["Apmv.m", [2004, 1, 5, 0, 9, 9], 29202, 227], ["SN_7.m", [2004, 1, 5, 0, 9, 8], 16999, 1024], [":-32.m", [2004, 1, 5, 0, 8, 57], 11321, 178], ["Smart.m", [2004, 1, 5, 0, 9, 8], 4980, 464]]

> z, x;

z, x

```
> CureLib("C:/rans/academy/libraries/svetlana", 'z', 'x');
    Warning, library file <C:/rans/academy/libraries/svetlana\Maple.lib> does not exist
    [["holdof.m", [2004, 1, 5, 0, 9, 9], 301636, 1030], ["DoF1.m", [2004, 1, 5, 0, 9, 9], 299450, 558],
    ["SSet.m", [2004, 1, 5, 0, 9, 8], 294829, 218], ["MAM.m", [2004, 1, 5, 0, 9, 7], 291563, 331],
    ============================================================================
    ["Apmv.m", [2004, 1, 5, 0, 9, 9], 29202, 227], ["SN_7.m", [2004, 1, 5, 0, 9, 8], 16999, 1024],
    [":-32.m", [2004, 1, 5, 0, 8, 57], 11321, 178], ["Smart.m", [2004, 1, 5, 0, 9, 8], 4980, 464]]
> z, x;
                                        z, x
> CureLib("C:/rans/academy/libraries/svetlana", 'z', 'x');
    Warning, library file <C:/rans/academy/libraries/svetlana\Maple.ind> does not exist
    Warning, Analysis of contents of library lib-file is being done. Please, wait!
    Warning, Analysis of contents of library lib-file has been completed!
    [["holdof.m", [2004, 1, 5, 0, 9, 9], 301636, 1030], ["DoF1.m", [2004, 1, 5, 0, 9, 9], 299450, 558],
    ["SSet.m", [2004, 1, 5, 0, 9, 8], 294829, 218], ["MAM.m", [2004, 1, 5, 0, 9, 7], 291563, 331],
    ============================================================================
    ["Apmv.m", [2004, 1, 5, 0, 9, 9], 29202, 227], ["SN_7.m", [2004, 1, 5, 0, 9, 8], 16999, 1024],
    [":-32.m", [2004, 1, 5, 0, 8, 57], 11321, 178], ["Smart.m", [2004, 1, 5, 0, 9, 8], 4980, 464]]
> z;
    [[":-13.m", 1024, 2319], [":-41.m", 3343, 217], ["Plib.m", 7517, 296], ["SSN.m", 5832, 1080],
    ["LSL.m", 6912, 605],["SQHD.m", 4365, 224], ["Smart.m", 4980, 464], ["LnFile.m", 5444, 388],
    ============================================================================
    ["memberL.m", 297991, 260], ["DirFT.m", 298251, 426], ["DoF1.m", 298677, 558], ["DAopen.m",
    299235, 595], ["MapleLib.m", 299830, 1033], ["holdof.m", 300863, 1030]]
> x;
    [[`:-13`, 1024, 2319], [`:-41`, 3343, 217], [`type/sequent`, 3560, 149], [`type/package`, 3709, 656],
    [T_SQHD, 4365, 224], [Nint, 4589, 391], [Smart, 4980, 464], [LnFile, 5444, 388], [SSN, 5832,
    ============================================================================
    [tabar, 296362, 711], [ReadProc, 297073, 918], [memberL, 297991, 260], [DirFT, 298251, 426],
    [DoF1, 298677, 558], [DAopen, 299235, 595], [MapleLib, 299830, 1033], [holdof, 300863, 1030]]
> _ProcMod;
    {"diffP", "mapTab", "Empty", "type/sequent", "swmpat", "Test", "null", "Rssl", ":-13", "extrS",
    "RTfile", "convert/ssll", "type/package", "pathtf", "subseqn", "LnFile", ":-7", "delf1", "NDP"
    ============================================================================
    "etf", "Close", "save1", "Search3", ":-36", "trConv", ":-8", "type/path", "FTabLine", "Is_Color",
    "Reduce_T", "EntS", "FFP", "FLL", "Subs_all2", "RmDir", "CorMod1", "Iddn", "Iddn1"},
    {"SoLists", "ACC", "DIRAX", "AlgLists", "SimpleStat", ":-19"},
    {":-46", ":-44", ":-14", ":-42", ":-23", ":-45", ":-26", ":-16", ":-17", ":-43", ":-21", ":-22"}
> CureLib("C:/rans/academy/libraries/vilnius/vgtu.lib", 'c', 'd');
    Warning, library file <c:\rans\academy\libraries\vilnius\vgtu.ind> does not exist
    Warning, Analysis of contents of library lib-file is being done. Please, wait!
    Warning, Analysis of contents of library lib-file has been completed!
> CureLib("C:/rans/academy\\grodno/tallinn/vilnius\\maple.lib");
    Warning, library <C:/rans/academy\grodno/tallinn/vilnius> is complete!
    Error, (in CureLib) one or both library files {".lib", ".ind"} are crucially damaged
```

> CureLib("C:/Temp/Book\\TestLib/Galina", 'z', 'x');

Warning, library file <C:/Temp/Book\TestLib/Galina\Maple.ind> does not exist
Warning, Analysis of contents of library lib-file is being done. Please, wait!

Warning, library contains multiple entries of the followingf means [Atr, CCM, Currentdir, DAclose, DAopen, DAread, DUlib, FSSF, F_atr2, FmF, Imaple, Is_Color, LibElem, LnFile, ModFile, NLP, ParProc1, RTfile, Reduce_T, SDF, SSF, Suffix, Vol, Vol_Free_Space, WD, WS, cdt, conSA, dslib, ewsc, gelist, mapTab, readdata1, redlt, sfd, sorttf, type/dir, type/file]

Warning, Analysis of contents of library lib-file has been completed!

[":-41.m", "type/package.m", "type/sequent.m", "extrS.m", "LnFile.m", "LSL.m", "MinMax.m", "NDP.m", "Nint.m", "pathtf.m", "Plib.m", "redss.m", "Rmf.m", "Smart.m", "SSN.m", "subSL.m",
==
"DAclose.m", "ModFile.m", "WD.m", "WS.m", "type/file.m", "type/file.m", "Vol.m", "conSA.m", "conSA.m", "conSA.m", "NLP.m", "DUlib.m", "dslib.m", "redlt.m", "redlt.m", "LibElem.m"]

> CureLib("C:/rans\\academy/libraries\\ArtKr", 'z', 'x'):

Warning, library file <C:/rans\academy/libraries\ArtKr\Maple.ind> does not exist
Warning, Analysis of contents of library lib-file is being done. Please, wait!

Warning, library contains multiple entries of the following means
[Atr, CCM, CureLib, Currentdir, DAclose, DAopen, DAread, DUlib, FSSF, F_atr1, F_atr2, FmF, Imaple, Is_Color, LibElem, LnFile, ModFile, NLP, ParProc1, RTfile, Reduce_T, SDF, SSF, Suffix, Uninstall, Vol, Vol_Free_Space, WD, WS, cdt, conSA, dslib, ewsc, gelist, sfd, mapTab, readdata1, redlt, sorttf, type/dir, type/...];
the sorted nested list of their names with multiplicities appropriate to them is in predefined variable `_mulvertools`

Warning, Analysis of contents of library lib-file has been completed!

> _mulvertools;

[[*F_atr2*, 6], [*Atr*, 4], [*CureLib*, 3], [*FSSF*, 3], [*SDF*, 3], [*SSF*, 3], [*Suffix*, 3], [*conSA*, 3], [*ewsc*, 3], [*gelist*, 3], [*sorttf*, 3], [*type/file*, 3], [CCM, 2], [*Currentdir*, 2], [*DAclose*, 2], [*DAopen*, 2], [*DAread*, 2], [*DUlib*, 2], [*F_atr1*, 2], [*FmF*, 2], [*Imaple*, 2], [*Is_Color*, 2], [*LibElem*, 2], [*LnFile*, 2], [*ModFile*, 2], [*NLP*, 2], [*ParProc1*, 2], [*RTfile*, 2], [*Reduce_*T, 2], [*Uninstall*, 2], [*Vol*, 2], [*Vol_Free_Space*, 2], [*WD*, 2], [*WS*, 2], [*cdt*, 2], [*dslib*, 2], [*sfd*, 2], [*mapTab*, 2], [*readdata1*, 2], [*redlt*, 2], [*type/dir*, 2]]

The above examples of procedure use well explain structure of the returned results. At operation with the user libraries having structural organization, similar to the main *Maple* library, in a series of cases the necessity of restoration of libraries arises.

---

cmlib – an operation with procedures and modules of the damaged user libraries

Call format of the procedure:

cmlib(F, M)

Formal arguments of the procedure:

F – symbol or string defining full path to a library datafile *".lib"*
M – symbol defining name of a procedure or module of an analyzed library datafile

Description of the procedure:

At operation with the user libraries having structural organization, similar to the main *Maple* library, in a series of cases the necessity of restoration of libraries arises, when the ***`march`*** utility appears powerless (for example, at a damage or absence of index file "Maple.ind" of a library). The more detailed information about possible methods of an analysis of damaged libraries structurally similar to the main *Maple* library has been represented above in description of the procedure **CureLib.** The procedure **cmlib** gives some additional means for such analysis.

The successful procedure call **cmlib(F, M)** creates in the root directory of *Maple* a subdirectory whose name is specified by the second actual argument **M.** In addition, if such directory already does exist, then the procedure call creates a new subdirectory with name **cat("_", M).** If an analyzed ***lib***-file does not contain a procedure or module with name **M,** the error arises and the above subdirectory will not be created. If **M** defines a procedure then the procedure call returns the *NULL* value, i.e. nothing, and creates in the above subdirectory the file with name **"M.m"** with definition of the procedure with output of an appropriate warning.

If **M** defines a module then the procedure call returns the table whose indices are names of *exports* of the module **M** and entries are definitions appropriate to them. In addition, the table is saved in the above subdirectory in datafile with name **"M.exp"** in the input *Maple* language format with output of an appropriate warning. Furthermore, in the same subdirectory a set of ***m***-files analogous to a case of the procedure **LibElem1** is created.

It is necessary to note, as a result of the successful procedure call through global variable ***_ProcMod*** the 3-element sequence is returned, namely: its first element defines a list of procedures names, the second - a list of modules names, and the third element - a list of names of *Maple* expressions of types different from types {*`procedure`*, *`module`*} which are contained in an analyzed datafile **F;** all names are represented as *strings*. However if a datafile **F** belongs to a library containing both basic files {***".lib"***, ***".ind"***}, the procedure call returns the global variable ***_ProcMod*** unevaluated. The procedure handles basic especial and erratic situations. The procedure **cmlib** can be considered as a rather useful tool of restoring of the damaged user libraries structurally analogous to the main *Maple* library.

```
cmlib:= proc(F::path, M::symbol)
local a, b, c, d, k, h, f, p, r, n, Tab, s, t, z;
  if type(F, file) and IsFtype(F, "lib") then
    if relml(F) then
      assign(b = cat(CDM(), "\\", M)), assign('b' = `if`(type(b, dir), cat(CDM(), "\\_", M), b));
      WARNING("Analysis of contents of library datafile <%1> is being done. Please, wait!",F);
      assign(a = interface(warnlevel)), interface(warnlevel = 0), LibElem1(F, [M], b);
      assign(f = Case(cat(b, "\\", M, ".m")), d = {op(DirF(b))}), interface(warnlevel = a);
      `if`(d={}, [rmdir(b), ERROR("<%1> does not contain procedure/module <%2>", F,M)],7);
      assign(r = Release(), Tab = table([]), 'h' = sstr(["\r" = NULL, "\n" = NULL],
      readbytes(f, TEXT, infinity)), 'a' = length(Iddn(M)) + length(M) + 6), close(f);
      if h[a .. a + 1] <> "`6" then RETURN(WARNING("file <%1> contains procedure <%2>,
        its definition is located in file <%3>", F, M, f))
      else
        for k in d do read k end do;
        assign('c' = [exports(M)], s = {seq(cat("", k), k = 0 .. 9)}, n = length(h), p = "");
        for k in c do
          if not swmpat(h, cat(Iddn(k), k, "*:-"), [1], "*", 'z') then
            ERROR("library <%1> has been crucially damaged", F)
```

```
          else
            for t from z[1][2] + 1 to n do
              if member(h[t], s) then p := cat(p, h[t]) else break end if
            end do;
            assign('Tab'[k] = eval(cat(``, ":-", SD(p))), 'p' = "")
          end if
        end do
      end if
    else ERROR("datafile <%1> is not compatible with the current Maple release", F)
    end if
  else ERROR("<%1> is not a lib-datafile", F)
  end if;
  eval(Tab), assign('a' = cat(b, "\\", M, ".exp")), (proc(x) save Tab, x end proc)(a),
    WARNING("table with exports of module <%1> has been saved in datafile <%2>", M, a)
end proc
```

Typical examples of the procedure use:

> cmlib("C:/Archive/Adverts/Noosphere/VGTU.lib", DIRAX);

Warning, Analysis of contents of library datafile <C:/Archive/Adverts/Noosphere/VGTU.lib> is being done. Please, wait!

Warning, table with exports of module <DIRAX> has been saved in datafile <C:\Program Files\Maple 8_DIRAX\DIRAX.exp>

$$\mathrm{table}\Big(\Big[reverse = \Big(L{::}dirax \to \text{`if`}\Big(L_2 < 2, L, \Big[L_1, L_2, \Big[\mathrm{seq}\Big(L_{3_{L_2 - k + 1}}, k = 1 \,..\, L_2\Big)\Big]\Big]\Big)\Big),$$

$$sortd = ((L{::}dirax, h) \to \text{`if`}(2 \le L_2, [L_1, L_2, \mathrm{sort}(L_3, h)], L)),\ new = (() \to [dirax, \mathrm{nargs}, [\mathrm{args}]]),\ insert = \Big($$

$$(L{::}dirax, n{::}posint, a{::}anything) \to$$

$$\text{`if`}\Big(1 \le n \text{ and } n \le L_2, \Big[L_1, L_2 + 1, \Big[\mathrm{op}(L_{3_{1\,..\,n}}), a, \mathrm{op}\Big(L_{3_{n+1\,..\,L_2}}\Big)\Big]\Big], Wrgn(L, n)\Big)\Big),\ size = (L{::}dirax \to L_2),\ delete$$

$$= ((L{::}dirax, n{::}posint) \to \text{`if`}(1 \le n \text{ and } n \le L_2, [L_1, L_2 - 1, [\mathrm{seq}(\text{`if`}(k \ne n, L_{3_k}, NULL), k = 1 \,..\, L_2)]], Wrgn(L, n)))$$

$$, empty = (L{::}dirax \to \text{`if`}(L_2 = 0, true, false)),\ printd = (L{::}dirax \to L_3),\ replace = (\mathbf{proc}(L{::}dirax, n{::}posint, a{::}anything$$

)

try

if $1 \le n$ **and** $n \le L[2]$ **then** $[L[1], L[2], [\mathrm{op}(L[3][1\,..\,n-1]), a, \mathrm{op}(L[3][n+1\,..\,L[2]])]]$

else ERROR("Member")

end if

catch "Member": $Wrgn(L, n)$

end try

end proc), *extract* = (**proc**($L{::}dirax, n{::}posint$)

try if $1 \le n$ **and** $n \le L[2]$ **then** $L[3][n]$ **else** ERROR("Member") **end if catch** "Member": $Wrgn(L, n)$ **end try**

end proc),

$conv = ((L{::}dirax, T{::}\mathrm{type}(\text{'set'}, list)) \to \mathrm{convert}(L_3, T))$

])

> cmlib("C:/Archive/Adverts/Noosphere/VGTU.lib", DIRAX); **# *Maple 6***

Error, (in **cmlib**) datafile <C:/Archive/Adverts/Noosphere/VGTU.lib> is not compatible with the current Maple release

> cmlib("C:/Archive/Adverts/Noosphere/VGTU.lib", MkDir);

Warning, Analysis of contents of library datafile <C:/Archive/Adverts/Noosphere/VGTU.lib> is being done. Please, wait!

Warning, file <C:/Archive/Adverts/Noosphere/VGTU.lib> contains procedure <MkDir>, its definition is located in file <c:\program files\maple 8\mkdir\mkdir.m>

Naturally, the procedure **cmlib** and the above means of work with the damaged libraries structurally similar to the main *Maple* library are the most effective tools under the condition of existence of the description of software of such libraries. In addition, above all, the question of contents revealing of a destroyed library is very topical. The procedure **ContLib** represented below allows to receive contents of a damaged user library under the condition of existence at least of one library datafile {*".lib", ".ind"*}.

ContLib - finding of contents of the damaged *Maple* libraries

Call format of the procedure:

ContLib(L)

Formal arguments of the procedure:

L - symbol or string defining full path to a directory with *Maple* libraries

Description of the procedure:

The procedure call **ContLib(L)** returns the nested list describing contents of the damaged *Maple* libraries located in a directory specified by the actual **L** argument. For example, the directory can contain files "Grodno.lib", Grodno.ind", and "Tallinn.lib" of two different *Maple* libraries **Grodno** and **Tallinn.** The argument **L** defines full path to a directory, and under the term *"Maple library"*, a library structurally analogous to the main *Maple* library is supposed. If a *Maple* library is complete, i.e. both its basic files **{*".lib", ".ind"*}** exist, the procedure call returns the *NULL* value with output of the appropriate warning. In other cases, the procedure uses the above procedure **CureLib** for a contents receiving of the corresponding *Maple* library from **L.**

In the general case the procedure call returns the nested list whose elements are lists describing contents (in format of the ***`march`*** utility or the above **CureLib** procedure) of a damaged *Maple* library specified by their first element of *string*-type. The procedure running is accompanied by a rather lucid protocol of the libraries processing. The procedure handles all basic especial and erratic situations. The examples represented below well illustrate the told.

```
ContLib:= proc(L::dir)
local a, b, c, d, h, t, p, k, psi, Res;
  assign(h = DirF(L, only)),
    `if`(h = [], ERROR("<%1> does not contain Maple libraries", L), NULL);
  h := [seq(`if`(member(Ftype(h[k]), {".lib", ".ind"}), h[k], NULL), k = 1 .. nops(h))];
  assign('h' = SLj(map(CCF, h), 2)), assign(t = nops(h), d=[], Res=[], a = cat(CDM(), "\\_61_"));
  psi:= a -> `if`(type(a, nestlist), map(procname, a), cat(a[1], "\\", a[2], ".", a[3]));
  for k to t - 1 do
    if h[k][2] = h[k + 1][2] then p := k; d := [op(d), [h[k], h[k + 1]]]; k := k + 1
    else d := [op(d), h[k]]
    end if
  end do;
```

```
`if`(p <> t - 1, assign('d' = [op(d), h[-1]]), NULL), assign('d' = psi(d));
 for k to nops(d) do `if`(type(d[k], list), assign('b' = CCF(d[k][1])[2]),
       assign('b' = CCF(d[k])[2])),
       WARNING("Analysis of library <%1> is being done. Please, wait!", b);
   if type(d[k], list) then
       WARNING("library <%1> is complete. The `march` utility can be used.", b)
   elif Ftype(d[k]) = ".lib" then
     try Res := [op(Res), [b, op(CureLib(d[k]))]]
     catch: close(d[k]), WARNING("library <%1> is crucially damaged or does not exist!", b)
     end try
   else mkdir(a), assign('c' = cat(a, "\\", CFF(d[k])[-1])),
        writebytes(c, readbytes(d[k], infinity)), close(d[k], c);
     try Res := [op(Res), [b, op(CureLib(c))]]
     catch:  WARNING("library <%1> is crucially damaged or does not exist!", b)
     finally  close(d[k]);  fremove(c);  rmdir(a)
     end try
   end if
 end do;
 WARNING("Below is analysis result of Maple libraries from directory <%1>:", L), Res
end proc
```

Typical examples of the procedure use:

```
> ContLib("C:/Temp/Galina");
   Warning, Analysis of library <grodno> is being done. Please, wait!
   Warning, library <grodno> is complete. The `march` utility can be used.
   Warning, Analysis of library <grsu> is being done. Please, wait!
   Warning, library file <c:\temp\galina\grsu.ind> does not exist
   Warning, Analysis of contents of library lib-file is being done. Please, wait!
   Warning, library contains multiple entries of the following means
        [Atr, CCM, CureLib, Currentdir, DAclose, DAopen, DAread, DUlib, FSSF,
        F_atr2, FmF, Imaple, Is_Color,      LibElem, LnFile, ModFile, NLP, ParProc1, RTfile,
        Reduce_T, SDF, SSF, Suffix, Vol, Vol_Free_Space, WD, WS, cdt, conSA, dslib, ewsc,
        gelist, mapTab, readdata1, redlt, sfd, sorttf, type/dir, type/file]
   Warning, Analysis of contents of library lib-file has been completed!
   Warning, Analysis of library <tallinn> is being done. Please, wait!
   Warning, library file <c:\temp\galina\tallinn.ind> does not exist
   Warning, Analysis of contents of library lib-file is being done. Please, wait!
   Warning, Analysis of contents of library lib-file has been completed!
   Warning, Below is analysis result of Maple libraries from directory <C:/temp/galina>:

   [["grsu", ["DestLm.m", [2004, 1, 8, 15, 50, 24], 1024, 1498], [":-6.m", [2004, 1, 8, 15, 50, 24],
   4999, 61], [":-17.m", [2004, 1, 8, 15, 50, 24], 5060, 48], [":-11.m", [2004, 1, 8, 15, 50, 24], 5231,
   62], [":-9.m", [2004, 1, 8, 15, 50, 24], 5293, 61], [":-14.m", [2004, 1, 8, 15, 50, 24], 5414, 48],
   [":-3.m", [2004, 1, 8, 15, 50, 24], 5827, 61], ["type/mlib.m", [2004, 1, 8, 15, 50, 24], 2522, 491],
   ["ContLib.m", [2004, 1, 8, 15, 50, 24], 3013, 1321], ["SimpleStat.m", [2004, 1, 8, 15, 50, 24], 4334,
   539], [":-12.m", [2004, 1, 8, 15, 50, 24], 4873, 66], [":-1.m", [2004, 1, 8, 15, 50, 24], 5108, 60],
   [":-18.m", [2004, 1, 8, 15, 50, 24], 5522, 50], [":-7.m", [2004, 1, 8, 15, 50, 24], 5572, 60], [":-15.m",
   [2004, 1, 8, 15, 50, 24], 5632, 48], [":-4.m", [2004, 1, 8, 15, 50, 24], 5680, 99], [":-8.m", [2004, 1, 8,
   15, 50, 24], 4939, 60], [":-19.m", [2004, 1, 8, 15, 50, 24], 5168, 63], [":-2.m", [2004, 1, 8, 15, 50,
```

24], 5354, 60], [":-5.m", [2004, 1, 8, 15, 50, 24], 5462, 60], [":-16.m", [2004, 1, 8, 15, 50, 24], 5779, 48], [":-13.m", [2004, 1, 8, 15, 50, 24], 5888, 66], [":-10.m", [2004, 1, 8, 15, 50, 24], 5954, 62]], ["tallinn", "Kr.m", "Art.m", "Gal.m", ":-1.m", "Vict.m", ":-2.m", "Sv.m"]]

> ContLib("C:/RANS/Academy/Libraries/Svetlana");

Error, (in **ContLib**) <C:/RANS/ Academy/ Libraries/ Svetlana> does not contain Maple libraries

---

Uninstall – cancellation of logical linking of the user libraries

Call format of the procedure:

Uninstall(F {, h})

Formal arguments of the procedure:

F - full path defining a directory with the user library
h - (optional) an arbitrary valid *Maple* expression

Description of the procedure:

The procedure call **Uninstall(F)** provides cancellation of logical linking of an user library whose full path the first actual **F** argument specifies, with the main *Maple* library. Successful call of the procedure returns a new value of the predefined ***libname***-variable with output of the appropriate warning. The logical links associated with the **F** library are deleted from the initialization file "Maple.ini" also. In addition, if the optional second actual **h** argument has been coded, a subdirectory with the library **F** will be removed from a file system of computer. However, in a series of cases (*details here are not discussed*), the procedure cannot delete file "Maple.ind" of the eliminated library. In this case, the procedure call outputs the appropriate warning. An example below gives an approach to processing of similar especial situation in the *Maple* environment. This procedure provides the effective cancellation of logical link of the libraries with the main *Maple* library.

```
Uninstall:= proc(F::mlib)
local a, k, G, h, Q, Z, S, L, p, t, omega;
global libname;
  Release('h'), assign(omega = (s -> Case(Subs_All("\\" = "/", s, 2)))),
    assign67(S = omega(F), G = "_62_", Q = cat(h, "/", KCM(), "/Maplesys.ini"));
  assign('Z' = iolib(4, Q, TEXT, infinity)), iolib(7, Q),
     assign('Z' = cat(Z[searchtext("UserD", Z) + 14 .. searchtext("\n", Z, searchtext("UserD", Z) ..
     length(Z)) + searchtext("UserD", Z) - 2], "/Maple.ini")),
    assign(L = [libname]), assign(p = fopen(Z, READ));
  libname := op([seq(`if`(omega(L[k]) = S, NULL, L[k]), k = 1 .. nops(L))]);
  assign(a = interface(warnlevel)), interface(warnlevel = 0);
  while not Fend(p) do writeline(G, sub_1([`"",` = NULL, `,""` = NULL, `, ""` = NULL],
    Red_n(Subs_All(S = NULL, omega(readline(p)), 2), ",", 2)))
  end do;
  close(G, Z), writebytes(Z, readbytes(G, infinity)), close(Z), fremove(G), interface(warnlevel = a);
  WARNING("logical linking of <%1> with the main Maple library has been canceled.", F);
  try `if`(1 < nargs, DUlib(F), NULL)
  catch: libname
  end try;
  libname
end proc
```

Typical examples of the procedure use:

> restart; libname;

"c:/program files/maple 9/lib/userlib", "C:\Program Files\Maple 9/lib",
"C:/Rans/Academy/Libraries/Svetlana"

> Uninstall("C:/Rans/Academy\\Libraries/Svetlana");

Warning, logical linking of <C:/Rans/Academy\Libraries/Svetlana> with the main Maple library has been canceled.

"c:/program files/maple 9/lib/userlib", "C:\Program Files\Maple 9/lib"

> libname;

"c:/program files/maple 8/lib/Tallinn", "c:/program files/maple 8/lib/userlib",
"C:\Program Files\Maple 8/lib"

> Uninstall("C:/Program Files/Maple 8/LIB/Tallinn", 157);

Warning, logical linking of <c:/program files/maple 8/lib/Tallinn> with the main Maple library has been canceled.

Warning, linkage of <c:/program files/maple 8/lib/Tallinn> with main Maple library has been canceled

Warning, DUlib could not remove all files of <c:/program files/maple 8/lib/Tallinn> – files {maple.ind} have not been removed; use other tools

"c:\program files\maple 8\lib\userlib", "c:\program files\maple 8\lib"

> restart; RmDir("C:/Program Files/Mmaple 8/LIB/Tallinn");

> type("C:/Program Files/Mmaple 8/LIB/Tallinn, dir);

false

> Uninstall("C:/Rans/Academy/Libraries/Svetlana", 7);

Warning, logical linking of <C:/Rans/Academy/Libraries/Svetlana> with the main Maple library has been canceled.

Warning, library <C:/Rans/Academy/Libraries/Svetlana> has been removed.

"c:/program files/maple 9/lib/userlib", "C:\Program Files\Maple 9/lib"

Having created own library of procedures with use of the above three procedures **User_pfl, User_pflM** and **User_pflMH,** or in a different way, quite naturally arises the problem of its optimization, in particular, with purpose of a frequency revealing of use of tools contained in it, and basic computer resources used by them. In this context, the problem of optimization of the user libraries is a rather important. For these purposes, the procedure **StatLib(L)** seems an useful enough, providing gathering of the basic statistics on a specified **L** library and the return of the statistics for the subsequent analysis. Above all, this procedure supposes that an analyzed **L** library is located in the **LIB** directory with the main *Maple* library. The **StatLib** procedure demands certain additional resources of memory and time. However, the procedure allows to receive very useful statistics allowing to estimate efficiency of the library tools usage, what is important at creation of the *Maple* libraries.

StatLib – a statistical analysis of tools use of the user libraries

Call format of the procedure:

StatLib(L {, *args*})

Formal arguments of the procedure:

L - string or symbol defining name of an user library
args - (optional) the remaining actual arguments

Description of the procedure:

The procedure **StatLib(L)** as the first obligatory **L** argument uses the name of an user library organized similarly to the main *Maple* library and located in **LIB** subdirectory of the package, and logically linked with it. The other arguments in quantity up to **3** are optional and the sense of tuples of their values is defined as follows:

StatLib(L) - initiation of gathering of statistics on a library **L** in a whole;

StatLib(L, 0) - deletion of all files with statistical information on the given library **L;**

StatLib(L, 1) - cancel of gathering of statistics on library **L** with saving of results;

StatLib(L, P) - initiation of gathering of statistics on **P** procedures of a library **L;**

StatLib(L, T, 2) - return of the gathered statistics on library **L** or its **P** procedures in the given context **T;**

StatLib(L, T, 2, m) - return of the gathered statistics on library **L** or its **P** procedures in the **T** context characterized by a number **m;**

StatLib(L, T, N) - return of the gathered statistics on procedure with name **N** in the demanded context {*calls, bytes, depth, maxdepth, time*};

StatLib(L, *abend*) - abort of the procedure **StatLib** at arising of critical errors.

The procedure call **StatLib(L)** with one actual **L** argument initiates process of gathering of statistics on calls of procedures located in an **L** library. The given process can be stopped by the call **StatLib(L, 1),** which is obligatory in a case of necessity of continuation of gathering of statistics in the next *Maple* sessions, because does saving of the gathered statistics in five special files located in the same subdirectory with library **L.** Names of these datafiles have the same "$@_"-prefix.

The procedure call **StatLib(L, 0)** with *zero* value of the second actual argument returns the *NULL* value with deletion of the above all statistical datafiles for the given library **L.** The given call is being done up to an initialization of a new process of profiling of procedures with intention of a saving of its results in subdirectory with the **L** library. The existing statistical datafiles can be preliminarily saved in the other subdirectory or remain in old under other names.

The procedure call **StatLib(L, P)** initiates a process of gathering of statistics on calls of procedures located in an **L** library and specified by list **P** of their names. In addition, the required procedures not necessarily should belong to the **L** library - they can be in any library logically linked with the main *Maple* library, or can be evaluated in a current *Maple* session. Otherwise, the erroneous situation arises with output of appropriate diagnostics. That allows to make the analysis of both the library as a whole, and in the context of procedures composing it (*without a combining of both these processes*), and of any active or accessible procedures of a current *Maple* session.

The procedure provides return of statistical information in five contexts specified by the following keywords **T** = {*calls, bytes, depth, maxdepth, time*}, which define accordingly quantity of calls, volume of used memory, depth and maximal depth of nesting, and also used time in secs. The procedure call **StatLib(L, T, 2)** returns an array of the numerical characteristics for all profiled procedures of a **L** library {or procedures specified by the **P** argument at call **StatLib(L, P)**} at the current moment in the given context **T** (for example, *calls* - a quantity of calls). Whereas the procedure call **StatLib(L, T, 2, m)** returns an array for a context **T** of the profiled procedures whose values of characteristics corresponding to **T,** less than **m.** At absence of such rows the procedure returns the *lack* value, i.e. at present any of profiled procedures has not of the **T** characteristic with a value not smaller, than **m.** The returned array is sorted by the row in decreasing order of values of characteristics; in addition, for equal values the rows are sorted in lexicographical order.

At last, the procedure call **StatLib(L, T, N)** (if **N** value of type {*symbol, string*} has been coded as the third actual argument of the procedure) returns the value of characteristic of a procedure with name **N**, if such does exist and earlier has been profiled, in the given context **T**. For instance, the procedure call **StatLib(ArtK, calls, RmDir)** returns a quantity of procedure calls of the **RmDir** from the **ArtK** library, if the given library or **RmDir** formerly have been profiled. On other tuples of values of actual arguments, the **StatLib** procedure returns the *NULL* value, i.e. nothing. The idle tuples is a rather good prerequisite of the subsequent development of the procedure. For an ensuring of reliability and safety of the gathered statistical information, we recommend periodically to save it by means of the calls **StatLib(L, 1)** with the subsequent recommencing of profiling by the call **StatLib(L).**

The procedure admits two modes of monitoring of results of procedures profiling - *dynamic* and *single*. The *dynamic* mode provides an opportunity of monitoring by means of the procedure calls, described above, in a process of profiling, i.e. between two procedure calls **StatLib(L)** and **StatLib(L, 1).** In addition, the process of profiling is being not broken. Whereas the *single* mode allows to make random inspection of results of the previous profiling, without renewing the process of profiling. Such approach allows to carry out the more flexible monitoring of process of profiling of procedures both inside, and outside.

In some versions of the package of releases **6 - 8** and **9/9.5** the use of the **StatLib** procedure in a series of cases can cause the critical errors linked first of all with a stack overflow. In this case we recommend to fulfill the procedure call **StatLib(L, *abend*)** with the subsequent execution of the clause **`restart`.** The given reception allows to save in statistical files the information, at least, about a quantity of calls of the profiled procedures onto the moment of an erratic situation. The subsequent procedure call **StatLib(L)** provides recommencing of a process of profiling from the interrupted moment. The above erratic situations can be substantially explicated as follows.

Our prolonged enough experience of use of package *Maple* of releases **4 - 9/9.5** in various appendices, including development of the means extending base means of the package, with all definiteness has revealed one rather essential circumstance, namely. Many of frequently-used standard means of *Maple* have been provided by an insufficiently advanced system of handling of special and erroneous situations, that in a lot of cases results in mistakes with obviously incorrect diagnostics, for example, "*Execution stopped: Stack limit reached*" with subsequent ABEND of a current session. Thus, or *Maple* has stack of insufficient depth, or the above situation has been caused (at absence of any cyclic calculations) by an error having some other ground. Unfortunately, increase of numbers of the *Maple* releases is accompanied by decrease of their robustness.

At use of the **StatLib** procedure, a delay of calculations and an increase of used memory whose sizes depend on quantity of the profiled procedures and frequency of their use first of all take place. However, because of the small enough user libraries (up to 500 procedures and frequency of their use no more than 300 during a current session, excepting cyclic constructions) the given circumstance does not lead to the critical situations linked with use of the basic resources of a computer and time of information processing.

The mechanism of use of the **StatLib** procedure consists in the following. In the beginning of each session a procedure call **StatLib(L {, P})** is carried out, where **L** - name of an analyzed user library satisfying to the above conditions (the **P**-argument can define a list of procedures names from an **L** library or outside of it). Before completion of a current *Maple* session, the procedure call **StatLib(L,1)** is carried out, providing a saving of gathered statistical information in five special text datafiles with names of the general view "**$@_h**", which are located in the same subdirectory with an analyzed **L** library (where **h** belongs to the set {*calls, bytes, maxdepth, depth, time*}). At any moment (in a *dynamic* or in a *single* mode) by means of calls of view **StatLib(L, T {, 2|, *name*} {, m})** it is possible to receive the information on characteristics of calls of procedures of the **L** library in the above contexts.

The most effective mode of the procedure use is the following. Each next *Maple* session begins with procedure call **StatLib(L),** after which a current work in the *Maple* environment is being carried out. During work is periodically recommended to make pairs of procedure calls {**StatLib(L, 1)** and **StatLib(L)**} for a providing of reliability of integrity of results of profiling of procedures of an **L** library. Before completion of a current session the procedure call **StatLib(L, 1)** is carried out, providing the termination of process of profiling and saving of its results in the above five files. The analysis of a gathered statistical information allows to improve both the organization of library as a whole, and the efficiency of its separate procedures having a high enough frequency of use or/and which essentially use the basic resources of a computer. Experience of appendices of the **StatLib** procedure with all definiteness speaks about its high efficiency in case of decision of optimization problems of the user libraries. The mechanism and methods of use of the **StatLib** procedure are surveyed in our previous books [29-33,39,43,44] enough in detail. Examples below rather well illustrate principles and results of appendices of the **StatLib** procedure.

```
StatLib:= proc(L::{string, symbol})
local a, k, h, S, P, K, G, H, t, pf, unp, u, V, T, R, W, Z, calls1, bytes1, depth1, maxdepth1, time1;
global profile_maxdepth, profile_calls, profile_bytes, profile_depth, profile_time, `$Art_Kr`,
        calls2, bytes2, depth2, maxdepth2, time2;
  unp:= () -> unassign("`$Art_Kr`", 'profile_proc', op(T), seq(cat(profile_, k), k = R));
  `if`(nargs = 2 and args[2] = abend, RETURN(unprofile(), unp(),
    "Abend! Execute `restart` command!"), NULL);
  W := table([1 = true, 2 = `if`(nargs = 2 and type(args[2], {list({symbol}), binary}), true, false),
    3 = `if`(nargs = 3 and member(args[2], {time, maxdepth, bytes, depth, calls}) and
    (args[3] = 2 or type(args[3], symbol)), true, false), 4 = `if`(nargs = 4 and
    member(args[2], {time, maxdepth, bytes, depth, calls}) and
    args[3]=2 and type(args[4], numeric), true, false)]);
  `if`(W[nargs] = true, unassign('W'),
    ERROR("invalid arguments %1 have been passed to the StatLib", [args]));
  assign(K = Plib(L)), assign(R = [bytes, calls, depth, maxdepth, time]),
    assign(calls1 = cat(K, "/$@_", R[2]), bytes1 = cat(K, "/$@_", R[1]),
    depth1 = cat(K, "/$@_", R[3]), maxdepth1 = cat(K, "/$@_", R[4]),
    time1 = cat(K, "/$@_", R[5])), assign(T = [seq(cat(R[h], `2`), h = 1 .. 5)],
    V = [bytes1, calls1, depth1, maxdepth1, time1],
    Z = [seq(cat(profile_, R[h]), h = 1 .. 5)], G = []);
  `if`(nargs = 2 and args[2] = 0, RETURN(WARNING("datafiles with statistics have been
    removed out of directory with library <%1>", L), unprofile(),
    op(map(Fremove, [bytes1, calls1, depth1, maxdepth1, time1])), unp()), NULL);
  if nargs = 2 and args[2] = 1 then `if`(type(eval(profile_calls), table),
      assign(calls2 = eval(cat(profile_, R[2])), bytes2 = eval(cat(profile_, R[1])),
      depth2 = eval(cat(profile_, R[3])), maxdepth2 = eval(cat(profile_, R[4])),
      time2 = eval(cat(profile_, R[5]))), ERROR("profiling does not exist"));
    (proc() save maxdepth2, maxdepth1; save calls2, calls1; save bytes2, bytes1;
```

```
        save depth2, depth1;  save time2, time1  end proc)(), RETURN(unprofile(), unp(),
        WARNING("profiling of library <%1> has been completed", L))
   else NULL
   end if;
   if nargs = 3 and belong(cat(``, args[2]), R) and
      member(whattype(args[3]), {string, symbol}) then
      assign(`$Art_Kr` = eval(cat(profile_, args[2])), pf = cat(K, "/$@_", args[2])),
        `if`(type(eval(`$Art_Kr`), table), NULL, `if`(DoF(pf) = file, [(proc() read pf end proc)(),
        assign(`$Art_Kr` = eval(cat(``, args[2], `2`)))],
        RETURN("a profiling information does not exist")));
      assign(a = [cat(``, args[2]), `$Art_Kr`[cat("", args[3])]]), RETURN(`if`(type(a[2], numeric), a,
        "procedure has been not profiled"), `if`(type(eval(profile_proc), symbol), unp(), NULL))
   else NULL
   end if;
   if 3 <= nargs and belong(cat(``, args[2]), R) and args[3] = 2 then
      `if`(nargs = 4 and type(args[4], numeric), assign(a = args[4]), assign(a = 0));
      assign(pf = cat(K, "/$@_", args[2]), `$Art_Kr` = eval(cat(profile_, args[2]))),
        `if`(type(eval(`$Art_Kr`), table), NULL, `if`(DoF(pf) = file, [(proc() read pf end proc)(),
        assign(`$Art_Kr` = eval(cat(``, args[2], `2`)))],
        RETURN("a profiling information does not exist")));
      RETURN(tabar(`$Art_Kr`, Procedures, cat(`Procedure's `, args[2]), a),
        `if`(type(eval(profile_proc), symbol), unp(), NULL))
   else NULL
   end if;
   `if`(K <> false, [assign(P = march('list', K)), assign('h' = nops(P))],
      ERROR("library <%1> does not exist or is not linked with the main Maple library", L));
   `if`(nargs = 2 and type(args[2], list({symbol})), assign(S = args[2]),
      assign(S = [seq(`if`(P[k][1][1 .. 2] <> ":-", cat(``, P[k][1][1 .. -3]), NULL), k = 1 .. h)]));
   for k in S do
      try `if`(type(eval(k), procedure), assign('G' = [op(G), k]), NULL)
      catch "":  NULL("Exception handling with program modules")
      end try
   end do;
   `if`(G = [], ERROR("procedures ordered for profiling do not exist both in library <%1> and in
      libraries logically linked with the main Maple library", L, unprofile(), unp()), NULL);
   try
      if DoF(calls1) = file then null((proc()  profile(op(G));
            seq((proc(x) read x end proc)(eval(V[h])), h = 1 .. 5);
            seq(_SL(Z, eval(T[h]), h), h = 1 .. 5)  end proc)())
      else profile(op(G))
      end if
   catch "%1 is already being profiled":  WARNING("profiling is already being executed!")
   end try
end proc
```

Typical examples of the procedure use:

> restart; StatLib(UserLib);

> MkDir("C:/Temp/Art/Kr"), type("C:/Temp/Art/Kr", dir), type("C:/Temp/Art/Kr", file);

"c:\temp\art\kr", *true*, *false*

> StatLib(UserLib, calls, 2, 7);

Procedures	*Procedure's calls*
Red_n	33
Case	31
Subs_All	16
Subs_all1	16
Search	10
holdof	9
CF	7

> StatLib(UserLib, time, Case);

[*time*, 0.020]

> StatLib(UserLib, bytes, StatLib);

[*bytes*, 17980]

> seq(StatLib(UserLib, calls, h), h = [belong, tabar, Case, CF]);

[*calls*, 4], [*calls*, 1], [*calls*, 61], [*calls*, 7]

> seq(StatLib(UserLib, calls, p), p = [frame_n, com_exe1, Kernels]);

[*calls*, 8], [*calls*, 3], [*calls*, 1]

> StatLib(UserLib, time, 2, 0.03);

Procedures	*Procedure's time*
Adrive	0.761
tabar	0.311
SLj	0.140
belong	0.070
Red_n	0.060
convert/list	0.060
Plib	0.040
StatLib	0.040

```
> StatLib(UserLib, bytes, 2, 20000);
```

Procedures	Procedure's bytes
SLj	2920484
tabar	1877432
Red_n	324348
Case	188348
type/nestlist	101096
StatLib	94508
Search	82980
Adrive	60944
Plib	54192
Subs_all1	37928
sub_1	35232
CF	34084
type/dir	32816
convert/list	27728
belong	26212
type/file	20236

```
> StatLib(UserLib, calls, 2, 15);
```

Procedures	Procedure's calls
Case	81
Red_n	33
sub_1	22
Search	20
Subs_All	16
Subs_all1	16

```
> StatLib(UserLib, calls, 2, 20);
```

Procedures	Procedure's calls
Case	86
Red_n	33
sub_1	24
Search	21

> StatLib(UserLib, 1);

Warning, profiling of library <**UserLib**> has been completed

[A new *Maple* session:

> StatLib(UserLib); CureLib("C:/rans\\academy/libraries\\ArtKr", x, y):

Warning, library file <C:/rans\academy/libraries\ArtKr\Maple.ind> does not exist

Warning, Analysis of contents of library lib-file is being done. Please, wait!

Warning, library contains multiple entries of the following means
[Atr, CCM, CureLib, Currentdir, DAclose, DAopen, DAread, DUlib, FSSF, F_atr1, F_atr2, FmF, Imaple, Is_Color, LibElem, LnFile, ModFile, NLP, ParProc1, RTfile, Reduce_T, SDF, SSF, Suffix, Uninstall, Vol, Vol_Free_Space, WD, WS, cdt, conSA, dslib, ewsc, gelist, sfd, mapTab, readdata1, redlt, sorttf, type/dir, type/...];
the sorted nested list of their names with multiplicities appropriate to them is in predefined variable `_mulvertools`

Warning, Analysis of contents of library lib-file has been completed!

> _mulvertools;

[[*F_atr2*, 6], [*Atr*, 4], [*CureLib*, 3], [*FSSF*, 3], [*SDF*, 3], [*SSF*, 3], [*Suffix*, 3], [*conSA*, 3], [*ewsc*, 3], [*gelist*, 3], [*sorttf*, 3], [*type/file*, 3], [*CCM*, 2], [*Currentdir*, 2], [*DAclose*, 2], [*DAopen*, 2], [*DAread*, 2], [*DUlib*, 2], [*F_atr1*, 2], [*FmF*, 2], [*Imaple*, 2], [*Is_Color*, 2], [*LibElem*, 2], [*LnFile*, 2], [*ModFile*, 2], [*NLP*, 2], [*ParProc1*, 2], [*RTfile*, 2], [*dslib*, 2], [*Reduce_T*, 2], [*Uninstall*, 2], [*Vol*, 2], [*Vol_Free_Space*, 2], [*WD*, 2], [*WS*, 2], [*cdt*, 2], [*mapTab*, 2], [*readdata1*, 2], [*redlt*, 2], [*sfd*, 2], [*type/dir*, 2]]

> SLj([[Vic, 62], [Gal, 57], [Sv, 37], [Arn, 41], [Art, 15], [Kr, 7]], 2);

[[*Kr*, 7], [*Art*, 15], [*Sv*, 37], [*Arn*, 41], [*Gal*, 57], [*Vic*, 62]]

> StatLib(UserLib, calls, 2, 15);

Procedures	*Procedure's calls*
Case	123
Red_n	75
Subs_all1	37
Subs_All	37
Search	36
sub_1	28
CF	20
holdof	17
belong	16

> StatLib(UserLib, bytes, 2, 62000);

Procedures	*Procedure's bytes*
SLj	5813452
tabar	4316044
Red_n	1109084
Case	319444
type/nestlist	294160
Search	144232
StatLib	113836
CF	96592
Adrive	96176
Subs_all1	89784
Plib	78000
CureLib	74284

> StatLib(UserLib, 0);

Warning, datafiles with statistics have been removed out of directory with library <**UserLib**>

In view of the explanations made above, the examples of the given fragment are transparent enough and do not demand any additional explanations. In particular, the given procedure has been used for improvement of operating characteristics of the user library containing all tools represented in the present monograph.

The procedures represented in the present section, provides the user by a number of tools for a handling of the user libraries having structural organization similar to the main *Maple* library. These tools provide a series of functions simplifying a restoring problem of damaged libraries. Along with that, they support also other structural organizations useful in a series of appendices. These and other prerequisites caused by the above procedures allow to essentially automate the user libraries handling and extend possibilities dealing with saving of procedures and program modules in external computer memory.

Summary

This book introduces students and professionals into the different aspects of the famous mathematical package *Maple* of releases 6 - 9/9.5. The book is immediate continuation of our previous books on the *Maple*-problematics published in Russia, Byelorussia, Estonia and Lithuania. The basic attention is given to additional features created in the process of exploring and approbation of the package, which in many cases essentially dilate potentialities of the package. On examples of embodying of the given facilities, the application of various useful receptions, including non-standard ones, is illustrated; these receptions allow in many cases essentially to simplify programming problems in environment of the package and to make them by more transparent from the mathematical point of view. The given features cover such components of the package as the basic software environment, graphic facilities, data processing, work with the user libraries, statistical facilities and access facilities to datafiles, etc.

The given software, carried out on an *innovation* level, has organized as the user *Library*, logical linking of which with the main *Maple* library allows to use the given features at a level of the built-in *Maple* facilities. The *Library* contains well-designed software (a set of more than **570** procedures and program modules), which supplements well the already available *Maple* software with the orientation towards the widest circle of the *Maple* users, greatly enhancing its usability and effectiveness. The experience of use of the given *Library* by both the single users and in a number of universities and scientific research institutes in Russia, Byelorussia, Latvia, Lithuania, the Ukraine, Germany, etc. has shown its good operating performances at the solution of manifold interesting mathematical problems. In many cases the represented additional procedures, functions and the program modules illustrate both useful receptions of programming, and elements of the principles and methodology of programming in environment of the package.

The *Library* is intended for *Maple* of releases 6 - 9/9.5 functioning on platforms *Windows* 95/98/98SE/ME/NT/XP/2000/2003, however the *txt*-datafile attached to the *Library* allows easily enough to adapt it for the operational platforms different from *Windows*. Furthermore, the sets of source codes of all *Library* tools and help-pages composing its Help database allow easily to update the *Library* or create on its basis own libraries. The attached datafile "ProcLib_6_7_8_9.mws" contains self-installing *Library* and is oriented on execution in the *Maple* environment of releases 6 - 9/9.5. It allows to automate installation of the *Library* for the above *Maple* releases.

Each software represented in the book is supplied with description and explanations, and contains the source code and the more typical examples of its application. As required, the description has supplied by the necessary considerations, concerning peculiarities of its execution in the *Maple* environment of one or another release, and of the current *Windows* platform.

Alongside with it, presence of source codes of procedures and program modules containing a whole series of both effective and non-standard receptions of programming, allows to use them in programming various applied tasks. In particular, rather effective method called by us as ***"disk transits"*** can be used for programming dynamically generated program constructions. This method is rather useful in conditions of advanced programming in *Maple*. The given method has been developed by us and successfully used for computers of 3rd generation (IBM/360, IBM/370, etc.) and appeared especially effective with appearance of the personal computers whose development is characterized (*including*) by enough quick decreasing of time of access to external memory on hard disks. Rather interesting receptions of creation of high-performance one-line extra-codes are represented by source codes of procedures.

Procedures and program modules of the given list have various complexity of organization and used algorithms; in many cases, they use both effective and non-standard receptions of programming in the *Maple* environment. At the same time, in a whole series of cases they allow to simplify essentially programming of various tasks in *Maple* and have well shown oneself in many applications of the package of releases **6 - 9/9.5.** The given means can be used as individually (*for the decision of various problems or for creation on their basis of new means*), and in structure of the user library extending standard means of the package, eliminating a number of defects and mistakes of the package, raising compatibility of the package relatively to its releases and raising efficiency of programming of problems in its environment. From our six-year experience of the package usage in various applications and from experience of our colleagues from universities and the academic institutes of the Baltic States, Belarus, Russia, Czechia, Poland, Germany, the USA, etc. a lot of the procedures represented here or their clones are rather expedient for including in set of standard means of the package of the subsequent releases.

The tools represented in the *Library* increase the range and efficiency of use of the package on the *Windows* platform owing to the innovations in three basic directions: (1) elimination of a series of basic defects and shortcomings, (2) extension of capabilities of a series of standard tools, and (3) replenishment of the package by new means which increase the capabilities of its program environment, including the features essentially improving the compatibility of releases 6 - 9/9.5. The basic attention has been devoted to additional tools created in the process of practical use and testing of the package of releases 4 - 9 that by some parameters considerably extend the capabilities of the package making the work with it much easier; a considerable attention is also devoted to the tools providing package compatibility of releases 6 - 9/9.5. The experience in using the above software for various applications has confirmed its valuable operational characteristics.

It should be noted that a series of our books and papers on *Maple*, representing tools developed by us and containing suggestions on further development of the package encouraged the development of such applications as package modules **FileTools, LibraryTools, ListTools** and **StringTools.** All this allows to hope, that the represented book and the software attached to it will appear useful enough to broad audience of the *Maple* users, both for the beginners and the skilled ones.

References

1. ***Aladjev V.Z., Hunt U.J., Shishakov M.L.*** Basics of Computer Science.- Gomel: Salcombe Eesti Publisher, 1997, 396 p., ISBN 5-14-064254-5 (*in Russian with extended English summary*)

2. ***Aladjev V.Z., Hunt U.J., Shishakov M.L.*** Basics of Computer Science: Textbook.- Moscow: Filin Press, 1998, 496 p., ISBN 5-89568-068-2 (*in Russian with extended English summary*)

3. ***Aladjev V.Z., Hunt U.J., Shishakov M.L.*** Basics of Computer Science. 2nd edition.- Moscow: Filin Press, 1999, 520 p., ISBN 5-89568-068-6 (*in Russian with extended English summary*)

4. ***Aladjev V.Z., Gershgorn N.A.*** Computing Problems on Personal Computer.- Kiev: Technics Press, 1991, 248 p. (*in Russian with extended English summary*)

5. ***Aladjev V.Z., Tupalo V.G.*** Algebraic Calculations on Computer.- Moscow: Mintopenergo Press, 1993, 251 p., ISBN 5-942-00456-8 (*in Russian with extended English summary*)

6. ***Aladjev V.Z., Hunt U.J., Shishakov M.L.*** Mathematics on Personal Computer.- Gomel: FORT Press, 1996, 498 p., ISBN 3-420-614023-3 (*in Russian with extended English summary*)

7. ***Aladjev V.Z., Shishakov M.L.*** Introduction into Environment of Package Mathematica 2.2.- Moscow: Filin Press, 1997, 362 p., ISBN 5-89568-004-6 (*in Russian with English summary*)

8. ***Aladjev V.Z., Vaganov V.A., Hunt U.J., Shishakov M.L.*** Introduction into Environment of Mathematical Package Maple V.- Minsk: IAN Press, 1998, 452 p., ISBN 14-064256-98

9. ***Aladjev V.Z., Vaganov V.A., Shishakov M.L., Hunt U.J.*** Programming in Environment of Mathematical Package Maple V.- Gomel: TRG & Salcombe, 1999, 470 p, ISBN 4-10-121298-2

10. ***Aladjev V.Z., Vaganov V.A., Shishakov M.L., Hunt U.J.*** Workstation for Mathematicians.- Gomel-Tallinn: International Academy of Noosphere, 1999, 605 p., ISBN 3-42061-402-3

11. ***Aladjev V.Z., Bogdevicius M.A.*** Solution of Physical, Technical and Mathematical Problems with Package Maple V.- Vilnius: Technics Press, 1999, 686 p., ISBN 9986-05-398-6

12. ***Aladjev V.Z., Shishakov M.L.*** Workstation for Mathematicians.- Moscow: Laboratory of Basic Knowledge, 2000, 751 p. + CD, ISBN 5-93208-052-3 (*in Russian with English summary*)

13. ***Aladjev V.Z., Bogdevicius M.A.*** Maple 6: Solution of Mathematical, Statistical, Engineering and PhysicalProblems.- Moscow: Binom Press, 2001, 850 p. + CD, ISBN 5-93308-085-X

14. ***Aladjev V.Z., Bogdevicius M.A.*** Interactive Maple: Solution of Mathematical, Engineering, Statistical and Physical Problems.- Tallinn-Vilnius, International Academy of Noosphere, the Baltic Branch, 2001 - 2002, CD, ISBN 9985-9277-1-0

15. ***Aladjev V.Z., Bogdevicius M.A.*** Use of package Maple V for solution of physical and engineering problems // Intern. Conf. TRANSBALTICA-99, Technics Press, 1999, Vilnius, Lithuania.

16. ***Aladjev V.Z., Hunt U.J.*** Workstation for mathematicians // International Conference TRANSBALTICA-99, Technics Press, April 1999, Vilnius, Lithuania.

17. ***Aladjev V.Z., Hunt U.J.*** Workstation for mathematicians // Intern. Conf. "Perfection of Mechanisms of Management", Institute of Modern Knowledge, April 1999, Grodno, Byelorussia.

18. ***Aladjev V.Z., Shishakov M.L.*** Programming in Package Maple V // **2**nd Int. Conf. "Computer Algebra in Fundamental and Applied Researches and Education".- Minsk: BGU Press, 1999.

19. ***Aladjev V.Z., Shishakov* M.L.** A Workstation for mathematicians // **2**nd Int. Conf. "Computer Algebra in Fundamental and Applied Researches and Education".- Minsk: BGU Press, 1999.

20. ***Aladjev V.Z., Shishakov M.L., Trokhova T.A.*** Educational computer laboratory of the engineer // Proc. **8**th Byelorussia Mathemat. Conf., vol. **3**, Minsk, Byelorussia, 2000.

21. ***Aladjev V.Z., Shishakov M.L., Trokhova T.A.*** Applied aspects of theory of homogeneous structures // Proc. **8**th Byelorussia Mathemat. Conf., vol. **4**, Minsk, Byelorussia, 2000.

22. ***Aladjev V.Z., Shishakov M.L., Trokhova T.A.*** Modelling in program environment of the mathematical package Maple V // Proc. Internat. Confer. on Mathematical Modelling MKMM-2000.- Herson, Ukraine, 2000.

23. ***Aladjev V.Z., Shishakov M.L., Trokhova T.A.*** A workstation for solution of systems of differential equations // Third Int. Conf. "Differential Equations and Applications".- Saint-Petersburg, Russia, 2000

24. ***Aladjev V.Z., Shishakov M.L., Trokhova T.A.*** Computer laboratory for engineering researches // Intern. Conf. ACA-2000.- Saint-Petersburg, Russia, 2000.

25. ***Aladjev V.Z., Bogdevicius M.A., Hunt U.J.*** A Workstation for mathematicians // Lithuanian Conference TRANSPORT-2000.- Vilnius: Technics Press, April 2000, Lithuania.

26. ***Aladjev V.Z.*** Computer Algebra // Alpha, № **1**, 2001, Grodno, GRSU, Byelorussia.

27. ***Aladjev V.Z.*** Modern computer algebra for modeling of the transport systems // Proc. Int. Conf. TRANSBALTICA-2001.- Vilnius: Technics Press, April 2001, Lithuania.

28. ***Aladjev V.Z., Shishakov M.L., Trokhova T.A.*** Workstation for the engineer-mathematician // Proc. of the GSTU, № **3**, 2000, pp. 42-47, Gomel State University, Gomel, Byelorussia.

29. ***Aladjev V.Z., Bogdevicius M.A.*** Special Questions of Operation in Environment of the Mathematical Maple Package.- Tallinn-Vilnius: International Academy of Noosphere & Vilnius Gediminas Technical University, 2001, 215 p., ISBN 9985-9277-2-9 (*with English summary*)

30. ***Aladjev V.Z., Vaganov V.A., Grishin E.*** Additional Functional Tools of Mathematical Package Maple 6/7.- Tallinn: International Academy of Noosphere, 2002, 325 p., ISBN 9985-9277-2-9

31. ***Aladjev V.Z.*** Effective Work in Maple 6/7.- Moscow: Laboratory of Basic Knowledge, 2002, 334 p. + CD, ISBN 5-93208-118-X (*in Russian with extended English summary*)

32. ***Aladjev V.Z., Liopo V.A., Nikitin A.V.*** Mathematical Package Maple in Physical Modelling.- Grodno: Grodno State University, 2002, 416 p., ISBN 3-093-31831-3 (*with English summary*)

33. ***Aladjev V.Z., Vaganov V.A.*** Computer Algebra System Maple: A New Software Library.- Tallinn: International Academy of Noosphere, 2002, 420 p.+ CD, ISBN 9985-9277-5-3

34. ***Aladjev V.Z., Veetousme R.A., Hunt U.J.*** General Theory of Statistics: Textbook.- Tallinn: TRG & SALCOMBE Eesti Ltd., 1995, 201 p., ISBN 1-995-14642-8 (*with English summary*)

35. ***Aladjev V.Z., Hunt U.J., Shishakov M.L.*** Course of General Theory of Statistics: Textbook.- Gomel: BELGUT Press, 1995, 201 p., ISBN 1-995-14642-9 (*in Russian with English summary*)

36. ***Aladjev V.Z., Hunt U.J., Shishakov M.L.*** Questions of Mathematical Theory of the Classical Homogeneous Structures.- Gomel: BELGUT Press, 1996, 151 p., ISBN 5-063-56078-5

37. ***Efimova M.O., Hunt U.J.*** Geometry of Drawing: Graphical Package AutoTouch (Ed. acad. **V. Z. Aladjev**).- Gomel: Russian Academy of Noosphere, 1997, 72 p., ISBN 7-14-064254-7

38. ***Aladjev V.Z., Hunt U.J., Shishakov M.L.*** Scientific-Research Activity of the Tallinn Research Group: Scientific Report during 1995 - 1998.- Tallinn - Gomel - Moscow: TRG & SALCOMBE & VASCO, 1998, 80 p., ISBN 14-064298-56 (*in Russian with extended English summary*)

39. ***Aladjev V.Z., Bogdevicius M.A., Prentkovskis O.V.*** New Software for Mathematical Package Maple of Releases 6, 7 and 8.- Vilnius: Vilnius Technical University & International Academy of Noosphere, the Baltic Branch, 2002, 404 p., ISBN 9985-9277-4-5, 9986-05-565-2

40. ***Aladjev V.Z., Hunt U.J., Shishakov M.L.*** Mathematical Theory of the Classical Homogeneous Structures.- Tallinn-Gomel: TRG & Salcombe Eesti Ltd., 1998, 300 p., ISBN 9-063-56078-9

41. ***Aladjev V.Z., Vaganov V.A., Hunt U.J., Shishakov M.L.*** Workstation for Mathematicians.- Tallinn-Minsk: Russian Academy of Natural Sciences, 1999, 608 p., ISBN 3-42061-402-3

42. ***Aladjev V.Z.*** Interactive Course of General Theory of Statistics.- Tallinn: International Academy of Noosphere, the Baltic Branch, 2001, CD with Booklet, ISBN 9985-60-866-6

43. ***Aladjev V.Z., Vaganov V.A.*** Systems of Computer Algebra: A New Software Toolbox for Maple.- Tallinn: International Academy of Noosphere, 2003, 270 p., ISBN 9985-9277-6-1

44. ***Aladjev V.Z., Bogdevicius M.A., Vaganov V.A.*** Systems of Computer Algebra: A New Software Toolbox for Maple. Second edition.- Tallinn: International Academy of Noosphere, 2004, 462 p., ISBN 9985-9277-8-8

45. ***Aladjev V.Z., Bogdevicius M.A.*** Computer algebra system Maple: A new software toolbox **// 4**th Int. Conf. TRANSBALTICA-03, Technics Press, April 2003, Vilnius, pp. 458-466.

46. ***Aladjev V.Z., Barzdaitis V., Bogdevicius M.A., Gecys S.*** The solution of the dynamic model of asynchronous engine by finite elements method **// 4**th Intern. Conf. TRANSBALTICA-03, Technics Press, April 2003, Vilnius, Lithuania, pp. 339-352.

47. ***Aladjev V.Z.*** Computer Algebra System Maple: A New Software Library **//** Intern. Conf. "Computer Algebra Systems and Their Applications", CASA-2003, Sankt-Petersburg, Russia, 2003.

48. ***Aladjev V., Bogdevicius M., Vaganov V.*** Systems of Computer Algebra: A New Software Toolbox for Maple / the 2004 International Conference on Software Engineering Research and Practice, SERP'04, 2004, Las Vegas, Nevada, USA.

49. ***Owen D.B.*** Handbook of Statistical Tables.- London: Addison-Wesley Publishing Co., 1963

50. ***Kelley T.L.*** The Kelley Statistical Tables.- Cambridge: Harvard University Press, 1948.

List of procedures and program modules represented in the monograph

&Shift, _0N, _1N, _nnn, _mMfile, _rd, _SL, A_Color, ACC:- acc, ACC:- Ds, ACC:- Sr, Aconv, ACP, Adrive, AFdes, AlgLists:- &*, AlgLists:- &+, AlgLists:- &-, AlgLists:- &/, AlgLists:- &^, All_Close, AnalT, Animate2D, Animate3D, andN, Aobj, Apmv, Aproc, Arobj, ArtMod, Assign, assign6, assign67, assign7, AtrRW, avm_VM, belong, Bit, Bit1, blank, Built-in, came, Case, Catch_Vir, CCF, CCM, CDiag, CDM, CF, CF1, CF2, cfdd, CFF, cfln, Chess, ChkPnt, Close, Clr, Cls, cdt, clsd, cmlib, cmtf, cnvtSL, CoefTaylor, com6_9, CompStr, Con_Mws, conSA, conSV, ContLib, convert/list, convert/listlist, convert/lowercase, convert/module, convert/rlb, convert/set, convert/ssll, convert/TEXT, convert/uppercase, conwf, Cookies, CorMod, CorMod1, CrNumMatrix, CS, CureLib, Currentdir, D_ren, DAclose, DagTag, DAopen, DAread, dcemod, DDT, DeCod, decode, DeCoder, DefOpt, DEL_F, delf, delf1, delsc, deltab, DestLm, Dialog, diffP, Dir, Dir_ren, dslib, DIRAX:- conv, DIRAX: -delete, DIRAX:- empty, DIRAX:- extract, DIRAX:- insert, DIRAX:- new, DIRAX:- printd, DIRAX:-replace, DIRAX:-reverse, DIRAX:-size, DIRAX:-sortd, DirE, DirF, DirFT, Dist_rand_n_2, DoF, DoF1, DT, DUlib, E_mail, email, Empty, EntS, Etest, etf, Evalf, ewsc, exeP, expLS, Extract, ExtrF, extrS, F_alias, F_Analys, F_atr, F_atr1, F_atr2, F_ren, F_T, F_test_Ds, F_Type, Fac, Fappend, FBBcopy, FBcopy, FBfile, FD, Fend, Fequal, FFP, filePM, FindFSK, FLib, FLib1, FLL, FmF, FNS, Fnull, FO_state, FOR_DO, Fopen, fpathdf, Fremove, frss, FSSF, FT_part, FT_part1, FT_restr, FT_subs, FTabLine, FTabLine1, FTcopy, FTmerge, ftpd, Ftype, Fword, gelist, Gener, GenFT, getable, GG, Heap, helpman, Histo, holdof, howAct, HS_1D, IAN_REA, Iddn, Iddn1, IdR, IDS, Images, Imaple, Indets, inel, Ins, Insert, insL, InstUlib, INT, InvL, InvList, InvT, IO_proc, IP, Is_Color, isDir, IsFempty, isFile, isflo, IsFtype, isLnkMws, ismLib, isMSDcom, IsOpen, IsOpen1, IsOpen2, IsOpenF, IsPtf, isRead, KCM, KL, KL1, Kr_Mesh, LGD, LibElem, LibElem1, LibElem2, LibLink, LibLink1, LibUser, Linear_Const, LnFile, LO, LocalnL, Lrare, LRM_NRM, LSF, LSL, LT, LTfile, M_Type, MA, MAM, map3, map4, map5, MapleLib, mapLS, mapN, mapTab, MatrSort, maxl, MEM, Mem, Mem1, memberL, MiniMax, minl, MinMax, MkDir, MmF, mmf,mmp, mod21, mod3, mod_proc, ModFile, ModProc, modproc, mpl_proc, MPL_txt, mPM, MSDcom, mtf, Mulel, mws789_6, Mwsin, MwsRtb, N_Cont, NDLN, NDP, NDT, Nint, nlcvector, nLine, NLP, NonaLP, Nstring, null, NumOfString, OP, open2, OpenLN, Open, orN, ParProc, ParProc1, Path, pathtf, perml, permlib, Pind, Plib, plotTab, Pnd, PNorm, Polyhedra, porf, Porshen, Pos, PP, PSubs, Pul_Bal, Pul_Bal_Cor, Pulsar, Q2plot, QFline, Qsubstr, Queue, rand_Histo, Read, Read1, read3, readdata1, ReadProc, reconf, Red_n, redL, RedList, redlt, redss, Reduce_T, RegMW, Release, Release1, relml, Remember_T, reprolib, Residue, Resl, ResRtb, Rev, Rfact, rID, Rlss, RmDir, rmdir1, Rmf, rName, Root, RS, RSLU, Rssl, Rt_s, RTab, Rtable, RTfile, RTsave, S_D, SA_text, Save, save1, Save1, save2, Save2, save3, SaveMP, SaveProc, SD, SD_S, SD_S1, SDF, Search, Search1, Search2, Search3, Search_D, Search_D1, Search_D2, SEQ, seq1, seq2, seqstr, sext, sextr, sextr1, sident, SimpleStat:- ACC, SimpleStat:- CC, SimpleStat:- CR, SimpleStat:- Ds, SimpleStat:- FD, SimpleStat:- LRM_NRM, SimpleStat:- LT, SimpleStat:- MAM, SimpleStat:- MCC, SimpleStat:- PCC, SimpleStat:- Sko, SimpleStat:- SR, SimpleStat:- Weights, SL, SLD, SLj, SLS, sls, Slss, Smart, Smart_Plot, sMf, SN, SN_6, SN_7, SN_8, SN_9, sof, SoLists:- intersect, SoLists:- minus, SoLists:- sublist, SoLists:- union, SortL, sorttf, Sproc, SQHD, SS, SSet, SSF, ssf, SSN, sspos, sstr, STACK, statf, StatLib, stpm, Sts, stype, sub_1, Sub_all, Sub_list, SUB_S, Sub_st, Subs_All, Subs_all1, Subs_all2, subseqn, Subset, subSL, subsLS, Suffix, swmpat, swmpat1, Sys_Env, T_Font, T_SQHD, T_SQHT, T_test_AV, tabar, TabList, Test, transmf7_6, trConv, TRm, ttable,

type/binary, type/boolproc, type/byte, type/digit, type/dir, type/dirax, type/file, type/heap, type/letter, type/Lower, type/lower, type/mlib, type/mod1, type/nestlist, type/package, type/fpath, type/path, type/realnum, type/rlb, type/sequent, type/setset, type/ssign, type/Table, type/Upper, type/upper, Type_D, typeseq, U_test_MW, UbF, Ulib, ulibpack, Uninstall, UpLib, User_pfl, User_pflM, User_pflMH, UserC, Users, V_Solve, VisM, Vol, Vol_Free_Space, VTest, Weight_LF, Weights, Weights_L, winver, writedata1, WD, WDS, WS, WT, X_test_VW, xbyte, xbyte1, xNB, xorN, xpack, XSN, XTfile, intaddr, minmax3d, plotvl, type/complex1, varsort, MSK, ExprOfString, com_exe1, com_exe2, mwsname, Kernels, frame_n.

Procedures and program modules of the given list have various complexity of organization and used algorithms; in many cases, they use both effective and non-standard receptions of programming in the *Maple* environment. At the same time, in a whole series of cases they allow to simplify essentially programming of various tasks in *Maple* and have well shown oneself in many applications of the package of releases **6 - 9.** The given means can be used as individually (*for the decision of various problems or for creation on their basis of new means*), and in structure of the user library extending standard means of the package, eliminating a number of defects and mistakes of the package, raising compatibility of the package relatively to its releases and raising efficiency of programming of problems in its environment. From our six-year experience of the package usage in various applications and from experience of our colleagues from universities and the academic institutes of the Baltic States, Belarus, Russia, Czechia, Poland, Germany, the USA, etc. a lot of the procedures represented here or their clones are rather expedient for including in set of standard means of the package of the subsequent releases.

The above-named **578** *procedures* and *program modules* have been organized into the user Library provided with Help database that are structurally analogous to the main *Maple* library. After logical linking of it with the main library, the tools contained in it can be used analogously to standard *Maple* tools. Archive with the Library versions for releases **6, 7, 8** and **9/9.5** with source codes of all tools contained in them can be downloaded from the web sites of the Publisher and Author:

`http://writers.fultus.com/aladjev/source/UserLib6789.zip`

`http://www.aladjev.newmail.ru/UserLib6789.zip`

Our work on computer algebra systems continues, therefore new software of various purpose and oriented on various appendices will appear. The library represented in the present monograph will be constantly updated by the most useful of these means. Therefore, the reader is recommended to check periodically presence on the above-mentioned sites of new versions of the library. Whereas all questions, remarks and offers can be directed to one of the following addresses

`valadjev@yahoo.com, aladjev@hotmail.com, aladjev@lenta.ru`

We welcome your feedback. Additionally, for direct inquiries, suggestions, or comments you can use

phone +(372) 635 6078 or **fax** +(372) 600 7969

Index

About the Author:
Victor Aladjev

Aladjev V. was born on June 14, 1942 in the town of Grodno (*Western Byelorussia*). After successful completion of the secondary school in 1959, he matriculated on the first course of physical and mathematical faculty of Grodno State University and in 1962 has been transferred to the Department of Mathematics of Tartu State University (*Estonia*). In 1966, he successfully finished Tartu State University the speciality of *Mathematician*. In 1969, **Aladjev V.** started post-graduate studies of the Estonian Academy of Science, and successfully finished in 1972 two specialities "*Theoretical Cybernetics*" and "*Technical Cybernetics*". The doctoral degree in mathematics has been assigned to him for the monograph "*Mathematical Theory of Homogeneous Structures and their Applications*". From 1972 to 1990, **Aladjev V.** has the respectable positions in a number of project-technological and research organizations of Tallinn (*Estonia*). In 1999, he has been appointed the president of the *Tallinn Research Group* (TRG), the scientific findings of which received international recognition, first, in the field of activities in mathematical theory of Homogeneous Structures (*Cellular Automata*). **Aladjev V.** is the author of more than 300 scientific and technological publications, including 60 monographies, books and collections of the articles, published in the former USSR, Estonia, Great Britain, Russia, Byelorussia, GDR, the Ukraine, Lithuania, Germany, Hungary, USA, Czechoslovakia, Japan and Holland. He founded the Estonian school of mathematical theory of Homogeneous Structures that has been recognized internationally and covered the fundamentals of a new section of modern mathematical cybernetics. **Aladjev V.** participates as a member of the organizing committee and/or a guest lecturer in many international scientific forums in mathematics and cybernetics. In April 1994, **Aladjev V.** has been elected as an *academician* of the Russian Academy of Cosmonautics in the section of "*Fundamental Researches*", and in September 1994, he has been elected as an *academician* of the Russian Academy of Noosphere in the section of "*Information Science*". In September 1995, he has been elected as an *academician* of the Russian Academy of Natural Sciences in the section of "*Noosphere Knowledge and Technologies*", and in June 1998 as an *honorary academician* of the Russian Ecological Academy. In November 1997 **Aladjev V.** has been elected as the *academician-secretary* of the Baltic Branch of the Russian Academy of Noosphere, integrating scientists and specialists of three Baltic countries and Byelorussia, who work in the field of a complex of scientific disciplines, which are included in the doctrine about *Noosphere* and areas, adjacent to it. Because of reorganization of the Russian Academy of Noosphere into the International Academy, **Aladjev V.** has been elected as its First vice-president in December 1998. At last, in December 1999 **Aladjev V.** has been elected as a *foreign member* of the Russian Academy of Natural Sciences in the section of "*Information Science and Cybernetics*". At present, **Aladjev V. Z.** is *academician-secretary* of the Baltic Branch of the International Academy of Noosphere. Most considerable scientific results of **Aladjev V.** are related to the mathematical theory of homogeneous structures and its applications. The sphere of his scientific interests includes mathematics, informatics, cybernetics, computer science, physics, cosmology, etc.

Lightning Source UK Ltd.
Milton Keynes UK
UKOW012001150313

207737UK00004B/344/A